WILDLIFE TOXICOLOGY and POPULATION MODELING

Integrated Studies of Agroecosystems

Edited by

Ronald J. Kendall, Ph.D.
TIWET/Clemson University
Thomas E. Lacher, Jr., Ph.D.
Archbold Tropical Research Center/Clemson University

Proceedings of the Ninth Pellston Workshop
Kiawah Island, South Carolina, July 22-27, 1990

Contributors
American Cyanamid Company, Battelle, Ciba-Geigy Corporation, Conservation Foundation, Dow-Elanco, E. I. du Pont de Nemours and Company, Mobay Corporation, National Wildlife Federation, Rhône-Poulenc, SETAC Foundation for Environmental Education, United States Environmental Protection Agency – Environmental Research Laboratory, Athens, GA, United States Environmental Protection Agency, Office of Policy and Planning, Washington, D.C., and the United States Navy.

SETAC Special Publications Series

Series Editor

Dr. T. W. La Point
The Institute of Wildlife and Environmental Toxicology
Clemson University

Publication sponsored by the Society of Environmental Toxicology and Chemistry (SETAC) and the SETAC Foundation for Environmental Education

LEWIS PUBLISHERS
Boca Raton Ann Arbor London Tokyo

Library of Congress Cataloging-in-Publication Data

Wildlife toxicology and population modeling : integrated studies of agroecosystems / edited by Ronald J. Kendall, Thomas E. Lacher.
p. cm. -- (SETAC special publications series)
Includes bibliographical references and index.
ISBN 0-87371-591-8
1. Pesticides and wildlife--Congresses. 2. Birds--Effect of pesticides on--Congresses. 3. Pesticides--Toxicology--Congresses. 4. Animal populations--Congresses. 5. Bird populations--Congresses. 6. Biological models--Congresses. I. Kendall, Ronald J. II. Lacher, Thomas E. III. Series.
QH545.P4W55 1993
598.252'22--dc20

93-25189
CIP

Direct all inquiries to CRC Press, Inc., 2000 Corporate Blvd., N.W., Boca Raton, Florida 33431.

No claim to original U.S. Government works
International Standard Book Number 0-87371-591-8
Library of Congress Card Number 93-25189
Printed in the United States of America 1 2 3 4 5 6 7 8 9 0
Printed on acid-free paper

The SETAC Special Publication Series

Wildlife Toxicology and Population Modeling: Integrated Studies of Agroecosystems is the eighth volume to be published in this series. It presents the proceedings of a conference convened by the "Avian Effects Dialogue Group" of the Conservation Foundation and SETAC. The conference, "The Population Ecology and Wildlife Toxicology of Agricultural Pesticide Use: A Modeling Initiative for Avian Species," was held on Kiawah Island, South Carolina, July 22–28, 1990. The purpose of the conference was to integrate fields of population ecology with wildlife toxicology through new approaches to modeling. The volume presents insights into modeling and large-scale field studies designed to assess the effects of pesticides on avian species, populations, and communities. The book represents a major contribution to risk assessments of avian exposure to xenobiotics.

Thomas W. LaPoint
Editor, SETAC Special Publications
The Institute of Wildlife and Environmental Toxicology
Clemson University

Foreword

The conference, "Population Ecology and Wildlife Toxicology of Agricultural Pesticide Use: A Modeling Initiative for Avian Species," was the tenth in a series of Pellston-style workshops where experts convened to assess the state-of-the-science of an environmental issue of international concern. Four of the workshops predated the formation and involvement of the Society of Environmental Toxicology and Chemistry (SETAC), but the six gatherings since 1982 all have been sponsored by SETAC. It is with pride that I commend to you this SETAC Special Publication. The information contained in it is relevant to the progress and needs of biologists, chemists, ecologists, modelers, toxicologists, and managers in understanding the effects of agricultural pesticides on avian populations, and the guidance provided will be invaluable to regulators, pesticide manufacturers, researchers, and the general public.

On behalf of SETAC, I thank the following: Dr. Ron Kendall, Conference Chair, and Dr. Tom Lacher, Conference Co-Chair, for their special efforts in organization; the staff from The Institute of Wildlife and Environmental Toxicology, Clemson University, for their help in facilitating the event; numerous corporate, public interest, and governmental groups for their financial support; and especially the group of professionals who gave their time and intellect to assure the success of the conference.

Rodney Parrish
Executive Director and 1990 President
Society of Environmental Toxicology and Chemistry

Executive Summary

On July 22, 1990, the Society of Environmental Toxicology and Chemistry (SETAC) and the Avian Effects Dialogue Group of the Conservation Foundation (now RESOLVE) convened a conference entitled "The Population Ecology and Wildlife Toxicology of Agricultural Pesticide Use: A Modeling Initiative for Avian Species," at Kiawah Island, South Carolina. The purpose of this 6-day conference was to integrate the fields of population ecology and wildlife toxicology through approaches to modeling. The goal was to achieve a more unified approach while gaining new insights on the effects of pesticides on avian populations as measured by field studies. The conference, in the usual spirit of SETAC, involved representatives from academia, government, industry, and environmental organizations. We achieved increased cooperation and communication among wildlife toxicologists, population biologists, and ecological modelers.

The most important product of the conference was the completion of this textbook, which conference co-chairs T. E. Lacher, Jr., Archbold Tropical Research Center (ATRC), Clemson University, and R. J. Kendall, The Institute of Wildlife and Environmental Toxicology (TIWET), Clemson University, had the task of editing. The book is published as part of the SETAC Special Publication Series. This was not the only product of the conference. A series of cooperative research ventures are underway as a result of the interactions among ecologists, modelers, toxicologists, members of the regulatory agencies, and industry representatives. A less tangible but equally important result was the amount of cross-disciplinary learning that took place, as was universally acknowledged by the participants. The interchange of basic and applied approaches will greatly improve the dialogue among all groups involved.

As noted by one of the attendees, Raymond O'Connor of the University of Maine, "A model is a tool, a caricature of reality, to help us understand issues otherwise difficult to comprehend." One of the major objectives of the conference was to be able to use models to provide the framework for the synthesis of field studies. Large-scale field research studies designed to assess the effects of pesticides on individual birds and populations are rapidly being developed. The conference attendees recognized that no single model could be built to serve all purposes related to environmental risk assessment of avian exposure to pesticides. Specific models must be constructed to meet specific needs. The conference attendees committed to developing initiatives to refine the interface of ecological modeling that integrates data and concepts from both population ecology and wildlife toxicology. A variety of simpler, separate models that can be updated as more information becomes available is preferred to one complex model. Attendee Peter Greig-Smith of the Ministry of Agriculture, Fisheries and Food of the United Kingdom suggested the following aspects required ecological modeling:

1. Fate and transport of pesticides
2. Exposure involving different species, chemicals, and routes of exposure
3. Uptake/metabolism/elimination models that show bioaccumulation, detoxification, and other aspects in relation to ecology
4. Toxic effects (i.e., dose-response relationships)
5. Extrapolation from one species (or chemical) to another
6. Thresholds of concern (e.g., mortality, reproduction, and sublethal effects)
7. Assessment of field trials for consequences of direct and indirect effects on local reduction and recovery
8. Prediction of effects of local usage to refine management tools
9. Long-term/regional patterns detected by exploratory models of monitoring data

10. Effects of total product utilization using models that integrate impact of single applications with area and frequency of proposed use
11. Community effects

Future effort toward the integration of population ecology with toxicology has been finalized as the *PET Initiative*. The first goal of the PET Initiative was to complete publication of this book. In addition, it established collaborations between research laboratories and instituted a scientific exchange program whose funding is to be shared by EPA and industry. The PET Initiative encourages the design of research projects that will use ecological modeling processes to integrate population ecology and wildlife toxicology of agricultural pesticide use in the area of wetlands, agricultural fields, and turf. For the years 1991 and 1992 the PET Initiative called for special symposia at SETACs annual meetings to encourage scientific exchange to promote future collaboration in scientific development of this field of research. As part of the PET Initiative, an ecological modeling conference will be convened in 1995 to update the book for a second edition, to be held again at Kiawah Island. We will also, through SETAC, communicate to U.S. EPA and other funding sources the need to move ahead with the research goals and recommendations made in this book. A newsletter has been established to support communication that aids the development of ecological modeling and the use of such data as applied to environmental risk assessment of avian exposure to pesticides.

The workshop participants agreed to work toward accomplishing the components of the PET Initiative within a 5-year plan. The goals of the 5-year plan are to: (1) obtain the field data necessary as input to predictive models, (2) develop an understanding of the relationship of avian exposure and resulting mortality for priority chemicals and species, (3) develop an understanding of the relationship of sublethal effects and biomarkers of avian exposure to avian population dynamics, and (4) encourage significant modeling activity as evidenced through scientific presentation at meetings and through scientific publications. A revised 5-year plan will succeed the first in completing the objectives of the PET Initiative. Our goal for the year 2000 is to be able to use predictive models in risk assessments of avian exposure to pesticides and other toxic substances.

The 54 attendees present at the Kiawah conference departed with a commitment to further integrate the fields of population ecology and wildlife toxicology in order to improve our ability to assess the risk of birds and their populations exposed to pesticides. It is only through such commitment, focused at personal levels of interaction, that we will be able to develop models necessary to enhance our ability to manage the use of pesticides and reduce any potential effects to birds and the environment. I believe that the 1990 Ecological Modeling Conference and this book are testimony to our success to date and should promote the future advancement of the science.

Ronald J. Kendall, Chairman
1990 Ecological Modeling Conference
Society of Environmental Toxicology and Chemistry

Acknowledgments

The conference entitled "The Population Ecology and Wildlife Toxocilogy of Agricultural Pesticide Use: A Modeling Initiative for Avian Species" was convened by The Society of Environmental Toxicology and Chemistry (SETAC) and The Avian Effects Dialogue Group (AEDG) of the Conservation Foundation (AEDG now associated with RESOLVE) from July 22–July 27, 1990, at Kiawah Island, South Carolina. We gratefully acknowledge the sponsors that made this meeting possible, including: American Cyanamid Company, Battelle, CIBA-GEIGY Corporation, Conservation Foundation, Dow-Elanco, E.I. DuPont de Nemours Company, Mobay Corporation, National Wildlife Federation, Rhône-Poulenc, United States Environmental Protection Agency–Environmental Research Laboratory, Athens, GA–Office of Policy and Planning, and the United States Navy. We also wish to thank the contributions of the Steering Committee for the conference, including: Richard Balcomb, CIBA-GEIGY Corporation; Anne Barton, U.S. EPA: Sandra Bird, U.S. EPA/ERL Athens; Ed Fite, U.S. EPA; James Gagne, American Cyanamid Company; Raymond O'Connor, University of Maine; Suzanne Orenstein, RESOLVE; Molly Whitworth, U.S. EPA; Bill Williams, ep & t, Corvallis, Oregon. The atmosphere and support for the meeting by the Kiawah Island Inn and conference facilities were extraordinary and greatly contributed to the creative and intellectual exchange that occurred.

We appreciate the attendance and substantial contribution to the meeting of the current and past presidents of SETAC: Rod Parrish, Executive Director of SETAC, and Jim Fava, Roy F. Weston, Inc., respectively. We extend special thanks to the staff of the Institute of Wildlife and Environmental Toxicology (TIWET) who attended and supported this meeting, in particular, Debbie Burrell, Linda Carlock, Jana Hofius, Claudia McLaughlin, Sherry Skipper, and Robin Stengel. All attendees to the conference appreciated the support of these personnel and made that clearly known to the Chair, Ronald J. Kendall, Clemson University, and to the Co-Chair, Thomas E. Lacher, Jr., Clemson University.

This conference would not have been possible without the patience, hard work, and dedication of our Program Coodinator, Suzanne (Swan) Williamson, Clemson University. We deeply appreciate, as did all of the attendees, the attention to detail and the smooth process before, during, and after the conference which attested to her organizational abilities. Her input really made the difference. A special word of thanks is expressed by the editors for her outstanding contribution to the organization of the workshop, and for her support during the early stages of the development of this book. We also would like to thank Anolyn Watkins for her help. Cam Hopkins and Beth Ann Cushman provided extremely valuable assistance in the concluding stages of this effort.

The publication of this book was made possible by financial support from American Cyanamid Company, Battelle, CIBA-GEIGY Corporation, Clemson University Foundation, Dow-Elanco, E.I. DuPont de Nemours Company, Mobay Corporation, National Wildlife Federation, Rhône-Poulenc, Society of Environmental Toxicology and Chemistry (SETAC) Foundation, Inc., U.S. EPA, Environmental Research Laboratory, Athens, GA, U.S. EPA, Office of Policy and Planning, and the U.S. Navy.

The editors acknowledge the excellent editorial support provided by Kent Premo and Elizabeth Rogers of White Water Associates and the editorial staff at CRC Press, Inc., particularly Sarah Fortener.

Editors

Ronald J. Kendall is a Wildlife Toxicologist and is currently a professor and the head of the Department of Environmental Toxicology and director of The Institute of Wildlife and Environmental Toxicology (TIWET) at Clemson University, Clemson, S.C. He is President of the Society of Environmental Toxicology and Chemistry (SETAC), has served on the Executive Committee and Board of Directors of SETAC, and is currently Editor of the Annual Review of *Environmental Toxicology and Chemistry.* His research in the field of wildlife toxicology has focused primarily on agricultural chemical and heavy metal ecotoxicology, ecological modeling, and ecological risk assessment.

Dr. Kendall continues to publish widely in the field of wildlife toxicology, to advise graduate students at both the M.S. and Ph.D. level, and has participated among committees and boards—including the Avian Effects Dialogue Group of the Conservation Foundation and the Ecological Risk Assessment Subcommittee of the Committee on Risk Assessment Methodologies of the National Academy of Sciences. He was awarded a Fulbright Fellowship to the University of Rajasthan in Jaipur, India, and currently is serving on the Expert Advisory Committee of the Canadian Network of Toxicology Centers.

He received his B.S. degree from the University of South Carolina, his M.S. from Clemson University, and a Ph.D. from Virginia Polytechnic Institute and State University.

Thomas E. Lacher, Jr. is a population and community ecologist with a specialty in tropical ecosystems. He was on the faculty of the Department of Zoology at the Universidade de Brasilia and at Huxley College of Environmental Studies, Western Washington University, before coming to Clemson University. He has also been a visiting professor at the Unversidade Federal de Minas Gerais in Brazil and a visiting scientist of the Brazilian National Research Council. Since 1989, he has been Director of the Archbold Tropical Research Center and a member of the Department of Aquaculture, Fisheries and Wildlife. He also holds an adjunct appointment in the Institute of Wildlife and Environmental Toxicology at Clemson.

Dr. Lacher has served as a member of committees and panels of the National Academy of Sciences/National Research Council, National Science Foundation, Council for International Exchange of Scholars (Fulbright), and Committee for the National Institutes of the Environment. He has published extensively on population and community ecology, tropical ecology and conservation, and wildlife toxicology. He teaches courses on the ecology and conservation of tropical wildlife.

Dr. Lacher received his B.S. in Biological Sciences and his Ph.D. in Biological Sciences, Section of Ecology, Evolution and Systematics, both from the University of Pittsburgh. His research interests focus on three areas: the population and community level effects of disturbances, such as agricultural chemicals, on wildlife populations; the structure and dynamics of communities of mammals in tropical forests and savannas; and the application of ecological theory and methods to conservation.

Contributors

M. Craig Barber, Ph.D.
Environmental Research Laboratory
U.S. Environmental Protection Agency
Athens, Georgia

Richard J. Barker, Ph.D.
Department of Statistics
Massey University
Palmerston North, New Zealand

Anne L. Barton, M.Sc.
Director, Environmental Fate
and Effects Division
Office of Pesticide Programs
U.S. Environmental Protection Agency
Washington, DC

Gary E. Belovsky, Ph.D.
Professor
Department of Fisheries and Wildlife
College of Natural Resources
Utah State University
Logan, Utah

Richard S. Bennett, Ph.D.
Environmental Research Laboratory
U.S. Environmental Protection Agency
Corvallis, Oregon

Sandra L. Bird, Ph.D.
Environmental Research Laboratory
U.S. Environmental Protection Agency
Athens, GA

Lawrence J. Blus, M.Sc.
Research Biologist
Patuxent Wildlife Research Center
Northwest Research Group
U.S. Fish and Wildlife Service
Corvallis, Oregon

Michael A. Bowers, Ph.D.
Associate Professor
Department of Environmental Sciences
and Blandy Experimental Farm
University of Virginia
Charlottesville, Virginia

Mark S. Boyce, Ph.D.
College of Natural Resources
University of Wisconsin
Stevens Point, Wisconsin

Larry W. Brewer, M.Sc.
President
EBA, Inc.
Sisters, Oregon

I. Lehr Brisbin, Jr., Ph.D.
Senior Ecologist
Savannah River Ecology Laboratory
Aiken, South Carolina

Michael Brody, Ph.D.
Staff Ecologist
Office of Policy, Planning, and Evaluation
U.S. Environmental Protection Agency
Washington, DC

Ronald K. Chesser, Ph.D.
Head, Wildlife Ecology and Toxicology Division
Savannah River Ecology Laboratory
Aiken, South Carolina

George P. Cobb, Ph.D.
Department of Environmental Toxicology
The Institute of Wildlife
and Environmental Toxicology
Clemson University
Pendleton, South Carolina

Donald L. DeAngelis, Ph.D.
Senior Scientist
Environmental Sciences Division
Oak Ridge National Laboratory
Oak Ridge, Tennessee

Richard T. Di Giulio, Ph.D.
Ecotoxicological Laboratory
School of the Environment
Duke University
Durham, North Carolina

Philip M. Dixon, Ph.D.
Savannah River Ecology Laboratory
Aiken, South Carolina

Andrew Dobson, Ph.D.
Ecology and Evolutionary Biology
Princeton University
Princeton, New Jersey

John Emlen, Ph.D.
U.S. Fish and Wildlife Service
National Fisheries Research Center
Seattle, Washington

Anne Fairbrother, D.V.M., Ph.D.
Environmental Research Laboratory
U.S. Environmental Protection Agency
Corvallis, Oregon

Ed Fite, M.Sc.
Wildlife Biologist
Office of Pesticide Programs
U.S. Environmental Protection Agency
Washington, D.C.

James A. Gagne, Ph.D.
Associate Research Fellow
Agricultural Research Division
American Cyanamid Company
Princeton, New Jersey

Evan P. Gallagher, Ph.D.
Department of Environmental Health
University of Washington
Seattle, Washington

Karen A. Garrett, M.Sc.
Department of Statistics
Colorado State University
Fort Collins, Colorado

Paul Geissler, Ph.D.
Chief, Branch of Surveys
Office of Migratory Bird Management
U.S. Fish and Wildlife Service
Laurel, Maryland

Sarah Gerould, Ph.D.
Office of Water Quality
U.S. Geological Survey
Reston, Virginia

Peter W. Greig-Smith, Ph.D.
Ministry of Agriculture, Fisheries and Food
Fisheries Laboratory
Lowestoft, Suffolk
United Kingdom

Louis J. Gross, Ph.D.
Professor of Mathematics and Ecology
Department of Mathematics
and Graduate Program in Ecology
University of Tennessee
Knoxville, Tennessee

Christian E. Grue, Ph.D.
Associate Professor and Leader
Washington Cooperative Fish
and Wildlife Research Unit
School of Fisheries
University of Washington
Seattle, Washington

Clifford Habig, Ph.D.
Jellineck, Schwartz, Connolly and Freshman, Inc.
Washington, D.C.

Thomas G. Hallam, Ph.D.
Department of Mathematics
Graduate Program in Ecology
Program in Environmental Toxicology
University of Tennessee
Knoxville, Tennessee

Gary H. Heinz, Ph.D.
Research Biologist
Environmental Contaminants Research Branch
Patuxent Wildlife Research Center
U.S. Fish and Wildlife Service
Laurel, Maryland

Elwood F. Hill, Ph.D.
Research Toxicologist
Patuxent Wildlife Research Center
U.S. Fish and Wildlife Service
Lauren, Maryland

David J. Hoffman, Ph.D.
Patuxent Wildlife Research Center
U.S. Fish and Wildlife Service
Lauren, Maryland

Michael J. Hooper, Ph.D.
Department of Environmental Toxicology
The Institute of Wildlife
and Environmental Toxicology
Clemson University
Pendleton, South Carolina

Peter Hudson, Ph.D.
Upland Research Group
Game Conservancy
Crubenmore Lodge
Newtonmore, Scotland

Roy A. Hummell, M.Sc.
URS Engineers, Inc.
Seattle, Washington

Ronald J. Kendall, Ph.D.
Director, Professor and Department Head
Department of Environmental Toxicology
The Institute of Wildlife
and Environmental Toxicology
Clemson University
Pendleton, South Carolina

Thomas E. Lacher, Jr., Ph.D.
Director
Archbold Tropical Research Center
Clemson University
Clemson, South Carolina

Ray R. Lassiter, Ph.D.
Environmental Research Laboratory
U.S. Environmental Research Agency
Athens, Georgia

Bruce G. Marcot, Ph.D.
Wildlife Ecologist
Pacific Northwest Experiment Station
U.S. Department of Agriculture Forest Service
Portland, Oregon

Nancy E. Mathews, Ph.D.
Assistant Professor
Department of Range and Wildlife Management
Texas Tech University
Lubbock, Texas

Karen McBee, Ph.D.
Associate Professor
Department of Zoology
Oklahoma State University
Stillwater, Oklahoma

Joseph S. Meyer, Ph.D.
Department of Fisheries
Humboldt State University
Arcata, California

Raymond J. O'Connor, Ph.D.
Professor
Department of Wildlife
University of Maine
Orono, Maine

H. Ronald Pulliam, Ph.D.
Institute of Ecology
University of Georgia
Athens, GA 30602

John R. Sauer, Ph.D.
Wildlife Biologist
Branch of Migratory Bird Research
Patuxent Wildlife Research Center
U.S. Fish and Wildlife Service
Laurel, Maryland

Edward W. Schafer, Jr., Ph.D.
Registration Liaison
Animal and Plant Health Inspection Service
U.S. Department of Agriculture
Denver Wildlife Research Center
Lakewood, Colorado

John R. Skalski, Ph.D.
Associate Professor, Biological Statistics
Center for Quantitative Science
School of Fisheries
University of Washington
Seattle, Washington

Norman A. Slade, Ph.D.
Museum of Natural History and
Department of Systematics and Ecology
University of Kansas
Lawrence, Kansas

Steven G. Smith, Ph.D.
Research Consultant
Center for Quantitative Science
School of Fisheries
University of Washington
Seattle, Washington

Donald W. Steffack, M.Sc.
Chief, Division of Environmental Contaminants
Ecological Services
U.S. Fish and Wildlife Service
Portland, Oregon

George A. Swanson, Ph.D.
Northern Prairie Wildlife Research Center
U.S. Fish and Wildlife Service
Jamestown, North Dakota

Michael Tome, Ph.D.
American Waterfowl and Wetlands Research Center
U.S. Fish and Wildlife Service
Arlington, Virginia

Richard E. Warner, Ph.D.
Department of Forestry
University of Illinois
Urbana, Illinois

Bill Williams, Ph.D.
ep and t
Corvallis, Oregon

Michael R. Willig, Ph.D.
Professor of Biology and Ecology Chair
Ecology Program
Department of Biological Sciences
and The Museum
Texas Tech University
Lubbock, Texas

Kevin Willis, M.Sc.
Conservation Biologist
American Association of Zoological Parks and Aquariums
Bethesda, Maryland

Vladimir M. Zakharov, Ph.D.
N. K. Koltzov Institute of Developmental Biology
Russian Academy of Sciences
Moscow, Russia

Contents

PART 9

DISCUSSION GROUPS AND SYNTHESIS RESULTS

1

Using Information Derived from Wildlife Toxicology to Model Ecological Effects of Agricultural Pesticides and Other Environmental Contaminants on Wildlife Populations

Ronald J. Kendall

ABSTRACT

Wildlife toxicology, the study of the effects of environmental contaminants on the reproduction, health, and well-being of wildlife, has greatly contributed to current initiatives in ecological risk assessment. Since the 1962 publication of Rachel Carson's *Silent Spring*, the science of wildlife toxicology has taken on new meaning. The public has a heightened concern for the environment and a demand for information on how chemicals affect our wildlife resources.

Wildlife toxicology generally focuses on birds, mammals, reptiles, and amphibians, although integration of activities in aquatic toxicology, plant toxicology, ecological modeling, and environmental chemistry facilitate its development. Prior to the 1980s, the science relied heavily on laboratory investigations, particularly data derived from LD_{50} and LC_{50} tests. The field studies conducted at this time were generally of a dimension and scale that failed to provide the necessary information to truly assess the risk of exposure of wildlife to various environmental contaminants.

During the early to mid-1980s, the science necessarily underwent rapid development with the increasing obligation of the United States Environmental Protection Agency (U.S. EPA) to strengthen the ecological risk assessment process for registration and reregistration of many of the pesticides used in the United States. Field studies that previously had little quantitative nature were developed into more comprehensive assessments of the effects of environmental chemicals on wildlife. These studies included not only mortality endpoints, but also various sublethal endpoints including reproduction, behavior, and disturbances in habitat and food quality and abundance.

Our challenge is to continually improve the ecological risk assessment process by refining our environmental exposure estimates for wildlife and by improving toxicity assessment by integrating laboratory and field research. As data from large-scale field studies of high quality assurance become increasingly available, our ability to use such information in ecological modeling will improve.

KEY WORDS

wildlife, toxicology, ecological, pesticides, modeling

INTRODUCTION

Wildlife toxicology has been reviewed and defined as the study of the effects of environmental contaminants on wildlife species as related to the animal's well-being, general health, and reproduction.[1] This material has been expanded by Hoffman et al.[2] who noted that three principal strategies for understanding xenobiotic effects on wildlife include chemical analysis, field ecology, and various control field studies. To further expand this observation, I propose that wildlife toxicology requires an interdisciplinary approach of three major emphasis areas. Initially, environmental chemistry and analytical toxicology are extremely critical to identifying the toxic substances of concern and their concentrations both in the environment and in vivo, respectively. Then, biochemical toxicology, encompassing the physiological and biochemical disturbances of contaminants and their toxicokinetics, provides information as to the "mechanism of action" and dynamics of certain toxic substances that may gain entry into the bodies of wildlife. Finally, we must consider ecotoxicology or the field effects of chemicals on both the aquatic and terrestrial environment concerning impact on wildlife and their populations. Through ecotoxicological testing methods in wildlife toxicology, we may conduct semicontrolled field studies, pen studies, and full-scale field studies to be discussed later in this chapter. Thus, we are integrating analytical chemistry, biochemical toxicology, and toxicological testing methods with field ecotoxicological assessments. From these data sets, we are able to construct information on the fate and effects of toxic chemicals so that, as through our studies in wildlife toxicology, we may be able to conduct ecological risk assessments.[3]

People may wonder what is so special about wildlife toxicology that it should be identified as a particular branch of toxicology. All toxicologists have in common a primary objective: to study the mechanisms and processes by which toxic substances produce adverse effects on biological systems. Such studies are based largely on the use of certain laboratory animals, animal-derived tissue, or cell models regardless of whether they are conducted to assess hazard to humans or other species. It is in risk assessment that a certain diversification among toxicologists becomes noticeable. Those whose primary concern is assessing chemical hazards to humans should be able to judge whether animal data are relevant to humans. They must understand the toxic substance, the comparative physiology of the test species and humans, and the potential for human exposure to the chemicals in question. Likewise, wildlife toxicologists must understand the toxic substance, the biological system, its response, and the potential for exposure.

Although it is sometimes possible in wildlife toxicology to use the actual species of interest in laboratory studies, generally data from one species must be extrapolated to another using comparative physiology. When risk assessments are conducted, wildlife toxicologists must consider not only hazards to one particular species, but those to a variety of species and their ecosystem. This requires sufficient understanding of the environmental chemistry of toxic chemicals (i.e., fate, activation, and degradation) as well as knowledge of the life cycles of wildlife, which may influence the likelihood of exposure.

In this area of risk assessment, wildlife toxicologists have one major advantage over human toxicologists: opportunities sometimes arise to conduct field studies to assess directly the impact of the chemical on certain species or group of species.[4] In most instances with humans, the effects of hazardous exposures to chemicals can be evaluated only indirectly.

The design of field experiments in wildlife toxicology requires the same combination of skills previously mentioned: environmental chemistry and analytical toxicology, biochemical toxicology, and wildlife biology and field ecotoxicology. Consequently, wildlife toxicology should be seen as a rather broad area in which scientists with different educational backgrounds and training can and should collaborate.

Wildlife Toxicology

There is widespread evidence that wildlife has suffered greatly from exposure to toxic substances. For instance, millions of waterfowl have died from lead poisoning after ingestion of spent lead shot.[5] In addition, brown pelican (*Pelecanus occidentalis*) populations declined as residual levels of the pesticide metabolite DDE (1,1-dichloro-2,2-*bis*(p-chlorophenyl)ethylene) in their eggs increased.[6] In an attempt to quantify the impacts of toxic chemicals on wildlife, scientists have used wildlife toxicology, which essentially develops and uses ecological data related to the acute and chronic toxicological information concerning the organisms being studied. One area that has been lacking considerably is the expansion of such data sets into wildlife population impact models. This lack was the basis for the development of the 1990 Ecological Modeling Symposium sponsored by the Society of Environmental Toxicology and Chemistry. Wildlife toxicology has already been defined as the study of the effects of environmental contamination on wildlife species as related to their well-being, general health, and reproduction. A state of well-being implies, for instance, that there is no significant increase in the probability of being preyed upon nor in aberrations in migratory behavior. A state of good general health means that the organism can maintain homeostasis and, therefore, survive in a variety of environmental situations. Because the reproductive process is often very sensitive to the influence of chemical contaminants in the environment, it has been accorded a high research priority.

The term *wildlife* usually pertains to vertebrate animals living in a natural undomesticated state, although there are no clear taxonomic guidelines for defining the word. Emphasis is placed in this volume on those species of wildlife, particularly birds, but also on mammals, reptiles, and amphibians, particularly those from which economic benefit may be derived. This emphasis is in line with the concept that the most often mentioned species of wildlife are those that provide benefit or detriment to society.[7] This is not to say, however, that other species are less important components of the ecosystem, and the discussion will be expanded to include many species that are relatively obscure (e.g., reptiles and amphibians) and not well studied from a toxicological perspective as reported in the literature.

History and Perspective

In the United States, wildlife experiments addressing the effects of DDT (1,1,1-trichloro-2,2-*bis*(p-chlorophenyl)ethane), in birds extend back to the early 1940s. In addition to this early DDT research, a proposal was made in the late 1950s to eradicate the imported fire ant (*Solenopsis invicta*) in the United States using heptochlor and dieldrin. This program was important for the future of pesticide/wildlife relationships.[8] The extensive wildlife mortality that resulted from the spraying program and the efforts initiated by the U.S. Fish and Wildlife Service (U.S. FWS) and others to evaluate these effects increased concern about pesticide safety. A great deal of research was undertaken to investigate chemical impact on wildlife and other organisms in the United States even before the publication of *Silent Spring*. Without question, however, as then-Secretary of the Interior Stuart Udall noted in 1963, we owe a great deal to Rachel Carson. Partly as a result of her concerns, wildlife research programs emphasizing the effects of chemicals on wildlife were developed at various locations in the United States, and many continue today. Recent statistics provided by the U.S. FWS indicate that more than half of Americans now participate in some form of wildlife-related recreation including feeding wild birds, preserving wildlife, fishing, photographing wildlife, hunting, and maintaining natural areas. Therefore, not only is it important to maintain ecosystem integrity by protecting wildlife populations, it is critical to understand that wildlife represents a substantial national and international economic interest and that toxic chemicals are a threat to this valuable resource. Wildlife species are also vital because their toxicological responses provide an indication of potential hazards to human health associated with chemical contaminants such as agricultural pesticides and industrial waste.

Many early pesticides were derived from naturally occurring materials in plant products and were neither very toxic nor environmentally persistent. These kinds of chemicals generally dominated American agriculture before World War II; however, after the war the variety and production of organic insecticides increased to significant levels. DDT is an excellent example of an organic insecticide that held great promise for the control of insects such as malaria-carrying mosquitoes, but this poison also was able to persist in the environment and enter the food chains of nontarget organisms, including fish and wildlife. Despite the expression of concern for DDT as early as the 1940s, its use continued even with the associated decline by the early 1950s of American robin (*Turdus migratorius*) populations which was linked to spraying for Dutch elm disease. It became evident, with research underway in the United States and evidence of the impact of DDT on raptors in Europe, that the American bald eagle (*Haliaeetus leucocephalus*) and osprey (*Pandion haliaetus*) were at risk.[9] DDT and a metabolite, DDE, impaired calcium metabolism in many avian systems, and eggs broke easily because of calcium-deficient shells, thereby adversely affecting reproduction.[10] The biochemical toxicology of well-studied compounds such as DDT is now more thoroughly characterized; however, even now we do not understand completely all the mechanisms associated with the phenomenon known as "raptor pesticide syndrome."

WILDLIFE AND PESTICIDE TESTING

With the establishment of the U.S. EPA in 1970 and its charge for protection of the environment, including our fish and wildlife resources, pesticides have come under much more scrutiny with regard to registration before their introduction into the marketplace. Under the Federal Insecticide, Fungicide, and Rodenticide Control Act (FIFRA) the U.S. EPA has the regulatory mandate of assessing the effects of pesticides on birds and other wildlife before these products are registered (Table 1).

Pesticide registration in the United States now requires acute oral toxicity tests on two species of birds, the mallard (*Anas platyrhynchos*) and the northern bobwhite (*Colinus virginianus*), as well as data provided on the chemistry of the product and the generation of dose-response curves from toxicity tests. Single-dose acute tests are required for adult bobwhite and mallard in which a single oral dose LD_{50} dose-response curve can be generated. In addition, a subacute feeding test in which a 5-day exposure to the pesticide in food of juvenile bobwhite and mallard is necessary to generate an LC_{50} determination (concentration resulting in lethality to 50% of the test group).[11]

Table 1. Tiered Framework for Pesticide Toxicity Testing in Birds under FIFRA

Tier I
Laboratory testing (acute)
LD_{50}—Adult mallard and northern bobwhite (single oral dose)
LC_{50}—2-week-old mallard and northern bobwhite (concentration in feed)
Songbirds may be included for comparative toxicology

Tier II
Laboratory testing (subchronic—chronic)
Reproductive toxicity where pairs of northern bobwhite and mallards are exposed in their food during breeding period; available data includes egg production, embryo survival, hatchability, and survival of hatched young to 2 weeks of age

Tier III
Field-pen testing
Field testing (acute–subacute)
Field pens are used to confirm wildlife exposure to pesticides in certain treatment areas; data acquired include mortality

Tier IV
Field-full-scale testing (acute–subacute–subchronic–chronic)
Field testing using Level I or II designs; Level I design or "screening study" generally focuses on assessment of pesticide residues plus mortality and bird use data of agricultural fields; Level II design seeks to quantitate responses of wildlife populations exposed to pesticides through radiotelemetry, nest box techniques, blood/fecal sampling, and extensive residue chemistry

These subacute testing methodologies allow evaluation of the feeding behavior and other responses, serving as indicators of the toxicity of a dietary exposure of the pesticide to these birds.

Protocols are also in place for examining the effects of pesticides on avian reproduction. Generally speaking, the exposure of adult breeding birds consists of pretreatment, treatment, and posttreatment evaluation phases that may extend for weeks or months in terms of experimental design. These costly investigations generate data on reproductive toxicity but also provide valuable sublethal effects data, which can be included in risk assessments.

WILDLIFE TOXICOLOGY AND FIELD TESTING

During the 1980s, wildlife toxicology saw a significant increase in the number of field studies conducted to rebut the presumption of risk of pesticides to wildlife through the U.S. EPA pesticide registration process. Field studies conducted in the early to mid-1980s on the effects of pesticides in birds generally were reported to include environmental chemistry data and response of birds (including mortality) to applied pesticides.[12] U.S. EPA has now evolved a set of criteria upon which current pesticide field trials can be conducted as either Level I or Level II studies.[13]

A Level I field study generally involves mortality as a key endpoint for a variety of sites treated with pesticides and assessed for response of birds and other wildlife possibly exposed to the chemical. Some residue analyses as well as behavioral observations may be employed. These Level I field studies or "screening trials" are more of a qualitative assessment of the potential hazard of the pesticide to birds that may utilize habitats receiving pesticide spray. Data acquired from Level I field trials make possible an assessment of the need for a Level II field study. A Level II field study seeks to quantify the response of wildlife populations to chemicals such as agricultural pesticides. This quantification can take the form of extensive chemical analyses, radiotelemetric monitoring techniques,[12] and starling nest box studies.[14] Conducting an experiment generally includes a number of field sites and a variety of observations that allow quantification of the response of wildlife to chemicals in terms of their reproduction, health, and well-being.[15] A Level II field study conducted in Iowa to assess the response of wildlife to the corn rootworm insecticide COUNTER® 15-G has been submitted to the U.S. EPA to meet the requirements of FIFRA guidelines. This is one of the only studies known to date that has been submitted to satisfy Level II requirements. Data generated from Level I and particularly Level II field studies will provide extremely useful information for the development of ecological models related to the effects of pesticides in avian and other wildlife species.

THE FUTURE OF WILDLIFE TOXICOLOGY

The assessment of the effects of environmental chemicals, particularly pesticides, on avian and other wildlife populations is a critical concern in wildlife toxicology. In the future, it will not be possible to test every pesticide for every geographic condition, crop, and wildlife species. The Iowa research program on COUNTER® 15-G alone represented approximately 6 years of effort, integrating both laboratory and field research to achieve a full-scale Level II study in just one geographic region on one crop.[15] The use of ecological modeling techniques will be extremely critical as we seek to better manage our environment, to protect valuable wildlife resources, and to adopt agricultural control techniques that allow cost-effective production of important food and fiber crops.

To generate the necessary data for ecological modeling and ecological risk assessment (Table 2), it is necessary to improve our design and conduct of field studies where actual response under natural conditions can be measured in wildlife exposed to pesticides. Included in the conduct of these field studies is an improved ability to assess the environmental chemistry of the compounds in question and their metabolites. The availability of such data allows for the improvement of our ecological risk assessments

Table 2. Future Needs in Wildlife Toxicology to Improve the Ecological Risk Assessment Process

Computer/simulation modeling of wildlife population response to agricultural pesticides
Improved design/conduct of field studies to acquire data for model development, application, and verification
More information on comparative toxicity (e.g., with songbirds) and toxicokinetics to allow broader extrapolation of models
Bird use/exposure assessment in pesticide treated agricultural areas to balance out risk (i.e., exposure times toxicity)
Sublethal sampling techniques, such as biomarkers, to assess exposure/effects that can be tied to health and survival of birds (immunotoxicity research holds much promise); these improve exposure assessments in models
More definitive data on actual nesting success vs pesticide-induced loss of successful nests of songbirds and other avian wildlife exposed to pesticides; these data help verify predictions of models as to reproductive success of birds exposed to pesticides

based on actual residue data obtained from the habitat occupied by wildlife under pesticide exposure conditions.

It is very important that we improve our ability to monitor and evaluate wildlife use of agricultural, pesticide-treated areas to determine wildlife exposure. Of course, our own investigations have proved that wildlife, particularly birds, utilize agricultural fields only for certain percentages of their overall lifetime. For instance, we have found that northern bobwhite utilize open corn fields to a lesser extent than do ring-necked pheasants (*Phasianus colchicus*). For this reason, the ring-necked pheasant may be a better model for the study of pesticide exposure in open corn fields, as northern bobwhite generally use only the edges. Because granular pesticides are now often incorporated into the soil in corn agroecosystems, birds that venture into the field and utilize open corn fields for feeding and other activities might be at greatest risk because of their potential exposure to such granular products. Again, species such as northern bobwhite may incur less risk from the granular products as compared to aerially applied pesticides, which may drift into hedgerows and expose a wide variety of songbirds as well as northern bobwhite and other game birds in hedgerow habitats.

An understanding of exposure increases our ability to conduct accurate and quantifiable ecological risk assessments. It is well understood that the risk of wildlife exposed to pesticides is a function of both the toxicity of the active ingredient and/or the formulated product and the exposure potential of the species concerned. For this reason, we must implement better techniques to assess exposure and conduct such analyses in a sublethal way so as to minimize wildlife mortality as a function of a pesticide field trial.[16] Sublethal techniques now utilized by our laboratory include blood sampling and fecal sampling for the assessment of pesticide-induced aberrations in the blood and for the determination of pesticide metabolites in the feces, respectively. Detecting reduced levels of cholinesterase (ChE) in the blood that can result from exposure to organophosphate (OP) pesticides such as methyl parathion[12] is useful for monitoring such exposure in wildlife, especially birds. The use of ChE reactivation techniques allows for individual birds to serve as their own "control" because OP-inhibited enzyme can be reactivated with various biochemical methodologies to allow the determination of exposure to ChE-inhibiting compounds.[17]

Knowledge of specific pesticide formulations and vehicles is important in evaluating hazards of agricultural chemicals to wildlife. The field hazard of granular pesticides appears as much dependent on the species and the feeding behavior of birds that may inhabit a treated area than on the application rate or toxicity of the compound. Aspects of contaminant interactions are little understood presently, although we have documented that exposure to a variety of applied pesticides occurs often in wildlife, particularly in birds using agroecosystems for feeding or breeding habitat. Possibilities exist to address the additivity of effects as well as synergism and antagonism. Of 13 pairs of chemicals fed to northern bobwhite and pheasants, 2 showed moderate synergism, whereas added effects were noted for others.[18] In addition to interactions between pesticides, which can result in additive, antagonistic, or synergistic effects, we must also consider other interacting parameters, including cold and heat stress, nutritional stress, reproductive stress, and other situations that may make the animal more prone to responding to the toxicity of a certain product.

Much more needs to be learned about the interaction of stress variables with pesticides and other toxicants. For instance, the toxicity of lead has been shown to result in increased mortality and higher blood and liver lead concentrations in ringed turtle doves (*Streptopelia risoria*) exposed to 6°C compared with birds maintained at 21°C.[19] Parathion toxicity doubled in quail chronically exposed to either heat or cold.[20,21] Methyl parathion has been shown to cause significant mortality in American kestrels

(*Falco sparverius*) at -5°C, but at 22°C comparable doses caused only sublethal effects.[22]

Advances in biochemical toxicology and toxicokinetics are important in assessing the "mechanism of action" of toxic substances, particularly pesticides, in wildlife as well as understanding their uptake in tissue, distribution, and elimination. Experiments on pesticide residue kinetics with DDT and other persistent organochlorines have shown that levels of DDT in the brains of birds could be diagnostically related to lethality.[23] The kinetics of both OP and carbamate insecticides, which are now being widely used, is little understood in wildlife. This lack of toxicokinetic data makes it very difficult to diagnose death from residues present or to design appropriate sampling schemes from field collected specimens and have a data set adequate for interpretation.

Bioaccumulation of pesticides in wildlife is related to and a function of metabolism via mixed function oxygenase (MFO) ability. Generally, bioaccumulation tends to be inversely related to the rate of xenobiotic metabolism. The bioaccumulation of pesticides in wildlife food chains is a critical component in understanding exposure. In addition, determining the actual bioavailability of pesticides to wildlife assists us in conducting appropriate and more accurate ecological risk assessments of the hazards pesticides pose to exposed wildlife.

The use of cell, tissue culture, and other in vitro techniques should prove useful for improving the database available on the "mechanism of action" and metabolism of pesticide products of interest. From a comparative toxicological viewpoint, it is critical that our database on the metabolism of pesticide products to their metabolites be improved so as to assess potential toxification and detoxification mechanisms. Adapting cell and tissue culture techniques to those for wildlife is valuable in predicting such outcomes and in identifying the most sensitive species of wildlife as this relates to the metabolism of the product.

The identification of sentinel species in wildlife still presents a great problem from the perspective of both toxicology and exposure. In concept, the use of sentinel species and sentinel bioassays[24] should provide signals of impending harm to wildlife in general. In this regard, we must continue to improve our ability to identify appropriate sentinel species which, by becoming our "early warning systems," may identify a threat or hazard from pesticide exposure. With the use of various biomarkers of exposure, including enzyme assays,[25] we can determine the sublethal response of birds and other wildlife to pesticide exposure. Other promising biomarker techniques include immunological quantification of P-450 isoenzymes, which can be adapted to avian studies. Oxyradical-mediated responses for biomonitoring have been used in environmental toxicology[26] because there are probably more xenobiotics occurring in contaminant mixtures that are "redoxed active" than in mixtures that are "MFO inducers."[27] Other promising biomarkers for wildlife include measurements of genotoxicity of chemicals to wildlife.

ECOLOGICAL RISK ASSESSMENT

The U.S. EPA Office of Pesticide Programs assesses risk to ecological resources using an ecotoxicological approach: (1) determination of hazard through laboratory toxicity bioassays, (2) determination of exposure either through monitoring data or predictions from models, and (3) comparison of exposure to hazard using the quotient (or ratio) method. In the quotient method, the exposure value is directly compared with a toxicity endpoint such as concentration in water to an LC_{50} value (10 ppm/100 ppm).

The closer the quotient is to 1 (or greater), the higher the probability that an adverse effect will occur.[3]

Interpreting this adverse effect (e.g., the likelihood that what is observed in the laboratory will actually occur in the field) brings up many questions.[3] The Office of Pesticide Programs follows four steps in a preliminary assessment of ecological risk: (1) review and evaluate hazard data to identify the nature of the hazards, (2) identify and evaluate the observed quantitative relationship between dose and response, (3) identify the conditions of exposure (such as intensity, frequency, and duration of exposure), and (4) combine the information on dose-response effects with that on exposure to estimate the probability that nontarget populations will be adversely affected by actual use of the pesticide.[3] Refer to Table 3, which can be reviewed for further clarification of ecological risk assessment at the U.S. EPA.

The Office of Pesticide Programs recognizes the numerous weaknesses inherent in the quotient method for assessing risk. For example, it does not adequately account for effects of incremental dosages nor does it compensate for differences between laboratory tests and field populations. The latter is a very significant weakness, and ecological modeling should provide much more insight as to field verification of laboratory-predicted results. In addition, the quotient method cannot be used for assessing indirect effects of toxicants, such as through food chain interactions. It also lacks the ability to quantify uncertainties, and does not adequately account for ecosystem-level effects including predator/prey relationships, community metabolism, and population structural shifts. From this process, sound ecological modeling clearly could add a great deal to improving ecological risk assessment at the EPA.[28] Without resolving these problems through ecological modeling or other means, these weaknesses will persist and render ecological risk assessments modestly useful in assessing the risk of exposure of birds to pesticides.

Table 3. Ecotoxicological Assessment Criteria for Pesticides[3]

Presumption of Minimum Hazard	Hazard that May Be Mitigated by Restricted Use	Level of Concern (LOC)
	Granular Formulations	
Mammals		
$EEC^a < 1/5\ LC_{50}$	$1/5\ LC_{50} \leq EEC < 1/2\ LC_{50}$	$EEC \geq 1/2\ LC_{50}$
$mg/kg/day < 1/5\ LD_{50}$	$1/5\ LD_{50} \leq mg/kg/day < 1/2\ LD_{50}$	$mg/kg/day \geq 1/2\ LD_{50}$
Birds		
$EEC < 1/5\ LC_{50}$	$1/5\ LC_{50} \leq EEC < LC_{50}$	$EEC \geq LC_{50}$
$mg/kg/day < 1/5\ LD_{50}$	$1/5\ LD_{50} \leq mg/kg/day < 1/2\ LD_{50}$	$mg/kg/day \geq 1/2\ LD_{50}$
Aquatic Organisms		
$EEC < 1/10\ LC_{50}$	$1/10\ LC_{50} \leq EEC < 1/2\ LC_{50}$	$EEC \geq 1/2\ LC_{50}$
	Chronic Toxicity	
EEC < Chronic No effect level	N/A	EEC ≥ effect level (including reproductive)

Note: Current EPA Guidelines provided by Edward Fite, Office of Pesticide Programs, Ecological Effects Branch, EPA Headquarters, Washington, DC.

[a]Estimated environmental concentration. This is typically calculated using a series of simple nomographs to complex exposure models.

THE NEED FOR WILDLIFE TOXICOLOGY

In the last decade, we have seen wildlife toxicology evolve from a relatively undeveloped science to one undergoing rapid growth and receiving intense pressure from a public demanding answers about potential chemical hazards to the environment, particularly to wildlife. The field is still in its infancy, and much work and international collaboration are needed before we are able to evaluate fully how chemicals affect the wildlife resources of the world.

REFERENCES

1. Kendall, R. J., Wildlife toxicology, *Environ. Sci. Technol.*, 16(8), 448A–453A, 1982.
2. Hoffman, D. J., B. A. Rattner and R. J. Hall, Wildlife toxicology, *Environ. Sci. Technol.*, 24, 276–283, 1990.
3. Bascietto, J., D. Hinckley, J. Plafkin and M. Slimak, Ecotoxicity and ecological risk assessment, *Environ. Sci. Technol.*, 24, 10–15, 1990.
4. Balk, F., and J. H. Koeman, Future hazards from pesticide use, *Environmentalist*, 4, 1–100, 1984.
5. Bellrose, F. C., Lead poisoning as a mortality factor in waterfowl populations, *Ill. Nat. Hist. Surv. Bull.*, 27, 235–288, 1959.
6. Blus, L. J., A. A. Belisle and R. M. Prouty, Relations of the brown pelican to certain environmental pollutants, *Pestic. Monitor. J.* 7, 181–194, 1974.
7. Giles, R. H., Jr., *Wildlife Management*, W.H. Freeman, San Francisco, 1978.
8. Hall, R. J., Impact of pesticides on bird populations, in *Silent Spring Revisited,* G. J. Marco, R. M. Hollingsworth and W. Durham, Eds., American Chemical Society, Washington, DC, 1987, pp. 85–111.
9. Carson, R., *Silent Spring*, Houghton Mifflin, Boston, 1962.
10. Ratcliffe, D. A., Decrease in eggshell weight in certain birds of prey, *Nature*, 215, 208–210, 1967.
11. Hudson, R. H., R. K. Tucker and M. A. Haegele, *Handbook of Toxicity of Pesticides to Wildlife*, Resource Publ. No. 153, U.S. Fish and Wildlife Service, Washington, DC, 1984.
12. Brewer, L. W., R. J. Kendall, C. J. Driver, C. Zenier and T. E. Lacher, Jr., Effects of methyl parathion on ducks and duck broods, *Environ. Toxicol. Chem.*, 8, 375–379, 1988.
13. Fite, E. C., L. W. Turner, N. J. Cook and C. Stunkard, Guidance Document for Conducting Terrestrial Field Studies, U.S. EPA, Washington, DC, 1988.
14. Kendall, R. J., L. W. Brewer, T. E. Lacher, B. T. Marden and M. L. Whitten, *The Use of Starling Nest Boxes for Field Reproductive Studies: Provisional Guidance Document and Technical Support Document*, EPA/600/8-89/056, NTIS No. PB89 195 028/AS, National Technical Information Services, Springfield, VA, 1989.
15. Tank, S. L., L. W. Brewer, G. P. Cobb, M. J. Hooper, R. A. Hummel, R. A Craft, J. A. Buck, W. McKenzie, C. M. Bens and R. J. Kendall, A Third Year Investigation of the Response of Selected Wildlife Populations to Planting Time Application of COUNTER Systemic Insecticide-Nematicide in an Iowa Corn Agroecosystem, Final report submitted to American Cyanamid Company, March 1990, and to Ecological Effects Branch, Office of Pesticide Programs, U.S. EPA, Washington, DC.
16. Cobb, G. P., and M. J. Hooper, Non-lethal wildlife monitoring to determine exposure to xenobiotics and resulting impacts. In *Wildlife Toxicology and Population Modeling: Integrated Studies of Agroecosystems*, R. J. Kendall and T. E. Lacher, Jr., Eds., Lewis Publishers, Chelsea, MI, 1993.
17. Hooper, M. J., P. J. Detrich, C. P. Weisskopf and B. W. Wilson, Organophosphorus insecticide exposure in hawks inhabiting orchards during winter dormant-spraying, *Bull. Environ. Contam. Toxicol.*, 42, 651–659, 1989.
18. Kreitzer, J. F., and J. W. Spann, Tests of pesticidal synergism with young pheasants and Japanese quail, *Bull. Environ. Contam. Toxicol.*, 9, 250–256, 1973.

19. Kendall, R. J., and P. F. Scanlon, The toxicology of lead shot ingestion in ringed turtle doves under conditions of cold exposure, *J. Environ. Pathol. Toxicol. Oncol.*, 5-4/5, 183-192, 1984.
20. Rattner, B. A., L. Sileo and C. G. Scanes, Hormonal responses and tolerance to cold in female quail following parathion ingestion, *Pestic. Biochem. Physiol.* 18, 132-138, 1982.
21. Rattner, B. A., J. M. Becher and T. Nakatsugawa, Enhancement of parathion toxicity to quail by heat and cold exposure, *Pestic. Biochem. Physiol.*, 27, 330-339, 1987.
22. Rattner, B. A., and J. C. Franson, Methyl parathion and fenvalerate toxicity in American Kestrels: acute physiological responses and effects of cold, *Can. J. Physiol. Pharmacol.*, 62, 787-792, 1984.
23. Stickel, L. F., Pesticide residues in birds and mammals, in *Environmental Pollution in Pesticides*, C. A. Edwards, Ed., Plenum Press, London, 1973, pp. 254-312.
24. Lower, W. R., and R. J. Kendall, Sentinel species and sentinel bioassays, in *Use of Biomarkers for Assessment of Environmental Contaminant Exposure*, L. Shugart and J. McCarthy, Eds., CRC Press, Boca Raton, FL, 1990, pp. 309-331.
25. Huggett, R. J., R. A Kimerle, P. M. Mehrle, Jr., and H. L. Bergman, *Biomarkers, Biochemical, and Histological Markers of Anthropogenic Stress*, Lewis Publishers, Inc., Chelsea, MI, 1992.
26. Di Giulio, R. T, Oxidant-mediated responses as indices of environmental contamination, ASTM, Spec. Tech. Publication (Annual Symposium on Aquatic Toxicology and Risk Assessment, April 1990, San Francisco, CA), 1991.
27. Di Giulio, R. T., P. C. Washburn, R. J. Wenning, G. W. Winston and C. S. Jewell, Biochemical responses in aquatic animals: determinates of oxidative stress, *Environ. Toxicol. Chem.*, 8, 1103-1123, 1989.
28. Tipton, A. R., R. J. Kendall, J. F. Coyle and P. F. Scanlon, A model of the impact of methyl parathion spraying on a quail population, *Bull. Environ. Contam. Toxicol.*, 25, 586-593, 1980.

2

Incorporating Concepts from Population and Behavioral Ecology into Models of Exposure to Toxins and Risk Assessment

H. Ronald Pulliam

ABSTRACT

A number of recent advances in population and behavioral ecology are relevant to attempts to predict the exposure of animals to toxic chemicals and the demographic consequences of such exposure. Recent experiments with birds, fish, and other animals have demonstrated that mobile organisms choose what to consume and where to feed based on the energetic and nutrient content of prey organisms and on the risk of mortality associated with exposure to predation. It is not surprising that animals make foraging decisions based on the reproductive and mortality costs associated with these factors, but ignore the mortality hazard due to toxins with which they have not evolved. Models developed by behavioral ecologists to predict habitat use and diet can be used to predict the rates of consumption of toxic substances if the concentrations of these substances in prey are known. The population consequences of exposure to toxins can only be assessed in the context of the habitat distribution of organisms and their exposure to toxins in each habitat type. A mobile population of organisms may occupy a variety of habitats in a local landscape; in the past few years, population ecologists have paid considerable attention to habitat distribution and habitat-specific demography. In some instances, small but highly productive source habitats may export surplus individuals and maintain large populations in nearby sink habitats, where within-habitat reproduction does not keep pace with mortality. In such cases, predicting the overall population dynamics requires a knowledge of habitat-specific demography as well as immigration and emigration rates. A case example is presented illustrating how the population consequences of exposure to toxins depend on landscape geometry, habitat-specific demographic parameters, and the dispersal behavior of the organisms. Finally, means to determine the sensitivity of population size and extinction probability to variations in reproductive, survival, and dispersal traits of mobile organisms in heterogeneous landscapes are discussed.

KEY WORDS

behavioral ecology, population dynamics, habitat-specific demography, exposure to contaminants, heterogeneous landscapes

INTRODUCTION

Although many well-documented cases directly link the use of pesticides and other environmental contaminants to vertebrate mortality, little is known about the population-level consequences of such exposure. Depending on such factors as the timing of exposure events, the behavior of individual organisms, life history characteristics of exposed populations, and the nature of density-dependent mortality compensation, substantial mortality may or may not have a substantial impact on the size of the reproducing population and on overall population viability. This chapter briefly outlines an approach to using models from behavioral and population ecology to predict the magnitude of animals' exposure to environmental contaminants and to predict the population-level consequences of such exposure.

There are several ways that ecologists and toxicologists can work together to solve ecotoxicological problems. Figure 1 presents a hierarchical structure of models and studies at different levels of organization. The first step is characterizing the fate and transport of toxic materials in the environment. The output of these studies is a prediction or description of where toxic materials occur in the environment. The second step includes models and field studies of target species concerning habitat use and foraging behavior of individual organisms. Combined with the output of fate and transport studies, these behavioral studies and models at the level of individual organisms describe and predict the exposure of target species to toxic substances. Behavioral ecologists can contribute substantially to understanding exposure of species to toxic substances by specifying where target species feed and what they consume.

A better understanding of the effects of toxins on wildlife populations can be obtained by combining studies of where animals feed and what they ingest with models and studies of the toxicological effects of exposure to describe and predict the magnitude of toxin-induced mortality. The population consequences of such mortality can only be assessed in the context of demographic models and field population studies. In this general scheme, ecologists contribute to ecotoxicological studies at three levels: fate and transport models and studies (e.g., food web analysis), behavioral studies of individuals to improve understanding of exposure, and demographic studies to help determine the population consequences of exposure and toxin-induced mortality. Because ecological principles and approaches have long been incorporated

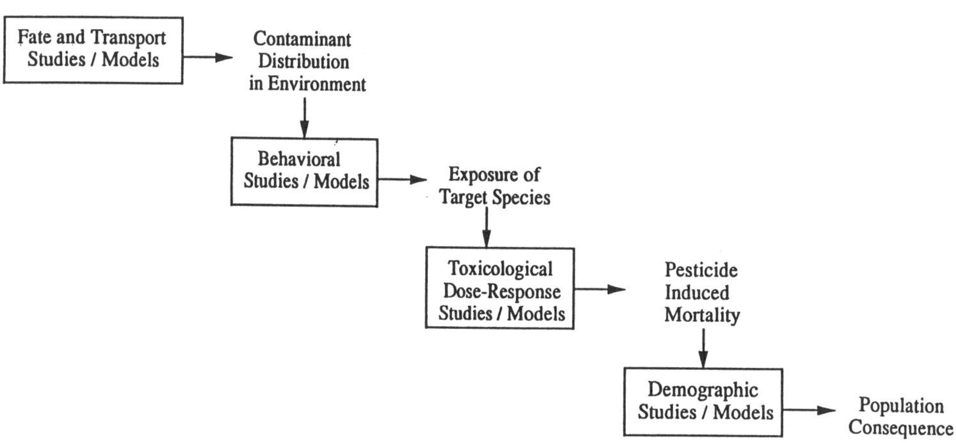

FIGURE 1. Hierarchical view of how ecologists and toxicologists can work together to solve ecotoxicological problems (see text for explanation).

into many fate and transport studies,[1] the author's comments are focused on what behavioral ecologists, studying foraging behavior and habitat selection, and population ecologists, studying population demography and dynamics, can contribute to ecotoxicology.

SOME RELEVANT CONCEPTS FROM BEHAVIORAL ECOLOGY

An important area in behavioral ecology is the study of how organisms acquire and allocate resources.[2] Much of the motivation of ecologists who study behavior is to understand better the mechanisms of resource acquisition and use in order to provide a better basis for understanding population dynamics and patterns of species diversity. Behavioral ecology provides a theoretical framework for predicting where animals will feed and what they will consume, of obvious relevance to ecological toxicology in predicting exposure to and ingestion of toxic materials.

Diet Models

One of the primary areas of concern to behavioral ecologists is predicting which foods an animal will eat among the many foods available. This problem is usually divided into two parts: (1) predicting what foods a mobile forager will encounter and (2) predicting which of those food items encountered will be pursued and eaten. Many simpler models ignore the first step by assuming that potential prey items are encountered in proportion to their relative abundance in the environment. Though this assumption often is not justified, it does sometimes provide a useful first step to predicting diets.

Most models of diet selection assume that an animal will choose the diet that maximizes its overall rate of intake.[3] The intake rate is usually measured as the energy (or caloric) intake per unit time or, occasionally, the rate of consumption of some essential nutrient. The procedure is to calculate the intake rates that could be achieved considering all possible diets and to find the combination of prey items accepted and rejected that maximizes intake rate. The information used in these models consists of, at a minimum, the encounter rate with each type of potential prey, its energetic or nutritional content, and the handling time of each prey type (time required to pursue and consume it).

The basic model of energy maximization ranks prey according to the ratio of energy content to handling time for each type of prey. The models predict that the highest ranked prey will always be pursued when encountered and that lower ranked prey will be consumed only if the overall rate of intake achieved by including them in the diet is higher than that achieved by excluding them. If the highest ranked prey is sufficiently abundant, the rate of energy intake during the time required to find and consume another of the highest ranked prey is greater than the rate of intake that can be achieved consuming a prey item of lower rank. Accordingly, if the highest ranked prey is not as abundant, prey of lower rank are added to the diet in rank order until a diet that maximizes intake rate is achieved.

Example: Prey Size Selection by Bluegill Sunfish in the Laboratory

Many investigators have sought to test the energy maximization hypothesis in the laboratory. For example, Werner and Hall[4] studied bluegill sunfish (*Lepomis macrochirus*) foraging on three size classes of *Daphnia magna* in laboratory aquaria. They estimated encounter rates, handling times, and biomass (as a surrogate for energy content) for prey of each size class. Their experiments consisted of manipulating prey densities and using a simple intake maximization model to predict which sizes of prey should be consumed at each experimental density of prey.

The results of the Werner and Hall experiments are in qualitative agreement with the predictions of the energy maximization model. At low prey densities, prey of all sizes were eaten in the same proportions as they were encountered. As prey density was increased, the smaller sizes of *Daphnia* were sequentially dropped from the diet, as predicted by the model. As with other tests of the model, some deviations exist between the exact proportions predicted and consumed; nonetheless, the qualitative predictions of the model were confirmed.

Example: Diets of Chipping Sparrows in the Field

Pulliam[5] examined the seed diets of chipping sparrows (*Spizella passerina*) feeding in a desert grassland in Arizona. Approximately 60 species of seeds were available in the soil, but only about 20 species were consumed by the sparrows. For all of the seed species common in either the soil or the diet, extensive laboratory tests were performed to estimate energy contents and handling times of the seeds. Soil samples were taken in the field to estimate the abundance of seeds in the soil, and it was assumed that encounter rates were proportional to seed abundances. The encounter rate, handling time, and energy content parameters were used to predict the diets that maximized intake rate. The actual diets were estimated by inducing regurgitation with an emetic administered to birds caught in the field.

Chipping sparrows require approximately 15 kcal/day of ingested food in mid-winter. A typical seed contains less than 0.8 cal, requiring the birds to consume on the order of 20,000 seeds per day just to meet energetic requirements. Given the number of hours in the day, this means that a chipping sparrow must find and consume a seed every 1 or 2 seconds during a mid-winter day of 10 daylight hours. Thus, the chipping sparrow seems to be a good candidate for a species that might feed so as to maximize its rate of intake while foraging.

Samples of the seeds available in the soil and those being consumed were taken in both January and March. On both dates the sparrows consumed only the seeds with higher ratios of energy content to handling time as predicted by the theory. However, they did not consume seeds in proportion to their availability as predicted, probably because the encounter rates were not proportional to relative seed abundance as assumed. Seed abundance in the soil declined by about 35% between the January and the March samples. As expected by the theory, the sparrows expanded their diets in March to include several species of seeds of lower rank that were not consumed in January. The overall results of this study are similar to the results of other field tests of the energy maximization model in that the theory fails to predict the exact proportions of prey items in the diet, but the qualitative features of the model predictions (for example, eating higher ranked prey and expanding the diet as prey abundance declines) were confirmed.

Laboratory tests of the energy maximization model have generally supported the qualitative predictions of the model; however, as in the previous example, the quantitative predictions of the simple versions of the model often do not accurately correspond to dietary choices in the field. What does seem to be a very general result, under both laboratory and field conditions, is that animals consume preferred prey items when these are abundant and expand their diets to include lower ranked (less preferred) prey when the preferred ones are uncommon. Even in cases for which the energy maximization model is inapplicable, this more robust general result can be useful.

If, for example, the less preferred prey items have higher concentrations of toxic pesticide residues, one can expect that the ingestion of highly toxic prey will increase over time if the more preferred prey items are depleted through consumption. In such a case, the ingestion of toxic prey might be reduced by artificial supplies of highly preferred prey items.

Microhabitat Selection

Much of the foraging theory has been built on the assumption that animals maximize their rates of food intake. According to this hypothesis, shifts from one habitat to another occur whenever the prey abundances change in such a manner as to reverse the relative intake rates associated with those habitats. Variations on this consider differences in nutritional content of prey in various habitats[6] and cognitive factors such as sampling prey availability and learning.

A large body of literature demonstrates that mobile organisms often shift their use of habitats in response to changes in both prey abundance and predation hazard. Gilliam and Fraser[7] argue that under a variety of circumstances, animals shift habitats in order to minimize the ratio of mortality rate (u) in a habitat to the feeding rate (f) experienced there. This hypothesis can be stated as minimizing the ratio of u to f (or u/f) or maximizing the ratio of f to u (or f/u). In the latter form, the hypothesis states that the amount of food received per unit of exposure to predation is maximized. If predation hazard does not vary among habitats, this hypothesis reduces to the familiar prediction of feeding in whichever habitat maximizes intake rate.

Example: Bluegill Sunfish Microhabitat Selection

Werner et al.[8] employed the optimal foraging theory to predict diets and microhabitat selection of bluegill sunfish in experimental ponds in Michigan. Three microhabitats were available in the ponds: (1) open water, (2) bare sediments, and (3) vegetation. Most (>90%) of the prey organisms were available only in one of the three microhabitats; therefore, by identifying prey from stomach samples, Werner et al. were able to estimate what fraction of all feeding was done in each microhabitat. Prey abundances were measured in each microhabitat and encounter rates, handling times, and energy content of prey were estimated from laboratory experiments.

The observed diets were very similar to the predicted diets in open water and vegetation microhabitats, except that diets were slightly broader than predicted. Two species of *Daphnia* were common prey in open water habitats and, as the season progressed and the abundance of *Daphnia* declined, the sunfish became less size-selective of *Daphnia*, as predicted by the model. The foraging model was less successful at predicting the actual diets of sunfish feeding in sediment microhabitats, apparently due to failure of the model to account for the burrowing behavior of some prey, which made them less available to the fish.

Werner et al. predicted that the sunfish would concentrate their foraging activity on the habitat yielding the highest intake rate. Furthermore, because the intake rate associated with each habitat changes as the prey abundances change seasonally, they expected the fish to shift their habitat use as the season progressed so as always to maximize intake rates. They measured bluegill energy intake rates in each habitat over the 3-month period of the study for small (35 mm), medium (53 mm), and large (73 mm) bluegills. The intake rates in vegetation never exceeded those achieved in either of the other two microhabitats; accordingly, as predicted, the bluegills never concentrated their foraging in the vegetation. The observations in the other two microhabitats also support the model predictions in establishing a close correspondence between where the fish were foraging and where the energy intake rate was highest. The bluegills shifted microhabitat use so as to maximize intake rates, as predicted.

Example: Creek Chubs Feeding Under Predation Hazard

Gilliam and Fraser[7] tested the "minimize u/f" hypothesis by allowing juvenile creek chubs (*Semotilus atromaculatus*) to choose foraging microhabitats in an experimental field stream. They manipulated feeding rates (f) by controlling the densities of tubificid worms and manipulated predation hazard (u) by controlling the number of cannibalistic adult chubs in each microsite. The adult chubs were too large to pass through the barriers between microsites, but the juveniles could move freely among the available microsites. The juvenile chubs could choose to remain in a site with no food and no predators or to feed in sites that had different levels of food and different numbers of predators. The experiments demonstrated that for each site the chubs could assess both the feeding rate that could be achieved and the predation hazard. The amount of time spent in each microsite corresponded closely with the predictions of the "minimize u/f" hypothesis. Furthermore, Gilliam and Fraser were able to predict the level of food enhancement necessary to induce the chubs to shift from one site into a more hazardous site with a higher feeding rate.

The results of the Gilliam and Fraser experiment are relevant to ecotoxicology inasmuch as different microhabitats, such as open water and bottom sediments, may contain different concentrations of contaminants. Studies combining the results of fate and transport studies that predict and measure where toxins accumulate with behavioral ecological studies that predict and determine where animals feed could significantly improve our understanding of exposure to and ingestion of toxic materials. Furthermore, a better understanding of what factors influence microhabitat and dietary choices may allow active intervention to reduce exposure.

Example: Sparrows Balancing Feeding Rate and Predation Exposure

Many small passerine birds feed near cover along the edge of open fields. This behavior can significantly influence the degree to which birds are exposed to pesticides or other toxic materials used in agricultural fields. The frequency with which birds feed in open fields is thought to be related to both the relative feeding rate that can be achieved and the exposure to predation in the open vs near cover. Field studies have quantified many of the factors that determine how frequently and to what extent birds will enter open fields.[9–11]

Schneider[12] demonstrated that white-throated sparrows (*Zonotrichia albicolis*) prefer to feed in areas near cover and will nearly exhaust food supplies near cover before feeding in open areas. As food is exhausted, these sparrows begin to feed farther from cover with the more dominant individuals occupying the favored microsites closer to cover. Pulliam and Dodd (unpublished manuscript) extended Schneider's result by quantifying how much reduction in feeding rate white-throated sparrows will accept in order to stay near cover. They videotaped a group of sparrows feeding in an aviary where seeds were available only on a 1-m square feeding board covered with brown artificial turf. The feeding board was divided into four 25 cm × 1 m feeding zones, and at the beginning of each experiment an equal quantity of seeds was scattered over each feeding zone. The feeding board was placed immediately adjacent to a brushpile (2 m in diameter and 1 m high). A frame-by-frame analysis of the videotapes could determine the feeding rates and positions of individual birds.

Figure 2 shows an example of how feeding rates declined near cover (zone 1) as seed supplies there were exhausted and how the birds progressively shifted their feeding activities to the feeding zones farther from cover as the seeds closer to cover were depleted. Based on detailed analyses of the feeding positions and the feeding rates, Pulliam and Dodd concluded that the sparrows preferred to feed adjacent to the brushpile rather than at a 20% faster rate half a meter farther from cover, but if the feeding rate closer to cover dropped by more than 20%, they preferred to feed farther from cover. This "titration" technique allows one to predict how far from the edge of a field the sparrows will feed depending on the initial concentrations of seeds, the density of birds, and the rates of depletion.

Other species of sparrows are less dependent on cover and will move farther into open fields with less food depletion near the edge. Watts[13] has also shown that the type of grass and weed cover in and along the edge of agricultural fields can strongly

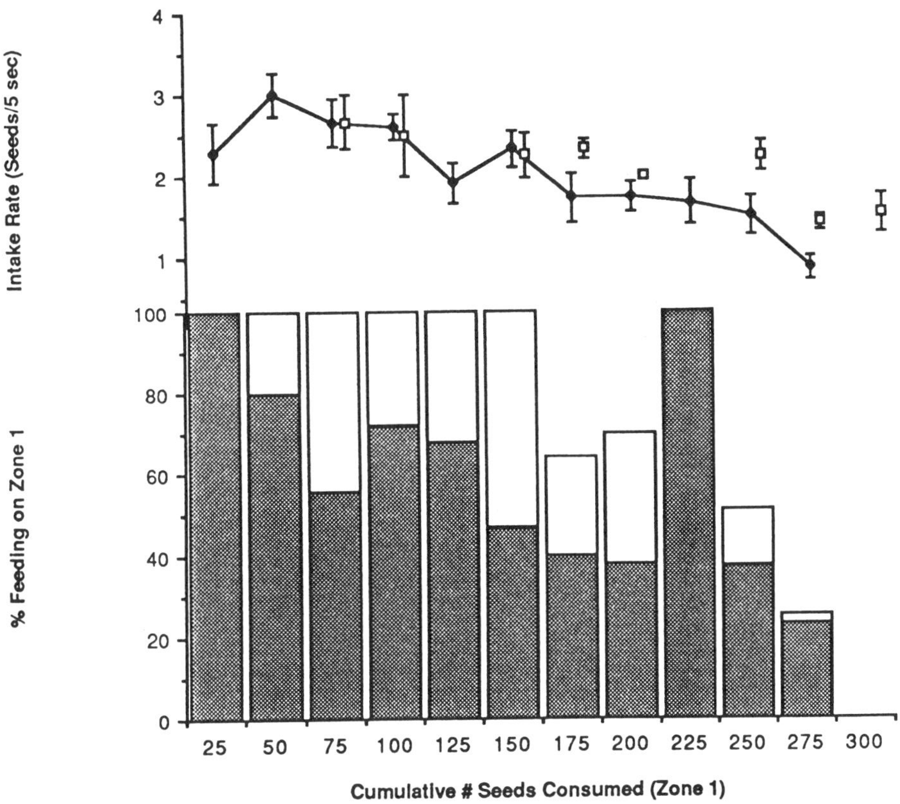

FIGURE 2. Feeding location of white-throated sparrows in relationship to food intake rate. The upper graph shows the feeding rate of those individuals feeding closest to cover (zone 1, represented by solid circles) and the feeding rate being simultaneously achieved by other individuals feeding farther from cover (represented by open circles). The lower graph for each date shows where the most dominant individual (open bar) and all other individuals (stippled bar) were feeding at the time. Feeding rates decline on all zones as seeds are consumed, but the decline is most rapid in zone 1. Initially most feeding is concentrated in zone 1, but as the seeds there are consumed, the feeding activity shifts to the more distant zone. These and similar results indicate that the sparrows prefer to feed at a microsite adjacent to cover rather than feeding 20% faster at a site 0.5 m from cover.

influence how far different species will venture into fields. These observations all suggest that by manipulating grassy borders around agricultural fields, seed supplies in the borders and in the fields, and the weed and stubble left in the fields, one can manipulate the foraging intensity in the fields and thereby control the exposure of birds to pesticide residues in the fields.

DEMOGRAPHIC MODELS

Population ecologists often study the factors regulating the local abundance of a species. This typically involves relating demographic parameters (such as age-specific birth and death rates) to environmental factors such as weather and the abundances of food, competitors, and predators. Most models of population dynamics can be modified to incorporate the impact of an environmental contaminant such as a pesticide, if the impact of the contaminant on birth and death rates can be quantified. However, a contaminant may well have a significant impact on individuals living on or adjacent to contaminated sites without having a significant impact on the population as a whole. This section discusses a class of landscape-level models that incorporate both habitat-specific demographics and dispersal between sites.[14–16] These models may be useful for evaluating the overall impact of a contaminant on a natural population.

Some ecotoxicological studies have employed a version of a basic Leslie matrix model for projecting the growth of an age-structured population.[17] This model employs a population projection matrix whose first row gives age-specific natality and whose subdiagonal contains survival probabilities from one age class to the next. A vector of the number of individuals in each age class for year t multiplied by this matrix gives the age distribution for year $t+1$. If the matrix is similarly multiplied by the resultant vector many times over, eventually a stable age distribution is reached after which each age class increases by the same factor λ each year. This factor λ is the dominant eigenvalue of the projection matrix, and the eigenvector associated with the dominant eigenvalue gives the stable age distribution.

The Leslie matrix model is well suited for ecotoxicological studies because the age classes are explicit. Thus, if a particular toxin differentially affects the survival or reproductive success of certain age classes, it is easy to determine the sensitivity of population growth to this effect. The survival and reproductive success parameters of the Leslie matrix are usually considered constants, but it is relatively straightforward to make them functions of population density or the density of either predators or prey.

Though the discrete-time Leslie matrix model is widely used by ecologists, many prefer to use the continuous-time, Lotka-Euler approach. In this case, the probability of survival to age x is given by l_x and the rate of production of offspring per female aged x is m_x. The lifetime production of offspring is then given by

$$R_o = \int_{\phi}^{\infty} l_x \, m_x \, dx,$$

and the intrinsic growth rate of the population (r) is found by solving the Lotka-Euler equation

$$\int_{0}^{\infty} \exp(-rx) \, l_x \, m_x \, dx = 1.$$

This approach was used recently to represent the dynamics of populations in a manner that could be particularly appropriate to ecotoxicology as discussed below.

Metapopulation Models: A Landscape Approach

Levins[18] introduced the concept of a metapopulation to describe a set of geographically separate subpopulations, each occupying a suitable patch of habitat in a landscape of otherwise unsuitable habitats. He demonstrated that the proportion of suitable habitat patches occupied at any given time results from a demographic equilibrium between the rate of extinction of local subpopulations and the rate of colonization of empty patches of habitat. This landscape approach to demography may prove particularly useful in ecotoxicological studies because it can represent the demographic consequences of having agricultural habitats where pesticides are used, embedded in otherwise natural landscapes. Most populations that may be affected by the use of pesticides in an agricultural landscape consist of numerous subpopulations, some of which exist in habitat adjacent to agricultural land where pesticides are applied and some of which do not.

To determine the demographic consequences of a pesticide application, one must consider not only the habitat patches adjacent to the affected agricultural land but the effect on the entire metapopulation. In some cases, a pesticide may cause extensive mortality in local patches but have little or no effect at the population level.

Lande[19] models a large landscape that is divided into discrete areas (or tracts), each the size of an individual territory. In his model, at most one female of reproductive age (adult) can occupy a single territory (suitable tract), and some suitable tracts may not be occupied at all. Lande assumes that suitable territories are randomly dispersed throughout the landscape and that, of all available tracts, a proportion h constitutes habitable tracts and a proportion $u = 1 - h$ constitutes uninhabitable tracts.

The model deals only with the demography of females, assuming that females can always find a mate. Juveniles have a probability (e) of inheriting their parents' territory; if they fail to inherit a territory, they must disperse, in which case they may search m tracts for a suitable territory before perishing. Under these assumptions, the probability of a juvenile finding or inheriting a suitable territory is given by

$$1 - (1 - e)(u + ph)^m$$

where p is the proportion of suitable territories already occupied. According to this model, as the population grows, the proportion of suitable territories that are actually occupied increases and juvenile survivorship declines because dispersing females are more likely to perish before finding an unoccupied suitable territory.

To calculate the equilibrium population size, Lande calculates the expected lifetime reproductive success (denoted as R_0') of those females that live to find a suitable territory. Note that R_0' times the probability that a juvenile finds a suitable unoccupied territory equals R_0, the expected lifetime reproductive success, i.e.,

$$R_o = R_0'(1 - (1 - e)(u + ph)^m).$$

Using the Lotka-Euler equation, which identifies R_0 with unity, one can calculate the equilibrium proportion of all suitable territories that are occupied. Because the proportion of the landscape that consists of potentially suitable territories is specified, one can then easily calculate the equilibrium population size for the equilibrium number of breeding females.

Table 1. Predictions of a Simple Metapopulation Model of the Population Impacts of Pesticide Use in Agricultural Fields Embedded in a Forested Landscape

	m		
d	2	5	10
0.00	0.817	0.922	0.960
0.01	0.808	0.918	0.958
0.03	0.790	0.910	0.949
0.05	0.769	0.900	0.949
0.07	0.745	0.889	0.943
0.09	0.717	0.875	0.936
0.11	0.685	0.859	0.927
0.13	0.647	0.840	0.917
0.15	0.601	0.816	0.903
0.17	0.546	0.785	0.889
0.19	0.458	0.742	0.862
0.21	0.380	0.679	0.823
0.23	0.228	0.553	0.744
0.25	0.000	0.000	0.000
0.27	0.000	0.000	0.000

Note: The table shows the fraction of suitable territories that are occupied as a function of two parameters. The parameter *d* is the fraction of the landscape that is agricultural and on which pesticides are applied. The parameter *m* is the number of potential territories searched by a dispersing juvenile female. It is assumed that this parameter is generally unknown, but, as shown in the table, the results may not be very sensitive to *m*.

To apply this approach to an agricultural landscape where the toxicological effects are only experienced by those individuals breeding in territories adjacent to agricultural fields sprayed with pesticide, assume the worst case scenario in which the females in such territories fail to produce any offspring. Suppose that a proportion d of the landscape consists of tracts that are sprayed annually. If the tracts are all square, then the probability that a habitable tract has no neighboring tracts that are sprayed agricultural fields is $(1 - d)^4$. In this case, the expected lifetime production of young by a female that successfully finds a territory is reduced to $R_0'(1 - d)^4$ and the Lotka-Euler equation becomes

$$R_0'(1 - d)^4(1 - (1 - e)(u + ph)^m) = 1. \tag{1}$$

For illustration, assume that all juvenile females must disperse from their natal site (i.e., $e = 0$) and that the juvenile females only search potentially suitable territories ($h = 1$). In this simple case, the equilibrium proportion of sites occupied is given by

$$p = (1 - 1/(R_0' (1 - d)^4))^{(1/m)}. \tag{2}$$

Table 1 shows the expected percentage of suitable territories occupied based on Equation 2 for an example in which $R_0' = 3.0$. All the parameters that go into the calculation of p are obtainable in field studies, except m. The parameters h and d are obtainable from descriptions of land-use patterns. R_0, based on standard l_x and m_x data, and e, the probability of inheriting the natal territory, are frequently measured in field demographic studies. Table 1 shows the expected equilibrium percentage of suitable habitats occupied for three values of m: 2, 5, and 10. Notice that the pattern is very similar for these three values: p stays reasonably high (mostly above 50%) as the percentage of sprayed agricultural land increases from 0 to about 20%, but if *d* increases toward 25%, the population rapidly becomes extinct. This result suggests that population size and extinction probability are not very sensitive to the actual value of *m*, as long as two or more potential territories can be evaluated before a juvenile female perishes.

MAP: A Spatially Explicit Model of Avian Population Dynamics

The metapopulation model discussed above is probably too general for use as a management tool. In addition to having one parameter that is difficult to measure, the model makes unrealistic assumptions about dispersal behavior. In particular, the model is not spatially explicit in that the actual locations of habitat types are not specified, and therefore assumes that an animal is equally likely to disperse from one tract to any other regardless of the actual locations of the tracts.

Mobile Avian Population (MAP), is a model that considers the actual locations of habitats on a landscape and simulates the dispersal behavior of birds on such an explicit landscape. The model was developed to predict the abundance and habitat distribution of Bachman's sparrow (*Aimophila aestivalis*) in managed pine plantations at the Savannah River Site (SRS) near Aiken, SC.[16] The structure of the model is quite general, permitting it to be run for a great variety of species with different life history characteristics and for any landscape, either hypothetical or real.

The Bachman's sparrow version of the model, which we refer to as BACHMAP, is run on a 2000-ha simulation landscape consisting of 800 hexagonal cells in a 20×40 array. Each cell represents 2.5 ha, the approximate average size of a Bachman's sparrow breeding territory. Pine plantations at the SRS vary in size, but on average occur in approximately 10 ha tracts of even-age trees. The simulation landscape mimics this by having 10 ha tracts, each consisting of four adjacent grid cells (potential territories), treated as a single vegetation type.

The vegetation types used in BACHMAP are chosen to represent the upland habitats at SRS, which mostly consist of pine plantations on an approximately 20-year rotation. There are 22 vegetation types considered in the model, 20 of which represent the 20 age classes of the pine rotations, and one represents clearcuts during the year of harvest. The 22nd vegetation type is mature or "old-growth" pines. Bachman's sparrows breed in mature pines (>80 years old) that are burned frequently. At SRS they also breed in young pine plantations (1 to 5 years old). Dunning and Watts[20] have shown that the structural characteristics of the vegetation in the first 3 m off the ground are very similar in mature pine stands and in 1- to 5-year-old plantation stands. The sparrows do not breed in the pine stands of intermediate age, presumedly because the foliage is too dense to permit light penetration to the forest floor; the resulting shade inhibits growth of grasses and forbs, which provide food and nesting sites.

BACHMAP has been run with a variety of demographic parameters, but based on field studies[20,21] and sensitivity analysis of model results, we use the following parameter values: adult females produce 3.0 female offspring per year; 60% of adults alive at the beginning of the breeding season survive until the beginning of the next breeding season; juvenile survival is 40%, exclusive of mortality that occurs during dispersal. Though estimates of these demographic parameters may change with further study, they will be used here to illustrate the model.

BACHMAP explicitly simulates the dispersal behavior of juveniles by incorporating what is known about dispersal in Bachman's sparrows and similar species. Dispersal is assumed to occur primarily in the spring just prior to reproduction. If the adult female on a territory dies, it is assumed that one of the juvenile females born on the territory

Table 2. Effects of Pesticide Application in Agricultural Fields Based on the BACHMAP Model

Agricultural Land (%)	$b = 3.0$	$b_1 = 3.0$ $b_2 = 1.5$
	Population Size	
0	24.3	24.3
10	21.5	17.6
20	16.6	12.6
30	16.0	11.3
	Extinction Probability	
0	0.00	0.00
10	0.00	0.00
20	0.00	0.13
30	0.00	0.33

Note: The results show the population sizes and extinction probabilities for the situation in which there is no impact of the agricultural practices on demographics as compared to the situation in which territories adjacent to agricultural field have half the reproductive success ($b_2 = 1.5$) of territories that are not adjacent to agricultural fields ($b_1 = 3.0$).

inherits the natal territory. Juveniles that do not inherit the natal territory must disperse in search of a suitable territory, in this case an unoccupied mature pine stand or 1- to 5-year-old pine plantation. Dispersing individuals remember where they have been and do not visit the same grid cell twice. If a dispersing individual occupies a site surrounded by several potential territories, she chooses at random among them. If the site chosen is already occupied, she moves on in a similar manner to search additional sites.

We modified the BACHMAP model to include agricultural habitat in the simulation landscape. We considered situations where 10, 20, and 30% of all tracts were agricultural land. We assumed that the impact of breeding on a territory next to an agricultural tract was on reproductive success (i.e., on the number of juvenile females successfully fledged). We consider two levels of effect: (1) no reduction in reproductive success (an average of 3.0 females fledged) and (2) a 50% reduction (an average of 1.5 females fledged) for birds breeding on a site adjacent to agricultural land.

As shown in Table 2, when there is no pesticide-induced mortality, increasing the fraction of agricultural land from 0 to 30% reduces breeding population size by about 34%. In this case, none of the simulated populations became extinct, even when as much as 30% of the landscape consisted of uninhabitable agricultural land. When pesticide-induced mortality reduced the production of fledged juveniles from 3.0 to 1.5 per nest, the impact of increasing the percentage of the landscape in agricultural land on the simulated avian population was much more substantial. An increase from 0 to 30% agricultural land decreased the breeding bird population an average of 54% and the extinction probability increased from 0 to 33%.

DISCUSSION

This chapter argues that models and concepts from behavioral and population ecology may improve our understanding of the factors that determine exposure of an animal population to toxic substances such as pesticides and the population-level consequences of such exposure. In both cases the quantitative models should be viewed as tools for understanding qualitative impacts rather than as tools for predicting quantitative details. As shown by the several examples presented, models of animal foraging often predict general patterns of where animals feed and what prey items they consume. Nonetheless, whenever one looks closely enough at the results, discrepancies between actual and predicted behavior are apparent.

Despite such quantitative discrepancies, behavioral and population models should prove useful tools for designing and interpreting field studies and application strategies. For example, as discussed earlier, many species of birds that feed around the edges of sprayed fields move farther into the fields (away from the shrub or tree cover surrounding the field) as they deplete the food near the edge. Field studies and models of such behavior could lead to a fairly accurate understanding of the factors controlling the rate of entry into sprayed fields compared to the rate of degradation of pesticides sprayed in the fields. This could in turn lead to planting and spraying strategies that minimize the exposure of birds to toxic pesticides.

Two approaches to modeling the population-level consequences of pesticide-induced mortality have been presented. The metapopulation models discussed are useful heuristic tools, but are not sufficiently realistic to be useful as management tools. More realistic models such as BACHMAP may become useful for the management of natural populations, particularly threatened and endangered species of vertebrates. However, these models require fairly detailed knowledge about habitat-specific demography and dispersal behavior, information not available for most species (see Reference 16 for a detailed example). Application of such models will require extensive field studies of the affected species.

Despite the obvious limitations of current behavioral and population models, field studies involving collaborations between ecologists and toxicologists hold much promise and should be encouraged. These studies will lead to an improved understanding of the individual and population consequences of pesticides and other contaminants in the environment and to improved technologies for their safe application.

REFERENCES

1. Lassiter, R. R., G. L. Baughman and L. A. Burns, Fate of toxic organic substances in the aquatic environment, in S. E. Jorgensen, Ed., *State-of-the-Art in Ecological Modeling*, Vol. 7, International Society of Ecological Modeling, Copenhagen, 1978, pp. 219-245.
2. Pulliam, H. R., Individual behavior and the procurement of essential resources, in J. Roughgarden, R. M. May and S. A. Levin, Eds., *Perspectives in Ecological Theory*. Princeton University Press, Princeton, NJ, 1989.
3. Pyke, G. H., H. R. Pulliam and E. L. Charnov, Optimal foraging: a selective review of theory and tests, *Q. Rev. Biol.*, 52, 137–154, 1977.
4. Werner, E. E. and D. J. Hall, Optimal foraging and the size selection of prey by bluegill sunfish (*Lepomis macrochirus*), *Ecology*, 55, 1042–1052, 1974.
5. Pulliam, H. R., Do chipping sparrows forage optimally?, *Ardea* 68, 75–82, 1980.
6. Belovsky, G. E., Food plant selection by a generalist herbivore: the moose, *Ecology*, 62, 1020–1030, 1981.
7. Gilliam, J. F. and D. F. Fraser, Habitat selection under predation hazard: test of a model with foraging minnows, *Ecology*, 68, 1856–1862, 1987.
8. Werner, E. E., G. G. Mittlebach, D. J. Hall and J. F. Gilliam, Experimental tests of optimal habitat use in fish: the role of habitat profitability, *Ecology*, 64, 1525–1539, 1983.
9. Pulliam, H. R. and G. S. Mills, The use of space by sparrows, *Ecology*, 58, 1393–1399, 1977.

10. Lima, S. L., Protective cover and the use of space: different strategies in finches, *Oikos* 58, 151–158, 1990.
11. Watts, B. D., Segregation of an old field sparrow community along a cover gradient: effect of cover dependency on spatial co-occurrence, *Ecology*, in press.
12. Schneider, K. J., Dominance, predation, and optimal foraging in white-throated sparrow flocks, *Ecology*, 65, 1820–1827, 1984.
13. Watts, B. D., Effects of predation risk on distribution within and between habitats in Savannah sparrows, *Ecology* 72, 1515–1519, 1991.
14. Pulliam, H. R., Sources, sinks and population regulation, *Am. Nat.*, 132, 652–661, 1988.
15. Pulliam, H. R., and B. J. Danielson, Sources, sinks and habitat selection: a landscape perspective on population dynamics, *Am. Nat.*, 137, S50-S66, 1991.
16. Pulliam, H. R., J. B. Dunning and J. Liu, Population dynamics in complex landscapes: a case study, *Ecol. Appl.*, 2, 165, 1992.
17. Grant, W. E., and S. O. Fraser, Effect of vertebrate pesticides on non-target wildlife populations: evaluation through modeling, *Ecol. Model.*, 21, 85–108, 1984.
18. Levins, R., Some demographic and genetic consequences of environmental heterogeneity for biological control, *Bull. Entomol. Soc. Am.*, 15, 237–240, 1969.
19. Lande, R., Extinction thresholds in demographic models of territorial populations, *Am. Nat.*, 130, 624–635, 1987.
20. Dunning, J. B., and B. D. Watts, Regional differences in habitat occupancy of Bachman's Sparrow, *Auk* 107, 463–472, 1990.
21. Haggerty, T. M., Reproductive Ecology of Bachman's Sparrow (*Aimophila aestivalis*) in Central Arkansas, Ph.D. dissertation, University of Arkansas, Fayetteville, 1986.

3

Ecological Risk Assessment in the Office of Pesticide Programs

Anne L. Barton

ABSTRACT

Pesticides are regulated under the Federal Insecticide, Fungicide, and Rodenticide Act (FIFRA), which requires that the benefits of a pesticide outweigh the effects on human health and the environment in order for it to be registered for use. Both health and ecological risk assessments are performed to support decisions at various points in a pesticide's history. Risk assessments are done on new pesticides to support decisions to register or reregister pesticides, or to allow a new use of a registered pesticide. More elaborate and thorough risk assessments are done when evidence of risk triggers a special review of an already registered pesticide. Ecological risk assessments are also done to protect endangered species.

To carry out its risk assessment responsibility, the Office of Pesticide Programs (OPP) has laid requirements on the pesticide industry to develop certain data. These data include studies of acute and reproductive toxicity to birds and terrestrial field studies.

The risk assessment procedures employ the Quotient (or Ratio) Method, using laboratory data. The Quotient Method may be supplemented by field data, either incident reports or specific field studies. Important endpoints include mortality; effects on reproduction, growth, or development; and population effects. Currently, the risk assessment process is changing to place greater emphasis on risk mitigation and monitoring, and less emphasis on field studies.

Within the decade, OPP expects to receive data on some 600 pesticide active ingredients as part of a massive reregistration effort. In order to be useful for regulation, any models developed must utilize information that OPP possesses (such as results of required studies) and must produce information that can be used in a regulatory decision (i.e., information specific to chemical and use).

KEY WORDS

FIFRA, tier testing, risk assessment, Quotient Method, weight of evidence

INTRODUCTION

Since 1970, the regulatory authority and basis for pesticide risk assessment has rested with the U.S. Environmental Protection Agency (U.S. EPA) under FIFRA. Under this

Act, U.S. EPA must determine whether a pesticide can be registered for a particular use. The test for whether a pesticide can be registered is that it will not "generally cause unreasonable adverse effects on the environment" (P.L.95396, Sec. 3(c)(5)(D)).

Unreasonable adverse effects on the environment means that both the risks and the benefits are to be taken into account. Thus, risks must be estimated and measured so they can be compared with the benefits of the pesticide. The risk assessments used for a pesticide decision include human health and ecological risk assessment. In the past, the adverse effects of greatest concern for ecological risk have been mortality to single species of nonhuman, nontarget organisms, including endangered or threatened species. The Ecological Effects Branch (EEB) in the Environmental Fate and Effects Division (EFED) in the OPP has primary responsibility for developing ecological risk assessments for pesticides. Ecological risk assessments are used by OPP for several purposes: (1) in the pesticide registration and reregistration processes, the EPA considers indications of ecological risk and whether the pesticide's use should be restricted to certified applicators; (2) evidence of ecological risk may trigger a special review of an already registered pesticide, which may lead to the suspension or cancellation of the pesticide's registration; and (3) when reviewing pesticides for their risks to endangered species, the EPA may place limitations on the use of a pesticide in certain areas. Because each use of a pesticide (e.g., specific crop) requires a separate registration, the risk assessment must be specific as to use as well as to substance. In recent years, EEB has performed an average of about 600 risk assessments per year. With the accelerated reregistration required by the 1988 amendments to FIFRA, this number is expected to increase greatly.

TESTING

Under FIFRA, the Agency is not responsible for producing the test data needed to make an ecological risk assessment. That burden is placed upon the applicants seeking registration. OPP has published regulations that specify the data required for registration (40 CFR Part 158) and guidelines for recommended testing methods needed to produce the required data (Pesticide Assessment Guidelines, Subdivision E). Testing for OPP is done in a tiered or hierarchical manner, involving three tiers of progressively more complicated tests. A hierarchical testing scheme provides an efficient method of arranging tests. The more complicated tests are performed much less often, and only when the results of the simpler tests indicate the need for further testing.

This tiered approach recognizes that time and resources are essential considerations in the testing and assessment processes and that both OPP and the chemical industry have finite resources. EPA's tiered testing system follows lines similar to OECD (Organization for Economic Cooperation and Development) guidelines.

Testing starts at Tier I with short-term acute and subchronic laboratory studies, which provide basic toxicity information. The tests most commonly done at this stage are LC_{50} and LD_{50}. At Tier II, additional studies may be required to estimate the potential for chronic effects when basic data and environmental conditions suggest possible problems. These studies commonly produce estimates of ecotoxicological hazard, expressed in the form of a NOEL (no observed effect level), NOEC (no observed effect concentration), or MATC (maximum acceptable toxicant concentration). The NOEL or NOEC indicates the highest level or concentration of the chemical at which no effect is observed (i.e., the NOEC). An even higher concentration at which effects do begin to be observed is expressed as the LOEC (i.e., the lowest observed effect concentration). The MATC is often expressed as a geometric mean.

TERRESTRIAL ECOTOXICITY TESTING

Tier 1

- Mammalian toxicity data
- Acute oral LD_{50} test—bird
- Dietary LC_{50} test—bird
- Seed germination/seedling emergence and vegetative vigor
- Acute contact LC_{50} test—honeybee

Tier II

- Wild mammal toxicity data
- Avian reproductive studies
- Special studies with avian or mammalian species (e.g., avian cholinesterase test, secondary toxicity)
- Seed germination/seedling emergence and vegetative vigor

Tier III

- Simulated and actual field testing with avian and/or mammalian species

FIGURE 1.

At Tier III, simulated or field data are used to examine acute and chronic adverse effects (such as mortality, reduction in growth, or impairment of reproduction) on captive or monitored fish and wildlife populations under field conditions. Such studies may be required to document the hazard identified by the earlier tiers of testing.

OPP requires testing for both aquatic and terrestrial risk. The tests required at each tier for terrestrial risk are shown in Figure 1.

RISK ASSESSMENT

Risk is a function of hazard and exposure. Similarly, ecological risk is a function of toxicological hazard and environmental exposure. Toxicological hazard is an intrinsic quality of a pesticide that causes an adverse effect under a particular set of circumstances. Toxicological hazard data would include, for example, LD_{50} values for laboratory fish, aquatic invertebrates, or birds, and effect levels for fish and avian reproduction tests. Environmental exposure is a function of two data components. The first is the estimated amount of the pesticide residue that will be in the environment and available to nontarget species. This component is known as the estimated environmental concentration (EEC). The second consists of the numbers, types, distributions, abundances, dynamics, and natural histories of nontarget organisms that will be in contact with these residues. Information on the proposed label use of the pesticide is essential for such exposure estimates, which are necessarily done on a use-by-use basis. OPP first estimates toxicological hazard and environmental exposure separately, and then compares them.

Quotient Method

To compare hazard and exposure, OPP uses the Quotient (or Ratio) Method. The Quotient Method is essentially a "common sense" method whereby the toxicological level of concern for the most sensitive species is compared to an estimated environmental concentration for the same chemical:

$$\frac{\text{Estimated Environmental Concentration}}{\text{Toxicological Level of Concern}} = \text{Quotient}$$

Quotients ≥ 1 represent a high risk, i.e., a strong likelihood exists that an ecotoxicological effect of concern will occur. Some type of regulatory response is usually needed to address the high risk, possibly including a requirement of additional testing to characterize the risk.

Quotients considerably less than 1, such as 0.01, represent a low risk, i.e., there is a strong likelihood that an ecotoxicological effect of concern will not occur. Usually little or no regulatory response is needed. Quotients approaching 1 represent an uncertain risk. Additional data are usually needed to further characterize the risk.

The Quotient Method is simple, can be understood intuitively, and permits clear decisions for chemicals posing high or low risk. Nevertheless, it does not adequately account for effects of incremental dosages, indirect effects (e.g., food chain interactions), or other ecosystem effects (e.g., changes in species diversity; community structure, function, and stability; and predator-prey relationships). Furthermore, it cannot compensate for differences between laboratory tests and field populations, nor can it quantify uncertainties or provide a known level of reliability. Presently, therefore, the Quotient Method does not allow for a complete characterization of the magnitude of risk or the degree of confidence associated with the characterization. Confirmatory data in the form of terrestrial field studies, aquatic field studies, and incident reports are used to corroborate or negate the ecological risk estimates.

Tier III field testing may be required when the quotient results in a prediction of high risk. Field studies can document adverse effects (usually mortality to birds, mammals, and/or fish) predicted in the ecological risk estimates using the Quotient Method. These studies can show whether these effects will occur in the field from the use of the pesticide according to label directions. However, because of many variable circumstances associated with field studies, the magnitude, frequency, and duration of adverse effects likely is greater than indicated by the actual number of bird, mammal, or fish carcasses found.

Incident reports of pesticide-related mortality are also useful as confirmatory data when available. Unfortunately, so many factors contribute to the great underreporting of incidents that the *absence* of incident reports cannot be used as evidence of little or no field effect.

Weight of Evidence

The final step in a pesticide risk assessment is a review and balancing of the weight of the evidence, including both the quotient and any additional information available from field tests and incident reports. Modeling results are also included when available, although they are presently rare for terrestrial risk assessment. The weight of evidence approach considers the quality and adequacy of the data, as well as the frequency and magnitude of the observed or estimated effect.

IMPLICATIONS FOR POPULATION MODELING

The avian risk endpoints with which OPP is now principally concerned are reflected in its 1985 Special Review Criteria, which lay out criteria for initiating a special review for significant risk. The specific criterion for ecological effects reads: "The uses of the pesticide may result in residues in the environment of non-target organisms at levels which equal or exceed concentrations *acutely* or *chronically toxic* to such organisms, or at levels which produce *adverse reproductive effects* in such organisms . . ." (emphasis added).

The effects singled out for mention in these criteria are those that are clearly of concern and correspond to the data that OPP has or is able to obtain. Population reduction is not mentioned because there is no way to measure directly population reduction from a specific pesticide use. Furthermore, in the opinion of OPP scientists, there are no validated models that would allow us to estimate or predict the degree of such reduction from specific pesticide uses.

Because population reduction is an endpoint of concern, models are needed that will be useful in pesticide risk assessment. In order to be useful, such models must be validated. They also need to be easily useable by wildlife toxicologists without extensive training. Information required as input should be routinely available (such as the results of the studies listed in Figure 1) or readily obtainable. Developing and validating such models will require a concerted effort by researchers in the various disciplines.

ACKNOWLEDGMENTS

This paper was almost entirely excerpted from References 2 and 3, authored by Douglas Urban and Norman Cook of the Office of Pesticide Programs, U.S. EPA; Donald Rodier of the Office of Toxic Substances, U.S. EPA; and Gilah Langner of Strettan Associates, Inc.

REFERENCES

1. Environmental Protection Agency, Special Reviews of Pesticides: Criteria and Procedures; Final Rule. 40 CFR Parts 154, 162, and 172, *Fed. Regist.* 50 (229); 1985.
2. Hazard Evaluation Division, *Standard Evaluation Procedure*, *Ecological Risk Assessment*, Office of Pesticide Programs, U.S. EPA, Washington, DC, 1986.
3. Office of Pesticides and Toxic Substances, *State of the Practice Ecological Risk Assessment Document*, U.S. EPA, March 14, 1990.

4

What is Measurable in Wildlife Toxicology?

Richard E. Warner

The second session of the 1990 Ecological Modeling Conference, "What is Measurable in Wildlife Toxicology?," was strategically placed after the plenary session and before presentations focusing on ecological modeling. The plenary session laid a foundation for this session because it (1) provided a historical background for the emergence of wildlife toxicology as a scientific discipline, (2) exposed a diverse group of scientists to a number of relevant concepts, and (3) proposed future needs for research. Early in the conference, participants became acutely aware—many for the first time—that legal mandates for pesticide regulation have not been developed concomitantly with the techniques for gaining reliable knowledge relative to the risks and benefits of pesticide use. Whereas pesticide regulation has been responsive to the documented loss of individuals, the science of wildlife toxicology appears more focused on demographic effects on populations and community composition over well-defined scales of space and time. Nonetheless, even as they presently exist, pesticide regulation and ecotoxicology share a considerable overlap in information gaps. The interchange between those involved in pesticide regulation—the legal apparatus—and wildlife toxicologists is likely to be dynamic for many years to come.

Questions asked at the end of the plenary session reflected what has been emphasized in wildlife toxicology studies to date and provided a lead-in for the second session with its overview of state-of-the-art monitoring of the effects of pesticides in ecosystems, especially wildlife. Among these questions were, why have studies of avians seemingly been preeminent, why have agricultural environments and a few key crops been given priority in monitoring efforts, and what is a model? The group discussions highlighted the need to draw distinctions between various kinds of models and established a first look at the potential interplay of various models as tools for the framing of conceptual questions, ecological simulation, and regulatory decisionmaking.

Whereas the plenary session was both historical and futuristic, the "What is Measurable in Wildlife Toxicology?" session focused on the present tools of the trade. Participants were given a quick overview of the availability of technologies and the infrastructures now in place to address wildlife toxicology. Two papers, one by Cobb and Hooper[1] and another by Hoffman,[2] provided background on determinations of LC_{50} and LD_{50} for selected avian species at various life stages, especially for insecticides widely used at present. Moreover, these scientists began to link state-of-the-art methods for measuring and detecting exposure with assessments of the ultimate effects of exposure in field studies.

Subsequent presentations focused more directly on field assessment. Crystal Driver of Battelle Northwest delivered a talk describing past and present radiotelemetry techniques for assessing responses by animals to pesticide applications and introduced emerging technologies that may have wide application for studying physiological and behavioral responses to chemical perturbations in the future. A paper by Brewer and Hummell[3] provided a conceptual framework for moving from the collection of exposure and toxicology data in highly controlled settings to field assessments under more realistic situations. They also illustrated the need and context for applying field assessment information to the development of ecological models.

By the end of the second session, participants were beginning to attach consistent definitions to various terminologies and to identify the need for integrating reliable toxicology information and ecological models. For example, differences between lethal/sublethal and direct/indirect effects were widely discussed throughout the workshop but were of particular concern in the session on the measuring of wildlife toxicology. Discussions of terminology also helped to identify what the end products of wildlife toxicology studies should be; i.e., what should be monitored, what inferences can be made from toxicological data, and what role may ecological models play?

REFERENCES

1. Cobb, G. P., and M. J. Hooper, Non-lethal wildlife monitoring to determine exposure to xenobiotics and resulting impacts, in *Wildlife Toxicology and Population Modeling: Integrated Studies of Agroecosystems,* R. J. Kendall and T. E. Lacher, Jr., Eds., Lewis Publishers, Chelsea, MI, 1993.
2. Hoffman, D. J., Measurements of toxicity and critical stages of development, in *Wildlife Toxicology and Population Modeling: Integrated Studies of Agroecosystems,* R. J. Kendall and T. E. Lacher, Jr., Eds., Lewis Publishers, Chelsea, MI, 1993.
3. Brewer, L. W., and R. A. Hummell, What is measurable in wildlife toxicology: field assessment, in *Wildlife Toxicology and Population Modeling: Integrated Studies of Agroecosystems,* R. J. Kendall and T. E. Lacher, Jr., Eds., Lewis Publishers, Chelsea, MI, 1993.

5

Nonlethal Wildlife Monitoring to Determine Exposure to Xenobiotics and Resulting Impacts

George P. Cobb and Michael J. Hooper

ABSTRACT

Severe limitations are encountered when using traditional methods to assess chemical exposure and impact in wildlife populations. This is primarily due to the lethality of such interrogation to the test organism. Newly implemented techniques exhibit utility in assessing wildlife exposure to xenobiotics. These techniques involve analyzing biological fluids and waste products.

Organophosphate (OP) exposure has traditionally been assessed using brain cholinesterase (ChE) activity determinations and gastrointestinal (GI) tract residue analysis. Plasma ChE activities and excreta residue levels provide alternative methods by which large numbers of individuals can be sampled without affecting the population level evaluation in a study. Multiple samplings from an individual accommodate survivability studies as well as evaluation of recovery periods following documented exposure.

The utility of plasma ChE evaluations frequently has been underestimated due to wide inter- and intraspecies variability. These perceived disadvantages are overcome when baseline values are determined for individuals prior to exposure and when large-scale sampling is performed, yielding improved variances of population means. Inclusion of both spatial and temporal control sampling can account for seasonal variation that may occur. The separate determination of acetyl- and butyrylcholinesterase (AChE, BChE) strengthens the method due to the particular sensitivity which butyrylcholinesterase exhibits for many OPs.

Avian fecal-urate samples have also been utilized to determine OP exposure. The method uses alkylphosphate analysis as the endpoint. This technique affords the opportunity to relate plasma ChE to excreted alkylphosphates in each individual. Data from different chemical treatments show promising correlations with residue concentrations in the environment. Discussions include residue trends and the applicability to species survivability.

KEY WORDS

biomarkers, nonlethal sample collection, organophosphate pesticides, chlorinated hydrocarbons.

INTRODUCTION

Results from modeling studies can be no better than the data used to develop the model. Thorough evaluations of toxicological impact require accurate and precise determination of chemical concentrations in the environment, chemical uptake by species residing in the study area, response to exposure, and survival of the species studied. Each aspect of an investigation complements the other aspects. Compilation of these data sets allows true determination of toxicological responses of animals to xenobiotics in the environment studied. Each parameter is necessary to assess chemical effects on wildlife. Particularly well-designed studies can accumulate the data necessary for true risk assessments in these areas. Such assessments are based on measured parameters as described above.

Significant obstacles have existed in risk assessments and subsequent modeling attempts. A primary impediment has been the availability of data that describe chemical concentrations, reactions in the environment, and uptake by test species. Uptake of pesticides by animals may result from direct exposure (dermal, inhalation, and ingestion) or by ingesting an organism that has previously ingested a pesticide. A secondary problem is encountered when simultaneously acquiring environmental, bioavailability, and survival data over a sufficient period. Both problems have been addressed by painstaking logistical planning and execution of interdisciplinary studies. Much to the dismay of traditional researchers, these steps are still not sufficient to achieve temporal evaluations of the toxic responses in a single habitat.

Many individuals must be monitored over extended periods of potential exposure. Monitoring individuals within the larger investigations produces data concerning individual variance within populations. Without the precision possible from such data, modeling efforts require inordinate numbers of samples to achieve an acceptable level of significance. Nonlethal sample collection is the only means of temporal, repetitive, individual sampling. Therefore, nonlethal monitoring must be implemented in large-scale environmental toxicology studies. These types of studies were undertaken recently to assess the toxicity of xenobiotics to test organisms in selected ecosystems.

Determination of a significant pesticide exposure requires two distinct lines of evidence. The first is the presence of the pesticide or its metabolites in tissues of the organism. The second is a biochemical and/or physiological response that reflects exposure to the chemical. The absence of either residue or biochemical response data weakens the case for a positive determination of significant exposure. New analytical methods that utilize the animal's biological fluids and waste products have been developed and implemented to satisfy the need for nonlethal accumulation of biochemical and residue data.

THE APPROACH

Traditional methods for exposure assessments primarily utilize tissue analyses.[1] Gastrointestinal (GI) tracts and abdominal organs have been emphasized in the chemical analysis of tissues. Biochemical evaluations have revolved around enzymatic assays of brain and liver tissue homogenates. Such practices hinder population modeling by removing individuals from the study population. Significant portions of a population must be sampled to obtain contaminant concentrations with any precision. Thus, with traditional methods, ecotoxicologists face a perplexing dilemma: to effectively monitor survival, or to precisely determine contaminant exposure.

Nonlethal sampling techniques can ease this dilemma. Each individual may be sampled for contaminant exposure and then returned intact to the study environment. Repeated interrogation of individuals is also feasible with these methods and may span one or more years. Differing approaches have been used for the collection of biological samples from species at risk of pesticide exposure. Extensive live-trapping of individuals on treatment sites can provide a broad assessment of exposure in a variety of species occupying the site. Another approach is the artificial enhancement of a species on the test site. Such "indicator" species can provide an extensive population of accessible individuals with quantifiable foraging, reproductive, and biochemical characteristics. Common to both of these approaches is the necessity for both temporal and spatial control data for all aspects of sampling. Nonlethal, low-impact sampling then provides the means by which large portions of both native and enhanced populations can be evaluated for pesticide exposure and impact.

Fecal-urate samples that are collected from birds recently exposed to OP pesticides will contain alkylphosphates. These compounds are hydrolysis products from OP pesticides. Presence of an alkylphosphate indicates exposure of the individual to OP compounds. Fecal-urates may be easily sampled by confining individuals to a small area with a collection media below the confinement area.

The chorioallantoic membrane (CAM) can be collected from the discarded eggs of most avian species. The CAM is the waste repository that separates metabolic byproducts from the embryo. These membranes appear in eggs during early development and grow throughout the incubation process. Therefore, contaminants present in eggs can be determined from analysis of CAMs. This type of analysis can be applied to lipid-soluble contaminants that are slowly metabolized in an organism and readily transferred from a mother to the egg. Membranes may be found inside eggshells that have been discarded beneath bird nests after chicks have hatched.

Certain enzymes respond specifically to particular contaminants, providing an indication of impact following exposure of the individual. Induction of hepatic mixed function oxygenases (MFOs) has been used in exposure assessments of organochlorine pesticides such as DDT, aldrin, and dieldrin;[2] prochloraz;[3] and many other environmental contaminants such as polynuclear aromatic hydrocarbons[4,5] and polychlorinated hydrocarbons.[6]

Biochemical assessment of exposure to OP and carbamate pesticides has been accomplished by the measurement of ChE enzymes in the blood or brain tissue from individuals suspected of exposure. In the latter case, methods for the nonlethal evaluation of ChE response using blood ChE were developed initially for human evaluation[7] and recently have been applied to wildlife species.[8-13] The target of OP insecticides, ChEs (particularly acetylcholinesterase, AChE), hydrolyzes the neurotransmitter acetylcholine in neuroneuronal and neuromuscular junctions. Inhibition of this enzyme results in the accumulation of acetylcholine in the synapse, leading to overstimulation and eventually desensitization of the postsynaptic tissue. Mortality follows disruption of respiratory and cardiovascular functions that are dependent on the compromised neurotransmission. Studies have shown that depression of brain ChE on the order of 20% below mean normal values is indicative of exposure to OPs, while death accompanied by ChE depressions of 50% or greater can be attributed to OP intoxication. Depression of brain ChE activity is thus a strong indicator of both exposure to and impact of OP and carbamate insecticides.

Plasma also contains ChE activity in the form of AChE and BChE. Although the function of these enzymes in plasma is unclear, their presence offers a nonlethal means by which exposure to OPs and carbamates can be evaluated. Depression of ChEs in the plasma occurs earlier than that in the brain, as the inhibitory compounds must travel via the blood to arrive in the brain. Sampling of plasma thus offers a sensitive and nonlethal approach to pesticide exposure evaluations.

Plasma ChE evaluations were recently applied to multiple species as well as indicator species assessments of OP exposure under field conditions. In the first case, extensive live-trapping provided plasma samples of numerous control birds from untreated sites. Evaluations are underway using methods such as diagnostic threshold (normal mean –2 S.D.) determinations from control populations for comparison to birds inhabiting treatment fields.[14] Evaluations of mean ChE activities over postapplication time intervals can also demonstrate pesticide-induced ChE depressions.

IMPLEMENTATION

Novel methods for assessing wildlife exposure to contaminants have been implemented in concert with environmental chemistry and species survivability determinations.[15-18] Test organisms have been evaluated for chemical uptake from the environment using chemical and biochemical analyses. Organochlorine residues have been quantified in CAMs from hatched and unhatched eggs to determine the correlations between residue values in the different matrices.[19] Fecal-urates have been analyzed for alkylphosphates as the nonlethal indication of OP exposure. Biochemical methods have heavily utilized plasma cholinesterase activity analyses for assessment of OP exposure.

Fecal-urate and cholinesterase data are used to model pesticide residue and avian survival data. Models of residue data from tissue samples with survival data are also discussed. CAM contaminant determinations are evaluated as an indicator of organochlorine and polychlorinated biphenyl (PCB) contaminants in great blue herons.

Environmental, fecal-urate, and blood plasma samples were collected using methods developed for large-scale toxicology evaluations.[9,10,15,16,20-23] Samples were collected at designated times during large-scale ecotoxicological investigations.[17,18] Analytical and biochemical determinations were conducted with recently developed methods.[15,16,24] Integrated sampling design, sample analysis, and complementary evaluation procedures are essential to obtain data of sufficient quality for risk assessment and modeling.

Great blue heron (*Ardea herodias fannini*) eggs and CAMs from discarded eggshells were collected in pilot studies in western Washington. Samples were collected from one colony. CAMs and egg contents from many whole eggs were analyzed using modifications of Canadian Wildlife Service methods[25] as described by Cobb et al.[15,19] Some eggs were analyzed in full, and CAMs collected from hatched eggs were analyzed individually.

TOXICOLOGICAL EVALUATIONS

Data collected during an ecotoxicity test of diazinon in apple orchards were modeled from application through ingestion and metabolism to organism recovery or death. This involved development of several rudimentary models. First, the mean degradation of diazinon on vegetative surfaces from all fields was fit to an exponential decay (Figure 1) as described by classic kinetics.[26] The correlation coefficient for these decay kinetics indicated factors other than time were involved. However, the decay constant was consistent for similar matrices.

Pesticide exposure is likely from ingestion of contaminated food. When aerial sprays are applied to perennial agricultural crops, food contamination can be estimated from

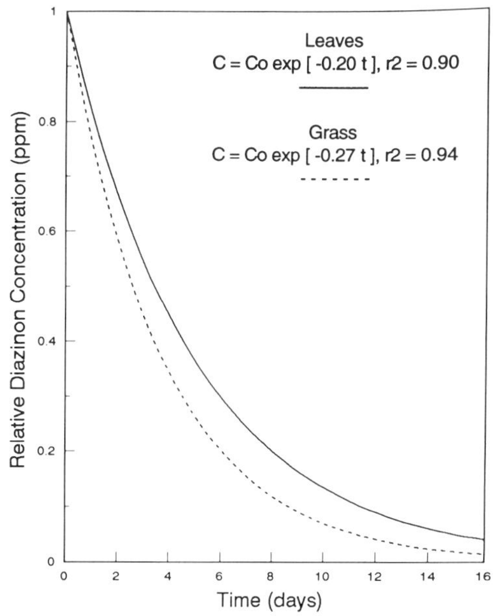

FIGURE 1. Diazinon concentration decrease from the surface of leaves and grass following liquid application.

chemical concentrations on vegetation. This can be the most abundant surface in such agroecosystems. Animals will receive exposure through contact with vegetation, direct ingestion of contaminated vegetation, or ingestion of prey items that have come into contact with this vegetation. Direct ingestion of pesticide will cause the maximum chemical concentration in the GI tract. Chemical concentrations in GI tracts as a function of time were fit to nonparametric curves (Figures 2 and 3). Kernel functions were used for this modeling. Mathematical approaches to this type of estimation are described by Eubank.[27] Curves were generated using the function:

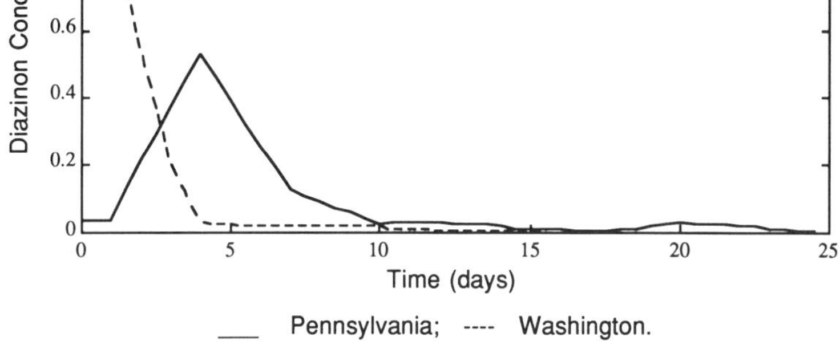

FIGURE 2. Diazinon concentrations in GI tracts of passerines from chemically treated fields.

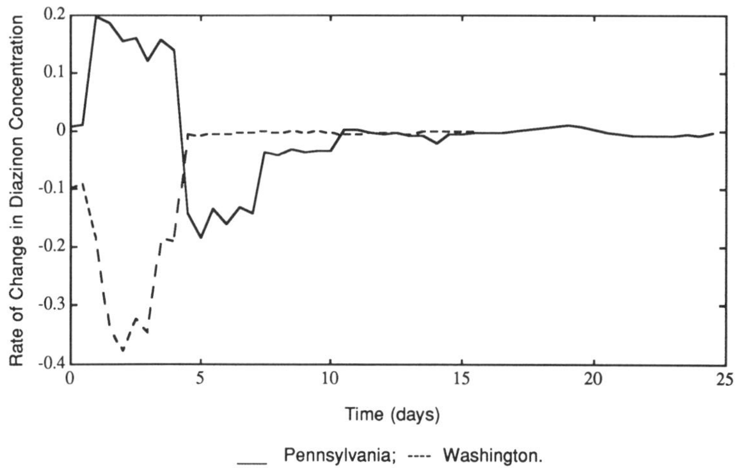

FIGURE 3. Change in diazinon concentrations within GI tracts.

$$\hat{y}_\lambda(d) = \frac{\sum_{j=1}^{n} K\left(\frac{d - d_j}{\lambda}\right) Y_j}{\sum_{j=1}^{n} K\left(\frac{d - d_j}{\lambda}\right)} \tag{1}$$

In this approach, determining the integer, λ, required repetitive calculations with incrementally changing λ values. The curve that best fits the experimental values was chosen as the proper model. Pesticide concentrations in GI tracts were evaluated for temporal patterns. As might be expected, an exponential decline (Figure 2) in pesticide concentration was found in the state of Washington. Here, pesticide uptake follows pesticide availability. However, dead birds from Pennsylvania orchards contained low concentrations during the first 2 days, after which time subsequent residue burdens rapidly increased and then similarly decreased. This was unexpected. Derivatives of chosen models were generated to determine the rate at which pesticide ingestion fluctuated as a function of environmental concentrations of these compounds. Exposure patterns were still unclear for the latter case (Figure 3).

The question remained, can mortality be predicted from environmental chemistry data? The final test was made to assess the feasibility of a model to predict mortality as a function of pesticide concentration in the environment. In these tests, data depicting temporal mortality were evaluated. These data show similar cumulative mortality patterns in the two study areas (Figure 4). Curve fits of the function:

$$M = K + P^{-c^q} \tag{2}$$

where M is mortality; c is concentration, and K, P, and q are constants

were used to evaluate death rate vs GI tract pesticide burden. Data are presented in Figures 5 and 6. The mortality curve from Pennsylvania is slightly sigmoidal. This shape indicates an increasing mortality rate following application with a subsequent reduction of mortality later in the study. Upper bounds were not known, however, so

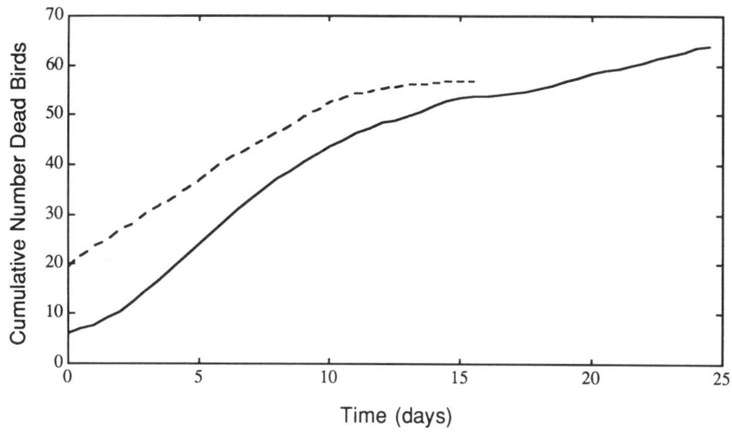

FIGURE 4. Cumulative passerine mortality in diazinon-treated fields.

simple functions were used in the plotting. These data illustrate the ability of well-planned investigations to predict mortality from environmental data.

Alkylphosphate data from fecal urate samples were analyzed in presence/absence format. The fraction of birds excreting alkylphosphate was determined and plotted over time. Exposure rates were used to develop temporal nonparametric curves (Figure 7). Excretion data from Washington orchards followed a logarithmic decay. Exposures in Pennsylvania followed a bimodal pattern with maxima at 4 and 12 days. The first apex corresponded to the time that maximum diazinon concentrations were found in GI tracts of dead birds. Patterns in both fecal urate data sets were quite similar to those found in GI tracts. These data hold promise for future evaluations of pesticide exposure and pesticide-induced mortality.

Data from heron eggs were simply tested for differences between contaminant concentrations in CAMs and eggs. Contaminant concentrations in CAMs were determined

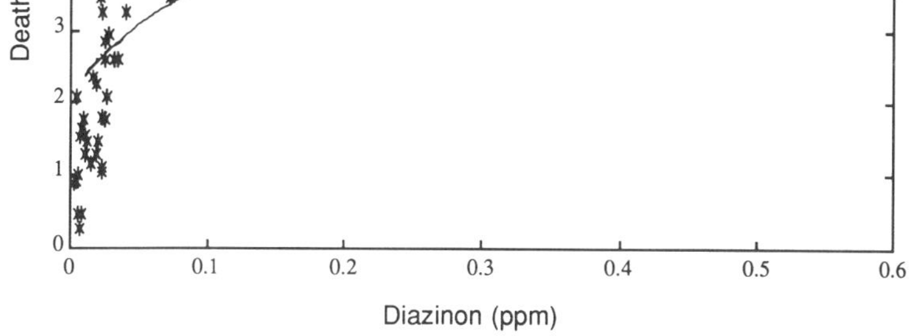

FIGURE 5. Passerine mortality rate vs pesticide concentrations in passerine GI tracts from Washington orchards.

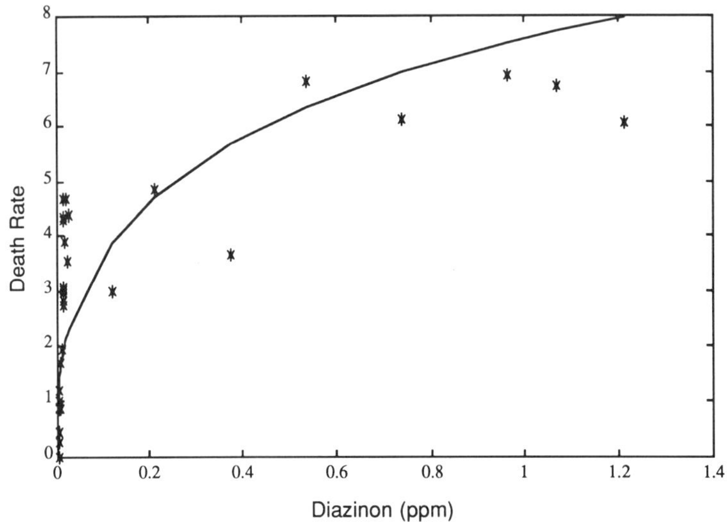

FIGURE 6. Passerine mortality rate vs pesticide concentrations in passerine GI tracts from Pennsylvania orchards.

to be equivalent to concentrations in whole eggs. CAMs from hatched eggs were then compared to data from whole eggs that had not been separated into CAM/egg tissues. Although sample sizes were insufficient for statistically significant comparisons, organochlorine contaminants from CAM tissues of hatched eggs were less than threefold different from concentrations in whole eggs. These data suggest an opportunity for nonlethal sampling to monitor contaminants that bioaccumulate in birds and may be transferred to their eggs.

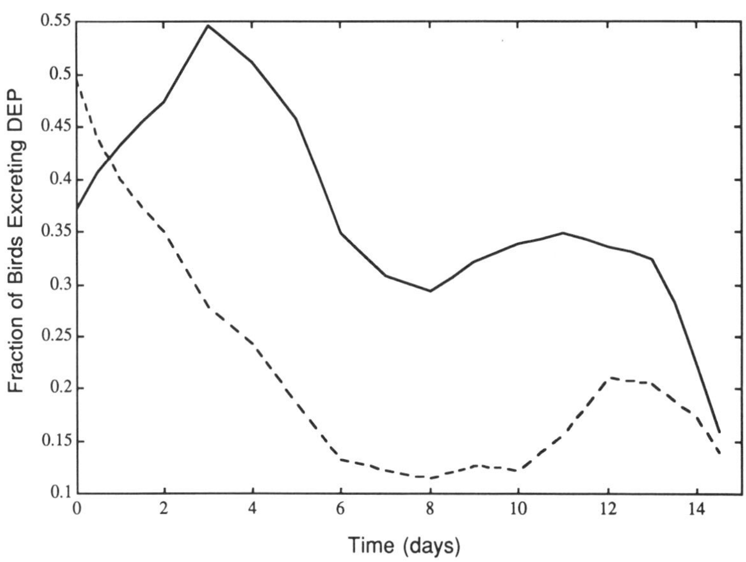

— Pennsylvania; ---- Washington.

FIGURE 7. Temporal alkylphosphate excretion by living passerines from diazinon-treated apple orchards.

ChE evaluations were made using data obtained following repeated treatments of apple orchards with the OP diazinon, or corn fields receiving a single application of COUNTER®, whose active ingredient is terbufos (*S*-[[(1,1-dimethylethyl)thio]methyl]-0,0-diethyl phosphorodithioate). Figure 8 demonstrates the effect of an OP on the plasma AChE from bluejay songbirds living on experimental corn fields. Birds living on control fields had AChE activities that fluctuated around normal levels. Birds inhabiting fields treated with either banded or infurrow applications had AChE activities depressed to approximately two thirds their normal activities. In this example, the birds reached their maximum depression during the 10- to 20-day interval following application. During the subsequent 10 days, activities began to return to normal levels.

Another approach to ChE evaluations is to compare individual plasma ChE activities from birds trapped on treatment fields with values obtained from a control, untreated location.

Utilizing a diagnostic threshold (one tailed, $p < 0.04$) for cholinesterase depression, temporal depression/recoveries were determined over the time of five distinct, sequential treatment events (Figure 9). The patterns for the five exposure/recovery processes proceed toward reduced exposure. ChEs were consistently impacted in the first few days following application and partially recovered during the period preceding the subsequent application. The initial increase in exposure during the last spray is unexplained. General patterns here compare well with the decline in residue concentrations determined by lethal and nonlethal sampling.

Although plasma ChE depression itself is not believed to decrease an animal's ability to survive, many studies have shown correlations between ChE depression and certain behavioral alterations. OP exposures capable of depressing ChE activity have been correlated with anorexia,[28,29] loss of predator aversion responses,[30,31] decreased tolerance to cold,[32] and inability to initiate nests or care for nestlings.[33,34] Much of the existing literature describes these correlations using brain ChE activity. Current research efforts need to be directed at extending these findings to plasma ChE activity depressions.

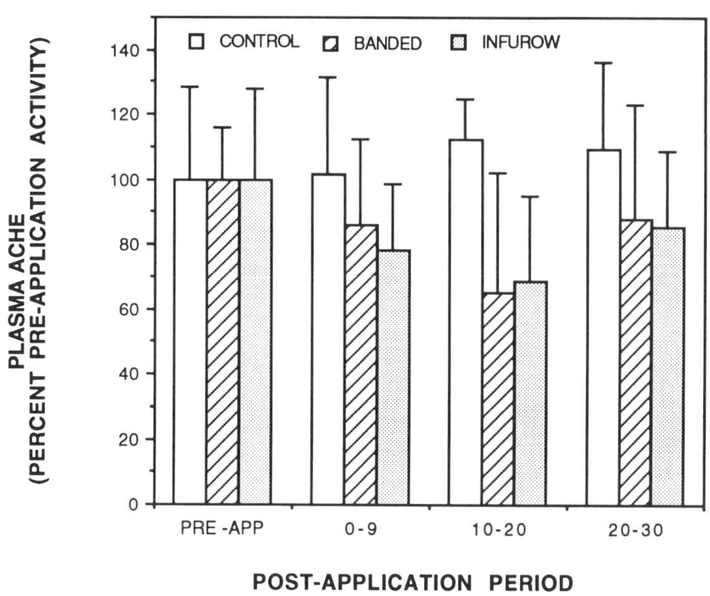

FIGURE 8. Plasma AChE response of bluejays during normal agricultural use of COUNTER®-15G.

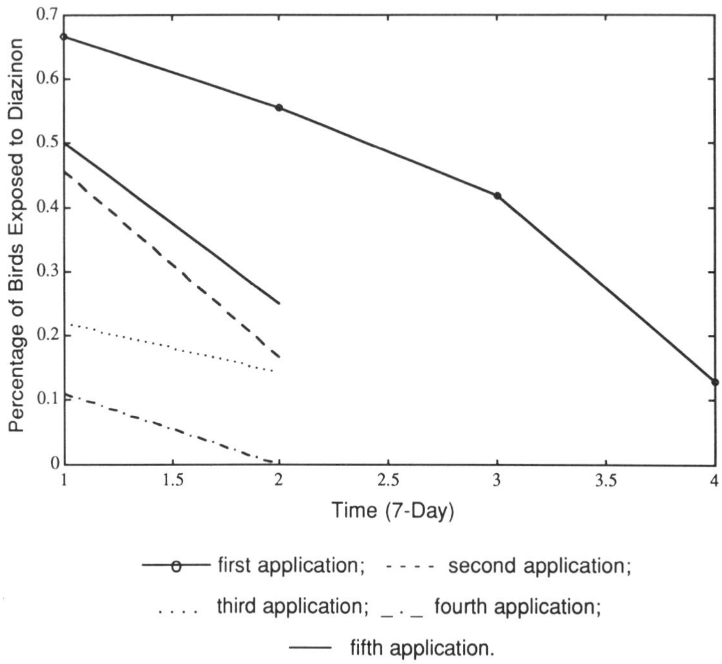

FIGURE 9. Fraction of passerines exhibiting ChE depression during 1-week intervals following diazinon application.

CONCLUSION

This review admittedly does not delve into the many complexities of chemical and biochemical toxicology. Rather, it explores a variety of nonlethal exposure evaluation techniques that can be used in concert with population modeling studies for correlations between population decline or perturbation and pesticide impact. Integration of nonlethal techniques discussed herein with those used for immunological function[35] and genetic aberration studies[36] can provide a broad basis for correlating population level impacts with chemical, biochemical, and physiological impacts in the individual.

ACKNOWLEDGMENTS

Funding for research described in this article was provided by the Washington State Department of Wildlife, U.S. Fish and Wildlife Service, CIBA-GEIGY Corporation, and American Cyanamid Company. Significant input into statistics was performed by Mr. Zhuojun Feng and Mr. E. H. H. Hol. Chemical analyses for CAM samples were performed by Mr. D. M. Norman.

REFERENCES

1. Fite, E. C., L. W. Turner, N. J. Cook, C. Stunkard and R. M. Lee, Guidance Document for Terrestrial Field Studies, Ecological Effects Branch, Hazard Evaluation Division, Office of Pesticide Programs, U.S. EPA, Washington, DC, 1987.

2. Bend, J. R., D. S. Miller, W. B. Kinter and D. B. Peakall, DDE induced microsomal mixed-function oxygenases in the puffin (*Fratercula arctica*), *Biochem. Pharmacol.*, 26, 1000–1001, 1977.
3. Riviere, J. L, Prochloraz, a potent inducer of the microsomal cytochrome P-450 system, *Pestic. Biochem. Physiol.*, 19, 44–52, 1983.
4. Payne, J. F., L. L. Fancey, A. D. Rahimtula and E. L. Porter, Review and a perspective on the use of mixed-function oxygenase enzymes in biological monitoring, *Comp. Biochem. Physiol.*, 86C(2), 233–245, 1987.
5. Walters, P., S. Kahn, P. J. O'Brien, J. F. Payne and A. D. Rahimtula, Effectiveness of a Prudhoe Bay crude oil and its aliphatic, aromatic and heterocyclic fractions in inducing mortality and aryl hydrocarbon hydroxylase in chick embryo in ovo, *Arch. Toxicol.*, 60, 454–459, 1987.
6. Rattner, B. A., D. J. Hoffman and C. M. Marn, Use of mixed-function oxygenases to monitor contaminant exposure in wildlife, *Environ. Toxicol. Chem.*, 8, 1093–1102, 1989.
7. Wills, J. H, The measurement and significance of changes in the cholinesterase activities of erythrocytes and plasma in man and animals, *Crit. Rev. Toxicol.*, 1, 153–202, 1972.
8. Hill, E. F., Sex and storage affect cholinesterase activity in blood plasma of Japanese quail, *J. Wildl. Dis.*, 25(4), 580–585, 1989.
9. Hooper, M. J., P. J. Dietrich, C. P. Weisskopf and B. W. Wilson, Organophosphorous insecticide exposure in hawks inhabiting orchards during after dormant-spraying, *Bull. Environ. Contam. and Toxicol.*, 42, 651–659, 1989.
10. Hooper, M. J., L. W. Brewer, G. P. Cobb and R. J. Kendall, An integrated laboratory and field approach for assessing hazards of pesticide exposure to wildlife, in *Pesticide Effects on Terrestrial Wildlife*, L. Somerville and C. H. Walker, Eds., Taylor & Francis, New York, 1990, pp. 271–283.
11. Fairbrother, A., B. T. Marden, J. K. Bennett and M. J. Hooper, Methods used in determination of cholinesterase activity, in *Cholinesterase-inhibiting Insecticides—Their Impact on Wildlife and the Environment*, P. Mineau, Ed., Elsevier, Amsterdam, 1991, pp. 35–72.
12. Fairbrother, A., and J. K. Bennett, The usefulness of cholinesterase measurements, *J. Wildl. Dis.*, 24, 587–590, 1988.
13. Westlake, G. E., P. J. Bunyan, A. D. Martin, P. I. Stanley and L. C. Steed, Organophosphate pesticides on plasma enzymes and brain esterases of Japanese quail, *J. Agric. Food Chem.*, 29, 772–778, 1981.
14. Hill, E. F., Brain cholinesterase activity of apparently normal birds, *J. Wildl. Dis.*, 24, 54–60, 1988.
15. Cobb, G. P., E. H. Hol, R. B. Lestina and R. J. Kendall, Wildlife sentinels as indicators of environmental contamination, paper presented at 199th Annual Meeting of the American Chemical Society, Boston, April 1990.
16. Hooper, M. J., L. L. Carlock, S. L. Tank, L. W. Brewer and R. S. Mellott, Biochemical impacts of organophosphate insecticides at the population level, paper presented at 199th Annual Meeting of the American Chemical Society, Boston, April 1990.
17. Mellott, R. S., L. W. Brewer, M. J. Hooper, G. P. Cobb III, R. A. Hummell, C. Crabtree and R. J. Kendall, The Response of Wildlife Exposed to Multiple Applications of Diazinon 50W in Apple Orchards of Eastern Washington and South Central Pennsylvania, The Institute of Wildlife and Environmental Toxicology, Clemson University, final research report submitted to CIBA-GEIGY Corp., Greensboro, NC, 1990.
18. Tank, S. L., L. W. Brewer, M. J. Hooper, G. P. Cobb III and R. J. Kendall, Third Year Investigation of the Effects of COUNTER® 15-G Insecticide-Nematicide on Selected Wildlife Species in a Corn Agroecosystem, The Institute of Wildlife and Environmental Toxicology, final research report submitted to American Cyanamid Company, Princeton, NJ, 1990.
19. Cobb, G. P., D. M. Norman and R. J. Kendall, Organochlorine Contaminants in Great Blue Herons Collected from the Puget Sound Ecosystem, U.S. Fish and Wildlife Service, Washington, DC, 1990.
20. Brewer, L. W., D. E. Dix, G. P. Cobb, M. J. Hooper, S. L. Tank, R. A. Craft, J. A. Buck, W. M. Buerger and C. M. Bens, Second Year Investigation of the Response of Selected

Wildlife Populations to Planting Time Application of COUNTER® 15-G Systemic Insecticide-Nematicide in an Iowa Corn Agroecosystem, Report submitted to American Cyanamid Company and U.S. EPA, 1989.

21. Weisskopf, C. P., J. N. Seiber, N. Maizlsh and M. Schenker, Personnel exposure to diazinon in a supervised pest eradication program, *Arch. Environ. Contam. Toxicol.*, 17, 201–212, 1988.
22. Cobb, G. P., L. W. Brewer and R. J. Kendall, Degradation of terbufos in soils during drought conditions, in Proc. Pesticides in Terrestrial and Aquatic Environments, Richmond, VA, May 1989.
23. Cobb, G. P., R. B. Lestina, E. H. Hol, S. L. Tank and R. S. Mellott, Determination of chemical movement in agroecosystems, in paper presented at 11th Annual Meeting of Society of Environmental Toxicology and Chemistry, Washington, DC, November 1990.
24. Weisskopf, C. P., and J. N. Seiber, Simple method for analysis of dialkylphosphate transformation products of organophosphorus pesticides at the population level, paper presented at 199th Annual Meeting of the American Chemical Society, Boston, April 1990.
25. Won, H., and R. Thurle, Method of Analysis, Can. Wildl. Serv. Rep. CWS-87-00, National Wildlife Research Center, Ottawa, Ontario, 1988.
26. Atkins, P. W, *Physical Chemistry*, W. H. Freeman, San Francisco, 1982.
27. Eubank, R. L, *Spline Smoothing and Nonparametric Regression*, Marcel Dekker, New York, 1988.
28. Bennett, R. S., Jr., and H. H. Prince, Influence of agricultural pesticides on food preference and consumption by ringnecked pheasants, *J. Wildl. Manage.*, 45(1), 74–82, 1981.
29. Grue, C. E, Response of common grackles to dietary concentrations of four organophosphate pesticides, *Arch. Environ. Contam. Toxicol.*, 11, 327–335, 1982.
30. Galindo, J., R. J. Kendall, C. D. Driver and T. E. Lacher, Jr., The effects on methyl parathion on susceptibility of bobwhite quail to domestic house cat predation, *Behav. Neural Biol.*, 43, 21–36, 1984.
31. Hunt, K. A., Fenthion as a Secondary Poisoning Hazard to American Kestrels, M.S. thesis, McGill University, Montreal, Quebec, Canada, 1990.
32. Rattner, B. A., and J. C. Franson, Methyl parathion and fenvalerate toxicity in American kestrels: acute physiological responses and effects of cold, *Can. J. Physiol. Pharmacol.*, 62, 787–792, 1984.
33. Brewer, L. W., C. J. Driver, R. J. Kendall, C. Zenier and T. E. Lacher, Jr., Effects of methyl parathion on ducks and duck broods, *Environ. Toxicol. Chem.*, 7, 375–379, 1988.
34. Robinson, S. C., R. J. Kendall, R. Robinson, C. J. Driver and T. E. Lacher, Effects of agricultural spraying of methyl parathion on cholinesterase activity and reproductive success in wild starlings (*Sturnus vulgaris*), *Environ. Toxicol. Chem.*, 7, 343–349, 1988.
35. Fairbrother, A., Immunotoxicology of captive and wild birds, in *Wildlife Toxicology and Population Modeling: Integrated Studies of Agroecosystems*, R. J. Kendall and T. E. Lacher, Jr., Eds., Lewis Publishers, Chelsea, MI, 1993.
36. McBee, K., Adaptation of mutagenicity and clastogenicity assays for avian species, in *Wildlife Toxicology and Population Modeling: Integrated Studies of Agroecosystems*, R. J. Kendall and T. E. Lacher, Jr., Eds., Lewis Publishers, Chelsea, MI, 1993.

6

Measurements of Toxicity and Critical Stages of Develoment

David J. Hoffman

ABSTRACT

First reports of anthropogenic environmental contaminants causing adverse effects in wildlife began to accumulate during the Industrial Revolution of the 1850s, when arsenic emissions, hydrogen sulfide fumes, and lead ingestion were implicated in wildlife mortality. Studies on the effects of modern pesticides on wildlife began soon after the introduction of DDT in 1943. Measured endpoints of toxicity in laboratory and field studies have included lethality, reproductive impairment, behavioral aberrations, alterations in growth and development, and changes in physiological indicators that may foreshadow effects crucial to reproduction and survival. In toxicity evaluation of agricultural and other environmental contaminants, critical stages of development in which vertebrate species are most vulnerable include embryogenesis and the neonatal or early posthatching period. This chapter focuses on effects during these critical periods of development. Measurements discussed include embryotoxicity, teratogenicity, and physiological perturbations in avian embryos and hatchlings. Examples are drawn upon from both laboratory and field studies including effects of petroleum, insecticides, herbicides, and agricultural drainwater contaminants. Interactive effects among toxicants and other environmental factors are discussed.

KEY WORDS

embryo, nestling, teratogenic, embryotoxic, pesticide, petroleum

INTRODUCTION

Since ancient times an awareness has existed of critical stages in avian development.[1] Artificial incubation of eggs was known in ancient Egypt, and Hippocrates and Aristotle recorded variability in morphological appearance and viability of bird embryos during the hatching of eggs. During the 18th and 19th centuries, pioneer experimental embryologists first became aware that environmental factors, including xenobiotics, could disrupt normal avian development.[2] Table 1 gives common and chemical names of some of the insecticides and herbicides reviewed in this chapter.

Reports of anthropogenic environmental contaminants affecting free-ranging wildlife first began to accumulate during the Industrial Revolution of the 1850s. Early reports included cases of arsenic poisoning and lead shot ingestion as well as industrial

Table 1. Common and Chemical Names of Pesticides Reviewed

Common Name	Chemical Name
	Insecticides
Acephate	Acetylphosphoramidothioic acid *O*, *S*-dimethyl ester
Azinphos methyl	Phosphorodithioic acid *O,O*-dimethyl *S*-[(4-oxo-1,2,3-benzotriazin-3(4*H*)-yl)methyl] ester
Carbaryl	1-Naphthalenol methylcarbamate
DDT	1,1,1-Trichloro-2,2-*bis*(p-chlorophenyl)ethane (by common usage, an isomeric mixture of dichlorodiphenyl-trichlorethane in which the p,p′ isomer is not less than 60 to 70%)
Decamethrin	(3-(2,2-dibromoethenyl)-2,2-Dimethyl cyclo-propane carboxylic ester with m-phenoxymandelonitrile
Diazinon	Phosphorothioic acid *O,O*-diethyl O-[6-methyl-2-(1-methylethyl)-4-pyrimidinyl] ester
Dichlorvos	Phosphoric acid 2,2-dichloroethenyl dimethyl ester
Dicrotophos	Phosphoric acid dimethyl ester, ester with *cis*-3-hydroxy-N,N-dimethylcrotonamide
Dimethoate	Phosphorodithioic acid *O,O*-dimethyl *S*-[2-(methylamino)-2-oxoethyl] ester
Endrin	1,2,3,4,10,10-Hexachloro-6,7-epoxy-1,4,4a,5,6,7,8,8a-octa-hydro-endo-endo-1,4:5,8-dimethanonaphthalene
EPN	*O*-Ethyl-*O*-(p-nitrophenyl) phenylphosphonothioate
Lindane	1,2,3,4,5,6-Hexachloro-cyclohexane, gamma isomer of not less than 99% purity
Malathion	[(dimethoxyphosphino-thioyl)thio]butanedioic acid diethyl ester
Methomyl	N-[[(methylamino)carbonoyl]oxy] Ethanimidothioic acid methyl ester
Methyl demeton	Phosphorothioic acid *O*-[2-(ethylthio)ethyl] *O,O*-dimethyl ester and *S*-[2-(ethylthio)ethyl] *O,O*-dimethyl phosphorothioate
Monocrotophos	(*E*)-Phosphoric acid dimethyl ester, ester with 3-hydroxy-N-methylcrotonamide
Naled	Phosphoric acid 1,2-dibromo-2,2-dichloroethyl dimethyl ester
Parathion	Phosphorothioic acid *O,O*-diethyl *O*-(4-nitrophenyl)ester

Table 1. Continued

Common Name	Chemical Name
Permethrin	3-(phenoxyphenyl)-methyl(±)-*cis,trans*-3-(2,2-dichloroethenyl)-2,2-dimethyl-cyclopropane-carboxylate (specify isomeric composition by ratios or percentages of active isomers)
Phosmet	Phosphorodithioic acid *S*-[(1,3-dihydro-1,3-dioxo-2H-isoindol-2-yl)methyl]*O,O*-dimethyl ester
Sulprofos	*O*-Ethyl *O*-[4-(methylthio)phenyl] phosphorodithioic acid *S*-propyl ester
Temephos	Phosphorothioic acid *O,O′*-dimethyl ester *O,O*-diester with 4,4′-thiodiphenol
Toxaphene	Chlorinated camphene containing 67–68% chlorine
Trichlorfon	(2,2,2-trichloro-1-hydroxyethyl)-Phosphonic acid dimethyl ester
	Herbicides
Amitrole	3-Amino-1H-1,2,3-triazole
Atrazine	6-Chloro-N-ethyl-N′-(1-methylethyl)-1,3,5-triazine-2,4-diamine
Bifenox	Methyl-5-(2,4-dichlorophenoxy)-2-nitrobenzoate
Bromoxynil; MCPA	3,5-Dibromo-4-hydroxybenzonitrile; 2-methyl-4 chlorophenoxyacetic acid
2,4-D	(2,4-dichlorophenoxy)-Acetic acid
Dalapon	2,2-Dichloropropanoic acid
Dicamba	3,6-Dichloro-2-methoxybenzoic acid
Diclofopmethyl	Methyl 2-[4-(2,4-dichlorophenoxy) phenoxy]propionate
Fosamine-ammonium	Ammonium ethyl carbamoylphosphonate
Glyphosate	N-(phosphonomethyl)-Glycine
Nitrofen	2,4-Dichlorophenyl-p-nitrophenylether
Oxyfluorfen	2-Chloro-1-(3-ethoxy-4-nitrophenoxy) 4-(trifluoromethyl)benzene
Paraquat	1,1′-Dimethyl-4,4′-bipyridinium
Picloram	4-Amino-3,5,6-trichloro-2-pyridine carboxylic acid
Prometon	6-Methoxy-N,N′-*bis*(1-methylethyl)-1,3,5-triazine-2,4-diamine
Propanil	N-(3,4-dichlorophenyl)-Propanamide
2,4,5-T	(2,4,5-trichlorophenoxy)-Acetic acid
Trifluralin	2,6-Dinitro-N-N-dipropyl-4-(trifluoromethyl) benzenamine

smokestack emission toxicity.[1] The use of synthetic organic insecticides became common after World War II. It soon became apparent that modern insecticides had considerable biological potency to poison and also to persist in the environment within nontarget organisms such as fish and wildlife.

As a consequence of wildlife losses due to modern pesticides, screening of pesticides for adverse effects became important, and standardized avian testing protocols were developed by the U.S. Fish and Wildlife Service (U.S. FWS).[3] Lethality is the principal endpoint in these basic protocols, which include acute tests—single oral LD_{50} with adult bobwhite (*Colinus virginianus*) and mallards (*Anas platyrhynchos*), and mallard embryo tests—and subacute tests (5 days feeding LC_{50}). Subchronic tests examine sublethal effects as well. Pen reproductive tests as well as *pen in field* and *field trials* have been utilized, but less frequently.

Population-level effects of pesticides may occur through reduction in survival at different stages of the life cycle. In order to model population-level effects of pesticides and other environmental contaminants, the potential impact on survival and reproduction must be established. In assessing toxicity at different stages of the life cycle, with experimental studies of both captive and free-ranging wildlife species, indicators of contaminant toxicity have included reproductive impairment; behavioral aberrations; alterations in growth and development, including teratogenesis; and physiological changes that may foreshadow effects crucial to reproduction, development, and survival. In toxicity evaluation of agricultural and other environmental contaminants, critical stages of development in which vertebrate species are most vulnerable include embryogenesis and the neonatal or early posthatching period. This chapter focuses on contaminant effects during these critical periods of development. Endpoints discussed include measurements of embryotoxicity, teratogenicity, and physiological perturbations in avian embryos and hatchlings. Examples are drawn upon from both laboratory and field studies. The effect(s) of multiple contaminants and other interactive factors are discussed.

MEASUREMENT OF EMBRYOTOXICITY AND TERATOGENICITY FOLLOWING TOPICAL EXPOSURE OF BIRD EGGS TO PESTICIDES

In evaluating the potential hazard of external exposure of mallard eggs to petroleum, pesticides, and industrial effluents, tests conducted on over 70 environmental contaminants established that 8 of 30 pesticides tested caused teratogenic effects at exposure levels well below the calculated median lethal levels.[4] Other measurements of embryotoxicity included presence of subcutaneous edema, blisters, hemorrhages, and stunted growth. Studies with different species but similar classes of pesticides often lead to similar conclusions.[2] Findings for insecticides and herbicides are discussed in the following sections.

Insecticides

Hoffman and Albers[4] determined the order of absolute toxicity of commercial formulations for 14 insecticides in aqueous emulsion as reflected by LC_{50} values (Table 2). Embryonated mallard eggs were exposed by immersion upon 3 days of incubation and allowed to incubate through 18 days. In these studies concentrations of insecticides were based on the active ingredient. The amount of aqueous emulsions remaining on eggs after immersion was determined by increase in weight and fell within the estimated amount expected after field spraying at a commonly used rate of 100 gal/acre

Table 2. Effects of Insecticides Applied to Bird Eggs[a]

Insecticide (Formulation)	Species	Effects	Ref.
Acephate (Orthene®)	Mallard	LC_{50} = 65 g/L; defects of bill, eye, and neck; stunted growth; edema	4
Azinophos methyl (Gusathion®)	Pheasant, gray partridge, red-legged partridge	0.7–1 L/ha caused mortality, lordosis of axial skeleton, altered sexual differentiation, paralysis	2
Carbaryl (Ortho liquid Sevin®)	Mallard	LC_{50}>259 g/L; no apparent adverse effects	4
DDT (in xylene and fuel oil)	11 species, mostly passerines	At 119 g/L; few effects apparent	2
DDT (unspecified commercial formulation)	Japanese quail	2.5 g/L caused gonad germ deficit	2
Decamethrin (unspecified commercial formulation containing xylene)	Japanese quail	0.5 g/L caused embryonic mortality and sterility due to excipient portion	2
Diazinon (Diazinon® AG 500)	Mallard	LC_{50} = 89 g/L; incomplete caudal ossification, stunted growths; moderate depression of brain AChE	4,5
Dichlorvos (Nogo®)	Ring-necked pheasant	0.1% decreased hatching success and caused paralysis	2
Dichlorvos (DDVP)	Japanese quail	0.00218 mg/m^3 air caused embryonic mortality, lordosis, and scoliosis of the axial skeleton	2
Dimethoate (Cygon® 2E)	Mallard	LC_{50} = 30 g/L; defects of bill, brain, and joints; gastroschisis; stunted growth	4
Endrin (Endrin 1.6 EC)	Mallard	LC_{50} = 30 g/L; defects of brain and limbs	4
EPN (Velsicol EPN)	Mallard	12 g EPN/egg caused mortality; stunted growth; axial scoliosis; edema; depressed brain AChE and NTE activity	2
Lindane (UNICO Lindane, EC)	Mallard	LC_{50} = 74 g/L; defects of bill, brain, eye, limb, liver; stunted growth	4
Malathion (UNICO Malathion 51 spray)	Mallard	LC_{50} = 142 g/L; defects of bill and limb; scoliosis of axial skeleton; stunted growth; slight depression of brain AChE and plasma ChE activities	4,5
Methomyl (Lannate® L)	Mallard	LC_{50}>215 g/L; defects of neck; edema; stunted growth	4
Methyl demeton (Metasystox®)	Ring-necked pheasant	Decreased hatching at receipt of 0.1%; paralysis; cervical scoliosis	2
Monochrotophos (Azodrin®)	Chicken, bobwhite, quail	2.1 μg/cm^2 of egg surface caused no apparent adverse effect	2
Parathion (unspecified commercial formulation)	Chicken, Japanese quail, duck, ring-necked pheasant	20 μg/L caused mortality, lordosis, anuria, defects of urogenital system, edema, sexual dysfunction	2

Table 2. Continued

Insecticide (Formulation)	Species	Effects	Ref.
Parathion (unspecified commercial formulation)	Japanese quail	10% active ingredient caused mortality and severe malformations of the axial skeleton; inhibited brain AChE and BChE activities	2
Parathion (Parathion 8 EC)	Mallard	LC_{50} = 49 g/L; scoliosis and lordosis of axial skeleton; edema; stunted growth; inhibited brain AChE and plasma ChE activities	4,5
Permethrin (Pounce® 3.2 EC)	Mallard	LC_{50}>48 g/L; not teratogenic	4
Phosmet (Imidan® 50 WP)	Mallard	LC_{50}>600 g/L; not teratogenic	4
Sulprofos (Bolstar®)	Mallard	LC_{50} = 41 g/L; defects of bill, eye, and neck; axial skeletal scoliosis; edema	4
Temephos (Abate® 4E)	Mallard	LC_{50} = 79 g/L; not teratogenic	4
Toxaphene (UNICO Toxaphene l.s.c.)	Mallard	LC_{50} = 130 g/L; defects of joints; stunted growth	4

[a]Modified from Hoffman.[2] Insecticides were applied by immersion or spraying of eggs in three studies.

(935 L/ha). The LC_{50} values for these insecticides ranged from 30 g/L to >600 g/L. The order of absolute toxicity of insecticides in aqueous emulsion (LC_{50} values) was endrin > sulprofos > parathion > acephate > lindane > temephos > diazinon > dimethoate > toxaphene > malathion > carbaryl, permethrin, phosmet, and methomyl. However, the order of potential hazard in terms of the highest permissible field level of application in the United States was toxaphene > malathion > endrin ≈ dimethoate > lindane > sulprofos > diazinon > parathion > acephate > temephos > carbaryl, phosmet, methomyl, and permethrin. Apparent differences between the ranking based on absolute toxicity and potential hazard were due to differences in maximum permissible levels of application, which, for example, were extremely high for toxaphene and malathion. Insecticides with LC_{50} values that occurred at approximately ten times the highest permissible field level of application or less included dimethoate, endrin, malathion, and toxaphene. Of these, all were teratogenic, but only dimethoate and endrin were teratogenic at levels below the LC_{50}.

Of all classes of insecticides studied, the greatest focus has probably been on organophosphates. At least four different investigators have examined the embryotoxic effects of parathion after immersion of eggs of the following species: chickens, Japanese quail (*Coturnix japonica*), mallards, and pheasants (*Phasianus colchicus*), as summarized in a recent review.[2] Lutz-Ostertag et al. (see Hoffman[2]) first reported that an unspecified commercial formulation of parathion in acetone solution at 20 g/L resulted in embryonic mortality, lordosis of the axial skeleton, anuria, altered sexual differentiation, and edema by 15 days of incubation. Hoffman and Albers[4] found an LC_{50} for parathion 8 EC of 49 g/L when mallard eggs were immersed upon 3 days of incubation and observed through 18 days of incubation. However, concentrations well below the LC_{50} were teratogenic, causing axial skeleton scoliosis and lordosis as well as edema and stunted growth.[4,5] Subsequent observations showed that many of these abnormal embryos were unable to hatch, and this would have significantly lowered the calculated LC_{50}. Many other organophosphate insecticides studied, including azinphos-methyl, methyl demeton, dichlorvos, acephate, EPN, malathion, and sulprofos resulted in defects characterized by a shortening and contortion (scoliosis and lordosis) of the axial skeleton (Figure 1).[2,4] On the other hand, several organophosphates, including phosmet and temephos resulted in few teratogenic effects in mallards,[4] and monocrotophos resulted in few effects in chickens and bobwhite.[2]

Herbicides

The order of absolute toxicity of commercial formulations for 14 herbicides in aqueous emulsion, as reflected by the LC_{50} values, was determined when embryonated mallard eggs were exposed by immersion upon 3 days of incubation.[4,6] The LC_{50} values spanned a range of 1.8 to 479 g/L (Table 3). The order of absolute toxicity of herbicides in aqueous emulsion (LC_{50} values) was paraquat ≈ trifluralin > propanil, bromoxynil with MCPA (Bronate®) > methyl-diclofop > prometon > picloram > 2,4,5-T > amitrole > glyphosate > 2,4-D > atrazine, dicamba, and dalapon. However, the order of potential hazard in terms of the highest permissible field level of application in the United States was trifluralin > paraquat > prometon > methyldiclofop ≈ propanil > bromoxynil with MCPA (Bronate®) > picloram > amitrole > 2,4,5-T > 2,4-D, glyphosate > atrazine and dicamba. Herbicides with LC_{50} values that occurred at ten times the field level of application or less included bromoxynil with MCPA, methyldiclofop, paraquat, prometon, propanil, and trifluralin. The phenoxy herbicides appear to be the most widely studied class of herbicides with respect to potential embryotoxicity of spray application to bird eggs. The earliest reported studies with 2,4-D and 2,4,5-T by Lutz and Lutz-Ostertag and Lutz-Ostertag and Didier (see Hoffman[2]) examined the effects of immersion on eggs of four different species [Japanese quail (*Coturnix japonica*), gray partridge (*Perdix perdix*), red-legged partridge (*Alectoris rufa*), and chicken]. At concentrations of 4000 to 10,000 ppm, mortality of 25 to 35% occurred with accompanying malformations which included abnormal gonads, stunted growth, and fused vertebrae. Unfortunately, neither the formulation nor the derivative used was provided by the investigators.

Kenaga,[7] in a review evaluating the safety of 2,4,5-T to birds, concluded that the formulation ingredients can be just as important as the herbicide. An extensive study conducted by Somers et al. (see Hoffman[2]) reported no adverse effects after spraying 2,4-D (Esteron®99) at 15 g/L (10 times the normal application rate) on chicken eggs. Similarly, Somers et al. reported no adverse effects for 2,4,5-T (Esteron® 245) at 15 g/L.

Four other studies that examined the effects of 2,4-D or 2,4,5-T did not detect any apparent adverse effects in Japanese quail, pheasants, or mallards at levels below 30 times the field level of application.[2,4] Paraquat appears to be the most highly embryotoxic and teratogenic herbicide identified by three different investigators in chicken, Japanese quail, and mallard embryos when eggs were tested by spraying or immersion.[2,4] Lutz-Ostertag and Henou (see Hoffman[2]) first demonstrated that spray applications of 4 g/L concentration on Japanese quail eggs and chicken eggs caused embryonic mortality and defects. Paulov (see Hoffman[2]) reported a decrease in hatching success in Japanese quail following immersion in a concentration of 0.5 g/L (500 ppm). Hoffman and Albers[4] determined the LC_{50} following immersion of mallard eggs to be 1.8 g/L as reflected by survival through 18 days of incubation. Additional testing[45] suggested that further mortality occurred at the time of hatching, resulting in an even lower LC_{50} in terms of hatching success. Overt anomalies of the brain as well as edema and stunted growth were apparent in mallards by 18 days of incubation.

FIGURE 1. (A) Control mallard embryo, 18 days old (left) and embryo, 18 days old, treated at 72 hr with EPN (right). Note shortening of axial skeleton, twisted neck, deformed foot, edema after EPN treatment. (B) Mallard embryo skeleton (18 days old) after treatment at 72 hr with EPN (left). Note scoliosis and incomplete ossification of caudal vertebrae after EPN treatment. Control embryo skeleton, 18 days old (right).

PHYSIOLOGICAL AND BIOCHEMICAL EFFECTS FOLLOWING TOPICAL EXPOSURE OF BIRD EGGS TO PESTICIDES

Parathion and dicrotophos resulted in severe axial deformities and strongly inhibited cholinesterase (ChE) activities, whereas malathion caused relatively few effects.[2] Similarly, Hoffman and Eastin[5] demonstrated an association between axial skeletal teratogenesis and cholinergic effects of parathion and diazinon in mallard embryos. When day-old hatchlings were examined, only parathion in oil vehicle had a significant residual effect on brain ChE activity. Plasma ChE activity in the hatchlings remained depressed by about 45% with parathion and by about 15% with diazinon.

Other neurotoxic and teratogenic effects in mallard embryos were seen following topical exposure at one to three times the field level of application of the organophosphate insecticide EPN, which has been used on a wide variety of crops including

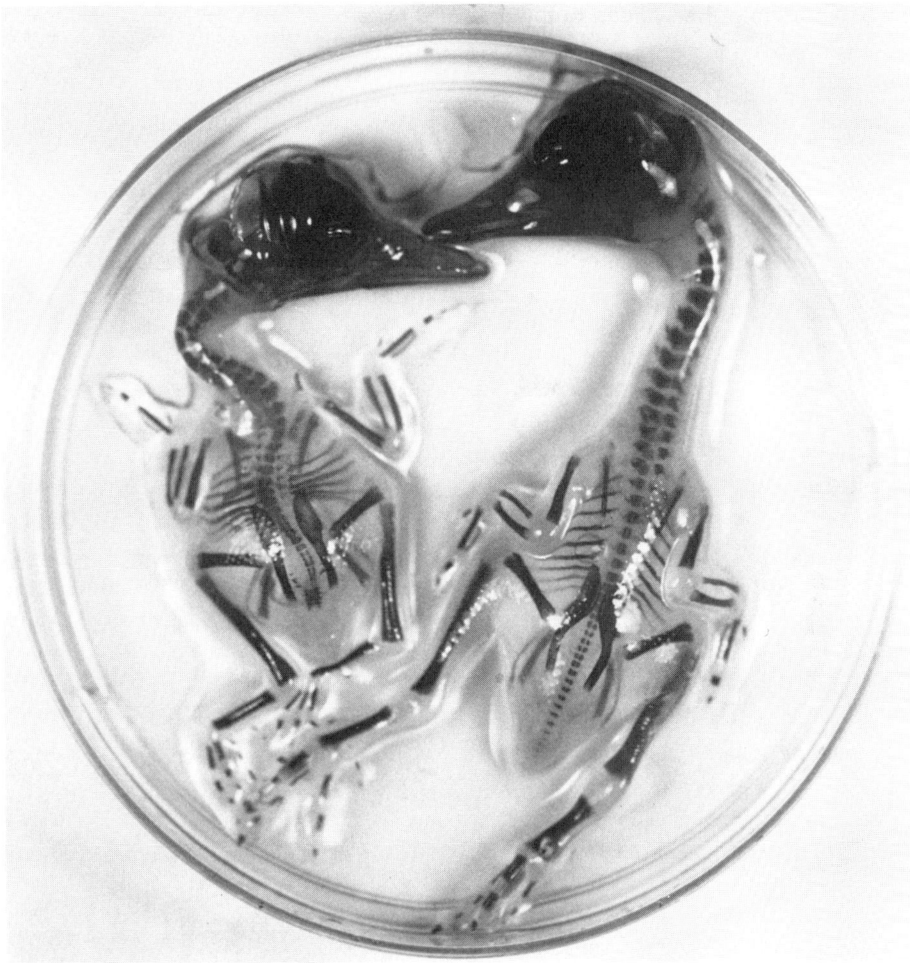

B

FIGURE 1. Continued

cotton, orchard fruit, corn, and soybeans.[2] Brain weights were significantly lower throughout development and brain neurotoxic esterase (NTE) and acetylcholinesterase (AChE) activities were inhibited. Duffard et al., as reviewed by Hoffman,[2] focused on the effects of 2,4-D on the central nervous system in developing chicken embryos, which exhibited postural and motor dysfunctions after the application of a commercial 2,4-D butylester in ethyl ether to eggs. Brain protein levels and total lipid levels were dose-dependently decreased by 2,4-D. Phospholipids were decreased in the highest dose group, gangliosides were increased, and cholesterol and glycolipids were diminished. Fosamine ammonium (Krenite®), a highly water-soluble phosphonate herbicide used to control woody brush, was found not to be very teratogenic to mallards and northern bobwhite after egg immersion.[6] Brain AChE activity was not depressed in embryos of either species 1 week after treatment or in hatchlings. Some alterations were seen but probably reflected developmental delays of physiological ontogeny rather than manifestations of severe toxicity.

External exposure of bobwhite eggs to Treflan® (trifluralin) by immersion at field-level application (2 g/L) reduced hatching success in mallard eggs,[4] but not in northern bobwhite.[2]. However, several physiological alterations as well as reduced hatching

Table 3. Effects of Herbicides Applied to Bird Eggs[a]

Herbicide (Formulation)	Species	Effects	Ref.
Amitrole (Amitrol T®)	Mallard	LC_{50} = 211 g/L; stunted growth	4
Atrazine (Aatrex® 4L)	Mallard	LC_{50} > 479 g/L; not teratogenic	4
Bromoyxnil + MCPA (Bronate®)	Mallard	LC_{50} = 9 g/L; eye defects; edema; stunted growth	4
2,4-D (isoctyl ester)	Chicken, ring-necked pheasant	No apparent effects at 6 g/L	4
2,4-D (unspecified formulation)	Chicken, Japanese quail, gray partridge	Abnormal gonads; stunted growth; fused vertebrae at 3.75 mL/L	2
2,4-D (Hedonal®)	Japanese quail	No apparent effects at 4 L/ha (conc. = 500 g/L)	2
2,4-D (U-46-D-Fluid)	Ring-necked pheasant, Japanese quail	Decreased hatching at 30X normal application; not teratogenic	2
2,4-D (Esteron® 99)	Chicken	No apparent effects at 15 g/L	2
2,4-D (butyl ester)	Chicken	LD_{50} = 5 mg/egg; motor dysfunction; altered brain biochemistry	2
2,4-D (dimethyl amine salt)	Mallard	LC_{50} = 230 g/L; not teratogenic	4
Dalapon (Dowpon® M)	Mallard	LC_{50} > 449 g/L; not teratogenic	4
Dicamba (Banvel®)	Mallard	LC_{50} > 240 g/L; eye defects; stunted growth	4
Fosamine ammonium (Krenite®)	Northern bobwhite, mallards	More reduced hatching in mallards than bobwhite at 65 g/L; not highly teratogenic	6
Glyphosate (Roundup®)	Mallard	LC_{50} = 213 g/L; not teratogenic	4
Methyldiclofop (Hoelon® 3EC)	Mallard	LC_{50} = 12 g/L; eye defects	4
Paraquat	Chicken, Japanese quail	4 g/L caused mortality and defects of the lung	2
Paraquat (Gramoxone®)	Japanese quail	0.5 g/L decreased hatching success	2
Paraquat (Ortho® paraquat CL)	Mallard	LC_{50} = 1.8 g/L; brain defects; edema; stunted growth	4

Table 3. Continued

Herbicide (Formulation)	Species	Effects	Ref.
Picloram (Tordon® 10K)	Mallard	LC_{50} = 120 g/L; stunted growth	4
Prometron (Pramitol® 25E)	Mallard	LC_{50} = 46 g/L; edema; stunted growth	4
Propanil (Stampede® 3E)	Mallard	LC_{50} = 8 g/L; limb and neck defects	4
2,4,5-T (unspecified formulation)	Chicken	4 g/L caused embryonic mortality, growth inhibition, abnormal gonads	2
2,4,5-T (Tormona® 80)	Japanese quail, ring-necked pheasant	Decreased hatching at 30X normal application; not teratogenic	2
2,4,5-T (Esteron® 245)	Chicken	No effects apparent at 15 g/L	2
2,4,5-T (isoctyl ester)	Mallard	LC_{50} = 127 g/L; not teratogenic	4
Trifluralin (Treflan®)	Mallard	LC_{50} = 2 g/L; bill defects; stunted growth	4
	Northern bobwhite	Hatching success not affected, but hatching weight lower at 38 g/L; decreased muscle and liver DNA to protein ratios	2

[a]Modification from Hoffman.[2] Herbicides were applied by immersion or spraying of eggs in these studies.

weight were reported for bobwhite. These included a decrease in the ratio of liver DNA to protein in embryos as well as a decrease in the ratio of muscle DNA and RNA to protein in hatchlings. It is difficult to speculate what long-term consequences such physiological alterations would have during growth and further development of the chicks.

MEASUREMENTS OF TOXICITY IN ALTRICIAL NESTLING BIRDS FOLLOWING CONTAMINANT INGESTION

Lead

The neonatal period in mammals is a highly critical period for accumulation and susceptibility to contaminant exposure, including lead. Similarly, lead (metallic) ingestion by nestling American kestrel (*Falco sparverius*)[8,9] was considerably more toxic than previously reported for adults and young precocial species, where LC_{50}s were greater than 5000 ppm.[10]

Overt toxicity was manifested by 40% mortality at the dose level 625 mg/kg, with significant impairment of growth occurring at 125 mg/kg (Table 4). Because nestling American kestrels consume approximately their own body weight in food per day, 625

Table 4. Effects of Contaminant Ingestion in Altricial Nestlings

Contaminant	Species	Exposure Method	Observation Period	Effects	Ref.
Bifenox	American kestrel	Daily oral	Days 1–10	500 mg/kg, high mortality, 250 mg/kg, reduced growth, altered physiology	Hoffman (unpublished)
Diazinon	European starling	Single oral	24 hr Postdosing	Day 1 LD_{50} = 13 mg/kg Fledgling LD_{50} = 145 mg/kg Adult LD_{50} = 630 mg/kg	20
Dicrotophos	European starling	Single oral (day 5 or day 15)	24 hr Postdosing	Day 5 LD_{50} = 4.9 mg/kg Day 15 LD_{50} = 9.0 mg/kg Depressed brain AChE and growth	21
Lead, metallic	American kestrel	Daily oral	Days 1–10	625 mg/kg, high mortality; 125 mg/kg, reduced growth; 25 mg/kg altered physiology	8,9
Lead, highway	Barn swallow	Nest sites along highway	Through day 18	3 mg/kg, lower ALAD activity	
	European starling	Nest boxes along highway	Through day 18	94 mg/kg, lower brain weight, altered physiology	37
Nitrofen	American kestrel	Daily oral	Days 1–10	500 mg/kg, complete mortality; 250 mg/kg, impaired growth; 50–250 mg/kg, altered physiology, including T4	Hoffman (unpublished)
Oxyfluorfen	American kestrel	Daily oral	Days 1–10	500 mg/kg, few effects	Hoffman (unpublished)
Paraquat	American kestrel	Daily oral	Days 1–10	60 mg/kg, high mortality; 10–25 mg/kg, reduced growth, altered physiology	12,13
Kuwait or South Louisiana crude oil	Herring gull	Single oral, 3–4 weeks old	9 days	0.3 mg/kg reduced growth, altered physiology	22
Weathered South Louisiana crude oil	Black Guillemot	Single oral, mixed ages	22 days	0.1–0.2 ml reduced growth, altered physiology	23

mg/kg of lead daily can be considered equivalent in daily lead intake to approximately 625 ppm of lead in the diet, and 125 mg/kg equivalent to 125 ppm in the diet.

Biochemical and hematological alterations in nestlings were also more severe than those reported in adult kestrels or precocial young birds exposed to lead. Lead poisoning due to ingestion of lead shot and food items containing lead would be expected to have a greater impact on nestlings of various altricial birds, including eagles, than previously reported for adults.

Bipyridinium Herbicides (Paraquat)

Paraquat ingestion by nestling American kestrels[11,12] was considerably more toxic than previously reported for adults and young precocial species. For northern bobwhite quail, Japanese quail, ring-necked pheasants, and mallards, the LC_{50}s ranged from nearly 1000 to >4000 ppm after a 5-day feeding trial.[10,13] In kestrels, 60 mg/kg of paraquat resulted in 44% mortality on days 4 to 8 of the experimental treatment period. The relatively high sensitivity of altricial nestlings to paraquat appears to parallel the sensitivity of embryos of precocial birds.

Paraquat altered growth in kestrels in a somewhat different manner than lead. For lead, overall growth curves were significantly different from controls over 10 days of exposure where a cubic polynomial was used to describe growth in each treatment group. Paraquat did not affect organ weights in the same manner as lead; although body weights were significantly less following treatment with paraquat, organ weights were not. Histologically, a localized focus of necrosis was observed in the liver of a high-dosed nestling and cellular degeneration and focal tubular dilation were present in the kidney of a second nestling. Congestion was apparent in the liver and kidney of one intermediate-dosed nestling. As histological examination was only conducted for nestlings that survived for 10 days, it is possible that a greater percentage of nestlings may have shown lesions if all nestlings, including those that died, had been examined. The lungs were histologically unremarkable, which is in contrast to the frequent occurrence of pulmonary lesions caused by either inhalation or ingestion of paraquat by mammals. Paraquat ingestion also caused a number of significant physiological and biochemical alterations apparent in the nestlings at 10 days. Plasma lactate dehydrogenase (LDH) activity and total protein concentration were elevated, and plasma alkaline phosphatase activity was lower in survivors of the 60 mg/kg group at 10 days. Lung total sulfhydryl (TSH) and protein-bound sulfhydryl (PBSH) concentrations were significantly higher in the 10, 25, or 60 mg/kg groups. Liver nonprotein sulfhydryl (NPSH) was lower in the 60 mg/kg group. Kidney DNA, RNA, and the ratio of RNA to protein concentration were higher in the 25 or 60 mg/kg groups. These findings suggest that altricial nestling kestrels are more sensitive to paraquat exposure than young or adult birds of precocial species. However, the lungs of nestling kestrels are less sensitive to paraquat than mammalian lungs.

Diphenyl Ether Herbicides (Bifenox, Nitrofen, Oxyfluorfen)

Bifenox ingestion by nestling American kestrels[45] was considerably more toxic than reported for precocial species; for ducks and pheasants, the 8-day LC_{50}s were reported to be greater than 5000 ppm in the diet.[14] Precocial young appear to be less sensitive than nestling kestrels to nitrofen ingestion. Bifenox at 500 mg/kg caused 66% mortality in nestling kestrels, while 500 mg/kg nitrofen caused 100% mortality within 5 days. Levels of bifenox or nitrofen at 250 mg/kg caused 14% mortality, and nitrofen at 50 mg/kg caused 9% mortality within 7 days. The higher sensitivity of altricial nestlings to diphenyl ether herbicides may parallel the developmental sensitivity of nitrofen in mammals exposed prenatally. However, bifenox has considerably less developmental toxicity than nitrofen in mammals.[15] No published avian reproductive studies exist for nitrofen or bifenox, so it is not possible to draw comparisons with avian species for reproductive effects or teratogenicity.

In addition to causing reduced growth and survival, nitrofen resulted in several signs suggestive of impending hepatoxicity, including an increase in liver weight relative to body weight. However, hepatic lesions were not apparent in kestrels or in ducklings, and only at dietary concentrations above 3000 ppm in rats.[16] Significant increases in plasma enzyme activities in the kestrels occurred for alanine aminotransferase (ALT), aspartate aminotransferase (AST), and LDH-L, which may be indicative of hepatic perturbations. Hepatic tissue glutathione (GSH) peroxidase activity was significantly higher in all nitrofen-treated groups and in the two treated with the highest doses of bifenox (50 and 250 mg/kg). Other alterations included an increase in plasma total thyroxine concentration (T4) by nitrofen, which may indicate thyromimetic activity, as suggested for mammals.[17,18] Because of nitrofen's teratogenicity and persistence in the aquatic ecosystem, it was withdrawn from commercial use.[19] Bifenox is currently in use, both as a pre- and postemergent herbicide to control a wide range of broadleaf weeds, certain grasses, and certain aquatic weeds. Bifenox is applied in the field at concentrations within 75% of the treatment administered to the kestrels in the present study (50 mg/mL at 250 mg/kg per nestling) that caused decreased survival and growth.

Therefore, these findings suggest some potential for bifenox to affect development of altricial nestlings in the wild that consume prey items contaminated internally and externally with bifenox. However, the potential of bifenox for nestling toxicity does not appear to be as great as that for paraquat. The potential for reproductive toxicity of bifenox in birds is unknown, but the threshold could occur at a lower concentration than for nestling toxicity.

Organophosphate Insecticides (Diazinon, Dicrotophos)

Hooper et al.[21] revealed that the 24-hour median lethal dose (LD_{50}) of the organophosphate insecticide diazinon for free-living, newly hatched European starling (*Sturnus vulgaris*) nestlings was < 1/10 that of their fledgling-aged counterparts and 1/50 that of adults; nestling sensitivity was supported by concurrent brain and plasma ChE depression. Grue and Shipley[20] reported that the LD_{50} of the organophosphate insecticide dicrotophos for free living 5-day-old nestling European starlings was about half that obtained for 15-day-old nestlings and adults. Brain ChE activity was severely depressed in all birds that died (74 to 94%), but degree of inhibition did not appear to vary with age. Greater body weight depression in dosed birds that died occurred in younger nestlings. Average weight loss in dosed survivors was greater for nestlings than adults. These findings suggest that young nestling songbirds may be nearly twice as sensitive as adults to organophosphate insecticides, and that growth of nestlings can be severely depressed.

Petroleum

Nestling birds are sensitive to exposure of petroleum hydrocarbons containing an aromatic fraction, but not as sensitive as embryos. A single small oral dose of Kuwait crude oil or South Louisiana crude oil at approximately 0.3 mL/kg caused reduced growth, osmoregulatory impairment, and hypertrophy of hepatic, adrenal, and nasal gland tissue in herring gull (*Larus argentatus*) chicks living in a simulated marine environment.[22] Weathered SLCO caused generally similar effects in black guillemot (*Cepphus grylle*) nestlings remaining within their natural habitat.[23] The authors concluded that these findings suggest oil ingestion by nestlings may cause long-lasting physiological effects that could reduce a young bird's ability to survive at sea.

FIELD EXPOSURE OF EMBRYOS AND NESTLINGS

Petroleum

The most complete evidence of embryotoxicity in the field due to direct topical exposure of bird eggs to contaminants comes from studies concerning oil pollution. Rittinghaus (see Hoffman[2]) first reported that contact exposure of bird eggs in the field was due to an environmental contaminant. During the breeding season, oil pollutants on the breast feathers, feet, and nesting material of sandwich terns (*Sterna sandvicencis*) were transferred to their eggs, impairing hatching ability. Birkhead et al. (see Hoffman[2]) reported a similar transfer of oil to eggs of great black-backed gulls (*Larus marinus*). Several controlled field studies confirmed these early reports. No. 2 fuel oil applied to the brood patch of laughing gulls (*Larus atricilla*) was transferred to the eggs, causing 41% embryonic mortality within 5 days.[24] Similar findings were reported for great black-backed gulls and herring gulls with this oil, and for herring gulls with Kuwait crude oil.[25] When artificial Prudhoe Bay oil slicks were created in water pans during the first week of incubation, as little as 5 ml/m^2 of water surface resulted in transfer of the oil to eggs with a marginal decrease ($p = 0.06$) in hatching success.[26]

Pesticides

Few attempts to measure field exposure of bird eggs to pesticides have been documented. Mitchell (see Hoffman[2]) sprayed the nests of 11 passerine species with typical field concentrations of DDT, but found virtually no adverse effects. Aerial spraying for western budworms (*Choristoneura occidentalis*) with relatively low volume trichlorfon and carbaryl in Montana forests did not alter the success of nests with eggs or young birds of various species.[27] However, application rates were lower than many pesticide field applications. Putera et al.[28] studied the effects of pesticides on mourning doves (*Zenaida macroura*) nesting in orchards in southern Illinois. The pesticides applied routinely by spraying usually consisted of a combination of an insecticide and a fungicide at intervals of approximately 10 days from March through September. Most adults incubating eggs left during the actual spraying but all returned within 30 minutes following application. Of the nests that contained eggs during spraying, 89% were unsuccessful, with most failures occurring 4 days or more after being sprayed.

Patnode and White[29] utilized a prototype experimental design to assess sublethal reproductive effects of multiple and varied organophosphate and carbamate operational exposures on avian productivity, allowing for classification of pesticide exposures according to toxicity of applied compounds and frequency of applications. Nest, egg, and nestling daily survival rates (DSRs) were determined for northern mockingbirds (*Mimus polyglottos*), brown thrashers (*Toxostoma rufum*), and northern cardinals (*Cardinalis cardinalis*) nesting along edges of pecan orchards and row crops in southern Georgia. Egg and nestling DSRs for all species combined varied inversely with exposure. Nestling growth was reduced with increasing exposure, but northern cardinal and northern mockingbird parenting behavior did not differ among the three exposure levels. Brain AChE activities were age dependent and substantiated adult, but not nestling, exposure. The authors concluded that increasing exposure to operational pesticide use may reduce songbird productivity.

PCB Related

Reproductive failure was noted among fish-eating birds in the lower Great Lakes in the early 1970s.[30] In the 1980s, subsequent examination of colonial fish-eating birds of Green Bay, WI, revealed a variety of deformities (e.g., crossed bills, reduced mandibles, and foot and eye problems) among nestlings of several species of terns (*Sturna* sp.) and double-crested cormorants (*Phalacrocorax auritus*). When tern eggs were collected from some of the colonies and artificially incubated, liver microsomal aryl hydrocarbon hydroxylase was found to be elevated more than threefold over control eggs of the same species collected from apparently normal colonies.[31] This finding, in combination with poor incubator hatching success of Green Bay eggs compared to control eggs, was suggestive of the presence of contaminants in the eggs themselves. Residue analysis has revealed high levels of several toxic polychlorinated biphenyl (PCB) congeners.[32]

An epidemic increase in the incidence of abnormalities in common and roseate terns (*Sturna dougalli*) was recorded on Long Island Sound, NY, in 1970, with the number of visible defects in young terns increasing to 1.3 from 0.1% in 1969.[33,34] The more prominent defects were feather loss among juveniles and eye, bill, and foot deformities among nestlings. Although the abnormalities were attributed to PCDD (polychlorinated dibenzo-p-dioxins), PCBs, and mercury, cause-effect relationships have not been verified.

Selenium

During Spring 1983, multiple malformations were encountered as well as poor hatching success in a natural population of aquatic birds exposed to agricultural drainwater ponds with food items containing high concentrations of selenium at Kesterson Reservoir in central California.[35,36] Embryonic death was generally high; approximately 17 to 60% of the nests of different species contained at least one dead embryo. The incidence of malformed embryos was also high; approximately 22 to 65% of the nests where at least two embryos were examined contained abnormal embryos. American coots (*Fulica americana*) and black-necked stilts (*Himantopus mexicanus*) experienced the highest incidence of malformed embryos. Multiple gross malformations of the eyes, brain, beak, and feet were often present. Based upon simultaneous examination of a control population of aquatic birds of the same species and published studies, the incidences of embryonic mortality and deformities were 9 to 30 times greater than expected.

Lead

Lead concentrations and reproductive success were studied in free-living European starlings (*Sturnus vulgaris*) nesting on the verge of two Maryland highways with different traffic volumes and in a nearby control area.[37] Concentrations of lead in the ingesta (84 to 94 mg/kg dry weight) of the Baltimore-Washington Parkway nestlings (18-day-old) and adults were 3 to 13 times those found in starlings from the control area. Parkway nestlings were more readily affected by lead ingestion than adults as reflected by a number of toxicologic responses including lower brain weights and depressed hemoglobin concentration, hematocrit, and red blood ALAD (Δ-aminolebulinic acid dehydrate) activity. These findings in free-living starling nestlings were similar to findings in experimentally lead-dosed nestling kestrels.[8,9]

INTERACTIVE EFFECTS AMONG CONTAMINANTS AND OTHER ENVIRONMENTAL FACTORS

Interactive effects of multiple contaminants can be additive, synergistic, or antagonistic. For example, toxicity of a mixture of aromatic hydrocarbons to mallard embryos was greater than the sum of the individual components, indicating moderate synergism.[4] Interactive effects of vehicle and formulation ingredients in toxicity studies with eggs may be important in determining the extent of toxicity. For example, David, as reviewed by Hoffman,[2] concluded that reduction in the gonadic germ population in 5-day-old Japanese quail embryos was primarily due to the excipient used in the formulation of commercial decamethrin, a pyrethroid insecticide. Kenaga[7] compared early reports of 2,4,5-T embryotoxicity, in which formulations were unspecified, to later reports in which known formulations showed much less embryotoxicity and concluded that the formulation ingredients of an herbicide can be just as important as the herbicide.

Hoffman and Albers[4] compared the effects of 13 pesticides applied by microliter pipette on mallard eggs in an oil vehicle (a nontoxic aliphatic mixture) to emulsifiable concentrate in water. Several pesticides, including lindane, paraquat, diazinon, parathion, and methomyl, were considerably more toxic in oil vehicle at 103 L/ha than in water at 935 L/ha, probably due to better penetration of the pesticide through the shell and its membranes. This contention is supported by studies comparing the degree of inhibition of brain and plasma ChE activity with organophosphate insecticides, including malathion, diazinon, and parathion applied in water and in aliphatic vehicle.[5]

Studies were conducted with day-old mallard ducklings examining the interactive effects of major mineral components (selenium, boron, and arsenic) of agricultural drainwater with one another as well as with dietary protein.[38] Selenium (Se) and boron (B) interacted to further decrease duckling growth during the first 3 weeks, and organ weights were lower at 4 weeks than when either element was fed alone. In combination, Se and B increased plasma GSH reductase activity, a measurement of stress on GSH metabolism. Arsenic (As) in combination with Se increased the survival rate of ducklings and also alleviated Se-related effects on body and organ growth. Physiological alterations caused by Se, including elevated GSH peroxidase activity, decreased hepatic thiol concentration, and increased hepatic lipid peroxidation, were less pronounced with 200 ppm As in the diet. When dietary protein was restricted to 7% instead of 22%, Se (60 ppm) caused 100% mortality within 4 weeks, and 15 ppm Se had a more pronounced effect on growth than in the 22% protein diet. Boron (1000 ppm) also had a more pronounced effect on growth when protein was reduced. Selenium accumulated to twice as high a concentration in the liver as with 22% protein. More severe interactive effects occurred between Se (15 ppm) and B (1000 ppm), including mortality (27%), further growth impairment, and still higher accumulation of Se in the liver. Arsenic (200 ppm) alone resulted in significant mortality with restricted protein and failed to offset the toxic effects of Se as it was able to do with 22% protein. These findings suggest that in locations in which quantity and quality of protein food sources for ducklings are diminished, ducklings could suffer more severe toxicological consequences due to Se, B, and As alone and in combination.

Other interactive factors influencing the toxicity of environmental contaminants to wildlife include naturally occurring stressors such as temperature extremes[39,40] and disease.[41,42] Data concerning these other factors are mainly based on interaction studies with adult birds in the laboratory and under fixed conditions, but could apply equally to critical stages of development.

CONCLUSIONS AND PERSPECTIVE

Field observations as well as controlled field, pen, and laboratory studies have provided convincing evidence of the hazard of oil pollution to eggs of wild birds during incubation. There are at least three routes in which external exposure of eggs to environmental contaminants could occur: (1) transfer to eggs by the plumage or feet of adult birds, (2) transfer to eggs by nesting materials, and (3) direct exposure of eggs following a spray application. Of insecticides, organophosphates have been the most widely studied with respect to potential for direct embryotoxicity and teratogenicity following spraying or immersion of eggs, and are most toxic to mallards when in oil vehicle. Several herbicides, including paraquat and trifluralin, were found to be more toxic than insecticides, with LC_{50} values occurring at normal use levels.

Altricial nestlings including American kestrels and European starlings have proven to be more sensitive to insecticides, herbicides, and lead ingestion than adult birds or precocial young as reflected by effects on survival, growth, and physiology.

Relatively few attempts have been made to measure the impact of direct exposure of bird eggs to pesticides following agricultural spraying in the field. However, at least 100 confirmed reports involving die-offs of adult and juvenile birds exposed to organophosphate and carbamate insecticides have been documented.[43] Such observations reflect highly visible situations, and there may be a multitude of undetected incidents for every one observed. Mourning dove nests in orchards that received routine spraying with organophosphate insecticides and fungicides were largely unsuccessful, with most failures occurring 4 days or more after being sprayed. The nests of species such as ring-necked pheasants and mourning doves that nest in wheat stubble may be exposed to spray operations during conservation tillage.[2]

Grue et al.[44] concluded that the potential for agricultural chemicals to enter prairie-pothole wetlands and impact-dependent wildlife appears to be great. Herbicides including bromoxynil with MCPA (Bronate®), diclofopmethyl (Hoelon®), and trifluralin (Treflan®) were frequently applied within the prairie-pothole region of North Dakota, a prime waterfowl nesting habitat. These herbicides were found to have LC_{50}s of one to ten times the field level of application in laboratory-treated mallard eggs,[4] and therefore could represent a potential hazard to waterfowl embryos in the field. Quite often, two or more herbicides or insecticides are sprayed in the same tank mixture, but few studies have examined the possible interactive effects of these combinations. Therefore, it is possible that two pesticides that may be only moderately toxic alone could result in additive or synergistic effects when applied together.

Future studies are needed to focus on field exposures in multiple avian species. Laboratory studies have shown ringed turtle dove (*Streptopelia risoria*) eggs to be more sensitive to aqueous pesticide exposure than mallard eggs, which are more sensitive than northern bobwhite. Under normal conditions of agricultural field application, one possible synergistic combination alone of two pesticides could have far-reaching effects if it were commonly used. Further laboratory and field studies examining the effects of pesticides most routinely applied in combinations of two or more are needed.

REFERENCES

1. Hoffman, D. J., B. A. Rattner and R. J. Hall, Wildlife toxicology, *Environ. Sci. Technol.*, 24, 276–283, 1990.
2. Hoffman, D. J., Embryotoxicity and teratogenicity of environmental contaminants to bird eggs, *Rev. Environ. Contam. Toxicol.*, 115, 40–89, 1990.
3. Hill, E. F., and D. J. Hoffman, Avian models for toxicity testing, *J. Am. Coll. Toxicol.*, 3, 357–376, 1984.
4. Hoffman, D. J., and P. H. Albers, Evaluation of potential embryotoxicity and teratogenicity of 42 herbicides, insecticides, and petroleum contaminants to mallard eggs, *Arch. Environ. Contam. Toxicol.*, 13, 15–27, 1984.
5. Hoffman, D. J., and W. C. Eastin, Jr., Effects of malathion, diazinon, and parathion on mallard embryo development and cholinesterase activity, *Environ. Res.*, 26, 472–485, 1981.
6. Hoffman, D. J., Effects of Krenite® brush control agent (fosamine ammonium) on embryonic development in mallards and bobwhite, *Environ. Toxicol. Chem.*, 7, 69–75, 1988.
7. Kenaga, E. E., The evaluation of the safety of 2,4,5-T to birds in areas treated for vegetation control, *Residue Rev.*, 59, 1–19, 1975.
8. Hoffman, D. J., J. C. Franson, O. H. Pattee, C. M. Bunck and A. Anderson, Survival, growth and accumulation of ingested lead in nestling American kestrels (*Falco sparverius*), *Arch. Environ. Contam. Toxicol.*, 14, 89–94, 1985.
9. Hoffman, D. J., J. C. Franson, O. H. Pattee, C. M. Bunck and H. C. Murray, Biochemical and hematological effects of lead ingestion in nestling American kestrels (*Falco sparverius*), *Comp. Biochem. Physiol.*, 80C, 431–439, 1985.
10. Hill, E. F., and M. B. Camardese, Lethal Dietary Toxicities of Environmental Contaminants and Pesticides to Coturnix, Tech. Rep. 2, U.S. Fish and Wildlife Service, Washington, DC, 1987.
11. Hoffman, D. J., J. C. Franson, O. H. Pattee and C. M. Bunck, Survival, growth, and histopathological effects of paraquat ingestion in nestling American kestrels (*Falco sparverius*). *Arch. Environ. Contam. Toxicol.*, 14, 495–500, 1985.
12. Hoffman, D. J., J. C. Franson, O. H. Pattee, C. M. Bunck and H. C. Murray, Toxicity of paraquat in nestling birds: effects on plasma and tissue biochemistry in American kestrels, *Arch. Environ. Contam. Toxicol.*, 16, 177–183, 1987.
13. Heath, R. G., J. W. Spann, E. F. Hill and J. F. Kreitzer, Comparative Toxicities of Pesticides to Birds, Spec. Sci. Rep. Wildl. No. 152, U.S. Fish and Wildlife Service, Washington, DC, 1972.
14. *Herbicide Handbook of the Weed Science Society of America*, 6th ed., WSSA, Champaign, IL, 1989.
15. Francis, B. J., Teratogenicity of bifenox and nitrofen in rodents, *J. Environ. Sci. Health*, B21(4), 303–317, 1986.
16. Burke-Hurt, S. B., J. M. Smith and A. W. Hayes, Nitrofen: a review and perspective, *Toxicology*, 29, 1–37, 1983.
17. Gray, L. E., Jr., and R. J. Kavlock, The effects of the herbicide 2,4-dichlorophenyl-p-nitrophenyl ether (NIT) on serum thyroid hormones in adult female mice, *Toxicol. Lett.*, 15, 231–235, 1983.
18. Manson, J. M., T. Brown and D. M. Baldwin, Teratogenicity of nitrofen and its effect on thyroid function in the rat, *Toxicol. Appl. Pharmacol.*, 73, 323–335, 1984.
19. Francis, B. M., Relative developmental toxicities of nine diphenyl ethers related to nitrofen, *Environ. Toxicol. Chem.*, 8, 681–688, 1989.
20. Grue, C. E., and B. K. Shipley, Sensitivity of nestling and adult starlings to dicrotophos, an organophosphate pesticide, *Environ. Res.*, 35, 454–465, 1984.
21. Hooper, M. J., L. W. Brewer, G. P. Cobb and R. J. Kendall, An integrated laboratory and field approach for assessing hazards of pesticide exposure to wildlife, in *Pesticide Effects on Terrestrial Wildlife*, L. Somerville and C. H. Walker, Eds., Taylor & Francis, New York, 1990.

22. Miller, D. S., D. B. Peakall and W. B. Kinter, Ingestion of crude oil: sublethal effects on herring gull chicks, *Science,* 199, 315–318, 1978.
23. Peakall, D. B., D. Hallett, D. S. Miller, R. G. Butler and W. B. Kinter, Effects of ingested crude oil on black guillemots: a combined field and laboratory study, *AMBIO*, 9, 28–30, 1980.
24. King, K. A., and C. A. Lefever, Effects of oil transferred from incubating gulls to their eggs, *Mar. Pollut. Bull.*, 10, 319–321, 1979.
25. Lewis, S. J., Effects of Oil on Avian Productivity and Population Dynamics. Ph.D. thesis, Cornell University, Ithaca, NY, 1982.
26. Albers, P. H., Transfer of crude oil from contaminated water to bird eggs, *Environ. Res.*, 22, 307-313, 1980.
27. DeWeese, L. R., C. J. Henny, R. L. Floyd, K. A. Bobal and A. W. Schultz, Response of Breeding Birds to Aerial Sprays of Trichlorfon (Dylox) and Carbaryl (Sevin-4-Oil) in Montana Forests, Fish Wildl. No. 224, U.S. Dept. of the Interior, Washington, DC, 1979.
28. Putera, J. A., A. Woolf and W. D. Klimstra, Mourning dove use of orchards in southern Illinois, *Wildl. Soc. Bull.*, 13, 496–501, 1985.
29. Patnode, K. A., and D. H. White, Effects of pesticides on songbird productivity in conjunction with pecan cultivation in southern Georgia: a multiple exposure experimental design, *Environ. Toxicol. Chem.*, 10, 1479–1486, 1991.
30. Gilbertson, M., and G. A. Fox, Pollutant-associated embryonic mortality of Great Lakes herring gulls, *Environ. Pollut.*, 12, 211–216, 1977.
31. Hoffman, D. J., B. A. Rattner, L. Sileo, D. Docherty and T. J. Kubiak, Embryotoxicity, teratogenicity and aryl hydrocarbon hydroxylase activity in Forster's terns on Green Bay, Lake Michigan, *Environ. Res.*, 42, 176–184, 1987.
32. Kubiak, T. J., H. J. Harris, L. M. Smith, T. R. Schwartz, P. L. Stalling, J. A. Trick, L. Sileo, D. E. Docherty and T. C. Erdman, Microcontaminants and reproductive impairment of the Forster's tern on Green Bay, Lake Michigan, *Arch. Environ. Contam. Toxicol.*, 18, 706–727, 1989.
33. Hays, H., and R. W. Risebrough, Pollutant concentrations in abnormal young terns from Long Island Sound, *Auk*, 89, 11–35, 1972.
34. Gochfield, M., Developmental defects in common terns of western Long Island, New York, *Auk*, 92, 58–65, 1975.
35. Ohlendorf, H. M., D. J. Hoffman, M. K. Saiki and T. W. Aldrich, Embryonic mortality and abnormality of aquatic birds: apparent impacts by selenium from irrigation drainwater, *Sci. Total Environ.*, 52, 49–63, 1986.
36. Hoffman, D. J., H. M. Ohlendorf and T. W. Aldrich, Selenium teratogenesis in natural populations of aquatic birds in central California, *Arch. Environ. Contam. Toxicol.*, 17, 519–525, 1988.
37. Grue, C. E., D. J. Hoffman, L. P. Franson and W. N. Beyer, Lead within highway rights-of-way and reproduction of a ground-foraging passerine, the starling (*Sturnus vulgaris*), *Environ. Pollut.*, 42, 157–182, 1986.
38. Hoffman, D. J., C. J. Sanderson, L. J. LeCaptain, E. Cromartie and G. W. Pendleton, Interactive effects of boron, selenium and dietary protein on survival, growth, and physiology in mallard ducklings, *Arch. Environ. Contam. Toxicol.*, 20, 288–294, 1990.
39. Rattner, B. A., Diagnosis of anticholinesterase poisoning in birds: effects of environmental temperature and underfeeding on cholinesterase activity, *Environ. Toxicol. Chem.*, 1, 329–335, 1982.
40. Kendall, R. J., and P. F. Scanlon, The toxicology of lead shot ingestion in ringed turtle doves under conditions of cold exposure, *J. Environ. Pathol. Toxicol. Oncol.*, 5, 183–192, 1984.
41. Franson, J. C., Immunosuppressive effects of lead, in *Lead Poisoning in Wild Waterfowl—A Workshop*, J. S. Feierabend and A. B. Russell, Eds., National Wildlife Federation, Washington, DC, 1986, pp. 106–109.
42. Fairbrother, A., T. M. Yuill and L. J. Olson, Effects of three plant growth regulators on the immune response of young and aged deer mice, *Arch. Environ. Contam. Toxicol.*, 15, 265–277, 1986.

43. Hill, E. F., and W. J. Fleming, Anticholinesterase poisoning of birds; field monitoring and diagnosis of acute poisoning, *Environ. Toxicol. Chem.*, 15, 265–277, 1982.
44. Grue, C. E., L. R. DeWeese, P. Mineau, G. A. Swanson, J. R. Foster, P. M. Arnold, J. N. Huckins, P. J. Sheehan, W. K. Marshall and A. P. Ludden, Potential impacts of agricultural chemicals on waterfowl and other wildlife inhabiting prairie wetlands: an evaluation of research needs and approaches, *Trans. N. Am. Wildl. Nat. Res. Conf.*, 51, 357–383, 1986.
45. Hoffman, D. J., unpublished data.

7

What is Measurable in Wildlife Toxicology: Field Assessment

Larry W. Brewer and Roy A. Hummell

ABSTRACT

The success and integrity of predictive ecological model development are related to the quantity and quality of empirical data available about the parameters that drive the models. High-quality empirical data are again essential when new models, or modified existing models, are to be verified. It is, therefore, appropriate to initially evaluate which ecological parameters associated with the fate and effects of contaminants in the environment are measurable, and how they are measured.

Tracking the movement and trophic partitioning of contaminants in biological systems and measuring the subsequent effects on individual animals and populations must rely heavily on the use of nonlethal sampling techniques. However, the use of certain important biomarkers as assessment endpoints will in some cases continue to require lethal sampling.

In situ vertebrate bioassays that are supported by in vitro and/or in vivo laboratory research have become the cutting edge in wildlife toxicology field assessments. The use of wildlife populations that are enhanced on study sites by techniques such as placement of nest boxes can solve statistical problems related to sample size. Adequate sample size in the experimental population allows for the measurement of multiple chemical and biochemical endpoints necessary to assess fate and effects of contaminants. Many population and biochemical parameters are now measurable in free-ranging wildlife populations and strong new statistical techniques exist for the quantitative analysis of the data produced from these measurements.

Parameters that are currently measurable and quantifiable in wildlife field studies include the distribution of contaminants in the soil, water, air, and vegetation; amount of contaminants in various trophic levels within food chains; amount of contaminant uptake by individual species; the lethal and sublethal response to contaminants of certain biochemical functions such as brain and plasma cholinesterase (ChE) activities, mixed function oxygenase (MFO) induction, electron transport alterations and genetic mutation; individual survivability related to both direct mortality and behavioral response; reproductive success and/or impairment; and comparative sensitivity to specific contaminants among species. Closely coordinated field and laboratory studies can, in some cases, establish correlations between the severity of response and exposure levels, between the degree of biochemical response and survival, or between the degree of biochemical response and survival or reproductive potential.

Research design, sampling techniques, and statistical methods play a key role in the overall utility of these types of data.

KEY WORDS

wildlife toxicology, field assessment, ecotoxicology

INTRODUCTION

There is a pressing need for ecological models that, when applied to real environmental problems, can be used as tools for environmental decisionmaking. Recent research efforts to quantify the impact of single chemical contaminants on wildlife[1] have demonstrated the enormous amount of effort, time, and expense required to complete such a study. More complex assessments of the impact of multiple chemical contamination will be far more costly in all of these respects. Conducting research on every toxic compound for every species in multiple geographic regions is clearly unrealistic as a means of assessing environmental risk and generating information on which to base environmental regulation and policy. One alternative involves studying chemical effects in sentinel systems and species[2] and utilizing the resultant data to predict, through ecological modeling, the fate and effects of contaminants in similar systems and species.

The success and integrity of predictive ecological modeling are directly correlated to the quality of the available empirical data concerning the parameters that can be utilized in model development and operation. Therefore, it is appropriate to ask initially which environmental parameters associated with the fate and effects of chemicals in the environment are measurable, and how these parameters are measured.

INTERDISCIPLINARY APPROACH

Wildlife biologists, analytical chemists, and biochemists are all essential to the success of a wildlife toxicology field study. Because of the interdisciplinary nature of ecotoxicological research, a discussion of measurable variables involves data generated in laboratories from measurements of samples collected in the field as well as from measurements taken directly in the field. Data measuring chemical concentrations throughout the ecosystem and used to assess exposure of organisms, such as temporal monitoring of plasma ChE activities in passerine species, are collected in the field by wildlife biologists, but the actual measurements are made subsequent to sample collection in the biochemistry laboratory. Reproductive success variables of a passerine species or daily survival rates of upland birds are examples of parameters that are actually measured in the field.

This discussion of what is measurable in wildlife toxicology field assessments incorporates parameters and methodologies for measurements taken in the field as well as those taken in the laboratory from samples collected in the field. Field data as well as chemical and biochemical laboratory analyses are inseparable aspects of toxicology field assessments and must be accepted as related integral components of a single study.

QUANTITATIVE CONSIDERATIONS WHEN CHOOSING TOXICOLOGICAL VARIABLES FOR RISK ASSESSMENT

Although an almost infinite array of ecological variables could be measured relative to chemical contaminants, sample size and repetition are necessary considerations when choosing variables that are statistically comparable and can provide dependable, mechanistic data on which to base the development and verification of predictive models.

Sample size is almost always a limiting factor when dealing with biological variables. Measuring variables that will not produce an adequate sample size for appropriate statistical analyses produces data of little use for risk assessment. Researchers must ensure that their experiments will yield significant statistical power so that tests will yield conclusive data regardless of acceptance or rejection of the null hypothesis.[3-6] A condensed overview of the importance of and relationship between sample size, power, and replication in terms of field study experimental design is presented by Fite et al.[7]

MEASURING CHEMICAL CONCENTRATIONS

Pollutants in a contaminated ecosystem can be identified through chemical component isolation. Research has demonstrated the capability of analytical chemists to isolate chemicals that contain specific functional groups.[8] Modern instrumentation has progressed to the point that most compounds in homologous series can be resolved and detected at environmentally significant concentrations.[9-13] Chemical classes can be isolated from environmentally important matrices and generally identified to a desired level of sophistication. These matrices include air, soil, water, vegetation, invertebrate animals, and various organs of vertebrate animals. Sampling the physical environment (i.e., soils, water, and air) is usually limited only by the physical ability of the researcher to collect and transport samples or by the funds available to conduct analyses. The repeatability of this type of sampling allows not only measurement of concentrations of chemicals and chemical degradation products, but the persistence and environmental behavior of the compounds over time.[14]

Sampling plants, both aquatic and terrestrial, in the field for measurement of chemical concentrations is often unlimited with regard to sample size and sample replication. Repeated measurements can provide data on uptake rates, metabolic rates, and persistence of parent compounds and metabolites in plant tissue. Chemicals can be washed from the surface of plants and their concentrations measured separately from those contained in the plant tissues. Different plant parts can be analyzed separately to determine how specific contaminants are partitioned in the plant (e.g., roots, leaves, fruit, and seeds), thus facilitating evaluation of potential risk to herbivorous animals.

Invertebrate animals are usually more difficult to sample randomly for contaminant analysis. Although generally plentiful in most ecosystems, they must be located and extracted from their medium, usually soil, sediment, or water. At this level of sampling, one may first encounter the problems associated with lethal collection as part of a toxicology study. Removing invertebrates from a study site each time sampling is conducted may produce population declines that result in insufficient numbers being available for the next scheduled sampling effort. This problem is relatively unlikely if study sites are large, but could occur if small plots are involved or invertebrate numbers are reduced by environmental factors.

An alternative to collecting invertebrates directly from their environmental medium involves the collection of food items gathered by selected bird species and brought to their nest to feed nestlings. By constricting the esophagus of individual nestlings for short periods such that they cannot swallow food items and by removing the items at regular intervals after parent birds have made several feeding forays, a substantial sample of invertebrates can be collected.[1,15] Measuring chemical concentrations in lower trophic levels provides information on how readily a chemical is accumulated in these organisms and how much is available for transport along the food chain. Additionally, those specific invertebrate species that are most involved in trophic movement of the compound of interest can be identified, and through knowledge of the feeding ecology of various vertebrate animals, inferences to those species of vertebrates most likely to be at risk can be identified.

Measuring degradation or metabolic products of chemicals in invertebrate samples is an important step in understanding the environmental behavior of a compound. Physical degradation products and metabolites of chemicals can be of equal or greater toxicity as compared to the parent compound.[16] Similarly, a degradation product may be more mobile or persistent in the environment than the parent compound and may have a greater tendency to accumulate at the upper end of the food chain.[17] If the analytical methods involve only the measurement of the parent compound, important aspects of the environmental risk posed by the compound may be overlooked.

Animals in the upper trophic levels of a food chain are fewer than in lower trophic levels,[18] the top predator population being the least dense. At the vertebrate animal level, measuring chemical concentrations (or any other variables) via lethal sampling methods can become a seriously limiting factor. Even when study sites are relatively large, vertebrate animal populations often are not sufficient to withstand repeated removal of individuals without severely limiting the sample population remaining on the study sites. Additionally, when the research design also includes measurements of survival, reproduction, behavior, and other density-dependent population parameters, lethal sampling may directly influence and change resulting data.

Lethal sampling techniques are valuable for the analysis of chemical concentrations in animals found dead at the study sites or for those species whose individuals are very numerous. Measurement of chemical concentrations in various tissues (i.e., gastrointestinal tract, liver, kidney, skin, and others) provides evidence of mode and degree of exposure. However, in complex ecotoxicological studies, nonlethal sampling techniques for verifying chemical exposure are essential.[19,20] Three nonlethal analytical chemistry procedures currently in use include the measurement of alkyl phosphates in fecal-urate samples collected from live-trapped birds, measurement of contaminants in the chorioallantoic membranes (CAMs) of hatched or addled bird eggs,[19,21] and the measurement of chemical concentrations in various food items removed directly from the crops of live nestlings, as mentioned previously. All of these methods verify chemical exposure in birds, and all are applicable to complex studies involving the measurement of multiple endpoints using the same animals (i.e., reproductive success, survival, blood ChE activity, and others).[1,22]

MEASUREMENT OF BIOCHEMICAL ENDPOINTS OF EXPOSURE

Biochemical responses of wildlife to chemical exposure can be measured for the purpose of verifying and measuring exposure, and determining, at least on a relative scale, the level of exposure in individual animals across multiple species. Historically, biochemical analyses have also required the sacrifice of the target animal. As previously discussed, this conflicts with many of the objectives of complex toxicology field studies. Studies of organophosphate insecticide effects on wildlife have for years relied on the measurement of brain ChE activity to assess exposure,[7,19,23] which requires sacrifice of the animal. The recent development of nonlethal field sampling and data handling techniques for measurement of blood plasma ChEs[19] has facilitated broad-scale temporal assessment of exposure in multiple avian species simultaneously at a given study site.[1,22] The nonlethal nature of the blood sampling allows for repeated capture of many individuals, and the trapping effort can also facilitate a mark-recapture study to determine survival rates.

Other biochemical assays use specific enzymes of oxidative metabolism, such as MFO, to determine wildlife exposure to aromatic hydrocarbons and chlorinated contaminants.[24-26] Wildlife exposure to metals also can be measured by enzymatic assays.[27] The avian immune system, as a means to assess exposure of birds to toxicants, has received considerable attention during the past 10 years.[28,29] Dr. Anne Fairbrother (U.S. EPA, Environmental Research Laboratory, Corvallis, OR) has been working to develop suitable bioassays involving immune systems of avian species, and she presents a discussion, "Immunotoxicology of Captive and Wild Birds," in a subsequent chapter. Such assays usually require the capture and sacrifice of the test animal in order to acquire appropriate tissues. Therefore, study designs must avoid limitations resulting from the removal of animals from the study sites. The development of nonlethal tissue sampling methods will be an important advance in increasing the applicability of these bioassays to complex toxicology field studies.

MEASUREMENT OF AVIAN SURVIVAL, REPRODUCTION, AND BEHAVIOR

The assessment of avian survival rates related to chemical exposure is an important part of a toxicology field assessment. One of the most definitive measurements of survival is obtained by radiotagging individual animals and relocating them at regular intervals to verify survival.[30,31] Other avian marking methods, such as leg banding, and several methods of trapping, such as mist-netting or funnel trapping, are applicable to toxicology field studies, particularly when study sites are very large. As previously mentioned, mark-recapture studies can incorporate nonlethal blood and fecal-urate sampling such that exposure information can be correlated with survival.[1,22,32] Survival rates also can be correlated with measurements of environmental concentrations of the chemical in question when appropriate statistical tests are applied.[32]

Another method frequently used in toxicology field studies to obtain a relative measure of mortality involves intensive searches for animal carcasses in conjunction with carcass disappearance tests.[7,33] These searches and tests are conducted on chemically treated study sites as well as control sites in order to make relative comparisons.

Reproductive success parameters such as egg survival rates, hatching success, and nestling survival rates can be measured by monitoring bird nests.[34-36] Very large study sites are usually needed to produce an adequate sample of nests of a given species.[7] The use of nest box-enhanced populations of certain cavity nesting species, such as the European starling (*Sturnus vulgaris*) and American kestrel (*Falco sparvarius*), can add substantially to the statistical power of the research design by increasing sample size on study sites. Approximately 40 to 50 starling nest boxes on a study site led to a considerably increased number of individual nestlings. The nest box study facilitates repeated blood and fecal-urate sampling and esophageal constriction for invertebrate sampling. It usually provides a large enough sample population to accommodate some lethal sampling if needed for immune system bioassays.

The nest box colony also provides an opportunity to measure certain behavioral aspects of adult birds. Using photographic techniques, nest attentiveness and food gathering behavior of birds on treated sites can be monitored and compared to those of birds on control sites.[1,22] A remote temperature-sensing transmitter mounted in a false egg can be used with automated recording devices to measure incubation temperature in the nest, incubation behavior patterns, and nest abandonment rates.[37]

Foraging behavior, home range, habitat use, and nesting success of radiotagged raptors, such as the great horned owl (*Bubo virginianus*), and upland bird species, such as the northern bobwhite (*Colinus virginianus*), can be measured using radiotelemetry techniques.[1,22,31] These studies also can be conducted in conjunction with nonlethal blood and fecal-urate sampling to correlate the measured parameters to chemical exposure and use of contaminated habitats. Radiotelemetry techniques may be used to monitor postfledging survival of young passerines and other birds as they leave the nest.[36]

This discussion should provide the reader with a general understanding of what types of data can be gathered from wildlife toxicology field studies and subsequently used in the development of chemical fate and effects models. The incorporation of measurable wildlife responses to chemical exposure, particularly survival and reproductive responses, into simulation models will enhance our ability to predict accurately the effects of chemical use on wildlife populations and will influence future regulatory decisionmaking regarding chemical use in the environment.

REFERENCES

1. Tank, S. L., L. W. Brewer, M. J. Hooper, G. P. Cobb, III and R. J. Kendall, Third Year Investigation of the Effects of COUNTER® 15-G Insecticide-Nematicide on Selected Wildlife Species in a Corn Agroecosystem, The Institute of Wildlife and Environmental Toxicology, final research report submitted to American Cyanamid Company, Princeton, NJ, 1990.
2. Lower, W. R., and R. J. Kendall, Sentinel species and sentinel bioasssays, in *Use of Biomarkers for Assessment of Environmental Contaminant Exposure*, L. Shugart and J. McCarty, Eds., CRC Prcss, Boca Raton, FL, 1990, pp. 306–331.
3. Eberhardt, L. L., Assessing the dynamics of wild populations, *J. Wildl. Manage.*, 49(4), 997-1012, 1985.
4. Eberhardt, L. L., Appraising variability in populations studies, *J. Wildl. Manage.*, 42(2), 207-237, 1978.
5. Eberhardt, L. L., Quantitative ecology and impact assessment, *J. Environ. Manage.*, 4, 27–70, 1976.
6. Hurlbert, S. H., Pseudoreplication and the design of ecological field experiments, *Ecol. Monogr.*, 54, 187-211, 1984.
7. Fite, E. C., L. W. Turner, N. J. Cook and C. Stunkard, Guidance Document for Conducting Terrestrial Field Studies, USEPA, Washington, DC, 1988.
8. Krahn, M. M., L. K. Moore, R. C. Bogar, C. A. Wigren, S. L. Chan and D. W. Brown, High-performance liquid chromatographic method for isolating organic contaminants from tissue and sediment extract, *J. Chromatogr.*, 437, 161–175, 1988.
9. Lee, R. C., L. J. O'Neil, E. J. Clairian, D. L. Brandon, R. J. Rhett, D. R. Sanders et al., Remedial Investigation of Contaminant Mobility at Concord Naval Weapons Station, Concord, Calif., Dept. of Army Waterways Experiment Station, Vicksburg, MS, 1986.
10. Krone, C. A., D. G. Burrows, D. W. Brown, P. A. Robisch, A. J. Friedman and D. C. Malins, Nitrogen-containing aromatic compounds in sediments from a polluted harbor in Puget Sound, *Environ. Sci. Technol.*, 20, 1144–1150, 1986.
11. Luffer, D. R., L. J. Galante, P. A. David, M. Novotny and G. M. Heiftje, Evaluation of a supercritical fluid chromatograph coupled to a surface-wave-sustained microwave induced plasma detector, *Anal. Chem.*, 60, 1365-1369, 1988.
12. Onuska, F. I., and F. W. Karasek, *Open Tubular Column Chromatography in Environmental Sciences*, Plenum Press, New York, 1984.
13. Yu, W. C., D. H. Fine, K. S. Chiu, and K. Biemann, Determination of nitrated polycyclic aromatic hydrocarbons in diesel detection, *Anal. Chem.*, 56, 1158-1162, 1984.
14. Cobb, G. P., L. W. Brewer, and R. J. Kendall, Degradation of terbufos in soils during

drought conditions, in Proc. Natl. Pestic. Res. Conf., Virginia Polytechnic Institute and State University, Blacksburg, 1989.

15. Johnson, E. J., L. B. Best and P. A. Heagy, Food sampling biases associated with the "ligature method," *Condor*, 82, 186-192, 1980.
16. Moriarity, F., *Ecotoxicology, The Study of Pollutants in Ecosystems,* 2nd ed., Academic Press, London, 1984.
17. Rudd, R. L., R. B. Craig, and W. S. Williams, Trophic accumulation of DDT in a terrestrial food web, *Environ. Pollut.*, 25A, 219-228, 1981.
18. Odum, E. P., *Ecology*, W. B. Saunders, Philadelphia, 1959.
19. Cobb, G. P., and M. J. Hooper, Non-lethal wildlife monitoring to determine exposure to xenobiotics and resulting impacts, in *Wildlife Toxicology and Population Modeling: Integrated Studies of Agroecosystems*, R. J. Kendall and T. E. Lacher, Jr., Eds., Lewis Publishing, Chelsea, MI, 1993.
20. Hooper, M. J., L. W. Brewer, G. P. Cobb and R. J. Kendall, An integrated laboratory and field approach for assessing hazards of pesticide exposure to wildlife, in *Pesticide Effects on Terrestrial Wildlife*, L. Somerville and C. H. Walker, Eds., Taylor & Francis, New York, 1990, pp. 272-283.
21. Norman, D., G. P. Cobb, L. W. Brewer, M. J. Hooper and R. J. Kendall, Organochlorine Contaminants in Great Blue Herons Collected from the Puget Sound, The Institute of Wildlife and Environmental Toxicology, Clemson University, final research report submitted to the U.S. Fish and Wildlife Service, Olympia, WA, 1990.
22. Mellott, R. S., L. W. Brewer, M. J. Hooper, G. P. Cobb III, R. A. Hummell, C. Crabtree and R. J. Kendall, The Response of Wildlife Exposed to Multiple Applications of Diazinon 50W in Apple Orchards of Eastern Washington and South Central Pennsylvania, The Institute of Wildlife and Environmental Toxicology, Clemson University, final research report submitted to CIBA-GEIGY Corp., Greensboro, NC, 1990.
23. Hill, E. F., and W. J. Fleming, Anticholinesterase poisoning of birds: field monitoring and diagnosis of acute poisoning, *J. Environ. Toxicol. Chem.*, 1, 27-38, 1982.
24. Payne, J. F., L. L. Fancey, A. D. Rahimtula, and E. L. Poter, Review and perspective on use of mixed-function oxygenase enzymes in biological monitoring, *Comp. Biochem. Physiol.*, 86C, 233-245, 1987.
25. Peakall, D. B., D. A. Jeffery and D. Boersma, Mixed-function oxidase activity in seabirds and its relationship to oil pollution, *Comp. Biochem. Physiol.*, 88C, 151-154, 1987.
26. Walters, P., S. Kahn, P. J. O'Brien, J. F. Payne and A. D. Rahimtula, Effectiveness of Prudoe Bay crude oil and its aliphatic, aromatic and heterocyclic fractions in inducing mortality and aryl hydrocarbon hydroxylase in chick embryo in ovo, *Arch. Toxicol.*, 60, 454-459, 1987.
27. Kendall, R. J., and P. F. Scanlon, The toxicology of lead shot ingestion in ringed turtle doves under conditions of cold exposure, *J. Environ. Pathol. Toxicol. Oncol.*, 5-4/5, 183-192, 1984.
28. Rocke, T. E., T. M. Yuill and R. D. Hindsdill, Oil and related toxicant effects on mallard immune defenses, *Environ. Res.*, 33, 343-352, 1984.
29. Toivanen, A., and P. Toivanen, *Avian Immunology, Basis and Practices*, Vol. 1 and 2, CRC Press, Boca Raton, FL, 1987.
30. Brewer, L. W., C. J. Driver, R. J. Kendall, C. Zenier and T. E. Lacher, Jr., Effects of methyl parathion on ducks and duck broods, *Environ. Toxicol. Chem.*, 7, 375-379, 1988.
31. Buck, J. A., L. W. Brewer, M. J. Hooper, G. P. Cobb, R. A. Craft and R. J. Kendall, Monitoring organophosphate insecticide exposure to great horned owls in Iowa, in Proc. 11th Annu. Meet. Society of Environmental Toxicology and Chemistry, Washington, DC, November 1990.
32. Skalski, J. R., and S. G. Smith, Assessing the fate of wild populations in an epidemiological framework, in Proc. Natl. Pestic. Res. Conf. Virginia Polytechnic Institute and State University, Blacksburg, 1989.
33. Balcomb, R., Songbird carcasses disappear rapidly from agricultural fields, *Auk*, 103, 817-820, 1986.

34. Mayfield, H. F., Nesting success calculated from exposure, *Wilson Bull.*, 73, 255–261, 1961.
35. Mayfield, H. F., Suggestions for calculating nest success, *Wilson Bull.*, 87, 456–466, 1975.
36. Kendall, R. J., L. W. Brewer, T. E. Lacher, Jr., M. L. Whitten and B. T. Marden, The Use of Starling Nest Boxes for Field Reproductive Studies: Provisional Guidance and Technical Support Document, NTIS Publ. No. PB 89 195 028/AS EPA/600/8-89/056.36, National Technical Information Service, Springfield, VA, 1989.
37. Varney, J. R., and D. H. Ellis, Telemetry egg for use in incubation and nest studies, *J. Wildl. Manage.*, 38(1), 142–148, 1974.

8

The Building Blocks for Ecological Modeling in Wildlife Toxicology

Richard E. Warner

ABSTRACT

Ecological models applicable to wildlife toxicology have been slow to emerge because (1) gaining reliable knowledge regarding the toxicity of various chemical perturbations has been tedious and fraught with economic, technological, and bureaucratic challenges; (2) the study of linkages between exposure and effects at individual, population, community, and ecosystem levels is in its infancy; (3) most widely applied models have assumptions often inappropriate for highly disturbed environments; and (4) no generic ecological model currently exists that will simulate these dynamics under well-defined scales of time, space, and species demography and no single approach to modeling will ever fit all situations. The session of the 1990 Ecological Modeling Conference, "What is Measurable in Wildlife Toxicology?," provided a forum whereby scientists from varied disciplines began to speak the same language, identify key stumbling blocks, and plan for potential avenues of integrated research in this area.

KEY WORDS

ecological models, research, wildlife toxicology

INTRODUCTION

Developing ecological models that can advance the science of wildlife toxicology and be of use to policymakers and regulators requires sustained effort by scientists in many disciplines. Better communication among these scientists is needed if data from laboratory and field studies is to be integrated and extended through ecological modeling. Thus far, the endeavors of wildlife toxicologists and ecologists have been too disjointed to permit the interactive processes necessary to assess the ecological effects of pesticides.

The fact that ecological models applicable to wildlife toxicology have been slow to emerge is at least in part a symptom of the disjointed disciplines as they presently exist. More specifically, such models have been slow to emerge for at least three reasons:

1. Reliable knowledge regarding the toxicity of various chemical perturbations to a variety of species has been tedious and fraught with economic, technological, and bureaucratic challenges.

2. The study of linkages between exposure and effects at individual, population, community, and ecosystem levels is in its infancy.
3. No generic ecological model currently exists that will simulate these dynamics under well-defined scales of time, space, and species demography, and no single approach to modeling may ever fit all situations. Further, most ecological models assume closed populations, a narrow range of environmental conditions, and density-dependent cues that operate over a considerable range of demographic conditions.

These assumptions clearly are not universally appropriate for highly disturbed environments.

The session "What is Measurable in Wildlife Toxicology?" provided a forum whereby scientists from varied disciplines began to speak the same language and to view from new vantage points the stumbling blocks and potential avenues of integrated research in this area. While conveying a wealth of scientific information, the second session also established a group dynamic and forged a common understanding of the tools of the trade in wildlife toxicology. Within this context, the four papers presented in this session were essential to understanding where the science of wildlife toxicology has been, where it is now, and where it may be headed.

DISCUSSION

Hoffman[1] illustrated standardized applications of LC_{50} and LD_{50} tests for several avian species and pesticides. He reviewed how well these indicators of contaminant toxicity have been linked to such endpoints as impaired reproduction, altered growth and development, and other behavioral and physiological parameters. He then connected findings from laboratory and field studies in order to assess the effects of contaminants on avian embryos and hatchlings. In extending laboratory findings to field situations, researchers must have information regarding absolute toxicity (effective concentrations) and other factors affecting the potential hazard, such as application levels and environmental conditions. For example, he noted a dramatic spread in the potential hazards of widely used herbicides, insecticides, and heavy metals based on the conditions under which they were applied and variations in toxicity among avian species and age cohorts.[1] Moreover, the likelihood of exposure to multiple contaminants is real, and exposure to multiple toxicants can be additive, synergistic, or antagonistic. Diet can also exacerbate the effects of single and multiple toxicants. Hoffman's review of the current use of LC_{50} and LD_{50} tests signals the need for improved linkages between standardized laboratory procedures and such biomarkers as cholinesterase (ChE), which can be measured in the field without sacrificing animals.[1]

Cobb and Hooper[2] provided contemporary examples of how biomarkers can be used to assess exposure. They noted that a "diagnostic" test of depressed ChE levels—a rule of thumb for detecting exposure to organophosphorus insecticides—refers to ChE levels 2 S.D. below an established $\overline{X}$ value for unexposed animals. Because of the inherent variability in such tests, diagnostic values should be determined from relatively large sample sizes under standardized conditions wherever possible. Because of variability, Cobb and Hooper[2] went on to explore the use of biomarkers to determine the probabilities of detecting lethal/sublethal and direct/indirect effects in field studies. For example, vulnerability to pesticides varies among species and age groups as well as individuals. In field situations, additional variability is introduced because of feeding behavior and the movement of pesticides and their metabolites through ecosystems. Cobb and Hooper[2] inferred that in general the ability to infer sublethal and indirect effects of pesticides from such biomarkers as ChE is promising but not yet established.

Future advances in the use of biomarkers to assess the effects of pesticides at population and community levels are in part dependent on more information regarding variability in uptake and subsequent effects. Radiotelemetry technologies offer a promising avenue for future research. At the 1990 conference, Crystal Driver of Battelle Northwest outlined emerging applications of radiotelemetry that may enable us to monitor the physiological responses of vertebrates to pesticides. This technique includes several attractive features for studying pesticide exposure: ingestion, inhalation, and dermal exposure singly or in combination. This technique also permits monitoring of real-time responses of vertebrates to chemical exposure. Further, by monitoring for such relatively fine-scale changes in physiological parameters as heart functions, respiration, body temperature, energetics, and behavior, the effect of seemingly sublethal doses may ultimately be assessed. Finally, radiotelemetry offers advantages in terms of study controls because the responses of each individual can be compared before and after exposure.

Preliminary results of toxicological studies in controlled (primarily laboratory) settings indicate that radiomonitoring of physiological functions appears to hold great promise, as pointed out by Driver. Except for body temperature and motor activity, however, these functions cannot be reliably monitored in free-ranging animals at present. Thus, if radiotelemetry is to contribute significantly to the study of wildlife toxicology, efforts must be made to improve radiotelemetry technologies to the point where they can track the numerous physiological effects of pesticide exposures in both laboratory and field situations.

Brewer and Hummell[3] underscored the tremendous time, energy, and, ultimately, dollars that have thus far been directed to determining the effects of single toxicants on wildlife, primarily in controlled settings. Thus, ecological models have proved to be compelling tools for studying and simulating wildlife toxicology, if such models can extend well-documented findings over a wide range of chemical compounds, species, and environments. In order to move forward with the design and execution of studies that can be expanded using ecological models, one must account for several factors:

1. Controlled laboratory and field situations may well benefit from improved standardization of methods, including the use of minimally adequate sample sizes and repeated measurements.
2. In field studies the movement of pesticides should be tracked through an array of environmentally important matrices including air, soil, water, vegetation, invertebrates, and vertebrates.
3. Consideration should be given to temporal and spatial events that affect the movement and degradation of toxicants in the environment.
4. Lethal/sublethal and direct/indirect effects must be simultaneously evaluated.
5. If field studies are to be the building blocks for widely applicable ecological models, careful selection must be made of the species, chemical perturbations, and environments to be studied.

CONCLUSION

A lively discussion followed the scientific presentations in the "What is Measurable in Wildlife Toxicology?" session, and the following areas of concern were identified.

Definitions of Terms

Clearly, participants were only beginning to speak the same language. The use of many terms such as *sublethal* and *indirect* were still confusing to many, and a glossary of terms was developed.

Biomarkers

Biomarkers associated with serum chemistry and immunotoxicology (for example, ChE) have been in a rapid development phase, with considerable progress made in detecting exposure. The use of biomarkers to associate exposure with sublethal and indirect effects, however, is only beginning. Participants were somewhat troubled by the variability found in "normal" ChE levels relative to species, age, sex, and environmental factors and by the inadequate sample sizes that have sometimes been used to establish and infer normal values. Thus, the reliable detection of normal/subnormal levels of such biomarkers as ChE presents one challenge for research; gaining the ability to infer the effects of exposure at the population level represents another. The need for development of other biomedical markers was evident as was the need to study various markers in concert in controlled settings. As in toxicological studies of humans, multiple measurements of the same individual over time would be a beneficial approach for wildlife toxicology. This technique would eliminate some of the concerns regarding individual variability and detection of exposure.

Comparability of Studies

Communication among researchers in the relevant disciplines must begin if in-depth studies are to yield information that can be used for multiple purposes. For example, studies designed primarily to determine the impact of toxicants on individuals may also provide information needed to assess the effect of chemical perturbations on population demography and community composition. Improved communication between scientists and agencies should lead to improved methodologies and in general facilitate an iterative process whereby each successive study improves the bank of reliable knowledge and useful techniques available to assess the effects of pesticide exposure.

REFERENCES

1. Hoffman, D. J., Measurements of toxicity and critical stages of development, in *Wildlife Toxicology and Population Modeling: Integrated Studies of Agroecosystems*, R. J. Kendall and T. E. Lacher, Jr., Eds., Lewis Publishers, Chelsea, MI, 1993.
2. Cobb, G. P., and M. J. Hooper, Nonlethal wildlife monitoring to determine exposure to xenobiotics and resulting impacts, in *Wildlife Toxicology and Population Modeling: Integrated Studies of Agroecosystems*, R. J. Kendall and T. E. Lacher, Jr., Eds., Lewis Publishers, Chelsea, MI, 1993.
3. Brewer, L. W., and R. A. Hummell, What is measurable in wildlife toxicology: field assessment, in *Wildlife Toxicology and Population Modeling: Integrated Studies of Agroecosystems*, R. J. Kendall and T. E. Lacher, Jr., Eds., Lewis Publishers, Chelsea, MI, 1993.

9

Critical Parameters in Population Studies

Thomas E. Lacher, Jr.

Ecologists have observed, quantified, and analyzed hundreds of different variables related to the population biology of organisms during the past century. Which set of variables is most important to a wildlife toxicologist will depend upon the compound of interest, the mechanism of action, and the species of concern. Several areas of population ecology now hold particular promise for use in studies on the effect of pesticides on critical parameters of wildlife populations because of recent theoretical and methodological advances. The four papers in this section present new developments in four of these areas: avian survival rates, use of space and habitat, avian foraging, and phenogenetics.

Dobson and Hudson[1] discussed the difficulty of estimating avian survival rates over long-term temporal and large spatial scales. Avian survival rates frequently show high variability in response to natural variability in environmental conditions, changing age-class distributions in populations, and differential mortality rates in different habitats. In order to carry out field studies that are capable of detecting, in a statistically significant manner, small changes in survival rates due to the effect of a chemical, avian toxicologists must have an understanding of this natural variability. They provide support for the utility of banding schemes carried out over the long term. Indeed, studies of many years duration will be required to detect depressed survival rates and declining populations whenever the effects are subtle or small in magnitude in any one year.

Bowers[2] addressed the importance of incorporating habitat selection and landscape perspective theories into modeling initiatives in the area of ecological risk assessment. He argued that whereas many risk assessment models attempt to incorporate demographic information, most overlook the importance of habitat selection in determining possible exposure and subsequent risk. Spatial heterogeneity in food resources or nesting sites presents organisms with habitats of different quality. Active selection of higher quality habitats will result in habitat-specific demographic rates and exposure probabilities. Simulations presented by Bowers show that habitat-structured populations are more stable and less susceptible to local extinction than age-structured models. There is a developing literature on habitat models based upon the concept of sources and sinks[3] that incorporate aspects of the model that Bowers presented and have intuitive appeal to wildlife toxicologists. This will likely become a profitable area of research in wildlife risk assessment modeling.

A number of methods and models are derived from optimal foraging theory that can be applied directly to wildlife toxicology with little additional development or modification. Belovsky has successfully applied linear programming methods to the study of foraging ecology,[4,5] and extends these methods to include constraints imposed by the application of agricultural chemicals to the environment.[6] Pesticides may have more subtle impact than direct mortality. Acting through indirect pathways, such as altering the composition of the diet, could have less obvious yet serious effects on reproduction and metabolism. The chronic effects of a persistent pesticide can be incorporated as an additional constraint in a linear programming model. All linear programming approaches require detailed information on the basic foraging biology of the bird species in question, the effect of the toxin on the prey population, and the effect of the toxin on the bird. When this information is available, novel approaches, such as incorporating the bird into an integrated pest management program, can be evaluated.

Zakharov[7] addresses the use of morphological characteristics to imply that developmental abnormalities are attributable to anthropogenic effects. Prior research on fishes[8] and marine mammals[9] has shown that certain developmental abnormalities, especially fluctuating asymmetry, can be related to environmental stresses. Fluctuating asymmetry frequently is related to the reduction of genetic variability in populations or through the destruction of genetic coadaptation when genetically distinct forms hybridize. In addition strong directional changes in phenotype can occur as the result of persistent stress. Morphological traits are much easier to measure than behavioral traits, and historical data can be obtained by examining museum specimens. Analyses of developmental stability can be an important tool in monitoring for effects of anthropogenic stress, especially when used for background monitoring of historical trends.

The papers discussed above review only a small selection of variables important in population ecology that are affected by toxic substances. Others, such as measures of reproductive success, are of equal importance (see chapters by Meyer and Boyce and Slade in this volume[10,11]). The chapters in this section of the book, however, do present challenging new approaches to the analysis of the impacts of agricultural chemicals on wildlife populations, and they will stimulate further research on these and other critical parameters.

REFERENCES

1. Dobson, A., and P. Hudson, Assessing the impact of toxic chemicals: temporal and spatial variation in avian survival rates, in *Wildlife Toxicology and Population Modeling: Integrated Studies of Agroecosystems*, R. J. Kendall and T. E. Lacher, Jr., Eds., Lewis Publishers, Chelsea, MI, 1993.
2. Bowers, M. A., Use of space and habitats by individuals and populations: dynamics and risk assessment, in *Wildlife Toxicology and Population Modeling: Integrated Studies of Agroecosystems*, R. J. Kendall and T. E. Lacher, Jr., Eds., Lewis Publishers, Chelsea, MI, 1993.
3. Pulliam, H. R., Sources, sinks and population regulation, *Am. Nat.*, 132, 652–661, 1988.
4. Belovsky, G. E., Herbivore optimal foraging: a comparative test of three models, *Am. Nat.*, 124, 97–115, 1984.
5. Belovsky, G. E., M. E. Ritchie and J. Moorehead, Foraging in complex environments: when prey availability varies over time and space, *Theor. Pop. Biol.*, 36, 144–160, 1989.
6. Belovsky, G. E., Modeling avian foraging: implications for assessing the ecological effects of pesticides, in *Wildlife Toxicology and Population Modeling: Integrated Studies of Agroecosystems*, R. J. Kendall and T. E. Lacher, Jr., Eds., Lewis Publishers, Chelsea, MI, 1993.
7. Zakharov, V. M., Population-phenogenetic approach for ecological monitoring (a biotest), in *Wildlife Toxicology and Population Modeling: Integrated Studies of Agroecosystems*, R. J. Kendall and T. E. Lacher, Jr., Eds., Lewis Publishers, Chelsea, MI, 1993.

8. Ames, L. J., J. D. Felley, and M. H. Smith, Amounts of asymmetry in Centrarchid fish inhabiting heated and non-heated reservoirs, *Trans. Am. Fish. Soc.*, 108, 489–495, 1979.
9. Zakharov, V. M., and A. V. Yablokov, Skull asymmetry in the Baltic grey seal: effects of environmental pollution, *Ambio*, 19, 266–269, 1990.
10. Meyer, J. S., and M. S. Boyce, Life historical consequences of pesticides and other insults to vital rates, in *Wildlife Toxicology and Population Modeling: Integrated Studies of Agroecosystems*, R. J. Kendall and T. E. Lacher, Jr., Eds., Lewis Publishers, Chelsea, MI, 1993.
11. Slade, N. A., Models of structured populations: age and mass transition matrices, in *Wildlife Toxicology and Population Modeling: Integrated Studies of Agroecosystems*, R. J. Kendall and T. E. Lacher, Jr., Eds., Lewis Publishers, Chelsea, MI, 1993.

10

Assessing the Impact of Toxic Chemicals: Temporal and Spatial Variation in Avian Survival Rates

Andrew Dobson and Peter Hudson

ABSTRACT

Avian survival varies on a number of temporal, spatial, and taxonomic scales. In order to determine the effect of human-induced environmental hazards on avian survival it is essential to have a complete understanding of the "natural" variation in avian survival. This chapter examines four major sources of variation: interspecific variation in adult survival, geographic variation in adult survival, age-dependent survival, and seasonal patterns of survival. The concentration is on examining these patterns using data collected from long-term banding studies of common British bird species. We conclude by discussing how these data might be used to discern changes in avian survival due to either a buildup of toxic pollutants or long-term changes in climate.

KEY WORDS

population ecology, avian ecology, population trends, survival rates, density dependence

INTRODUCTION

The approximately 8700 species of birds that exist in the world today live in every habitat from tropical rain forests to the high arctic. All of these habitats have been subjected to direct and indirect effects of human-induced environmental insults, including the effects of toxic chemicals. Monitoring the survival rates of a wide variety of bird species provides a very important mechanism for monitoring the health of whole ecosystems, not only with regard to the buildup of potential toxicants, but also in terms of their response to other environmental perturbations, such as long-term climate change. This chapter discusses the temporal and spatial patterns of variation observed in avian mortality rates using data collected in the British Isles. The analyses we present suggest a variety of methods for monitoring the potential effects of ecotoxicants on bird populations.

Potential Sources of Data

Because most birds are relatively easy to see, their presence is constantly monitored by a huge multitude of amateur and professional bird watchers. In many countries, organizations coordinate the activities of these ornithologists so that data on the demography of avian populations may be collected in a systematic fashion. The huge amounts of data collected annually by the national bird-banding (or ringing) schemes provide possibly the most powerful tool available to these organizations for monitoring the health of a wide variety of bird species.

Several types of data can be used to look at spatial and temporal variations in survival rates and subsequently assess the effects of toxicants. First, detailed population studies coupled with experiments and modeling can demonstrate the actual effects of toxicants. We initially consider an example of this approach. The remainder of the chapter concentrates on describing ways in which bird-banding data may be analyzed. Rather than concentrate on the statistical techniques used in these analyses, we shall focus on describing the observed spatial and temporal patterns of variation in survival. These descriptions will then be used as the basis for a final discussion on the ways in which banding data might be more widely used as a powerful tool for a broader scale of environmental monitoring and, more specifically, for identifying the potential effects of ecotoxicants on bird populations.

Detailed Population Studies

Detailed population studies of potential indicator species are necessary if we are to understand how toxicants influence the population dynamics of individual species within a community. In principle this approach involves the intensive collection of basic population data from a series of study areas. Analyses are then undertaken to identify key changes in the mortality rates for the population. Manipulative experiments are conducted to test associations between toxicant levels and changes in mortality (or fecundity); models may then be constructed to aid in understanding interactions between factors and to predict potential effects of changes in the use of ecotoxicants in other areas. This understanding can be used to develop management procedures to reduce the effects of toxicants and to construct more effective monitoring programs.

Case Study: The Grey Partridge

Detailed studies on the survival of grey partridges (*Perdix perdix*) have demonstrated that both the direct and indirect effects of pesticides can reduce survival.[1,2] While cases have been reported of partridges dying as a direct result of pesticide exposure, this problem is declining as biodegradable pesticides, such as the synthetic pyrethroids, have been developed and more toxic pesticides have been removed from the market. Population studies, experiments, and modeling have demonstrated that the most important effects of pesticides have been indirect ecological effects. Chick survival is considerably reduced in areas in which pesticides have reduced broad-leaved weeds, such as *Polygonum* and their associated insect faunas, particularly the insect species preferred by partridge chicks (sawfly larvae, leaf beetles, and plant bugs). Reductions in the density of insect-rich plant species appear to be the most important effect of the application of insecticides in habitats used by partridges.

While the application of herbicides and insecticides have reduced chick survival dramatically within the past 40 years, mathematical models of partridge populations indicate that the impact of this effect at the population level depends on the availability of suitable nesting cover and predation pressure during the nesting season. When extensive nesting cover is available and predators are controlled, chick survival and population size can be greatly increased by reducing the use of selective pesticides on field margins. This experimental approach has been refined into a highly successful management procedure known as "conservation headlands."[3]

INTENSIVE BANDING SCHEMES

Detailed banding schemes where color bands or numbered tags are used to individually identify birds can provide good estimates of annual survival. Perrins[4] suggested that data from such studies provide better estimates of survival than analysis of extensive banding data that depend on the recovery of dead birds. It is interesting to note that discrepancies between the two techniques tend to occur mainly in long-lived species where age-related changes in survival rate may bias the technique.[5]

Generally, ornithologists agree that the mortality rates of young birds are initially high, decline as the bird approaches sexual maturity, and then either remain constant or increase as the bird ages. However, attempts to quantify age-dependent variation in mortality using banding data are frequently frustrated by age-dependent variation in the recovery rates of birds; this can lead to significant biases in the estimates of mortality.[6,7] Even with large initial sample sizes, estimates of age-dependent survival in the older age classes are ineluctably influenced by decreasing sample sizes and increasing errors.

Case Study: Short-Tailed Shearwater

Age-dependent survival has been examined in a number of seabird species. One of the best examples is the analysis of survival data from a 40-year study on the short-tailed shearwater (*Puffinus tenuirostris*), initiated by the late D. L. Serventy on Fisher Island, Australia.[8] Problems of band wear and loss were avoided by double-banding birds and replacing all bands showing signs of abrasion. Mortality was measured from the time of first breeding. Initially, rates of mortality were high and then decreased and remained low for about 10 years before increasing until all birds were dead 27 years after first breeding. Application of two models showed survival was neither constant with age nor fit any simple relationship with age.

EXTENSIVE BANDING SCHEMES

National banding schemes are run in many countries to collect extensive data on the movement and survival of a wide variety of bird species. Such data are particularly useful in describing temporal and spatial variations in survival of common species, while at the same time providing an important monitoring scheme that may help detect the effects of ecotoxicants. In Britain, the British Trust for Ornithology (BTO) runs an extensive banding program and a number of schemes to monitor breeding production (the nest record scheme) and changes in the size of breeding populations (the common bird census). In this section we examine ways of analyzing these data.

Interspecific Variation in Avian Mortality

The original techniques developed by Lack[9,10] and Haldane[11] still provide the basis for the more sophisticated techniques available today. The most comprehensive review of these methods is given in the volume edited by Brownie et al.;[12] the works by Taylor[13] and Seber[14] provide additional information. Although Lack's and Haldane's methods provide the most straightforward way of obtaining estimates of the survival of adult birds for a wide range of species, both techniques assume that the mortality rates of adult birds are constant and that the recovery rate of banded birds is independent of age. These assumptions are only likely to hold for shorter-lived species; band-wear and senescent mortality of older birds may cause serious biases in longer-lived groups such as many sea birds. Nevertheless, the techniques provide a powerful tool for comparing survival between a range of species for which comparable sets of banding data are available.

Interspecific variation in survival for 50 species of common British birds were estimated from birds ringed in the period 1963 to 1975 (Figure 1).[5] Most of the variation in these data may be explained by body size (75% among the 38 passerine species or 63% if all species are considered). Two species are significant outliers from the general trend: whitethroat (*Sylvia communis*) and kingfisher (*Alcedo athis*) lie well below the line.

Reduced survival in the whitethroat appeared dramatically in 1969 when 77% of the previous year's breeding stock failed to reappear. This population decline has been attributed to high mortality during a period of severe drought in its overwintering areas in West Africa.[15] Kingfisher populations appear susceptible to both severe winter weather and the effects of pollutants that probably act to reduce food availability in streams. Reduced survival may also be due to pesticide runoff from agricultural areas into rivers.[16] In the survey of British breeding birds, the absence of kingfishers from the river system around Leeds and Bradford was attributed to industrial pollutants.[17]

Comparative Methods

A whole suite of comparative methods have recently been developed to address questions on the evolutionary significance of life history variation.[18] These tech-

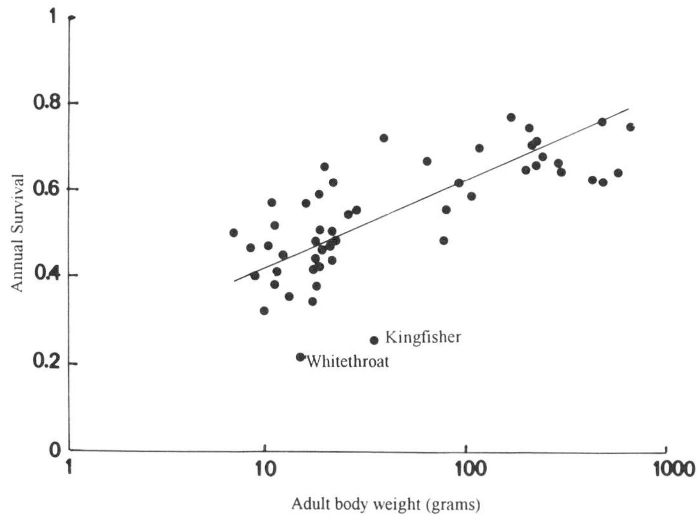

FIGURE 1. The relationship between adult survival rate (estimated using Lack's method) and body weight for 50 species of common British birds (after Dobson[5]). The banding data used are for birds ringed as adults between the years 1963–1975. A major axis regression has been fitted to these data by averaging the observed data for each family. This reduces potential taxonomic biases and produces a line with slope of 0.204. The data for whitethroat and kingfisher were excluded from the analysis as they are clear outliers.

niques seek to remove taxonomic biases from comparative analyses and to allow comparison of life history traits between species with different body size. Their application to data sets, such as those illustrated in Figure 1, would provide a more rigorous method of identifying species that exhibit "abnormal" life-history variation, possibly resulting from the influence of environmental change or pollution. Although they are unlikely to be able to identify the source of the pollutant, comparative studies may act as an important first filter for identifying species in need of more detailed examination.

Age-Dependent Patterns of Mortality

Methods that seek to determine age-dependent variation in the survival rates of individual bird species have received by far the greatest attention from biostatisticians. This interest stems largely from evolutionary biology's fascination with the potential importance of senescence as a life history trait. (A fascination that is as worthy of Jungian analysis as it is of Darwinian!) Unfortunately, attempts to quantify age-dependent variation in mortality using banding data are frequently frustrated by variation in the rates of recovery of birds of different ages. These variations can lead to significant biases in the estimates of mortality.[6,7] Despite these problems of quantification, most ornithologists would agree that the mortality rates of young birds are initially high, then decline to a lower level as the bird approaches sexual maturity. They then either remain constant, as assumed for the short-lived species mentioned above, or increase as the bird ages.

Figure 2 illustrates the instantaneous mortality rate functions for two common British birds of prey: sparrowhawk (*Accipiter nisus*) and kestrel (*Falco tinnunculus*).[19] In both cases the greatest change in mortality rate occurs during the first 2 years of life. It is then equivocal whether mortality remains constant or increases as the birds become older. To a large extent our ability to estimate rates of senescent mortality always will be constrained by the small sample sizes we obtain in the tail of the age distribution. If we naively assume that recovery rates of banded birds are constant, we can fit a fairly simple function of the form $\mu_t = a\,e^{-bt} + c$ to the instantaneous mortality rate data. Integration of this function provides the expected survival curve (Figure 3). Although this technique provides biased estimates of age-dependent survival rates (the survival estimates for birds of any age are ineluctably biased by the unknown recovery rates for birds of that and previous ages), it does provide an important "null-model" for the expected pattern of recoveries obtained from a cohort of birds banded as chicks.[20] This potentially allows past cohorts of ringing data to be used to estimate future expected rates of recovery. Significant differences between observed and expected recoveries may indicate that a species is suffering from the effects of environmental change, or that the null model's parameter estimates are flawed. As a potential example of this effect, Figure 3 contains data for years of high and low pesticide use. The null survival curve is fitted to data for all years; the raw survival data are presented for years with and without pesticides; the differences between this curve and the data from the pesticide era suggests pesticides may well have reduced survival rates in both species.

Density-Dependent Variation in the Survival of Immature Birds

An important factor that is rarely considered in evolutionary discussions of age-dependent avian mortality is the importance of population density in determining the survival of individuals through their first year of life. This is by far the most critical period of a bird's life history (more than 80% of passerines die before reaching age 1).

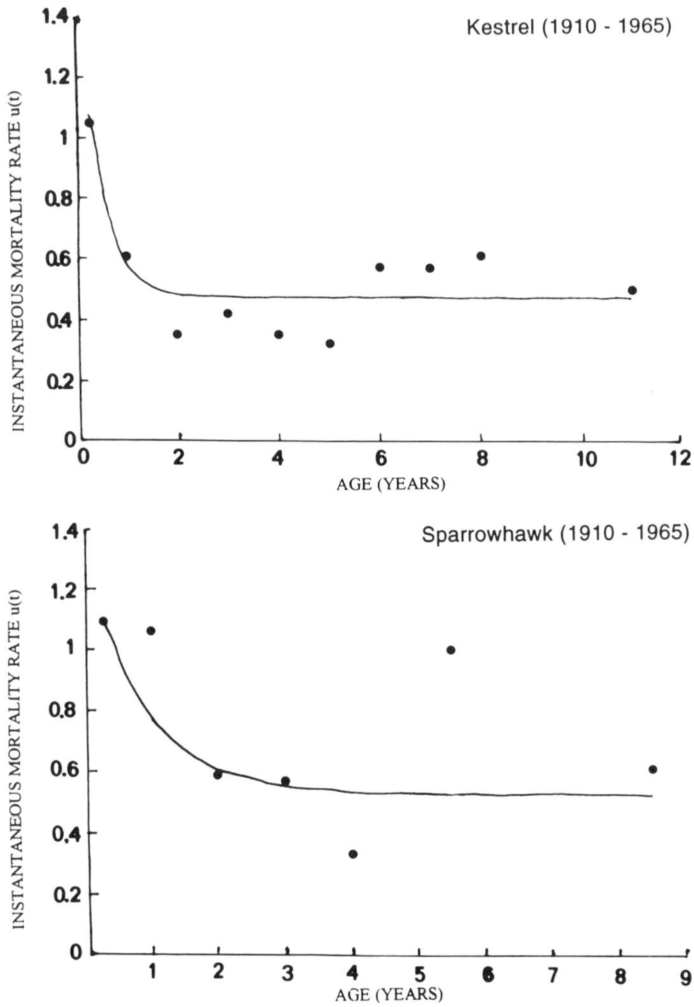

FIGURE 2. The relationship between instantaneous mortality rate and age for British kestrel and sparrowhawk banded as chicks and recovered at known ages. The function, $\mu_t = a\,e^{-bt} + c$ has been fitted to each set of observed data; parameter estimates for each species are kestrel $a = 1.1$, $b = 2.4$, $c = 0.47$; sparrowhawk $a = 0.77$, $b = 1.1$, $c = 0.52$ (after Dobson[19]).

Unfortunately, banding data alone is not particularly helpful in examining possible density-dependent sources of mortality, as it is gathered on a much larger spatial scale than that in which most density-dependent processes operate.

Determining the magnitude of these density-dependent effects usually requires detailed long-term studies of clearly defined populations that have exhibited some variation in population size. The two bird-of-prey species discussed above have been subject to such studies; Figure 4 illustrates the fit of a simple density-dependent recruitment function to data from the studies of Newton et al.[21] on sparrowhawks (*Accipiter gentilis*) and Southern[22] on tawny owls (*Strix aluco*). Between 37 and 51%, respectively, of the variation in recruitment may be explained by density-dependent processes. Any attempt to monitor or model these populations must consider this important source of age-dependent variation in survival.

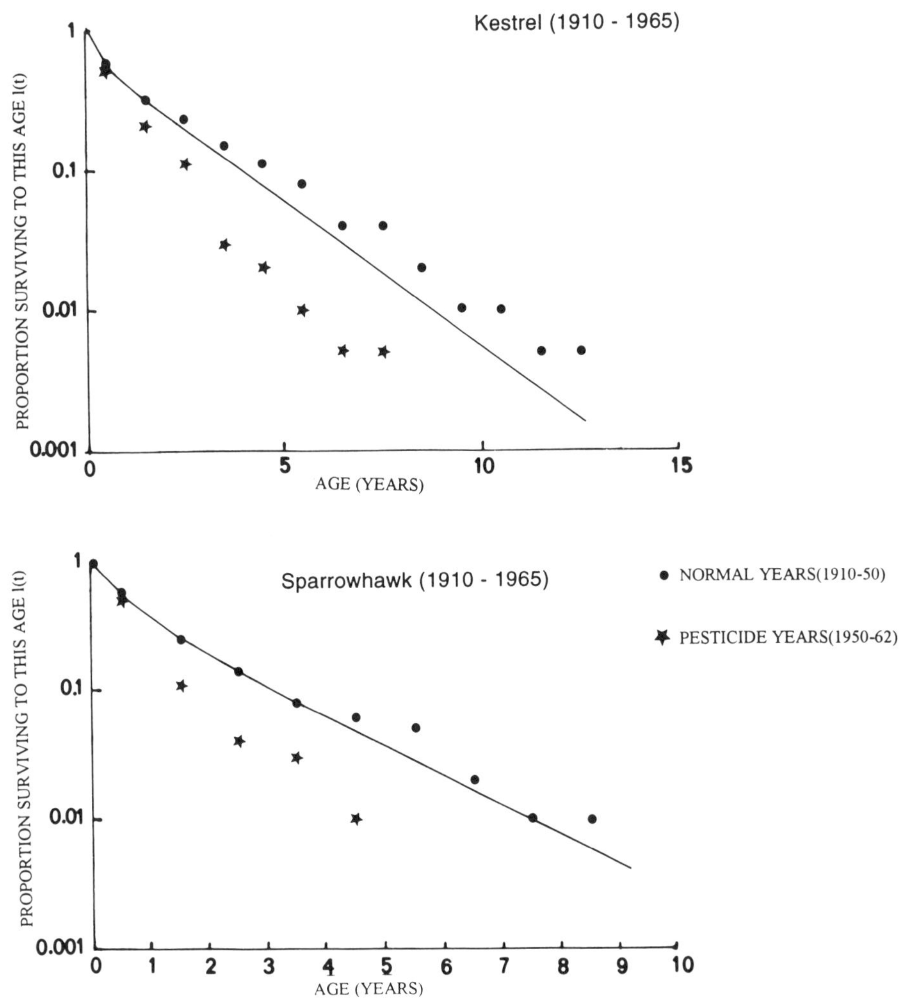

FIGURE 3. The observed survival data and fitted survival functions for the two species illustrated in Figure 2. The fitted curve has the form $N_t = N_0 \, e^{a/b} \, (e^{-bt} - 1) - ct$, with the parameters *a*, *b*, and *c* estimated from the fitted instantaneous mortality rate function. Raw survival data are presented for years of low or no pesticide use (1915 to 1950) and high pesticide use (1951 to 1965) (after Dobson[19]).

Annual Variation in Adult Survival Rates

It is also possible that temporal variation may occur in the survival rates of adult birds. Such variation may be due in part to variation in the density of birds in different years or in different locations; it may also be due to weather variations between years or the buildup of environmental pollutants. The analysis of temporal and geographical variation in survival rates requires very large sets of data; these are only available for the most common species. Nevertheless, their ubiquitous nature makes common species powerful indices of environmental quality.

Annual variations in adult mortality can be estimated from national banding data using a technique described by Coulson.[23] Significant variations in mortality were found for four species of common British birds (Figure 5). None of the sets of data

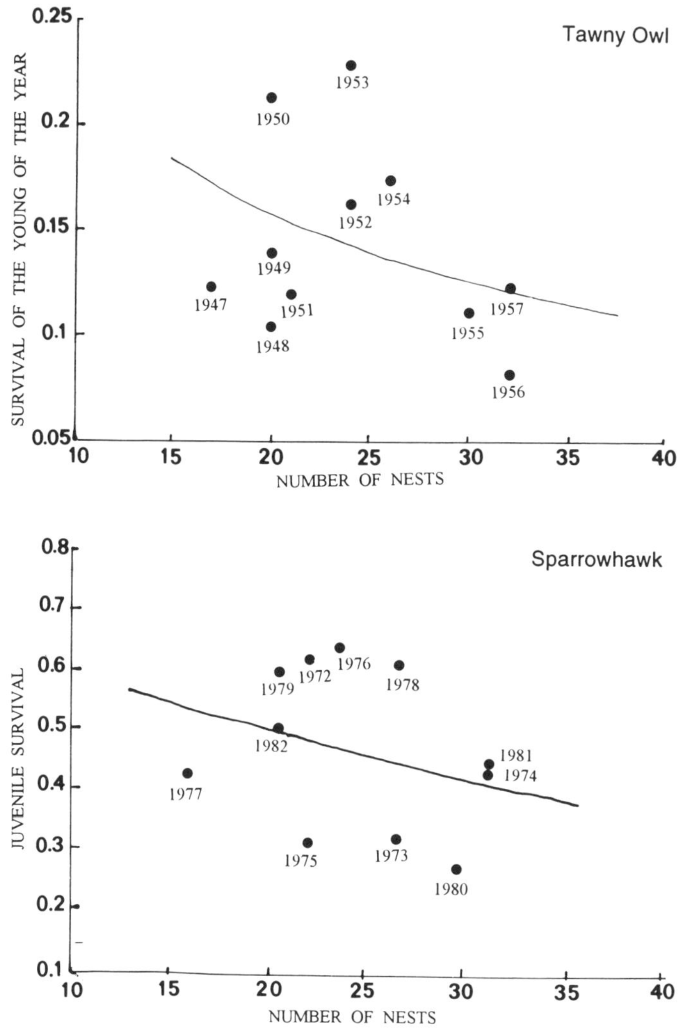

FIGURE 4. Density-dependent recruitment functions fitted to data from long-term studies of sparrowhawks[21] and tawny owls.[22] The function used is $f(Nt + 1) = 1/(1 + (aNt - 1)\hat{}\, b)$. Parameter estimates for sparrowhawks are $a = 1/31.5$, $b = 1.7$; estimates for tawny owls are $a = 1/1.5$, $b = 0.65$ (after Dobson[27]).

exhibit any significant long-term trend, although in all cases the frequency distribution of survival rates is leptokurtic, with a few years of low survival forming the tail of the distribution.

The variation between years could be due to a number of different causes: adult survival may vary with the weather in any year, particularly if unusually harsh conditions prevail. Similarly, survival could vary with density, although this would be very hard to test in data collected over a broad geographic scale. An analysis of the observed inter-year variations in survival in terms of average climatological data for Britain suggest that around 70% of the variation in adult survival may be explained by temperature variation in just 2 months of the year.[27] Figure 6 illustrates the relationship between the average temperature in one winter and one summer month for two of the species, both figures suggest that a 1°C change in temperature in these months corre-

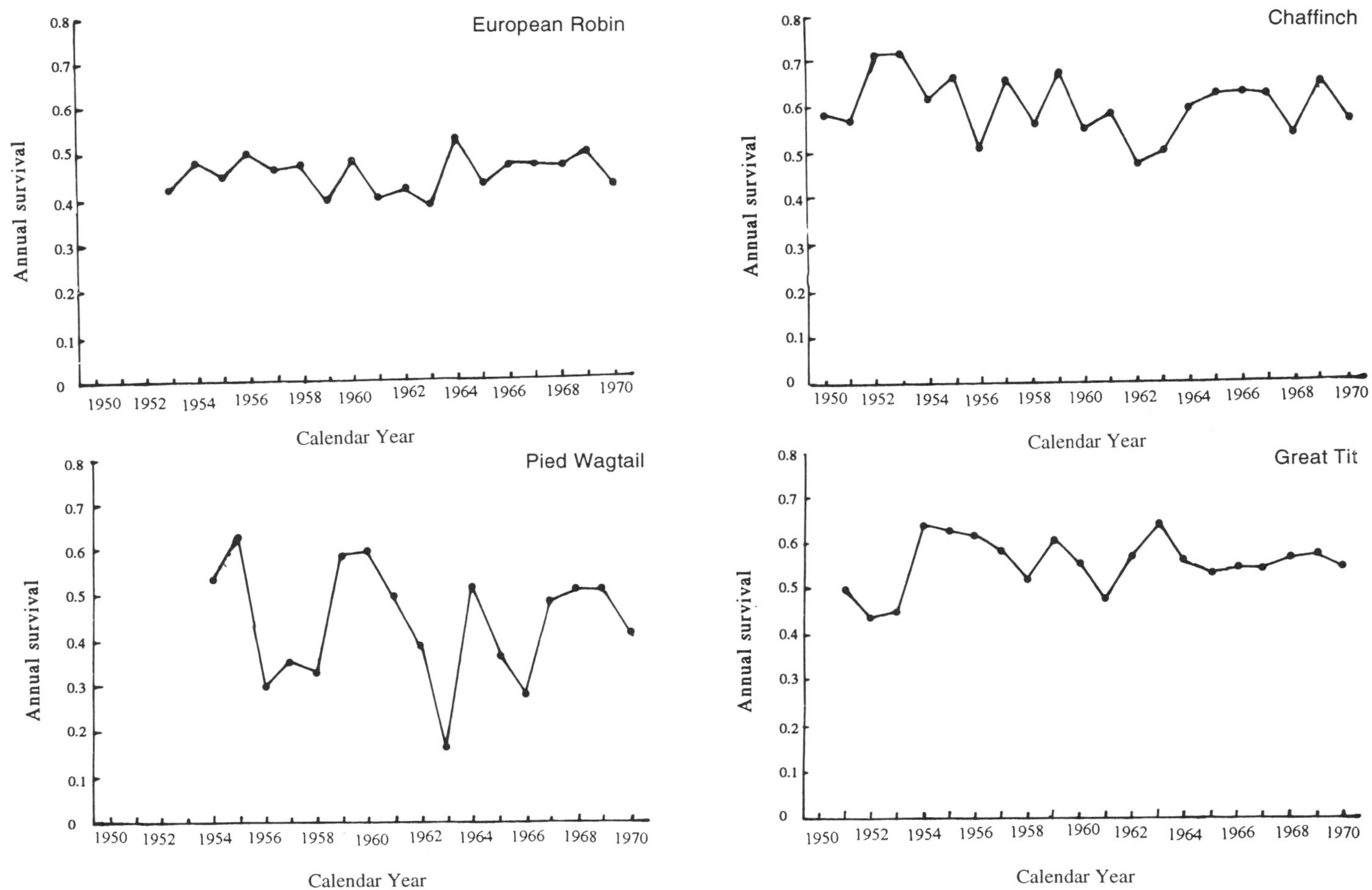

FIGURE 5. Estimates of calendar-year specific survival rates for four species of British birds for the years 1953 to 1970.

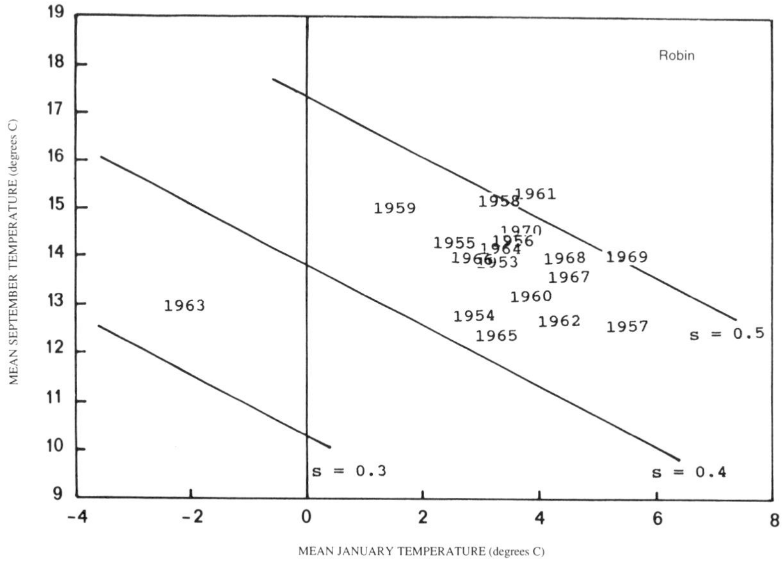

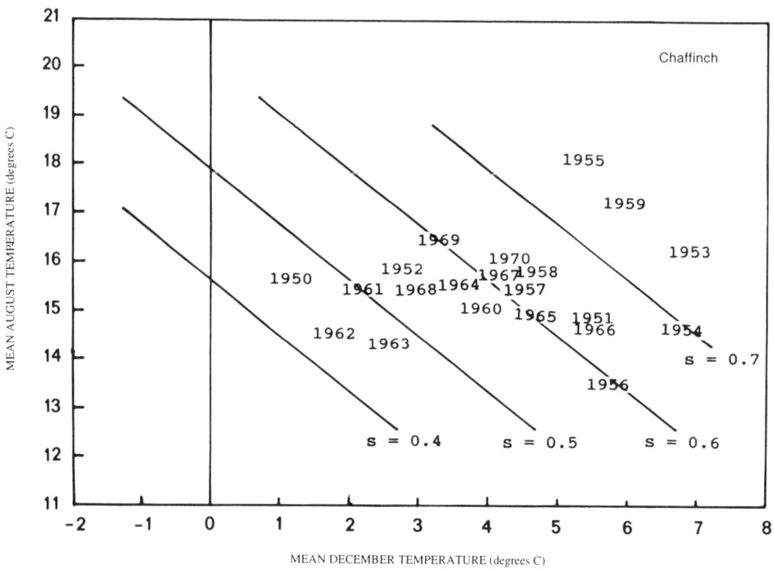

FIGURE 6. The relationship between calendar year specific survival and temperature in the 2 months that correlated most strongly with observed variation. The relationship for the robin accounts for 68% of the variation in survival (s = 0.46 + 0.018 mean Jan. + 0.028 mean Sept.), that for the chaffinches accounts for 72% of the observed variation (s = 0.65 + 0.050 mean Dec. + 0.044 mean Aug.). The years marked on each of the two graphs represent the *expected* survival rate in this year.

lates with a 3 to 4% change in annual survival. Because of strong correlations between the weather in adjacent months of most years, adding weather data for additional months into the analysis does not lead to a great increase in the amount of variation explained.

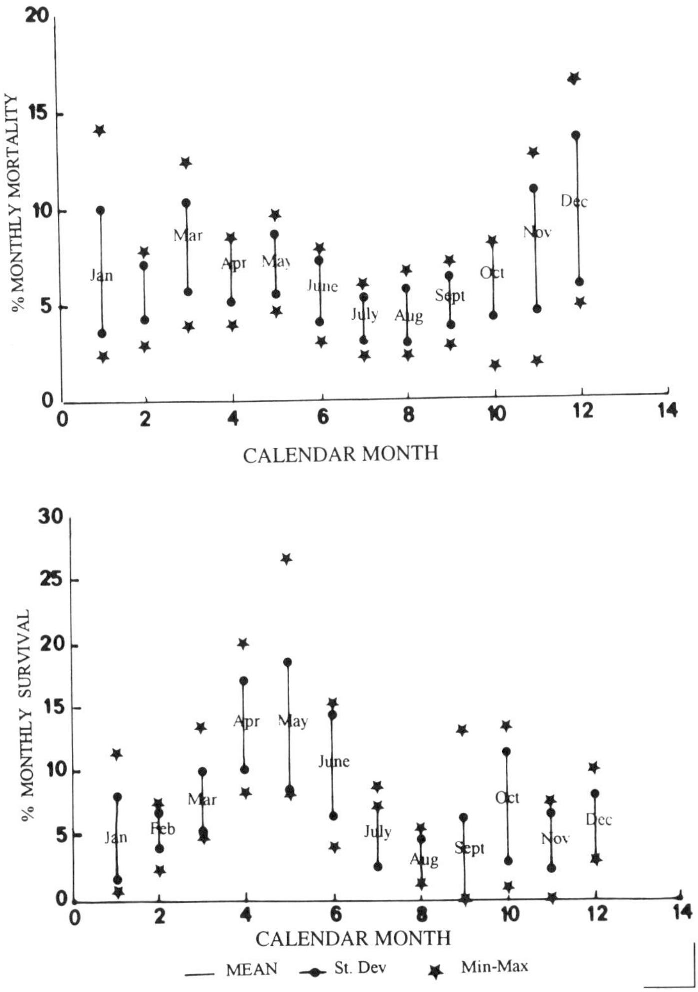

FIGURE 7. Box and whisker diagrams illustrating estimates of percentage monthly mortality rate for robins and chaffinches between the years 1960 to 1970, inclusive. Mean monthly mortality (± 1 S.D.) is given for each month, and the minimum and maximum estimates are also illustrated.

Seasonal Variation in Adult Survival

The data for chaffinches (*Fringilla coelobs*) and robins (*Erithacus tubecula)* are sufficiently numerous to be analyzed on a month-by-month basis for this time period. Figure 7 illustrates the estimated average monthly mortality and the variation in mortality observed over the period of the analysis. These data illustrate a number of interesting patterns: both species show increases in mortality rates during the winter and during the breeding season. This analysis again suggests that winter and late summer are times when prevailing weather is important in determining survival. There is also pronounced mortality during the breeding season which may reflect some "cost of reproduction."

Table 1. Geographical Variation in Survival of Four Common British Birds from Birds Banded and Recovered Dead within the Same Region

	Region				
Species	**East**	**West**	**South**	**Mid**	**North**
Pied wagtail (*Motacilla alba*)	0.48	0.53	0.47	0.52	0.52
Robin (*Erithacus rubecula*)	0.48	0.38	0.50	0.43	0.48
Great tit (*Parus major*)	0.58	0.45	0.55	0.57	0.55
Bullfinch (*Pyrrhula pyrrhula*)	0.44	0.44	0.51	0.38	0.53

[a]Estimated after Lack;[9,10] further details in Dobson.[5]

Geographical Variation in Survival

If weather is important in determining inter-annual variations in survival, it is likely that geographical variations in climate may lead to regional variation in avian survival rates. Dobson[5] failed to detect any significant geographical variation in the survival rates of several of the species of common British birds discussed above (Table 1). However, a recent study of *Turdus* sp. using ratios of old to young birds from museum collections detected significant geographical variation in survival of different species in the genera.[28] More studies of this type are obviously important in addressing questions about geographic variation in demographic parameters.

DISCUSSION

This chapter has described a variety of ways in which avian survival rates are likely to vary both spatially and temporally. Many of the results presented are preliminary and are presented as examples of the types of analyses that need to be undertaken if we wish to use bird populations and bird-banding data as means of monitoring the environment for both the possible effects of ecotoxins and the possible climatic changes that may occur given present concerns about global warming. In contrast the intensive population studies, as illustrated by the work on the grey partridge, provides the only definite method of demonstrating the effects of pesticides and how they influence population size.

In this final discussion we describe a number of factors that will be important in further quantifying variation in survival. As with any discussion in population dynamics, the most difficult problems arise when an attempt is made to disentangle density-independent factors from density-dependent effects, which may occur in either a compensatory or depensatory manner. Although the simplest assumption about avian population dynamics is that adult survival is density independent while juvenile survival and hence recruitment is density dependent, this approximation is only likely to be true for short-lived species. Even in the absence of ecotoxins, a full appreciation of avian population dynamics still requires a better understanding of the effects of population density on avian demography.

Although ecotoxins may have a direct and quantifiable effect on the demography of controlled laboratory populations, their effects on the demography of free-living populations are still not clearly understood.[24] Many studies have illustrated the effect of pesticides on avian fecundity, but only a few have illustrated their effect on survival. This may be because interactions between fecundity and survival may confound attempts to quantify the full demographic impact of ecotoxins on avian populations. Similarly, interactions between high reproductive effort in low density years and high potential costs to reproduction that reduce survival at these times may be important in masking density-dependent effects on survival.[25,26]

The integration of studies that examine the local effects of weather variation with the geographical effects of climate variation present a very powerful method of monitoring avian survival rates on a range of temporal and spatial scales. Although the analyses described above are still preliminary, they suggest a very important mechanism for monitoring the effects of potential long-term climate changes on a variety of bird species.

REFERENCES

1. Potts, G. R., *The Partridge. Pesticides, Predation and Conservation*, Collins, London, 1986.
2. Potts, G. R., and N. J. Aebischer, Control of population size in birds: the grey partridge as a case study, in *Towards a More Exact Ecology*, P. J. Grubb and J. B. Whittaker, Eds., Blackwell Scientific, Oxford, 1989.
3. Sotherton, N. W., N. D. Boatman and M. R. W. Rands, The "Conservation Headland" experiment in cereal ecosystems, *Entomologist*, 108, 135-143, 1989.
4. Perrins, C. M., Age of first breeding and adult survival rate in the swift, *Bird Study*, 18, 61-70, 1971.
5. Dobson, A. P., Survival rates and their relationship to life-history traits in common British birds, *Curr. Ornithol.*, 7, 115-146, 1990.
6. Cave, A. J., Pitfalls in the estimation of age-dependent survival rates of birds from ringing and recovery data, *Vogelwarte*, 29, 160-171, 1977.
7. Lakhani, K. H., and I. Newton, Estimating age-specific bird survival rates from ring recoveries—can it be done?, *J. Anim. Ecol.*, 52, 83-91, 1983.
8. Bradley, J. S., R. D. Wooler, I. J. Skira and D. L. Serventy, Age-dependent survival of breeding short-tailed shearwaters *Puffinus tenuirostris*. *J. Anim. Ecol.*, 58, 175-188, 1989.
9. Lack, D., The age of the blackbird, *Br. Birds*, 36, 161-175, 1943.
10. Lack, D., The age of some more British birds, *Br. Birds*, 36, 193-197, 1943.
11. Haldane, J. B. S., The calculation of mortality rates from ringing data, *Proc. Int. Ornithol. Cong.*, 11, 454-458, 1955.
12. Brownie, C., D. R. Anderson, K. P. Burnham and D. S. Robson, Statistical Inference from Band Recovery Data—A Handbook, Fish and Wildlife Service Resource Publ. No. 131, U.S. Dept. of the Interior, Washington, DC, 1978.
13. Taylor, S. M., Recent quantitative work on British bird populations, *Statistician*, 16, 119-132, 1966.
14. Seber, G. A. F., *The Estimation of Animal Abundance and Related Parameters*, 2nd ed., Griffin & Co., London, 1982.
15. Winstanley, D., R. Spencer and K. Williamson, Where have all the whitethroats gone?, *Bird Study*, 21, 1-14, 1974.
16. Meadows, B. S., Kingfisher numbers and stream pollution, *Ibis*, 110, 443-451, 1972.
17. Sharrock, J. T. R., *The Atlas of Breeding Birds in Britain and Ireland*, Poyser, Calton, U.K., 1976.
18. Harvey, P. H., and M. D. Pagel, *The Comparative Method in Evolutionary Biology*, Oxford University Press, New York, 1990.

19. Dobson, A. P., Mortality Rates of British Birds, D.Phil. thesis, University of Oxford, Oxford, U.K., 1982.
20. Dobson, A. P., Age-dependent mortality rates of some common British birds, in *Statistics in Ornithology*, Lecture Notes in Statistics 29, B. J. T. Morgan and P. M. North, Eds., Springer Verlag, 1985, pp. 275–288.
21. Newton, I., M. Marquiss and P. Rothery, Age structure and survival in a sparrowhawk population, *J. Anim. Ecol.*, 52, 591-602, 1983.
22. Southern, H. N., The natural control of a population of tawny owls (*Strix aluco*), *J. Zool.*, 162, 197-285, 1970.
23. Coulson, J. C., A study of the mortality of the Starling based on ringing recoveries, *J. Anim. Ecol.*, 29, 251-271, 1960.
24. Walker, C. H., Pesticides and birds – mechanisms of selective toxicity, *Agric. Ecosyst. Environ.*, 9, 211-226, 1983.
25. Dobson, A. P., P. J. Hudson and E. R. Carper, Population biology and life-history variation in gamebirds, in *Ecology and Management of Gamebirds*, P. J. Hudson and M. R. Rands, Eds., Blackwell Scientific, Oxford, 1988, pp. 48–71.
26. Partridge, L., and P. H. Harvey, The ecological context of life history evolution, *Science*, 241, 1449-1455, 1988.
27. Dobson, A. P., manuscript in preparation.
28. Ricklefs, R. E., manuscript in preparation.

11

Population-Phenogenetic Approach for Ecological Monitoring (a Biotest)

Vladimir M. Zakharov

ABSTRACT

A procedure termed *biotest*, defined as an estimation of the quality of the environment through the assessment of the condition of natural populations, is proposed as a useful tool for ecological monitoring. A population-phenogenetic approach is developed as a method of biomonitoring for estimating the condition of populations of a spectrum of species in the wild. Analysis of phenotypic variability in natural populations from the viewpoint of general developmental stability or homeorhesis forms the basis of the procedure. Analysis of the natural populations of different species suggests intra- and interpopulation and interspecific differences in developmental stability, which allows us to consider the latter as a population characteristic. The analysis of developmental stability is essential for estimating the optimal conditions of development and genetic coadaptation. The application of a variety of approaches for characterizing homeostasis (morphogenetic, genetic, biochemical, physiological, and immunological) to biomonitoring is discussed.

KEY WORDS

biotest, developmental homeostasis, fluctuating asymmetry, phenotypic variability, homeorhesis, biomonitoring

INTRODUCTION

Organizing a Biotest at the Population Level

Various general approaches are being used to assess the quality of the environment. One of the possible (and perhaps most important) ways to evaluate environmental quality is to assess the condition of living organisms. This is the case for the following reasons. A strict necessity exists to identify global and local changes in the environment as well as to detect the content of different pollutants in various abiotic and biotic components of the environment. In spite of our efforts to decrease anthropogenic impacts on nature, it is not possible to stop entirely the pressure of our activities; therefore, the effects of any impact can become progressively complex. In these cases the identification of the consequences of complex impacts on the condition of living organisms of various species, including man, is of primary importance.

Biotests, described as the assessment of the quality of an environment through the estimation of the condition of living systems, can be realized at several levels of organization.[1] Their application at the population level, however, seems to be of particular significance at present because increasing attention is being paid to the analysis of natural populations. The population level is key to understanding the mechanisms of a number of important biological processes, and is emerging as the main unit of evolution. Investigations at this level of organization are important because the condition of particular populations eventually determines the survival and breeding success of individual species as well as the normal functioning of the ecosystem as a whole.[2]

THE BIOTEST

Requirements for a Biotest

In order to detect the quality of the environment through a biotest in any particular area one needs to be provided with basic data about the diversity of different species and possible changes in diversity. However, some detectable changes in a set of species in an area take place as a result of both global and local successional processes derived from different, nonhuman sources. Often it is not easy to distinguish changes of natural or artificial (i.e., human induced) origins. Nevertheless, in such a case the assessment of the condition of natural populations of species inhabiting a particular area seems to be the main and most important task for a biotest.

A number of various methods and approaches can be applied in this respect. They have the following requirements:

- *Significance.* To detect real changes in a condition of organisms and their populations, one needs to focus attention on the most general and essential biological parameters.
- *Sensitivity.* As nature conservation measures are developed and new, less dangerous technologies are being used, rather sensitive methods allowing one to distinguish even small, reversible changes in the environment become more important.
- *Universality.* The procedure should be broadly applicable in relation to both the assemblage of species tested and the type of impact evaluated. The best opportunity to obtain a general estimate of the quality of an environment through a biotest occurs when the condition of populations of various species in an ecosystem is assessed. This requires that methods be applicable for a number of species. Populations of any species now face a complicated mixture of various types of impact. An attempt to test any kind of human impact and to find a specific measure of the biological consequence for each species seems to be a nearly impossible task. Biotest methods should be focused on assessment of many kinds of impact and various combinations of impact as well; i.e., such methods should be available as a nonspecific measure of many kinds of environmental stresses.
- *Wide availability.* Methods should be inexpensive and not too complicated for wide application.
- *Applicability to natural populations.* Many sensitive modern methods can be used to model the effects of different kinds of impact, but a majority of them are appropriate only for laboratory experiments. Admittedly, basic information about the effect of various kinds of impact obtained via experimental models is needed. However, such data alone often are not sufficient to explain the situation in nature. Moreover, experimental modeling of the effect of the impact of various factors in different combinations seems to be impossible. Integrating the assessment of the condition of natural populations of different species living in the wild under a complex set of impacts of various factors appears to be the most informative and attractive approach to the problem.

In addition, all such methods should be able to detect possible differences among populations (spatial control) and to identify changes in a condition of the same population before and after any particular impact, during both the increase and decrease of the degree of influence (temporal control). Biotest systems need to be applicable for the assessment of possible changes in a condition of populations of various species in natural environments (background monitoring) as well as under the impact of various kinds of human activity.[2]

POPULATION PHENOGENETICS AND THE BIOTEST

The Place and Role of Population Phenogenetics in a Biotest

The first requirement for detecting the quality of the environment through the assessment of the condition of natural populations is to use the most general and important characteristics. If we want to study the ontogenetic processes of living organisms, one important characteristic is the presence of a general or a developmental homeostasis. We need to find appropriate ways to estimate these most general ontogenetic characteristics in the natural populations of various species. This specific field of study, the analysis of developmental homeostasis in natural populations, is the subject of population phenogenetics.[3,4] Various methods are used to study this general characteristic from different points of view (e.g., physiological, biochemical, genetic, and morphogenetic).

Consider an opportunity to apply the general concept of developmental homeostasis for a biotest, with a particular interest in characterizing the morphogenetic point of view. Developmental homeostasis can be defined as a stabilized form or as a developmental trajectory.[5] When analyzing one developmental stage rather than the whole developmental trajectory, homeorhesis is interpreted as the ability to develop a similar phenotypic effect. It has been shown, both theoretically and in practice, that it is necessary to distinguish and analyze two aspects of general developmental homeostasis. Stability (measured by the level of developmental noise) is the capacity for stable development under given conditions. Canalization or plasticity (measured by possible directional changes) is the capacity for stable development under different conditions.[4,6] Thus, an analysis of phenotypic variability in natural populations from the viewpoint of overall developmental homeostasis includes the characterization of both random disturbances and directional changes.

Developmental stability seems to be of primary importance for a biotest, while the applicability of canalization and plasticity of development is more narrow and rather specific (see below). Phenotypic similarity corresponds to a high level of developmental stability. To evaluate developmental stability, however, it is necessary to take into account only those phenotypic features that result from developmental disturbances; i.e., features that cannot be reduced to genotypic or environmental differences. Such phenotypic differences can be revealed between monozygotic twins or individuals of one genetic clone developed under similar environmental conditions. A more universal approach to detect this kind of variation is through the study of homologous structures of the same organism developing on the basis of the same genotype under identical external conditions. Such differences can be considered a manifestation of chance variability of development or as developmental noise.[5,7] The manifestation of such variability, which is most common and accessible to analysis, is fluctuating asymmetry; i.e., minor nondirectional deviations from perfect bilateral symmetry, as well as differences in the value of morphological traits on the left and right sides of the body.[8-10]

Fluctuating asymmetry presents clear analytical advantages mainly for many species of animals, while for plants this kind of variation as developmental noise can be easily detected by comparing many homologous structures, such as leaves on a plant.

Another measure of developmental stability is the frequency of phenodeviants as abnormal phenotypes or as small deviations from the conditionally normal structures of various morphological characters.[11-14] In general, any other approaches allowing one to distinguish deviations from the genetically determined norm, such as intraclutch variation in different parameters of eggs as a measure of female condition in birds, could be useful for the measurement of developmental noise.

Normally, in order to evaluate the level of developmental stability from the morphogenetic point of view, a limited number of morphological characters is used. Because a disturbance in general developmental stability is manifested by an increase in developmental noise of different, even uncorrelated, characters,[3] such an approach can be considered justifiable. Then it requires only a precise assessment of these certain morphological characters. Analysis in this case can be limited to characters of external morphology, which allows acquisition of the necessary information from living specimens. For example, intraclutch variation in egg parameters and fluctuating asymmetry in characters of leg pholidosis for nestlings can provide researchers with needed information about developmental stability, even for strictly protected species of birds.

The approach is universal in that it can be applied to various groups of living organisms. This is because the causes for the changes in developmental stability measured by the level of developmental noise are similar in principle in relation to different morphological structures in various species. A high level of developmental stability corresponds only to a certain value (or narrow range) of the environmental parameter under study, which can be characterized as optimal.[14] Disturbances of developmental stability are a nonspecific reaction to the deviation from the optimal values in environmental parameters. The results of the experimental study of the dependence of developmental stability on incubation temperature measured by fluctuating asymmetry for the sand lizard (*Lacerta agilis*) are presented as an example (Figure 1). The data show unambiguously that the maximum level of developmental stability (minimum level of fluctuating asymmetry) is observed at a temperature of about 25°C. Developmental stability is disturbed by any deviations from these optimal values.[3,4]

The interest and importance in estimating optimal conditions by the criterion of developmental stability in natural populations is determined by the following. First, optimal conditions are similar for many populations of the species. Second, the range of these conditions is narrower than in nature. Third, some changes in phenotypic characters or other parameters are consequences of generation under nonoptimal conditions and may disappear under optimal ones. Fourth, some small, environmentally induced disturbances of developmental stability appear to be within the limits of a certain drift allowed by natural selection and may not adversely affect viability.[3,4]

This opens up the prospects for discovering populations existing under suboptimal conditions. Moreover, this permits the elucidation of differences not only between populations, but even between generations in the same population if developmental conditions were different. As regards the species with overlapping generations, such information could be obtained even from a study of a single sample by comparing individuals of different ages.[3] If developmental conditions for different generations were themselves different, individuals characterized by high and low levels of developmental stability could be found in the same population. Thus, it is possible to monitor a population for a long period, before and after the onset of a certain human impact. This means that the dynamics of developmental stability can be monitored as the impact declines or increases.

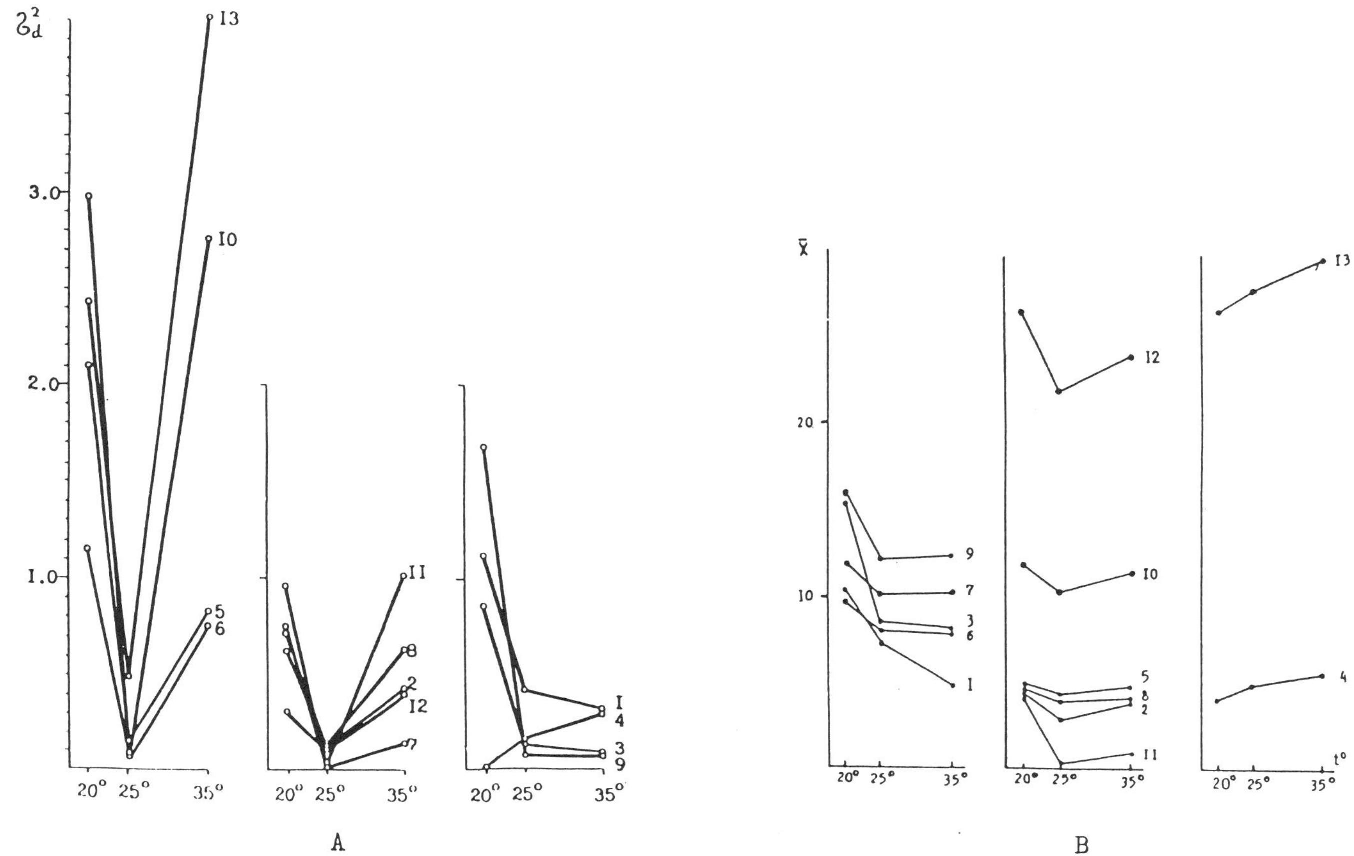

FIGURE 1. Variance of fluctuating asymmetry (δ_d^2) and mean ($\bar{x}$) of pholidosis characters (1–13) in the sand lizard (*Lacerta agilis*) as a function of the temperature of egg incubation. Key for number of scales: (1) postnasal and antejugular, (2) preorbital, (3) upperlabial in front of and (4) behind suborbital, (5) orbitotemporal, (6) supraorbital, (7) upper ciliary, (8) upper temporal, (9) lower labial, (10) mandibular, (11) granules, (12) lower ciliary, and (13) femoral pores.

The results of a comparison of two samples of the grey seal, *Halichoerus grypus*, collected from the same population in the Baltic Sea in pre- and postpollution periods are presented as an illustration of possible temporal changes in the level of developmental stability (Figure 2). The findings suggest a dramatic change in developmental stability of this species as measured by asymmetry and phenodeviants of various characters of skull morphology during the period of heavy pollution.[15]

One peculiarity of the genetic control of developmental stability that has been established in a number of special studies is the observation that it mainly depends on the general genetic coadaptation.[6,16-18] A disturbance of developmental stability can take place when the gene balance is disturbed, which has been demonstrated in crosses of essentially genetically different forms adapted to different environmental conditions.[10,19] This presents us with an opportunity to reveal changes in the condition of a population in the zones of secondary intergradation, especially when caused by human activities.

Another cause of disturbance in genetic coadaptation manifested in changes of developmental stability could be strict directional selection that can essentially change gene arrangement. Such an effect has been modeled in laboratory experiments,[20] but it is less likely in natural populations. It is probable that even directional natural selection will always be focused on two processes: to create the desirable effect and to keep general genetic coadaptation at a certain required level. Decreased genetic coadaptation as a result of selection under extremely severe conditions has been observed in only one particular case for a rather short time (corresponding to the longevity of several generations). Such a decrease in developmental stability has been registered for the first generations of flies selected for resistance to an insecticide.[21] Even in this situation developmental stability provides us with needed information about changes in the condition of a population after a human impact. Thus, the characterization of the

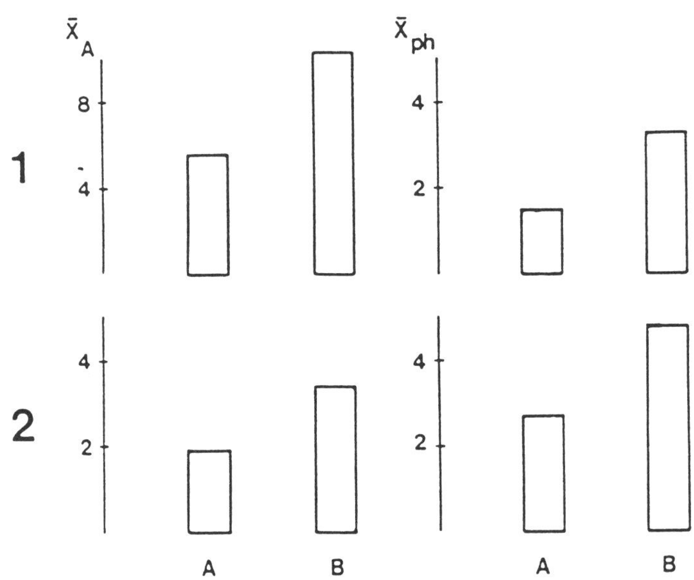

FIGURE 2. The mean number of asymmetric characters ($\bar{x}_A$) and the mean number of phenodeviants ($\bar{x}_{ph}$) for the two sets of skull characters (1 and 2) in two sample groups (A and B) of the Baltic grey seal (*Halichoerus grypus*). Set 1, 24 quantitative meristic characters as a number of foramina; set 2, 12 qualitative characters as different variants of shape and position of foramina and bones. Group A, prepollution sample of animals born before 1940 (n = 21); group B, postpollution sample of animals born after 1960 (n = 29).

condition of natural populations by the level of developmental stability meets the principal modern requirements for biomonitoring.

The disturbance of developmental stability is an integral nonspecific measure of many kinds of environmental stress and that is one of its advantages. In some cases it may be necessary to detect the main cause of such a deterioration in developmental stability. Note that a combination of the factors affecting developmental stability occurs relatively rarely. Genetically, these characteristics depend on general genetic coadaptation, disturbances of which are seldom detected in natural populations, and then mainly in zones of secondary intergradation between different forms adapted to essentially different environmental conditions at species and subspecies levels. The same is true for environmental factors.

Moreover, the effect of a suspected main factor can be modeled in an experiment and the result can be compared with an observed situation in natural populations under a certain level of a particular factor (such as a concentration of a particular pollutant). If this result corresponds to the experimental data obtained for the same level of the studied factor, it means that this factor probably is responsible for such an effect. If the degree of deterioration of stability in natural populations is higher than in a laboratory experiment under a similar level of the factor, then it is likely that other factors are adversely affecting developmental stability in nature.

In some cases the experimental analysis of another aspect of general homeostasis, canalization, or plasticity of development estimated by directional changes in development can reveal which environmental parameter has caused a disturbance of developmental stability. These changes are often specific and their appearance (for example, an increase in the magnitude of a character) is observed only under a strictly defined range of levels of a given environmental parameter. A comparison of specific marker traits to the actual phenotypic properties of the population under study would elucidate the causative agent of the disturbance of developmental stability.[3]

One example of how to get information about both aspects of homeostasis for the same set of morphological characters during the same experiment is presented in Figure 1. In spite of the fact that the level of developmental stability nonspecifically decreased at low and high temperatures for practically all characters under study, the pattern for the dependence of the mean value for the same characters on temperature is very different (positive or negative correlation, or lack of any interaction) for different characters. Thus, a particular phenotypic change that corresponds to the certain environmental conditions can be considered as only a marker of them.

CONCLUSIONS

Future Prospects for Biotest at the Population Level

This study has mainly focused on an illustration of some opportunities to detect the level of developmental homeostasis in natural populations of various species from the morphogenetic point of view. If these characteristics are in fact the most general measures of the condition of living organisms it should be possible to register changes by some other approaches as well. An important and challenging task for the future is to combine various methods for a more complete characterization of the effects of an impact. This is of interest for both theoretical and practical aspects of the biotest. If we can identify changes in all different measures used it means that the effect we have detected is real and we should act to mitigate any further change in the condition of the population in spite of the absence of any obvious changes in viability.

Such a complex approach will allow researchers to characterize developmental homeostasis in different functional aspects of an individual at different levels, ranging from the molecular to the organismic. The following approaches are needed for such a characterization:

- *Morphogenetic.* This approach has been specifically described in this study.
- *Genetic.* Concern here is specifically with changes at the chromosomal and DNA level that correspond to two main requirements: (1) they should not adversely affect the viability of an organism, and (2) they should be a more or less nonspecific reaction to various kinds of stress. Thus, the cytogenetic approach allows one to characterize developmental homeostasis by some structural and numerical changes in the chromosomes in somatic cells (e.g., aneuploidy, sister chromatid exchanges, and lateral asymmetry of the chromosome). It was demonstrated that an increase in the frequency of aneuploid cells and an appearance of their clones mainly reflect a condition of the immune system of an organism.[22-25]
- *Physiological/biochemical.* Many physiological and biochemical tests can be used to characterize homeostasis and the "condition" of an organism. It has been established for various animal species that in spite of a well-known exponential dependence of the rate of oxygen consumption on temperature, total oxygen consumption for a stage of individual development is minimal at a certain range of the temperatures only, which can be characterized as the optimal range, and nonspecifically increases under the lower and higher temperature. This means that the most economical energetic metabolism can be considered as a criterion to detect optimal temperature conditions for development. It has been shown that such an energetic approach can be used for any other environmental parameters and in general as a criterion of developmental stability.[26-28] Additionally, some biochemical tests, such as the registration of a change in activity of some enzymes as a specific reaction to certain impacts or as a nonspecific consequence of any kinds of stress, such as a modulation of superoxide dismutase activity under oxidative stress, can be used to measure the condition of an organism and its possible alterations from the norm.[29]
- *Immunological.* Various molecular and histological tests of the efficiency of the immune response could prove useful in characterizing the homeostatic condition of an organism. It has been established that parasitic infection depends on a complex of various parameters, but mainly reflects the physiological condition of a host organism and its homeostatic ability.

Events that can be registered at different levels are probably not directly related to each other, at least in the sense that their interactions are not necessarily cause-effect. In such a case it is of particular interest to distinguish the degree of noise at different levels (from molecule to organism as a whole). Thus, an estimation of the quality of the environment through an assessment of the condition of natural populations of various species of living organisms by the level of their developmental homeostasis seems to have great potential for the wide application for a biotest. This level can be measured by an integrated and sensitive complex corresponding to the main requirements mentioned in this chapter.

REFERENCES

1. Yablokov, A. V., and S. A. Ostroumov, *Levels of Nature Protection*, Nauka, Moscow, 1985 (in Russian).
2. Zakharov, V. M., Analysis of fluctuating asymmetry as a method of biomonitoring at the population level, in *Bioindications of Chemical and Radioactive Pollution*, Advances in Science and Technology in the USSR, Biology Series, D. A. Krivolutsky, Ed., Mir, Moscow, 1990, pp. 187–198.

3. Zakharov, V. M., *Animal Asymmetry: A Population-Phenogenetic Approach*, Nauka, Moscow, 1987 (in Russian).
4. Zakharov, V. M., Future prospects for population phenogenetics. *Sov. Sci. Rev. F. Physiol. Gen. Biol.*, 4, 1-79, 1989.
5. Waddington, C. H., *The Strategy of the Genes*, Allen & Unwin, London, 1957.
6. Mather, K., Genetical control of stability in development, *Heredity*, 7, 297-336, 1953.
7. Astauroff, B. L., Analyze der erblichen Storungsfalle der bilateralen Symmetrie im Zusammenhang mit selbstständigen Variabilität ähnicher Strukturen, *Z. Indukt. Abstamm. Vererb.*, 55, 183-262, 1930.
8. Van Valen, L., A study of fluctuating asymmetry, *Evolution*, 16, 125-142, 1962.
9. Soule, M., Phenetics of natural populations. II. Asymmetry and evolution in a lizard, *Am. Nat.*, 101, 141-160, 1967.
10. Zakharov, V. M., Fluctuating asymmetry as an index of developmental homeostasis, *Genetica (Beograd)*, 13, 241-256, 1981.
11. Lerner, I. M., *Genetic Homeostasis*, Oliver & Boyd, Edinburgh, 1954.
12. Kincaid, H. L., Effects of inbreeding on rainbow trout populations, *Trans. Am. Fish. Soc.*, 105, 273-280, 1976.
13. Kirpichnikov, V. S., *Genetic Bases of Fish Selection*, Nauka, Leningrad, 1979 (in Russian).
14. Zakharov, V. M., Analysis of homeorhesis in its ontogenetic, population and evolutionary aspects, in *Thermodynamics and Regulation of Biological Processes*, I. Lamprecht and A. I. Zotin, Eds., Walter de Gruyter, Berlin, 1985, pp. 497-508.
15. Zakharov, V. M., and A. V. Yablokov, Skull asymmetry in the Baltic grey seal: effect of environmental pollution, *Ambio*, 19, 266-269, 1990.
16. Goldshmidt, R. B., *Theoretical Genetics*, University of California Press, Berkeley, 1955.
17. Adams, A. S., and J. D. Niswander, Developmental "noise" and congenital malformation, *Genet. Res.*, 10, 313-317, 1967.
18. Ferguson, M. M., Developmental stability of rainbow trout hybrids: genomic coadaptation or heterozygosity?, *Evolution*, 40, 323-330, 1986.
19. Graham, J. H., and J. D. Felley, Genomic coadaptation and developmental stability within introgressed populations of *Enneacanthus gloriosus* and *E. obesus* (Pisces, Centrarchidae), *Evolution*, 39, 104-114, 1985.
20. Thoday, J. M., Homeostasis in a selection experiment, *Heredity*, 12, 401-415, 1958.
21. Clarke, G. M., and J. A. McKenzie, Developmental stability of insecticide-resistant phenotypes in blowfly; a result of canalizing natural selection, *Nature*, 325, 345-346, 1987.
22. Shah, J., K. Mayeda, F. Koppitch, S. Mahmood and D. Nemitz, Karyotypic polymorphism in acute myelofibrosis, *Blood*, 60, 841-844, 1982.
23. Cimino, M., R. R. Tice, and J. C. Liang, Aneuploidy in mammalian somatic cells in vivo, *Mutat. Res.*, 176, 107-122, 1986.
24. Deen, D. F., W. F. Morgan, P. J. Tofilon and M. H. Barcellos-Hoff, Measurement of sister chromatid exchanges and their relationship to DNA damage, repair and cell killing, *Pharmacol. Ther.*, 42, 349-360, 1989.
25. Ilynskikh, N. N., M. A. Medvedev, S. S. Bessudnova and I. N. Ilynskikh, *Mutagenesis in an Organism being in Different Condition*, Tomsk University Press, Tomsk, 1990 (in Russian).
26. Winberg, G. G., The temperature optimum of development, *Adv. Mod. Biol.*, 5, 650-651, 1936 (in Russian).
27. Zotin, A. I., and N. D. Ozernyuk, The effect of temperature on respiration and ATP level at the period of cleavage of *Misgurnus fossilis*, *Rep. Acad. Sci. USSR*, 171, 1002-1004, 1966 (in Russian).
28. Zotin, A. I., and T. A. Alekseeva, Stability of ontogenesis, in *Thermodynamics and Regulation of Biological Processes*, I. Lamprecht and A. I. Zotin, Eds., Walter de Gruyter, Berlin, 1985, pp. 485-496.
29. Perke, D. V., Molecular mechanisms of chemical toxicity, *Pol. J. Occup. Med.*, 1, 18-38, 1988.

12

Use of Space and Habitats by Individuals and Populations: Dynamics and Risk Assessment

Michael A. Bowers

ABSTRACT

Habitat selection has the effect of increasing fitness differences among individuals within a population. Most models used in ecological risk assessment, however, assume that the only source of demographic variability is attributable to age, and that populations occupy homogeneous environments. Such an approach ignores individual differences in performance and is more epiphenomenal than process oriented. A source of potential error in such models is that demographic rates may vary more between individuals living in different habitats than between demographic classes. One issue addressed in this chapter is whether demographic status can legitimately be used as a first-order approximation of demographic performance. Another is whether age-structured populations exhibit different population dynamics than those structured by habitat. Computer simulations showed age-structured populations to be less stable and to experience extinction more frequently than habitat-structured populations. This reflects differences in the spatiotemporal mode of population regulation: under habitat-structure individuals have the ability to change their reproductive status simply by changing habitats, while in age-structured populations fitness can only be changed through aging. One means of incorporating habitat structure into demographic analyses is through the use of submodels that predict the distribution of individuals among habitats, followed by habitat-specific analyses of demographics. Parameterization of such models with field data will be time and effort intensive, however, and will not be feasible in most situations. As a shortcut, it is suggested that studying dynamics in suboptimal habitats can provide much information concerning the maximal effect of most disturbances (including that of toxic compounds of human origin) on population processes.

KEY WORDS

habitat selection, population modeling, risk assessment, wildlife toxicology, source/sink populations

INTRODUCTION

There is much current interest in the development of models that predict population-level responses to human-caused disturbances, particularly those that involve contamination of the environment by xenobiotic chemicals.[1,2] Research detailing in vitro physiological responses of organisms to toxicants can be used to predict the level of contamination at which death is likely, but experience has shown that extrapolating such results to account for phenomena at or above the population level is not simple. Problems are (1) depending on physiological condition, age, sex, and where they live, organisms vary in both exposure and susceptibility to toxins; and (2) because of variation in rates of survival and reproduction, individuals contribute differentially to long-term population dynamics and regulation. One method of accounting for these shortcomings is through the separate estimation of age-specific demographic rates, and the incorporation of these into compartmental models of population dynamics. Submodels can then be used to alter birth, death, and survivorship under different contamination regimes. The appeal of such models is that they can be easily generalized to most populations inhabiting a wide range of sites.

Habitat occupancy is a much overlooked component impinging on whether organisms are risk-averse or risk-prone to contamination. A basic premise of past risk assessment models is that some individuals are more susceptible than others to contamination; a widely used method of accounting for this variation has been the demographically based Leslie-matrix approach (for discussion, see Reference 1). However, in some cases fecundity and survivorship may vary more between individuals living in different habitats than between age classes,[3-6] indicating one potential source of error in this approach.

Habitat selection is a process involving the interaction between individuals and their environments. At some spatial scale, most environments are comprised of habitats of different quality. The success or ability of individuals to evaluate and occupy the better habitats represents a source of variability to individual performance that has been largely ignored in models of population dynamics. One question the author addresses here is whether differences in ability of individuals to select habitats of overall high quality should be considered when the goal is to predict an overall population dynamic. The role that interspecific competition plays in habitat selection (choosing to lump such influences under the rubric of "the environment") is not directly addressed, but the author does acknowledge that in some cases competition with other species is largely responsible for delimiting which habitats are available for species to select.[7-9]

While the focus of this symposium is on the assessment of ecological damage accompanying the application of toxic chemicals, xenobiotic compounds represent just one type of ecological disturbance. Pickett and White[10] define disturbance as "any relatively discrete event in time that disrupts ecosystem, community, or population structure and changes resources, substrate availability, or the physical environment." While a number of studies[10,11] have focused on disturbances in the generic sense, and a number have considered the ecological importance of disturbance in heterogeneous environments,[12] the focus of most studies of this genre have been on an ecosystem or a landscape scale. However, much evidence suggests that spatial heterogeneity can strongly influence processes occurring on the scale of individuals and populations.

This chapter asks whether habitat selection should be included as a basic parameter in population dynamical models, particularly those models that are oriented toward ecological risk assessment. Working assumptions acting as a base for this paper are

1. Individuals within most populations have the potential of occupying habitats of different quality.
2. Habitat selection, which the author defines as the active selection of habitats of relatively high quality and the rejection of habitats of low quality, is the most important

mechanism creating nonrandom patterns of habitat occupancy, either directly or indirectly.

3. The differential occupancy of habitats by individuals within a population creates a spatial template over which demographic rates may vary widely.

BASICS OF HABITAT OCCUPANCY AND SELECTION

Ecologists have always had difficulty distinguishing between the habits of species and the habitats they occupy.[13,14] Thus, in most cases variation in individual performance will be totally inseparable from variation in habitat quality. There always will be high premiums associated with the occupancy of the "best" habitats. By selecting some habitats and avoiding others, organisms may use their environments in a highly nonrandom (coarse-grained) manner.[15] For birds, in particular, natural selection appears to have exerted an especially strong influence on where individuals choose to live.[7,9,16]

Individuals should always choose the best habitat available to them at a given time,[15,17] but how is quality assessed? Factors that are of ultimate importance are food, protection from predators and climate, breeding sites, and more idiosyncratic factors related to the morphology or behavior of a particular species. In some cases organisms appear to select habitats on the basis of proximal cues that are correlated with factors of biological relevance. For example, catbirds (*Dumetella carolinensis*) select sites according to the size and shape of leaves,[18] and bobolinks (*Dolichonyx oryzivorus*) choose sites where ground litter is sparse.[19] In both cases habitats appear to be selected according to proximal factors that are correlated with prey availability and foraging efficiency. In other cases, species appear to respond to ultimate factors such as nest hole availability in the case of the pied flycatcher (*Ficedula hypoleuca*),[20] and lemming abundance for jaegers, hawks, and owls in the arctic.[21]

It is rare, however, that any habitat is selected according to single features. Demands on organisms are multifactorial, and it is probably exceedingly rare that sites can be found that simultaneously maximize survivorship, growth, and reproduction. Therefore, in an absolute sense most habitats will be suboptimal, and organisms must consider tradeoffs when habitats are evaluated. For example, while nest hole availability is the crucial factor determining suitable habitats for the pied flycatcher (see above), deciduous forest with a mix of appropriate tree species is also an important, albeit more ancillary, habitat requirement.[20] A similar conclusion was reached by Martin,[22] who showed that while diversity of bird species in forested drainage basins in Arizona was correlated with features characteristic of both foraging and nesting habitats, the strongest correlations involved features related to nest sites.

The occupancy of a particular site will depend on the time and energy that an organism allocates to searching, as well as the probability that a better site will be found. Leaving a satisfactory site for one that may be better is a risky proposition at best. Individuals are under constant pressure to improve their fitness, but must evaluate the risks involved in doing so. In this sense improving fitness by changing habitats is not that different from using dispersal to a similar end. As pointed out by Gadgil,[23] ". . . the factor favoring the evolution of dispersal would be the chance of colonizing a site more favorable than the one that is presently inhabited."

Historical factors and/or prior experience also play a role in habitat selection. Mature individuals that have successfully fledged young at a site may return to that site in the future,[24] even if the quality of such sites is reduced.[25] Less site tenacity apparently exists for younger birds without ties to a particular location. Consequently, younger individuals play an important role filling vacancies in established habitats, adjusting to changes in habitat quality, or exploiting sites in newly created habitats.[26]

STUDIES OF HABITAT SELECTION

Studies of avian habitat selection typically identify structural features correlated with species distributions rather than ultimate factors, or the means, by which habitats are actually selected.[7] Hence, much of our understanding of habitat selection is based on correlative relationships. Moreover, habitat associations may result from habitat selection per se or from external influences that produce habitat associations (i.e., differential predation leading to differential survivorship among habitats). Experimentally identifying the causes of habitat associations has proved to be difficult.[27] In many cases, distributions are best described using multivariate statistical techniques,[7] reflecting both the diverse requirements of organisms and the inability of ecologists to accurately identify factors of overriding importance.

There is a range of scales in time and space over which habitat associations can be studied:[28] from the use of habitats by single individuals, the use of habitat composites by whole populations, the distribution of metapopulations among habitat patches within regions, to habitat associations measured over entire geographic ranges. While there are processes that cut across all these scales, some are more important at one level than others. Interspecific interactions (i.e., competition and predatory risk) or variation in the availability of critical resources may create habitat associations that can only be understood at the level of individual organisms. At larger spatial scales, variation in gross habitat characteristics, productivity, or climatic conditions may create habitat mosaics that are best studied at the metapopulation or population level.[28] At these different scales are fundamental differences in what habitat associations indicate, and what one might infer from them. At the local scale the focus may rest on relative differences in habitat quality; e.g., given that a species occurs locally, what are the underlying factors that cause individuals to discriminate among habitats? At larger scales, the primary question is whether a particular species occurs at a particular site, and why. It is possible that the answer could be as simple as the presence or absence of suitable habitats, or alternatively, there could be historical reasons why species are absent. With an increase in scale, the focus of inquiry shifts from relative to absolute measures of fitness, and from local processes that influence individual choice to those that influence population persistence (i.e., from foraging to extinction and dispersal vectors). A number of recent reviews[28-31] discuss responses of organisms to environmental heterogeneity at different spatial scales.

SPATIAL DEMOGRAPHY

A central premise of habitat selection theory is that animals occupying habitats of different quality will survive and reproduce at different rates, a relationship largely ignored in population models where the standard approach is to use age-specific estimates of natural, undisturbed birth and death rates, and to predict population dynamics via matrix projection models. The goal of ecorisk (ecological risk assessment) models of this genre is to first establish baseline estimates of demographic parameters, and then to include the influence of some specified external agent on immediate and future population dynamics (using a range of submodels to vary the matrix elements). How adaptable is such an approach to situations where demographic parameters vary more between habitats than between age classes? If habitat occupancy is density dependent, and habitats vary in quality, then some allowance must be made for habitat-specific demographic rates. Habitat structuring may be particularly important when social interactions and dominance are involved in habitat occupancy,[32,33] as may be the case with birds.

The social system of many species strongly influences their spatial distribution patterns. For example, while the preferred habitat of the American redstart (*Setohphaga ruticilla*) is mature deciduous forests, a large proportion of first-year birds are relegated by adults to live in immature deciduous and coniferous forests.[34,35] Similarly, it is typically young English great tits (*Parus major*) from secondary habitats (mostly hedgerows) that fill sites vacated by mature birds in preferred forest habitats.[36] A number of other studies have documented both age-[37-39] and sex-related[26] differences in foraging sites.

In flocks of wintering black-capped chickadees (*Parus atricapillus*), females forage high and on the outermost branches, while males forage low and close to the trunks. There is a strict dominance hierarchy with males having higher rank than females and older birds dominating younger birds of the same sex.[37] Experimental manipulation showed that females occupied the lower foraging territories when males were removed,[40] suggesting that females are relegated by males to poorer habitats. The feeding rates of females in the two habitats were similar, suggesting that differences in the quality of habitats was due to predatory risk. In a similar case, Ekman[41] showed that adult willow tits (*Parus montanus*) used sites with greater protection from pygmy owls (*Glaucidium passerinum*) than did yearlings, and that as a result of these habitat differences older birds were more efficient in foraging and also suffered less mortality.

Anderson[42] distinguished between high fitness "survival habitats" and low fitness "colonization habitats." In the former, reproductive success is high and population density is relatively constant from year to year, while in the latter densities fluctuate widely and fitness is relatively low (comparable to the "source" and "sink habitats" of Pulliam[43]). In most years densities in colonization habitats are determined by the emigration of excess individuals from survival habitats, because in colonization habitats it is rare that local reproduction will compensate for mortality.

It is generally recognized that, at some spatial scale, source and sink habitats can be found for most populations,[42-45] and that the capacity of one habitat to produce and the other to absorb excess individuals is crucial to understanding population dynamics.[44,45] This fact notwithstanding, models of population dynamics in spatially subdivided systems typically treat process and state variables averaged over space. See, for example, References 23, 46, and 47, but especially review Reference 48.

DYNAMICS OF AGE- AND HABITAT-STRUCTURED POPULATIONS: RESULTS OF SIMPLE COMPUTER SIMULATIONS

When developing models of population dynamics, ecologists walk a fine line between including sufficient detail over which accurate estimates of demographic parameters are likely and including so much detail that overall generality is sacrificed.[15] One means of achieving both objectives is by careful separation of processes that occur on different temporal or spatial scales, and by averaging population processes over homogeneous subunits.[49] The standard demographic approach identifies age-structure as a logical grouping variable over which population dynamics can be modeled. An alternate method for partitioning the population into workable subunits for modeling is to consider where organisms live and the social processes that determine habitat use.[33,44]

Critical questions lie in determining how population dynamics differ between age- and habitat-structured populations. Specifically, this paper addresses whether the ability of individuals to select better habitats affects overall population dynamics, and whether populations in which habitat occupancy is based on social dominance (where older, intrinsically more fit individuals select better habitats) have different dynamics than those in which habitat accessibility is equal for individuals of all ages. A factorial approach and computer simulations were used to answer these questions. The first factor contrasted intrinsically determined vs extrinsically influenced sources of demographic success, i.e., where survivorship and fecundity were set according to age, habitat, or a age-by-habitat interaction. The second factor represented various ways that individuals might respond to differences in habitat quality, i.e., no habitat selection, age-independent habitat selection, and age-despotic habitat selection (in which adults selected the better habitat and young were relegated to the poorer habitat). Dynamics were simulated using the standard Leslie-matrix approach for a population comprised of two age classes (first-year individuals and adults) and two habitat types (better and poorer). In all simulations there was an upper limit of how many individuals could select the better habitat; the remainder were relegated to the poorer habitat, with no limit as to how many individuals could occupy the poorer habitat. The different habitat selection models determined whether individuals of different ages had priority in selecting the better habitat: under the no-selection model, individuals did not distinguish between the two habitat types; under age-independent habitat selection, adults and first-year individuals selected the better habitat with equal probabilities; and under despotic habitat selection, both age classes preferred but adults had higher priority in selecting the better habitat (excess adults were forced to use the poorer habitat).

Steps in the simulations included setting the initial densities and calculating population dynamics over time using the population projection algorithm standard for Leslie matrices. External (environmental) influences were included by allowing survivorship to vary stochastically; i.e., both age- and habitat-specific survivorship rates were allowed to vary randomly generation to generation within ± 0.50 of the set value (survivorship for adult and first year, or for individuals in better and poorer habitats, were increased or decreased at the same random increment).

Preliminary simulations were used to set demographic values. Parameters that resulted in unlimited growth or promptly led to extinction were substituted until a range of acceptable values was identified. Within this range, fecundity and survivorship rates were set 50% higher for adults (or individuals in the better habitat, depending on the model) than for first-year individuals (or those in the poorer habitat), respectively (Table 1).

One hundred simulations of 100 generations each were performed for each of the six factor level combinations. The same series of random numbers were used (in the same order) for each set of 100 simulations.

The models produced very different population trajectories. A series of runs over 20 generations (Figure 1) showed population fluctuations to be less for models in which demographic success was habitat dependent and habitats were selected, and much greater for simulations in which performance was determined by age alone or where habitats were not selected. The similar dynamics for nonselecting and pure age-determined simulations suggests that the factor responsible for creating these differences was not the presence of habitats of different quality per se, but the tendency for individuals to select the better habitat when given a choice.

Single simulation results (i.e., similar to those in Figure 1) suggested a number of hypotheses that could be tested using the full complement of 100 computer runs.

Table 1. Three Demographic Schedules Used in the Various Simulation Models

		Age Determined		
Habitat	F_f	F_a	P_f	P_a
Better	1.0	1.5	0.50	0.75
Poorer	1.0	1.5	0.50	0.75
		Habitat Determined		
Habitat	F_f	F_a	P_f	P_a
Better	1.5	1.5	0.75	0.75
Poorer	1.0	1.0	0.50	0.50
		Age-Habitat Determined		
Habitat	F_f	F_a	P_f	P_a
Better	1.25	1.50	0.625	0.75
Poorer	1.00	1.25	0.50	0.625

Note: F_x is the number of offspring born to an individual of age *x*; P_x the survival rate of an individual of age *x*. In all cases there were two age classes: first-year individuals (with *f* as the subscript) and adults (denoted by *a*), and two habitat types: better and poorer.

Specifically, because the age-structure and nonselecting models appeared to produce greater fluctuations in density than those incorporating habitat selection, the author predicted higher coefficients of variation in density, and as a related phenomenon, higher rates of extinction in the former when compared to the latter.

Both hypotheses were confirmed by the simulations. Using generation 30 as the point at which coefficients of variation were calculated (beyond which extinctions occurred; see below), the simulations showed that all age-based and nonselecting models predicted population sizes that were two to three times more variable than those in which habitats were selected (Figure 2). Moreover, extinction occurred in 12, 14, and 9 runs of the pure age-based model, and in 10, 9, and 9 runs for the model without habitat selection, but no extinctions were recorded for simulations that included habitat selection. However, simulations allowing individual performance to vary with both age and habitat produced densities that were more variable than when performance was deter-

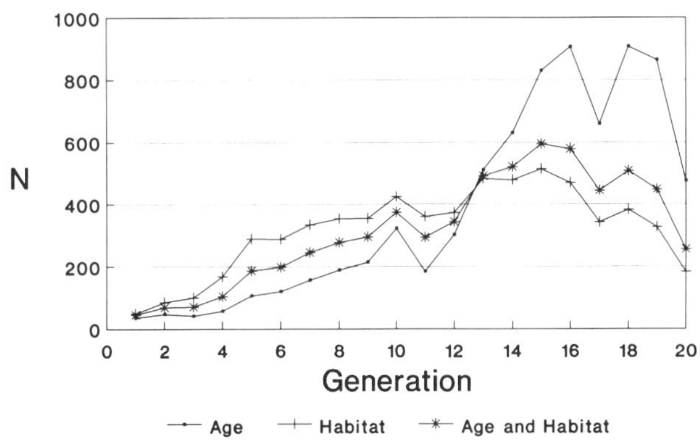

FIGURE 1. Population trajectories over a 20-generation run for age-determined, habitat-determined, and age-by-habitat models. While each simulation experienced the same schedule of stochastic mortality, the degree to which this mortality depressed population density varied markedly. All simulations graphed here were based on the model of despotic-habitat selection (see text for more detail).

Surv/Fecundity Schedule	Mode of Habitat Selection: None	Age-independent	Despotic
Age	CV= 62 EXTINC= 9	CV= 56 EXTINC= 12	CV= 61 EXTINC= 14
Habitat	CV= 52 EXTINC= 9	CV= 34 EXTINC= 0	CV= 26 EXTINC= 0
Age-by-habitat	CV= 53 EXTINC= 10	CV= 44 EXTINC= 0	CV= 37 EXTINC= 0

FIGURE 2. Computer simulation results listed factorially for each schedule of survivorship or fecundity (down) and mode of habitat selection (across). Within each cell are listed the coefficient of variation in density (i.e., CV; calculated at generation 30 over 100 simulations), and the number of extinctions recorded over the full complement of 100 simulations for that combination of factors.

mined by habitat alone. Densities under the age-independent habitat selection model were more variable than those for habitat selection that was despotic.

This model assumes that stochastic environmental influences affect survivorship but not fecundity. However, several runs of the model with fecundity as the stochastic variable showed the same qualitative result as those for which survivorship was allowed to vary, i.e., fluctuations were greater in age-structured and less in habitat-structured models. Hence, in this case the generality of the results do not appear to depend on which parameters are allowed to vary and which are fixed.

While the main simulations treat the case in which both first-year and adult survivorship or fecundity are similarly affected by environmental stochasticity, it is possible that individuals living in one or the other habitat might be better buffered from environmental variability than the other. However, this is unlikely when environmental perturbations take the form of chemical toxicants applied at large spatial scales and affecting physiological processes independent of habitat type. More specific modeling efforts are needed to treat the frequent situation where contaminants are applied to just one of several habitats used by a species.

These simple simulations were designed to be more heuristic than prescriptive. They identify some potential errors of pure age-specific models rather than recommend how such models can be improved. The simulations do suggest that:

1. Habitat selection tends to stabilize short- and long-term population dynamics. It has the effect of dampening density fluctuations and reduces the probability of extinction.
2. Populations where individual performance is determined more by age than habitat will be less stable (i.e., they will have more violent population fluctuations) than populations where performance is determined predominantly by habitat.
3. Populations where older, more intrinsically fit individuals select the better habitats, and relegate younger, less fit individuals to secondary habitats will be more stable than those where habitat selection is non-despotic.

POPULATION CONSEQUENCES OF HABITAT SELECTION: IMPLICATIONS FOR MODELING DYNAMICS

Demographic analyses treat dynamics as epiphenomena of populations, while habitat-selection models presented here identify the individual as the basic unit—the former is based more on pattern, the latter on process. It is rare that the two approaches will yield predictions that are comparable. Under a habitat-structured model, individuals have the ability to change their reproductive success simply by changing habitats, while in age-structured populations fitness can only be changed through aging. Thus, habitat selection may occur on a faster time scale than do demographic events.[43,48,50]

The issue is whether demographic status can be used as a first-order approximation of demographic performance. It is almost a maxim in ecology that either age or its proxy, individual size, is correlated with demographic performance. While a multitude of factors contribute to this relationship (e.g., experience), the one the author emphasizes is habitat selection or relegation. If age determines social status, and social ranking determines where individuals live, then demographic variability may have as much a spatial (habitat) as a temporal (demographic) component. Then the question becomes whether it is age or habitat that ultimately determines fitness. Where dominant individuals occupy better habitats and subdominant individuals occupy poorer habitats, the effects of age and habitat on fitness will be totally confounded. In the rare case in which the correlation between habitat quality and demographic composition is high, age and habitat structured models will yield similar results.

What if, however, there is not a one-to-one match between the number of individuals in particular age classes and the capacity of habitats of different quality to support them? Inequalities in habitat capacity-to-individual ratios create potential for error in strict demographic approaches, and increases the necessity of employing habitat structure as the basis for dynamics. This is especially true when habitat selection is density dependent.[51,52] For example, at low densities all individuals regardless of age should select the best habitat available. As densities increase, dominance hierarchies will play an increasingly important role in segregating individuals of different ages among habitats of different quality. At very high densities only a small fraction of the dominant individuals will occupy the better sites, with a large fraction inhabiting poorer habitats.

Density-dependent habitat occupancy represents one mechanism of density-dependent population regulation. Most models of population dynamics include terms for density-dependent influences. In many cases, these terms seem to be included because ecologists believe such effects are important and thus reflect something about the known biology of particular species. Adjustments in model parameters are usually made by assuming a linear impact of density on either birth rates, death rates, or both. However, it is rare that any mechanistic explanation is offered that connects density per se to density-dependent terms in the models.

Habitat selection and occupancy represent one clear case in which the nature of density dependence can be specified, although the effects may not be linear. Consider the case in which individuals occupy a limited number of home ranges or territories in adjacent habitats of different quality. Further assume that these individuals occupy sites whose quality is not reduced by other individuals in that habitat. As long as the total number of individuals is less than the number of sites in the better habitat, all individuals should select the better habitat type. Only when population size exceeds the capacity of the better habitat will individuals begin to occupy the poorer habitat type. In this case, density-dependent effects measured at the population level would be zero up to the switchpoint, and would increase in magnitude with an increase in the fraction of individuals occupying the poorer habitat. Moreover, if habitat occupancy is socially based, subdominant individuals will be relegated to poorer habitats at lower densities than the dominant individuals, producing density effects that vary among demographic classes.

This simple case illustrates that asymmetries in individual fitness are to be expected when habitat occupancy is based on a combination of pure habitat selection and habitat relegation mediated through social dominance. For an individual living in the better habitat, fitness will always be the maximal attainable for that habitat regardless of how many individuals are living in the poorer habitat. By contrast, the ultimate factor determining fitness for subdominant individuals is not habitat per se, but the appropriation of better habitats by socially dominant individuals. In such cases fitness is based on both social rank (affecting which habitats an individual uses) and features of the habitat occupied. Most current population-based models homogenize demographic rates across individuals irrespective of habitat or social position, an approach that is prone to inaccurately represent population dynamics.[31]

The author suggests that density-dependent habitat selection provides one context for analyzing demographic variability among individuals within a population. Under high mortality, surviving individuals (regardless of their age) have the ability to improve individual performance simply by moving to the better habitat—the result being that a large fraction of the population reproduced and survived at the maximal rates. The greater the mortality rate, the more dramatic the shift of individuals into the better habitat and the steeper the population recovery rate. Under low mortality, population size increases and better habitats become saturated, forcing an increasingly large fraction of the population into the poorer habitat, hence, decreasing the per capita reproductive rate to a point at high densities in which the average rate of increase for the population will approach that set by the poorer habitat.

The inclusion of habitat in population dynamic models is partially a scaling problem. The smaller the spatial scale of inquiry the more likely that only one of several interconnected habitat types will be considered. Treatments of population dynamics in single habitats (i.e., in spatially homogeneous systems) typically assume the system to be closed, and the dynamics determined by local rates of fecundity and survivorship. Such models offer a sharp contrast to those incorporating spatial heterogeneity, where emigration of individuals becomes an increasingly important factor in population regulation.[31,32,44] For populations whose individuals move between source and sink habitats or between habitats of different quality, it is important to know whether a habitat is a net consumer or a net producer of dispersers.[43,48] These are not dispersers in the biogeographical sense (i.e., those that move far distances across uninhabitable areas with low probability of finding a suitable site), but are individuals that move freely and with high transition probabilities among habitats occupied by the same population. It is these "local dispersers" that affect habitat selection and through it have strong effects on local population dynamics.[31]

HABITAT SELECTION AND ECORISK MODELS

Ecological risk assessment considers either local extinction or depression in mean abundance, in one or over several years, as important endpoints in a continuum of ecological damage.[1] How does one evaluate whether a population is depressed when under unperturbed, natural conditions populations fluctuate widely? Attempts to evaluate whether depression has occurred quickly become more statistical than biological; i.e., what level of statistical confidence is required before evidence for damage is sufficient. If the criteria for significance are high, ecological damage may proceed to its limit (i.e., local extinction). At the other endpoint, unsubstantiated conclusions about statistically (not necessarily biologically) insignificant effects creates other sorts of problems. Because of these limitations, there is much interest in the development of models that can be used to evaluate ecological damage, optimally before the damage progresses too far. In order for such models to be useful they must be accurate, general, and predictive.

Most models used in ecological risk assessment are constructed around some variant of the demographically structured, Leslie-projection matrix. Such models are pattern oriented. Bookkeeping methods are used to estimate matrices of demographic parameters, and these parameters are used to project present population structure into the future. Because most populations can be characterized in terms of age classes, life history stages, or some closely related proxy variable (i.e., size), these models are very general. Such models, however, can provide little information concerning the processes affecting population dynamics; i.e., they are more phenomenological in nature than mechanistic.

Habitat selection represents a process by which individuals improve their reproductive success. Habitat selection occurs in most populations on some spatial scale. One of its consequences is to create demographic rates that vary markedly among individuals. One means of accounting for this variation has been to explicitly include habitat selection in submodels.[43,53,54] In these approaches separate models are constructed for populations in different habitat types, and the selection of habitats by individuals provides the diffusive element that creates an overall population dynamic. The habitat selection submodel determines the distribution of individuals among habitats, and the habitat-specific schedule of demographics determines the relative contribution of that habitat to overall population dynamics.

Models with this structure can be used to predict population responses to a wide range of perturbations, including increased mortality as a result of contamination or destruction of habitat. For example, in the case of elevated mortality rates the model would first predict a redistribution of individuals (so that all spaces in the better habitat were filled) and then predict the demographic consequences of the redistribution.[54]

Applying the model to "real world" situations requires the estimation of: (1) demographic rates in each habitat; (2) abundance of each habitat type; (3) initial distribution of individuals among habitats; and (4) transition probabilities for dispersal among connected habitats. While the simplest case will be for individuals that select habitats that are discrete and of uniform quality, with some effort it should be possible to model cases in which habitats are more continuous.[53] Such models will allow for variation in habitat distribution or abundance, and the manner and extent to which individuals select the better habitat type.

Habitat-structured population models (such as those developed in this chapter) are based on the premise that population dynamics are not epiphenomena, but are the by-product of individuals attempting to maximize their fitness in a heterogeneous environment. The nature of such models dictates that they be habitat, site, and species specific. As May[55] argues, the average dynamical behavior of a nonlinear system can be very different from the dynamical behavior of the average. Hence, differences among individuals may play an important role in population regulation and dynamics.

For those with applied or basic interests in modeling populations in the field, the ideas presented here will no doubt cause some concern. Specifically, not only are tremendous technical and logistic difficulties involved with quantifying habitats in terms of availability and demographic suitability, but also with defining the appropriate scale over which habitat mosaics should be studied. Measuring all these parameters in one system may take years; when environmental stochasticity and interactions with other species are considered, the task of developing and validating such a model may prove impossible.

There is, however, a real need to identify systems in which habitat structure can be studied in all the requisite detail. Such systems could then be used as models to test general predictions concerning the importance of habitat structure to population dynamics and, by extension, to establish the appropriate temporal and spatial bounds for ecological risk assessment.

Such studies would further indicate the necessity of understanding processes occurring both within and between connected habitats. The author expects that under detailed analysis individuals comprising most populations will exhibit differences in fitness attributable to habitat quality, that habitats will differ in the rate at which "surplus" individuals are produced, and that some habitats will be net producers and others net consumers of emigrants. Modeling approaches that fail to recognize that emigrants from one habitat may be immigrants for another are susceptible to a number of errors. For example, the application of a harmful compound in a source habitat may not significantly or measurably decrease densities in that habitat, but may decrease or totally eliminate the number of emigrants (i.e., surplus individuals filling vacancies of those that died instead of creating a pool of dispersers). The consequence of reducing the number of emigrants, however, could be local extinction in the poorer habitat, particularly if its population is regulated largely by immigration. For this reason populations in poorer (sink) habitats will be particularly sensitive to ecological damage.[43] The differential susceptibility of populations in habitats of contrasting qualities actually could be used as a tool in ecological risk assessment. In short, population depression in suboptimal habitats could be used as a measure of maximal effect, but overall risk assessment needs to consider the fate of residents in both habitat types and that of locally dispersing individuals.

ACKNOWLEDGMENTS

Computational facilities were provided by the Blandy Experimental Farm. Discussions with E. F. Connor and C. A. Flanagan were of great help in channeling the ideas and designing the simulations.

REFERENCES

1. Emlen, J. M., Terrestrial population models for ecological risk assessment: a state-of-the-art review, *Environ. Toxicol. Chem.*, 8, 831–842, 1989.
2. Hallam, T. G., C. E. Clark and R. R. Lassiter, Effects of toxicants on populations: a qualitative approach. I. Equilibrium environmental exposure, *Ecol. Model.*, 18, 13–20, 1983.
3. Martin, K., S. J. Hannon and R. F. Rockwell, Clutch size variation and patterns of attrition in fecundity of willow ptarmigan, *Ecology*, 70, 1788–1799, 1989.
4. Verner, J., and M. F. Willson, The influence of habitats on mating systems of North American passerine birds, *Ecology*, 47, 143–147, 1966.
5. Lack, D. L., *Ecological Adaptations for Breeding in Birds*, Methuen, London, 1968.
6. Harper, J. L., *Population Biology of Plants*, Academic Press, London, 1977.
7. Cody, M. L., Habitat selection in birds: the role of vegetation structure, competitors, and productivity, *Bioscience*, 31, 107–113, 1981.
8. Pimm, S. L., and M. L. Rosenzweig, Competition and habitat use, *Oikos*, 37, 1–6, 1981.
9. Hilden, O., Habitat selection in birds: a review, *Ann. Zool. Fenn.*, 2, 53–75, 1965.

10. Pickett, S. T. A., and P. S. White, Eds., *The Ecology of Natural Disturbance and Patch Dynamics*, Academic Press, New York, 1985.
11. Mooney, H. A., and M. Godron, Eds., *Disturbance and Ecosystems*. Springer-Verlag, New York, 1983.
12. Turner, M. G., R. H. Gardner, V. H. Dale and R. V. O'Neill, Predicting the spread of disturbance across heterogeneous landscapes, *Oikos*, 55, 121–129, 1989.
13. Elton, C., *Animal Ecology*, Sidgwick & Jackson, London, 1927.
14. Uvardy, M. D. F., Notes on the ecological concepts of habitat, biotype, and niche, *Ecology*, 40, 725–728, 1959.
15. Levins, R., *Evolution in Changing Environments*, Princeton University Press, Princeton, NJ, 1968.
16. Lack, D., *Ecological Isolation in Birds*, Blackwell Scientific, Oxford, 1971.
17. Fretwell, S. D., *Populations in a Seasonal Environment*, Princeton University Press, Princeton, NJ, 1972.
18. Oelke, H., and P. M. Klopfer, Licht als Stimulationsfaktor in des biotop von Katzendrosseln (*Dumetella carolinensis*, Mimidae), *J. Ornithol.*, 111, 357-361, 1970.
19. Cody, M. L., On the methods of resource division in grassland bird communities, *Am. Nat.*, 102, 107–147, 1968.
20. Merikallio, E., Uber regionale Verbreitung und Anzahl der Landvogel in Sud and Mittelfinland, *Ann. Zool. Soc. Bot. Fenn. Vanaimo*, 12, 34, 1946.
21. Pitelka, F. A., P. Q. Tomich and G. W. Treichel, Ecological relations of jaegers and owls as lemming predators near Barrow, Alaska, *Ecol. Monogr.*, 25, 85–117, 1955.
22. Martin, T. E., Habitat and area effects on forest bird assemblages: is nest predation an influence?, *Ecology*, 69, 74–84, 1988.
23. Gadgil, M., Dispersal: population consequences and evolution, *Ecology*, 52, 253-261, 1971.
24. Coulson, J. C., The influence of the pair-bond and age on the breeding biology of the kittiwake gull *Rissa tridactyla*, *J. Anim. Ecol.*, 35, 269–279, 1966.
25. Austin, O. L., Site tenacity, a behavioral trait in the common tern, *Bird Banding*, 20, 1–39, 1949.
26. Morse, D. H., *Behavioral Mechanisms in Ecology*, Harvard University Press, Cambridge, MA, 1980.
27. Weins, J. A., Population responses to patchy environments, *Annu. Rev. Ecol. Syst.*, 7, 81–120, 1976.
28. Weins, J. A., Spatial scale and temporal variation in studies of shrubsteppe birds, in *Community Ecology*, J. Diamond and T. J. Case, Eds., Harper & Row, New York, 1986, pp. 154–172.
29. Dayton, P. K., and M. J. Tegner, The importance of scale in community ecology: a kelp forest example with terrestrial analogs, in *A New Ecology: Novel Approaches to Interactive Systems*, P. W. Price, C. N. Slobodchikoff and W. S. Gaud, Eds., John Wiley & Sons, New York, 1984, pp. 457–481.
30. O'Neill, R. V., Perspectives in hierarchy and scale, in *Perspectives in Ecology Theory*, J. Roughgarden, R. M. May and S. A. Levin, Eds., Princeton University Press, Princeton, NJ, 1989, pp. 140–156.
31. Lomnicki, A., Regulation of population density due to individual differences and patchy environment, *Oikos*, 35, 185–193, 1980.
32. Lidicker, W. Z., Jr., Emigration as a possible mechanism permitting the regulation of population density below carrying capacity, *Am. Nat.*, 96, 29–33, 1962.
33. Cockburn, A., and W. Z. Lidicker, Jr., Microhabitat heterogeneity and population ecology of an herbivorous rodent, *Microtus californicus*, *Oecologia*, 59, 167–177, 1983.
34. Ficken, M. S., and R. W. Ficken, Age specific differences in the breeding behavior of the American redstart, *Wilson Bull.*, 79, 188–199, 1967.
35. Morse, D. H., The foraging of small populations of yellow warblers and American redstarts, *Ecology*, 54, 346-355, 1973.
36. Krebs, J. L., Territory and breeding density in the great tit, *Parus major* L., *Ecology*, 52, 2–22, 1971.

37. Glase, J. C., Ecology of social organization in the black-capped chickadees, *Living Bird*, 12, 235–267, 1973.
38. Ekman, J., and C. Askenmo, Social rank and habitat use in willow tit groups, *Anim. Behav.*, 35, 445–452, 1984.
39. Gustafsson, L., Foraging behavior of individual coal tits, *Parus ater*, in relation to their age, sex, and morphology, *Anim. Behav.*, 36, 696–704, 1988.
40. Desrochers, A., Sex, dominance, and microhabitat use in wintering black-capped chickadees: a field experiment, *Ecology*, 70, 636–645, 1989.
41. Ekman, J., Exposure time use in willow tit flocks: the cost of subordination, *Anim. Behav.*, 32, 508–514, 1987.
42. Anderson, P. K., Ecological structure and gene flow in small mammals, *Symp. Zool. Soc. London*, 26, 299–325, 1970.
43. Pulliam, H. R., Sources, sinks and population regulation, *Am. Nat.*, 132, 652–661, 1988.
44. Lidicker, W. Z., Jr., The role of dispersal in the demography of small mammals, in *Small Mammals: Their Productivity and Population Dynamics*, F. B. Golley, K. Petrusewicz and L. Ryszkowski, Eds., Cambridge University Press, New York, 1975, pp. 103–128.
45. van Horne, B., Density as a misleading indicator of habitat quality, *J. Wildl. Manage.*, 47, 893–901, 1983.
46. Vance, R. R., The effect of dispersal on one species, discrete space population growth models, *Am. Nat.*, 123, 230–254, 1984.
47. Levin, S. A., Population dynamics in heterogeneous environments, *Annu. Rev. Ecol. Syst.*, 7, 287–310, 1976.
48. Holt, R. D., Population dynamics in two-patch environments. Some anomalous consequences of an optimal habitat distribution, *Theor. Pop. Biol.*, 28, 181–208, 1985.
49. Koehl, M. A. R., From individuals to populations, in *Perspectives in Ecological Theory*, J. Roughgarden, R. M. May and S. A. Levin, Eds., Princeton University Press, Princeton, NJ, 1989, pp. 39–53.
50. Rosenzweig, M. L., A theory of habitat selection, *Ecology*, 62, 327–335, 1981.
51. Fretwell, S. D., and H. L. Lucus, Jr., On territorial behavior and other factors influencing habitat distribution in birds. I. Theoretical development, *Acta Biotheor.*, 19, 16–36, 1970.
52. Morris, D. W., Density-dependent habitat selection: testing the theory with fitness data, *Evol. Ecol.*, 3, 80–94, 1989.
53. Pacala, S. W., and J. A. Silander, Neighborhood models of plant population dynamics. I. Single species models of annuals, *Am. Nat.*, 125, 385–411, 1985.
54. Pulliam, H. R., and B. J. Danielson, Sources, sinks and habitat selection: a landscape perspective on population dynamics, *Am. Nat.*, 137, S50-S66, 1991.
55. May, R. M., The search for patterns in the balance of nature: advances and retreats, *Ecology*, 67, 1115–1126, 1986.

13

Modeling Avian Foraging: Implications for Assessing the Ecological Effects of Pesticides

Gary E. Belovsky

ABSTRACT

The impact of many pesticides on nontarget species occurs by chronic exposure through the food chain and reduction of their prey populations. This makes an understanding of diet choice by nontarget species critical to predicting potential effects. A conceptual method to assess these impacts and to adopt better pesticide application strategies is developed in a series of steps:

1. A model of foraging is developed based upon the ecological theory of optimal foraging. The model is tested using the cowbird (*Molothrus ater ater* [Boddaert]) and demonstrates that these birds maximize their daily energy intake during the summer by consuming a mixed diet of grasshoppers and seeds.
2. Assuming a pesticide is applied to control grasshoppers, the model can be used to determine the conditions under which the birds would experience chronic exposure to the pesticide and/or population reductions due to the pesticide's reduction of their prey.
3. The foraging model indicates that a level of pesticide application that is less than maximal for grasshopper control might lead to optimal pest control when combined with bird predation.

KEY WORDS

pest control, foraging, grasshoppers, birds

INTRODUCTION

The immediate poisoning (acute effects) of animals by environmental toxins may require careful experimental studies, but it is a straightforward problem. However, the assessment of chronic effects of exposure to environmental toxins in the food chain and/or reduction in prey populations is not as straightforward. To assess these potential effects on nontarget species requires a better understanding of diet choice by nontarget species.[1] While diet choice models from ecology occasionally have been employed in developing pest control strategies,[2] their application in assessing the impact on nontarget species of pesticides in the food chain has not been addressed. Therefore, risk assessment of pesticide applications and the development of integrated pest management strategies that combine pesticide use with biological control agents (e.g., predators and parasitoids) may benefit from the use of diet choice models.

Mathematical models of animal diet choice have been employed by ecologists for more than 20 years[3] (optimal foraging theory, or OFT). The majority of these models have employed principles of mathematical optimization that are widely used in economics to model consumer choices. These models are applied to animal diet choices based upon the assumption that individuals that select better diets than other individuals in the species population will be favored by natural selection (greater survival and reproductive success). This implies that foraging is related to individual survival and reproductive output, and individuals are "plastic" in their foraging behavior.

The mathematics of optimization often permits ecologists to make specific predictions about the ways that animals should select their diets under different environmental conditions. These predictions can be compared then with the observed foraging behavior of animals to assess how closely their behavior approaches the predicted optima. Birds have been most frequently employed to test these models.[3] Birds appear to be able to select diets very similar to the predicted optima; however, most of these tests have been conducted in the laboratory rather than in the field.

One set of OFT studies has employed the mathematical optimization technique of linear programming. Almost every deterministic diet choice model in OFT can be written as a linear programming model.[4] In addition, these studies have successfully predicted diet choices for a wide range of animals feeding on very different foods and in very different field environments.[4-10] Therefore, the linear programming method represents a very flexible and useful modeling format to address animal diet choices.

The linear programming studies have been cited as examples that may be applicable in developing better pest control strategies[2] and in assessing the risk of persistent pesticides on nontarget species.[1] This modeling approach also has been employed to address the effects on consumer diet choice of naturally occurring toxins in foods.[11] Therefore, it is a small step to adapt this modeling technique to an examination of how pesticides in the food chain affect consumers.

This chapter presents an example of how a linear programming model of diet choice can be used to examine the potential effects of a pesticide in the food chain on a nontarget species. If the nontarget species is a predator of the pest, the diet choice model can be used to help develop an optimal integrated pest management strategy. This example is developed by:

1. constructing the diet choice model;
2. examining the effects of the pesticide through the food chain by imposing its ingestion as a constraint in the model, and including its reduction of the nontarget species' prey populations; and
3. contrasting the reduction of the pest by the pesticide and predator at different levels of pest abundance, as a potential integrated pest management program.

The example is developed using cowbirds (*Molothrus ater ater* [Boddaert]) as the nontarget species and grasshoppers (Orthoptera: Acrididae) as the pest.

THE FORAGING MODEL

A linear programming diet choice model for the cowbird in a Montana prairie environment has been published.[12] The cowbird is an ideal example because its foraging behavior is determined solely by its individual nutritional requirements, as the cowbird is a nest parasite that does not feed its young. The cowbirds in this environment consume only two major food categories during the summer: seeds and insects, mostly grasshoppers. Also, the cowbird is second only to the western meadowlark (*Sturnella neglecta*) as an avian predator of grasshoppers in the western United States[13] and is very abundant, making it a potential control agent of grasshoppers.

The linear programming model for cowbirds includes three constraints that might influence their decision on how much seed vs grasshoppers to include in their diet:[12] daily feeding time, daily digestive capacity, and daily energy requirements. Each constraint must be written as a linear inequality:

$$C \geq \text{ or } \leq aG + bS \tag{1}$$

where C is the constraint value, a converts consumed grasshopper mass, G (grams dry mass per day), into the constraint units, and b converts consumed seed mass, S (grams dry mass per day), into the constraint units. Which inequality is applicable depends upon whether the constraint cannot be exceeded or must be exceeded by the forager. Average observed cowbird foraging constraints are presented in Table 1, and the observations used to determine them are summarized below from the original study[12].

Daily Feeding Time

The cowbird's foraging time during summer daylight hours was observed to be limited by the thermal environment with the mid-day being too hot for foraging. The daily time the birds have available to harvest seeds is greater than that to hunt grasshoppers, because grasshoppers can only be caught when they are active, and grasshoppers' activity is constrained more by the thermal environment than is the birds' activity. This means that the birds are active in the cooler hours of the day (morning and evening) when grasshoppers are inactive, and the grasshoppers are active during some of the warmer hours when the birds are inactive. Even when grasshoppers are active, the cowbirds cannot search for grasshoppers and seeds at the same instant, because each requires a different behavioral pattern of search.

Therefore, the daily feeding time constraint is composed of two equations:

1. temporal non-simultaneous search, when seeds are available but grasshoppers are not, and
2. spatial non-simultaneous search, when seeds and grasshoppers are both available but cannot be searched for using the same behavioral mode.

The first constraint sets an upper limit to grasshopper consumption that is less than the amount that could be consumed given total daily feeding time, because grasshoppers are not available for the entire feeding time. The second constraint converts the total daily feeding time into the consumption of seeds versus grasshoppers given a trade-off in the consumption of the two food types because they require different search modes.

The conversion of feeding time into consumption was measured using captive birds in an aviary, where either grasshoppers or seeds were available for consumption and their abundances could be controlled. The feeding observations on grasshoppers were further differentiated between grasshoppers of different sexes and body sizes. These measurements provided a measure of how much time on average a cowbird requires to encounter a food item at different food densities (functional response: Figure 1).[12] In addition, the time it takes for a bird to ingest a food item once captured was measured by feeding the captive birds different grasshoppers and seeds (handling time: Figure 2).[12]

Knowing abundances of different grasshoppers and seeds in the environment, the functional response and handling time data can be converted into the expected time

Table 1. The Linear Programming Model's Constraints for Cowbird Foraging

Digestive Constraint

	Physiological Measures	Observed
SEEDS:	$\left(\frac{17\text{ hours/day}}{4.4\text{ hours/fill}} + 1\text{ fill/day}\right) \times 0.94\text{ g/fill} = 4.57\text{ g/day}$	4.57 + 0.07 g/day (N = 6)
HOPPERS:	$\left(\frac{17\text{ hours/day}}{0.7\text{ hours/fill}} + 1\text{ fill/day}\right) \times 0.94\text{ g/fill} = 23.77\text{ g/day}$	24.80 + 4.13 g/day (N = 6)

Based on Activity Over a 12-H Period Each Day in the Field

$$\left(\frac{1.03\text{ g-wet/g-dry}}{3.71\text{ fills/day}} \times \text{SEEDS g-dry}\right) + \left(\frac{2.70\text{ g-wet/g-dry}}{17.44\text{ fills/day}} \times \text{HOPPERS g-dry}\right) < 0.94\text{ g/fill}$$

Feeding Time Constraint

Determinants of daily feeding time in the field

Observed feeding from 0700–1900 hours
No feeding when air temperature >30°C
Actual feeding for only 7 hours/day
Grasshoppers only available (active) for 3.23 hours/7 hours

$$(41.90\text{ min/g-dry} \times \text{SEEDS g-dry}) + (60.27\text{ min/g-dry} \times \text{HOPPERS g-dry}) < 420\text{ min/day}$$

$$60.27\text{ min/g-dry} \times \text{HOPPERS g-dry} < 194\text{ min/day}$$

Energy Constraint

$$(5\text{ kcal/g-dry} \times 0.61\text{ digestibility} \times \text{SEEDS g-dry}) + (6.3\text{ kcal/g-dry} \times 0.78\text{ digestibility} \times \text{HOPPERS g-dry}) > 12.6\text{ kcal/day}$$

Note: The methods used to measure the parameters and construct the constraint equations are described in Belovsky.[12]

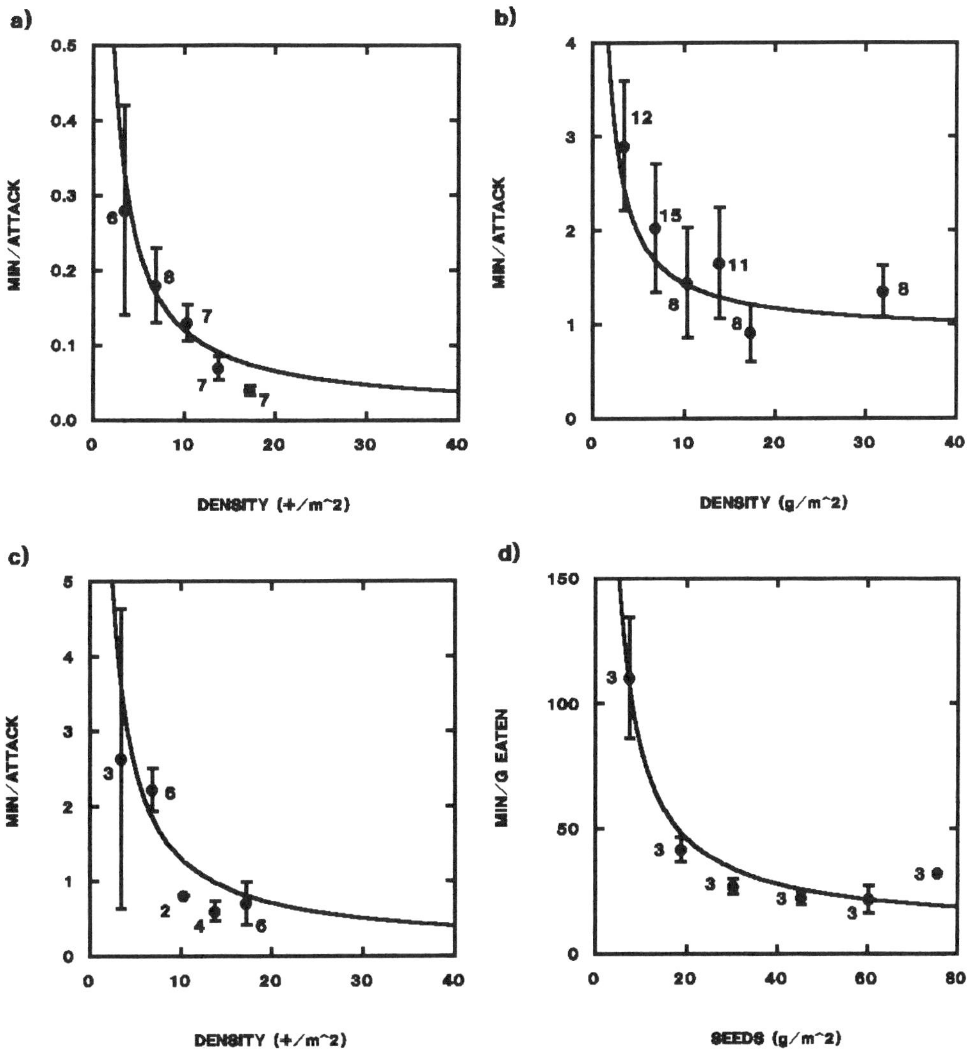

FIGURE 1. The functional response curves for cowbirds, as reported in Belovsky[12] for (a) small grasshoppers (<100 mg), (b) intermediate-sized grasshoppers (100–400 mg), (c) large grasshoppers (>400 mg), and (d) seeds. These curves represent the time between encounters with potential food items.

required to ingest a food item by a free-roaming bird. These values were computed for an average grasshopper and seed in the environment, and were used to convert the daily feeding time into potential consumption. Therefore, the feeding time constraint reflects the maximum intake of different combinations of the two foods (diets), given the cowbird's daily feeding time.

Daily Digestive Capacity

A bird's daily digestive capacity depends upon the amount of food that can be processed through its alimentary tract in a day. This quantity will be limited by the capacity of the crop, the organ in which food is stored while awaiting digestion. Crop capacity was measured as the maximum mass of grasshoppers or seeds that a captive

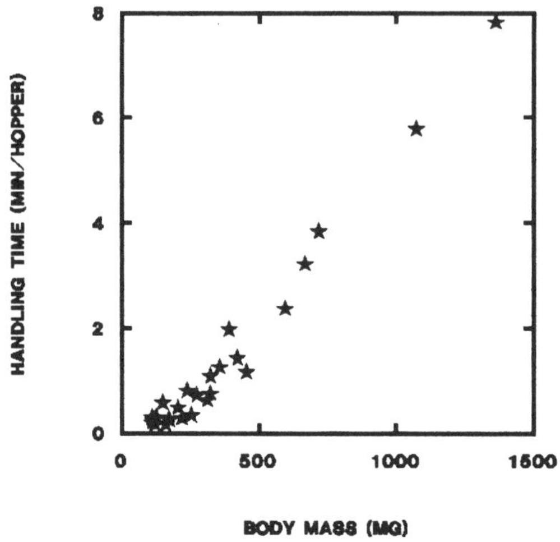

FIGURE 2. The cowbird's handling times for grasshoppers of different body sizes, as reported by Belovsky.[12] This is the time necessary to subdue and ingest a grasshopper.

bird could ingest before it becomes satiated (stopped feeding), after it has been starved for 3 hours to ensure that its crop is empty.

Different foods pass through the alimentary tract at different rates, determining how fast the crop is emptied. This passage rate was measured by feeding birds that had been starved for 3 hours, either grasshoppers or seeds that had been colored with an indigestible dye, and determining how much time was required before the dye was excreted. Dividing this passage rate into daylight hours (the time period over which feeding is distributed) and adding 1 for the bird's ability to empty the crop overnight provides an estimate of the number of times the crop can be emptied in a day for each food type.

The digestive constraint reflects the cowbird's crop capacity, and its conversion into daily ingestion of different amounts of grasshoppers and seeds (grams of dry mass per day) based upon each food's turnover and bulk (wet mass/dry mass). Therefore, this constraint, like the feeding time constraint, reflects a maximum ingestion of the two foods.

Energy Requirements

As mentioned above, a cowbird's daily energy requirements do not depend upon the provisioning of offspring because cowbirds are nest parasites. Cowbird metabolic requirements (kilocalories per day) are available in the literature,[14] as are the gross energetic values for seeds and grasshoppers (kilocalories per grams of dry mass).[15] However, gross energetic value is not as important as net value, which depends upon the bird's ability to assimilate the energy; this was measured by feeding captive birds known amounts of each food and measuring cloacal output.

The bird's energy requirements must be satisfied by the ingestion of grasshoppers and seeds, which differ in their net energy value. Therefore, this constraint defines the minimum ingestion of the two foods.

Predicting the Bird's Diet

Using the constraint equations in Table 1, the linear programming model of cowbird diet choice can be depicted graphically (Figure 3). Each constraint eliminates the con-

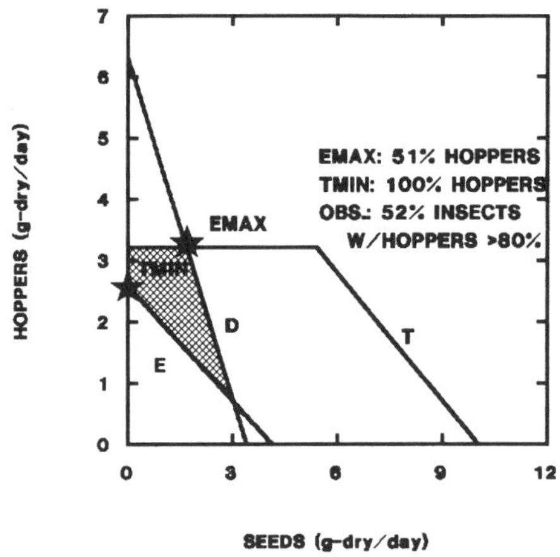

FIGURE 3. The graphic solution to the linear programming model of cowbird diet choice. Each constraint equation is labeled (see Table 1): D = digestive capacity, T = feeding time, and E = energy requirement. The cross-hatched region (feasible region) contains all diet combinations of seeds and grasshoppers that satisfy the constraints. The stars represent solutions to the energy-maximizing (EMAX) and time-minimizing (TMIN) goals. Also, the observed diet is presented, indicating that cowbirds appear to forage as energy maximizers.[12]

sumption of certain combinations of grasshoppers and seeds, and in total, the constraints define a region of diets that is feasible for the bird to consume. Using the Simplex algorithm of linear programming,[16] we know that if an optimal solution to the problem exists, it must occur at a corner of the feasible region. The corners are defined by the intersection of two constraint equations or a constraint equation and an axis. Finally, the optimal solution depends upon the goal being sought by the forager.

OFT has traditionally assumed that foragers have two dichotomous goals: feeding time minimization and nutrient maximization.[9] Feeding time minimization should occur if the forager's evolutionary fitness (survival and reproduction) increases when the forager spends less time feeding after minimum nutritional requirements are attained. This goal may be sought if the forager's fitness increases when it has more time to look for mates, care for young, or hide from predators. Nutrient maximization should occur if the forager's evolutionary fitness increases as its intake of some nutrient increases. This goal may be sought if the forager's survival and reproductive output continually increase with its nutritional status.

The solutions for the two foraging goals are presented in Figure 3, along with the cowbird's observed diet. Given observed cowbird diets, it appears that they select a nutrient-maximizing diet and, in particular, a diet that maximizes their daily energy intake in summer. Finally, the diet solutions are very robust (i.e., insensitive to parameter changes within the standard errors of the parameter estimates).

THE TOXIC IMPACTS OF A PERSISTENT PESTICIDE

The foraging model developed above provides a baseline description of this bird's position in the food chain. Because cowbirds are common inhabitants of prairie environments and consume large quantities of grasshoppers, and because grasshoppers are a frequently controlled pest in prairie environments, cowbirds could easily be impacted by a pesticide applied to control grasshoppers. How can the foraging model help assess potential impact?

Several realistic assumptions must be made to begin the analysis. First, assume that the pesticide is never applied at levels causing immediate toxic effects on the birds (i.e., acute effects are not important). This would make the pesticide's impact apparent and extreme. Second, assume that the pesticide is accumulated at greater concentrations in the grasshoppers than in seeds. This assumption is realistic because grasshoppers, as consumers of plants, may biologically amplify the pesticides that they ingest with the plants (or with the poison bait used to apply some grasshopper pesticides).

The chronic effects of a persistent pesticide can be viewed as an additional constraint added to the foraging model. This constraint will require knowledge of the amount of pesticide (or a derivative from environmental degradation) that must be accumulated in the bird before detrimental effects (i.e., reduced survival, reduced fecundity, etc.) will be observed. If different detrimental effects emerge as the accumulation of toxin increases, multiple constraints can be built into the model. However, for simplicity assume that the bird simply dies when a threshold level (T) of the toxin is accumulated (e.g., parts per million).

Accumulation of the toxin will depend upon:

1. the differential concentration (in parts per million) of the toxin in grasshoppers (L_g) and seeds (L_s),
2. the ability of the birds to degrade the toxin metabolically and excrete it (R: ppm/day), and
3. the life expectancy of a fledged bird in the absence of the toxin (E: days).

The constraint for the foraging model can be written as:

$$T/E + R \geq L_g G + L_s S \tag{2}$$

where the subscripts g and s, respectively, refer to grasshoppers and seeds. The term for life expectancy is restricted to fledged birds for several reasons. First, they feed themselves. Second, we are concerned with the population effects of the pesticide and, thus, the potentially reproductive individuals maintaining the population. Third, nestling mortality or failure to reproduce would be incorporated in a term converting food intake into nestling production.

The constraint for the persistent pesticide can be superimposed upon the foraging model (Figure 4). A conservative assumption (i.e., worst case) is that the bird will not avoid foods containing the toxic compound simply because they contain the toxin. This can be assumed because the toxin is not naturally occurring, as are plant secondary chemicals.[11] Therefore, natural selection would not have operated over time on the foragers to evolve an aversion or an ability to learn to avoid the food, and thus they will consume the contaminated foods. If the energy-maximizing diet now falls above the toxin constraint, the pesticide will exhibit a chronic effect on the birds.

The chronic effect of the pesticide will reduce life expectancy to a new level E':

$$E' = (T + R)/(L_g G + L_s S).$$

Depending upon whether T reduces reproduction or survival, the chronic effect of the pesticide could reduce population density, if the population is stationary, or population growth rate, if it is not stationary, by as much as E'/E. If both survival and reproduction are reduced, then population density or growth rate can be reduced by as much as $(E'/E)^2$.

The pesticide, however, will exhibit an effect on the foraging model in addition to its toxicity on the nontarget species. The pesticide will decrease the abundance of the pest

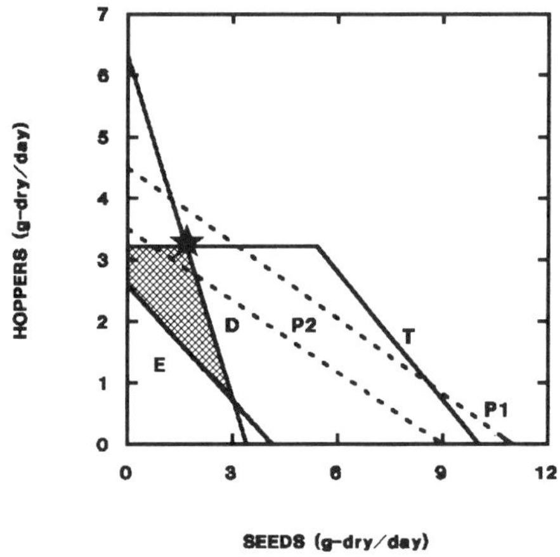

FIGURE 4. The linear programming model of cowbird diet choice modified to include the toxic effects of a persistent pesticide in the food chain. The constraint equations and feasible region are denoted as in Figure 3, while the star only represents the energy-maximizing solution, as cowbirds appear to be energy maximizers. The toxin constraints are represented by the dashed lines, P1 and P2, and their mathematical formulation is presented in the text. In the case of P1, the toxin has no impact on the cowbird because the cowbird's diet (star) falls below the constraint. For P2, the toxin would affect the cowbird because the diet (star) falls above the constraint.

(grasshoppers). This will decrease the rate with which birds encounter grasshoppers (functional response: Figure 1), which will change the feeding time constraint in the foraging model and the energy-maximizing diet. This could change the birds' diet; i.e., the consumption of grasshoppers will remain the same or decrease. If the consumption of grasshoppers declines enough, it is possible that the effects of chronic exposure to the pesticide may not materialize. Therefore, evaluating whether the pesticide will elicit chronic effects requires specific values for the toxin constraint, the baseline foraging model, and the reduction of prey via pesticide application.

NONTOXIC IMPACT OF THE PESTICIDE

The foraging model indicates that when both grasshoppers and seeds are available, cowbirds use all of their feeding time to consume only grasshoppers. This never changes unless grasshoppers are unavailable, which explains why cowbirds are considered to be such important avian predators on grasshoppers.[13] Because of this relationship, a general solution to the cowbird's energy-maximizing diet within the feasible region (Figure 3) can be written:

$$G = T_{gs}/C_g, \text{ and}$$

$$S = (D - G\,B_g/P_g)/(B_s/P_s),$$

where T_{gs} is the daily feeding time when grasshoppers and seeds are both available, C_g is the time required to ingest grasshoppers (grams of dry mass), D is the crop capacity, P_* (* stands for either g or s) is the number of times the crop can be filled each day with food, and B_* is the wet mass/dry mass of food.

The above result indicates that the energy-maximizing solution for cowbirds at a site will only depend upon C_g, which in turn is dependent on the abundance of grasshoppers and distribution of grasshopper body masses. Abundance affects the functional response (Figure 1), and grasshopper body mass affects the functional response (Figure 1) and handling time (Figure 2). Given the cowbird functional response curves (Figure 1), changes in grasshopper abundance would dramatically change the feeding time constraint when grasshopper densities are below $10/m^2$, because the time to encounter a grasshopper increases dramatically when grasshopper densities fall below this value. When intermediate sizes of grasshoppers (100 to 400 mg) are abundant, the encounter rate declines even more, because these grasshoppers appear to be better able to avoid birds (Figure 1).[12]

In prairie environments, grasshoppers are considered pests when their total densities exceed $9.6/m^2$;[17] densities in the cowbird functional responses are expressed per square meter of vegetation, which would make the defined pest density approximately $16/m^2$ after bare ground is eliminated. A successful treatment with grasshopper pesticides at the usual application levels can reduce grasshopper densities by 70 to 90% with spray-applied pesticides (e.g., malathion), and 30 to 50% with bait-applied pesticides (e.g., carbaryl).[17] Therefore, when grasshopper densities are less than $50/m^2$ of vegetation and spray pesticides are applied, or less than $17/m^2$ of vegetation and bait pesticides are applied, the birds' intake of grasshoppers will decrease. This means that spray pesticides will generally impact the birds' diet selection, while bait pesticides will not.

To complicate the effects of pesticides on bird diets further, different pesticides and application methods differentially impact grasshopper species. This differential effect on grasshopper species may be related to body mass and taxon, which also influence a bird's ability to capture grasshoppers.[18] It appears that the grasshoppers more easily captured by birds (small and large body mass, and Oedopodinae)[18] may be further reduced by pesticides.[17]

Because cowbirds are energy maximizers prior to pesticide application, any reduction in their consumption of grasshoppers will result in the average bird ingesting less energy. Animals tend to be energy maximizers when their survival or reproductive success or both depend upon nutritional intake. Thus, pesticide-induced dietary changes should reduce survival and/or reproduction for the birds. This will ultimately reduce the population density of the birds. Given the foraging model, each gram of dry mass decrease in grasshopper consumption will reduce the bird's daily energy intake by nearly 14% (3.22 kcal/day).

The baseline model was constructed for an observed grasshopper density of $18.6/m^2$ of vegetation, slightly above the USDA/APHIS (Animal and Plant Health Inspection Service) definition of pest levels. Assume that cowbirds do not substitute other arthropod prey for grasshoppers when pesticides reduce grasshopper densities, or the pesticide indiscriminately reduces all arthropod prey for the bird, which is likely. The application of a spray pesticide would reduce the cowbird's energy intake above maintenance by 74%, and the bait pesticide would reduce surplus energy by 39%. It is this surplus energy that should be related to cowbird reproductive output and survival over the winter and hence population density. If surplus energy is monotonically related to survival and reproduction, these declines in energy intake could have dramatic effects on the bird population.

Birds that are feeding time minimizers also may experience reduction in survival and reproduction via pesticide application. This would occur if more time is required to acquire a minimum intake of grasshoppers, and feeding increases vulnerability to predators and leaves less time for care of young. If cowbirds were feeding time minimizers, the application of a spray pesticide would increase feeding time by only 19% above baseline, while grasshopper densities decline by 80%. Therefore, the baseline foraging model indicates that there would be less effect on feeding time minimizers.

Data comparing bird populations in areas treated with grasshopper pesticides and nontreated controls[19] provide some indication that bird populations decline in the treated areas. However, whether the population decline reflects the death and reproductive failure of birds from either exposure to the pesticide or reduction in food, or their movement to areas where grasshopper densities are higher, cannot be ascertained. Also, the controls are not necessarily comparable to the treated areas (i.e., areas with grasshoppers at pest densities are treated, while the control areas never had pest densities).

If the bird encounters toxic effects of the pesticide, even at the reduced prey densities, the new relative population density or growth rate might be as low as $(E'/E)^2(1 - M)$, where M is the reduction in energy intake due to pesticide application. This portrays a worse-case scenario, assuming the pesticide effects are linearly related to survival and reproduction. However, this estimate would at least provide an *a priori* limit to expected occurrence with pesticide application. A better understanding of how reductions in food intake affect survival and reproduction, and how changes in survival and reproduction in turn affect population growth rate or density or both, would lead to more accurate estimation of the pesticide effects.

USING THE FORAGING MODEL FOR INTEGRATED PEST MANAGEMENT

The above analysis indicates how the foraging model can be used to address the impact of pesticide use on a nontarget species. The first avenue is through chronic exposure to toxic effects of a persistent pesticide that is transmitted through the food chain and accumulated by the nontarget species. The second way is through the reduction via pesticide of the pest that is a food resource for the nontarget species. One can argue that neither of these effects is environmentally acceptable. Nonetheless, the control of pests at times may be economically and politically mandated, and pesticides are often the most effective means of control. However, pesticide use may be moderated by the use of integrated pest management; pesticide use might actually increase the effectiveness of the avian predator as a control agent. The cowbird-grasshopper example developed above may illustrate how this can be envisioned.

Using the cowbird functional response curves (Figure 1), the foraging model for cowbirds can be solved for different hypothetical distributions of grasshopper abundances and body sizes. These solutions show ways in which the model can be used to predict how the bird's diet changes in different environments by varying the feeding time constraint.

For pest management, diet changes themselves are not critical; rather, it is the number of grasshoppers consumed per day relative to grasshopper abundance. Even without knowing the area covered by the bird in its foraging, or bird density, the ratio of grasshoppers consumed per bird to grasshopper abundance will indicate the environmental conditions (grasshopper abundance and relative abundance of body sizes) in which each bird will have the greatest predatory impact (Figure 5a). An individual bird has the greatest impact at low grasshopper densities and when grasshoppers of small body size are relatively the most abundant.

The predatory impact of an individual bird in different environments (functional response) is not the sole determinant of the impact on the pest; predator density is also important (numerical response). Assume that energy intake above metabolic requirements limits the bird's reproductive success and survival, which limits population density. This is likely for cowbirds, because as with domestic chickens, they continuously lay eggs as long as their nutritional level is above a minimum.[20] The same assumption may be valid for bird species that are not nest parasites, as survival of their young may

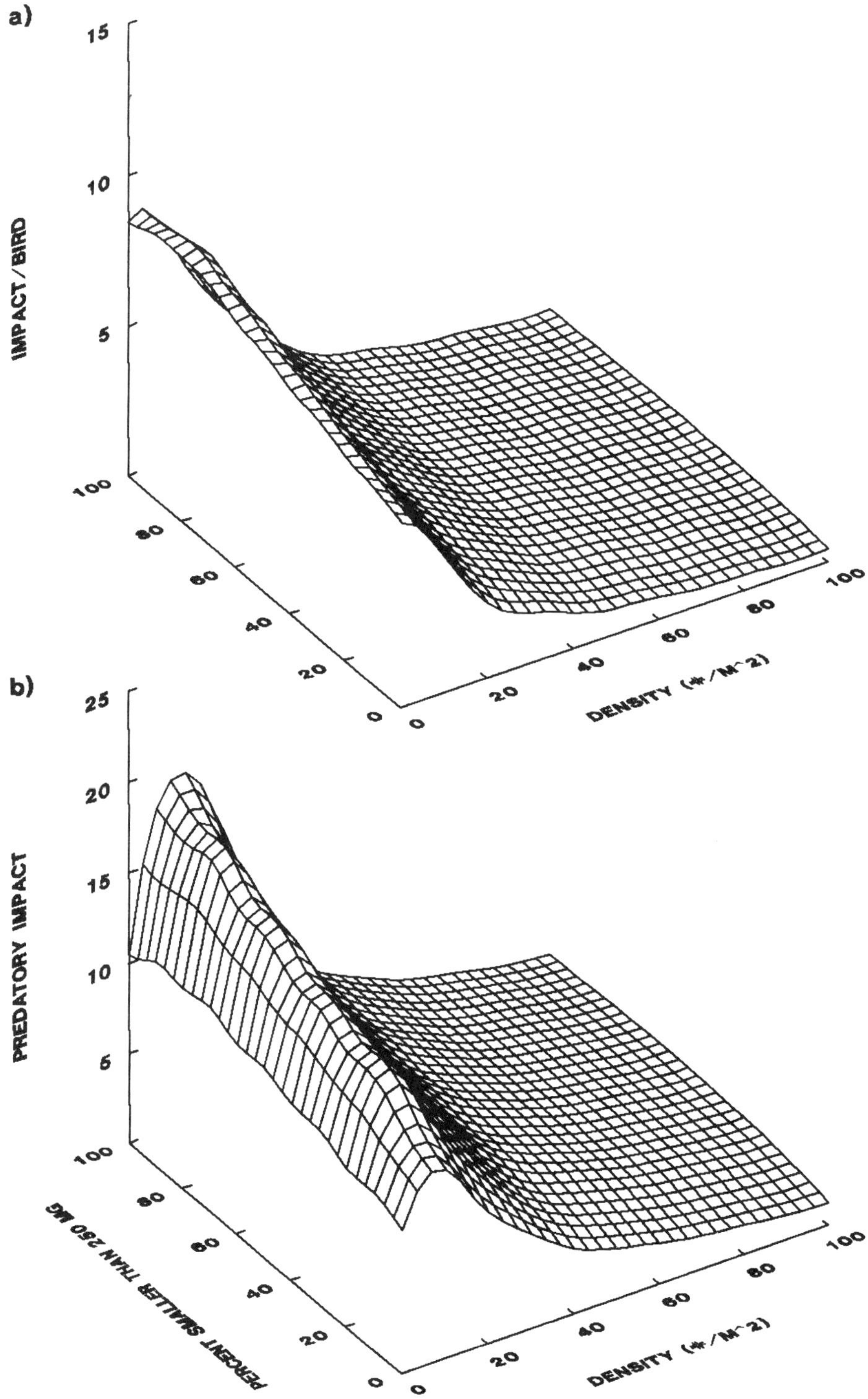

FIGURE 5. The linear programming model of cowbird diet choice used to estimate the predatory impact of the bird on grasshoppers at different densities and with different relative abundances of body sizes: (a) the impact of an individual bird, and (b) the impact of the entire population at different grasshopper abundances. The values of predatory impact are indices (see text) that reflect relative values at different grasshopper abundances. Therefore, the graphs only indicate under what conditions the birds will have the greatest predatory impact.

be a function of the parents' ability to provide food to the young. Therefore, the numerical response is proportional to energy intake above requirements, and this can be computed for different environments (grasshopper abundances and relative abundances of body sizes) using the foraging model.

The product of the functional and numerical response provides the overall predatory impact of the bird. Assuming that the grasshoppers eaten relative to their abundance is proportional to the functional response, and energy intake above requirements is proportional to the numerical response, one can compute the predatory impact from the foraging model, as presented in Figure 5b. The bird would have the greatest potential impact on grasshoppers at moderate densities ($5/m^2$ of vegetation) and when the grasshoppers are represented by species of small body size. Therefore, birds are unlikely to reduce grasshoppers at pest levels ($16/m^2$ of vegetation), and when the grasshoppers are dominated by large species. Furthermore, as grasshopper densities increase, the predatory impact of birds rapidly becomes negligible.

While birds may not reduce grasshoppers from outbreak levels, they may be able to prevent them from achieving outbreak densities. This would occur in environments in which the grasshoppers generally are at densities near the level at which birds could have a large predatory impact (low to moderate densities), and the conditions that permit the grasshoppers to exceed these densities and achieve pest status occur infrequently.

More importantly, birds may be used in an integrated pest management plan. This might be useful for regions where grasshoppers frequently achieve densities greater than those where birds have an appreciable predatory impact and the rate at which grasshoppers are able to increase is high. Here, the application of pesticides or other control agents (e.g., fungal pathogens) could be used to reduce the grasshopper densities to levels at which birds have a potential predatory impact. In this way, further reduction of the grasshoppers and retardation of their return to outbreak densities could be achieved by the birds. Integrating bird predation with the application of control agents would reduce the level and frequency of application, which would reduce the deleterious environmental impact from pesticides as well as costs.

Because predator effectiveness depends upon functional and numerical responses, the analysis presented in Figure 5b will not be adequate to develop an integrated pest management plan, if the plan uses a pesticide that is persistent in the food chain and causes chronic toxic effects on the bird. In this case, the pesticide's chronic effect could reduce the bird's numerical response and predatory impact on the pest by as much as $(E'/E)^2$. Therefore, Figure 5b must be recomputed with the toxin constraint included in the foraging model and its effect on population dynamics incorporated. The bird's reduced numerical response might negate the use of integrated pest management or require a reduced use of pesticide; the option chosen will depend upon the effectiveness of pesticides and avian predation, economic costs, and environmental hazards.

The foraging model illustrates for pest control that the ability of birds to limit grasshopper densities depends upon initial grasshopper densities, the relative abundance of grasshoppers of different body sizes, and the ability of grasshopper populations to increase. This is evident from studies of the impact of avian predation on grasshoppers that illustrate the presence of predatory impact,[19,21] and the absence of impact.[18,22] Therefore, one should not expect a single statement about the ability to use birds as a control agent for grasshoppers (e.g., birds limit grasshopper densities). Rather, the potential to use birds, and how to use them in an integrated fashion with other control measures, will vary between environments.

CONCLUSION

Optimal foraging models, such as the one developed here for a cowbird, using the mathematical technique of linear programming, may provide useful insights into understanding the potential impacts of pesticides on birds and how birds may be integrated into pest control programs. Evident from the analysis presented here, the models do not provide answers in the absence of detailed biological data. However, they aid in defining the critical elements of information that need to be supplied. Given the model and parameter values, projections can be made to assess the impact of pesticide application: toxic effects of persistent pesticides that are amplified in the food chain and their effect on the bird through reduction of potential prey. Finally, the model can be used to define the environmental conditions under which the bird may be built into an integrated control program for pests, thereby reducing the use of costly and potentially environmentally hazardous pesticides.

ACKNOWLEDGMENTS

I wish to thank J. B. Slade for reading and commenting on this manuscript. This work was supported with a grant from the United States Department of Agriculture.

REFERENCES

1. Emlen, J. M., Terrestrial population models for ecological risk assessment: a state-of-the-art review, *Environ. Toxicol. Chem.*, 8, 831–842, 1989.
2. Gosling, L. M., The role of wild plants in the ecology of mammalian crop pests, in *Pests, Pathogens and Vegetation*, J. M. Threth, Ed., Pitman, London, 1981, pp. 341–364.
3. Stephens, D. W., and J. R. Krebs, *Foraging Theory*, Princeton University Press, Princeton, NJ, 1986.
4. Belovsky, G. E., M. E. Ritchie and J. Moorehead, Foraging in complex environments: when prey availability varies over time and space, *Theor. Pop. Biol.*, 36, 144–160, 1989.
5. Belovsky, G. E., Herbivore optimal foraging: a comparative test of three models, *Am. Nat.*, 124, 97–115, 1984.
6. Belovsky, G. E., Optimal foraging and community structure: implications for a guild of generalist grassland herbivores, *Oecologia*, 70, 35–52, 1986.
7. Belovsky, G. E., Hunter-gatherer foraging: a linear programming model, *J. Arch. Anthropol.*, 6, 29–76, 1987.
8. Ritchie, M. E., Optimal foraging and fitness in Columbian ground squirrels, *Oecologia*, 82, 56–67, 1990.
9. Schmitz, O. J., Management implications of foraging theory: evaluating deer supplemental feeding, *J. Wildl. Manage.*, 54, 522–532, 1990.
10. Ball, J. P., Active diet selection or passive reflection of changing food availability: the underwater foraging behaviour of canvasback ducks, in *Behavioural Mechanisms of Food Selection*, NATO ASI Ser., R. N. Hughes, Ed., Springer-Verlag, Berlin, 1990, pp. 95–110.
11. Belovsky, G. E., and O. J. Schmitz, Mammalian herbivore optimal foraging and the role of plant defenses, in *Plant Defenses Against Mammalian Herbivores*, R. T. Palo and C. T. Robbins, Eds., CRC Press, Boca Raton, FL, 1991, pp. 1–28.
12. Belovsky, G. E., How important are nutrient constraints in optimal foraging models or are spatial/temporal factors more important?, in *Behavioural Mechanisms of Food Selection*, NATO ASI Ser., R. N. Hughes, Ed., Springer-Verlag, Berlin, 1990, pp. 255–280.
13. Bent, A. C., Life Histories of North American Blackbirds, Orioles, Tanagers, and Allies, Bull. No. 211, Smithsonian Institution/U.S. Natural History Museum, Washington, DC, 1958.

14. Lustick, S., Energetics and water regulation in the cowbird, *Physiol. Zool.*, 43, 270–287, 1970.
15. Bailey, C. G., and M. K. Mukerji, Energy dynamics of *Melanoplus bivittatus* and *M. femurrubrum* in a grassland ecosystem, *Can. Entomol.*, 109, 605–614, 1977.
16. Intriligator, M. D., *Mathematical Optimization and Economic Theory*, Columbia University Press, New York, 1971.
17. *USDA-APHIS-PPQ Grasshopper Integrated Pest Management Annual Report* (*GHIPM*), USDA-APHIS-PPQ, Boise, ID, 1989.
18. Belovsky, G. E., J. B. Slade and B. A. Stockhoff, Susceptibility to predation for different grasshoppers: an experimental study, *Ecology*, 71, 624–634, 1990.
19. McEwen, L. C., T. L. George, B. E. Petersen, D. W. Beyers, A. C. Fowler, and F. P. Howe, Environmental monitoring and evaluation of the grasshopper integrated pest management program, in *USDA-APHIS-PPQ Grasshopper Integrated Pest Management Annual Report* (*GHIPM*), USDA-APHIS-PPQ, Boise, ID, 1989, pp. 93–101.
20. Friedman, H., *The Cowbirds: A Study in the Biology of Social Parasitism*, Charles C Thomas, Springfield, IL, 1929.
21. Joern, A., Experimental study of avian predation on coexisting grasshopper populations (Orthoptera: Acrididae) in a sandhills grassland, *Oikos*, 46, 243–249, 1986.
22. Belovsky, G. E., and J. B. Slade, personal communication.

14

Direct and Indirect Pathways in Spatially Structured Habitats

Thomas E. Lacher, Jr.

ABSTRACT

No review of critical parameters in studies of ecotoxicology can be comprehensive. Parameters of importance will vary among studies; however, several consistent themes emerged during discussion. It is important to distinguish between lethal and sublethal effects and between direct and indirect pathways. Sublethal effects by indirect pathways traditionally have been neglected and may be of importance; future research should place more emphasis on the sublethal effects of agricultural chemicals on wildlife via indirect pathways. In addition, agricultural fields occur with relatively undisturbed habitats in a complex mosaic. Models that incorporate habitat-specific parameters for species under consideration are necessary in order to estimate the overall effects of chemicals on regional populations. The recent literature on source and sink habitats can provide insight in this area.

KEY WORDS

population parameters, sublethal effects, indirect pathways, habitat selection, sources, sinks

INTRODUCTION

The history of population ecology[1] involves the constant modification of ideas concerning population-level processes. The development of mathematical models of natural populations is such an ongoing process. Darwin[2] presented some simple calculations on the population growth rate of a plant in the *Origin of Species* in 1859. The logistic equation was first used to describe population growth in 1838,[3] and was not popularized until 1920.[4] It was during the 1920s that a formal, mathematical discussion of population biology was published.[5] The logistic equation has been modified extensively during the 20th century to incorporate a number of different parameters considered to be of importance. These include time lags, various kinds of intraspecific competition, and other factors affecting carrying capacity.[6] This equation has been used in multispecies models of interspecific competition, predator-prey interactions, and host-parasite relationships. Parameters have been added and new parameters defined. We continue to modify these equations to attempt to assess the effects of compounds foreign to nature and released into the environment by humans with little prior knowledge of their potential impact.

The modeling of the effects of chemicals on wildlife will require an evolution of theory and method; it is presumptuous to think that all critical parameters can be defined by the participants of this symposium (the 1990 Ecological Modeling Conference). Indeed, this is a job that will fall largely into the hands of ecotoxicologists. The discussion following the presentations of the preceding four chapters at the meeting focused less on which parameters to measure (as these will likely vary from species to species, habitat to habitat, and chemical to chemical) and more on some general concerns related to the measurement of any parameter. Two major themes emerged from the open discussion: the importance of considering effects acting both directly and indirectly on wildlife populations over an appropriate temporal scale, and the importance of considering the ecological heterogeneity of the landscape.

DIRECT AND INDIRECT PATHWAYS OF AGRICULTURAL CHEMICALS

Mortality can be one of the easier parameters to observe and measure, although differentiating between mortality and emigration on finite sites with artificial boundaries is a difficult task.[7] Public opinion is more quickly aroused by direct mortality, and this influences the regulatory process. More subtle effects such as shifts in diet, changes in patterns of habitat use, and concomitant changes in community structure and function are more difficult to observe and measure. They also occur over longer time periods and wider geographic areas than studies investigating acute mortality of a single species over a limited number of sites.

Several terms are used when discussing the effects of a perturbation on a population. These include lethal vs sublethal and direct vs indirect. These terms were frequently interchanged during the discussion at the meeting and merit clarification. It is perhaps better to associate the word "effect" with lethal/sublethal and the word "pathway" with direct/indirect. Lethal effects result in the death of the organism; lethal effects, however, can occur via direct or indirect pathways. A chemical can result in acute mortality of individual birds upon contact or ingestion (a direct pathway), or can devastate an important prey population, resulting in the death by starvation of those birds dependent upon the prey species (an indirect lethal effect). In the latter case, attempts to relate the mortality of birds to chemical exposure by looking for residues in the tissues may reveal nothing, if the chemical was degraded by the prey species, or the dead prey were ignored by the birds. The long-term ramifications of these more indirect effects can be greater than direct mortality. Many agricultural chemicals now have low acute toxicity and much mortality may be indirect.

Sublethal effects result in impaired biochemical processes, depressed reproductive potential, altered behavior, or other nonlethal effects. Sublethal effects can also occur via both direct and indirect pathways. Birds can be exposed to a pesticide by ingestion or dermal contact and exhibit a depressed metabolism (a sublethal effect) through the actual contact with a compound (a direct pathway). Birds can show shifts in dietary composition or altered use of preferred habitats in response to prey levels altered by a pesticide application, in turn altering nutritional status and reproductive success.[8] This is a sublethal effect caused by an indirect pathway.

Ecotoxicology has to a large degree emphasized the study of lethal effects via direct pathways. This is partially related to the LD/LC_{50}, and the evolution of extensive field trials from tightly designed laboratory experiments. The author expects directly acting lethal effects to be one of the least common outcomes of a pesticide application, especially with the formulations now available. If mortality is rare, and field trials are designed with mortality as the dependent variable, then the incremental mortality attributable to the application of a pesticide would require a design with extremely large sample sizes.[9] Because the sampling unit in studies of this type is usually an agricultural field site[10] in order to avoid problems of pseudoreplication,[11] logistics would prohibit many studies from detecting a significant effect.

Recent research has shown that sublethal effects are far more common than lethal, and numerous techniques are available to measure them.[12-14] A recently published guidance document emphasizes the importance of sublethal effects in the study of pesticide effects on passerine populations,[15] and the technique presented in this document is becoming widely adopted. Indirect pathways are rarely addressed, especially the combination of sublethal effects via indirect pathways (for exceptions, see References 16 through 18), yet this is an area in which future research should be concentrated. A number of fields of theoretical ecology are well suited to address indirectly caused sublethal effects of pesticide application. Cooperative initiatives between ecologists and toxicologists can result in the design of field experiments and research protocols that will focus attention on sublethal effects occurring indirectly.

HABITAT-STRUCTURED POPULATIONS AND RISK ASSESSMENT

Throughout most of the world, agricultural fields exist in a complex mosaic with undisturbed or natural vegetation. In forested areas, agricultural fields represent suboptimal habitat for most of the native species of the region. In areas formerly covered by prairies or grassland, certain kinds of agricultural fields can serve as surrogate grasslands for many species and mimic the original optimal habitat.

Pulliam[19] and Bowers[20] argue that insufficient attention has been paid to differential local birth and death rates among habitats; in fact, such habitat-specific demographic rates may be ecologically more important than the traditional age- or stage-structured demographic rates. As Bowers states, this is not a trivial point in the ecological risk assessment of agricultural pesticides. The reproductive success of a particular avian species from a source habitat cannot be used as a standard for comparison to treated agricultural fields, if these fields function as sinks for the species in question. The consideration of habitat-specific demographic rates in the context of the ecological suitability of the altered habitat in relation to native species of birds should be a fundamental aspect of virtually all studies.

There was a great deal of debate during the discussion section of the symposium concerning the relative importance of monitoring source vs sink habitats in large-scale field studies. Several participants, in particular Raymond O'Connor and Ron Pulliam, presented reasons for monitoring both. It will not be possible to estimate regional trends in population numbers unless the effect of agricultural chemicals on both sink and source habitats is known. Estimates of pesticide-induced mortality (or other more subtle indirect sublethal effects) should be obtained for birds in both agricultural lands subjected to direct application of the pesticide and nearby undisturbed vegetation. When agricultural lands represent suboptimal habitat, local reproduction may not compensate for local mortality, even in the absence of induced mortality above background levels. If reproductive success in optimal habitat is not depressed, then pesticide-induced mortality in agricultural fields may have little effect on the long-term viability of the population.

Agricultural fields located in former grasslands or savannahs may be more problematic because birds may view crop fields as little different from native vegetation, and pesticide-induced mortality in agricultural fields would represent mortality in substantial areas of source habitat, especially where little native grassland remains. In many cases it will be difficult to determine if agricultural land serves as a source or a sink habitat for a particular species of bird because there are few regions where reproductive success in the agroecosystem is not confounded by some kind of chemical application. These areas may only appear to be sinks because of chronic effects of chemical application or other kinds of disturbance. This emphasizes the value of long-term and geographically broad monitoring programs, such as proposed by Dobson and Hudson.[21] The work of Pulliam[19] and Bowers[20] emphasizes, however, that in cases in which agricultural fields are suboptimal or unsuitable for birds, the large-scale conversion of lands to agricultural uses will reduce population viability far more than any incremental mortality due to pesticide use.

Large-scale field studies of agricultural chemical use will undoubtedly become more complicated and more sophisticated in the coming years. The effects of the application of agricultural chemicals on the landscape will only be known if large areas with replicates for both controls and treatments are carefully studied for a long time.

REFERENCES

1. Kingsland, S. E., *Modeling Nature*, University of Chicago Press, Chicago, 1985.
2. Darwin, C., *On the Origin of Species by Means of Natural Selection, or the Preservation of Favoured Races in the Struggle for Life*, John Murray, London, 1859.
3. Verhulst, P. F., Notice sur la loi que la population suit dans son accroissement, *Corresp. Math. Phys.*, 10, 113–121, 1838.
4. Pearl, R., and L. J. Reed, On the rate of growth of the population of the United States since 1790 and its mathematical representation, *Proc. Natl. Acad. Sci. U.S.A.*, 6, 275–288, 1920.
5. Lotka, A. J., *Elements of Physical Biology*, Williams & Wilkins, Baltimore, 1925.
6. Begon, M., and M. Mortimer, *Population Ecology*, 2nd ed., Sinauer & Associates, Sunderland, MA, 1986.
7. Seber, G. A. F., *The Estimation of Animal Abundance and Related Parameters*, 2nd ed., Griffin, London, 1982.
8. Belovsky, G. E., Modeling avian foraging: implications for assessing the ecological effects of pesticides, in *Wildlife Toxicology and Population Modeling: Integrated Studies of Agroecosystems*, R. J. Kendall and T. E. Lacher, Jr., Eds., Lewis Publishers, Chelsea, MI, 1993.
9. Gould, A., Comments on statistical aspects of the "guidance document for contrasting terrestrial field studies," in *Pesticide Effects on Terrestrial Wildlife*, L. Somerville and C. H. Walker, Eds., Taylor & Francis, London, 1990, pp. 207–215,
10. Fite, E. C., L. W. Turner, N. J. Cook, C. Stinkard and R. M. Lee, Guidance Document for Conducting Terrestrial Field Studies, Ecological Effects Branch, Hazard Evaluation Division, Office of Pesticide Programs, U.S. Environmental Protection Agency, Washington, DC, 1988.
11. Hurlbert, S. H., Pseudoreplication and the design of ecological field experiments, *Ecol. Monogr.*, 54, 187–211, 1984.
12. Hooper, M. J., L. W. Brewer, G. P. Cobb and R. J. Kendall, An integrated laboratory and field approach for assessing hazards of pesticide exposure to wildlife, in *Pesticide Effects on Terrestrial Wildlife*, L. Somerville and C. H. Walker, Eds., Taylor & Francis, London, 1990.
13. Weisskopf, C. P., J. N. Seiber, N. Maizlish and M. Schenker, Personnel exposure to Diazinon in a supervised pest eradication program, *Arch. Environ. Contam. Toxicol.*, 17, 201–212, 1988.

14. Cobb, G. P., R. J. Kendall and D. M. Norman, Organochlorine Contaminants in Great Blue Herons Collected from the Puget Sound, Report submitted to the United States Fish and Wildlife Service, Olympia, WA, 1990.
15. Kendall, R. J., L. W. Brewer, T. E. Lacher, Jr., B. T. Marden and M. L. Whitten, *The Use of Starling Nest Boxes for Field Reproductive Studies: Provisional Guidance Document and Technical Support Document,* PB89 195 028/AS, National Technical Information Services, Springfield, VA, 1989.
16. Galindo, J., R. J. Kendall, C. Driver and T. E. Lacher, Jr., The effect of methyl parathion on the susceptibility of bobwhite quail (*Colinus virginianus*) to domestic cat predation, *Behav. Neur. Biol.*, 43, 21-36, 1985.
17. Buerger, T. T., R. J. Kendall, B. S. Mueller, T. DeVos and B. A. Williams, Effects of Methyl Parathion on northern bobwhite survivability, *Environ. Toxicol. Chem.*, 10, 527-532, 1991.
18. Hunt, K. A., Fenthion as a Secondary Poisoning Hazard to American Kestrels, M.S. thesis, McGill University, Montreal, Canada, 1990.
19. Pulliam, H. R., Sources, sinks, and population regulation, *Am. Nat.*, 132, 652-661, 1988.
20. Bowers, M. A., Use of space and habitats by individuals and populations: dynamics and risk assessment, in *Wildlife Toxicology and Population Modeling: Integrated Studies of Agroecosystems*, R. J. Kendall and T. E. Lacher, Jr., Eds., Lewis Publishers, Chelsea, MI, 1993.
21. Dobson, A., and P. Hudson, Assessing the impact of toxic chemicals: temporal and spatial variation in avian survival rates, in *Wildlife Toxicology and Population Modeling: Integrated Studies of Agroecosystems*, R. J. Kendall and T. E. Lacher, Jr., Eds., Lewis Publishers, Chelsea, MI, 1993.

15

Approaches to Modeling

M. Craig Barber

Emlen[1] recently reviewed the use of models appropriate for assessing the risk of chemicals to wildlife populations. In this review, Emlen recognized three general categories of population models: (1) bioenergetic models that relate individual growth, foraging, territoriality, and mating to survival and reproduction; (2) demographic models that predict things such as population size, age class distribution; and (3) genetic structure models. In this chapter, examples of each of these general model types are presented and discussed.

Bioenergetics relevant to population dynamics can be modeled in a variety of ways. Models for describing these processes generally represent a continuum from the purely empirical to the process based. In the Chapter 17 by Brisbin,[2] the author discusses the application of the Richards growth model to describe the growth of organisms and the bioaccumulation of chemicals. Many important ecological and toxicological responses of birds are dependent on their body size. The sizes of territories and home ranges of birds, for example, correlate well with their body size.[3-7] Similarly, Juanes[8] showed that population densities of birds are weakly but significantly correlated with their mean body weights. The body size dependence of the metabolic demands of birds is also well known.[9-17] Dietary and respiratory exposures of birds to pesticides likewise will be determined by their body size. In particular, because the size and amount of prey that birds consume generally depend on their body weight,[16,18-20] and because xenobiotic residues in both terrestrial and aquatic organisms generally are correlated with their body size, weight dependent predator-prey interactions may be extremely important in accurately assessing the dietary exposures of birds. Respiratory exposures to birds also will be size dependent because ventilation volumes and lung morphometrics that determine such exposures are size dependent.[21-27] Lastly, the lipid to water composition and lipid dynamics of birds that influence reproduction, migration, daily thermoregulation, and the accumulation of pesticides and other xenobiotics also depend on the bird's body size.[28-30] Considering these diverse size dependencies, it is clear that growth should be an endpoint of interest in any comprehensive avian risk assessment; the Richards growth model provides one method of analysis.

Most ecological effects of pesticides on bird populations are assessed by referencing the expected environmental concentrations (EEC) of the pesticides. For example, acute mortality often is assessed using a dietary LC_{50} which is the pesticide concentration in food that kills 50% of a population during a fixed duration of exposure. Although using a toxicological benchmark such as a dietary LC_{50} often does not cause serious errors in the risk assessment process, its use does obscure the fact that the effects of pesticides on birds or any organism are actually caused by the internal accumulation of those chemicals and not by their environmental concentrations per se. Several authors[31-34] have discussed the need to address chemical bioaccumulation explicitly when predicting ecological effects of organic pollutants on aquatic organisms and these arguments pertain to avian risk assessments as well. It should be clear, therefore, that the Richards growth model may also provide pesticide researchers and regulators with a useful methodology that can be used for residue-based risk assessments.

The Leslie-Lewis matrix models[35,36] are some of the most commonly used demographic models in population ecology. Although these models are traditionally formulated to project population age classes, Lefkovitch[37] reformulated these models to simulate size class dynamics. In Chapter 19, Slade[38] discusses both age- and size-dependent versions of these models. As indicated above, the importance of size-dependent versions of the Leslie matrix models is based on the fact that the size of an organism is often more useful than its age in characterizing and determining its ecological and toxicological responses.

Chesser et al.[39] demonstrate in Chapter 18 how another traditional population demographic model can be used to analyze the population dynamics of birds exposed to a generalized toxicant. They use the logistic equation to describe the nominal population dynamics of a bird species of interest. The authors then reformulate the parameters of this model to analyze pre- and postmating mortality and reduced fecundity that would be expected from chronic pesticide exposures. These reformulated logistic models are then analyzed within the context of genetic structure models that delineate a population's genetic variability and coancestry. Although this analysis may appear at first to be too esoteric for practical use, it provides a theoretical framework from which the recovery potential of populations could be analyzed for ecological risk assessments.

In Chapter 16, Bird[40] does not describe a traditional population model, but the Terrestrial Ecosystem Exposure Assessment Model (TEEAM) that simulates the fate and transport of pesticides in agricultural fields and their resulting bioaccumulation in soil, plants, and soil invertebrates. Consequently, TEEAM provides a framework by which the dietary exposures to granivorous and insectivorous birds can be assessed within a community context. Traditionally, most avian exposure assessments have assumed that the food chain is the predominant route of pesticide poisoning in birds. Although one could question the validity of this assumption in general,[41] even when this assumption is valid, the methods that often have been used to assess these exposures leave something to be desired. Avian dietary exposures, for example, often have been predicted using very simple descriptive models such as the Hoerger-Kenaga monograms[42] and Quantitative Structure Activity Relationships (QSAR).[43-47] Although the Hoerger-Kenaga monograms do address some important biological properties that influence dietary residues for birds, they do not address physicochemical properties of pesticides that influence these exposures. QSARs, on the other hand, ignore the biological properties of the food items in question and address only the influence the pesticide's physicochemical properties. An additional short coming of both of these methods is that they only predict steady state dietary residues which may be significantly different from the transient exposure levels to which birds are also subject. The TEEAM model attempts to overcome these shortcomings and provide a state-of-the-science methodology to predict realistic dietary exposures to granivorous and insectivorous birds under both acute and chronic field conditions.

REFERENCES

1. Emlen, J. M., Terrestrial population models for ecological risk assessment: a state-of-the-art review, *Environ. Toxicol. Chem.*, 8, 831–842, 1989.

2. Brisbin, I. L., Jr., Applications of a modified Richards sigmoid model to assess the uptake and effects of environmental contaminants upon birds, in *Wildlife Toxicology and Population Modeling: Integrated Studies of Agroecosystems*, R. J. Kendall and T. E. Lacher, Jr., Eds., Lewis Publishers, Chelsea, MI, 1993.
3. Schoener, T. W., Sizes of feeding territories among birds, *Ecology*, 49, 123–141, 1968.
4. Jenkins, S. H., Common patterns in home range-body size relationships of birds and mammals, *Am. Nat.*, 118, 126–128, 1981.
5. Mace, G. M., and P. H. Harvey, Energetic constraints on home-range, *Am. Nat.*, 121, 120–132, 1983.
6. Shank, C. C., Territory size, energetics, and breeding strategy in the Corvidae, *Am. Nat.*, 128, 642–652, 1986.
7. Calder, W. A., The scaling of sound output and territory size, *Ecology*, 71, 1810–1816, 1990.
8. Juanes, F., Population density and body size in birds, *Am. Nat.*, 128, 921–929, 1986.
9. Lasiewski, R. C., and W. R. Dawson, A reexamination of the relation between standard metabolic rate and body weight in birds, *Condor*, 69, 13–23, 1967.
10. Zar, J. H., Standard metabolism comparisons between orders of birds, *Condor*, 70, 278–279, 1968.
11. Hainsworth, F. R., and L. L. Wolf, Power for hovering flight in relation to body size in hummingbirds, *Am. Nat.*, 106, 589–596, 1972.
12. Kendeigh, S. C., Energy control of size limits in birds. *Am. Nat.*, 106, 79–88, 1972.
13. MacMillen, R. E., and F. L. Carpenter, Daily energy cost and body weight in nectarivorous birds, *Comp. Biochem. Physiol.*, 56A, 439–441, 1977.
14. Hails, C. J., Metabolic rate of tropical birds, *Condor*, 85, 61–65, 1983.
15. Adams, N. J., and C. R. Brown, Metabolic rate of sub-Antarctic Procellariiformes: a comparative study, *Comp. Biochem. Physiol.*, 77A, 169–173, 1984.
16. Nagy, K. A., Field metabolic rate and food requirement scaling in mammals and birds, *Ecol. Monogr.*, 57, 111–128, 1987.
17. Birt-Friesen, V. L., W. A. Montevecchi, D. K. Cairns and S. A. Macko, Activity specific metabolic rates of free-living northern gannets and other seabirds, *Ecology*, 70, 357–367, 1989.
18. Vezina, A. F., Empirical relationships between predator and prey inference, *Oecologia*, 75, 456–458, 1985.
19. Schramm, H. L., Jr., M. W. Collopy and E. A. Okrah, Potential problems of bird predation for fish culture in Florida, *Progr. Fish-Culturist*, 49, 44–49, 1987.
20. Bozinovic, F., and R. G. Medel, Body size, energetics and foraging mode of raptors in central Chile: an inference, *Oecologia*, 75, 456–458, 1988.
21. Lasiewski, R. C., and W. A. Calder, Jr., A preliminary allometric analysis of respiratory variables in resting birds, *Resp. Physiol.*, 11, 152–166, 1971.
22. Dubach, M., Quantitative analysis of the respiratory system of the house sparrow, budgerigar, and violet-eared hummingbird, *Resp. Physiol.*, 46, 43–60, 1981.
23. Abdalla, M. A., J. N. Maina, A. S. King, D. Z. King and J. Henry, Morphometrics of the avian lung. I. The domestic fowl (*Gallus gallus* variant *domesticus*), *Resp. Physiol.*, 47, 267–278, 1982.
24. Maina, J. N. and A. S. King, Morphometrics of the avian lung. II. The wild mallard (*Anas platyrhynchos*) and graylag goose (*Anas anser*), *Resp. Physiol.*, 50, 299–310, 1982.
25. Maina, J. N., and A. S. King, The thickness of the avian blood-gas barrier: quantitative and qualitative observations, *J. Anat.*, 134, 553–562, 1982.
26. Maina, J. N., Morphometrics of the avian lung. III. The structural design of the passerine lung, *Resp. Physiol.*, 55, 291–307, 1984.
27. Maina, J. N., Morphometrics of the avian lung. IV. The structural design of the charadriiform lung, *Resp. Physiol.*, 68, 99–119, 1987.
28. Walsberg, G. E., Evaluation of a nondestructive method for determining fat stores in small birds and mammals, *Physiol. Zool.*, 61, 153–159, 1988.

29. Ellis, H. I., and J. R. Jehl, Jr., Total body water and composition in phalaropes and other birds, *Physiol. Zool.*, 64, 973-984, 1991.
30. Lindstrom, A., Maximum fat deposition rates in migrating birds, *Ornis Scand.*, 22, 12-19, 1991.
31. Neely, W. B., An analysis of aquatic toxicity data: water solubility and acute LC_{50} fish data, *Chemosphere*, 13, 813-819, 1984.
32. Friant, S. L., and L. Henry, Relationship between toxicity of certain organic compounds and their concentrations in tissues of aquatic organisms: a perspective, *Chemosphere*, 14, 1897-1907, 1985.
33. McCarty, L. S., P. V. Hodson, G. C. Craig and K. L. E. Kaiser, The use of quantitative structure-activity relationships to predict the acute and chronic toxicities of organic chemicals to fish, *Environ. Toxicol. Chem.*, 4, 595-606, 1985.
34. McCarty, L. S., The relationship between aquatic toxicity QSARs and bioconcentration for some organic chemicals. *Environ. Toxicol. Chem.*, 5, 1071-1080, 1986.
35. Lewis, E. G., On the generation and growth of a population, *Sankya*, 4, 93-96, 1942.
36. Leslie, P. H., The use of matrices in certain population mathematics, *Biometrics*, 33, 183-212, 1945.
37. Lefkovitch, L. P., The study of population growth in organisms grouped by stages, *Biometrics*, 21, 1-18, 1965.
38. Slade, N. A., Models of structured populations: age and mass transition matrices, in *Wildlife Toxicology and Population Modeling: Integrated Studies of Agroecosystems*, R. J. Kendall and T. E. Lacher, Jr., Eds., Lewis Publishers, Chelsea, MI, 1993.
39. Chesser, R. K., K. B. Willis and N. E. Mathews, Impacts of toxicants on population dynamics and gene diversity in avian species, in *Wildlife Toxicology and Population Modeling: Integrated Studies of Agroecosystems*, R. J. Kendall and T. E. Lacher, Jr., Eds., Lewis Publishers, Chelsea, MI, 1993.
40. Bird, S. L., Fate and exposure in terrestrial ecosystems: a process approach, in *Wildlife Toxicology and Population Modeling: Integrated Studies of Agroecosystems*, R. J. Kendall and T. E. Lacher, Jr., Eds., Lewis Publishers, Chelsea, MI, 1993.
41. Driver, C. J., M. W. Ligotke, P. Van Voris, B. D. McVeety, B. J. Greenspan and D. B. Drown, Routes of uptake and their relative contribution to the toxicological response of Northern bobwhite quail (*Colinus virginianus*) to an organophosphate pesticide, *Environ. Toxicol. Chem.*, 10, 21-34, 1991.
42. Hoerger, F., and E. E. Kenaga, Pesticide residues on plants: Correlation of representative data as a basis of their magnitude in the environment, *Environ. Qual. Safety*, 1, 9-28, 1972.
43. Lord, K. A., G. G. Briggs, M. C. Neale, and R. Manlove, Uptake of pesticides from water and soil by earthworms, *Pestic. Sci.*, 11, 401-408, 1980.
44. Briggs, G. G., R. H. Bromilow, and A. A. Evans, Relationships between lipophilicity and root uptake and translocation of non-ionized chemicals by barley, *Pestic. Sci.*, 13, 495-504, 1982.
45. Briggs, G. G., R. H. Bromilow, A. A. Evans, and M. Williams, Relationships between lipophilicity and the distribution of non-ionized chemicals in barley shoots following uptake by the roots, *Pestic. Sci.*, 14, 492-500, 1983.
46. Bacci, E., M. A. Cerejeira, C. Gaggi, G. Chemello, D. Calameri and M. Vighi, Bioconcentration of organic chemical vapors in plant leaves: the azalea model, *Chemosphere*, 21, 525-535, 1990.
47. Connell, D. W., and R. D. Markwell, Bioaccumulation in the soil to earthworm systems, *Chemosphere*, 20, 91-100, 1990.

16

Fate and Exposure Modeling in Terrestrial Ecosystems: A Process Approach

Sandra L. Bird

ABSTRACT

Pathways for exposure of birds to pesticides include soil, water, air, soil-dwelling organisms, and insects. A process approach to avian exposure calculates transport and transformation of agricultural chemicals through each of the exposure media. Differential equations calculating the time rate of change of chemical in each exposure medium are solved in this approach. Development of terrestrial exposure algorithms at the United States Environmental Protection Agency (U.S. EPA), Environmental Research Laboratory, Athens, GA, draws on validated technology where well-established methodologies do not exist.

Multimedia process models are data intensive. An integral part of developing a usable and useful exposure calculation framework is the incorporation of supporting databases in the system. Supporting data for soils, meteorology, crops and cropping scenarios, and species distribution data are being developed on a regional scale based on the 186 major land resource areas (MLRA) defined by the United States Department of Agriculture (USDA).

This chapter further describes the process-oriented, mechanistically based approach to avian exposure calculations, the supporting databases required for doing regional analyses, and the interactive system design for accessing this information.

KEY WORDS

exposure assessment, pesticide transport, process modeling

INTRODUCTION

Relating the pesticide exposure pattern of birds to environmental characteristics of the toxic chemicals involved is a prerequisite for predicting the long-term response patterns of avian populations. Direct measurement of the exposure of birds in a field situation is costly, and only a limited number of exposure scenarios will likely ever be evaluated in this manner. Extrapolation techniques in the form of mathematical models are required to supplement field and laboratory studies in performing comprehensive environmental exposure assessments and assisting in the choice of critical field tests.

Pathways for exposure of birds to pesticides include soil, water, air, soil-dwelling organisms, plants, and insects. Transport and transformation of agricultural chemicals through each of the exposure media, coupled with the behavioral and physiological attributes of the bird, determine the ultimate exposure patterns in a field situation. TEEAM, described by Dean et al.,[1] used a process-based approach in looking at the fate of pesticides in the physical environment and their ultimate movement into the biota. Processes originally represented in TEEAM included behavior of the spray, water, and pesticide movement in the soil; evapotranspiration and volatilization; plant growth and uptake; bioconcentration in soil organisms; and ingestion and inhalation by terrestrial vertebrates. The TEEAM project provided the initial prototype for a terrestrial exposure system that will be incorporated into the Pesticide and Industrial Chemical Risk Analysis and Hazard Assessment (PIRANHA)[2] modeling system. The current components of the PIRANHA system provide an exposure assessment methodology for aquatic systems.

The general approach of the PIRANHA system models is to combine the loadings, transport, and transformation of a chemical into a set of differential equations using the law of conservation of mass as an accounting principle. Generally, the process descriptions for transport of chemicals between environmental compartments and transformation of chemicals within those compartments are based on process-oriented mechanistic constructs or accepted empirical relationships. The discussion in this chapter describes the components to be incorporated into the PIRANHA system as part of the U.S. EPA EcoRisk program, which will expand the methodology to terrestrial exposure assessments.

SYSTEM ARCHITECTURE

Exposure assessment within PIRANHA is structured as a series of stand-alone programs linked together by input and output files. The modules are based on well-developed and tested technology. The terrestrial exposure algorithm structure is a series of four stand-alone components: (1) spray drift and deposition, (2) terrestrial exposure media, (3) surface water, and (4) avian exposure, which are linked as illustrated in Figure 1. The spray drift and deposition component provides the spray deposition on habitats and spray day inhalation concentrations. The terrestrial exposure media com-

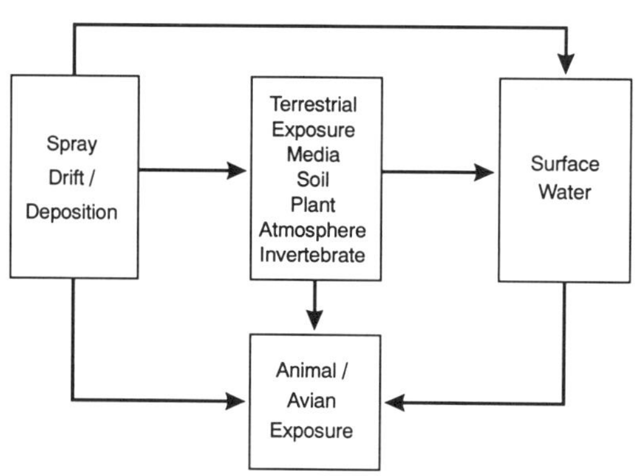

FIGURE 1. Architecture of a terrestrial exposure modeling system depicting the interactions between spray drift and deposition, terrestrial exposure media (soil, plants, invertebrates), surface water, and avian uptake.

ponent calculates the residues (soil, plants, invertebrates) to which the avian species may be exposed. The surface water component provides exposure concentrations via the drinking water.

There are several advantages to developing the terrestrial exposure system in this modular format. The primary advantage is the ability to incorporate updated versions of components easily and with a reduced risk of introducing errors in the integration process. Thus, the user can be confident that the validity of the components is not compromised in the transfer process and that the most recent versions of each component are included in the package. Finally, a major emphasis in the PIRANHA development is the linkage of the models to input databases. The mechanistic approach is data intensive and use of the process modeling approach outside of a research environment requires direct user access to parameter databases in developing input scenarios.

APPLICATION AND DEPOSITION

Evaluating the dynamics of toxicant application and deposition is the first step in determining the fate of pesticides. Pesticides are introduced in several ways: spray application to foliage or soil surfaces, incorporation into the soil, or deposition as time release granules or on seeds treated prior to planting. The type of initial application plays a large role in the type of environmental threat a pesticide may represent. Spray applications may contaminate offsite wildlife habitat; treated seeds or granules may provide a highly concentrated source of pesticide if ingested.

TEEAM used the USDA-Forest Service Spray Drift Model, FSCBG, to estimate deposition and offsite drift from aerial spray applications,[3] and the recent release, FSCBG 4.0, is a candidate for inclusion in PIRANHA. Offsite drift is a function of a variety of factors including aircraft type, speed, and application altitude; type and configuration of spray nozzles; weather conditions; and chemical formulation. Spray drift model algorithms calculate the spatial distribution of pesticide in the canopy and on the soil surface and, in addition, provide an estimate of the aerosol concentration that can be used to estimate inhalation dosage that could occur during the spray.

Wind tunnel tests provide the droplet size spectra, the most important equipment parameter, for spray nozzles under different application conditions. One of the biggest limitations in parameterizing spray drift models is estimating local meteorological conditions. Turbulence (i.e., atmospheric stability) is the most difficult model feature to accurately parameterize and lends the greatest uncertainty in modeling pesticide drift.

Two candidate spray drift models for inclusion in the PIRANHA system are FSCBG 4.0, the updated USDA-Forest Service model, and the Dow-Elanco spray drift model. Model testing efforts for FSCBG 4.0 have been reported previously,[4–6] but have not been oriented to performing exposure assessments. The Dow-Elanco model contains a more simplified algorithm for near-aircraft spray behavior and is simpler to parameterize. Both models require additional testing for use within a regulatory context. The spray drift model selection for incorporation within the PIRANHA system is coordinated with the spray drift task force, an industry coalition developing drift analysis tools for regulatory evaluations.

In addition to simulating aerial spray drift, pesticide application in the soil/plant component can be specified by the user as a direct application to the canopy, application to the soil surface, incorporation to a specified depth of the soil, or release from granules or treated seeds.

TRANSPORT AND TRANSFORMATION IN SOIL

Following application, a combination of chemical properties, soil properties, cropping practices, and meteorological conditions interact to determine whether a chemical will move into surface water via runoff and erosion, leach into the groundwater, volatilize into the atmosphere, move into the food chain via plants and soil-dwelling organisms, or degrade in the soil surface layers.

The Pesticide Root Zone Model (PRZM) forms the basis in TEEAM for calculating soil-associated movement of the pesticide.[7] PRZM is a one-dimensional compartmental model for use in simulating vertical chemical movement in unsaturated soil systems within and immediately below the plant root zone. PRZM processes are illustrated in Figure 2. PRZM allows the user to simulate movement of potentially toxic chemicals, particularly pesticides, that are applied to the soil or to plant foliage. Dynamic simulations allow the consideration of pulse loads, the prediction of peak events, and the estimation of time-varying mass emission or concentration profiles.

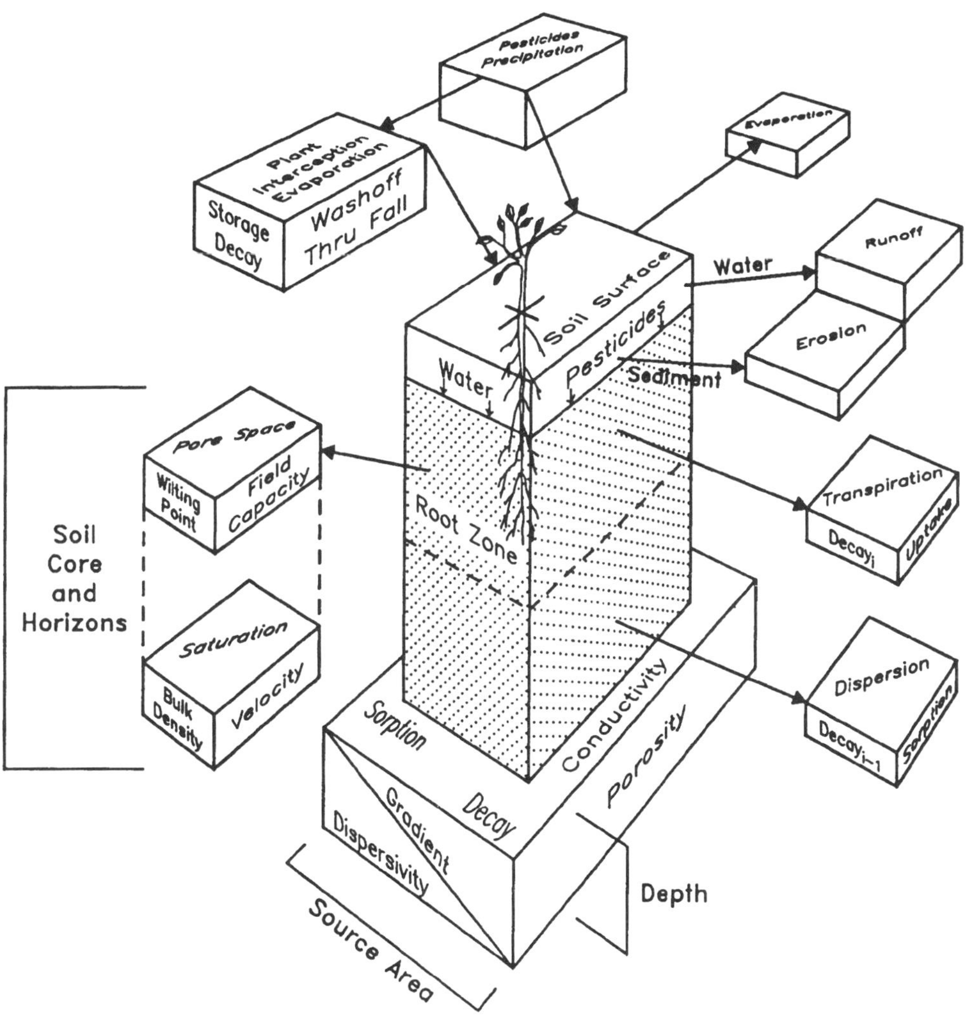

FIGURE 2. Process included in the pesticide root zone model (PRZM).

PRZM has two major components: hydrology and chemical transport. The hydrologic component for calculating runoff and erosion is based on the Soil Conservation Service curve number technique and the universal soil loss equation. Evapotranspiration is estimated from pan evaporation data, or by an empirical formula if input pan data are unavailable. Evapotranspiration is divided among evaporation from crop interception, evaporation from soil, and transpiration by the crop. Water movement is simulated by the use of generalized soil parameters, including field capacity, wilting point, and saturation water content. Irrigation also may be considered. Dissolved, adsorbed, and vapor-phase concentrations in the soil are estimated by simultaneously considering the processes of pesticide uptake by plants, surface runoff, erosion, decay, volatilization, foliar washoff, advection, dispersion, and retardation. Detailed descriptions of equation development, numerical solution techniques, and input variables may be found in the PRZM user's manual.[7]

PRZM performs calculations on a daily time step returning daily values of adsorbed, dissolved, and vapor phase pesticide concentrations in vertical soil layers; runoff volume and pesticide concentration in the runoff; eroded soil volume and pesticide concentration in the eroded particles; and volatilization loss from the soil surface and under canopy vapor concentrations. PRZM has been applied extensively and tested against field data.[8-10]

TRANSPORT AND TRANSFORMATION IN PLANTS

Exposure of plants to pesticide can occur either through direct application to the foliage or uptake of pesticide from the soil. On plants, the pesticide may wash off the leaf surface, photodegrade, or move into the plant cells and be translocated throughout the plant. Pesticides in soil water adsorb to root surfaces and move into the aboveground plant parts in the transpiration stream. The chemical may be either degraded by the plant or lost through the stomata to the atmosphere. Following application, pesticide concentrations in plants decline not only due to washoff, degradation, and volatilization but also due to growth dilution.

The plant growth simulations are based on the plant growth algorithm found in the USDA model EPIC.[11] Crop growth is calculated using crop-specific factors such as maximum leaf area index, heat units required for crop maturity, energy conversion efficiency, and optimal growth temperature along with environmental parameters such as solar radiation, temperature, and available water. Crop growth models for corn, wheat, soybeans, and other major field crops are relatively well developed and parameterized.

Uptake of pesticide from the soil and translocation into aboveground plant parts are simulated using a simple compartment model (Figure 3) applicable to neutral organic molecules passively transported by the plant. Uptake by the plant roots is parameterized based on studies[12] for a series of organic molecules with log K_{ow} values ranging from 1.2 to 5.0.

Individual components of the soil-plant transport system have been tested independently, and the combined system has been recently tested using field data.[10]

DRINKING WATER EXPOSURE

Drinking water exposure of birds to pesticides can be considered through two routes. When the soil surface layer is saturated following a rainstorm, this surface soil water is

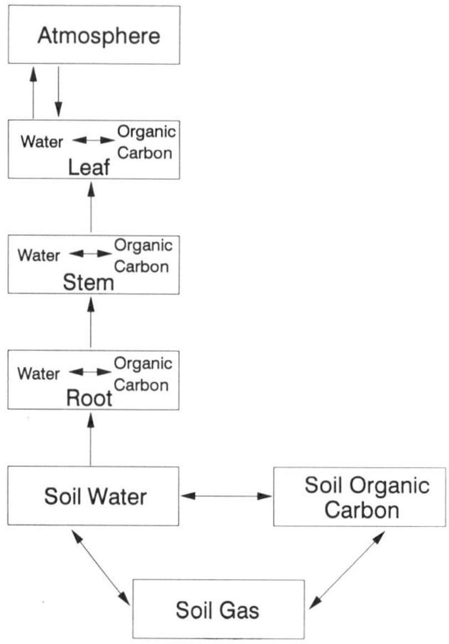

FIGURE 3. A simple compartment model simulating uptake of xylem-transported pesticides from the soil and translocation into aboveground plant parts.

assumed to be available to the bird. Additionally, they may drink from ponds or streams adjacent to treated areas.

EXAMS[13] is used to calculate pesticide concentration in streams or ponds. This model combines the loadings, transport, and transformations of a chemical in the body of water into a set of differential equations using the laws of mass balance. Loadings to the water body are measured from the runoff/erosion calculations of PRZM and direct deposition to the body of water. The chemical kinetics expressions in EXAMS are second order descriptions for transformations attributable to direct photolysis, hydrolysis, biolysis, and oxidation reactions. The model input has been designed to accept standard water quality parameters and system characteristics along with chemical data sets required by EPA regulatory procedures.

TERRESTRIAL FOOD CHAIN

The simulation of pesticide movement in soil, plant, air, and water provide the media concentrations for uptake by ecosystem fauna. Once environmental concentrations of toxicants are known, the behavior and physiology of the individual species interact to determine uptake and accumulation of the pesticide.

A simple Markov transition matrix is used to specify movement of animals between environmental compartments. Soil organisms are allowed to move vertically between soil horizons while aboveground dwellers move among the laterally defined habitats. For each organism, an $M \times M$ transition matrix is specified where M is the number of possible locations (horizons or habitats), and the transition probability specified within the matrix is the probability that an animal in a given location will move to a specific location given its current location.

Formulation of equations to describe the uptake and accumulation of chemicals in birds and other aboveground dwellers based on easily defined chemical properties and organism physiology is the ultimate goal for the development of uptake and exposure algorithms. The pharmacokinetic approach used in the development of the Food and Gill Exchange of Toxic Substances (FGETS) model[14] to simulate accumulation of neutral organics in fish serves as a prototype for estimating accumulation in terrestrial species. An exposure model must be able to calculate internal concentrations to evaluate impacts of time-varying dosages of the contaminant before extrapolations can be made to population-level impacts. The original approach to modeling the uptake of pesticides in TEEAM incorporated a calibration factor, often referred to as an assimilation efficiency factor, to parameterize the amount of material retained in an organism following ingestion. A soil organism module recently developed for inclusion uses this physiological approach.

Soil-dwelling organisms are assumed to remain in a single lateral habitat but may move vertically between soil horizons. Initial development and testing of the uptake by soil-dwelling organisms has centered on earthworms. Uptake of pesticide may be through ingestion of soil material or diffusion of dissolved pesticide though the integument.

The two routes of uptake by the earthworm are illustrated in Figure 4. Chemical dissolved in the soil water solution diffuses through the cuticle. Chemical ingested with soil and litter is assumed to equilibrate between the gut and body of the worm as the material passes through the worm's digestive tract. Equilibration is controlled by the organic matter content of the soil and of the worm. Concentration in the organism may decrease due to internal degradation and growth dilution. As of this writing, the formulation is applicable only to neutral hydrophobic molecules, but plans are to expand it to accommodate polar and ionic compounds.

A substantial database exists on accumulation of toxicants in earthworms because they are an economically important species in agroecosystems as well as being an important food source for many species of birds. Both model and field data indicate that uptake of chemicals by earthworms is relatively rapid, with concentrations equili-

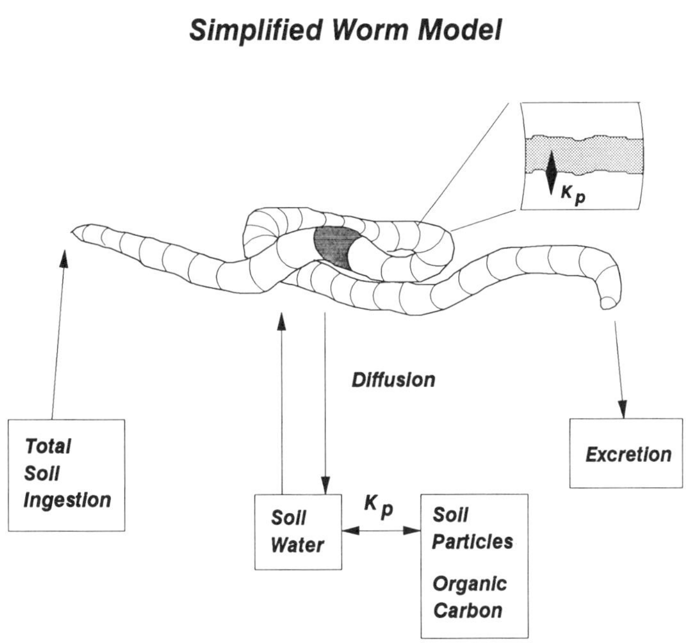

FIGURE 4. Routes of uptake by the earthworm where K_p is the partitioning coefficient of chemical between organic carbon and water.

brating within the worm in 10 to 20 days with a three- to tenfold organism concentration relative to soil concentration.

Aboveground organisms may move between different lateral habitats and take in pesticide from a variety of sources including water, air, plants, soil, pesticide granules, and prey. Daily drinking volumes and respiratory volumes are combined with water and air concentrations, respectively, to calculate dosage via the water and inhalation routes.

Total daily feeding rates along with preference factors for each food source are combined with each food type to calculate dosage via ingestion. In addition to food (plants, soil organisms, and prey animals), soil, pesticide granules, and coated seeds may be ingested. Capture of prey by a predator is simulated by basing capture on the probability that predator and prey are in the same habitat and a probability of capture for each predator/prey pair.

Development of a pharmacokinetic approach to estimating internal organism concentrations for bird is a precondition for adequately predicting impacts of field exposures on birds. Previous bioaccumulation modeling for birds[1] used an empirical assimilation factor, which has limited use in extrapolating between conditions, individuals, and species. The pharmacokinetic approach analogous to that used in FGETS, based on physiology of the organism and the properties of the chemical, is necessary to adequately predict the impact on a mixed population under multiple stresses.

INPUT DATA GENERATION

The input data requirements for performing mechanistic-based exposure calculations are extensive and include soil characteristics, chemical properties, crop and cropping practice information, meteorological conditions, animal behavior and physiology, and food chain interactions. Development of a utilitarian tool requires linkage of the model to databases for development of user input sequences. The PRZM Input Collator, Version 1.0 (PIC:V1)[15] is the first step in developing a comprehensive database support system for all elements of a terrestrial exposure system.

The geographical data in PIC:V1 are organized based on the 186 MLRAs defined by the U.S. Soil Conservation Service.[16] The delineation of MLRAs is based primarily on soils, climate, and hydrology of individual regions. The county-scale soil and crop data contained in NSSAD/SIRS (National Soil Survey Area Database/Soil Interpretations Record Database)[17] was overlaid onto the MLRA scale data.

PIC:V1 allows the user to enter the data generator by selecting either a crop or an MLRA. If a crop is selected, the program returns a list of MLRAs where this crop is potentially grown for user selection. Alternately, if an MLRA entry is selected, the program returns the list of potential crops grown in that area.

Based on crop and region selected, a list of soil names sorted by areal extent in the MLRA is returned to the user along with the number of hectares of the particular soil, hydrologic group, and soil textural class. The user then simply selects the soil of interest for input generation.

Selection of the MLRA also identifies the appropriate first order National Oceanic and Atmospheric Association (NOAA) meteorological station for the area. PIC:V1 contains a utility that allows the user to review 20 to 30 years of rainfall records to easily identify a specified 30-day period of interest. The user can then select the application date that represents the critical period for a particular evaluation.

Selection of the crop, MLRA, and soil specifies all PRZM input variables except the chemical properties and application scenarios. PIC:V1 requests input of the chemical properties and application parameters. Following specification of these parameters, a PRZM input sequence is generated.

Incorporation of additional databases including information such as plant growth parameters, feeding rates and food preferences, physiological parameters, and aircraft and spray nozzle characteristics will expand preprocessing to exposure calculation for terrestrial species. Development of typical farm pond scenarios within each MLRA for use with the EXAMS model is an additional component of the PIRANHA system.

Specific scenario generation is another approach to increasing model accessibility. This approach has been successfully used in the inclusion of the canonical series developed for use with the EXAMS model.

MODEL VALIDATION

Mathematical models and computer simulation codes designed to aid in risk assessment must be verified and validated before they can be used with confidence in a decisionmaking context. The builders and users of ecological models are rightfully concerned that the models be valid, because decisions based on them have significant economic, public health, and ecological consequences. Structuring a system such as PIRANHA to incorporate, where available, developed technological components (e.g., EXAMS and PRZM), each characterized by a substantial history of validation testing and maintained to the extent possible in a stand-alone context—allows the package to be updated easily as the components are updated, with reduced risk of introducing errors in the process.

Individual process models developed for incorporation in PIRANHA must be subjected to validation testing. In addition to process-level testing, the model as a whole must be evaluated. This overall system testing includes use of the databases and parameter estimation or measurement techniques when the model is used in a risk assessment context. This level of validation will be an integral part of the continuing development of the terrestrial exposure model.

One of the difficulties in evaluating terrestrial exposure, whether through models or through field-based evaluations, is the lack of local homogeneity of the soil medium itself, of the pesticide distribution in the system, and of the population of organisms. The potential variability of pesticide concentration must be considered as it moves through all the components of the system. Model results must be evaluated within the context of this concentration variability. Field studies for use in testing model validity must be carefully designed with consideration for this sample variability.[10]

Objective criteria for model validity must be defined. Models are, due to inherent assumptions, incapable of predicting exactly the "true" values and can only be expected to get close to them. A model's validity must be considered in the context of producing values that are sufficiently close to true values. Hence, defining what is meant by "sufficiently close" in the context of the user's application is the essence of model validity. One way to pose the model validity question is using a criterion statement of the form of whether the model is capable of predicting within a factor (e.g., 2) of the true value in a given application context. This type of statement can be easily translated into a hypothesis statement that can be subjected to statistical tests.

SUMMARY

Calculating the exposure of terrestrial biota to pesticides is the first step in evaluating possible ecological impacts from the application of these pesticides in terrestrial systems. A series of models that calculate transport of pesticides in the exposure media and into the terrestrial biota is necessary in developing population effects modeling of field applications. In addition to algorithm development, parameter estimation utilities and validation testing are key requirements for success of a terrestrial exposure model in a risk assessment framework and use in a regulatory setting.

REFERENCES

1. Dean, J. D., K. A. Voos, R. W. Schang and B. P. Popenuck, Terrestrial Ecosystem Exposure Assessment Model (TEEAM), EPA/600/3-88/038, U.S. Environmental Protection Agency, Athens, GA, 1989.
2. Burns, L. A., Ed., PIRANHA: Pesticide and Industrial Chemical Risk Analysis and Hazard Assessment, Version 2.0, U.S. Environmental Protection Agency, Athens, GA, 1992.
3. Bjorklund, J. R., C. R. Bowman and G. E. Dodd, User Manual, Forest Service Aerial Spray Computer Model, FPM 89-1, USDA, Forest Service, Forest Pest Management, Davis, CA, 1989.
4. Rafferty, J. E., J. M. White, J. F. Bowers and J. W. Barry, Comparison of Program Wind Phase III Deposition Measurements with FSCBG2 Model Predictions, U.S. Army Material Command, Technical Analysis and Information Office, U.S. Army Dugway Proving Ground, Project IM465710DO49, Joint Chemical/Biological (CB) Contact Point and Test (Project DO49), 1988.
5. Teske, M. E., J. W. Barry and R. B. Ekblad, Canopy Penetration and Deposition in a Douglas-Fir Seed Orchard, Paper No. 901019, Proc. American Society Of Agricultural Engineers, Columbus, OH, June 24-27, 1990.
6. Rafferty, J. E., and J. F. Bowers, Comparison of FSCBG2 and FSCBG3 Aerial Spray Model Predictions with Field Measurements, FPM 90-2, USDA Forest Service, Davis, CA, 1990.
7. Carsel, R. F., C. N. Smith, L. A. Mulkey, J. D. Dean and P. Jowise, User's Manual for the Pesticide Root Zone Model (PRZM), EPA/600/3-84/109, U.S. Environmental Protection Agency, Athens, GA, 1984.
8. Carsel, R. F., W. B. Nixon and L.G. Ballantine, Comparison of pesticide root model predictions with observed concentrations for the tobacco pesticide metalaxyl in unsaturated zone soils, *Environ. Toxicol. Chem.*, 5, 345-353, 1986.
9. Jones, R. L., G. W. Black and T. L. Estes, Comparison of computer model predictions with unsaturated zone field data for aldicarb and aldoxycarb, *Environ. Toxicol. Chem.*, 5, 1027-1037, 1986.
10. Smith, C. N., R. S. Parrish and D. S. Brown, Conducting field studies for testing pesticide leaching models, *Int. J. Environ. Anal. Chem.*, 39, 3-12, 1989.
11. Williams, J. R., C. A. Jones and P. T. Dyke, A model for assessing the effects of erosion on soil productivity, in *Analysis of Ecological Systems: State-of-the-Art in Ecological Modeling*, W. K. Lauenroth, G. V. Skogerboe and M. Flug, Eds., Elsevier, New York, 1983.
12. Briggs, G. G., R. H. Bromilow and A. A. Evans, Relationship between lipophilicity and root uptake and translocation of non-ionised chemicals by barley, *Pestic. Sci.*, 13, 495-504, 1982.
13. Burns, L. A., and D. M. Cline, Exposure Analysis Modeling System (EXAMS): Reference Manual for EXAMS II, EPA/600/3-85/038, U.S. Environmental Protection Agency, Athens, GA, 1985.
14. Barber, M. C., L. A. Suarez and R. R. Lassiter, Modeling bioaccumulation of organic pollutants in fish with an application to PCBs in Great Lakes salmonids, *Can. J. Fish. Aq. Sci.*, 48, 318-337, 1991.
15. Bird, S. L., PRZM Input Collator, Version 1 (PIC:V1), in PIRANHA: Pesticide and Industrial Chemical Risk Analysis and Hazard Assessment, Version 1.0, U.S. Environmental Protection Agency, Athens, GA, 1990.

16. U.S. Soil Conservation Service, Land Resource Regions and Major Land Resource Areas of the United States Agriculture Handbook 26, U.S. Department of Agriculture, Washington, DC, 1981.
17. U.S. Soil Conservation Service, User Manual for Interactive Soils Databases: National Soil Survey Area Database, Soil Interpretations Record Database and Plant Name Database, U.S. Department of Agriculture, Fort Collins, CO, 1985.

17

Applications of a Modified Richards Sigmoid Model to Assess the Uptake and Effects of Environmental Contaminants upon Birds

I. Lehr Brisbin, Jr.

ABSTRACT

The uptake of environmental contaminants by birds and other consumer organisms traditionally has been studied under controlled laboratory conditions and has been described with a fixed-shape monomolecular model. Under free-living field conditions, however, several species of birds utilizing both aquatic and terrestrial food webs have been shown to have a significant time lag in early contaminant uptake rates, with the resulting data being more accurately described by a flexibly shaped modified Richards sigmoid model. This same modified Richards model has also proved useful for describing changes in the growth of a variety of organisms, including birds, in response to exposures to a variety of stressors, such as uptake of environmental contaminants. The latter studies have shown that the shape of a bird's growth curve is more likely to change, and thus shape provides a more sensitive indicator of response to contaminant uptake than either the asymptotic size or rate at which overall growth occurs. Fixed-shape sigmoid models that fail to conform to the proper curve shape of the data being described can produce misleading or erroneous conclusions concerning the uptake or effects of contaminants upon birds. These errors can be particularly important in comparing contaminant uptake under field vs laboratory conditions and in assessing the effects of contaminant uptake upon growth.

KEY WORDS

contaminants, birds, growth, sigmoid models, radionuclides

INTRODUCTION

Designers of wildlife toxicology studies are invariably faced with the need to better describe both the pattern and rate at which free-living organisms take up and accumulate contaminants from the environment. Equally important is the need to develop more sensitive means of detecting the effects of such contaminants on the animals in which they have become concentrated. Both of these needs have been addressed recently through the development and application of a modified nonlinear three-parameter Richards sigmoid model.[1,2] Although never applied to data for pesticide toxicology in birds, this model has proved extremely robust in describing data for a variety of other forms of contaminants, including heavy metals, radionuclides, and polychlorinated biphenyls (PCBs), and for a variety of organisms, including birds.[3] In these applications, this model has proved particularly useful in comparing the results of laboratory vs field studies of contaminant uptake and concentration, using data from free-ranging recapturable "sentinel" birds residing in contaminated habitats.[4] This chapter describes the modified Richards sigmoid model and reviews its application to pertinent aspects in the field of wildlife toxicology. A particular emphasis is placed on those findings that either would not have been possible or might have produced contrary conclusions without the use of this particular model. Such situations should be cause for particular concern for any attempts to draw conclusions about the fate and effects of agricultural pesticides in birds.

THE RICHARDS SIGMOID GROWTH MODEL

As originally developed for the analysis of growth,[1] the Richards sigmoid model is characterized by its flexibility in describing changes in curve shape through changes in the numerical value of a curve "shape parameter" (m). Changes in the value of this parameter can generate any number of widely used, fixed-shape sigmoid models. When, for example, $m = 0, 0.667$, or 2.00, the Richards model becomes the monomolecular, von Bertalanffy, or logistic models, respectively; when m approaches a value of 1.00, the Richards model becomes the Gompertz model.[1] Thus, it is not irrelevant to ask whether the Richards model is inferior or superior to one of these fixed-shape models in its ability to describe a given set of data. If the data warrants such a shape, curve-fitting procedures will adjust the value of the Richards curve shape parameter until it becomes identical to that of the appropriate fixed-shape model as just described. However, intermediate values also may be assumed by this parameter, and thus a curve described by an equation with $m = 1.48$, for example, would have a shape intermediate between that of the Gompertz and logistic models. As the value of m increases, the point of inflection of the curve tends to occur progressively later in the growing period (Figure 1).

As originally formulated,[1] the Richards model contained several features that rendered it both mathematically intractable and statistically unsuitable for use in curve-fitting procedures and the testing of hypotheses concerning growth. Later changes in the formulation of the model[2] and in the way that it has been fit to actual data,[5,6] however, have made considerable improvements in these areas. Of particular importance has been the reparameterization of the model[2] so as to remove the original growth rate constant K and replace it with the parameter T. Although not actually used in the original formulation of the model, this T parameter was described by Richards[1] as a quantity that "may loosely be considered to measure the time required for the major part of growth, and for comparative purposes probably represents the period of development as well as any other figure derivable from an asymptotic function." In this formulation, T may be generally considered the amount of time required by the growing organism to attain about 95% of final asymptote (A).

In its reparameterized and integrated form, the Richards model is represented by the equation:

$$W_t = \left[A^{(1-m)} - \left(A^{(1-m)} - \mathrm{W}_0^{(1-m)} \right) \exp\left(\frac{-2t}{T}(m+1) \right) \right]^{\frac{1}{1-m}} \tag{1}$$

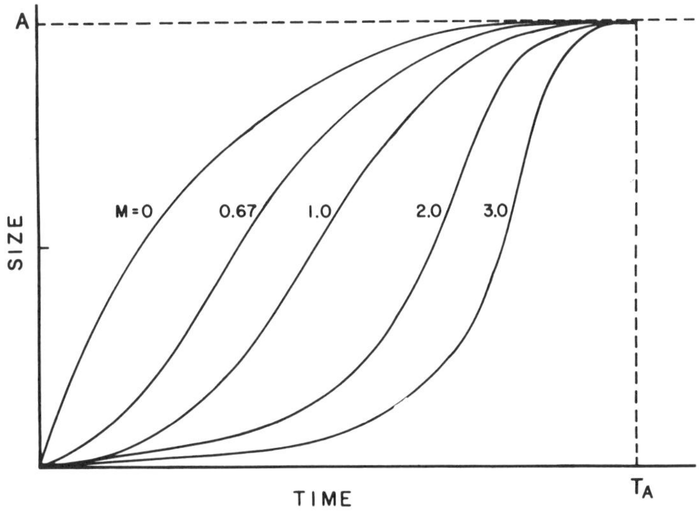

FIGURE 1. Diagrammatic representation of the effects of changing the value of the Richards sigmoid model's shape parameter (*m*) upon the shape of the curve produced. As explained in the text, *m* values of 0, 0.667, ±1.00, and 2.00 produce the fixed-shape monomolecular, von Bertalanffy, Gompertz, and logistic models, respectively.

where W_t and W_0 represent body weight (or some other measure of quantity or size) at time t and at time $t = 0$, respectively. As previously explained, the remaining three parameters represent asymptote (A), the approximate time (T) required to attain that asymptote, and the particular shape (m) of the path or trajectory taken in attaining final size within that amount of growing time. In formulating Equation 1, Richards' original growth rate parameter, K, has been transformed to T by substitution, using the relationship:

$$K = \frac{2(m + 1)}{T(1 - m)} \tag{2}$$

as explained by Brisbin et al.[2]

APPLYING THE RICHARDS MODEL TO CONTAMINANT UPTAKE PROCESSES

The Richards growth model may be applied to the uptake of an environmental contaminant by a free-living consumer organism by letting W_t from Equation 1 be represented by C_t, the concentration of the contaminant in the organism at time t. Time now represents the amount of elapsed time that the organism has been exposed to the contaminated habitat. Under these conditions, A of Equation 1 would now represent the final contaminant concentration in the organism at steady state or equilibrium, and T would represent the amount of exposure time required to attain approximately 95% of that equilibrium concentration.[4] The uptake of environmental contaminants by an organism classically has been described by the equation:

$$C_t = A(1 - e^{-k_d \cdot t}) \tag{3}$$

where k_d is the depuration or elimination rate constant, and the other parameters are as already described.[7] Equation 3 may be modified then by substituting the T parameter for k_d, using the relationship:

$$T = \frac{2(m + 1)}{k_d(1 - m)} \tag{4}$$

as explained by Brisbin et al.[4] Equation 4 represents a correction of the erroneous presentation of Equation 5 of Reference 4. This produces a Richards contaminant uptake model which in its integrated form is represented by the equation:

$$C_t = \left[A^{(1-m)} - \left(A^{(1-m)} - C_o^{(1-m)}\right) \exp\left(\frac{-2t}{T}(m + 1)\right)\right]^{\frac{1}{1-m}} \tag{5}$$

where C_o represents the concentration of the contaminant in the body of the organism at the beginning of the exposure period and the other parameters are as previously explained.

Substituting $m = 0$ in Equation 5 and reversing the substitution of T for k_d as given by Equation 4 will produce Equation 3. Thus, the classic uptake model described by Equation 3 is really the monomolecular form of the Richards uptake model, with $m = 0$. In order to determine whether a given set of contaminant uptake data is better described by the classic model vs the Richards uptake model with $m \neq 0$, the data are first fit to Equation 5 and best-fit estimates produced for the parameters A, T, and m. The procedure is then repeated with $m = 0$, producing estimates only for the parameters A and T. The resulting error sums of squares are then compared with an F-test using procedures described by White and Brisbin.[5] An example of the use of these procedures to describe actual field data for contaminant uptake by free-living birds has been provided by Brisbin et al.[4] and is described in detail in the next section.

APPLYING THE MODEL TO CONTAMINANT UPTAKE DATA FOR FREE-LIVING BIRDS

As in the case of growth, contaminant uptake can be described by either longitudinal, cross-sectional, or mixed data sets.[8] All else being equal, longitudinal data sets in which repeated measures are taken over time from the same individual are preferable due to their more favorable statistical properties in the testing of hypotheses.[5,8] However, collecting longitudinal data for the uptake of pesticides by individual free-living birds is generally not possible because destructive sampling is usually required to determine contaminant concentration levels. Gamma-emitting radionuclides, however, can be easily quantified in living birds by nondestructive whole-body counting techniques.[9-12] Data for these radioactive contaminants and the use of gamma-emitting radionuclide "tracers" have thus been used to produce field data for testing the use of uptake models such as those described here.[4,13] In the following example, however, cross-sectional data are analyzed for radionuclide uptake using a method that would be readily adaptable for describing the uptake of agricultural pesticides by birds under field conditions.

In a series of field experiments described in detail by Potter[13] and Brisbin et al.,[4] American coots (*Fulica americana*) were captured from portions of a nuclear reactor cooling reservoir system on the U.S. Department of Energy's Savannah River Site near Aiken, S.C. Portions of this reservoir system had been contaminated years earlier with several radionuclides, particularly the gamma-emitter radiocesium (^{137}Cs), as described by Ashley and Ziegler.[14] Birds from uncontaminated portions of the reservoir system, which showed little or no initial body burdens of ^{137}Cs ($\bar{x}\ C_o$ = 0.062 Bq/g live weight), were wing-scissored and released on more contaminated portions of the reservoir. Periodically recapturing these "sentinel" birds for whole-body burden determinations of ^{137}Cs and then re-releasing them into contaminated habitat for future recapture generated radiocesium uptake values, which were analyzed as a cross-sectional data set using both the Richards and classic uptake models (Equations 5 and 3, respectively). These analyses (Figure 2) indicated that the free-ranging coots on the reservoir reached an average ^{137}Cs equilibrium whole body burden (A) of 4.73 Bq/g live weight in approximately 14 days (T), with an uptake curve shape parameter m = 1.2.[4] More importantly, the fit of the data to the Richards uptake model was significantly better (F = 5.34; df = 1,11; $p<0.05$) than that provided by the classic model of Equation 3.

In subsequent studies, still further confirmation has been obtained in support of the superior ability of the Richards model to describe the uptake of radioactive contaminants by free-ranging birds compared to the classic model. This support has included qualitative analyses of the uptake of ^{137}Cs by feral bantam chickens (*Gallus gallus*) foraging in a contaminated floodplain forest[12] and by mallard ducks (*Anas platyrhynchos*) inhabiting the same contaminated reservoir habitat as previously described for the coots.[15]

In all analyses using the Richards model there was a slight but statistically significant time lag in early contaminant uptake compared to the predictions of the classic model (Figure 2). The importance of this often seemingly inconsequential difference between model predictions is magnified for the case of migratory birds, whose stopover times in contaminated areas are generally of only a short duration and within a time frame when the predictions of the Richards and classic models differ to the greatest extent

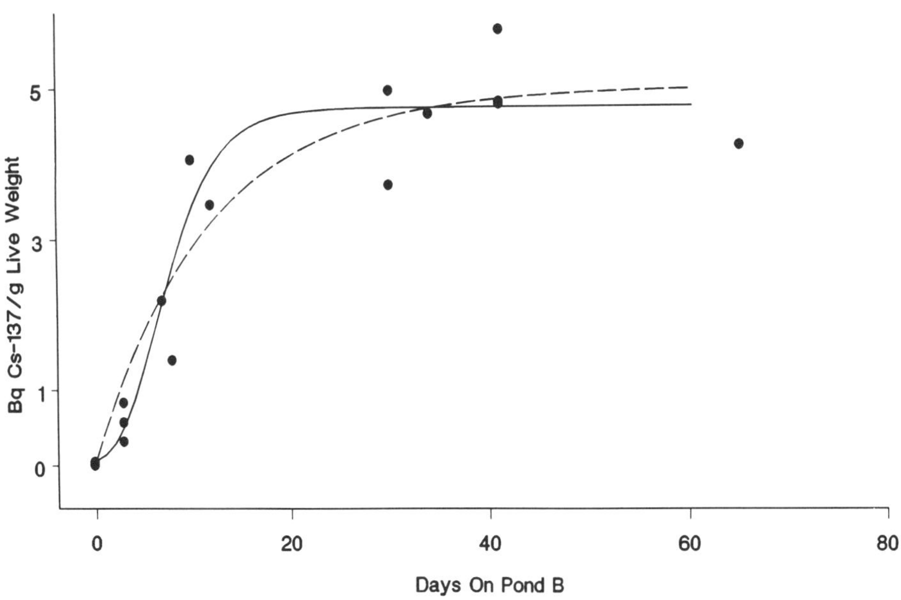

FIGURE 2. Uptake of ^{137}Cs by free-living American coots inhabiting a contaminated reactor cooling reservoir on the U.S. Department of Energy's Savannah River Site. A Richards contaminant uptake model (solid line), as described by Equation 5 in the text, provided a significantly better description of the data than did the classic uptake model of Equation 3 (dotted line).[4]

(Figure 2). This difference in model predictions would have particularly notable consequences in large-scale events of environmental contamination such as occurred in 1986 at the site of the Chernobyl nuclear reactor accident in the former Soviet Union.[12]

APPLICATIONS TO ANALYSES OF GROWTH DATA AS INDICATORS OF CONTAMINANT STRESS

When growth characteristics are used as indicators of an organism's response to environmental stress, such as exposure to a contaminant, parameters quantifying asymptotic size or the rate of growth are usually chosen for analysis.[16,17] However, the Richards growth model also offers the opportunity to quantify and thus study changes in a growth curve's shape as an additional response to such stress. Studies using this approach[2,3,18] have shown that changes in sigmoid growth curve shape (i.e., changes in the Richards m parameter) are up to three times more likely to occur following exposure to stressors than are changes in either asymptotic size or the time required to reach asymptote (the Richards model's A and T parameters, respectively); see Table 1. Furthermore, in every case in which changes in growth curve shape were detected in response to environmental stress, the changes were always from a lower to a higher numerical value of m.[3] This would indicate that stress exposure was causing the point of inflection (time of most rapid growth) of the growth curve to occur relatively later in the total growing period (Figure 3), even though, in some cases, neither final asymptote nor the time required to reach that asymptote changed significantly.[2] In the latter study, for example, captive mallard ducks were exposed to varying levels of PCBs in their diet throughout the growing period, and if conventional fixed-shape sigmoid models had

Table 1. Summary of Variables Shown to Have Significant Effects on Parameters of the Richards Sigmoid Growth Model

Treatment Variable	Species	Growth Parameter[a]	Level of Effect[b]	Ref.
Stressors				
Dietary PCBs	Mallard duck	m	0.04	2
Dietary cadmium	Wood duck	m	<0.001	18
		T	0.008	18
Heated water	Cypress tree	m	0.007	3
		A	0.002	3
Nonstressors				
Sex	Wood duck	m	0.05	2
		A	<0.001	2
		A	<0.001	18
		A	<0.001	19
Breeding location	Wood duck	A	<0.001	10

[a]Growth parameters, as used in Equation 1, include A = final asymptotic size, T = length of growing period, and m = Richards curve shape parameter.

[b]p-values derived from univariate analyses of variance for each parameter. Univariate analyses were conducted for all parameters only if a significant effect ($p \leq 0.05$) was first demonstrated for the vector (A, T, m) using multivariate analysis of variance. Parameters not shown in this table showed no significant treatment effects in the univariate analyses.

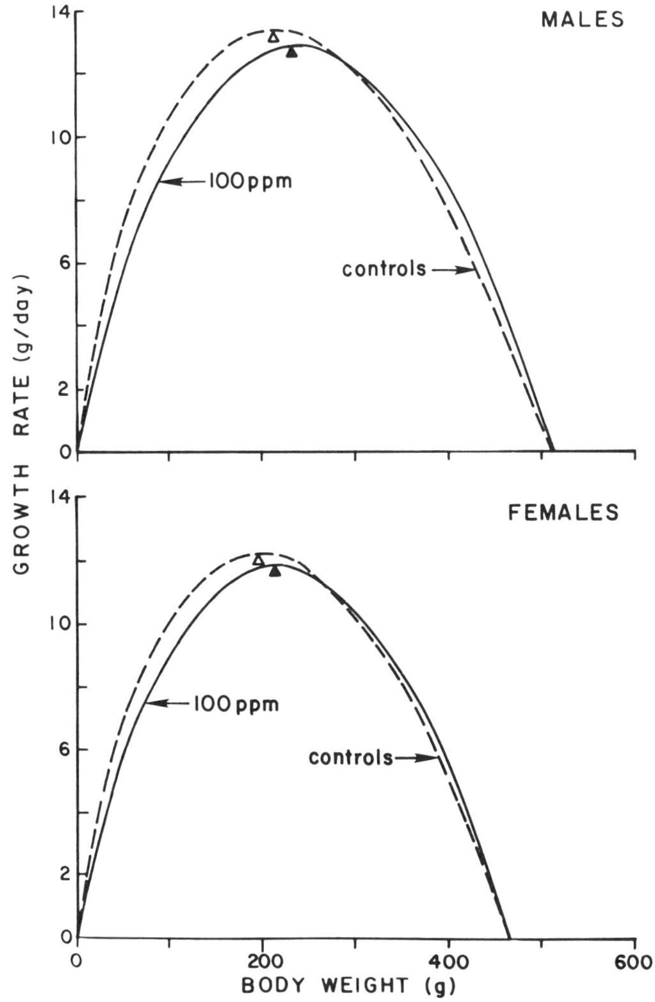

FIGURE 3. Changes in the shape of the sigmoid growth curves of wood ducks fed a diet contaminated with cadmium throughout their growing period. Changes in growth curve shape were quantified as changes in the curve shape parameter (*m*), using a process-error form of the Richards model.[18]

been used to analyze the resulting data, it would have been concluded that the levels of contaminant used had no significant effects on any aspect of the birds' growth. Only analyses using the Richards model showed a response to increasing levels of contaminant in the birds' diet, manifested as an increase in the value of *m* at the higher dietary contaminant levels tested.[2]

Analyses of the effects of other "nonstressor" variables on parameters of the Richards model have frequently failed to show significant changes in curve shape (*m*) even though in some cases the other parameters (*A* and *T*) changed significantly (Table 1). With birds, these studies have included analyses of the effects of sex,[2,18,19] dietary composition,[18] and differences between birds from different parts of the species' range.[19] There thus appears to be no inherent property of the Richards model itself which causes an unusual sensitivity of curve shape to environmental stress exposures. Moreover, the frequency and direction of response in *m* seem to be robust across a variety of species ranging from waterfowl[2,18] to cypress trees (*Taxodium distichum*)[3] as well as across a variety of stresses, including exposure to increased ambient temperature[3] and such contaminants as heavy metals[18] and PCBs.[2] Although the Richards model has not yet been used to analyze the growth of birds exposed to agricultural pesticides, the studies that have been conducted to date suggest that similar response patterns should be shown and that growth curve analyses based on parameters of the Richards model should also prove useful in this area of study.

Finally, the studies reported above suggest fruitful areas for future investigations into the more basic aspects of growth responses to environmental stress. The frequency of stress-related increases in the value of the Richards m parameter (termed a curve shape "decretion" by Brisbin et al.[18]) suggests that growth curve shape may be a "first line of defense" by which an organism responds to a lag in early growth with a compensatory increase in growth rate later in the growing period. The result would be an enhanced ability to attain the same asymptote (A) in the same amount of time (T). To the extent that both A and T are genetically determined parameters whose values may confer specific advantages to the organism, such nongenetic flexibility in growth curve shape (m) should be advantageous. Such a system, however, would require the ability to overcompensate for early growth depression by accelerating growth rate beyond that of the controls, especially during the latter part of the growing period. The studies conducted to date offer intriguing support for the ability of at least some growing organisms to do just this, although they may continue to be exposed to constant levels of contaminant stress throughout the entire growing period.[2]

CONCLUSIONS

The flexibly shaped Richards sigmoid model has proved in initial studies to be extremely robust and capable of providing unique insight into the nature of a variety of biological processes, ranging from contaminant uptake to the response of organismal growth in the face of various environmental stress. Of particular importance has been the ability of the Richards model's curve shape parameter to provide a degree of sensitivity in data analysis not possible with linear, exponential, or fixed-shape sigmoid models. This increased sensitivity in data analysis has provided a corresponding increase in the ability to detect organismal growth responses to contaminants, often at exposure levels so low as to produce no other detectable responses with other methods of data analysis. Similarly, the Richards model has also shown that the pattern of contaminant uptake by free-ranging birds under field conditions may not follow the predictions of other conventional models, which are based on laboratory studies of contaminant exposure and the many necessary assumptions inherent therein.

These findings suggest that with proper attention to the statistical procedures used for curve fitting[5,6] and hypothesis testing,[2,5,8,20] the Richards sigmoid model could offer similar advantages and insight into the nature of other functional processes involved in the exposure of birds and their populations to agricultural pesticides under both field and laboratory conditions. The finding that such processes may operate very differently under laboratory conditions compared to those in the field has important implications and injects a cautionary note regarding extrapolations between the two situations, regardless of the particular model or analytical technique used in their assessment.

ACKNOWLEDGMENTS

The work described in this chapter was supported by a contract (DE-AC09-76SROO-819) between the United States Department of Energy and the University of Georgia. Conference participation was made possible by support from the Society of Environmental Toxicology and Chemistry as well as the federal agencies and corporations that sponsored the 1990 Ecological Modeling Conference. Analyses of effects data were undertaken while the author was a visiting scientist at the Duke University Marine Laboratory. Gary White, G. Archibald McCallum, Phillip Dixon, Clark McCreedy, and James Matis have provided many helpful ideas and discussions, which have contributed to my thinking on these issues over the years. Ron Chesser and Cham E. Dallas provided helpful comments on the present manuscript.

REFERENCES

1. Richards, R. J., A flexible growth function for empirical use, *J. Exp. Bot.*, 10, 290-300, 1959.
2. Brisbin, I. L., Jr., G. C. White and P. B. Bush, PCB intake and the growth of waterfowl: multivariate analyses based on a reparameterized Richards sigmoid model, *Growth*, 50, 1-11, 1986.
3. Brisbin, I. L., Jr., K. W. McLeod and G. C. White, Sigmoid growth and the assessment of hormesis: a case for caution, *Health Phys.*, 52, 553-559, 1986.
4. Brisbin, I. L., Jr., M. C. Newman, S. G. McDowell and E. L. Peters, Prediction of contaminant accumulation by free-living organisms: applications of a sigmoidal model, *Environ. Toxicol. Chem.*, 9, 141-149, 1990.
5. White, G. C., and I. L. Brisbin, Jr., Estimation and comparison of parameters in stochastic growth models for barn owls, *Growth*, 44, 97-111, 1980.
6. McCallum, D. A., and P. M. Dixon, Reducing bias in estimates of the Richards growth function shape parameter, *Growth, Dev. Aging*, 54, 135-141, 1990.
7. Davis, J. J., and R. F. Foster, Bioaccumulation of radioisotopes through aquatic food chains, *Ecology*, 39, 530-535, 1958.
8. Brisbin, I. L., Jr., C. T. Collins, G. C. White and D. A. McCallum, A new paradigm for the analysis and interpretation of growth data: the shape of things to come, *Auk*, 104, 552-554, 1987.
9. Brisbin, I. L., Jr., R. A. Geiger and M. H. Smith, Accumulation and redistribution of radiocesium by migratory waterfowl inhabiting a reactor cooling reservoir, in *Proc. Int. Symp. Environmental Behavior of Radionuclides Released in the Nuclear Industry*, International Atomic Energy Agency, Vienna, 1973, pp. 373-384.
10. Straney, D. O., B. Beaman, I. L. Brisbin, Jr., and M. H. Smith, Radiocesium in birds of the Savannah River Plant, *Health Phys.*, 28, 341-345, 1975.
11. Brisbin, I. L., Jr., and M. J. Vargo, Four-year declines in radiocesium concentrations of American coots inhabiting a nuclear reactor cooling reservoir, *Health Phys.*, 43, 266-269, 1982.
12. Brisbin, I. L. Jr., Avian radioecology, in *Current Ornithology*, Vol. 8, D. M. Power, Ed., Plenum Press, New York, 1991, pp. 69-140.
13. Potter, C., Use of Reactor Cooling Reservoirs and Cesium-137 Uptake in the American Coot (*Fulica americana*), M.S. thesis, Colorado State University, Fort Collins, CO, 1987.
14. Ashley, C., and C. C. Ziegler, Releases of Radioactivity at Savannah River Plant 1954 through 1978, Rep. DPSU-75-25-1, E.I. DuPont de Nemours and Co., Aiken, SC, 1980.
15. George, L. S., C. E. Dallas, I. L. Brisbin, Jr., and D. L. Evans, Radiocesium (^{137}Cs) uptake in mallards at the Savannah River Site (SRS) and effects on DNA cell cycle in red blood cells, *Toxicologist*, 10, 248, 1990.

16. Fendley, T. T. and I. L. Brisbin, Jr., Growth curve analyses: investigations of a new tool for studying the effects of environmental stress upon wildlife populations, in *Proc. 13th Int. Congr. Game Biologists*, Atlanta, 1977, pp. 337–350.
17. Millard, J. B., F. W. Whicker and O. D. Markham, Radionuclide uptake and growth of barn swallows nesting by radioactive leaching ponds, *Health Phys.*, 58, 429–439, 1990.
18. Brisbin, I. L., Jr., G. C. White, P. B. Bush and L. A. Mayack, Sigmoid growth analyses of wood ducks: the effects of sex, dietary protein and cadmium on parameters of the Richards model, *Growth*, 50, 41–50, 1986.
19. Brisbin, I. L., Jr., and R. A. Kennamer, Growth characteristics of wood ducks from two southeastern breeding locations, *Wilson Bull.*, 99, 91–94, 1987.
20. Leberg, P. L., I. L. Brisbin, Jr., M. H. Smith and G. C. White, Factors affecting the analysis of growth patterns of large mammals, *J. Mammals*, 70, 275–283, 1989.

18

Impacts of Toxicants on Population Dynamics and Gene Diversity in Avian Species

Ronald K. Chesser, Kevin B. Willis, and Nancy E. Mathews

ABSTRACT

Computer simulation and mathematical modeling approaches were used to assess: (1) the relative effects of gradual vs rapid bioconcentration on rates of loss of genetic variation; (2) the recovery rates of avian populations after acute and chronic pesticide applications and the impact on loss of gene diversity; (3) the importance of timing (pre- and postmating) and mode (mortality vs fertility) of toxicant impacts on recovery rates and loss of genetic variation; and (4) the effectiveness of avian breeding and dispersal tactics in partitioning genetic variation within and among populations. Computer simulations of toxicant loads provided insight on cumulative effects of toxicants on population size and genetic variation. Gene diversity was shown to be more sensitive to rates of population recovery than to actual population sizes, and fertility reduction was much more important than mortality in maintenance of genetic variance and population size. Logistic models demonstrated that application schedules may have substantial impact on the recovery of population size and loss of genetic variation and that repeated acute impacts are of equal concern to chronic effects on population dynamics. Recovery rates are much quicker when impact is made primarily on adults rather than adults and juveniles concurrently. Additionally, toxicant impact that affects adult mortality rather than fertility will allow a more rapid rate of population recovery. Loss of gene diversity is inversely correlated to the recovery rate of the population. Thus, scenarios that prolonged the recovery rates of population sizes resulted in substantial losses in genetic variation. Genetic variance within populations was shown to conform to classical demic models. Results of the models are applicable to many avian species and document the need for careful consideration of the timing of toxicant applications to ensure population viability and variability.

KEY WORDS

genetic variation, chronic exposure, fertility, mortality, population recovery

INTRODUCTION

Although the varying impacts of agricultural pesticides on populations of avian species have been widely recognized since the early 1950s,[1,2] little is known about the potential impacts on population demography and genetic diversity. Historically, intensive pesticide application has been linked with declines in both survivorship and reproductive success for a number of avian species. Mass avian mortalities occurred in the 1950s,[3] when DDT was used for a variety of pest eradication efforts.[4] At Clear Lake, California, an estimated population of 100 breeding pairs of western grebes (*Aehmophorus occidentalis*) was reduced to fewer than 50 individuals in 1958.[5] Over 1100 birds of 10 species at Tule Lake National Wildlife Refuge died during 1960 to 1962 from agricultural runoff,[6] while dramatic local population reductions occurred in species such as bald eagle, brown pelican, Bermuda petrel, California condor, western grebe, peregrine falcon, osprey, and Cooper's hawk. Despite the extensive documentation of the classic impacts of DDE and DDT on reproductive success of these species,[7,8] we are just beginning to understand the potential impact of these massive population reductions on the genetic diversity of local populations.

To understand the mechanism involved in pesticide-related population reductions, we must consider the modes of action of those classes of chemicals likely to affect avian species. Agricultural pesticides may be grouped into two general categories, the organochlorines (OC) and organophosphates (OP). Most OCs are relatively stable and persist in the environment because of their lipophilic nature. They also can move through the food chain and bioaccumulate in higher level organisms. Organochlorines are generally not excreted to a significant degree before metabolism.[7,9,10] Persistence of DDE, for example, is high—up to 229 days in grackles[11] and 250 days in rock doves;[12] other half-lives range between 57 and 74 days for oxychlordane in brown-headed cowbirds, grackles, starlings, and red-winged blackbirds.[13]

OCs can affect avian species either directly or indirectly through acute or chronic effects. While OCs undoubtedly have had some acute effects on avian species, chronic or indirect effects are most frequently reported. In the classic DDT and DDE poisoning cases, eggshell thinning was credited as the primary reason for a decline in breeding success. Blus et al.[8] found that after direct application of endrin to orchards, birds generally died within 1 to 2 months after application. Some mortalities occurred 4 to 6 months after application and probably resulted from mobilization of residues from lipids.[8] Heptachlor has been found in Canada goose eggs on Umatilla National Wildlife Refuge as recently as 1978; concentrations are negatively correlated with nesting success. After the ban on heptachlor, nesting success of geese increased from 52 to 84%, with the number of eggs hatched increasing from less than 3 to greater than 4 (Blus, unpublished data; see Reference 1). One of the chronic, indirect effects that may have an important impact on fecundity is behavioral aberrations. Experimental data suggest a correlation between exposure to OCs and behaviors such as courtship, vigilance, incubation, and brooding. Nest attentiveness declined in herring gulls on Lake Ontario.[14] McArthur et al.[15] found a decline in incubation and brooding in rock doves after exposure to a mixture of chemicals. Koeman[16] documented a reduction in the courtship behavior in American kestrels.

In contrast, OPs generally have a relatively short environmental persistence and short half-life in body tissues.[17] OPs act by inhibiting enzyme cholinesterase (ChE), causing an accumulation of acetylcholine (ACh) at nerve synapses and consequent disruption of nerve function. The extent of action varies with chemical, species, age, level of exposure, and sex. Acute effects of OPs most frequently include loss of body weight,[18] which is apparently due to pesticide-induced anorexia, with losses up to 40% in adult birds.[19] Egg production was significantly inhibited in ring-necked pheasants and bobwhite quail[20] fed diets containing diazinon. Loss of body weight may be associated with susceptibility to environmental factors. This may lead to low fledgling weights, which have been associated with decreased survival. While chronic effects are not well documented, OPs, like OCs, may also affect behavioral traits associated with fecundity (nest attendance, courtship, territoriality, defense, and incubation). Consequently, pesticides may function either acutely or chronically to reduce population size.

Timing of pesticide applications is of critical importance in determining potential impact. Seasonally, pesticides are generally applied at the time of greatest insect abundance. This period usually coincides with the reproductive period of birds. Potential consequences of the application timing to birds include direct effects through consumption of contaminated prey and indirect effects such as severe declines in prey resources. Pesticide-induced reductions in insect abundance were hypothesized to have caused the dispersal of various bird species.[7] Survival of grey partridge chicks decreased following a reduction in insect abundance, the result of a pesticide application.[21]

Our objective is to determine the role that toxic substances may play in the rate of loss of genetic variability in avian taxa. Because loss of genetic variation is a function of population size, population growth rate, and the distributions of genes within affected populations, we necessarily address functions of toxic exposure on mortality, fertility reduction, and population recovery rates, as well as the role played by typical avian breeding structures. We take particular interest in the effects of timing of chronic or repetitive toxic exposures on population size and gene diversity. To achieve our objectives, we employ both computer simulations and logistic models that provide a deterministic solution for the fate of genetic variation and population recovery times.

SIMULATION MODELS OF TOXIC ACCUMULATION

Computer simulation models were implemented to investigate the impact of toxic exposure on natural avian populations when toxic effects on mortality and fecundity were cumulative, when the distributions of affected individuals were varied, and when variables such as sex, genotypes, and toxic exposure were subjected to sampling variance. As seen below, the results of the simulation and mathematical models are not directly comparable, but rather were designed to address a separate set of objectives.

The computer simulation model was written in FORTRAN-77 code on a Digital VAX 6320. The purpose of the program was to track the genotype, toxic load, reproduction, and mortality of individuals and to monitor population size and the loss of genetic variation over an indeterminant number of iterations. The program comprised five subroutines, *SETUP*, *BREEDING*, *FEEDING*, *MORTALITY*, and *AGING*, of which the last four constituted a repeated iterative loop.

In the *SETUP* subroutine a fixed number of individuals was created and the sex, age, and genotype were randomly assigned to each. The initial sex ratio was approximately 1:1, and the age distribution was approximately the asymptotically expected age distribution for a stable, unimpacted population (determined from trial runs of the model). The genotypes were assigned such that the allele frequency of an allele was approximately 0.5 and the population complied with Hardy-Weinberg expectations. All values were approximate because of binomial sampling error due to random assignment of individual characteristics. The *SETUP* subroutine created a demographically stable population at the assigned carrying capacity for a particular run of the model.

The lifetime accumulation of the toxin was determined in the *FEEDING* subroutine. Three separate toxic load distributions (constant, binomial, and Poisson) were considered. For the constant distribution, the toxic load for all individuals was incremented by 2.5; thus, there was no variance in the rate of toxin accumulation. For the binomial distribution five random numbers from a 0 to 1 uniform distribution were selected for each individual. These numbers were added to lifetime accumulation of the toxin for the individual. For the Poisson distribution a random number was selected from a Poisson distribution and added to the lifetime accumulation. The average increase in toxic load per individual was identical for each distribution. An individual's lifetime accumulation of the toxin was used in both the *BREEDING* and *MORTALITY* subroutines where the potential effects of the toxin were manifested.

In the *BREEDING* subroutine a mate was randomly selected for each female. In instances of more females than males, the excess females would not breed during that iteration of the model. Each female initially produced four eggs; however, hatching and fledgling success were set at 50%, reducing the female production of viable offspring at between zero and four with a mean of two. If either the male or the female of a pair had a lifetime accumulation of the toxin exceeding the critical value for that run of the model, no viable offspring were produced.

The *MORTALITY* subroutine removed individuals from the data matrix based on two criteria. First, if an individual had a lifetime accumulation of the toxin that exceeded the maximum allowable for that run of the model, the individual was removed. Second, there was a random probability of being removed from the data matrix, which was considered to be due to natural mortality or emigration. For juveniles the probability of random mortality was fixed. For adults, natural mortality was dependent on population size. The minimum probability of being removed was 0.3, a value comparable to the intrinsic mortality. If the population rose above the preset carrying capacity, however, the probability of natural mortality was equal to one minus the ratio of carrying capacity to population size, or $1 - (k/n)$.

The *AGING* subroutine simply incremented the age of each individual by one unit. After this routine the data for population size and genetic variation were checked. If the genetic variance was greater than 90% of the original value, then another iteration of the model was performed; however, if at least 10% of the original genetic variation had been lost, then data regarding the average population size over all iterations, final population size, and number of generation intervals performed were output and the program ceased operation.

The model was run a total of 5040 times: 7 critical levels for toxin-induced mortality, 7 critical levels for toxin-induced infertility, 3 load distributions, 2 carrying capacity levels (100 and 200), and 30 replicates of each set of conditions. The same seven critical levels were used for fertility and mortality, and therefore a full factorial design could not be used. For example, it does not make biological sense to have the critical value for mortality be less than the critical level for infertility; death likewise invokes infertility. This is why the 7×7 design yields 28 ($[7^2 + 7]/2$) cells rather than the full 49. Analysis of variance was then performed on both iteration number and average population size, with toxin distribution, mortality critical levels, carrying capacity, and fertility critical levels as factors.

RESULTS

The results for the full ANOVA model are presented in Table 1. Carrying capacity, load distribution, and fertility were significantly associated with the maintenance of genetic variation and population size. Mortality, however, was relatively unimportant as a factor to either genetic variation or population size. The differences between mortality and fertility effects are emphasized also when the analyses are performed for distribution and carrying capacity separately (Table 2). Mortality had a consistently greater effect on genetic variation and population size when the carrying capacity was at the lower level. It is expected that mortality effects would become a significant factor

Table 1. Results of Analysis of Variance on the Full Model from Computer Simulations

Source(s)	Genetic Variance	Population Size
Carrying capacity (*k*)	***	***
Load distribution (*D*)	***	***
$k \times D$	***	***
Mortality (*M*)	ns	ns
$k \times M$	*	ns
$D \times M$	ns	ns
$k \times D \times M$	ns	ns
Fertility (*F*)	***	***
$k \times F$	ns	***
$D \times F$	**	***
$k \times D \times F$	ns	***
$M \times F$	ns	ns
$k \times M \times F$	ns	ns
$D \times M \times F$	ns	ns
$k \times D \times M \times F$	ns	ns

Note: Carrying capacities were either 100 or 200; load distributions were constant, Poisson, and binomial; and toxic effects were manifested as either fertility reductions or mortality. Asterisks indicate significant effects on either the maintenance of genetic variance or population size: * = $p<0.05$, ** = $p<0.01$, *** = $p<0.001$, ns = not significant.

Table 2. Separate ANOVA Results from the Computer Simulation Models of Toxin Accumulation via Mortality or Fertility Reduction

Dependent Variable	Mortality	Fertility	Mortality × Fertility
Constant distribution			
$k = 100$			
Genetic variance	ns	**	ns
Population size	ns	***	ns
$k = 200$			
Genetic variance	ns	***	ns
Population size	ns	***	ns
Poisson distribution			
$k = 100$			
Genetic variance	*	***	ns
Population size	ns	***	ns
$k = 200$			
Genetic variance	ns	***	ns
Population size	ns	***	ns
Binomial distribution			
$k = 100$			
Genetic variance	ns	***	ns
Population size	ns	***	ns
$k = 200$			
Genetic variance	ns	***	ns
Population size	ns	***	ns

Note: Carrying capacity is indicated by *k*. Asterisks indicate significance as in Table 1.

when carrying capacities are very low (<50), but most populations at this value became extinct in our models. Interestingly, maintenance of genetic variance and population size were affected differently by the independent variables and their interactions (Table 1). Thus, genetic variance retention is not a simple function of population size; rather, maintenance of genetic variance is strongly influenced by reproductive rate (numbers of progeny). In our models highest mean population size and longest retention time for genetic variance were evident for the constant load distribution, followed by the Poisson, and finally the binomial. When carrying capacity was 100, the maintenance time for genetic variation for the constant and Poisson distributions were not significantly different from one another ($p > 0.05$). All other comparisons showed significant differences; obviously, values for population size and gene diversity were significantly higher when carrying capacity was 200 rather than 100.

TOXICANT IMPACT AND POPULATION RECOVERY

The computer simulations documented that both population size and maintenance of genetic variation were affected by pesticide-induced fertility reduction and, to a lesser extent, mortality. The lack of concordance between fertility and mortality effects suggests that rates of recovery of populations may be important to either ameliorating (fast recovery) or exacerbating (slow recovery) the toxic impacts. The simulations, however, did not directly assess the rates of change but rather the ultimate values of population size and genetic variance. To assess the impact of lethal or sublethal toxicants on population recovery rates we next employed a series of mathematical models. Although a mathematical approach is more definitive regarding the factors responsible for population changes, several simplifying assumptions are necessary. We first assume that the avian species conform to the constraints of logistic growth.[22] The change in population size over time for the logistic model is expressed by the differential equation

$$\frac{\partial n}{\partial t} = r\,n \left[\frac{k - n}{k}\right] \tag{1}$$

where n is population size, t is time or interval between breeding periods, r is intrinsic rate of increase (unconstrained rate of growth), and k is carrying capacity or maximal value of n for a given environment. Because k represents the asymptotic or equilibrial population size, the expression $(k - n)/k$ denotes the potential for growth. Obviously, for any positive values of r and n

$$\lim_{t \to \infty} [n] = k \tag{2}$$

The logistic equation, Equation 1, is sufficient only for determining *acute* effects of toxicants on population size and rates of recovery. The integral form

$$\int \frac{k}{n\,(k - n)}\,dn = \int r\,dn \tag{3}$$

is resolved as

$$\ln\left[\frac{n}{k - n}\right] = rt + C \tag{4}$$

where C is a constant of integration and ln denotes the natural logarithm.

We also assume that before toxicant exposure the population size was at carrying capacity ($n = k$). An acute exposure to the toxicant at $t = 0$ results in a reduction of the population by a proportion x such that $n_0 = xk$. Therefore, in the initial generation

$$\ln\left[\frac{xk}{k - xk}\right] = rt + C \text{ (and because } t = 0)$$

$$C = \ln\left[\frac{x}{1 - x}\right] \tag{5}$$

Substituting the result of Equation 5 for C in Equation 4 we determine

$$rt = \ln\left[\frac{(1 - x)n}{x\,(k - n)}\right] \tag{6}$$

and the population at any time t can be determined by taking the derivative of Equation 6 with respect to n

$$n_t = \frac{k}{1 + e^{C - rt}} = \frac{k}{1 + \dfrac{x\,e^{-rt}}{1 - x}} = \frac{k\,(1 - x)}{1 - x\,(1 - e^{-rt})} \tag{7}$$

The time necessary for a population to recover some percentage, P, of its original population size (k) following an acute toxicant exposure is determined from Equation 6 as

$$rt = \ln\left[\frac{(1 - x)\,Pk}{x\,(k - Pk)}\right] = \ln\left[\frac{P\,(1 - x)}{x\,(1 - P)}\right] \tag{8}$$

and finally

$$t = \frac{1}{r}\ln\left[\frac{P\,(1 - x)}{x\,(1 - P)}\right] \tag{9}$$

In the above expressions the population size for a given generation is carried forward to the subsequent generation to determine the population growth rate. Thus, as emphasized above, those equations pertain only to solitary exposures to toxicants. We shall consider chronic exposures as repetitive acute exposures, as would be the case when pesticides are applied at the same time each year or breeding cycle. Chronic exposures potentially affect the population size for each generation via mortality or reduced fecundity of exposed animals. Therefore, we must consider toxicant exposures to reduce population size on a continuous basis. The potential chronic impacts are considered to affect (1) adult and juvenile mortality (postmating), (2) adult mortality (premating), and (3) fecundity reduction.

Exposure of adults and juveniles to toxicants is assumed to reduce the population size by a constant proportion, M. Thus, the logistic equation for population growth rate with chronic, postmating mortality becomes

$$\frac{\partial n}{\partial t} = r\,n\left[\frac{k - n}{k}\right] - Mn \tag{10}$$

Rearranging and integrating this expression yields

$$\int \frac{k}{nk\,(r - M) - rn^2}\,dn = \int dn \tag{11}$$

and therefore,

$$\frac{k \ln\,[kn\,(r - M) - nr]}{k\,(r - M) - r} = t + C \tag{12}$$

Assuming that the population size prior to toxicant exposure was at carrying capacity (k), the initial population size (n_0) is equal to $k(1 - M)$. The constant of integration is resolved then as

$$C = \frac{k \ln\,[k(1 - M)(k\,(r - M) - r)]}{k\,(r - M) - r} \tag{13}$$

Upon substitution of Equation 13 into 12,

$$t = \frac{k}{k\,(r - M) - r} \ln \left[\frac{kn\,(r - M) - nr}{k(1 - M)(k\,(r - M) - \mathrm{r})}\right] \tag{14}$$

To determine the time necessary for the population to recover a proportion, P, of its original size (k), the variable n in Equation 14 is replaced by Pk, giving

$$t = \frac{k}{k\,(r - M) - r} \ln \left[\frac{kPk\,(r - M) - Pkr}{k\,(1 - M)(k\,(r - M) - r)}\right] = \frac{k \ln \left[\dfrac{P}{1 - M}\right]}{k\,(r - M) - r} \tag{15}$$

When $M \geq r$, a positive value for recovery time t can be obtained only if $P < 1 - M$, indicating a declining population size. Under this condition, ultimate extinction of the population will result (Figure 1). The equilibrium population size, n_E, will be evident when the derivative of n with respect to t is 0; thus, $n_E = k(1 - M/r)$. When postmating toxicant mortality is less than the intrinsic growth rate, recovery of the population is expected (Figure 1).

To assess premating adult mortality due to toxicant exposure we will assume that a proportion M of the adults are removed just prior to mating. Hence, the logistic model for premating adult mortality is

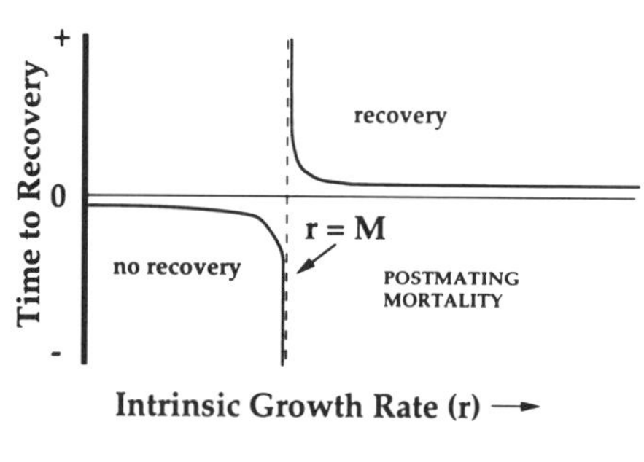

FIGURE 1. Graphic depiction of the time to 90% recovery of the original population size for populations with different intrinsic growth rates repetitively exposed to pesticides. In this example, pesticides induce a constant proportion of mortality (M) on adults and progeny (postmating effects). Negative recovery rates are indicative of eventual extinction of the population.

$$\frac{\partial n}{\partial t} = r\,n\,(1 - M)\left[\frac{k - n\,(1 - M)}{k}\right] \tag{16}$$

Note that toxicant effects are manifested before the reproductive events. Integrating Equation 16 produces

$$\int \frac{k}{rn\,(1 - M)[k - n\,(1 - M)]}\,dn = \int dn \tag{17}$$

Assigning $n_0 = k(1 - M)$ at $t = 0$, the constant of integration is resolved as

$$C = \frac{\ln\left[\frac{1 - M}{1 - (1 - M)^2}\right]}{r\,(1 - M)} \tag{18}$$

and t becomes

$$t = \frac{\ln\left[\frac{n}{k - n\,(1 - M)}\right]}{r\,(1 - M)} - \frac{\ln\left[\frac{1 - M}{1 - (1 - M)^2}\right]}{r\,(1 - M)} \tag{19}$$

The time necessary to recover a percentage, P, of the original population size is determined by assigning $n = Pk$

$$t = \frac{\ln\left[\frac{P[1 - (1 - M)^2]}{(1 - M)[1 - P\,(1 - M)]}\right]}{r\,(1 - M)} \tag{20}$$

Premating toxicant mortality does not suffer from the constraint that M must be less than r to ensure population recovery, as was evident in the postmating model. In fact, recovery of the population should occur even with rather substantial toxicant mortality; however, the time required for such recovery will be extended with increasing values of M (Figure 2).

Finally, we shall assess the potential impact of toxicants on fertility. In this model toxicant exposure will be considered to reduce the number of reproductive individuals by a percentage, R. However, because mortality is not induced by the toxicant, the potential for population growth is not affected. The logistic equation for fertility effects is

$$\frac{\partial n}{\partial t} = r\,(1 - R)n\left[\frac{k - n}{k}\right] \tag{21}$$

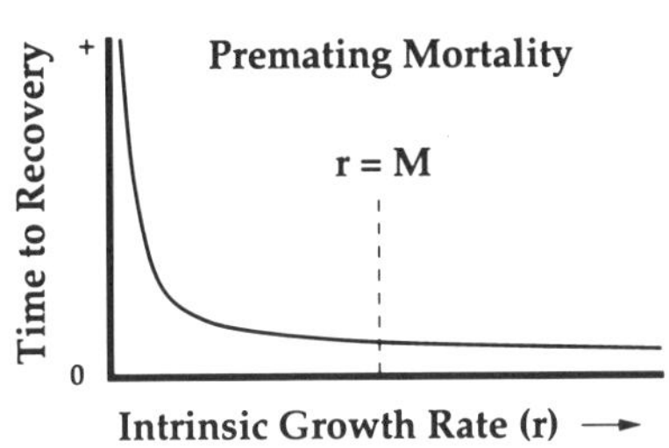

FIGURE 2. Graphic depiction of the time to 90% recovery of the original population size for populations with different intrinsic growth rates repetitively exposed to pesticides. In this example, pesticides cause mortality of adults only (premating effects).

The integral

$$\int \frac{k}{n\,(k - n)(1 - R)}\,dn = \int r\,dn \tag{22}$$

is resolved as

$$r\,t + C = \frac{1}{1 - R}\ln\left[\frac{n}{k - n}\right] \tag{23}$$

It would be trivial to assume that the population prior to toxicant exposure is at carrying capacity because the potential for population growth at that point is zero. For this situation only two scenarios are possible: (1) no change in the population results if $R < 1$; or (2) extinction of the population occurs if $R \geq 1$. Assuming that the initial population, n_0, is a percentage, X, of the carrying capacity ($n_0 = Xk$), the constant of integration is determined as

$$C = \frac{\ln\left[\dfrac{X}{1 - X}\right]}{1 - R} \tag{24}$$

The time necessary for recovery of a percentage, P, of the carrying capacity is determined by substituting Equation 24 into 23 giving

$$t = \frac{\ln\left[\dfrac{P(1 - X)}{X(1 - P)}\right]}{r\,(1 - R)} \tag{25}$$

Obviously, population extinction (or infinite recovery time) would result only if $R = 1$ and/or $X = 0$; otherwise relatively rapid recovery of populations will be feasible for a broad array of toxicant impact. The response curve for fertility reduction is very similar to that for premating mortality (Figure 2), but recovery times are slightly slower.

It is clear from the different recovery models that the timing and mode of impact of toxicant exposures can have very different results on the time necessary to restore population sizes. If the initial population size, n_0, was identical for the three models (using $n_0 = Xk$) and the magnitudes of toxic effects were the same ($M_{Post} = M_{Pre} = R_{Fert}$), then the recovery times for the three models are (from Equations 15, 20, and 25)

$$t_{Post} = \frac{k\ln\left[\dfrac{P}{X}\right]}{k\,(r - M) - r}$$

$$t_{Pre} = \frac{\ln\left[\dfrac{P[1 - X(1 - M)]}{X[1 - P(1 - M)]}\right]}{r\,(1 - M)}$$

$$t_{Fert} = \frac{\ln\left[\dfrac{P(1 - X)}{X(1 - P)}\right]}{r\,(1 - R)} \tag{26}$$

Numerical comparisons of these models demonstrate that the recovery times are expected to be shortest for premating effects, with fertility reduction having slightly slower times to population restoration (Figure 3). Chronic toxic exposures that affect

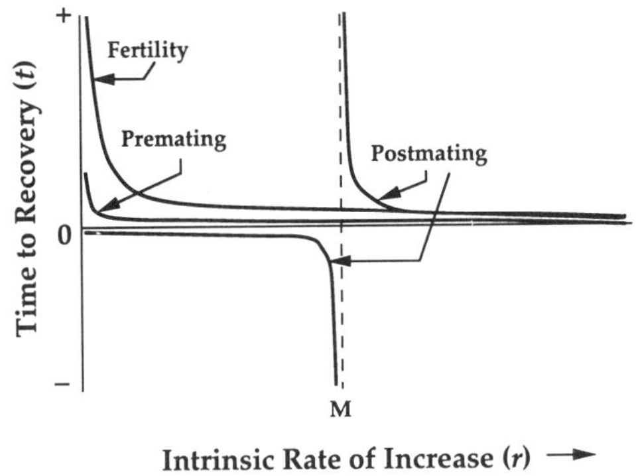

FIGURE 3. General comparisons of times to 90% recovery of the original population sizes for populations with different intrinsic growth rates. Comparisons are for toxic effects on premating and postmating mortality and for reduction in fertility. In this example, the magnitudes of effects for models were considered equal.

the mortality of adults and juveniles (postmating) will permit population recovery only when $M < r$, but rates of recovery are usually slower than for the other two models.

GENETIC VARIANCE PARTITIONING IN AVIAN SPECIES

To model rates of loss of gene diversity from avian populations it is necessary to determine whether there is significant genetic structure at the intrapopulational level. If the assumption of random mating within populations cannot be applied to avian species, models of rates of loss of gene diversity can become very complex, depending on whether the toxin affects the population subunits equally. Since their inception, fixation indices classically have been used to assess genetic partitioning among population units.[23-27] The three major components are F_{ST}, the proportion of the extant genetic variation that is found among the populations or groups; F_{IS}, the correlation of genes of individuals relative to the genes found within groups; and F_{IT}, the correlation of genes within individuals relative to those of the total array of groups. Cockerham's[26,27] modifications of genetic variance components were

$$F_{ST} = \frac{\Theta - \alpha}{1 - \alpha} \qquad F_{IS} = \frac{F - \Theta}{1 - \Theta} \qquad F_{IT} = \frac{F - \alpha}{1 - \alpha} \tag{27}$$

where Θ is the coancestry (correlation) of genes within groups, α is the correlation of genes of random individuals in the total population, and F is the average inbreeding coefficient, or the correlation of genes within individuals. Cockerham[26] considered α to be zero and showed[27] that it can usually be considered zero; therefore, the variance components can be simplified to

$$F_{ST} = \Theta \qquad F_{IS} = \frac{F - \Theta}{1 - \Theta} \qquad F_{IT} = F \tag{28}$$

The quantities for Θ and F are determined by the breeding and dispersal tactics within and among the breeding groups of a population.

Greenwood's[28] review of the dispersal and mating tactics of avian and mammalian taxa concluded that for birds, female dispersal was predominant while male philopatry was common, and polygyny was relatively rare in birds, with monogamy being prevalent. Mammal mating systems, on the other hand, typically involved female philopatry, male-biased dispersal, and predominant polygyny. Obviously, many exceptions exist and the relative magnitude of such strategies varies among taxa. We assume that dispersal is sufficient to prevent inbreeding; thus, the genetic variance components, F_{ST} and F_{IS}, are simple functions of Θ.

To illustrate the potential impact of social structure, consider a single population comprising a very large number of breeding groups (s), and n females within each breeding group. Further, assume that progeny replace their parents and therefore generations are discrete. The coefficient of coancestry (Θ) between any pair of individuals, x and y, is[29]

$$\Theta_{x,y} = \frac{1}{4}\left|\Theta_{Sx,Sy} + \Theta_{Sx,Dy} + \Theta_{Dx,Sy} + \Theta_{Dx,Sy}\right| \tag{29}$$

where S and D refer to the sire and dam, respectively, for individuals x and y. All individuals in the initial generation are assumed unrelated (all $\Theta_{x,y} = 0$). The coancestry of two parents is equal to the inbreeding coefficient (F) of their offspring, and the coancestry of an individual to itself is $(1 + F)/2$.[29,30]

Monogamy is assumed for most avian taxa with females dispersing among the breeding groups. Males shall be assumed to inherit their father's territory within a breeding group. For any arbitrary breeding group with n breeding females the coancestry among progeny resulting from n monogamous matings is

$$\Theta_{progeny} = \frac{1}{4}\left[\frac{1}{2}\,\frac{1}{n} + 0 + 0 + \frac{1}{2}\,\frac{1}{n}\right] = \frac{1}{4n} \tag{30}$$

Thus, Equation 30 represents the average coancestry within groups after mating, yet before female dispersal. Again, assuming a large number of breeding groups, after random dispersal by females to find mates in other breeding groups, the average coancestry within groups is zero. Therefore, with monogamy and random female movement, male philopatry does not promote any effective genetic variance partitioning among breeding groups. This model is equivalent to panmixia, and breeding group identification is admittedly arbitrary.

If a single male is now considered to mate with the n females within a breeding group (complete polygyny), the coancestry of progeny born within groups is

$$\begin{aligned}\Theta_{Progeny} &= \frac{1}{n}\,\frac{1}{4}\left[\frac{1}{2} + 0 + 0 + \frac{1}{2}\right] + \frac{n-1}{n}\,\frac{1}{4}\left[\frac{1}{2} + 0 + 0 + 0\right] \\ &= \frac{1}{4n} + \frac{n-1}{8n} = \frac{n+1}{8n}\end{aligned} \tag{31}$$

because the frequency is $1/n$ for full siblings and $(n - 1)/n$ for half siblings (identical fathers but different mothers). After female dispersal the coancestry of breeding individuals within groups returns to zero. Male polygyny with female dispersal does increase gene correlation among progeny born within groups but does not promote continuity of gene correlations among adults within groups.

The resultant within-group coancestry for typical mammalian breeding scenarios may differ substantially from that of birds. If females are philopatric and males are polygynous and disperse randomly among breeding groups, then in the initial generation the coancestry among progeny within groups will be

$$\Theta_{Progeny} = \frac{1}{4n}\left[\frac{1}{2} + 0 + 0 + \frac{1}{2}\right] + \frac{n-1}{4n}\left[\frac{1}{2} + 0 + 0 + 0\right] = \frac{n+1}{8n} \qquad (32)$$

which is identical to that for progeny born in the previous model. However, subsequent to male dispersal the coancestry among breeders for the next generation is

$$\Theta_{Breeders} = \frac{\frac{n^2 - n}{2}\,\frac{1}{8}}{\frac{(n+1)^2 - (n+1)}{2}} = \frac{n-1}{8(n+1)} \qquad (33)$$

(coancestry of females is 1/8, males are unrelated to females). Because the females remaining within the group are related, there is now a continued correlation of breeders within the mating groups. The coancestry of females within breeding groups becomes

$$\Theta_{D,D(t)} = \frac{1}{4}\left[\frac{1}{2} + 0 + 0 + \Theta_{D,D\,(t-1)}\right] = \frac{1}{8} + \frac{\Theta_{D,\,D\,(t-1)}}{4} \qquad (34)$$

which is a first order difference equation resolved as

$$\Theta_{\mathrm{D,D}(t)} = \frac{\Theta_{D,D\,(0)}}{8} + \frac{1}{8}\,\frac{1 - \frac{1}{4^t}}{1 - \frac{1}{4}} = \frac{1 - \frac{1}{4^t}}{6} \qquad (35)$$

The numerator of Equation 35 quickly approaches unity resulting in an asymptotic coancestry of females within lineages of 1/6. Using this result, a general equation to replace Equation 32 yields the asymptotic coancestry among progeny within groups

$$\Theta_{Progeny} = \frac{1}{4n}\left[\frac{1}{2} + 0 + 0 + \frac{1}{2}\right] + \frac{n-1}{4n}\left[\frac{1}{2} + 0 + 0 + \Theta_{D,D}\right] = \qquad (36)$$

$$= \frac{1}{4n} + \frac{4(n-1)}{24n} = \frac{2n-1}{12n}$$

Clearly, typical mammalian breeding and dispersal tactics may promote considerable genetic variance partitioning within populations for both adult and offspring generations. Most scenarios for avian taxa, however, will result in appreciable values for progeny only. Because little variance is maintained among subunits of avian populations, models of gene diversity can be based on the classic assumption of random breedings within entire populations.

GENETIC VARIABILITY AND POPULATION RECOVERY

The relationship between loss of genetic variation and finite population size has been recognized for many years.[23] The random loss of genetic variance (genetic drift) occurs due to the finite number of independent mates available. The concept of effective population size (N_e) was developed to assess the number of effective gametes and to determine the rates of loss of genetic variation within populations.[31,32] These concepts are easily applied to populations impacted by toxicants.

In the following models n will refer to the total population size, and males and females will be assumed to contribute equally to the progeny born each breeding period. Sex ratio is assumed to be 1:1. Loss of gene diversity will be considered for the scenarios of death of adults and segregation during mating. The mortality of adults per generation will be assumed to be a constant proportion, m, of the population size.

In a finite population of adults, the sampling variance (V_p) for allele frequency is

$$V_p = \frac{p(1 - p)}{2n} \tag{37}$$

where p is the average allele frequency for a genetic locus.[32] It is important to note that if there is no adult mortality there is no net change in the genetic variance in adults. With percent adult mortality (m) and assigning $V_0 = p_0(1 - p_0)$[26] the change in variance is

$$\begin{aligned} V_0 - V_t &= V_0 \left[\frac{1}{2n} - \frac{1}{2n(1 - m)}\right] \\ \frac{\partial V}{\partial t} &= -\frac{V}{2n}\left[\frac{m}{1 - m}\right] \end{aligned} \tag{38}$$

Thus, the rate of loss in genetic variance in the adults is the ratio of the variance times adult mortality and the number of gametes remaining.

The effective population size is approximately

$$N_e \approx \frac{2nb}{1 + \frac{\sigma_b^2}{2b}} \tag{39}$$

where b is the birth rate ($2b$ = the average number of progeny per breeding pair) and σ_b^2 is the variance in offspring number.[32] If we assume that the variance in progeny is distributed as Poisson,[32] then

$$\begin{aligned} \sigma_b^2 &= 2b \\ N_e &\approx nb \end{aligned} \tag{40}$$

and the genetic variance represented in progeny can be represented by the recursive equation

$$V_t = V_{t-1}\left(1 - \frac{1}{2nb}\right) \tag{41}$$

The rate of change in genetic variance from adult to offspring is

$$\frac{\partial V}{\partial t} = -\frac{V}{2nb} \tag{42}$$

The total change in genetic variance over time is determined from the weighted average of the differentials in Equations 40 and 42. Therefore, the rate of change of genetic variance is

$$\frac{\partial V}{\partial t} = -\frac{Vm}{2n(1 - m)}\frac{n(1 - m)}{n} - \frac{V}{2nb}\frac{nb}{n} = -\frac{Vm - V}{2n} = -\frac{V(1 + m)}{2n} \tag{43}$$

Importantly, the population size (n) and the genetic variance (V) change through time. The integrations of the logistic growth curves in the previous section were performed to determine the time necessary to recover a proportion of the maximal population. The integrals also can be used to calculate the size of the population at any time t. It can be shown that for the models of chronic effects

$$n_t^{(Post)} = \frac{k\,(r - M)\,n_0}{r\,n_0 + [k\,(r - M) + r\,n_0]\,e^{t\,(r - M)}}$$

$$n_t^{(Pre)} = \frac{k}{1 - M + \dfrac{k - n_0\,(1 - M)}{n_0}\,e^{-r\,t\,(1 - M)}}$$

$$n_t^{(Fert)} = \frac{k}{1 + \dfrac{k - n_0}{n_0}\,e^{-r\,t(1 - R)}} \quad (44)$$

Numerical solutions to the genetic variance associated with chronic toxicant exposure for postmating, premating, and fertility reduction for different life history scenarios are depicted in Figure 4 using the recursive equation

$$V_t = V_{t-1}\left[1 - \frac{1 + m}{2n_t}\right] \quad (45)$$

While loss of genetic variation was evident in all models, the rate of loss was slowest for the premating adult mortality model. Very slow recovery rates, such as in the postmating models, were typified by substantial losses in genetic variation.

DISCUSSION

Of the unintentional poisonings in North America, birds are the most frequently reported wildlife group.[17] As indicator species, birds represent a variety of feeding guilds[33] subject to the effects of agricultural pesticides. As granivores, insectivores, and herbivores, birds potentially provide a sensitive monitoring tool in agricultural habitats because they are generally more sensitive to delayed neurotoxicity than other vertebrates.[34] Young birds appear more sensitive to OPs than adults, but incidents of poisoning are rare, perhaps because nests are frequently away from direct contamination. Long-term effects of OP-related mortality on wildlife populations are not known. No known permanent declines in overall populations have been attributed to their use. Raptors (low fecundity species) may be more sensitive. Body size and physiological condition also may affect susceptibility to pesticides. Small birds may be more suscepti-

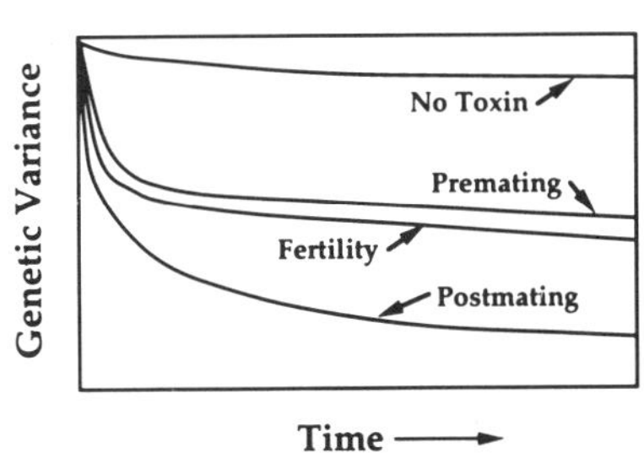

FIGURE 4. Graphic depictions of the loss of genetic variability for conditions of no toxin, premating mortality, fertility reduction, and postmating mortality.

ble due to higher metabolic rates and greater quantity of food ingested per unit body mass. Physiological condition may influence post-molt individuals as animals with low fat reserves may be more vulnerable. If this is true, then we may also expect that migratory birds, upon their return to summer or winter range, would be particularly vulnerable to the influence of pesticides.

Although OPs have not been shown to have chronic effects on avian taxa, our models demonstrate that chronic impact is not necessary to influence population size and gene diversity adversely. The logistic models did not consider the accumulation of toxicants; rather, the effects were manifested as repeated acute effects, with repetitions transpiring at particular stages of the avian life cycle. Therefore, even chemicals that have relatively short duration in the environment may have substantial negative impact on the population dynamics and genetic variation when populations are repeatedly exposed.

The models presented in this chapter further demonstrate that the potential impact of repeated pesticide applications on avian population size and gene diversity may be substantial. The magnitude of impact is dependent on the application time schedules, mode of action of the pesticide on fecundity and/or mortality, the toxicity of the chemical, and the distribution of individuals affected. Toxic compounds that affect fertility, either directly or indirectly, are expected to have greater impact on the rates of recovery and ultimate size of bird populations. Loss of genetic variation within impacted populations will be greatest when pesticide exposure either reduces fertility of adults or causes mortality of progeny. Rapid recovery rates of population size will ameliorate loss of gene diversity because for most models loss of genetic variance was greatest in the early generations.

Repeated pesticide applications are likely to coincide with the postmating periods of our models, i.e., brood production and pesticide use usually occur at times of peak insect abundance. Therefore, the consequences of pesticide-induced mortality will likely affect adults as well as nestlings. Reduction in prey resources may also have an adverse effect on viability of progeny.[21] Both the mathematical and computer simulation models demonstrated that such application schedules (reducing realized fertility) will have the greatest impact on population recovery rates, population size, and loss of genetic variation. If the toxic effects persist at levels greater than the intrinsic growth rate of the population, then recovery will be impossible.

Breeding tactics in avian species will have minimal effects on intrapopulational genetic variance partitioning. Predominant female dispersal in birds prohibits predictable genetic differentiation among breeding groups within a population. Although relatedness among offspring will result in within-group gene correlations, such effects are dissipated upon subsequent dispersal. In contrast, female philopatry will result in a maintenance of gene correlations within breeding groups subsequent to male dispersal. Genetic differentiation among breeding groups has been reported for many mammalian taxa that have socially organized populations and female philopatry.[35-38] Genetic subdivision produces an extra level of complexity not included in these models. The models do not imply that genetic subdivision will not be present in avian populations; rather, that the genetic structure usually will be due to factors other than sex-biased dispersal and breeding tactics.

Clearly, factors controlling the population dynamics and genetic variation in avian taxa are more complex than those considered in our models. We have regarded populations as relatively isolated and genetic characters as essentially neutral. Dispersal among populations, differing habitat topology, and natural selection for alleles and gene combinations may alter the results reported here. Our models, however, demonstrate the sensitivity of the genetics and sizes of bird populations to repeated toxic applications and indicate that care should be taken in the design of pesticide applications to minimize detrimental impact. Future studies should be designed to include specific population characteristics of target species. Also, the quantification of many variables used in our models is poorly understood. Toxicological assessment of chemical applications on gene diversity in natural populations is still in its infancy, and vigorous investigations are needed to predict population and genetic responses accurately.

ACKNOWLEDGMENTS

This work was conducted under contract DE-AC09-76SR00-819 between the U.S. Department of Energy and the University of Georgia.

REFERENCES

1. Fleming, W. J. and D. R. Clark, Jr., Organochlorine pesticides and PCBs: a continuing problem for the 1980s, *Trans. North Am. Wildl. Nat. Res. Conf.*, 48, 186-199, 1983.
2. Kendall, R. J., Wildlife toxicology: a reflection on the past and the challenge of the future, *Environ. Toxicol. Chem.*, 7, 337-338, 1988.
3. Rudd, R. L., *Pesticides and the Living Landscape*, University of Wisconsin Press, Madison, 1964.
4. Rudd, R. L., and R. E. Genelly, Avian mortality from DDT in California rice fields, *Condor*, 57, 117-118, 1955.
5. Hunt, E. G., and A. I. Bischoff, Inimical effects on wildlife of periodic DDD applications to Clear Lake, Calif., *Fish Game*, 46, 91-106, 1960.
6. Keith, J. A., Insecticide contaminants in wetland habitats and their effects on fish-eating birds, *J. Appl. Ecol.*, 3, 71-85, 1966.
7. Risebrough, R.W., Pesticides and bird populations, in *Current Ornithology*, Vol. 3, R. F. Johnston, Ed., Plenum Press, London, 1983.
8. Blus, L. J., C. J. Henny, T. E. Kaiser and R. A. Grove, Effects on wildlife from use of endrin in Washington state orchards, *Trans. North Am. Wildl. Nat. Res. Conf.*, 48, 159-174, 1983.
9. Bedford, C. T., R. K. Harrod, E. C. Hoadley and D. H. Hutson, The metabolic fate of endrin in the rabbit, *Xenobiotica*, 5, 485-500, 1975.
10. Matthews, H. B., Excretion of insecticides, *Pharmacol. Ther.*, 4, 657-675, 1979.
11. Stickel, L. F., W. H. Stickel, R. A. Dyrland and D. L. Hughes, DDE in birds: lethal residues and loss rates, *Arch. Environ. Contam. Toxicol.*, 13, 1-6, 1984.
12. Bailey, S., P. J. Bunyan, B. D. Rennisen and A. Taylor, The metabolism of 1,1,-di(p-chlorophenyl)-2,2-dicholoroethylene and 1,1-di(p-chlorophenyl)-2-chloroethylene in the pigeon, *Toxicol. Appl. Pharmacol.*, 14, 23-32, 1969.
13. Stickel, L. F., W. H. Stickel, R. D. McArthur and D. L. Hughes, Chlordane in birds. A study of lethal residues and loss rates, in *Toxicology and Occupational Medicine*, W. A. Deichmann, organizer, Elsevier, New York, 1979, pp. 387-396.
14. Fox, G. A., A. P. Gilman, D. B. Peakall and F. W. Anderka, Behavioral abnormalities of nesting Lake Ontario Herring Gulls, *J. Wildl. Manage.*, 42, 477-483, 1978.
15. McArthur, M. L. B., G. A. Fox, D. B. Peakall and B. J. R. Philogene, Ecological significance of behavioral and hormonal abnormalities in breeding ring doves fed an organochlorine chemical mixture, *Arch. Environ. Contam. Toxicol.*, 12, 324-353, 1983.
16. Koeman, J. H., Chemicals in the environment and their effects on ecosystems, in *Advances in Pesticide Science*, H. Geisshuhler, Ed., Pergamon Press, Oxford, 1972, pp. 25-38.
17. Grue, C. E., W. J. Fleming, D. G. Busby and E. F. Hill, Assessing hazards of organophosphate pesticides to wildlife, *Trans. North Am. Wildl. Nat. Res. Conf.*, 48, 200-220, 1983.

18. Costa, L. G., and S. D. Murphy, Passive avoidance retention in mice tolerant to the organophosphorus insecticide disulfoton, *Toxicol. Appl. Pharmacol.*, 65, 451-458, 1982.
19. Grue, C. E., Response of common grackles to dietary concentrations of four organophosphate pesticides, *Arch. Environ. Contam. Toxicol.*, 11, 617-626, 1982.
20. Stromborg, K. L., Reproductive tests of diazinon in bobwhite quail, in *Avian and Mammalian Wildlife Toxicology: Second Conference.* ASTM STP 757, D. W. Lamb and E. E. Kenaga, Eds., American Society for Testing and Materials, Philadelphia, 1981, pp. 19-30.
21. Potts, G. R., Population dynamics of the grey partridge: overall effects of herbicides and insecticides on chick survival rates, *Proc. Int. Congr. Game Biol.*, 13, 203-211, 1977.
22. Krebs, C. J., *Ecology: The Experimental Analysis of Distribution and Abundance,* 2nd ed., Harper & Row, New York, 1978, p. 186.
23. Wright, S., The genetical structure of populations, *Ann. Eugen.*, 15, 323-354, 1951.
24. Wright, S., *Evolution and the Genetics of Populations, Vol. 2, The Theory of Gene Frequencies*, University of Chicago Press, Chicago, 1969.
25. Wright, S., *Evolution and the Genetics of Populations, Vol. 4, Variability Within and Among Natural Populations*, University of Chicago Press, Chicago, 1978.
26. Cockerham, C. C., Variance of gene frequencies, *Evolution*, 23, 72-84, 1969.
27. Cockerham, C. C., Analysis of gene frequencies, *Genetics*, 74, 679-700, 1973.
28. Greenwood, P. J., Mating systems, philopatry and dispersal in birds and mammals, *Anim. Behav.*, 28, 1140-1162, 1980.
29. Jacquard, A., *The Genetic Structure of Populations*, Springer-Verlag, New York, 1974.
30. Falconer, D. S., *Introduction to Quantitative Genetics*, Longmans, New York, 1981.
31. Wright, S., Isolation by distance, *Genetics*, 28, 114-138, 1943.
32. Crow, J. F., and M. Kimura, *An Introduction to Population Genetics Theory*, Brugess Press, Minneapolis, MN, 1970.
33. Landres, P. B., Use of the guild concept in environmental impact assessment, *Environ. Manage.*, 7, 393-398, 1983.
34. Bursian, S. J., J. S. Brewster and R. K. Ringer, Differential sensitivity to the delayed neurotoxin tri-o-tolyl phosphate in several avian species, *J. Toxicol. Environ. Health*, 11, 907-916, 1983.
35. Chesser, R. K., Genetic variability within and among populations of the black-tailed prairie dog, *Evolution*, 40, 616-624, 1983.
36. Chesser, R. K., Gene diversity and female philopatry, *Genetics*, 127, 437-447, 1991.
37. Chesser, R. K., Influence of gene flow and breeding tactics on gene diversity within populations, *Genetics*, 129, 573-583, 1991.
38. Melnick, D. J., The genetic consequences of primate social organization: a review of macaques, baboons and vervet monkeys, *Genetica*, 73, 117-135, 1987.

19

Models of Structured Populations: Age and Mass Transition Matrices

Norman A. Slade

ABSTRACT

Models for studying toxicological impact on vertebrate populations should recognize demographic heterogeneities within populations. Simple matrix models incorporate class-specific probabilities of death and reproduction and allow differential impact on classes to be assessed for a net effect on the entire population. Models that partition the population into size classes also allow incorporation of nonlethal effects on somatic growth into models of population growth. The basic data to construct matrices of estimated vital statistics can be readily obtained with extant field methods. Projections of numbers from one census period to the next can be compared to observations to assess the accuracy of estimates and validity of assumptions. After a valid transition matrix has been constructed, the dominant eigenvalue is an estimate of the hypothetical multiplicative rate of population growth. Sensitivity analyses can identify which transitions are the most influential on the growth rates of a population, and the impact of small changes in class-specific birth and death rates can be predicted. Realistic models of population growth must incorporate environmental fluctuations, both seasonal and stochastic. The difference in predictions of deterministic and stochastic models depends on how far into the future populations are projected and the variance-covariance structure of vital rates. Estimation of the parameters necessary for realistic models of population growth is difficult, but a worthwhile goal for toxicological testing.

KEY WORDS

body size, age, structured populations, demography, growth, environmental stochasticity

INTRODUCTION

Much toxicological work has focused on identifying effects of chemical agents on physiology, behavior, survival, and reproduction in the laboratory. However, ultimately, regulatory agencies and others are interested in the impact on free-ranging populations. Thus, one goal of toxicological studies should be expressing results as the potential growth rate or probability of extinction of an exposed population. The actual impact will depend on myriad factors such as the physical environment, physiological condition of the animals, spatial pattern or degree of contagion, home range size, and migratory movements, as well as probabilities of reproduction and survival. This chapter focuses on the latter by providing an introduction to matrix representations of vital statistics and discussing recent developments in theory and practice relevant to toxicological studies of birds. Caswell[1] has recently published an excellent and far more comprehensive treatment of these matrix methods.

Most populations of higher vertebrates show demographic structure; i.e., they are composed of subgroups with differing probabilities of survival and reproductive success. To date most models of structured populations have treated birth and death rates as functions of age. Lotka[2] presented models for estimating instantaneous rates of birth and population growth from fecundity and survivorship curves that were continuous functions of age.

The simplest equation for population growth at discrete intervals is:

$$N_{t+1} = \lambda N_t \tag{1}$$

where N_t represents the number of animals in the population at time t and λ is the simple geometric growth multiplier.

AGE STRUCTURE

Lewis[3] and Leslie[4] independently developed matrix equations for projecting numbers of animals in discrete age classes from one time to the next. The Lewis-Leslie equation for this simple growth process is

$$\mathbf{n}_{t+1} = \mathbf{L}\mathbf{n}_t \tag{2}$$

where $\mathbf{n}_t$ is a vector of numbers of animals in each age class at time t and $\mathbf{L}$ is a matrix of age-specific fecundities and survivorships. When expanded, this is

$$\begin{bmatrix} n_1 \\ n_2 \\ n_3 \\ \cdot \\ \cdot \\ \cdot \\ n_k \end{bmatrix}_{t+1} = \begin{bmatrix} F_1 & F_2 & F_3 & \cdots & F_k \\ P_{21} & 0 & 0 & \cdots & 0 \\ 0 & P_{32} & 0 & \cdots & 0 \\ \cdot & \cdot & \cdot & \cdots & \cdot \\ \cdot & \cdot & \cdot & \cdots & \cdot \\ \cdot & \cdot & \cdot & \cdots & \cdot \\ 0 & 0 & 0 & P_{k,k-1} & 0 \end{bmatrix} \begin{bmatrix} n_1 \\ n_2 \\ n_3 \\ \cdot \\ \cdot \\ \cdot \\ n_k \end{bmatrix}_t \tag{3}$$

where n_i is the number of animals in age class i, F_i is the average number of offspring surviving to time $t+1$ produced by each animal in age class i at time t, and P_{ij} is the probability of an animal in age class j at time t surviving to time $t+1$. Probabilities of survival are straightforward and can be estimated from capture-recapture data, observations of marked animals, or band returns by many different methods.[5,6] Because they are projections from potential breeders at one time to hatchlings or fledglings at another, fecundity terms combine survival of the breeders through fledging, clutch size, viability of the eggs, and survival of the hatchlings until census. If, as is usually the case, populations are projected as numbers of females, the sex ratio of fledglings must also be considered. The precise method of combining all these factors into a single fecundity term depends on whether breeding is reasonably continuous (birth flow) or occurs in discrete bursts (birth pulse) and when the census is taken relative to the time of hatching.[1] Once the transition matrix, $\mathbf{L}$, has been estimated, projections into the future can be made from any observed vector, assuming the vital rates remain unchanged. For most feasible matrices describing avian populations, vectors of age distributions will converge to fixed ratios of numbers in each age class (stable age distributions). If a population is in a stable age distribution, the growth of the entire population and of any age class is given by Equation 1, and the dominant eigenvalue of **L** is the geometric growth constant, λ. The vector of numbers in a stable age distribution conforms to the general equation for a right eigenvector, **u**, associated with the eigenvalue, λ, by the equation

$$\mathbf{L}\mathbf{u} = \lambda\mathbf{u} \tag{4}$$

and so is an eigenvector of the matrix **L**. Any vector **v** satisfying the equation

$$\mathbf{v}'\mathbf{L} = \mathbf{v}'\lambda \tag{5}$$

is a left eigenvector and can be normalized to a vector of reproductive values[1] by dividing all elements in the vector by the first.

Because Leslie matrices have many elements equal to zero, eigenvalues and eigenvectors can be found with rather simple equations,[7] but many computer programs are limited to extracting eigenvalues from symmetric matrices and so will not analyze Leslie matrices. Caswell[1] lists several commonly available subroutines for eigenstructure of nonsymmetric matrices and recommends their use. Most packages yield only the right eigenvector of **L**, but the left eigenvector may be obtained as the eigenvector of the transpose of **L**. The calculations are quite easy, however, given access to a computer. As the matrix **L** is multiplied by itself m times, the product matrix, $\mathbf{L}^m$, will come to have rows that are proportional to one another, as will be the columns. The rows will converge on left eigenvectors (reproductive values), and the columns of $\mathbf{L}^m$ will become right eigenvectors (stable age distributions)[1]. The ratio of any single element of $\mathbf{L}^{m+1}$ to the corresponding element of $\mathbf{L}^m$ is the dominant eigenvalue. Matrix multiplication is available in many computer packages (e.g., SAS, MINITAB, IMSL), or a listing of a simple BASIC program to produce eigenvalues and eigenvectors of demographic matrices is available from the author.

SIZE STRUCTURE

Age is only one of several choices for state variables to describe structure in populations. Williamson[8] and Lefkovitch[9] generalized population transition matrices to allow any recognizable stage in a life history to be used as a state variable. In fact, the age categories of juvenile, subadult, and adult, in many field studies, are stages rather than age classes as required in Leslie matrices. Lefkovitch[9] applied stage-transition matrices to insect populations grouped into instars, and the techniques have been applied to plants grouped into size classes (see References 1 and 10 for reviews). Recently, Sauer and Slade[10–13] and Nichols[14] have advocated the application of size-based transition matrix models to terrestrial vertebrates. Substituting size classes or any other life history characteristics for age in the construction of transition matrices simply involves substituting a more general transition matrix, **A**, for the matrix **L** in Equation 2. The vector $\mathbf{n}_t$ still is the number of animals in the population, but they are stratified into size classes or other stages. The principal difference is that whereas surviving animals must advance one age interval with the Leslie matrix, they may advance, regress, or remain in the same stage in the more general case. Thus, the matrix **A** has the form

$$\mathbf{A} = \begin{bmatrix} F_1+P_{11} & F_2+P_{12} & F_3+P_{13} & \cdots & F_k+P_{1k} \\ P_{21} & P_{22} & P_{23} & \cdots & P_{2k} \\ P_{31} & P_{32} & P_{33} & \cdots & P_{3k} \\ \cdot & \cdot & \cdot & \cdots & \cdot \\ \cdot & \cdot & \cdot & \cdots & \cdot \\ \cdot & \cdot & \cdot & \cdots & \cdot \\ P_{k1} & P_{k2} & P_{k3} & \cdots & P_{kk} \end{bmatrix}$$

where P_{ij} is the probability that a surviving animal moves from stage j to stage i and F_i is the fecundity of animals in stage i, as before. Because survivors may be apportioned among several subsequent stages, stage-based matrices require estimation of more parameters than do most age-based matrices. Newly captured animals are more easily placed in stage classes than in age classes, however, so more data are available to estimate transition probabilities, especially in short-term studies. In **A**, all recruits to the population were assigned to the first stage (all fecundity terms were entered in the first row of **A**). It is also possible to assign recruits to more than one stage (e.g., small and large young[13]); in such instances, the fecundity terms must be divided among two or more classes. As with Leslie matrices, eigenvalues are geometric growth rates and eigenvectors are stable stage distributions and reproductive values.

Leslie matrix and related models have been called "bookkeeping" models[15] and were derived as a means of summarizing age-specific vital statistics,[1] but they can yield numerous insights into the dynamics of populations. They are models of closed population growth that yield projections of structured populations based on observed survival and reproductive rates. Comparing observed and projected population trajectories for single time intervals reveals the importance of immigration in determining population density. In field studies, it is difficult to separate losses from death and from emigration. Therefore, survival rates will generally be underestimated. Because both emigration and death are losses from the target population, however, projections may be relatively unaffected. If immigration is an appreciable demographic force, projections from **A** will consistently underestimate observed population densities.

Comparing projected numbers from **A** or **L** with projections from λ tests the demographic impact of departures from the stable age or stage distribution. Because populations are expected to reach stable stage distributions only if vital rates remain constant for several generations, departures from stable distributions indicate recent or frequent changes in structure; this is useful background information in planning toxicological experiments or in assessing impacts of treatments on populations. Discrepancies between observed numbers and projections over longer periods indicate cumulative departures from stable structure and population closure or from the assumption of fixed transition matrices. The assumption of time-invariant rates can be tested for individual elements of the transition matrix (for example, Sauer et al.[16]) or for entire matrices.[1]

Comparison of matrices representing different conditions, be they results of temporal variation or experimental intervention, fall under the heading of life table response experiments.[17] The variable of interest in these studies is usually the geometric growth rate, and Caswell[17] gives methods for decomposing experimental effects into differences in individual vital rates. Large differences in individual survivorships or fecundities may not have large impact on λ. Sensitivity analysis[1,18] is one of the most useful tools in identifying individual elements of **A** which are most influential on potential population growth rate. The impact of small changes in the transition from stage j to stage i, a_{ij}, is proportional to the product of the fraction of the population in stage j in the stable stage distribution, u_j, and the reproductive value (future reproductive output) of animals in stage j, v_i. If the stable stage vector, **u**, is multiplied by the vector of stage-specific reproductive values written as a row vector, $\mathbf{v}'$, the resulting matrix of sensitivities can be examined to identify critical stages and transitions. Alternatively, one could systematically vary each element of $\mathbf{A}$ and quantify variation in the dominant eigenvalue. Adding or subtracting a fixed amount from each element, in turn, and comparing eigenvalues will produce the same result as the sensitivity analysis of Caswell. However, survivorships are constrained between zero and one, whereas fecundity terms are frequently greater than one, so adding or subtracting a constant does not produce equal proportional changes in all elements.[19] There are few data on the form of variation in stage-specific vital rates, but it seems reasonable that they might vary in proportion to their magnitude. The relative impact of small proportional changes in elements of $\mathbf{A}$ can be compared by constructing a matrix of elasticities.[19] Sensitivities and elasticities are important in toxicological analyses because they identify critical stages and transitions in life histories, which if decreased would have the greatest consequence on subsequent population levels. In our analyses of mammalian life histories using size classes as stages,[11-13] growth rates of immature animals have been most influential, especially during the breeding seasons. Thus, chemicals that slow individual growth rates might have severe impact on population growth rates, even though they do not alter survival rates or size-specific fecundity rates substantially.

BODY MASS AS SIZE

Several choices can be made for measures of size, and Sauer and Slade[11,12] initially chose body mass because it was easy to measure without harming or anesthetizing living animals. In many vertebrates, both reproductive competence and probability of survival are functionally related to body mass.[1,10] In addition, the incorporation of somatic growth into transition matrices offers promise for integrating physiological models of energy or nutrient allocation into models of population growth. One can model the demographic impact of investing calories in reproductive tissue or maternal body mass. Thus, models of mass-based population dynamics are intermediate between statistical models, which incorporate statistical correlations among demographic rates and categorical variables, and process models, which incorporate biological processes that are causally linked to higher order phenomena.

Application of size- or mass-based matrix transition models to vertebrate populations is a relatively recent development, and the author knows of no application to avian populations. Nevertheless, there are numerous examples of the influence of body size on intraspecific variation in avian population dynamics.[10] Models incorporating body mass seem particularly well suited to avian toxicology because body size scales not only with metabolic rate and physiological parameters, but also diet composition, social dominance, habitat selection, and numerous other environmental factors.

Body mass is correlated with other measures of size,[20] but body mass, unlike measurements of body or appendage lengths, fluctuates with reproductive status, general body condition (e.g., fat stores), and availability of food. The fact that body mass varies more than other measures of size is an advantage in demographic studies, because such variation reflects general physiological status. For example, just before egg-laying, heavier maternal mass may identify more fecund females; increased body mass may reflect a larger investment in reproduction, either in more numerous or more massive eggs. Even among birds of equal skeletal size, heavier birds may be viewed as farther advanced in egg development and therefore more fecund because they are closer to laying. (Recall that fecundity involves the probability of maternal survival to laying.) Of course, a lighter bird, weighed immediately after laying, actually may be more likely to have living offspring in the next period than any female that had not yet laid, so fecundity is not a perfect correlate of size. Nothing in the use of body mass as a state variable in demographic studies requires that vital statistics increase or decrease as a monotonic function of size or that the relationship between size and vital rates remains constant through time or space; the only requirement is that fecundity or survival is related to mass at the time and place represented by the transition matrix. Thus, simple linear correlations between mass and vital rates support the hypothesis of mass specificity, but so do analyses of variance and other more general linear or log-linear models.

Variance of estimates of mass-specific survival and fecundity rates varies directly with variation among animals within classes and inversely with the sample size. Thus, choice of mass class intervals is a compromise between sample size and intraclass variation in vital statistics.[21,22] Wider intervals permit larger sample sizes within classes, hence more precise estimates of vital statistics, but increased intraclass variation in body mass will be accompanied by increased variation in survival and fecundity rates. Similarly, uncertainty in assigning animals to classes will decrease both the precision of intraclass estimates and the distinctions among classes. In contrast to seasonal and other longer term changes in body mass, variation in body mass associated with daily uptake and loss of food and water is a source of potential error in assigning birds to mass classes. Haramis et al.[23] and Johnson ct al.[24] suggested that daily variation in body mass of birds is a significant source of error in using body mass as a measure of size. In small mammals, coefficients of variation for body mass on a daily basis are on the order of 5%,[25] thus mass does not seem to be so variable as to preclude its use as a state variable in population dynamics. Daily variation is not as well quantified for birds as for mammals, but before adopting mass class intervals, it is important to quantify daily variation. Mass classes should be at least 4 to 6 S.D. in width[25] so birds in the middle of a class will be correctly classified.

VARIATION IN VITAL RATES

Simple, time-invariant, transition matrices are not realistic models of long-term population growth; no rate of geometric increase or decrease can be maintained indefinitely in a finite world. Nevertheless, eigenvalues of transition matrices are useful descriptors of sets of vital rates and can be reasonable estimates of short-term growth, especially for comparing and integrating the impacts of pesticides or other toxins on bird populations. By allowing, $\mathbf{A}$ or $\mathbf{L}$ to vary with time, the model of Equation 2 can be made quite general. Variation in $\mathbf{A}$ through time can be deterministic and incorporate seasonality,[12,13] population density,[1] or any other predictable pattern. Such variation is relatively easy to incorporate into matrix models.[1] An initial vector, $\mathbf{n}_o$, will lead to a single vector, $\mathbf{n}_t$, after t time units, following a fixed trajectory determined by the product of the matrices $\mathbf{A}_o$ through $\mathbf{A}_{t-1}$. Variation in vital rates can also result from random environmental fluctuations. Truly stochastic models of structured populations are more complex mathematically and heuristically,[1,26-28] but are also more realistic. Instead of projecting a single vector resulting from a given $\mathbf{n}_o$, models of environmental stochasticity yield a distribution of values of $\mathbf{n}_t$, which are determined by the distributions of sequences of $\mathbf{A}$ matrices. If each $\mathbf{A}$ is an independent observation drawn from a single statistical population, the distribution of total population size, N_t, will tend toward log normal as t increases. The expected value (arithmetic mean) of N_t, is given by the product of N_o and the t-th power of the eigenvalue of a matrix composed of expected values of each of the vital statistics. Lognormal distributions are skewed with a median and mode less than the mean. Because most life tables and transition matrices are reported as mean values of vital statistics, population projections and estimated growth rates generally overestimate actual growth rates of most populations following identical birth and death schedules.[1,29-32]

When projections of stochastic geometric growth, with the mean growth rate near one, are extended for many time intervals, the distribution of N_t includes an ever-diminishing proportion of populations projected to levels above the mean, and an increasing proportion approaching extinction.[29-32] If the expected value of the geometric growth rate is greater than unity, the expected value of N_t can approach infinity while the probability of extinction approaches one.[29,30] The degree to which any percentile of the distribution of N_t departs from the expected value increases with t and the variance of vital statistics or the population growth rate. Because probabilities of survival are restricted to the interval between zero and one, variances are likely to be small. With several age or size classes, independent variances will have a cumulative effect on the variance of the eigenvalue (geometric growth rate), but differences between deterministic and stochastic predictions will probably be small even over as many as 20 time intervals. If significant differences between deterministic and stochastic projections arise only after many intervals, then the practical applications of stochastic models to formulating management plans or assessing toxicological effects are limited.

Short-term deviations from expected numbers can be appreciable when more statistically complex models apply. If variation in vital rates is temporally autocorrelated, departures from expectation can accumulate rapidly.[28,33] Similarly, covariation among vital rates of different stages is critical in assessing the short-term effects of environmental stochasticity in structured populations. Positive covariation among vital rates can induce large effects on population growth rates, whereas negative covariation among fecundities and survivorships or among survival-growth rates of different age or size classes can diminish the impact on overall population growth. As variance of population growth rates increases, median population density decreases and probability of extinction increases. The author currently knows of no analyses of natural populations that approach this level of statistical complexity, but in attempting to assess the impacts of pesticides, it will be quite important to estimate covariation among stages and time intervals. In addition, both means and variances of vital statistics should be reported and considered. A pesticide with highly variable effects on individual birds would have quite different impact if it affected only a single stage or age, affected all stages in a similar fashion, or negatively impacted small birds while actually benefiting larger birds.

AN AVIAN EXAMPLE

This example, based loosely on Naik and Mistry[34] and Mirza,[35] is used to illustrate the analytical methods discussed in this chapter. The author has attempted to use realistic values for vital rates of house sparrows (*Passer domesticus*), but the original papers did not stratify the population and did not include estimates of adult survival. Thus, this is only a demonstration of techniques.

Suppose a house sparrow population is stratified into four mass classes, (0 to 9 g, 10 to 19 g, 20 to 25 g, and >25 g). A monthly transition matrix, **A**, and associated eigenvectors, **u** and **v**, from the peak of the breeding season might be

$$\mathbf{A} = \begin{bmatrix} 0 & 0.714 & 0.952 & 1.19 \\ 0.3 & 0.1 & 0.1 & 0.0 \\ 0.3 & 0.6 & 0.4 & 0.9 \\ 0 & 0.2 & 0.4 & 0 \end{bmatrix} \quad \mathbf{u} = \begin{bmatrix} 0.417 \\ 0.129 \\ 0.334 \\ 0.120 \end{bmatrix} \quad \mathbf{v} = \begin{bmatrix} 1.00 \\ 2.11 \\ 2.31 \\ 2.46 \end{bmatrix}$$

reflecting the following life history description. Hatchlings weigh about 3 g each and grow to at least 18 g in the 17 days before fledging, with survival over the first month of life being 60%. I have assumed about half of the survivors exceed 20 g, but none exceed 25 g. Thereafter, survival is 90%/month, with most of the mass class 2 birds exceeding 20 g. Mass class 4 females are assumed to be large birds with fully developed but unlaid clutches, so all survivors will regress one mass class when recaptured. Because all hatchlings fall in mass class 1 and do not breed in their first month and no survivors remain in that class beyond 1 month, the sums of elements in rows 2 through 4 are survivals for mass classes represented by columns 1 through 4, respectively. The fecundity term for mass class 2, 0.714, is the product of a clutch size of 3, proportion of female eggs of 0.5, maternal survival rate of 0.8, egg survival to hatching of 0.7, and viability at hatching of 0.85. Fecundity terms for mass classes 3 and 4 are calculated from the same values but assume clutch sizes of 4 and 5, respectively. The dominant eigenvalue of **A** is 1.33, so the population would be expected to increase rapidly during the breeding season. If the population were near a stable mass distribution, it could more than double in a 3-month breeding season, but would decline near previous levels if the monthly survival rate of 0.9 persisted through the year (see matrix **C**).

The proportions in mass classes 1 to 4 in **u**, the stable mass distribution, reflect the high fecundity and somatic growth rates that result in most sparrows being in the nest or medium-sized adults during the breeding season. Because survival of adults and nesting success of breeding birds are relatively high, reproductive values, **v**, are highest for the largest females. Population growth rate is most sensitive to changes in transitions from classes represented by the largest elements of **u** to those represented by the largest elements in **v**. Of the possible transitions, a small change in transition rate from mass class 3 to 4 ($u_3v_4 = 0.824$) would have the greatest impact on population growth rate, whereas similar increases in reproduction ($u_3v_1 = 0.334$) have relatively little impact. Perhaps this difference reflects the relatively high prereproductive mortality of hatchlings, but a small increase in a survivorship is a much larger proportional change than is the same increase in one of the fecundity terms. Elasticities, or sensitivities to proportional change for elements of the same matrix, can be compared by multiplying the sensitivities by the corresponding element of the transition matrix.[18] This is especially valuable in our example, in which several transitions are impossible and, therefore, have elasticities of zero, despite their high sensitivities. The elasticity for the transition from mass class 3 to 4 is 0.33, whereas that for mass class 3 to 1 (fecundity of mass class 3) is 0.32. The effects of proportional changes on somatic growth and on fecundity of mass class 3 are much more comparable than may be inferred from sensitivity analysis.

Stochastic effects can be illustrated with any transition matrix, but the matrix, **A**, from mid-breeding season has a growth rate too high to be sustained for any appreciable time. By assuming that **A** applies to 3 months and that a matrix, **B**, with the same growth-survival elements but with zeros for fecundities, describes the remaining 9 months of the year, it is possible to derive a yearly transition matrix, $\mathbf{C} = \mathbf{A}^3\mathbf{B}^9$, with a dominant eigenvalue of 1.042:

$$\mathbf{C} = \begin{bmatrix} 0.340 & 0.509 & 0.509 & 0.509 \\ 0.100 & 0.150 & 0.150 & 0.150 \\ 0.266 & 0.398 & 0.399 & 0.398 \\ 0.102 & 0.154 & 0.153 & 0.154 \end{bmatrix} \quad \mathbf{u} = \begin{bmatrix} 0.420 \\ 0.124 \\ 0.329 \\ 0.127 \end{bmatrix} \quad \mathbf{v} = \begin{bmatrix} 1.00 \\ 1.50 \\ 1.50 \\ 1.50 \end{bmatrix}$$

Note that although **C** is not the 12th power of a single matrix, the matrices **A** and **B** are similar enough that the rows and columns of **C** are proportional and are quite close to vectors of reproductive values and stable size distributions. Projecting into the future from an initial vector of 100 sparrows in the stable size distribution (42, 12, 33, and 13, in mass classes 1 through 4, respectively) yields a predicted population of 228 birds after 20 years and 6100 birds after 100 years.

The simplest form of environmental stochasticity can be demonstrated by allowing c_{13}, the contribution of mass class 3 to mass class 1 in the matrix **C**, to vary uniformly between 0.06 to 0.96. This is high variation in a transition element of intermediate importance (intermediate sensitivity value). The author simulated 500 replicate population trajectories using the pseudorandom number generator supplied with GW-BASIC (Microsoft Corporation). After 100 generations, the sample mean was 5663 ± 165 (SE) and the median was 4724, both significantly less than the expected value of 6100. Clearly, after many intervals even small sources of variation can cause appreciable differences between deterministic and stochastic predictions.[29,30] After 20 years, the average population from 500 random projections was 229 ± 2.86 and the median was 226; neither of these was a significant departure from the deterministic projection. The latter result is probably more relevant to studies of pesticides, because both the ability to test for toxic effects and the need to project toxic effects are more likely to be limited to shorter periods. Substantial stochastic effects in the short term depend on patterns of covariation among several vital rates. When all 16 elements of **C** were allowed to vary uniformly and independently over a range ±0.1 units around their means, the mean of a sample of 500 simulations of 20 years' length was 223 ± 3.09 and the median was 216. With even a modest amount of variation in each vital rate, the deterministic projection based on mean vital rates was a reasonable estimate of mean numbers but overestimated the median over a span of 20 time intervals. If the vital rates each varied over the same range, but with perfect positive correlation with one another (the same random value was added to all elements), the mean level was unchanged (223 ± 12.3), but the median population was only 145.4 (64% of the deterministic result). With the same range of variation, but making the random parts of transitions to mass classes 1 and 3 equal the negative of those to 2 and 4, both mean and median numbers after 20 time intervals were 228. Thus, negative covariances among vital rates negated the stochastic effects. Admittedly, these examples are extreme cases of covariation among elements of transition matrices. They are intended only as a demonstration of the potential impact of environmental stochasticity, but they do illustrate the need for population studies over several generations, which will yield estimates of vital rates and their patterns of variation and covariation over time.

CONCLUSIONS

The difficulty with estimating all of the required parameters for structured populations is compounded if vital rates must be estimated for several individual periods. Studies need to be planned with the expectation that vital rates will be estimated for each census period and for each of several stage or age groups. Even if expected values of vital rates are estimated correctly, observed rates may be quite different within small populations. Variation in observed rates as a result of low numbers, or demographic stochasticity, generally is negligible in animal ecology because studies are designed to yield large samples. Models of rare species, which may be prime concerns in formulating environmental policy, also must consider demographic stochasticity in making projections.[1,36] When studying relatively rare species the demographic concern should shift from short- or long-term population growth rates to probability of extinction to which stochastic models are particularly suited.[1,37]

It is probably not constructive to demand higher standards of data collection and analysis for toxicological studies than are met by most ecological studies; nevertheless, the recognition of demographic structure, especially the physiological and behavioral consequences of body size, and the incorporation of the role of environmental stochasticity into studies of pesticide impacts of birds are worthy goals. Many of the models and techniques discussed in this chapter are relatively novel, even to population ecologists; their incorporation into avian toxicological studies should increase the value of field trials of agricultural chemicals.

REFERENCES

1. Caswell, H., *Matrix Population Models: Construction, Analysis, and Interpretation*, Sinauer Associates, Sunderland, MA, 1989.
2. Lotka, A. J., *Elements of Physical Biology*, reprinted in 1956 by Dover Publishing, New York (originally published in 1924).
3. Lewis, E. G., On the generation and growth of a population, *Sankya*, 4, 93-96, 1942.
4. Leslie, P. H., The use of matrices in certain population mathematics, *Biometrika*, 33, 183-212, 1945.
5. Brownie, C., D. R. Anderson, K. P. Burnham and D. S. Robson, Statistical Inference from Band Recovery Data—A Handbook, Resource Publ. 156, U.S. Fish and Wildlife Service, Washington, DC, 1985.
6. Pollock, K. H., J. D. Nichols, C. Brownie and J. E. Hines, Statistical inference for capture-recapture experiments, *Wildl. Monogr.*, 107, 1-97, 1990.
7. Pielou, E. C., *Mathematical Ecology*, Wiley-Interscience, New York, 1977.
8. Williamson, M. H., Some extensions of the use of matrices in population theory, *Bull. Math. Biophys.*, 21, 13-17, 1959.
9. Lefkovitch, L. P., The study of population growth in organisms grouped by stages, *Biometrics*, 21, 1-18, 1965.
10. Sauer, J. R., and N. A. Slade, Size-based demography of vertebrates, *Annu. Rev. Ecol. Syst.*, 18, 71-90, 1987.
11. Sauer, J. R., and N. A. Slade, Mass-based demography of a hispid cotton rat (*Sigmodon hispidus*) population, *J. Mamm.*, 66, 316-328, 1985.
12. Sauer, J. R., and N. A. Slade, Size-dependent population dynamics of *Microtus ochrogaster*, *Am. Nat.*, 127, 902-908, 1986.
13. Sauer, J. R., and N. A. Slade, Uinta ground squirrel demography: is body mass a better categorical variable than age?, *Ecology*, 68, 642-650, 1987.
14. Nichols, J. D., Population models and crocodile management, in *Wildlife Management: Crocodiles and Alligators*, G. J. Webb, S. C. Manolis and P. J. Whitehead, Eds., Surrey Beatty & Sons, London, 1987, pp. 177-187.
15. Emlen, J. M., Terrestrial population models for ecological risk assessment: a state-of-the-art review, *Environ. Toxicol. Chem.*, 8, 831-842, 1989.
16. Sauer, J. R., R. J. Barker and P. H. Geissler, Statistical aspects of modeling population changes from population size data, in *Wildlife Toxicology and Population Modeling: Integrated Studies of Agroecosystems*, R. J. Kendall and T. E. Lacher, Jr., Eds., Lewis Publishers, Chelsea, MI, 1993.

17. Caswell, H., Analysis of life table response experiments. I. Decomposition of effects on population growth rate, *Ecol. Model.*, 46, 221–237, 1989.
18. Caswell, H., A general formula for the sensitivity of population growth rate to changes in life history parameters, *Theor. Pop. Biol.*, 14, 215–230, 1978.
19. De Kroon, H., A. Plaisier, J. M. van Groenendael and H. Caswell, Elasticity: the relative contribution of demographic parameters to population growth rate, *Ecology*, 67, 1427–1431, 1986.
20. Iskjaer, C., N. A. Slade, J. E. Childs, G. E. Glass and G. W. Korch, Body mass as a measure of body size in small mammals, *J. Mamm.*, 70, 662–667, 1989.
21. Vandemeer, J., Choosing category size in a stage projection matrix, *Oecologia*, 32, 79–84, 1978.
22. Moloney, K. A., A generalized algorithm for determining category size, *Oecologia*, 69, 176–180, 1986.
23. Haramis, G. M., J. D. Nichols, K. H. Pollock and J. E. Hines, The relationship between body mass and survival of wintering canvasbacks, *Auk*, 103, 506–514, 1986.
24. Johnson, D. H., G. L. Krapu, K. J. Reinecke and D. G. Jorde, An evaluation of condition indices for birds, *J. Wildl. Manage.*, 49, 569–575, 1985.
25. Slade, N. A., and C. Iskjaer, Daily variation in body mass of free living rodents and its significance for mass-based population dynamics, *J. Mamm.*, 71, 357–363, 1990.
26. Furstenberg, H., and H. Kesten, Products of random matrices, *Ann. Math. Stat.*, 31, 457–469, 1960.
27. Cohen, J. E., Long-run growth rates of discrete multiplicative processes in Markovian environments, *J. Math. Anal. Appl.*, 69, 243–251, 1979.
28. Tuljapurkar, S. D., An uncertain life: demography in random environments, *Theor. Pop. Biol.*, 35, 227–294, 1989.
29. Lewontin, R. C., and D. Cohen, On population growth in a randomly varying environment, *Proc. Natl. Acad. Sci. U.S.A.*, 62, 1056–1060, 1969.
30. Boyce, M. S., Population growth with stochastic fluctuations in the life table, *Theor. Pop. Biol.*, 12, 366–373, 1977.
31. Slade, N. A., and H. Levenson, Estimating population growth rates from stochastic Leslie matrices, *Theor. Pop. Biol.*, 22, 299–308, 1982.
32. Tuljapurkar, S. D., and S. Orzack, Population dynamics in variable environments. I. Long-run growth and extinction, *Theor. Pop. Biol.*, 18, 314–342, 1980.
33. Tuljapurkar, S. D., Population dynamics in variable environments. II. Correlated environments, sensitivity analysis and dynamics, *Theor. Pop. Biol.*, 21, 114–140, 1982.
34. Naik, R. M., and L. Mistry, Breeding season and reproductive rate of *Passer domesticus* (L.) in Baroda, India, in *Proc. Gen. Meet. IBP Working Group on Granivorous Birds*, The Hague, Netherlands, September 6–8, 1970, pp. 133–140.
35. Mirza, Z. B., Study on the fecundity, mortality, numbers, biomass and food of a population of house sparrows in Lahore, Pakistan, in *Proc. Gen. Meet. IBP Working Group on Granivorous Birds*, The Hague, Netherlands, September 6–8, 1970, pp. 141–150.
36. Pollard, J. H., *Mathematical Models for the Growth of Human Populations*, Cambridge University Press, Cambridge, U.K., 1973.
37. Lande, R., and S. H. Orzack, Extinction dynamics of age-structured populations in a fluctuating environment, *Proc. Natl. Acad. Sci. U.S.A.*, 85, 7418–7421, 1988.

20

Modeling Ecological Impacts of Pesticides on Avian Populations: Synthesis

M. Craig Barber

ABSTRACT

A goal of ecotoxicology is to develop models that can be used in risk assessment and the regulatory process. Many population level models are currently available that can be useful in this process. Few are designed explicitly to address chemical exposure. The TEEAM model presented by Bird is one of the few that meets this requirement and is linkable to other exposure models. Other models presented can be useful, but their utility is related to the broader issue of how models should be used in ecological risk assessment. The distinction between theoretical and empirical based models is used to evaluate their utility in risk assessment. Generally, theoretical models that can be parameterized objectively and independently from the data under analysis will be more appropriate, since the model can be used not only for comparative purposes, but in extrapolations as well. The inappropriate use of models in risk assessment must be cautiously avoided.

KEY WORDS

ecological modeling, risk assessment, population biology, population dynamics, pesticides, physiology, toxicology

INTRODUCTION

In order for models to be useful in the population risk assessment process, Emlen[1] concluded that they should:

1. directly address specific endpoints of regulatory interest,
2. use population data that are generally available or reasonably obtainable by extrapolation,
3. be linkable to models of chemical exposure, and
4. easily incorporate toxicological information.

Assuming that endpoints of general regulatory interest include (1) local extinctions of populations, (2) altered temporal or spatial population dynamics, (3) altered recovery capacities, and (4) altered population structure (e.g., sex ratios, age and size class distributions), then the Richards growth model, Leslie matrix models, and the logistic equation described in Chapters 17 to 19 clearly could be useful tools for assessing the risk of pesticides to bird populations. Because they are traditional population models, their data requirements are indeed consistent with the type of data that could be compiled during most field studies investigating the ecological effects of pesticide use. Of the models described in this chapter, only the TEEAM model discussed by Bird in Chapter 16 is designed to address explicitly chemical exposure and to be linkable to other exposure models. Although Brisbin discusses how to use the Richards growth model to describe chemical bioaccumulation, its utility for this purpose must be considered conditional because the model does not explicitly address the exposure scenarios that cause organisms to bioaccumulate the chemicals of concern. Although it may be argued that the "growth" coefficient (K) of the Richards model is implicitly a function of the bird's exposure, because this coefficient is time invariant, one must conclude that only constant or nearly constant exposure scenarios should be analyzed using this model. Depending on the pesticide's persistence and the bird's primary route of exposure, however, this may not be a serious limitation. For example, if the pesticide is principally ingested as a granule, the Richards model could be an appropriate tool to analyze bioaccumulation. Likewise, if dermal contact with nearly constant pesticide residues in soil is the bird's primary route of exposure, then the Richards model again may be an appropriate analytical tool.

A consideration of under what conditions the Richards model provides an appropriate methodology to analyze chemical bioaccumulation is actually a broader issue: how should models be used or applied in ecological risk assessments. In general, a pesticide researcher/regulator is faced with two quantitative problems. First, the researcher/regulator must analyze and interpret results from specific laboratory and field studies. This phase is basically one of data interpolation. The second problem facing the researcher/regulator, however, is one of data extrapolation. For example, how do the results from a field study concerning the effects of a pesticide in a Iowa corn field apply to other proposed uses in regions with different physiographic and biological characteristics? Although all mathematical models are implicitly designed to interpolate between observed data, only theoretical, process-based models are constructed explicitly to extrapolate outside of the range of observation. Therefore, to use models similiar to those described in this chapter appropriately, one must consider from the onset the empirical/theoretical nature of the models in question.

In general, theoretical and empirical modes can be distinguished from one another based on two fundamental characteristics. The first of these concerns the level of organization used to describe the object of interest. In theoretical-based models, the behavior of an object is predicted by integrating behaviors of objects or processes that function at a level of resolution lower than the object of interest. Empirical models, on the other hand, predict the behavior of an object as a function of objects or processes that operate at a level of resolution equal to or higher than the object of interest. Consider, for example, the different ways one could model a bird's growth. For an appropriate time interval, one could model this process using the simple differential equation

$$\frac{dW}{dt} = \gamma W \tag{1}$$

where γ is the bird's specific growth rate and W is the bird's body weight. If the bird's specific growth rate is not explicitly treated as a function of the bird's physiological processes (i.e., feeding, respiration, excretion, etc.), this equation would represent a purely descriptive or empirical model of a bird's growth. If, however, the bird's growth was modeled using even the overly simplified mass balance equation

$$\frac{dW}{dt} = A - R \tag{2}$$

where A and R denote the bird's dietary assimilation and routine respiratory demand, respectively, then this equation satisfies the first criterion for being a theoretical model.

The second characteristic that distinguishes a theoretical model from an empirical model concerns the parameterization and predictability of the functional forms used to describe the lower level processes of the object. In particular, not only can these functions be parameterized independently of the particular data set that the model is attempting to emulate, but these functions are capable of robust extrapolations regardless of their own descriptive/theoretical basis. For example, in Equation 2 one could describe the physiological processes of interest, either empirically as power functions of the organism's body size or theoretically as functions of biochemical reactions. In the former case, the theoretical-based model for the bird's growth now assumes the form

$$\frac{dW}{dt} = \alpha\, W^{\beta} - \gamma\, W^{\delta} \tag{3}$$

where α and β are empirical constants that describe a bird's dietary assimilation as a function of its body weight and γ and δ are empirical constants that describe its routine respiration. If this equation is indeed a theoretical process-based model, then power functions used to describe the bird's assimilation and respiration must be robust extrapolators, i.e., given these parameters for birds weighing 1 to 250 g, one could accurately quantify these processes for birds weighing >250 g.

The TEEAM model described by Bird clearly satisfies the above criteria for being a theoretical model. Similarly, although a population's size class structure could be modeled using complex partial differential equations that address physiological and ecological processes in fine detail such as those described by Hallam elsewhere in this volume, most researchers would also consider the Leslie matrix models described by Slade to be theoretically based. Given computerized databases for the parameters of these models, pesticide regulators and researchers could use these models not only to analyze avian population dynamics and dietary exposures in specific field studies, but, more importantly, to extrapolate these results to other intended uses. The utility of these models would be greatly increased if their supporting databases were structured to facilitate regional analyses using geographical information system (GIS) technology.

In general, however, the distinction between empirical and theoretical models may not be so clear. To illustrate this situation, consider, for example, the differential equation form of Richards growth model, i.e.,

$$\frac{dW}{dt} = \pm\, k\, W \mp l\, W^{m} = \pm\, P_1 \mp P_2 \tag{4}$$

where W denotes either the organism's body weight or chemical burden and k and l are positive rate coefficients. If P_1 and P_2 are interpreted as the competing processes of assimilation and metabolism (in the case of growth) or of uptake and excretion (in the case of pesticide accumulation) the Richards model clearly could be interpreted as being theoretically based. This interpretation, however, is justified only if the bird's assimilation is directly proportional to its body weight and its respiration proportional to W^m or if its assimilation is proportional to W^m and its respiration directly proportional to its body weight. If these allometric relationships are satisfied, then the Richards model indeed can be considered a theoretical model and consequently can be used for extrapolation. Failure to satisfy such allometric relationships implies that the Richards model should be used only for data interpolation such as the analysis treatment effects on growth and bioaccumulation within a particular field or laboratory study. Brisbin et al.[2] reached a similar conclusion when they wrote that ". . . although the Richards model can adequately describe accumulation data in an empirical sense, it does not provide information concerning the mechanisms responsible for the particular accumulation pattern observed."

Because the logistic model that Chesser et al. described in Chapter 18 to analyze population dynamics is actually a special case of the Richards growth model, it should be obvious from the preceding discussion that this model can be treated either as a theoretical or as an empirical construct, depending on circumstances. If the model can be parameterized independently from the data set that it is being used to analyze, clearly it should be treated as a theoretical construct. Under such conditions, pesticide regulators and researchers could use it not only to compare and contrast treatment effects of a pesticide in a field study, but also to extrapolate results to other field conditions.

Mathematical models are presently, to a greater or lesser extent, an integral part of the study and regulation of the ecological effects of pesticides. They provide an objective, scientifically defensible framework to plan and to analyze laboratory and field studies and, in some cases, to extrapolate to unstudied conditions. Inappropriate use of these models, however, must be avoided. Model developers must make it clear when their models can or cannot be used. Model users must not use a model simply to obtain a "number" but use the right model to obtain the correct "number." Although theoretically based process models allow pesticide researchers and regulators to extrapolate outside of the range of observation, unless these models can be parameterized objectively, they will be of little use in evaluating the ecological effects of pesticides on avian populations. Developers of process-based models, therefore, must consider the development of databases for model parameters as an integral and necessary part of their research activities. If coordinated model construction and database compilation is undertaken, not only will researchers be able to address important questions concerning model sensitivity and uncertainty, but regulators will be able to incorporate such models as a routine part of their ecological risk assessments.

REFERENCES

1. Emlen, J. M., Terrestrial population models for risk assessment: a state-of-the-art review, *Environ. Toxicol. Chem.*, 8, 831–842, 1989.
2. Brisbin, I. L., Jr., M. C. Newman, S. G. McDowell and E. L. Peters, Prediction of contaminant accumulation by free-living organisms: application of a sigmoidal model, *Environ. Toxicol. Chem.*, 9, 141–149, 1990.

21

Extending Laboratory Results to the Field

Christian E. Grue

Agencies that regulate pesticide use and environmental contaminants in general have relied heavily on data collected on selected species of captive animals in controlled environmental conditions to assess the biological significance of chemical concentrations in the environment. This is true for both aquatic and terrestrial hazard assessments.[1] There are several reasons for this dependence on standard laboratory toxicity tests instead of field data. First, standardized laboratory protocols facilitate comparisons among chemicals in terms of their relative toxicities. Second, some species adapt well to laboratory conditions. Third, field studies are considerably more costly than laboratory studies and, because of the number of potential confounding variables, have a greater likelihood of resulting in insufficient or inconclusive data. The laboratory data most frequently used in these ecological hazard assessments include measures of acute toxicity (LD_{50}s in birds and mammals), subacute toxicity (LC_{50}s in aquatic invertebrates, fish, birds, and mammals), and chronic toxicity (reproduction in birds and mammals, carcinogenicity in mammals).[2-4] For example, one approach currently used by the U.S. EPA to assess the ecological hazards of pesticides to birds and other nontarget wildlife is to compare subacute toxicity data for captive species (e.g., northern bobwhite, *Colinus virginianus*) with actual or estimated field concentrations on food items.[1,5] A major assumption in this approach is that the sensitivity of other avian species to a particular pesticide is similar to that for the northern bobwhite. In addition to sensitivity (acute oral toxicity), wild species may also differ in the chemical exposure they receive because of their food habits or other behaviors. A comprehensive review of the factors that may affect the vulnerability of birds to pesticides is provided by Buttiker.[6]

Despite the many assumptions in current assessments of the hazards pesticides pose to wildlife, relatively few cases of recurring adverse effects on wildlife populations from pesticides have been reported. The most notable exception is the effect of DDT and its metabolites on raptorial and fish-eating birds.[7] More recent, but less drastic examples include the decline in numbers of pink-footed geese (*Anser brachyrhynchus*) following the use of carbophenothion as a seed treatment on wheat in Great Britain,[8] and the population decline of black-billed magpies (*Pica pica*) associated with the use of the organophosphorus insecticide famphur on cattle in the Pacific Northwest.[9] Do these few cases accurately reflect the magnitude of effects of current pesticide use on wildlife populations, or are effects simply undetected?

Models may, in the short term, be an effective means of providing a framework guiding future research, both in the laboratory and in the field, to address this question and, in the long-term, predicting the effects of pesticides on wildlife populations.[10] If laboratory data are to continue to be used to assess the hazards of pesticides to wildlife, we must know to what extent these data can be extrapolated to the field. To do this, we must critically evaluate the methods and assumptions on which these hazard assessments are based.

The four chapters that follow review some of the laboratory tests and techniques currently used to assess the toxicity of pesticides and other contaminants to wildlife, as well as the effects of pesticides on behavior and disease resistance of captive animals that may ultimately adversely affect reproduction and survival of their free-living counterparts. These assessments of the state of the art also identify: (1) ways in which methods (data collection, analysis, and interpretation) may be improved; (2) existing data gaps; and (3) approaches by which needed data may be obtained. The synthesis that follows these chapters addresses the need for further evaluation of the assumptions upon which the majority of our hazard assessments have been based.

ACKNOWLEDGMENTS

The author is an employee of the U.S. Fish and Wildlife Service. The Washington Cooperative Fish and Wildlife Research Unit is jointly supported by the U.S. Fish and Wildlife Service, University of Washington, Washington Department of Ecology, Washington Department of Fisheries, Washington Department of Natural Resources, Washington Department of Wildlife, and the Wildlife Management Institute.

REFERENCES

1. Urban, D. J., and N. J. Cook, Hazard Evaluation Division Standard Evaluation Procedure—Ecological Risk Assessment, EPA 540/9-85-001, U.S. Environmental Protection Agency, Washington, DC, 1986.
2. Johnson, W. W. and M. T. Finley, Handbook of Acute Toxicity of Chemicals to Fish and Aquatic Invertebrates, Fish and Wildlife Service Resource Publ. 137, U.S. Department of the Interior, Washington, DC, 1980.
3. Hudson, R. H., R. K. Tucker and M. A. Haegele, Handbook of Toxicity of Pesticides to Wildlife, 2nd ed., Fish and Wildlife Service Resource Publ. 153, U.S. Department of the Interior, Washington, DC, 1984.
4. Hill, E. F., and M. B. Camardese, Lethal Dietary Toxicities of Environmental Contaminants to *Coturnix*, Fish and Wildlife Service Tech. Rep. 2, U.S. Department of the Interior, Washington, DC, 1986.
5. Kenaga, E. E., Factors to be considered in the evaluation of the toxicity of pesticides to birds in their environment, *Environ. Qual. Safety*, 11, 166–181, 1973.
6. Buttiker, W., Ecological effects of insect control on bird populations, in *The Ecological Effects of Biological and Chemical Control of Undesirable Plants and Animals*, D. J. Kuenen, Ed., E.J. Brill, Leiden, The Netherlands, 1961, pp. 48–60.
7. Stickel, W. H., Some effects of pollutants in terrestrial ecosystems, in *Ecological Toxicology Research*, A. D. McIntyre and C. F. Mills, Eds., Plenum Press, New York, 1975, pp. 25–74.
8. Hamilton, G. A., and P. I. Stanley, Further cases of poisoning of wild geese by an organophosphorus winter seed treatment, *Wildfowl*, 26, 49–54, 1975.
9. Henny, C. J., L. J. Blus, E. J. Kolbe and R. E. Fitzner, Organophosphate insecticide (famphur) topically applied to cattle kills magpies and hawks, *J. Wildl. Manage.*, 49, 648–658, 1985.
10. Emlen, J. M., Terrestrial population models for ecological risk assessment: a state of the art review, *Environ. Toxicol. Chem.*, 8, 831–842, 1989.

22

Acute and Subacute Toxicology in Evaluation of Pesticide Hazard to Avian Wildlife

Elwood F. Hill

ABSTRACT

Single-dose acute oral and short-term subacute dietary toxicity tests with captive birds provide critical information on the potential hazard of pesticides to wild populations. The two tests have similar experimental designs and both generate a lethality curve and estimation of its midpoint, the median lethal dosage (LD_{50}) or concentration (LC_{50}). Although LD_{50}s and LC_{50}s are widely used to characterize pesticide toxicity, the lethality curve and critical observation of animal response to chemical challenge provide necessary insight for hazard evaluation. The highly controlled acute test is based on graded dosage by body mass and provides a sound method of comparing naive sensitivity to toxicant and a means of detecting pesticides that may cause large-scale field kills. In contrast, the subacute test presents graded concentrations of a chemical in the diet for a specified duration, usually 5 days. This feeding trial provides an evaluation of response to repeated chemical exposures as may be encountered in the field. This chapter is an appraisal of the two basic tests of lethality with an emphasis on factors that may affect interpretation of potential hazard.

KEY WORDS

birds, pesticides, lethal toxicity, hazard

INTRODUCTION

The single-dose acute oral toxicity test is used in preliminary evaluation of virtually all substances of suspected biological activity. The test is based on administration of graded dosage of chemical in relation to body mass. The primary objective is to generate estimates of the dose-response or lethality curve and its midpoint, the median lethal dosage or LD_{50}.[1] Once these statistical parameters and their associated errors are properly determined, this test of lethality provides a proven means of quantifying chemical potency and comparing substances of different mechanisms and sites of action.[2] The value of an acute test is greatly enhanced by detailed observation of each animal from the time of dosage to its death or recovery. Too often, however, comparisons and interpretation of acute tests are focused on the LD_{50} exclusive of its statistical reliability and without reference to the lethality curve or other supplemental observations that provide important clues about acute toxicity and hazard evaluation. The

LD_{50}, per se, is simply a convenient index of toxicity that is subject to error, and its indiscriminate use can be misleading.[3].

In wildlife toxicology, two tests of lethality are routinely required on birds for pesticide registration in the United States.[4] The first is a standardized acute test of captive reared adult mallards (*Anas platyrhynchos*) or northern bobwhites (*Colinus virginianus*).[5] The second test is similar to the acute test except graded concentrations of chemical are presented ad libitum in the feed for 5 days to young mallards or northern bobwhites of specified ages, and the midpoint of the lethality curve is quantified as the median lethal concentration or LC_{50}.[6] This subacute feeding trial is intended to augment the acute test by measuring response to repeated exposures and accumulative effects. Whereas the acute test provides a measure of a species' naive sensitivity to a toxic substance and a convenient index for rating its potency, the subacute test provides a measure of the species' ability to cope with a contaminated diet for a specified duration, allowing for the metabolic changes that occur over time.[7] Careful observation for changes in behavior and rate of feeding and for onset and course of toxic signs is especially important during subacute tests because the subjects voluntarily eat the potentially lethal diets. These two tests of lethality must never be viewed casually because they are often the only required avian tests for pesticide registration.[4,8]

This chapter is an appraisal of avian single-dose acute oral and 5-day dietary subacute toxicity tests as they are used in the evaluation of pesticide hazard. The basic tests of lethality, their toxicologic rationale, and key statistical treatments are described. Data are presented to illustrate experimental factors that affect toxicologic interpretation. The focus of the examples is on contemporary pesticides, many of which work through the same toxic mechanisms but often yield profound differences in response and potential environmental hazard.

THE BASIC TESTS

Classical acute toxicity tests are designed to determine exposures that cause death under a prescribed protocol with treatment levels that are based on animal response rather than practical residues. When treatments are properly arranged, however, the resultant lethality curve provides estimates of the LD_{50} and other dose-response coordinates that may be used in hazard assessment. Once the basic lethality curve and response to a substance are determined for several appropriate species, determination of only the general order of the substance's toxicity by approximate tests[9,10] with alternative species or finished product formulations may then be adequate. The choice between use of a full-scale or an approximate test depends on the purpose of the study. Although one should always strive to use the smallest number of animals, good science that is supported by sound statistical analysis must never be compromised.

Toxicologic Rationale

Toxic response is graded by the concentration of the substance that penetrates the target and remains in contact for a sufficient time to elicit change. The concentration of substance that penetrates the target is usually correlated directly with the dosage that is received by the organism. However, various biological, chemical, and physical factors influence translocation and penetration of substances, and individuals may not be equally sensitive to a chemical. Therefore, response will vary even within a homogeneous population.[11] This natural diversity is approximated by a normal Gaussian distribution with about one third of the population divided equally between hyper- and hyposensitive individuals. When individual responses are described quantitatively, the frequency-response curve tends to be skewed toward hypersensitive respondents because their arithmetic range of tolerance is smaller than that of hyposensitive individuals.[1] Because the representation of hyper- and hyposensitive individuals is assumed to be equal in a homogeneous population, a series of groups may be randomly selected from the population and gradation of dose-related responses between groups may be generated if dosages of test substance are properly spaced. Responses can be quantified as qualitative changes by a preselected all or nothing (binary) endpoint. In acute testing of lethality, the endpoint is alive or dead, and the responses can be evaluated quantitatively because the percentage of respondents increases with dosage. This concept and the factors responsible for diversity of response among individuals are well documented.[1,2,9-14]

Dose-Response or Lethality Curve

The percentage of respondents in a lethality test is related to the composite tolerances of the population.[1,13] The pattern of response to graded dosages of substance is analogous to the graded tolerances of individual specimens and gives a frequency distribution skewed toward hypersensitivity and an asymmetric sigmoid curve when percentage response is plotted against dosage. The resultant dose-response curve is quite steep from its origin to the inflection point (at about the 30% response level) and then becomes gradual until virtually asymptotic. Because skewed data are difficult to analyze statistically, test dosages are usually arranged logarithmically to normalize the distribution of responses.[1,12] Normalization gives a symmetric sigmoid dose-response curve with the inflection point at the exact midpoint, the 50% response level.

The symmetric dose-response curve represents a cumulative normal distribution of log-tolerances. Steepness of the curve is similar for many substances but may become significantly steeper or shallower depending on the substance's mechanism of action, route or method of exposure, or shift of tolerance in the population. Thus, the dose-response curve has interpretive value in addition to determination of probable dose-response coordinates. However, the linear portion of the curve is limited to a range of only 30 to 35 percentage points on either side of the 50% response level. The entire curve can be made linear by transforming the percentage response for log-dosage to probits.[1,12] Responses can then be analyzed by probit analysis, a method of calculating maximum likelihood fit of a probit-log-dose line by an iterative weighted regression analysis. The analysis provides critical interpretive statistics such as the median response level and its 95% confidence interval, and the slope of the weighted linear regression of probits on log-dose and its error. A systematic probit analysis, including calculation of all relevant toxicity statistics, is presented by Finney.[1] Although probit analysis or shortcut procedures by probit analysis[15] are traditionally used in statistical evaluation of acute-type lethality tests, the movement is toward use of logit analysis as a more convenient computational method.[12]

Toxicity Comparisons

Comparison of toxicity between chemicals is possible with data generated by probit analyses if the level of tolerance of test populations is the same and the probit regression lines are parallel.[1]. The level of tolerance can be assumed comparable if the test subjects are selected randomly from a single population and are tested concurrently in a completely randomized experiment.[1] In hazard evaluation of pesticides, data sets from many laboratories usually provide the basis of comparison, and such restrictive criteria cannot often be met. Even when tests are conducted in one laboratory, problems as indicated by Finney,[13] may arise: "One feature possessed by all biological assays is the variability in the reaction of the test subjects and the consequent impossibility of reproducing at will the same results in successive trials, however carefully the experimental conditions are controlled." This variability can be corrected statistically by concurrent testing of a standard preparation that has the same biologically active principle as the test preparation.[13] This too is impractical because even pesticides that act on the same physiologic system may do so in different ways; e.g., central nervous system (CNS) stimulation by chlorinated cyclodiene insecticides or cholinesterase (ChE) inhibition by organophosphorus (OP) insecticides. Nonetheless, the researchers who generated most of the early avian subacute lethality data on pesticides believed that the test of a general standard substance should accompany all tests irrespective of mechanism of action.[16,17] Dieldrin was used as the standard and results have been summarized.[17-19] Even though the basic data from these reports have been widely used in hazard evaluation, a literature search failed to reveal evidence that the dieldrin standard was ever used as suggested for correction of LC_{50}s. Such specific corrections may best not be made on the basis of the dieldrin standard because consensus presently favors use of a nonspecific standard primarily for intralaboratory quality control rather than routine adjustment of LD_{50}s or LC_{50}s.[19-21].

Statistical techniques for comparison of potency among chemicals, including median response levels and slope of the probit regression curves, have been described.[1] A simplified method for separation of LD_{50}s or LC_{50}s is to compare the 95% confidence intervals for overlap; if they do not overlap, the median response levels may be considered different at $p<0.05$. Other methods such as the two-tailed t test and Bonferroni's t statistics[22] are also used for comparison of median response levels. Median response levels must be statistically separable ($p<0.05$) before quantitative comparison is credible. Toxicologic literature is replete with conclusions from comparison of LD_{50}s that are obviously not different or the data are inconclusive because of omission of the 95% confidence interval or other estimate of variation. Even when the median response levels are statistically different, the same relationship cannot be assumed at different response levels without testing the slopes of the dose-response curves for parallelism.[1,17] When the slope of the dose-response curve and the median (50%) response level are known, any derived response level can be estimated.[1,17-19] Although response levels other than the 50% response may be desired, estimates of this type must be used cautiously because extrapolation from a standard probit regression line can be misleading if the true regression equation has some curvature.[1] In wildlife toxicology, the historical focus of acute toxicity testing has been on estimation and general comparison of LD_{50}s with approximate statistical procedures that do not provide for statistical estimation of the dose-response curve.[23,24]

Test Protocols

Single-Dose Acute Oral Toxicity Test

Optimal use of the acute test in hazard evaluation requires statistical estimation of the lethality curve and its midpoint and descriptive information on toxic response. The test for birds is basically the same as that described for laboratory animals.[3,10] The test involves dosage of test substance as a proportion of body mass and detailed observation of response until death or recovery. Ideally, a statistically adequate number of adult nonbreeding birds are drawn from a homogeneous population, weighed, and randomly assigned to individual test pens in a controlled environment room about 2 weeks prior to testing. A few extra birds are provided in case substitution is necessary. Room temperature and photoperiod are maintained at about 24° to 28°C and 10L:14D. The short day ensures reproductive quiescence to minimize sex differences. After 1 week the birds are evaluated and any that appear obviously substandard are replaced. On the morning of the day prior to testing, birds are weighed in order to calculate dosage and are given a general health check. That evening, feed is removed in preparation for dosing the next morning.

Overnight-fasted birds receive a single dose of the test substance at midmorning. Feed is provided immediately after dosing, and observations for signs of intoxication are continued throughout the day. Special attention is given to the time of first evidence of toxicity, recovery, or death. Observations are continued twice daily or more often as indicated for 2 weeks after treatment or as long as toxic signs persist. Excellent summaries of observed toxic signs in acute tests of birds are available.[3,25] Gross necropsy should be performed on all birds that die and on a subsample of survivors to document significant toxic lesions.

Test substance is usually administered to the proventriculus in gelatin capsule or by gavage in water or suitable organic solvent. About five birds per sex are tested at each of five or six geometrically arranged dosage levels spanning the expected 10 to 90% mortality levels. Dosage levels are determined from a preliminary study of three widely spaced dosages administered to three to five birds each. Three kinds of controls (negative or sham, vehicle, and positive) may accompany each test; negative and vehicle are mandatory. The size of negative and vehicle control groups must each be equal to at least one dosage level; e.g., five birds per sex, with individuals integrated into the initial experimental design and treated exactly the same as those on test substance. Negative controls receive sham treatment—insertion of empty dosing apparatus. Vehicle controls receive vehicle minus test substance. Positive controls, if used, receive a standard substance of known potency with the same biological action as the test substance. Use of the standard substance requires a full test to compare the slope of the dose-response curve and LD_{50}.[13] The LD_{50} and its 95% confidence interval, expressed as milligram of active ingredient per kilogram of body mass, and the slope and error of the dose-response curve are derived by probit,[1] logit,[12] or other appropriate analysis.[3,10,15]

When only the general order of acute toxicity is desired, (e.g., to compare many species or finished product formulations), an approximate test of lethality may be used.[9,10,25,26] The treatment of test animals and post-dosage observations in these studies are the same as described for the full-scale acute test. The difference is that as few as three groups of three to five subjects are tested against a series of prearranged dosages, with LD_{50} and its 95% confidence interval calculated from published tables.[9,24]

Five-Day Subacute Dietary Toxicity Test

The design of the subacute test is based on the single-dose acute oral test.[8] The test was developed to quantify the toxicity of contaminants for which the diet was considered an important source of exposure.[16] The subacute test was optimized with young precocial birds, such as ducks and quail, but virtually any species can be tested under the protocol if it can be maintained in captivity in good health and cannot survive for 5 days without eating.[21,27,28] If a portion of the test population can fast for 5 days, the results are erratic and not easily reproduced. Thus, the species of choice must be susceptible to the test protocol. This condition of susceptibility has been questioned because death by starvation does not represent the direct toxicity of a chemical.[29] Others have demonstrated that susceptible birds eventually eat rather than starve,[30] and even though death is undoubtedly influenced by nutritional status, it remains primarily a chemical effect.[28]

Like the acute test, the subacute test generates a lethality curve and its midpoint as well as descriptive information on toxic response. The basic design uses the same number of animals, treatment levels, and control groups as the full-scale acute test. However, when testing very young precocial species, birds must be maintained in groups in heated brooder units with at least 14 hours of light.[6,18] Therefore, only one pen of equal-aged birds is usually tested at each concentration of test substance. To ensure susceptibility to the 5-day test, the recommended test ages for the most common model species are 5 days for mallard, 10 days for ring-necked pheasant (*Phasianus colchicus*), and 14 days for northern bobwhite and Japanese quail (*Coturnix japonica*).[6,18,21] Because of the young age at start, randomization to test pen is usually 2 days prior to testing. Any apparently substandard birds are replaced by surplus hatchmates.

Test substance is presented midmorning in an ad libitum diet to birds of the prescribed age and is continued for 5 days. Mortality and signs of intoxication are monitored at least twice daily. Food consumption is measured at 24-hour intervals. Fresh feed is added to all pens each day. After the fifth day, all feed, including that of control groups, is replaced with untreated feed and the study is continued for at least 3 days. When toxic signs persist, observation is continued through complete remission. The LC_{50} and its 95% confidence interval, expressed as milligram of active ingredient per kilogram of feed (or parts per million) in a 5-day ad libitum diet, and the slope and error of the dose-response curve are derived by probit analysis or other suitable method exactly as acute tests.

COMPARATIVE TOXICOLOGY

Birds vs Laboratory Rats

Acute tests of laboratory rodents are the most readily available toxicologic data on vertebrates and often serve as the primary factor in decisions on pesticide hazard to wildlife. For example, a rat LD_{50} above 200 mg/kg is generally considered only moderately toxic; if the pesticide also has poor affinity for lipids and is therefore not likely to bioaccumulate, the pesticide use may be considered low risk for general purposes of environmental impact, and often no additional attention is paid to potential wildlife hazard. However, such a conclusion may be inappropriate because the pesticide may be applied many times during the year, with its fate influenced by widely diverse factors, and the sensitivity to acute exposure may be quite different in birds than in laboratory rats.

Acute sensitivity to pesticides is not the same in birds as in laboratory rats. In Table 1, LD_{50}s for ring-necked pheasants and red-winged blackbirds (*Agelaius phoeniceus*) are compared to LD_{50}s for laboratory rats for OP insecticides of widely variable toxicity. All tests of each species were conducted at a single laboratory. Pheasants and blackbirds are presented because both species have general feeding habits, but represent extreme body mass compared to rats. The pesticides are all anticholinesterases that require metabolic activation for maximum potency, but whose extreme mammalian toxicity (i.e., rat LD_{50} for phorate or temephos) varies over 4000-fold. By most criteria for ranking acute toxicity, phorate is classed highly or extremely toxic and temephos is practically nontoxic.[2,10,18] Phorate is also highly toxic to ring-necked pheasants, but it is about three times more toxic to rats than pheasants whereas temephos is about 250

Table 1. Avian Sensitivity to Organophosphorus Pesticides of Widely Variable Toxicity in Mammals

	Rat[a]		Pheasant[b]		Blackbird[c]	
Pesticide	**Rank**	**LD_{50}[d,e]**	**Rank**	**LD_{50}[d]**	**Rank**	**LD_{50}[d]**
Phorate	1	2	1	7	1	1
Disulfoton	2	7	2	12	2	3
Azinphos methyl	3	13	7	75	5	8
EPN	4	36	6	53	2	3
Ethion	5	65	10	1297	9	45
Phosmet	6	113	9	237	6	18
Dimethoate	7	215	3	20	4	7
Fenitrothion	8	740	4	26	7	25
Malathion	9	1375	5	167	10	>100
Temephos	10	8600	8	35	8	42

[a]Sherman strain male laboratory rats, 3 months old, n = 5–60 per test; dosage by gavage in peanut oil.[31,32]
[b]Farm-reared male and female ring-necked pheasants, 3 to 4 months old, n = 8–28 per test; dosage by gelatin capsule.[25]
[c]Wild-captured pen-conditioned male and female red-winged blackbirds, adult, n = 8–28 per test; dosage by gavage in propylene glycol.[26,33]
[d]LD_{50} = mg active ingredient (technical grade) per kg of body mass calculated to kill 50% of test population.
[e]All rat LD_{50}s are statistically separable ($p<0.05$).

times more toxic to pheasants than rats. The blackbirds are consistently most sensitive to OP exposure, possibly because of influences of differential metabolic rate, but more likely because red-winged blackbirds are especially deficient in hepatic microsomal monooxygenase activity that is often essential for detoxication.[34,35]

Beyond phorate and disulfoton, the rank of the individual pesticides is quite variable among the species, but the real importance to acute hazard evaluation is in comparison of the compounds with rat LD_{50}s above 200 mg/kg. As mentioned, this level implies only moderate toxicity to rats and therefore little acute field hazard would be expected from dimethoate, fenitrothion, malathion, or temephos. However, of the four pesticides, only malathion is not classed as extremely toxic (i.e., $LD_{50}<40$ mg/kg) to both pheasants and blackbirds, and field application of fenitrothion has killed wild birds.[36] All insecticides listed in Table 1 elicit primary toxicity through the same mechanism, yet produce marked differences in toxicologic relationships between birds and rats; birds are much more sensitive than rats to the less toxic anticholinesterases. The differential sensitivity of birds and mammals to anticholinesterases is reviewed elsewhere.[37] This remarkably different response by birds and rats in response to chemicals of like action suggests that equal or greater differences should be expected for dissimilar pesticides and therefore reliance on rat data for prediction of hazard to birds is not adequate.

Interspecies sensitivity

LD_{50}

Avian species vary widely in sensitivity to acute pesticide exposure.[25,26,33,38] Table 2 presents LD_{50}s for ten anticholinesterase pesticides tested at a single laboratory on an array of species that weigh between 25 g (house sparrow, *Passer domesticus*) and 1.2 kg (ring-necked pheasant). Anticholinesterases are again presented because chemicals of

Table 2. Sensitivity of Seven Avian Species to Diverse Anticholinesterase Pesticides[a,b]

	House sparrow		Red-winged blackbird		European starling		Rock dove		Chukar		Mallard		Ring-necked pheasant	
Pesticide	**Rank**	**LD_{50}**	**Rank**	**LD_{50}**	**Rank**	**LD_{50}**	**Rank**	**LD_{50}**	**Rank**	**LD_{50}**	**Rank**	**LD_{50}**	**Rank**	**LD_{50}**
Monocrotophos	1	1.6	1	1.0	2	3.3	3	2.8	2	6.5	4	4.8	1	2.8
Dicrotophos	2	3.0	2	1.6	1	2.7	1	2.4	3	10	3	4.2	3	3.2
Parathion	3	3.4	4	2.4	5	5.6	2	2.5	5	24	1	2.1	6	12
EPN	4	13	5	3.2	6	7.5	5	5.9	4	14	8	53	2	3.1
Propoxur	4	13	6	3.8	7	15	9	60	5	24	6	12	8	20
Chlorpyrifos	6	21	8	13	3	5.0	7	27	9	61	9	76	5	8.4
Fenthion	7	23	3	1.8	4	5.3	4	4.6	7	26	5	5.9	7	18
Temephos	8	35	9	42	9	>100	8	50	10	270	10	79	9	32
Landrin	9	46	7	10	9	>100	10	168	8	60	7	22	10	52
Mexacarbate	10	50	7	10	8	32	6	6.5	1	5.2	2	3.0	4	4.5
Sensitivity rank[c]	3		1		6		3		7		5		2	

[a]Toxicity as LD_{50} = mg active ingredient (technical grade) per kg of body mass calculated to kill 50% of test population.

[b]Table reconstructed from Tucker and Haegele[38] with red-winged blackbird and European starling data from Schafer[33] and Schafer et al.[26] All studies were conducted at the Denver Wildlife Research Center (Denver, CO) by the same protocol. Mallards and gallinaceous species were farm-reared males and females, 2 to 4 months old; rock doves and passerine species were wild-captured pen-conditioned male and female adults. Eight to 28 birds were dosed per test either by gavage in propylene glycol (blackbirds and starlings) or by gelatin capsule.

[c]Sensitivity rank is based on the mean of across-species order of sensitivity to each pesticide.

the same toxic mechanism should yield the most conservative results. In contrast to OP compounds (Table 1), all of which require metabolic activation for maximum potency, examples (Table 2) include compounds that are direct ChE inhibitors; i.e., monocrotophos, dicrotophos, and the three carbamates. Monocrotophos and dicrotophos, whose primary structural difference is a single methyl group, rank as the most or second most toxic compound to all species except mallard, and both yield the most consistent results across the seven species. The extreme LD_{50}s differ by factors of about 6 to 7× for dicrotophos and monocrotophos with a median difference of 15× across species for all ten compounds. In contrast, the carbamates give highly variable results across species and among compounds. Extreme carbamate LD_{50}s differ across species by about 16 to 17×.

The red-winged blackbird is either the most or second most sensitive species to seven to ten compounds, whereas the chukar (*Alectoris chukar*) is either the most or second most tolerant species of eight of ten compounds (Table 2). The other five species are from four taxonomic orders and each species is either most or least sensitive of the seven species to at least one compound. When the seven species are compared in all possible combinations, LD_{50}s of the ten compounds correlated well between species in 18 of 21 comparisons ($r = 0.74$, $p<0.05$ to $r = 0.99$, $p<0.01$). The three exceptions ($0.05<p<0.1$) are mallard compared with chukar ($r = 0.68$), ring-necked pheasant ($r = 0.58$), and European starling (*Sturnus vulgaris*, $r = 0.59$). These data suggest any of the test species, except possibly mallard, represent the acute sensitivity of birds to anticholinesterase pesticides, but the response of one species cannot be used to predict the sensitivity of another species to a specific pesticide. The same conclusions are also reported for pesticides with other toxic mechanisms.[38]

Neither body mass nor close taxonomic relation can be consistently used to predict the sensitivity of birds to pesticides. A list of species in ascending size reveals no apparent trend in sensitivity (Table 2). The largest (ring-necked pheasant) and smallest (house sparrow) are ranked second and third in across-species sensitivity, whereas the chukar, a Phasianidae, is ranked seventh. LD_{50} is lower for pheasants than for chukars for listed pesticides, but the difference varies from 1.2 (NS) to 8.4× ($p<0.05$). It may be significant that the pesticides yielding the least difference between chukar and pheasants are the three carbamates and the two yielding the largest difference of 7.3 and 8.4× are the least toxic OP pesticides, chlorpyrifos and temephos.

LC_{50}

Species response to the subacute protocol has been thoroughly studied only for young of the precocial northern bobwhite, Japanese quail, ring-necked pheasant, and mallard.[18,19,21,30] The differences in LC_{50}s usually are not as large among the young as among adults of the same species.[37] When the subacute tests are conducted on birds of about the same level of susceptibility to the 5-day trial (i.e., recommended ages for regulatory purposes[6]), the order of response most often negatively correlates with body mass: bobwhite = Japanese quail > ring-necked pheasant > mallard.[18] This is probably an interactive function of differential maturation of detoxicating processes and rate of feeding and subsequent exposure in relation to body mass. Even though all combinations of species order of response occurred during tests of more than 100 pesticides, a typical species order tends to prevail within each class of chemicals and LC_{50}s for any two of the test species strongly correlate.[18] Nonetheless, tests of multiple species are always desirable.

LD_{50} vs LC_{50}

Acute and subacute tests yield different toxicologic relationships.[7,37] The differences are exemplified by listing a series of diverse pesticides in ascending order of LD_{50} for young adult mallards and comparing to LC_{50}s for 5-day-old ducklings (Table 3). All studies of each type were conducted at a single laboratory[18,25] with birds of the preferred age for regulation purposes.[5,6] The pesticides represent a near continuum of acute toxicities by overlapping confidence intervals for successive LD_{50}s that result in clusters of several consecutive inseparable LD_{50}s. When the subacute toxicities are compared for pesticides within a cluster of LD_{50}s (e.g., parathion through endrin), the LC_{50}s are almost always statistically separable. The disparity of response to the two tests is indicated by the arithmetic difference between LD_{50}s of little more than 2× for parathion and endrin, monocrotophos and methyl parathion, and endrin and methiocarb. In contrast, the difference in subacute toxicities within each of these LD_{50} clusters is about 60× between LC_{50}s for monocrotophos and aldicarb, 130× for monocrotophos and DDVP (dichlorvos), and 70× for endrin and DDVP. Each of the clusters of four or five pesticides contains both latent and direct ChE inhibiting OP compounds, a carbamate, and a chlorinated hydrocarbon. When the pesticides are ranked by ascending LC_{50}, no more than two successive compounds have overlapping confidence intervals. Overall, no statistically significant correlation exists between the paired LD_{50}s and LC_{50}s.

Some Factors Affecting Interpretation of LD_{50} and LC_{50}

LD_{50}s and LC_{50}s change significantly during growth and development of precocial birds.[21,30,39,40] The direction and amount of change often differ widely between the two tests of lethality. In the acute test, change is believed to be primarily influenced by developing metabolic processes that affect both toxication and detoxication of xenobiotics and an immature immune system. The subacute test is influenced by these same processes and by the highly individualistic response of the experimental animal to the ad libitum toxic diet. Changes in sensitivity as reflected by the oral LD_{50} often follow different patterns depending on the basic toxic mechanism of the pesticide (Table 4). For example, mallard LD_{50}s for anticholinesterases that require activation for maximum potency (i.e., latent cholinesterase inhibitors) tend to decrease between hatch and 7 days and then increase with maturation to adulthood, whereas the opposite pattern occurs for direct acting OP and carbamate anticholinesterases. LD_{50}s for both CNS stimulating chlorinated hydrocarbons follow the pattern of the latent ChE inhibitors. Significant change in LD_{50} occurs between successive ages at least once for each of the pesticides, but little change is evident in the overall order of toxicity among the compounds at the different test ages.

In contrast to the dichotomy of change between successive LD_{50}s during early avian maturation, LC_{50}s typically increase in variable degrees with age during early growth of precocial species.[21,30] The increase occurs across chemical class and is assumed to be primarily due to a change in the ability to cope with the toxic diet for the duration of the subacute protocol; i.e., larger (= older) chicks that eat less proportional to body mass are better able to survive a 5-day trial by reducing food consumption and, therefore, toxic exposure. This is demonstrated by a series of subacute tests with Japanese quail from a single hatch.[30] Food consumption of controls in proportion to body mass averaged 48 g/100 g at 3 days of age, 31 g at 10 days, 24 g at 17 days, and 19 g at 24 days, which is a reduction of about 35, 23, and 21%/week from hatch to 3 weeks of age. During this period, the average increase in LC_{50} for nine pesticides (three

Table 3. Comparative Toxicity of Diverse Pesticides to Mallards Tested Acutely and Subacutely

Pesticide	Class[c]	Acute[a] Rank	LD_{50}	(95% CI[d])	Subacute[b] Rank	LC_{50}	(95% CI)
Fensulfothion	OP-L	1	0.7	(0.6–0.9)	3	41	(32–55)
Parathion	OP-L	2	2.4	(1.7–4.0)	5	76	(61–93)
Aldicarb	CB	3	3.4	(2.7–4.3)	10	594	(507–695)
Monocrotophos	OP-D	4	4.8	(3.4–6.6)	1	10	(8–12)
Endrin	CH	5	5.6	(2.7–11.7)	2	18	(15–21)
DDVP	OP-D	6	7.8	(6.0–10.1)	12	1317	(1043–1674)
Methyl parathion	OP-L	7	10	(6.1–16.3)	8	336	(269–413)
Ethoprop	OP-D	8	13	(11–15)	7	287	(215–382)
Methiocarb	CB	8	13	(7–22)	11	1071	(808–1405)
Morsodren	Hg	10	53	(32–89)	4	51	(43–60)
Toxaphene	CH	11	71	(38–133)	9	538	(474–614)
Dieldrin	CH	12	381	(141–1030)	6	153	(123–196)

[a]Single-dose oral toxicity: LD_{50} as mg active ingredient (technical grade) per kg of body mass calculated to kill 50% of test population. Farm-reared male and female, 3 to 7 months old, n = 8–28 per test; dosage by gelatin capsule.[25]

[b]Five-day dietary toxicity: LC_{50} as mg active ingredient (technical grade) per kg of feed in ad libitum diet calculated to kill 50% of test population. Five groups of 10 unsexed ducklings (5 days old) were tested per pesticide.[18]

[c]Pesticide class: CB, carbamate; CH, chlorinated hydrocarbon; Hg, organic mercury; OP-D, organophosphorus-direct cholinesterase inhibitor; OP-L, organophosphorus-latent cholinesterase inhibitor.

[d]CI = confidence interval.

organophosphorus and two each of carbamate, chlorinated hydrocarbon, and methyl mercury) is 36% between 1 and 7 days, 43% between 7 and 14 days, and 28% between 14 and 21 days. In an acute study with mallards,[39] eight pesticides are compared and the LD_{50}s increase between 1 and 7 days for two compounds by an average of 70%, decrease for three compounds by an average of 80%, and are unchanged for three compounds (Table 4).

LC_{50}s must be used cautiously in comparison of pesticide toxicity among species because the species may not be equally challenged by the test protocol. However, as discussed previously, a reproducible LC_{50} can probably be obtained for any species that cannot survive for 5 days without eating.[27,28] When a portion of the population can survive severe food reductions for the duration of the test, responses tend to be erratic and produce an expanded 95% confidence interval for LC_{50} and a shallow lethality curve that may be a product of factors other than sensitivity. These relationships are demonstrated by subacute tests conducted at a single laboratory with 5- and 10-day-old mallards.[18,41] (*Note*: About 50% of 10-day-old mallards can fast for 5 days, whereas 5-day-old ducklings cannot.[21]) Comparable data sets for nine pesticides indicate variable degrees of increase between LC_{50}s at 5 and 10 days of age (Table 5). LC_{50}s for five of six anticholinesterases increase by an average of 180% while the sixth, fensulfothion, the two chlorinated hydrocarbons, and the methyl mercury are essentially unchanged. Overall, the proportional size of the 95% confidence interval (division of upper by lower bound) averages about 20% smaller and the slope of the lethality curve about 25% steeper for 5-day-old than 10-day-old ducklings. Methiocarb, the only carbamate, has the largest difference in LC_{50}s between ages, extremely wide confidence intervals at both ages, and the steepest lethality curve at 10 days. Carbamates typically yield the

Table 4. Acute Oral Toxicity of Anticholinesterase and CNS Stimulating Pesticides to Mallards from Hatch through Adulthood[39]

Pesticide	LD_{50}[a] (95% CI) 1.5 days	1 week	1 month	6 months
Carbofuran[b]	0.4 (0.3–0.5)	0.6 (0.5–0.7)	0.5 (0.4–0.6)	0.4 (0.3–0.5)
Aldicarb[b]	1.9 (1.6–2.4)	3.6 (2.9–4.5)	6.7 (5.3–8.6)	4.4 (3.5–5.6)
Monocrotophos[c]	5.9 (4.7–7.3)	7.2 (5.8–9.0)	5.1 (4.4–5.9)	3.4 (2.8–4.1)
Demeton[c]	13 (11–16)	15 (13–18)	15 (12–19)	8.2 (6.6–10.2)
Parathion[d]	1.6 (1.4–2.0)	1.4 (1.1–1.8)	1.6 (1.4–2.0)	2.3 (2.0–2.8)
Chlorpyrifos[d]	145 (56–377)	29 (19–47)	50 (32–78)	83 (44–158)
Endrin[e]	22 (10–50)	3.4 (2.4–4.8)	2.9 (2.2–3.9)	5.3 (3.7–7.7)
Endosulfan[e]	28 (23–34)	6.5 (5.2–8.1)	7.9 (5.8–10.8)	34 (26–45)

[a]Toxicity as LD_{50} = mg active ingredient (technical grade) per kg of body mass calculated to kill 50% of test population.
[b]Carbamate (direct ChE inhibitor).
[c]Organophosphorus (direct ChE inhibitor).
[d]Organophosphorus (latent ChE inhibitor).
[e]Chlorinated hydrocarbon (CNS stimulator).

Table 5. Subacute Dietary Toxicity[a] of Widely Diverse Pesticides to 5- and 10-Day Old Mallards[18]

Pesticide	5-Day Old			10-Day-Old		
	LC_{50}	(95% CI)	Slope[b]	LC_{50}	(95% CI)	Slope[b]
Monocrotophos[c]	10	(8–12)	5.4	32*	(19–57)	1.7
Endrin[d]	18	(15–21)	5.7	22	(17–31)	3.4
Fensulfothion[e]	41	(32–55)	5.1	43	(36–51)	4.4
Morsodren[f]	51	(43–60)	8.2	60	(47–76)	7.5
Parathion[e]	76	(61–93)	4.4	275*	(183–373)	3.7
Dicrotophos[c]	94	(80–111)	3.9	144*	(110–185)	3.3
Dieldrin[d]	153	(123–196)	5.4	169	(131–217)	4.9
Methyl parathion[c]	336	(269–413)	5.3	682*	(541–892)	3.2
Methiocarb[g]	1071	(808–1405)	2.5	4113*	(2817–7504)	5.1

[a]Five-day dietary toxicity: LC_{50} as mg active ingredient (technical grade) per kg of feed in ad libitum diet calculated to kill 50% of test population. Asterisk indicates paired LC_{50}s are statistically separable ($p<0.05$).
[b]Slope: probit on log concentration.
[c]Organophosphorus (direct cholinesterase inhibitor).
[d]Chlorinated hydrocarbon (CNS stimulator).
[e]Organophosphorus (latent cholinesterase inhibitor).
[f]Organic mercury.
[g]Carbamate (direct cholinesterase inhibitor).

most erratic response by birds to both acute (controlled dosage) and subacute (uncontrolled dosage) toxicity tests.[19,25,30,41]

Sex, reproductive condition, genetic lineage, nutritional status, and exogenous and endogenous stress may have variable effects on LD_{50} and LC_{50} determinations, but the importance of the factors is not well established for birds. Historically, most acute avian studies tested nonbreeding subadult game birds or adult passerines of both sexes.[25,26,33] This was done to reduce sex effect and thereby conserve the number of birds required for testing species sensitivity and ranking the acute toxicity of pesticides. The legitimacy of pooling sexes of reproductively quiescent birds has been validated for acute toxicity testing.[27,33,38,42] However, beyond general comparisons, this narrow focus may not be adequate for hazard assessment because pesticides are intensively applied in nature during avian breeding seasons and knowledge of sex differences in sensitivity is essential. The importance of this variable is indicated by an acute test of fenthion toxicity that showed female northern bobwhite to be 2.3 times ($p < 0.05$) as sensitive as males.[43]

Research on birds usually is with captive-reared specimens from haphazardly outbred stocks or wild-captured birds of unknown origin. Reproducibility of acute toxicity tests with birds of such vague genetic lineage is not known. However, in a study with equal-aged farm-reared northern bobwhites of both sexes from eight commercial breeders, extreme LD_{50}s for technical grade diazinon were 13 and 17 mg/kg body mass.[44] These two extremes are statistically inseparable, although the eight stocks differed in apparent vigor and body mass at dosing. Both factors are known to affect acute response,[45] but genetic variability from outbreeding could obscure detection of minor differences based on LD_{50} alone.

Adequate methods are not available to evaluate the suitability of a wild-captured individual or species for acute toxicity testing. Simple survival and weight maintenance for a few weeks in captivity may not reflect subtleties such as nutritional imbalance or stress response to confinement, isolation, or crowding. Whether captive specimens, either wild or farm hatched and reared, truly represent their free-living counterparts is not known. For example, DDT and several organophosphorus insecticides were tested subacutely on wild bluejays (*Cyanocitta cristata*), house sparrows, northern cardinals (*Cardinalis cardinalis*), and wild and farm northern bobwhites.[27] All birds were at their capture weight and believed to be adequately conditioned to captivity at the time of testing. Bluejays were the most sensitive species to all compounds and farm bobwhites the most tolerant. Bluejays are adaptable generalized feeders that are reputed to be quite resilient in contaminated environments[46] and are easily kept in captivity, yet based on LC_{50}s they are about 1.5 to 50 times as sensitive as the other species to the various insecticides. Wild bobwhites had much less subcutaneous and visceral fat than their farm counterparts, weighed about 25% less, and consistently gave lower LC_{50}s. The difference is attributed in large part to consumption of significantly more toxic feed proportional to body mass by the wild birds during the 5-day trial rather than to differential sensitivity. Neither body mass nor rate of feeding explains the unexpected bluejay sensitivity because they are nearly twice as heavy and eat proportionally less than either house sparrows or cardinals.

HAZARD EVALUATION

It is clear from the foregoing that the most often used criteria of toxicity, the single-dose acute oral LD_{50}, varies unpredictably among avian species, and responses by laboratory rats to acute tests do not adequately represent avian response. When feeding for 5 days is substituted for controlled dosage, the resultant subacute LC_{50} often produces relationships among species and chemicals that are quite different from those for LD_{50}s. Acute and subacute tests provide complementary measures of relative potency for the identification of chemical substances of potential lethal toxicity to wildlife. Although neither the LD_{50} nor LC_{50} per se is more than a convenient statistical reference point, evaluation of associated dose-response curves and observations of toxic responses enhance the utility of acute-type lethality tests in hazard assessment. These tests are meager considering that avian habitat is routinely treated with a variety of formulations and combinations of pesticides and that many factors alter the chemical fate and availability of a pesticide. However, ingestion is believed to be the most common route of pesticidal exposure in birds,[46] and therefore these oral tests of lethality provide a sound basis for preliminary screening.

LD_{50} and LC_{50} provide a statistical measurement that can be used to classify pesticides by an established scale of toxicity.[5,6,18,36] This criterion provides simplistic guidance in first-line reviews of any array of pesticides for lethal hazard. Caution must be exercised to ensure that comparisons are based on test subjects that are equally susceptible to the experimental protocol (e.g., special attention to age, body mass, and feeding habits) and that the median response level is supported by its 95% confidence interval. LD_{50} is derived by controlled dosage and therefore provides a tangible measure of naive sensitivity to toxic challenge that can be used for direct comparison of species, life stages, and chemicals. Although the emphasis herein is on oral dosage, the basic acute test can also be used to evaluate percutaneous toxicity. In comparative studies with mallards and several passerines, oral LD_{50}s were consistently lower ($p<$ 0.05) than percutaneous LD_{50}s for an array of pesticides[47,48]. An LD_{50} is difficult to relate to a field application of pesticide because some combination of inhalation, percutaneous, and ingestive exposure is probably the rule.

LC_{50} provides a basis for comparison of the ability of the test population to cope with chemically contaminated feed for 5 days. This subacute test is believed by some to be more practical than its acute predecessor because the birds must voluntarily ingest the pesticide and are then subject to the effects of repeated dosage as might be experienced in nature. However, subacute studies usually use technical grade pesticide mixed into dry feed, whereas natural ingestion of the finished product formulation may be from varied sources such as water, seeds, foliage, invertebrates, vertebrates, and granular pesticides,[46] and the toxicity of the pesticide may be different in each matrix because of its form or availability. In a realistic sense, except for some carbamates, a field residue equivalent to an LC_{50} in a specific food matrix may not be especially hazardous to a mobile population if the birds choose to emigrate. Emigration is more likely due to food deprivation (i.e., reduced arthropod population) than toxicity.[49-51]

Some insight into potential hazard associated with a specific level of 5-day subacute toxicity is provided by comparison of cumulative mortality patterns during exposure to LC_{50} concentration of carbamate, OP, chlorinated hydrocarbon, and organic mercury (Figure 1). The response curves are based on studies of 14-day-old Japanese quail and are typical for most compounds in the represented pesticidal classes.[19,30] (Comparable mortality patterns occur for 5-day-old mallards and 10-day-old ring-necked pheasants.[55]) LC_{50} is presented because it is the focus of the experimental design, and therefore responses are least variable, but lower or higher response levels produce the same characteristic pattern, with the sigmoid response beginning about 1 day later at lower levels and 1 day earlier at higher levels.

The mortality pattern for dicrotophos is consistent with the cumulative response theoretically necessary to kill a portion of the test population during 5-day exposure to a nonaccumulative toxicant. Mortality from OP compounds is rare after withdrawal of

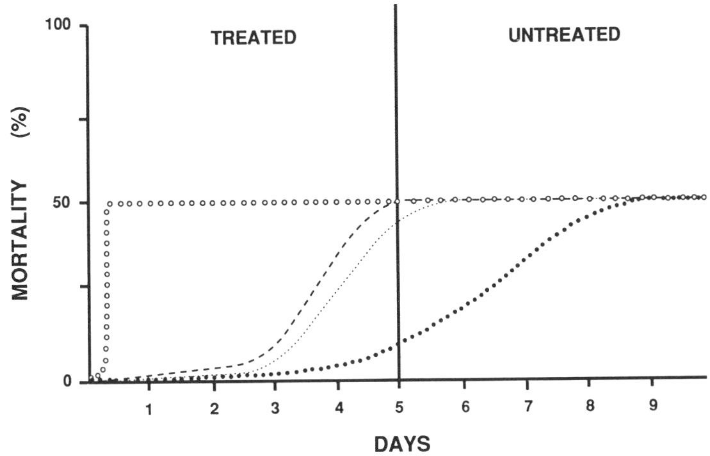

FIGURE 1. Cumulative mortality patterns for 14-day-old Japanese quail fed LC_{50} concentration of carbofuran (open circle), dicrotophos (dash), dieldrin (dot), and Ceresan M® (closed circle) for 5 days followed by untreated feed.

treated feed.[19] A typical response to OP exposure occurred with dicrotophos. Consumption decreased by 30% compared with controls during the first day of exposure, by 55% during the second and third days, and by 60% during the fourth and fifth days.[55] Feeding at lower and higher response levels is described in detail elsewhere for many species.[19,27,28,30,41] Dieldrin produced essentially the same cumulative response pattern as dicrotophos but some mortality occurred during the first day on untreated feed. Although dieldrin is lipophilic and accumulative, latent mortality is not common, provided ad libitum untreated feed is available.[19,30] Consumption of dieldrin-treated feed decreased compared with controls by about 15, 30, 40, 45, and 45% during the first through fifth days.[55] Quail fed Ceresan M® showed little evidence of toxicity preceding the first death on the last day of exposure, then toxic signs began to intensify and deaths ensued through the fourth day of untreated feed; all toxic signs remised in survivors by day 13.[30] Consumption of Ceresan M®-treated feed was consistently about 5 to 15% less than control consumption, but daily differences were not significant. A detailed account of subacute response to mercury is presented elsewhere.[40] In contrast to each of the above patterns, all deaths from carbofuran occurred during the first few hours of feed presentation. After an initial decrease of about 60%, feed consumption was reduced by only 25% on the second day and comparable to or in excess of controls thereafter.[55] This temporal pattern also occurs at higher and lower response levels and is generally representative of other carbamates.[19] The OP fensulfothion produced a carbamate-type response pattern with mallards,[17] but a typical OP pattern with Japanese quail.[30]

When the subacute response patterns depicted in Figure 1 are considered with their corresponding rates of consumed toxic feed, many different exposure scenarios can be developed to enhance the evaluation of the potential hazard. For example, potential effects on migrants can be compared to resident populations, and mobile residents to breeders, and so on. Certainly, from these patterns it would not have been difficult to predict that carbofuran poses an acute hazard to birds, which it does;[52,53] or that Ceresan M® is much more hazardous than indicated by its single-dose LD_{50} of 668 mg/kg (95% confidence interval, 530 to 842 mg/kg) for adult Japanese quail.[25] Nonetheless, caution must be used when projecting results of subacute studies to the field because in the laboratory, reasonably consistent exposure can be provided over time, whereas field exposure is erratic because pesticide is naturally degraded and translocated. Care must also be used in the interpretation of experimental feed consumption because subacute trials usually test technical grade chemical mixed into dry mash. Pesticide presented in this way may be easily sensed and consumption reduced; in the field, finished product formulation may be less easily detected when present in natural matrices including plant and animal tissues. Thus, different factors may render a pesticide either more or less toxic in the field than predicted from laboratory studies.

The dose-response or lethality curve calculated from acute and subacute toxicity tests is critical to the evaluation of potential pesticide hazard to wildlife. The curve is used in the same general way for both tests, but their interpretive implications are somewhat different because of the method of exposure. The most important concept applicable to both tests is that a steep lethality curve indicates increased hazard if for no reason other than proportionally less chemical increases effect; thus, applicator precision is essential. However, chemicals that produce shallow curves may be even more hazardous if the slope is not known. These somewhat contradictory notions are explained by comparison of hypothetical pesticides A and B with slopes (probit on log dose) of 8.0 and 2.0 and both with an arbitrary LD_{50} of 10 mg/kg (Figure 2). Assume the slope is known for pesticide A and the expected exposure is 6 mg/kg which may kill about 5% of the population; if treatment is accidentally doubled and results in exposure of 12 mg/kg, it would kill about 75% of the population, a 15-fold increase. In contrast, assume the slope is not known for pesticide B, but its LD_{50} of 10 mg/kg is the same as for pesticide A, and this time the target exposure of 6 mg/kg is met. The shallow slope indicates that about 35% of the population would be killed. Pesticides such as carbofuran tend to yield shallow slopes[30,42] and have been implicated in numerous avian die-offs.[54]

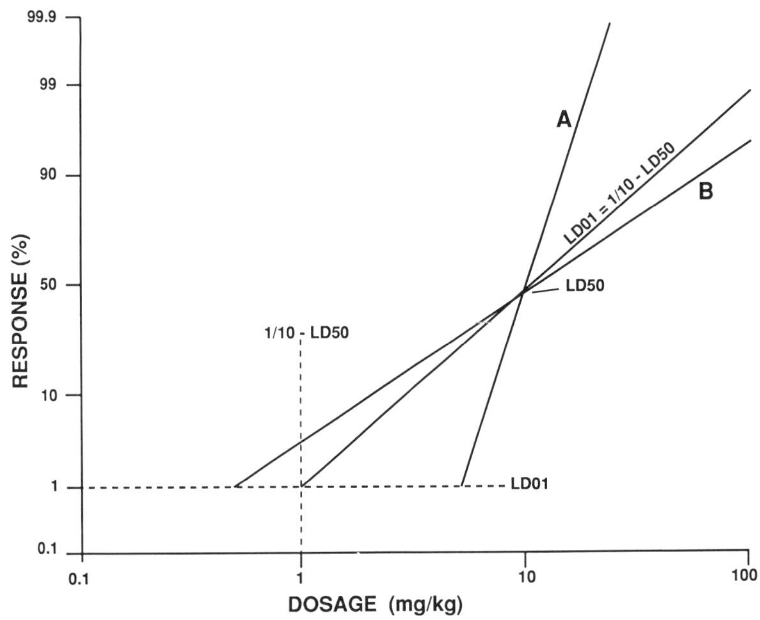

FIGURE 2. Dose-response curves of hypothetical pesticides A (slope 8.0) and B (slope 2.0) and a line (slope 2.3) intercepting the coordinates of the LD_{01} and 1/10 LD_{50}.

For regulatory purposes, a popular method is to use some fraction of the LD_{50} or LC_{50} to denote hazard and restrict use of treatments that probably yield an exposure potential to wildlife. Suppose the acceptable residue in the equivalent of one feeding bout is set at 1/10 of the LD_{50}, or 1 mg/kg. In this example, pesticide A would appear safe and pesticide B lethal to about 5% of the exposed population (Figure 2). In Figure 2 the 1/10 LD_{50} is arbitrarily intercepted with the calculated LD_{01} for reference. The resultant slope is about 2.5, which is much more shallow than that calculated for most pesticides tested either acutely or subacutely with birds.[18,19,42] Therefore, the 1/10 LD_{50} or LC_{50} criterion appears to be a reasonably conservative parameter for most purposes when the slope of the dose-response curve is not known.[42] Even when the dose-response curve is known, use of coordinates outside the linear limits (i.e., ±1 S.D. of the midpoint of the curve or the 16 and 84% response level) is discouraged.[1,17]

In a practical sense, the steepness of the dose-response curve can be reduced to a qualitative index based on the ratio between two constant response levels; e.g., LD_{10} and LD_{50}. The smaller the ratio, the more hazardous the substance because proportionally smaller amounts increase effect and thereby reduce the acceptable margin of error in a pesticidal application. In contrast, shallow slopes indicate greater inherent safety because it takes proportionally more chemical to increase effect; however, low levels may cause unacceptable effects.

CONCLUSIONS

Single-dose acute oral and 5-day subacute dietary toxicity studies are the preponderance of available data for preliminary assessment of pesticidal hazard to wildlife. Properly designed, these tests provide a method of comparing pesticides by lethality from one (acute) or multiple (subacute) exposures that generate statistical estimates of the dose-response curve and its midpoint, LD_{50} or LC_{50}. When these tests are supplemented with detailed observations of individual responses and food consumption through remission of toxicity, a meaningful appraisal of potential lethal hazard is possible.

Historically, only LD_{50} or LC_{50} has received extensive use, and often without consideration of its statistical validity. This approach is inappropriate because both LD_{50}s and LC_{50}s vary widely in unpredictable ways between chemicals, species, and the life stage of the test subjects. Therefore, careful review of test compatibility is essential before any comparisons are attempted. However, once the credibility of the study is ascertained, LD_{50} and LC_{50} provide useful guides to chemical potency for comparing pesticides of different mechanisms of toxic action. Specifically, LD_{50} provides a direct measure of sensitivity, whereas LC_{50} yields information on sensitivity to the chemical and the ability of birds to cope with toxic feed for a specified duration. A review of the responses indicated from mortality patterns and slopes of dose-response curves gives insight into potential hazards of both an acute and chronic nature.

However, literal projection of either acute or subacute tests to nature is not possible. Most laboratory tests use a technical grade chemical, either administered directly to the bird or in a dry feed. Field application almost always uses a finished product formulation of pesticide, and formulations may vary in toxicity and availability depending on the use and factors of environmental degradation. Therefore, extreme care is recommended in the use of acute and subacute toxicity tests; when used in combination and judiciously, the two tests of lethality are invaluable tools for preliminary evaluation of potential hazard of pesticides to wild birds.

REFERENCES

1. Finney, D. J., *Probit Analysis*, 3rd ed., Cambridge University Press, Cambridge, U.K., 1971.
2. Casarett, L. J., Toxicologic evaluation, in *Toxicology: the Basic Science of Poisons*, L.J. Casarett and J. Doull, Eds., Macmillan, New York, 1975, pp. 11-25.
3. Chan, P. K., and A. W. Hayes, Principles and methods for acute toxicity and eye irritancy, in *Principles and Methods of Toxicology*, 2nd ed., A. W. Hayes, Ed., Raven Press, New York, 1989, pp. 69-220.
4. Pesticide Assessment Guidelines, Subdivision E, Hazard Evaluation: Wildlife and Aquatic Organisms, EPA-540/9-82-024, Ecological Effects Branch, Hazard Evaluation Division, Office of Pesticide Programs, Washington, DC, 1982.
5. Bascietto, J., Hazard Evaluation Division Standard Evaluation Procedure Avian Single-dose Oral LD_{50}, EPA-540/9-85-007, U.S. Environmental Protection Agency, Washington, DC, 1985.
6. Bascietto, J., Hazard Evaluation Division Standard Evaluation Procedure Avian Dietary LC_{50} Test, EPA-540/9-85-008, U.S. Environmental Protection Agency, Washington, DC, 1985.
7. Heinz, G. H., E. F. Hill, W. H. Stickel and L. F. Stickel, Environmental contaminant studies by the Patuxent Wildlife Research Center, in *Avian and Mammalian Wildlife Toxicology*, STP 693, E. E. Kenaga, Ed., American Society for Testing and Materials, Philadelphia, 1979, pp. 9-35.
8. Hill, E. F., and D. J. Hoffman, Avian models for toxicity testing, *J. Am. Coll. Toxicol.*, 3, 357-376, 1984.
9. Gad, S., and C. Weil, *Statistics and Experimental Design for Toxicologists*, Telford Press, Caldwell, NJ, 1986.
10. Gad, S. C., and C. P. Chengelis, *Acute Toxicology Testing*, Telford Press, Caldwell, NJ, 1988.
11. Albert, A., *Selective Toxicity*, 4th ed., Butler & Tanner, London, 1965.
12. Hewlett, P. S., and R. L. Plackett, *The Interpretation of Quantal Responses in Biology*, University Park Press, Baltimore, MD, 1979.
13. Finney, D. J., *Statistical Methods in Bioassay*, 2nd ed., Hafner, New York, 1964.
14. Klaassen, C. D., Principles of toxicology, in *Casarett and Doull's Toxicology: The Basic Science of Poisons*, 3rd ed., C. D. Klaassen, M. O. Amdur and J. Doull, Eds., Macmillan, New York, 1986, pp. 11-32.
15. Litchfield, J., and F. Wilcoxon, A simplified method of evaluating dose-effect experiments, *J. Pharmacol. Exp. Ther.*, 96, 99-113, 1949.
16. Heath, R. G., and L. F. Stickel, Protocol for testing the acute and relative toxicity of pesticides to penned birds, in *The Effects of Pesticides on Wildlife*, Circular 226, U.S. Fish and Wildlife Service, Washington, DC, 1965, pp. 18-24.
17. Heath, R. G., J. W. Spann, E. F. Hill and J. F. Kreitzer, *Comparative Dietary Toxicities of Pesticides to Birds*, Spec. Sci. Rep. 152, U.S. Fish and Wildlife Service, Washington, DC, 1972.
18. Hill, E. F., R. G. Heath, J. W. Spann and J. D. Williams, *Lethal Dietary Toxicities of Environmental Pollutants to Birds*, Spec. Sci. Rep. 191, U.S. Fish and Wildlife Service, Washington, DC, 1975.
19. Hill, E. F., and M. B. Camardese, *Lethal Dietary Toxicities of Environmental Contaminants and Pesticides to Coturnix*, Fish and Wildlife Tech. Rep. 2, U.S. Fish and Wildlife Service, Washington, DC, 1986.
20. Tucker, R. K., and J. S. Leitzke, Comparative toxicology of insecticides for vertebrate wildlife and fish, *Pharmacol. Ther.*, 6, 167-220, 1979.
21. Hill, E. F., J. W. Spann and J. D. Williams, Responsiveness of 6 to 14 generations of birds to dietary dieldrin toxicity, *Toxicol. Appl. Pharmacol.*, 42, 425-431, 1977.
22. Miller, R. G., Jr., *Simultaneous Statistical Inference*, 2nd ed., Springer-Verlag, New York, 1981.

23. Thompson, W. R., Use of moving averages and interpolation to estimate median effective dose. I. Fundamental formulas estimation of error, and relation to other methods, *Bacteriol. Rev.*, 11, 115–145, 1947.
24. Weil, C. S., Tables for convenient calculation of median effective dose (LD_{50} or ED_{50}) and instruction in their use, *Biometrics*, 8, 249–263, 1952.
25. Hudson, R. H., R. K. Tucker and M. A. Haegele, *Handbook of Toxicity of Pesticides to Wildlife*, Resource Publ. 153, U.S. Fish and Wildlife Service, Washington, DC, 1984.
26. Schafer, E. W., W. A. Bowles, Jr., and J. Hurlbut, The acute oral toxicity, repellency, and hazard potential of 998 chemicals to one or more species of wild and domestic birds, *Arch. Environ. Contam. Toxicol.*, 12, 355–382, 1983.
27. Hill, E. F., Toxicity of selected mosquito larvicides to some common avian species, *J. Wildl. Manage.*, 35, 757–762, 1971.
28. Grue, C. E., Response of common grackles to dietary concentrations of four organophosphate pesticides, *Arch. Environ. Contam. Toxicol.*, 11, 617–626, 1982.
29. Turner, L. W., Development of an avian dietary LC_{50} toxicity test for potential use under the Toxic Substances Control Act, in *Avian and Mammalian Wildlife Toxicology: Second Conference*, STP 757, D. W. Lamb and E. E. Kenaga, Eds., American Society for Testing and Materials, Philadelphia, 1981, pp. 98–104.
30. Hill, E. F., and M. B. Camardese, Subacute toxicity testing with young birds: response in relation to age and intertest variability of LC_{50} estimates, in *Avian and Mammalian Wildlife Toxicology: Second Conference*, STP 757, D. W. Lamb and E. E. Kenaga, Eds., American Society for Testing and Materials, Philadelphia, 1981, pp. 41–75.
31. Gaines, T. B., The acute toxicity of pesticides to rats, *Toxicol. Appl. Pharmacol.*, 2, 88–99, 1960.
32. Gaines, T. B., Acute toxicity of pesticides, *Toxicol. Appl. Pharmacol.*, 14, 515–534, 1969.
33. Schafer, E. W., Jr., The acute oral toxicity of 369 pesticidal, pharmaceutical and other chemicals to birds, *Toxicol. Appl. Pharamcol.*, 21, 315–330, 1972.
34. Pan, H. P., J. R. Fouts and T. R. Devereaux, Hepatic microsomal N-hydroxylation of p-chloroaniline and p-chloro-N-methylaniline in red-winged blackbird compared with rat, *Xenobiotica*, 9, 441–446, 1979.
35. Walker, C. H., Species variation in some hepatic microsomal enzymes, *Prog. Drug. Metab.*, 5, 113–164, 1980.
36. Smith, G. J., Pesticide Use and Toxicology in Relation to Wildlife: Organophosphorus and Carbamate Compounds, Resource Publ. 170, U.S. Fish and Wildlife Service, Washington, DC, 1987.
37. Hill, E. F., Avian toxicology of anticholinesterases, in *Clinical and Experimental Toxicology of Anticholinesterases*, B. Ballantyne and T. C. Marrs, Eds., Butterworth, London, 1991, pp. 272–294.
38. Tucker, R. K., and M. A. Haegele, Comparative acute oral toxicity of pesticides to six species of birds, *Toxicol. Appl. Pharmacol.*, 20, 57–65, 1971.
39. Hudson, R. H., R. K. Tucker and M. A. Haegele, Effect of age on sensitivity: acute oral toxicity of 14 pesticides to mallard ducks of several ages, *Toxicol. Appl. Pharmacol.*, 22, 556–561, 1972.
40. Hill, E. F., and J. H. Soares, Jr., Oral and intramuscular toxicity of inorganic and organic mercury chloride to growing quail, *J. Toxicol. Environ. Health*, 20, 105–116, 1987.
41. Hill, E. F., Subacute dietary toxicities of dicrotophos and dieldrin in time-replicated trials with young ring-necked pheasants and mallards, in *Avian and Mammalian Wildlife Toxicology: Second Conference*, STP 757, D. W. Lamb and E. E. Kenaga, Eds., American Society for Testing and Materials, Philadelphia, 1982, pp. 105–120.
42. Hill, E. F., and M. B. Camardese, Toxicity of anticholinesterase insecticides to birds: technical grade versus granular formulations, *Ecotoxicol. Environ. Safety*, 8, 551–563, 1984.
43. Wiemeyer, S. N., and D. W. Sparling, Acute toxicity of four anticholinesterase insecticides to American kestrels, eastern screech-owls, and northern bobwhite, *Environ. Toxicol. Chem.*, 10, 1139–1148, 1991.

44. Hill, E. F., M. B. Camardese, G. H. Heinz, J. W. Spann and A. B. BeBevac, Acute toxicity of diazinon is similar to eight stocks of bobwhite, *Environ. Toxicol. Chem.*, 3, 61-66, 1984.
45. Weil, C. S., C. P. Carpenter, J. S. West and H. F. Smyth, Jr., Reproducibility of single oral dose toxicity testing, *Am. Indust. Hyg. Assoc.*, 27, 483-487, 1966.
46. Grue, C. E., W. J. Fleming, D. G. Busby and E. F. Hill, Assessing hazards of organophosphate pesticide to wildlife, *Trans. North Am. Wildl. Nat. Res. Conf.*, 48, 200-220, 1983.
47. Schafer, E. W., Jr., R. B. Bunton, N. F. Lockyer and J. W. DeGrazio, Comparative toxicity of seventeen pesticides to the *Quelea*, house sparrow, and red-winged blackbird, *Toxicol. Appl. Pharmacol.*, 26, 154-157, 1973.
48. Hudson, R. H., M. A. Haegele and R. K. Tucker, Acute oral and percutaneous toxicity of pesticides to mallards: correlations with mammalian toxicity data, *Toxicol. Appl. Pharmacol.*, 47, 451-460, 1979.
49. DeWeese, L. R., C. J. Henny, R. L. Ford, K. A. Bobal and A. W. Shultz, Response of Breeding Birds to Aerial Sprays of Triachlorfon (Dylox) and Carbaryl (Sevin-4-Oil) in Montana Forests, Spec. Sci. Rep. 224, U.S. Fish and Wildlife Service, Washington, DC, 1979.
50. McEwen, L. C., L. R. DeWeese and P. Schaladweiler, Bird predation on cutworms (Lepidoptera:Noctuidae) in wheat fields and chlorpyrifos effects on brain cholinesterase activity, *Environ. Entomol.*, 15, 147-151, 1986.
51. Stromborg, K. L., C. E. Grue, J. D. Nichols, G. H. Hepp, J. E. Hines and H. C. Bourne, Postfledging survival of European starlings exposed to an organophosphorus insecticide, *Ecology*, 69, 590-601, 1988.
52. Flickinger, E. L., C. A. Mitchell, D. H. White and E. J. Kolbe, Bird poisoning from misuse of the carbamate Furadan in a Texas rice field, *Wildl. Soc. Bull.*, 14, 59-62, 1986.
53. Littrell, E. E., Waterfowl mortality in rice fields treated with the carbamate, carbofuran, *Calif. Fish Game*, 74, 226-231, 1988.
54. Office of Pesticides and Toxic Substances, Carbofuran Special Review Technical Support Document, U.S. Environmental Protection Agency, Washington, DC, 1989.
55. Hill, E. F., unpublished data.

23

The Role of In Vitro Studies in Ecological Hazard Assessments

Richard T. Di Giulio, Evan P. Gallagher, and Clifford Habig

ABSTRACT

While perhaps not playing as clearly a fundamental role in ecological hazard assessments of contaminants as activities such as toxicity tests, field studies, and population modeling, in vitro studies can provide very useful information that aids in these assessments. In vitro methodologies provide rapid, cost-effective means for elucidating mechanisms of action, determining target sites, measuring kinetics of activation and deactivation of toxicants, comparing the abilities of different species to activate or detoxify toxicants, and developing techniques for biomonitoring. This chapter discusses the use of in vitro studies for accomplishing these purposes and thereby enhancing ecological hazard assessments. Relevant studies from our laboratory are described and include studies of mechanisms for and species comparisons of oxyradical production by redox-active compounds (e.g., paraquat, nitroaromatics), comparative studies of organophosphorus (OP) ester activation and cholinesterase (ChE) inhibition, and the role of glutathione in detoxification, both via glutathione S-transferase activity and as an antioxidant. Collectively, these studies and others support the use of in vitro studies in ecological hazard assessments. However, this role could be much more fully exploited if selected in vitro studies were designed specifically to support such assessments, rather than used randomly and posteriorly, as is typically the case.

KEY WORDS

in vitro, hazard assessment, pesticides

INTRODUCTION

"In vitro" refers to studies with cells or tissues that are "isolated from living tissue and artificially maintained" (Webster). In toxicology, in vitro approaches play fundamental roles, particularly in terms of elucidating pathways of xenobiotic metabolism and mechanisms of toxic action. Such information is critical to determinations of target sites (cell or organ specificities), health hazards, and potential remedial actions (i.e., therapeutic drugs). These issues clearly are central to the realm of human toxicology, where the individual comprises the level of biological organization of ultimate concern. In ecological toxicology and associated hazard assessments, however, higher levels of organization (populations, communities, and ecosystems) are the focal points.

For understanding or predicting hazards arising from inputs of xenobiotics, including pesticides, at these levels, in vivo toxicity tests, field studies, and ecological modeling provide key approaches. In vitro studies, therefore, may appear somewhat esoteric relative to the goals of ecological hazard assessments. By further considering, for example, simply the number of species of concern to "traditional" toxicology (one, the human) vs those of ecological concern (many), it is clear that the potential utility of the detailed information available through in vitro studies is not feasibly realized in the ecological setting.

These issues and limitations notwithstanding, in vitro approaches can provide highly useful techniques and information for ecological hazard assessments. This utility reasonably can be expected to be maximized when in vitro studies are carefully selected and designed into a specific assessment a priori, which rarely appears to have been the case in the albeit brief history of ecological hazard assessments. The purpose of this paper is to describe roles by which in vitro approaches can enhance these assessments; examples, including recent or current studies in our laboratory, will be employed to illustrate some of these roles.

KEY ROLES

Important roles in vitro studies may serve in ecological hazard assessments include screening compounds for various biological activities (such as biochemical effects associated with certain toxicities), providing fundamental species-comparative information (including relative species or tissue sensitivities), elucidating metabolic pathways for xenobiotics, and determining mechanisms of action. In these roles, in vitro approaches provide advantages related to speed and level of detail; in terms of the latter, in vitro studies are not only advantageous, but often required, as in the case of mechanistic studies.

In toxicity testing in general, the use of in vitro techniques for screening compounds for various cellular and biochemical perturbations has become widespread, and this trend is likely to continue. Motivations for this trend include the speed and economy afforded by in vitro tests vs in vivo tests and the substantial reduction in numbers of animals sacrificed for in vitro studies.[1] Testing for potential human carcinogens comprises an example of an area where the use of in vitro approaches is greatly increasing. In vivo-based rodent bioassays for carcinogen testing cost approximately $1 million per test chemical and require hundreds of animals. Therefore, in vitro testing with both bacterial and mammalian tissue/cell culture is increasingly employed. While at this time in vitro methodologies cannot completely supplant in vivo bioassays (and ignoring here the considerable uncertainties associated with these bioassays), in vitro techniques can at the least provide an efficient means for prioritizing chemicals for the lengthy bioassays.

Similarly, in vitro approaches can provide time- and cost-effective mechanisms for prioritizing pesticides for further study, as well as assisting with comparative studies of species or tissue sensitivity. This aspect of prioritization may be particularly true in the case of many insecticides (such as OP and carbamates) where the mechanism of action is understood and amenable to routine analysis (i.e., acetylcholinesterase activities). In view of the large number of pesticides, both registered and in development, requiring examination for ecological hazards, and the much larger number of nontarget species potentially at risk from routine applications, the role that in vitro studies could play in prioritizing compounds, species, and tissues within species, appears to be much greater than is presently exploited.

Closely related to these practical issues of screening and prioritizing is a more fundamental issue related to the generation of basic, comparative information that is woefully inadequate in ecotoxicology presently. In vitro studies could play an important role in elucidating comparative aspects of species (and where relevant, tissue) metabolism of, and sensitivity to, various xenobiotics including pesticides. As such studies become more common, and techniques more standardized, the resulting database will be of great use not only in the context of the basic research issues for which it was largely intended, but also to the process of ecological hazard assessments. For example, while the great number of species of concern to ecotoxicology may appear to limit the feasibility of gathering the detailed information that in vitro studies can provide, these studies could play a key role in underpinning interspecies extrapolations inherent to ecological assessments but presently based largely on guesswork.

While the roles in vitro studies can play in hazard assessments in terms of prioritizations and comparative information are compelling, it is in the realm of providing information concerning metabolism and mechanisms of action of particular compounds where these approaches are most fundamental and, indeed, indispensable. An understanding of the metabolism of a xenobiotic such as an insecticide is critical to the hazard assessment process. Information concerning the sites (e.g., tissues) and rates of metabolism, and the identity of metabolites formed, is essential to quantification of the exposure component of the hazard equation (i.e., hazard = f[exposure + toxicity]), and can also be necessary for elucidating mechanisms of action. Most organic compounds are metabolized by exposed organisms, and understanding this process for a given compound is required in order to interpret the tissue residue data that underlie exposure assessments. Understanding metabolism will provide answers to key questions such as: Which tissues will accumulate and/or metabolize the compound in question? At what rates will various tissues metabolize the compound? What are the metabolites formed? Answers to these questions will direct exposure assessments by indicating the proper tissues to analyze, elucidating the temporal patterns of degradation in these tissues, and most fundamentally, determining what compound(s) to analyze (i.e., parent compound or certain metabolites). In vitro studies are central to metabolism studies and are required to assess key variables such as cofactor and other biochemical/cellular requirements, rates of degradation, and identities of metabolites formed.

Additionally, understanding metabolism is often crucial to the final key role for in vitro studies to be considered here—that of elucidating mechanism of action. While metabolism is classically considered a mechanism by which organisms degrade organic compounds to less toxic and/or more excretable products, metabolism can also play essential roles in enhancing toxicities of many xenobiotics ("activations"). For example, the microsomal desulfuration of many OP pesticides from phosphorothionates (where sulfuration increases stability and hence shelf and environmental longevity) to the corresponding phosphates results in compounds of considerably greater potencies as anticholinergic agents, the key mode of acute toxicity of these compounds.[2] Another well-established example is provided by the microsomal oxidation of polycyclic aromatic hydrocarbon (PAH) procarcinogens to ultimately carcinogenic metabolites (e.g., diol epoxides).[3] For many compounds, including both preceding examples, various metabolic processes comprising both activations and detoxification occur simultaneously, and the degree of toxicity resulting from exposures to these compounds reflects the balance of these opposing phenomena. Another illustration of the importance and complexity of xenobiotic metabolism is provided by the dehydrohalogenation of DDT to form DDE.[2] In this case, the neurotoxicity of the parent compound (DDT) has been diminished; however, the product (DDE) can impair avian reproduction through a variety of other, more subtle mechanisms, such as eggshell thinning.

Elucidating mechanisms of action, including the roles of metabolism, comprises perhaps the most fundamental research activity of modern toxicology. This is in large part due to the assumption that these elucidations are critical to reducing uncertainties surrounding the risks posed by environmental agents to human health.[4] As discussed earlier, ecotoxicology focuses upon higher levels of biological organization and precludes as great an emphasis on mechanisms compared to medical toxicology. As the science of ecotoxicology evolves and develops a sounder theoretical basis, however, the utility of mechanistic studies will increase substantially; linkages between mechanisms and population effects, for example, may become quantifiable.

At the present time, the elucidation of mechanisms of action with in vitro studies can contribute substantially to ecological hazard assessments. This information, in concert with tissue distribution and metabolism data, can help identify key targets (tissues, cells, organelles) of the compound in question. Additionally, as ecological hazard assessments become more refined, there is growing emphasis on sublethal effects (for example, effects on reproduction, behavior, and energy metabolism) that can greatly influence population-level impacts. Understanding mechanisms of action plays a central role in the development of biochemical and physiological tools ("biomarkers") for monitoring contaminant exposures and/or sublethal effects in exposed animal populations. In terms of sensitivity and specificity, biomarkers associated with mechanisms of action generally will be superior to nonspecific indicators (of stress, for example).[5] A "model" biomarker is comprised by the enzyme acetylcholinesterase (AChE) as an index of exposure to OP pesticides and their neurotoxic effects. As mentioned earlier, this enzyme is very sensitive to OP compounds, and its inhibition is the mechanism underlying OP neurotoxicity. These properties, and its ease of measurement, have made AChE a very powerful tool in ecological hazard assessments of OPs.[6] The development of effective biomarkers for environmental contaminants comprises a vigorous research arena in ecotoxicology today, and an area wherein in vitro studies clearly can benefit ecological hazard assessments. Recent in vitro studies, largely from our laboratory, are described in the following that illustrate some of the ways in which these studies can enhance assessments. As is generally the case, however, these studies were not performed with assessments in mind, but were motivated by concerns for basic information concerning xenobiotic metabolism and mechanisms of action, including comparative aspects.

ILLUSTRATIVE EXAMPLES

Xenobiotic-Mediated Oxyradical Production in Aquatic Animals

A fundamental mechanism of action underlying the toxicities of a number of compounds that has received considerable attention by toxicologists in recent years is referred to as "oxygen toxicity" or "oxidative stress." This toxicity is due to production of oxygen-free radicals (oxyradicals) that overwhelm cellular antioxidant defense systems.[7,8] Oxyradicals are reactive, partially reduced oxygen species such as the superoxide anion radical (O_2–), hydrogen peroxide (H_2O_2), and the hydroxyl radical (·OH). These compounds are produced at limited rates during the normal course of oxygen-associated metabolic activities such as mitochondrial respiration, microsomal metabolism, photosynthesis, phagocytosis by leukocytes, and various enzymatic activities. In order to offset the potential deleterious effects of oxyradicals, all aerobic organisms possess antioxidant defenses comprised of certain enzymes including superoxide dismutase (SOD), catalase, and peroxidases, and various low molecular weight scavengers such as glutathione, ascorbic acid, and tocopherols.

A number of perturbations can lead to situations in which the production of oxyradicals is greater than the capacity of antioxidants to inactivate them. Of particular interest in the context of environmental toxicology is the ability of a variety of contaminants to enhance oxyradical production. This effect is often realized through the process of redox cycling, in which an organic xenobiotic is univalently reduced, often via microsomal P-450 reductase activity with the concomitant oxidation of NADPH, to an organic radical. The unshared electron of the organic radical is rapidly transferred to O_2, yielding O_2– and the parent compound, which can undergo another redox cycle.[7] This process can perturb cellular redox status (via NADPH oxidations, for example) and, through the production of oxyradicals, enhance oxidations of various biomolecules including lipids, proteins, and nucleic acids. These biochemical disturbances underlie the toxicities of structurally diverse compounds including nitroaromatics and quinones (both components of many pesticides), and bipyridilium herbicides (such as paraquat and diquat).

In recent years, our laboratory has been involved in examinations of this phenomenon in several organisms, particularly aquatic animals. This research has been motivated by interests in basic elucidations of mechanisms of action, as well as in the development of biomarkers for oxidative stress (such as antioxidant enzyme activities, redox status, and oxidations of selected macromolecules). In these studies, in vitro techniques have played essential roles in determining mechanisms of action, comparing different compounds, and performing species comparisons in a group of animals largely unstudied from this perspective previously. A key in vitro approach employed was the quantification of O_2– produced as a result of xenobiotic redox cycling; two techniques were employed: cyanide-insensitive oxygen (O_2) consumption[9] and cytochrome *c* reduction.[10] In the O_2 consumption assay, subcellular fractions (such as microsomes) are incubated with a potential redox cycling compound, cofactors (such as NADPH, required as an electron source), and cyanide to quench O_2 consumption by contaminating mitochondria. The rate of O_2 disappearance is carefully monitored, and this rate above control values is assumed to be due to reduction to O_2–. This is checked quantitatively by examining the effects of additions of SOD and catalase. SOD catalyzes the dismutation of O_2– to O_2 and H_2O_2, while catalase catalyzes the conversion of H_2O_2 to O_2 and H_2O. Thus, if O_2– production is responsible for observed increases in O_2 consumption rates, these rate increases will be reduced by either SOD (50%) or SOD + catalase (75%). For the cytochrome *c* reduction assay, similar incubations are employed, with the addition of cytochrome *c*, which readily accepts the unpaired electron from O_2–. The univalent reduction of cytochrome *c* can be readily followed spectrophotometrically; the role of O_2– in this reduction is again substantiated by quantifying the inhibition of the reduction by added SOD.

Using these techniques, we recently examined the abilities of three nitroaromatic compounds (p-nitrobenzoic acid or PNBA, nitrofurantoin or NF, and m-dinitrobenzene or MDNB) to stimulate O_2– production in hepatic microsomal fractions from three freshwater fish species: channel catfish (*Ictalurus punctatus*), largemouth bass (*Micropterus salmoides*), and rainbow trout (*Oncorhynchus mykiss*).[11] Each compound strongly enhanced O_2– production in all three species, with an observed potency ranking of NF > MDNB > PNBA. Also, for all compounds, greater enhancements were observed with trout microsomal fractions than with either catfish or bass microsomal fractions (which produced very similar results). Results from oxygen consumption experiments with three species using MDNB, a common environmental contaminant that is toxic to fish,[12] are displayed in Figure 1. In vitro studies such as these clearly can assist in selecting compounds and/or species for more labor- and time-intensive in vivo studies.

Concomitant with these comparative experiments, we performed detailed in vitro studies[13] using catfish hepatic fractions (microsomal and soluble fractions) to elucidate mechanisms underlying enhanced O_2- production mediated by NF, a model nitroaromatic well studied in mammals. Through the use of selected enzyme cofactors and inhibitors, it was concluded that redox cycling of NF was mediated by the activity of NADPH cytochrome P-450 reductase in microsomal fractions and by the activity of xanthine oxidase in soluble fractions. The similarity in the mechanism of metabolism of nitroaromatic compounds by fish microsome and cytosol to that of mammalian systems suggests that the severe health consequences of exposure to these environmental contaminants in mammals[14] might also have implications for the health of aquatic animals.

We have also examined xenobiotic-mediated O_2- production in bivalves using *Geukensia demissa*, the ribbed mussel. We compared this invertebrate to channel catfish and to a standard mammalian model, the rat.[15] In these experiments, we employed the cytochrome *c* assay in hepatic microsomal preparations to examine NF and the bipyridilium herbicide paraquat, a well-characterized redox cycling compound. Results from these investigations (Table 1) have proven very interesting and have motivated continued research on these phenomena in invertebrates. Bivalves are considered to have relatively very low levels of microsomal activity, and control rates of O_2- production bear this out. These rates show marked species differences in the order: rat> fish>mussel. Both NF and paraquat resulted in marked enhancements of O_2- produc-

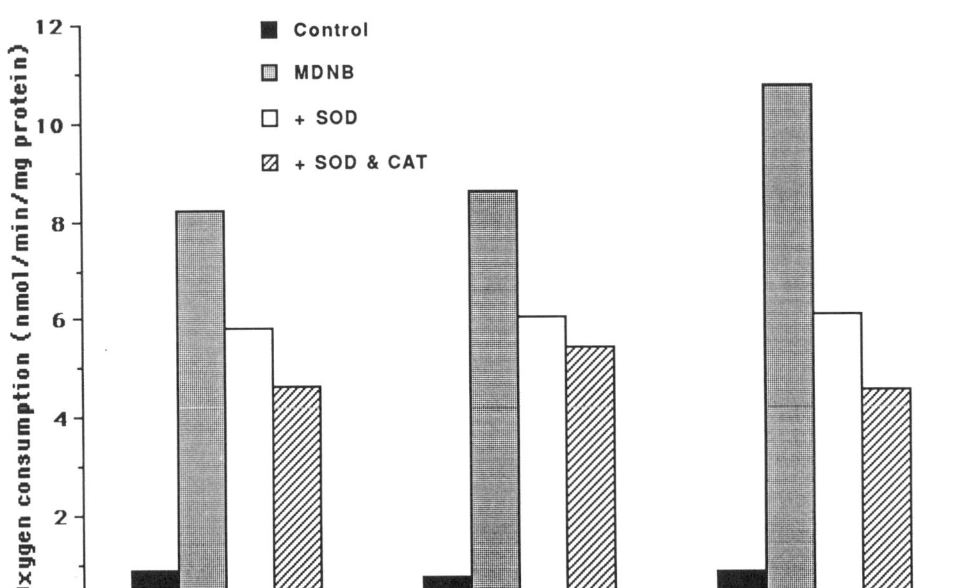

FIGURE 1. Stimulation of cyanide-insensitive, NADPH-dependent oxygen consumption by MDNB (2.5 m*M*) and inhibitory effects of added SOD and catalase (CAT) in hepatic microsomes from channel catfish, largemouth bass, and rainbow trout. (Adapted from Washburn and Di Giulio.[11])

Table 1. Effects of Paraquat and Nitrofurantoin on Cytochrome *c* Reduction (nmol/min/mg protein) in Hepatic Microsomal Preparations from Rat, Channel Catfish (*Ictalurus punctatus*), and Ribbed Mussel (*Geukensia demissa*)

	Rat	Catfish	Mussel
Baseline	32.06 ± 0.32	13.33 ± 0.58	1.90 ± 0.00
Paraquat (1 m*M*)	91.43 ± 1.31	23.81 ± 0.34	4.66 ± 1.16
Change	185%	79%	145%
Nitrofurantoin (1 m*M*)	120.43 ± 7.67	23.19 ± 0.89	3.81 ± 0.62
Change	276%	74%	100%

Adapted from Wenning and Di Giulio.[15]

tion in all species. While as expected, differences in absolute rates in xenobiotic-amended preparations were marked among the species and followed the order observed for baseline rates, these differences were greatly attenuated when considered as a percentage of baseline rates for each species (Table 1). These results suggest that while various biochemical activities, such as contaminant-mediated oxyradical generation, may be considerably reduced on an absolute basis in lower vertebrates and invertebrates compared to "standard" models, perturbations in activities due to xenobiotics may be of similar magnitudes and toxicological significance. Ecotoxicologists working with nontraditional models must be very careful to avoid being overly influenced by mammalian toxicology in the design of studies and interpretation of data. Very elegant in vitro studies in the laboratories of Winston[16] and Livingstone[17] have indicated markedly different mechanisms of microsomal oxyradical production in crustaceans and bivalves than have been described for mammals. Such differences can have important ramifications related to the hazards associated with contaminant exposures in different species, as well as to the development of sophisticated tools for environmental monitoring.

Glutathione-Mediated Detoxification

Our interest in the roles and mechanisms of glutathione in contaminant metabolism were initially motivated by the importance of this endogenous tripeptide in protecting organisms from oxidative stress.[7] Glutathione can nonenzymatically scavenge oxyradicals, but perhaps its most important antioxidant function in animals is as a cofactor for the enzyme glutathione peroxidase; this enzyme catalyzes the reduction of peroxides to their corresponding alcohols, at the expense of reduced glutathione (GSH) that is oxidized (to GSSG):

$$\mathrm{ROOH} + 2\,\mathrm{GSH} \rightarrow \mathrm{ROH} + \mathrm{GSSG} \qquad (1)$$

This reaction halts lipid peroxidizing chain reactions and catalyzes the reduction of H_2O_2 to H_2O. GSH is regenerated from GSSG by the activity of glutathione reductase, which employs NADPH as a reductant. In short, glutathione participates in a beneficial redox cycle to offset potentially deleterious oxyradical-generating redox cycles stimulated by xenobiotics. Responses of the glutathione system in aquatic animals to redox-active, or "pro-oxidant" xenobiotics, such as the herbicide diquat are under investigation presently, employing integrated in vitro and in vivo approaches.

In addition to the role GSH plays in ameliorating oxidative stress, it also functions as the principal conjugate in the phase II metabolism of electrophilic compounds (particularly halogenated, nitro- and epoxide-containing aromatics).[18] These conjugations of

GSH to such substrates are generally catalyzed by a family of enzymes referred to as the glutathione S-transferases (GSTs). As a key part of our investigations of the roles of GSH in xenobiotic detoxification in fish, we are investigating the GST-mediated metabolism of the fungicide chlorothalonil (2,4,5,6-tetrachloroisophthalonitrile) and a model GST substrate, CDNB (chlorodinitrobenzene). The use of a model compound such as CDNB greatly facilitates comparisons of results from our laboratory with those from other laboratories and with other organisms. We have been particularly interested in the importance of GST-mediated metabolism in gills which, relative to the liver, has received little attention in this regard. The use of in vitro studies has been essential to the mechanistic aspects of this research, as well as to the direction of associated in vivo studies. Recent highlights from this ongoing project are described below.

Chlorothalonil Conjugation in Channel Catfish

Chlorothalonil is a broad spectrum fungicide that is highly toxic to channel catfish (static 96 h LC_{50} = 52 μg/L, unpublished data), as well as other fish species (static 96 h LC_{50} for rainbow trout = 18 μg/L).[19] Due to the presence of four halogenated, electrophilic carbon centers, it readily undergoes nucleophilic substitution with GSH and other low molecular weight cellular thiols.[20] The fungicidal action of this compound has been directly attributed to its ability to inhibit glucose oxidation following depletion of GSH.[20] Although the exact mechanism of chlorothalonil toxicity to fish is unknown, covalent binding and inactivation of respiratory proteins may account for at least part of its toxicity.[19] This observation is also supported by a report demonstrating that short-term in vivo exposure to chlorothalonil reduces lamellar diffusive capacity and increases ventilatory frequency in rainbow trout.[21] In that study, GST-mediated conjugation of chlorothalonil with GSH played a protective role by limiting the amount of chlorothalonil available for protein binding.[22] In vivo studies in our laboratory have demonstrated that depletion of hepatic GSH by chemical agents greatly increases the toxicity of chlorothalonil to channel catfish.[23] We have also observed increased concentrations of gill and liver GSH in catfish exposed to sublethal levels of chlorothalonil.[23] Similarly, in vivo increases in hepatic GST activities and GSH concentrations in rainbow trout following exposure to chlorothalonil have been reported,[21] and Davies[24] reported that LC_{50} values for chlorothalonil were positively correlated with total hepatic GST activity for five species of fish. Since GSTs represent a significant pathway for the detoxification of chlorothalonil, assessments and comparisons of reaction rates of chlorothalonil metabolism by key tissues in channel catfish provide valuable information regarding its hazard.

Because chlorothalonil does not undergo a spectral shift upon conjugation with GSH, it is not possible to follow its metabolism by spectrophotometric methods commonly employed for measuring GST activities. Therefore, a high performance liquid chromatographic (HPLC) assay was developed to assess the in vitro metabolism of chlorothalonil.[25] Assay conditions for chlorothalonil are limited by its low aqueous solubility (2.25 μM), necessitating the use of low protein concentrations (<200 μg/mL) to avoid excess substrate depletion. Under the conditions established, the practical limit of assay sensitivity was 1 to 2 nmol chlorothalonil conjugated per minute. This assay was employed to examine GST-mediated chlorothalonil metabolism in cytosolic and microsomal fractions of liver and gill tissue of the catfish.

The results of the GST-mediated conjugation of chlorothalonil in liver and gill fractions are presented in Table 2. Chlorothalonil metabolism occurred in all four fractions, with the highest rate of metabolism catalyzed by hepatic cytosols. The rate of chlorothalonil metabolism by gill cytosolic fractions was approximately 33% of that

Table 2. Metabolism of the Fungicide Chlorothalonil by Channel Catfish Liver and Gill Subcellular Fractions (mean ± SE)

Fraction	Additions/Omissions			Chlorothalonil Metabolism (nmol/min/mg)
	GSH	NADPH	n	
Liver				
Cytosol	+	–	4	388 ± 77
	–	–	3	27 ± 3
Microsomes	+	–	6	51 ± 10
	–	–	3	<1.0
	–	+		<0.050
Gill				
Cytosol	+	–	4	115 ± 22
	–	–	4	5.7 ± 0.8
Microsomes	+	–	4	6.9 ± 2.6
	–	–	3	<1.0
	–	+		<0.050

Adapted from Gallagher et al.[25]

catalyzed by hepatic cytosolic fractions. For fish, the primary route of uptake of waterborne pollutants is through the gill. As a result, the gill, more so than the liver, may be continuously exposed to toxic compounds and therefore should be considered in assessing the metabolic routes of detoxification. Furthermore, the detoxification ability of the gill can have a profound effect on the level of a toxicant reaching other organs. For example, chlorothalonil accumulates over 1000-fold in the trout gill, and remains elevated after the fish are transferred to toxicant-free water.[19] Our data suggest that in channel catfish exposed to chlorothalonil, a significant percentage of the compound is conjugated by gill GSTs prior to entering the general circulation.

Liver microsome catalyzed the formation of chlorothalonil-GSH conjugates at a rate approximately sevenfold higher than observed for gill microsome. Microsomal GSTs comprise 2.5 to 3% of the total microsomal protein recovered from rat liver and approximately 2% of the microsomal protein in northern pike (*Esox lucius*) and rainbow trout.[26] These enzymes have been shown to play a critical role in the conjugation of hexachloro-1:3-butadiene[27] and mitrozantrone.[28] While microsomal GST activity may appear minor relative to cytosolic activity, it can be extremely important because, as described earlier, other (phase I) microsomal activities can activate xenobiotics to reactive, toxic intermediates.[3] In our study, the absence of chlorothalonil metabolism in microsomal preparations lacking GSH (and incubated with NADPH or an NADPH-generating system) indicated that a GST-catalyzed, GSH-dependent reaction is the dominant pathway for chlorothalonil metabolism for channel catfish liver and gill microsomes.

CDNB Conjugation in Channel Catfish

The rates of GST-catalyzed CDNB conjugations by channel catfish cytosolic and microsomal fractions are presented in Table 3. As with chlorothalonil, GST-catalyzed conjugations of CDNB with GSH occurred in all four fractions in the order: liver cytosol > gill cytosol > liver microsome > gill microsome. In addition, the rate of CDNB metabolism exceeded that for chlorothalonil in both tissues and fractions. The

Table 3. Activities of GST Toward the Model Substrate CDNB in Channel Catfish Liver and Gill Subcellular Fractions (mean ± SE)

Tissue	Fraction	n	Activity
Liver	Cytosol	4	1311 ± 120
	Microsomes	4	174 ± 38
Gill	Cytosol	4	711 ± 139
	Microsomes	4	25 ± 1

Adapted from Gallagher et al.[25]

rate of GST catalysis of GSH-CDNB conjugation reported here in channel catfish hepatic cytosol is almost fourfold greater than those reported for rainbow trout.[29] Channel catfish hepatic microsomal GSTs conjugate CDNB at a similar rate to that reported for northern pike,[26] but over 2.5-fold the same activity reported for rainbow trout.[29] Additionally, in contrast to our results and those for pike, gill microsomal GST activity toward CDNB was not detected in rainbow trout. These observations suggest that channel catfish and northern pike have an advantage over the rainbow trout in detoxifying certain electrophilic compounds or reactive intermediates.

These in vitro studies suggest that channel catfish are well equipped to detoxify chlorothalonil via liver and gill GSTs. However, environmental conditions that impose physiological demands upon the glutathione system on channel catfish (nutritional stress, co-exposure to other electrophilic compounds or to pro-oxidants) may compromise the organism's ability to detoxify chlorothalonil. For example, depletion of hepatic GSH by chemical-inducing agents results in a threefold increase in the acute toxicity of chlorothalonil.[23] Conversely, because cytochrome P-450-mediated metabolism appears to play a minor role in the detoxification of this compound in channel catfish,[25] exposure to agents that induce (e.g., polyhalogenated biphenyls, polyaromatic hydrocarbons) or inhibit (e.g., heavy metals, solvents) cytochrome P-450 activities may not have a direct effect on conjugation of chlorothalonil. The presence of microsomal GST activity in channel catfish gill and liver probably affords additional protection against chlorothalonil toxicity. However, as with cytochrome P-450-mediated biotransformation, it must be considered that conjugation with GSH may also bioactivate certain compounds to toxic products.[30] In particular, halogenated compounds such as hexachloro-1:3-butadiene[31] and bromobenzene[32] are activated to nephrotoxic products after initial enzymatic conjugation with GSH. The careful integration of in vitro studies with in vivo pharmacokinetic, pathological, and acute toxicity studies are essential to determinations of the biological significance of metabolic processes such as conjugations.

Cholinesterase Characterization and Inhibition Studies

In contrast to most compounds, the mechanism of action of OP and carbamate pesticides (inhibition of AChE) is relatively well understood.[33] However, much less is understood concerning species differences in AChE properties, such as subcellular localizations, substrate preferences, kinetics of inhibition and recovery, and assay optima. In vitro methods are indispensable for characterizing these properties of target enzymes such as AChE. A number of investigators, including ourselves, have employed in vitro techniques to characterize and compare properties of AChEs in aquatic vertebrates and invertebrates.

The first step in determining enzyme distribution involves isolating subcellular components by differential centrifugation. Differential centrifugation of crude tissue homogenates results in cellular organelles that are fractionated into components such as nuclear, plasma membrane and synaptosomes, peroxisomes, microsomes, and cytosol. This type of work with channel catfish brain tissue and blue crab (*Callinectes sapidus*) ganglion tissue demonstrated that AChE is primarily found in the synaptosomal and microsomal fractions with these species, with a small amount of soluble or cytosolic AChE.[34]

Following subcellular fractionation, specific enzyme inhibitors and substrates can be employed to characterize the ChEs associated with each fraction. Enzyme characterization of ChEs is particularly important from a biomonitoring standpoint because it is essential to know the substrate preference for assaying a cholinesterase being used as marker enzyme. ChEs are generally subdivided into two major classes: true or AChEs and pseudo- or butyrylcholinesterases (BChEs).[33] As examples, plasma ChEs from mammals and birds generally are BChEs, while neural and muscle ChEs from mammals, bird, fish, and crabs have been characterized as AChEs.[33-35] However, in some other crustaceans (such as lobsters, *Homarus*) the neural tissue has substantial BChE, and a single AChE has also been isolated from molluscan (*Mytilus edulis*) hemolymph.[36,37] Research on the cholinesterase of the earthworm (*Eisenia foetida*) indicated that an alternative substrate, proprionylthiocholine, is preferred.[38] In vitro approaches are also particularly suited for determining kinetic parameters of enzymes, such as the K_m and V_{max}. With AChE, this work can also include determining whether interspecific differences in sensitivity to OPs or carbamates result principally from differences in affinity or rates of phosphorylation/carbamylation.

The use of selected inhibitors and substrates in our laboratory indicated differences between catfish brain and crab ganglion ChEs although both are classified as "true" cholinesterases (i.e., AChE).[34] The crab ganglion AChE was more susceptible to eserine (a specific inhibitor of AChE) and had a higher affinity for acetylthiocholine as a substrate than fish brain AChE. On the other hand, fish brain AChE, particularly microsomal AChE, showed a much greater substrate preference for acetylthiocholine than crab ganglion AChE. Work on these different cellular pools of AChE has indicated differences among these AChE pools in both fish and crabs in terms of preferential hydrolysis of substrates, KI_{50} for eserine, and the effectiveness of Triton® X-100 in solubilizing AChE from synaptosomes and microsome. Additionally, distinct differences were observed in the effectiveness of Triton® X-100 in solubilizing AChE from synaptosomes and microsomes of crabs compared to those of fish. Elucidating the distribution of cellular pools of AChE also encourages examining specific pools following in vivo exposures to OPs or carbamates to determine if these pools may be selectively inhibited, which provides additional insight on the effects of these compounds.

A fundamental role of in vitro procedures in the context of hazard assessments is that of determining optimal conditions for assaying an enzyme activity in various species.[39] This approach has been used to determine preferred buffers and buffer ionic strength, pH, temperature, and substrate concentrations for assaying AChE activities in a number of fish species.[34,39] For example, although AChEs from various fishes have substantial activity over a broad pH range of about 6.8 to 8.5, differences in pH optima, such as 7.2, 7.8, and 8.2, have been reported for different fish species.[35,40] This type of information is critical to developing a biomonitoring program by increasing the sensitivity of the assay, as well as enhancing protocol standardization among laboratories. In vitro assays can assist in identifying sensitive or potential indicator species. These assays can readily determine differences in the sensitivity of AChEs from different species to a given inhibitor, differential inhibitory potentials of a series of inhibitors toward a chosen species, or a combination of these objectives. These methods are relatively straightforward and cost effective, and can contribute significantly as a screening tool to help identify potentially high-risk species and/or compounds. This approach has been successfully used to obtain comparative data for diisopropyl fluorophosphate, paraoxon, methyl paraoxon, gutoxon, and ethyl gutoxon in fish, amphibians, chickens, and three species of mammals.[41] Results from this study indicate that varying sensitivities of AChEs from these species to inhibitors arose both from differences in affinity and in phosphorylation rates. Similarly, we found crab ganglion synaptosomal AChE to be four times as sensitive to maloxon and ten times as sensitive to paraoxon as catfish brain synaptosomal AChE.[34] Such information, when integrated with use and exposure data, can assist with prioritizing both test compounds and species that should be examined in more detail using in vivo toxicity testing procedures. Although standard species are usually tested in vivo for regulatory purposes, interspecific comparisons using in vitro methods and exposure data may indicate whether additional species should be examined in certain situations.

CONCLUSION

While perhaps not as fundamental to ecological hazard assessments as other activities (in vivo studies, field studies, population modeling), in vitro studies play an important ancillary function. They serve presently in xenobiotic metabolite identification, the development and optimization of biomarkers, and the acquisition of information on comparative species and compounds. As the science of ecotoxicology develops, and these assessments concomitantly become more sophisticated, the importance of in vitro approaches will likely increase. This is due principally to the fundamental importance of understanding cause-effect linkages between exposures and biological responses. In vitro approaches are critical to the elucidation of these linkages and to building a more solid theoretical base for the young science of ecotoxicology and the critical decisions this science undergirds. Finally, the utility of in vitro approaches in hazard assessments would be considerably enhanced at the present time if appropriate studies were designed into specific assessments, rather than used posteriorally, as generally appears to be the case.

REFERENCES

1. Mehlman, M. A., Ed., *Benchmarks: Alternative Methods in Toxicology*, Princeton Scientific Publishing, Princeton, NJ, 1989.
2. Matsamura, F., *Toxicology of Insecticides*, 2nd ed., Plenum Press, New York, 1985.
3. Guengerich, E. P., and D. C. Liebler, Enzymatic activation of chemicals to toxic metabolites, *Crit. Rev. Toxicol.*, 14, 259–307, 1985.
4. Hildebrandt, A. G., Overdose toxicity studies versus threshold: elements of biology must be incorporated into risk assessment, *Arch Toxicol.*, 60, 217–233, 1987.
5. Di Giulio, R. T., *Biomarkers*, in *Ecological Assessments of Hazardous Waste Sites: a Field and Laboratory Reference Document*, W. Warren-Hicks, B. R. Parkhurst and S. S. Baker, Jr., Eds., U.S. EPA Report No. EPA 600/3-89/013, Corvallis, OR, 1989, pp. 7-1–7-34.
6. Cobb, G. P., and M. J. Hooper, Non-lethal wildlife monitoring to determine exposure to xenobiotics and resulting impacts, in *Wildlife Toxicology and Population Modeling: Integrated Studies of Agroecosystems*, R. J. Kendall and T. E. Lacher, Jr., Eds., Lewis Publishers, Chelsea, MI, 1993.
7. Kappus, H., Oxidative stress in chemical toxicity, *Arch. Toxicol.*, 60, 144–149, 1987.

8. Di Giulio, R. T., P. C. Washburn, R. J. Wenning, G. W. Winston and C. S. Jewell, Biochemical responses in aquatic animals: a review of determinants of oxidative stress, *Environ. Toxicol. Chem.*, 8, 1103–1123, 1989.
9. Mason, R. P., and J. L. Holtzman, The role of catalytic superoxide formation in the O_2 inhibition of nitroreductase, *Biochem. Biophys. Res. Commun.*, 67, 1267–1274, 1975.
10. McCord, J. M., and I. Fridovich, Superoxide dismutase: an enzymatic function of erythrocuprein (hemocuprein), *J. Biol. Chem.*, 244, 6049–6055, 1969.
11. Washburn, P. C., and R. T. Di Giulio, Stimulation of superoxide production by nitrofurantoin, *p*-nitrobenzoic acid and *m*-dinitrobenzene in hepatic microsome of three species of freshwater fish, *Environ. Toxicol. Chem.*, 8, 171–180, 1989.
12. Kumar, N. J., K. P. Krishnamoorthi and D. M. R. Rao, Toxicity and biochemical responses of carp to dinitrobenzene plant effluent, Water Air Soil Pollut., 28, 117–126, 1986.
13. Washburn, P. C., and R. T. Di Giulio, Nitrofurantoin-stimulated superoxide production by channel catfish (*Ictalurus punctatus*) hepatic microsomal and soluble fractions, *Toxicol. Appl. Pharmacol.*, 95, 363–377, 1988.
14. Rickert, D. R., Ed., *Toxicity of Nitroaromatic Compounds*, Hemisphere, New York, 1985.
15. Wenning, R. J., and R. T. Di Giulio, Microsomal enzyme activities, superoxide production, and antioxidant defenses in ribbed mussels (*Geukensia demissa*) and wedge clams (*Rangia cuneata*), *Comp. Biochem. Physiol.*, 90C, 21–28, 1988.
16. Jewell C. S. E., and G. W. Winston, Oxyradical production by hepatopancreas microsome from the red swamp crayfish, *Procambarus clarkii, Aquat. Toxicol.*, 14, 27–46, 1989.
17. Livingstone, D. R., P. G. Martinez and G. W. Winston, Menadione-stimulated oxyradical formation in digestive gland microsome of the common mussel, *Mytilus edulis, L. Aquat. Toxicol.*, 15, 213–236, 1989.
18. Chasseaud, L. F., The role of glutathione and glutathione S-transferases in the metabolism of chemical carcinogens and other electrophilic agents, *Adv. Cancer Res.*, 29, 175–274, 1979.
19. Davies, P. E., and R. W. G. White, The toxicology and metabolism of chlorothalonil in fish. I. Lethal levels for *Salmo gairdneri*, *Galaxias maculatus*, *G. truttaceus* and *G. auraratus* and the fate of ^{14}C-TCIN in *S. gairdneri*, *Aquat. Toxicol.*, 7, 93–105, 1985.
20. Vincent, P. G., and H. D. Sisler, Mechanisms of antifungal action of 2,4,5,6-tetrachloroisophthalonitrile, *Plant Physiol.*, 21, 1249–1264, 1968.
21. Davies, P. E., Physiological, anatomic and behavioral changes in the respiratory system of *Salmo gairdneri* Rich on acute and chronic exposure to chlorothalonil, *Comp. Biochem. Physiol.*, 88C, 113–119, 1987.
22. Davies, P. E., The toxicology and metabolism of chlorothalonil. II. Glutathione conjugates and protein binding, *Aquat. Toxicol.*, 7, 265–275, 1985.
23. Gallagher, E. P., A. T. Canada and R. T. Di Giulio, The protective role of glutathione in chlorothalonil-induced toxicity to channel catfish, *Aquat. Toxicol.*, 23, 155–168, 1992.
24. Davies, P. E., The toxicology and metabolism of chlorothalonil. III. Metabolism, enzymatics and detoxification in *Salmo* spp. and *Galaxias* spp., *Aquat. Toxicol.*, 7, 277–299, 1985.
25. Gallagher, E. P., G. L. Kedderis and R. T. Di Giulio, Glutathione S-transferase-mediated chlorothalonil metabolism in liver and gill subcellular fractions of channel catfish, *Biochem. Pharmacol.*, 42, 139–145, 1991.
26. Morgenstern, R., G. Lundqvist, G. Andersson, L. Balk and J. W. DePierre, The distribution of microsomal glutathione transferase among different organelles, different organs, and different organisms, *Biochem. Pharmacol.*, 33, 3609–3614, 1984.
27. Wolf, C. R., P. N. Berry, J. A. Nash, T. Green and E. A. Lock, Role of microsomal and cytosolic glutathione S-transferases in the conjugation of hexachloro-1:3-butadiene and its possible relevance to toxicity, *J. Pharmacol. Exp. Ther.*, 228, 202–208, 1984.
28. Wolf, C. R., J. S. Macpherson and J. F. Smyth, Evidence for the metabolism of mitrozantrone by microsomal glutathione transferase and 3-methylcholanthrene-inducible glucuronosyl transferases, *Biochem. Pharmacol.*, 35, 1577–1581, 1986.

29. Lauren, D. J., P. P. Halarnkar, B. D. Hammock and D. E. Hinton, Microsomal and cytosolic epoxide hydrolase and glutathione S-transferase activities in the gill, liver and kidney of the rainbow trout, *Salmo gairdneri*, *Biochem. Pharmacol.*, 38, 881-887, 1989.
30. Anders, M. W., L. Lash, W. Dekant, A. A. Elfarra and D. R. Dohn, Biosynthesis and biotransformation of glutathione S-conjugates to toxic metabolites, *Crit. Rev. Toxicol.*, 18, 311-341, 1988.
31. Lock, E. A., and J. Ishmael, Hepatic and renal non-protein sulfhydryl concentration following toxic doses of hexachloro-1:3-butadiene in the rat: the effect of Aroclor 1254, phenobarbitone or SKF-525A treatment, *Toxicol. Appl. Pharmacol.*, 47, 79-87, 1981.
32. Monks, T. J., S. S. Lau, R. J. Highet and J. R. Gillette, Glutathione conjugates of 2-bromohydroquinone are nephrotoxic, *Drug Metab. Disposition*, 13, 553-559, 1985.
33. Main, A. R., Mode of action of cholinesterases, in *Toxicities of Insecticides and Halogenated Aromatics*, F. Matsamura, Ed., Pergamon Press, New York, 1984, pp. 351-400.
34. Habig, C., R. T. Di Giulio and M. B. Abou-Donia, Comparative properties of channel catfish (*Ictalurus punctatus*) and blue crab (*Callinectes sapidus*) acetylcholinesterases, *Comp. Biochem. Physiol.*, 91C, 293-300, 1988.
35. Coppage, D. L., Characterization of fish brain acetylcholinesterase with an automated pH stat for inhibition studies, *Bull. Environ. Contam. Toxicol.*, 6, 304-310, 1971.
36. Maynard, E. A., Esterases in crustacean nervous system. I. Electrophoretic studies in lobsters, *J. Exp. Zool.*, 157, 251-266, 1964.
37. von Wachtendonk, D., and J. Neff, Isolation, purification, and molecular properties of an acetylcholinesterase (E.C. 3.1.1.7) from the haemolymph of the sea mussel *Mytilus edulis*, *Comp. Biochem. Physiol.*, 63C, 279-286, 1979.
38. Andersen, R. A., T. Aune and J. A. B. Barstad, Characteristics of cholinesterase of the earthworm *Eisenia foetida*, *Comp. Biochem. Physiol.*, 63C, 279-286, 1978.
39. Abou-Donia, M. B., and D. B. Menzel, Fish brain cholinesterase: its inhibition by carbamates and automatic assay, *Comp. Biochem. Physiol.*, 21, 99-108, 1967.
40. Weiss, C. M., The determination of cholinesterase in the brain tissue of three species of fresh water fish and its inactivation in vivo, *Ecology*, 39, 194-199, 1958.
41. Wang, C., and S. D. Murphy, Kinetic analysis of species difference in acetylcholinesterase sensitivity to organophosphate insecticides, *Toxicol. Appl. Pharmacol.*, 66, 409-419, 1982.

24

Do Behavioral Responses to Pesticide Exposure Affect Wildlife Population Parameters?

Richard S. Bennett

ABSTRACT

Avian behavioral responses to exposure to agricultural pesticides are reviewed in relation to ultimate effects on survival and reproduction. Pesticide-induced changes in foraging behaviors may lead to the formation of conditioned aversions that affect subsequent dietary exposure and, consequently, survival. Although laboratory birds can readily detect the presence of pesticides in food and prefer untreated alternatives if present, there are several other factors in the field that affect the ability of birds to modify their exposure to pesticides. Pesticides also can affect reproductive behaviors of birds at several periods, including pairing and nest building, egg laying, incubation, and rearing of young. Behavioral effects at these periods have been observed to reduce reproductive success. Research is needed to identify how pesticide effects on avian population parameters are influenced by their behavioral responses.

KEY WORDS

behavior, foraging patterns, reproduction, organophosphorus insecticide, carbamate

INTRODUCTION

Risk assessments of the potential effects of agricultural pesticides on wildlife species usually assume that after pesticides are applied no changes occur in the typical behavioral patterns of those species, such as foraging or reproductive behaviors. Because so little is known about the behavioral responses of wildlife to pesticide exposures in the field, it is very difficult to predict how an animal may behave in a given scenario of pesticide use. To what extent do wildlife species react to the presence of pesticides and change their behavioral patterns? How could changes in behavioral responses affect population parameters such as survival and reproduction?

This chapter reviews the literature on pesticide-related changes in avian behavioral patterns in the laboratory and field that could affect survival and reproduction. The emphasis is on organophosphorus (OP) and carbamate insecticides. For a review of the effects of other environmental contaminants on avian behavior, see Peakall.[1]

CHANGES IN FORAGING BEHAVIOR

Birds make choices daily concerning what to eat and where to search for food to meet energetic and nutrient requirements using the available resources. The variety of these choices is dictated by environmental and life history constraints. The choices available to a bird are partly a function of dietary specialization, territorial constraints, food availability, and nutrient content and palatability of foods. The palatability of prey items can be significantly affected by the presence of naturally occurring toxic chemicals. Many animals and plants have evolved chemical mechanisms for protecting themselves from possible predators, while animal consumers have learned to identify and reduce consumption of prey containing these naturally occurring toxins.[2,3]

Pesticide occurrence on or in prey items may also elicit changes in foraging behavior. The greatest difference from natural toxins may be that pesticides can suddenly appear on food items that were previously preferred foods. Pesticides also may be available from other routes of exposure besides diet, such as inhalation and dermal exposure.[4] These routes of exposure may add to the total pesticide exposure of birds and affect their ability to behaviorally respond to the presence of a chemical.

Much of the information on avian responses to dietary chemical exposure has come from the field of animal damage control, which for many years has been developing chemical methods for protecting crops from animal depredation. Hundreds of chemicals, including many with insecticidal and fungicidal properties, have been tested as potential bird repellents.[5] Observations of avian responses to repellents for crop protection can also contribute to an understanding of animal responses to pesticides used primarily to protect the crops from targeted insect pests.

It is commonly observed in avian subacute dietary tests using OP or carbamate insecticides that there is a dose-related reduction in food consumption in the range of dietary concentrations that affect survival.[6,7] In subacute dietary tests with OPs, mortality occurs most frequently near the end of the treatment period with birds in extremely emaciated condition. Grue[8] described this phenomenon as pesticide-induced anorexia. As food consumption decreases with increasing concentration, it has also been observed that the amount of active ingredient ingested during the test period remains relatively unchanged over the concentrations producing mortality.[8-10] Consequently, the observed effects on body weight and mortality are more a function of food consumption than the ingestion of pesticide.[9,10] Because birds in these studies had a choice of eating treated food or not eating, Hill and Camardese[6,7] described this as a measure of vulnerability rather than toxicity.

Dietary Discrimination

It is unclear to what extent laboratory measures of pesticide vulnerability reflect responses in the field. Do free-living birds succumb to pesticide-induced anorexia or can they respond to the presence of pesticides and modify their subsequent exposure? If they can respond to the presence of chemicals, under what field scenarios does this occur? Several studies have addressed these questions with a variety of test species and environmental contaminants. A common test of the ability of birds to detect and avoid contaminated food compares the amounts of food consumed when presented with feeders containing pesticide-treated food and untreated food for a specified period. Kenaga et al.[11] found that mallards (*Anas platyrhynchos*) readily chose untreated food over food treated with chlorpyrifos, and thus avoided ill effects. Juvenile bobwhite (*Colinus virginianus*) and mallards could detect the presence of OP and carbamate insecticides in food at dietary concentrations below those demonstrated to produce signs of toxicity.[9-13] For all chemicals there was a range of concentrations in which birds did not respond to the presence of pesticide by showing a preference. Above a concentration defined as the discrimination threshold,[14] birds increasingly consumed a greater proportion of untreated food, such that the amount of pesticide ingested often reached a plateau.[9,10] Most of these insecticides were not overtly repellent, but produced a change in feeding response through the formation of a conditioned aversion. Conditioned aversion agents may not initially repel an animal from the food, but may cause an illness that is subsequently associated with characteristic taste, smell, or visual cues of the food.[15]

Other Factors Affecting Discrimination

Tests using two food choices represent the simplest scenario for an animal making foraging decisions and may reflect the animal's greatest potential for detecting and responding to the presence of treated food in a way that reduces pesticide ingestion. However, the potential for changing pesticide ingestion under laboratory conditions may not be a good reflection of responses of birds in the field. Several other behavioral, physiological, or environmental factors can influence the expression of dietary discrimination behavior.

Environmental factors that affect dietary discrimination include the amount and relative proportion of treated and untreated food available to birds. Increasing the number of choices of feeders from two to ten, while keeping the number of treated and untreated choices equal, did not significantly affect dietary discrimination by bobwhite, except with birds exposed to methyl parathion.[9,10] However, given nine feeders containing treated food and one with untreated, the ability to discriminate was reduced for all pesticides tested, except chlorpyrifos. Bobwhite were able to detect and avoid chlorpyrifos-treated food in most treatment groups, but some birds were unable to find sufficient untreated food to avoid starvation. Consequently, birds in environments containing both pesticide-contaminated and uncontaminated foods may be able to detect the presence of pesticides, but their total food and chemical intake will be affected by the absolute and relative abundance of each food type. If untreated or less toxic alternative foods are readily available, a conditioned aversion may help a bird avoid lethal ingestions of pesticide, yet find sufficient food to avoid starvation. Also, the strength of available sensory cues is an important factor in the development of a conditioned aversion. If a treated food is associated with a readily detectable sensory cue, animals form aversions to that cue; if the cues are weak or undetectable, an aversion may form for that food type whether contaminated or not.[15]

Exposure to a pesticide through nonfood sources may confound the ability of a bird to discriminate between treated and untreated foods. Bobwhite initially given a single sublethal dose (6 mg/kg) of methyl parathion did not discriminate against methyl parathion-treated food and, in fact, initially favored the treated food.[16] In the field, birds may also receive a preliminary inhalation or dermal exposure that confounds subsequent dietary discrimination. Driver et al.[4] found that inhalation and dermal routes of exposure were a significant proportion of total pesticide exposure to bobwhite during a simulated spray application.

Visual cues can be more important to birds than taste or olfactory cues in the formation of conditioned aversions.[17-19] When treated food is visually distinct (e.g., different color cues or food types) from untreated food, birds develop an aversion to treated food more readily and switch to alternatives whether or not taste or olfactory cues are detectable.[15] Ring-necked pheasants (*Phasianus colchicus*) will switch their consumption to less preferred food types when their most preferred foods are contaminated with pesticides.[20] Color cues often improve the efficacy of methiocarb as an avian repellent on crops.[17,18]

Extending Laboratory Results to the Field

The potential exists for a pesticide to be detected and avoided by birds in some situations and not in others. These scenarios are much more definable in the laboratory than in the field. In the laboratory, response can be measured as food consumption or time spent foraging, endpoints much more difficult to measure in the field. Consequently, the role of pesticide-induced changes in foraging behaviors in the field is poorly understood. There are many incidences of acute poisoning by OPs where lethal amounts of insecticide were consumed before a conditioned aversion could be formed (see review by Grue et al.[21]). Some of the largest incidences of insecticide-related avian mortality occur when flocking behaviors congregate birds in treated habitats.[22,23]

There are also incidences in which behavioral responses to an insecticide application led to an increased exposure when birds and mammals were attracted to dead and dying invertebrates.[24-26] For example, pied kingfishers (*Ceryle rudis*) were attracted to fish kills following insecticide applications.[27] In these studies, no effects were observed on birds or mammals, and the residue concentrations in prey items were below those expected to be lethal.

Much less is known about the situations in which birds form conditioned aversions in the field and reduce their subsequent exposure to pesticide. Several studies have been conducted with methiocarb, which has been shown to be effective at reducing crop depredation by birds in corn,[28] rice,[29] cherries,[30] blueberries,[31] and grapes.[32,33] Methiocarb also has been shown to reduce feeding by free-ranging Canada geese (*Branta canadensis*) on turf for 4 to 8 weeks, even though little residue remained after the first week.[34]

Conditioned aversions also have been observed in field studies with other pesticides. Bobwhite exposed to corn seedlings in fields treated with bendiocarb consumed fewer seedlings than in control fields.[35] Granular bendiocarb was taken up systemically into seedlings, so the avoidance of seedlings increased over time. Japanese quail (*Coturnix japonica*) also developed conditioned aversions to seedlings grown in soil with aldicarb granules.[36] Canada geese, given a choice between untreated or diazinon-treated turf, avoided the treated areas beginning on the third day after treatment and grazed more heavily on untreated areas during the 3-day test.[37]

In field studies with insecticides, effects on survival and reproduction are often measured, but not feeding behaviors. Consequently, it is unknown if birds surviving a pesticide application have responded to the application by changing feeding behaviors. In cases in which pesticide-induced changes in feeding behavior were documented, it is unknown if changes were due to a learned aversion to the insecticide or to insecticide-induced changes in food resources. Understanding avian exposure to pesticides will require more information about the factors influencing changes in foraging patterns. Much of the observed variability of pesticide effects on birds in field studies may be related to the many variables affecting exposure, such as the overall availability of food or the temporal and spatial distribution of chemical residues. In laboratory tests, birds may detect and avoid a given concentration of an insecticide in some feeding scenarios and not in others. Our ability to predict risks of insecticides to birds will improve as these exposure scenarios are better understood.

CHANGES IN REPRODUCTIVE BEHAVIORS

The United States Environmental Protection Agency (U.S. EPA) may require that testing of a pesticide be conducted prior to registration to determine the potential for adverse effects on avian reproduction.[38] The current avian reproduction test guidelines were originally designed for chemicals with continuous or repeated exposure patterns.[38] Tests are conducted with captive birds primarily using artificial incubation and brooding. Important measurements in these tests are the occurrence of embryotoxicity; reductions in egg production, fertility, and hatchability; and chick size and survival.

Pesticides may affect parental reproductive behaviors that reduce the production and survival of the young. Many of the behavioral effects observed with OP or carbamate insecticides may not be evaluated with the current reproduction test methods. This is especially important because the initial contact with these insecticides may come at any time during reproduction (e.g., pairing and nest building, egg laying, incubation, and rearing of young), depending on the geographical location and the associated crop type. The next section reviews the effects of pesticides on parental behaviors at several periods during reproduction.

Breeding Display Behavior

Relatively little is known about the effects of pesticides during pairing and nest building. Forsyth[39] found that OP exposure altered the singing behavior of white-throated sparrows (*Zonotrichia albicollis*). Movement and vocalization by male starlings (*Sturnus vulgaris*) were reduced by OP exposure for up to 28 hours,[40] potentially affecting courtship and territorial defense. Aerial applications of fenitrothion reduced the ability of white-throated sparrows to defend territories or resulted in territorial abandonment.[41] McEwen and Brown[42] observed an increase of intense aggression in sharp-tailed grouse (*Tympanuchus phasianellus*) at breeding grounds when a portion of the males were given a sublethal dose of malathion and released.

Egg Laying

Insecticide exposure during egg laying can affect feeding behaviors, and consequently, dietary intake. Dose-related reductions in food consumption have been observed in avian reproduction tests with OP and carbamate insecticides just as they are in subacute dietary tests. Stromborg[43,44] found that reductions in bobwhite egg production were directly related to pesticide-induced reductions in food consumption. Dietary exposure of bobwhite to methyl parathion produced no evidence of an effect on egg production or eggshell quality other than those caused by pesticide-induced anorexia.[45] However, diazinon produced a greater effect on bobwhite egg production than could be explained by reduced food consumption alone.[46] Reduced bobwhite egg production also was observed during a 10-day dietary exposure to parathion without a pesticide-induced reduction in food consumption, and may have been related to a direct alteration in the secretion of reproductive hormones.[47] Although the causes of reduced egg production by OPs is not completely understood, the energetic requirements for a laying bobwhite are approximately double that of a nonlaying bird,[48] so reduced food consumption must affect sustained egg production.

Mallard egg production dropped rapidly after 2 days of exposure to methyl parathion-treated food in conjunction with an 82% decrease in food consumption.[49] Of ten hens, three abandoned nests during laying, with two birds subsequently initiating new clutches, while five birds incubated smaller than normal clutches. Franson et al.[50] reported that the average interval between eggs was increased in mallards fed 10 ppm temephos compared to controls. Reduced clutch size, extended period for completing a clutch, and clutch abandonment are also possible effects from pesticide exposure to free-ranging birds. However, it is unknown how pesticide effects on food consumption and egg production in laboratory tests relate to effects on laying birds in field environments.

Incubation

Initial exposure of birds to pesticides may occur during incubation. Captive mallard hens exposed to methyl parathion in the diet either early or late in incubation demonstrated a variety of behavioral effects.[49] Of 23 treated hens, 7 abandoned their nests compared to one of ten control hens. Of 23 hens, 6 exhibited reduced nest attentiveness for one or more days during treatment. The treatment groups produced 43 to 58% as many ducklings per hen as the control group.

Similar changes in incubation behavior have been observed in the field. Laughing gulls (*Larus atricilla*) dosed with parathion spent significantly less time incubating during the following 2 or 3 days than control birds,[51] although no effects on nest defense behavior or hatching success were found.[52] Nest attentiveness also was reduced for several hours in red-winged blackbirds (*Agelaius phoeniceus*) dosed with methyl parathion, although no changes in incubation behavior were observed after 24 hours.[53]

Few studies have documented effects on incubating birds from operational application of pesticides. Forest spraying with fenitrothion resulted in the disruption of incubation and nest abandonment by white-throated sparrows.[41] Several birds temporarily abandoned their nests for several days, resulting in a prolonged incubation period. Reproductive success in the sprayed area was only one third of that observed in the control area. In a similar study on fenitrothion, Spray et al.[54] did not observe behavioral changes in coal tits (*Parus ater*) during incubation nor effects on hatchability. Busby et al.[41] discussed some of the differences between these studies that might explain differences in observed results.

Brood-Rearing Behavior

Researchers using single pesticide exposures to either parents or juveniles during the brood-rearing period have observed behavioral changes that could affect the growth and survival of the entire brood. Female starlings dosed with dicrotophos made significantly fewer sorties to feed their broods and were absent for longer periods of time than control females.[55] This led to greater weight loss in young of treated females. Fairbrother et al.[56] studied the effects of methyl parathion on mallard hen and brood interactions in pond enclosures. In broods in which half of the ducklings were sublethally dosed with methyl parathion and released into ponds, the hen and all ducklings spent most of the time on land near intoxicated ducklings during the first 24 hours, while control broods were observed foraging or swimming on the water 60% of the time. Brood behavior also can be affected by indirect effects on available food resources. Black duck (*Anas rubripes*) and mallard ducklings spent more time searching for food and less time resting on ponds sprayed with carbaryl than ducklings on control ponds.[57]

Meyers and Gile[58] found that mallards breeding on ponds supplemented with food treated with 80 ppm chlorpyrifos did not produce any ducklings surviving to 7 days old, partially because the hens were not attentive to the broods, while control ponds produced six to eight ducklings per hen. This concentration is approximately 9% of the

LC_{50} for 10-day-old mallards.[59] Franson et al.[50] found a 40% reduction in survival of mallard ducklings fed 1 or 10 ppm temephos compared to controls, although there was no reduction in brain cholinesterase. It was not possible to determine whether mortalities resulted from ingestion of temephos, changes in maternal behavior, duckling vigor, or other factors. Concentrations used were significantly below the temephos LC_{50} of 894 ppm for 17-day-old mallards,[59] although the dietary toxicity of temephos is markedly enhanced by cold temperatures.[60] In the field, Brewer et al.[61] observed brood abandonment by wood duck (*Aix sponsa*) and blue-winged teal (*Anas discors*) hens after agricultural applications of methyl parathion.

Importance of Behavioral Changes to Reproduction

The reported pesticide-induced changes in avian reproductive behaviors have potential ramifications to the overall reproductive success of birds. It is not possible at this time to assess the relative importance of behavioral effects compared to direct pesticide effects on fertility, embryotoxicity, and chick vigor or indirect effects of food resource availability for chicks or egg-laying females. However, parental changes in behavior following pesticide exposure can reduce reproductive success under certain conditions. The abandonment of nests and broods by birds sublethally exposed to OP and carbamate insecticides cannot be predicted by current standardized test methods, but may represent a significant reduction in reproductive success because renesting attempts are usually less successful.[62]

CONCLUSIONS

Various behavioral responses to pesticide exposure that could ultimately influence survival or reproduction parameters for individuals have been reviewed in this chapter. The occurrence of these behavioral responses is dependent on a variety of factors that often are difficult to predict. Probably more important than determining the kinds of behavioral changes that take place is determining the set of field conditions under which behavioral responses would take place. The response observed to a given concentration of an insecticide in one exposure scenario may be quite different in another. Making generalizations on the effect of behavioral responses on survival or reproduction parameters will require further research to characterize differences of avian exposure patterns among pesticide use scenarios. This characterization is extremely important to the larger question of how pesticide effects observed in individuals relate to effects on population size.

REFERENCES

1. Peakall, D. B., Behavioral responses of birds to pesticides and other contaminants, *Residue Rev.*, 96, 45–77, 1985.
2. Freeland, W. J., and D. H. Janzen, Strategies in herbivory by mammals: the role of plant secondary compounds, *Am. Nat.*, 108, 269–289, 1974.
3. Sorensen, A. E., Taste aversion and frugivore preference, *Oecologia*, 56, 117–120, 1983.
4. Driver, C. J., M. W. Ligotke, P. Van Voris, B. D. McVeety, B. J. Greenspan and D. B. Drown, Routes of uptake and their relative contribution to the toxicologic response of northern bobwhite (*Colinus virginianus*) to an organophosphate pesticide, *Environ. Toxicol. Chem.*, 10, 21–33, 1983.

5. Schafer, E. W., Jr., W. A. Bowles, Jr., and J. Hurlbut, The acute oral toxicity, repellency, and hazard potential of 998 chemicals to one or more species of wild and domestic birds, *Arch. Environ. Contam. Toxicol.*, 12, 355–382, 1983.
6. Hill, E. F., and M. B. Camardese, Subacute toxicity testing with young birds: response in relation to age and intertest variability of LC_{50} estimates, in *Avian and Mammalian Wildlife Toxicology: Second Conference*, STP 757, D. W. Lamb and E. E. Kenaga, Eds., American Society for Testing and Materials, Philadelphia, 1981, pp. 41–65.
7. Hill, E. F., and M. B. Camardese, Lethal Dietary Toxicities of Environmental Contaminants and Pesticides to Coturnix, Fish and Wildlife Tech. Rep. 2, U.S. Fish and Wildlife Service, Washington, DC, 1986.
8. Grue, C. E., Response of common grackles to dietary concentrations of four organophosphate pesticides, *Arch. Environ. Contam. Toxicol.*, 11, 617–626, 1982.
9. Bennett, R. S., Role of dietary choices in the ability of bobwhite to discriminate between insecticide-treated and untreated food, *Environ. Toxicol. Chem.*, 8, 731–738, 1989.
10. Bennett, R. S., Factors influencing discrimination between insecticide-treated and untreated foods by northern bobwhite, *Arch. Environ. Contam. Toxicol.*, 18, 697–705, 1989.
11. Kenaga, E. E., R. J. Fink and J. B. Beavers, Dietary toxicity tests with mallards, simulating residue decline of chlorpyrifos and avoidance of treated food, in *Avian and Mammalian Wildlife Toxicology*, E. E. Kenaga, Ed., American Society for Testing and Materials, Philadelphia, 1979, pp. 36–44.
12. Kononen, D. W., J. R. Hochstein and R. K. Ringer, A quantitative method for evaluating avian food avoidance behavior, *Environ. Toxicol. Chem.*, 5, 823–830, 1986.
13. Kononen, D. W., J. R. Hochstein and R. K. Ringer, Avoidance behavior of mallards and northern bobwhite exposed to carbofuran-contaminated food and water, *Environ. Toxicol. Chem.*, 6, 41–50, 1987.
14. Bennett, R. S., and D. W. Schafer, Procedure for evaluating the potential of birds to avoid chemically contaminated food, *Environ. Toxicol. Chem.*, 7, 359–362, 1988.
15. Conover, M. R., Response of birds to different types of food repellents, *J. Appl. Ecol.*, 21, 437–443, 1984.
16. Bussiere, J. L., R. J. Kendall, T. E. Lacher, Jr., and R. S. Bennett, Effect of methyl parathion on food discrimination in northern bobwhite (*Colinus virginianus*), *Environ. Toxicol. Chem.*, 8, 1125–1131, 1989.
17. Avery, M. L., Relative importance of taste and vision in reducing bird damage to crops with methiocarb, a chemical repellent, *Agric. Ecosyst. Environ.*, 11, 299–308, 1984.
18. Mason, J. R., and R. F. Reidinger, Jr., Importance of color for methiocarb-induced food aversions in red-winged blackbirds, *J. Wildl. Manage.*, 47, 383–393, 1983.
19. Wilcoxon, H. C., W. B. Dragoin and P. A. Kral, Illness-induced aversions in rat and quail: relative salience of visual and gustatory cues, *Science*, 171, 826–828, 1971.
20. Bennett, R. S., Jr., and H. H. Prince, Influence of agricultural pesticides on food preference and consumption by ring-necked pheasants, *J. Wildl. Manage.*, 45, 74–82, 1981.
21. Grue, C. E., W. J. Fleming, D. G. Busby and E. F. Hill, Assessing hazards of organophosphorus pesticides to wildlife, *Trans. North Am. Wildl. Nat. Res. Conf.*, 48, 200–220, 1983.
22. Seabloom, R. W., G. L. Pearson, L. W. Oring and J. R. Reilly, An incident of fenthion mosquito control and subsequent avian mortality, *J. Wildl. Dis.*, 9, 18–20, 1973.
23. White, D. H., C. A. Mitchell, E. J. Kolbe and J. M. Williams, Parathion poisoning of wild geese in Texas, *J. Wildl. Dis.*, 18, 389–391, 1982.
24. Brewer, L. W., C. J. Driver, R. J. Kendall, T. E. Lacher, Jr., and J. C. Galindo, Avian response to a turf application of Triumph® 4E, *Environ. Toxicol. Chem.*, 7, 391–401, 1988.
25. Clements, R. O., and J. S. Bale, The short-term effects on birds and mammals of the use of chlorpyrifos to control leatherjackets in grassland, *Ann. Appl. Biol.*, 112, 41–47, 1988.
26. Stehn, R. A., J. A. Stone and M. E. Richmond, Feeding response of small mammal scavengers to pesticide-killed arthropod prey, *Am. Midl. Nat.*, 95, 253–256, 1976.
27. Douthwaite, R. J., Changes in pied kingfisher (*Ceryle rudis*) feeding related to endosulfan pollution from tsetse fly control operations in the Okavango Delta, Botswana, *J. Appl. Ecol.*, 19, 133–141, 1982.

28. Rogers, J. G., Jr., and J. T. Linehan, Some aspects of grackle feeding behavior in newly planted corn, *J. Wildl. Manage.*, 41, 444-447, 1977.
29. Besser, J. F., Protecting seeded rice from blackbirds with methiocarb, *Int. Rice Comm. Newsl.*, 22, 9-12, 1973.
30. DeHaven, R. W., D. F. Mott, J. L. Guarino, J. F. Besser, C. E. Knittle and E. W. Schafer, Jr., Methiocarb for repelling birds from ripening sweet cherries, *Int. Pest Cont.*, 21, 12-14, 1979.
31. Stone, C. P., W. F. Shake and D. J. Langowski, Reducing bird damage to highbush blueberries with a carbamate repellent, *Wildl. Soc. Bull.*, 2, 135-139, 1974.
32. Hothem, R. J., D. F. Mott, R. W. DeHaven and J. L. Guarino, Mesurol as a bird repellent on wine grapes in Oregon and California, *Am. J. Enol. Viticult.*, 32, 150-154, 1981.
33. Rooke, I. J., Methiocarb-induced aversion to grapes by gray-backed white-eyes, *J. Wildl. Manage.*, 48, 444-449, 1984.
34. Conover, M. R., Alleviating nuisance Canada goose problems through methiocarb-induced aversion conditioning, *J. Wildl. Manage.*, 49, 631-636, 1985.
35. Robel, R. J., and M. E. Morrow, Feeding by bobwhites on sprouting corn reduced by bendiocarb, *Trans. Kan. Acad. Sci.*, 90, 35-40, 1988.
36. Benjamini, L., The efficacy of seed treatment with aldicarb to establish conditioned aversion in birds to sprouting sugar beets, *Phytoparasitica*, 9, 3-9, 1981.
37. Dingledine, J., An Active Avoidance Study with Canada Goose on Turf, presented at 8th Annual Meeting of the Society of Environmental Toxicology and Chemistry, Pensacola, FL, November 9-12, 1987.
38. Pesticide Assessment Guidelines. Subdivision E. Hazard Evaluation: Wildlife and Aquatic Organisms, EPA-540/9-82-024 (NTIS PB83-153908), U.S. Environmental Protection Agency, Washington, DC, 1982.
39. Forsyth, D. J., Effects of dietary fenitrothion on the behavior and survival of captive white-throated sparrows, in Environmental Surveillance in New Brunswick, 1978-79, in *Effects of Spray Operations for Forest Protection Against Spruce Budworm*, I. W. Varty, Ed., Committee for Environmental Monitoring of Forest Insect Control Operations, Dept. Forestry Research, University of New Brunswick, Fredericton, 1980, pp. 27-28.
40. Grue, C. E., and B. K. Shipley, Interpreting population estimates of birds following pesticide applications—behavior of male starlings exposed to an organophosphate pesticide, *Stud. Avian Biol.*, 6, 292-296, 1981.
41. Busby, D. G., L. M. White and P. A. Pearce, Effects of aerial spraying of fenitrothion on breeding white-throated sparrows, *J. Appl. Ecol.*, 27, 743-755, 1990.
42. McEwen, L. C., and R. L. Brown, Acute toxicity of dieldrin and malathion to wild sharp-tailed grouse, *J. Wildl. Manage.*, 30, 604-611, 1966.
43. Stromborg, K. L., Reproduction of bobwhites fed different dietary concentrations of an organophosphate insecticide, methamidophos, *Arch. Environ. Contam. Toxicol.*, 15, 143-147, 1986.
44. Stromborg, K. L., Reproductive toxicity of monocrotophos to bobwhite quail, *Poult. Sci.*, 65, 51-57, 1986.
45. Bennett, J. K., and R. S. Bennett, Effects of dietary methyl parathion on northern bobwhite egg production and eggshell quality, *Environ. Toxicol. Chem.*, 9, 1481-1485, 1990.
46. Stromborg, K. L., Reproductive tests of diazinon on bobwhite quail, in *Avian and Mammalian Wildlife Toxicology: Second Conference*, ASTM STP 757, D. W. Lamb and E. E. Kenaga, Eds., American Society for Testing and Materials, Philadelphia, 1981, pp. 19-30.
47. Rattner, B. A., L. Sileo and C. G. Scanes, Oviposition and the plasma concentration of LH, progesterone and corticosterone in bobwhite quail (*Colinus virginianus*) fed parathion, *J. Reprod. Fert.*, 66, 147-155, 1982.
48. Case, R. M., Energetic requirements for egg-laying bobwhites, in Proc. 1st Natl. Bobwhite Quail Symposium, Stillwater, OK, April 23-26, 1972, pp. 205-212.
49. Bennett, R. S., B. A. Williams, D. W. Schmedding and J. K. Bennett, Effects of dietary exposure to methyl parathion on egg laying and incubation in mallards, *Environ. Toxicol. Chem.*, 10, 501-507, 1972.

50. Franson, J. C., J. W. Spann, G. H. Heinz, C. Bunck and T. Lamont, Effects of dietary Abate® on reproductive success, duckling survival, behavior, and clinical pathology in game-farm mallards, *Arch. Environ. Contam. Toxicol.*, 12, 529–534, 1983.
51. White, D. H., C. A. Mitchell and E. F. Hill, Parathion alters incubation behavior of laughing gulls, *Bull. Environ. Contam. Toxicol.*, 31, 93–97, 1983.
52. King, K. A., D. H. White and C. A. Mitchell, Nest defense behavior and reproductive success of laughing gulls sublethally dosed with parathion, *Bull. Environ. Contam. Toxicol.*, 33, 499–504, 1984.
53. Meyers, S. M., J. L. Cummings and R. S. Bennett, Effects of methyl parathion on red-winged blackbird (*Agelaius phoeniceus*) incubation behavior and nesting success, *Environ. Toxicol. Chem.*, 9, 807–813, 1990.
54. Spray, C. J., H. Q. P. Crick and A. D. M. Hart, Effects of aerial applications of fenitrothion on bird populations of a Scottish pine plantation, *J. Appl. Ecol.*, 24, 29–47, 1987.
55. Grue, C. E., G. V. N. Powell and M. J. McChesney, Care of nestlings by wild female starlings exposed to an organophosphate pesticide, *J. Appl. Ecol.*, 19, 327–335, 1982.
56. Fairbrother, A., S. M. Meyers and R. S. Bennett, Changes in mallard hen and brood behaviors in response to methyl parathion-induced illness of ducklings, *Environ. Toxicol. Chem.*, 7, 499–503, 1988.
57. Hunter, M. L., Jr., J. W. Witham and H. Dow, Effects of a carbaryl-induced depression in invertebrate abundance on the growth and behavior of American black duck and mallard ducklings, *Can. J. Zool.*, 62, 452–456, 1984.
58. Meyers, S. M., and J. D. Gile, Mallard reproductive testing in a pond environment: a preliminary study, *Arch. Environ. Contam. Toxicol.*, 15, 757–761, 1986.
59. Hill, E. F., R. G. Heath, J. W. Spann and J. D. Williams, Lethal Dietary Toxicities of Environmental Pollutants to Birds, Sci. Rep. Wildlife No. 191, U.S. Fish and Wildlife Service, Washington, DC, 1975.
60. Fleming, W. J., G. H. Heinz, J. C. Franson and B. A. Rattner, Toxicity of Abate® 4E (temephos) in mallard ducklings and the influence of cold, *Environ. Toxicol. Chem.*, 4, 193–199, 1985.
61. Brewer, L. W., C. J. Driver, R. J. Kendall, C. Zenier and T. E. Lacher, Jr., Effects of methyl parathion in ducks and duck broods, *Environ. Toxicol. Chem.*, 7, 375–379, 1988.
62. Lack, D., *The Natural Regulation of Animal Numbers*, Oxford University Press, London, 1970.

25

Immunotoxicology of Captive and Wild Birds

Anne Fairbrother

ABSTRACT

Environmental chemicals, including pesticides, have the potential to alter the immune response of laboratory or free-ranging animals. As a consequence, wild animals may become more susceptible to microbial or parasitic diseases; there is ample evidence that free-ranging wildlife frequently are diseased. Mathematical models predict that disease and parasitism could alter host population growth rates and that host susceptibility is an important component of the growth rate function. Laboratory studies have shown that all classes of insecticides and fungicides have immunosuppressive properties. Only a few studies have been done with wildlife species (fish, mallards, wild mice, and earthworms). A battery of immune function tests developed recently for captive birds is now being used in free-ranging birds to determine if pesticide use or chemical disposal may be at least partially responsible for recent increases in the incidence of disease outbreaks in free-ranging wildlife.

KEY WORDS

immunology, immunotoxicology, avian, pesticides

INTRODUCTION

Immunotoxicology is a subdiscipline of toxicology and is defined as the study of the effects of poisons on the immune system of humans and animals. Immunotoxicology emerged as a discipline separate from immunopharmacology (the study of the effects of drugs and therapeutics on the immune system) in the early 1970s. Since then, the science of immunotoxicology has progressed rapidly in conjunction with an increased understanding of the fundamental cellular and molecular mechanisms of immune function.

The immune system is the body's defense mechanism against invasion of foreign substances. It is an extremely complex interaction of many different cell types (e.g., macrophages, lymphocytes, granulocytes) as well as soluble circulating proteins (e.g., lymphokines, complement factors).[1] This system recognizes "nonself" particles and cells, and either breaks them down and eliminates them from the body, or walls them off behind protective barriers of fibrous tissue. When the immune system fails in its job of surveillance, the individual becomes sick through replication of pathogens, through physical trauma to tissues, by overreaction of the immune system itself (e.g., allergy), or by the proliferation of cancerous cells. Such a result often has ramifications throughout the entire body, resulting in problems such as reproductive or nervous system malfunction, chronic debilitating disease, or death. Failure of the immune system is strikingly illustrated by acquired immune deficiency syndrome (AIDS), the autoimmune disease caused by the human immunodeficiency virus (HIV). The virus itself does not cause death; rather, it invades and destroys certain immune cells (T cells), resulting in compromised immune function. The infected individual then is unable to withstand infections by opportunistic bacteria such as *Pneumocystis* spp., resulting in a fatal pneumonia, or to inhibit cancer cells such as those causing Kaposi's sarcoma.[2]

The immunotoxicity of environmental chemicals to free-living wild animals is reviewed in the following sections, with particular emphasis on pesticides. Techniques used in laboratory and field studies to evaluate the immune status of birds exposed to pesticides are briefly discussed. A review of population models that have incorporated pathogen- and parasite-induced disease effects on demographic parameters will demonstrate the potential for environmental immunotoxicants to affect growth rates of wild animal populations.

IMPORTANCE OF WILDLIFE IMMUNOTOXICOLOGY

Exposure to toxicants that cause immune dysfunction in wildlife species is of importance only if these animals are exposed to infectious and/or parasitic agents. There is, of course, ample evidence that this is the case.[3-6] Wildlife species are infected with many of the same pathogens and parasites that infect domestic livestock and humans. Indeed, free-ranging wildlife often are reservoirs for diseases that are of public health and economic importance. For example, bison in and around Yellowstone National Park are infected with brucellosis, which is readily transmissible to domestic cattle. Abortion and sterility caused by brucellosis is of major economic significance in the cattle industry. Similarly, mycoplasmas and avian influenza viruses harbored by many galliformes and waterfowl, respectively, may cause significant economic damage to the domestic poultry industry. Rabies, plague, Lyme disease, ornithosis, and viral encephalitides are all examples of significant zoonotic diseases harbored by various wildlife species.

During the last 50 years, the number and size of waterfowl epidemics have increased alarmingly due to avian cholera, duck plague, and other pathogens.[5] These epidemics are due in part to changing management practices[7] that force birds into closer contact on ever-smaller refuge areas. However, it is possible that disease outbreaks also may be a result of immune suppression caused by the concomitant increases in pesticide use and industrial waste production. Additionally, concern is increasing for preservation of endangered species; chronic exposure to immunomodulatory chemicals resulting in opportunistic infections may provide the final stress to many fragile natural ecosystems.

Historically, some of the most insightful thinkers in game management and ecology have recognized the potential for disease to influence the distribution and abundance of populations. Darwin, in *Origin of Species*,[8] noted that when a species reaches high densities epidemics often occur, providing a "limiting check" on the host population. Aldo Leopold devoted a whole chapter to disease effects in his book, *Game Management*.[9] Andrewartha and Birch,[10] in their classic treatise *The Distribution and Abundance of Animals*, classified disease with predation, suggesting that disease be considered separately from predator-prey relationships only when it differed in the methods of study.

Although ecologists no longer believe that disease or parasitism are the sole causes of population cycles of small mammals and gallinaceous birds, a great deal of controversy currently exists about what role disease and parasitism play in population regulation. Many ecologists[11,12] believe that parasites co-evolve with their hosts and eventually attain a commensalistic, if not mutualistic, relationship. Consequently, disease can affect a host population only if either the pathogen or the host is newly introduced in a region or if the host is abnormally immunosuppressed.

Some ecologists believe that a host population can compensate for the increased mortality or decreased natality caused by disease. Wild populations often have a surplus of individuals for breeding relative to the amount of habitat or food available. Thus, the breeding population is a subset of the population as a whole. If disease kills some individuals in the breeding population, others take their place. While this compensatory change in the breeding population may produce the same number of young, it also represents a shift in the genetic base of the species that may alter the long-term fitness of the population.

Predation is often a major mortality factor of young animals. If disease kills many in this cohort, then: (1) the remaining young are preyed upon more heavily than before, reducing the number reaching sexual maturity; or (2) predators faced with a declining number of this prey species would begin relying more heavily on other species (termed "buffer species"), leaving approximately the same number of young to mature as in the absence of disease. Additionally, predators may scavenge on those dying from disease, relieving predation pressure on the remainder of the prey population. This type of scavenging occurred in a population of foxes feeding on muskrat that died of a "hemorrhagic disease".[13]

Although parasites and pathogens may evolve to the point at which they do not kill their hosts, they still may have significant sublethal effects on the host population. Figure 1 illustrates how sublethal infections may have physiological effects impacting one or more organ systems and how immune dysfunction fits into this relationship. These organ system effects, in turn, could cause a decrease in the birth rate or indirectly contribute to the death rate by compromising the immune system of the host and increasing its susceptibility to future infections. Physiological effects can also cause behavioral changes that result in altered mass and energy balances, changes in movement patterns, or increased or decreased predation.

Others have suggested that mate selection can be influenced by the disease state of the mate being selected. Freeland[14] showed that the nematode *Heligmosomoides polygyrus* decreased the development of dominance among male mice and suggested that female sex characteristics evolved in part to induce competition by males so as to reveal their disease state.[15] Hamilton and Zuk[16] suggested that birds select nonparasitized mates based on phenotypic characteristics such as brightness of plumage. These types of sublethal influences could markedly change the genetic base of a population and decrease its overall fitness.

MATHEMATICAL MODELS

Interest in mathematical modeling of ecological consequences of disease has increased, based in part on the theories developed to describe predator-prey interactions. Although the literature on mathematical models of predator-prey systems is quite extensive, these models are not directly applicable to host-parasite relationships. The

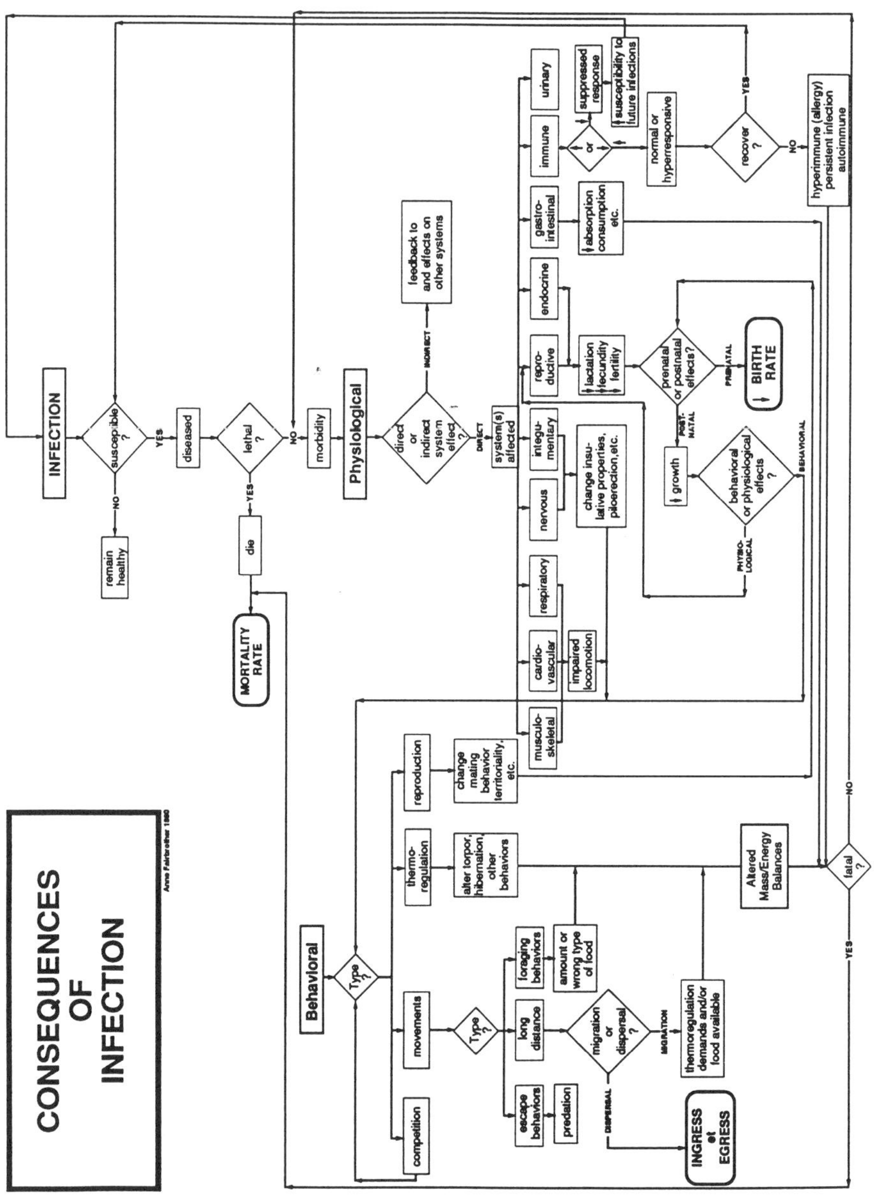

FIGURE 1. A simplified flow diagram of the relationships between disease-induced physiological and behavioral changes and population demographic factors.

models are usually based on Lotka and Volterra's equations,[17] which assume that parasites (predators) alter only the death rate of the host population. As discussed above, parasites and infectious disease agents may also influence host birth rate and movements. In a series of papers, Anderson and May[18-29] have developed new models describing changes in host population growth rate by microparasites (viruses, bacteria, and protozoa) and macroparasites (helminths). These are general models that give insight into how parasites might regulate the host population—through stable limit cycles, oscillations, or stable equilibria. The models indicate that both types of parasites can cause the host population to grow at a lower rate than if the parasites were not present.

Many of the mathematical models discussed above predict that pathogens and parasites can decrease the growth rate of a population. This prediction has been borne out in a study conducted by Anderson et al.,[24] which showed a decrease in the population of red foxes in Europe due to the spread of endemic rabies virus. Introduction of myxomatosis in Australia in the 1950s caused an initial dramatic decrease in numbers of rabbits (*Oryctolagus cuniculus*) followed by adaptation of the host-virus system and maintenance of a stable rabbit population at densities lower than predisease values.[30] On a smaller scale, the "hemorrhagic disease" described by Errington[13] nearly eliminated muskrats from the Iowa marsh in which the epidemic occurred. Although avian cholera and botulism kill thousands of waterfowl yearly,[5,6] the cumulative impact of such mortality on the entire North American population of waterfowl species is not known. Nevertheless, government agencies and private organizations concerned with the health and maintenance of wildlife populations, whether for purposes of hunting, the fur trade, or aesthetics, should be cognizant of the potential for environmental chemicals to reduce animal populations by decreasing the animals' abilities to resist disease caused by pathogens and parasites, resulting in decreased fecundity and increased mortality.

IMMUNOTOXICANTS

Most of the data available on immunotoxic properties of chemicals comes from controlled laboratory studies using rabbits, rodents, or chickens as test animals. A large percentage of these data are derived from studies of nonpesticide environmental contaminants such as heavy metals, pentachlorophenol (PCP), nonpesticide organochlorines (OCs) (e.g., polychlorinated biphenyls, PCBs), polycyclic aromatic hydrocarbons (PAHs), and dioxins (e.g., 2,3,7,8-TCDD).[31,32] However, as the primary purpose of this chapter is to examine pesticide effects on wildlife, this extensive body of literature will be reviewed only as it provides information about the immunotoxicity of pesticides with an emphasis on wildlife studies.

Dieldrin, lindane, and DDT have been shown to increase animals' susceptibility to infection. Ducks exposed to dieldrin had increased mortality rates following challenge with duck hepatitis virus;[33] lindane and hexachlorobenzene increased the susceptibility of mice to malaria infections;[34] and DDT increased the percent mortality of mice exposed to encephalomyocarditis (EMC) virus[35] and susceptibility of chickens to *Histomonas meleagridis*.[36] Circulating antibody titers were depressed in chickens by mirex and DDT,[37] while heptachlor increased the weights of chicken bursa of fabricius (suggestive of increased potential for antibody production).[38] Nonspecific immunity as measured by the phagocytic ability of macrophages was also depressed by p,p'-DDT in rats and rabbits,[39,40] while DDT and endrin, another OC insecticide, have been shown to reduce the ability of human lymphocytes to multiply and mature into functional B and T cells.[41]

Organophosphorus (OP) insecticides also have shown immunosuppressive effects in laboratory studies. Dimethoate, dichlorvos, methyl parathion, and malathion decreased antibody titers in mice, rats, or rabbits.[42-43] Methyl parathion has also been shown to increase the susceptibility of mice to infection by *Staphylococcus typhimurium*,[42] and to decrease murine lymphocyte proliferation.[41] Trichlorphon decreased phagocytic ability of neutrophils in carp (*Cyprinus carpio*).[44] Ronnel increases bursal weights in chickens.[38] Carbaryl and carbofuran, two carbamate insecticides, are also immunoreactive and depressed antibody production, decreased T cell responses, and decreased resistance to *Giardia muris* and *Escherichia coli* or *Staphylococcus* infections in rats and mice.[31]

Synthetic pyrethroids are relatively new insecticides that are being used in increasing quantities. A recent study by Desi et al.[45] demonstrated the potential for immunosuppression by this compound class. Rabbits exposed to cypermethrin had a dose-dependent decrease in antibody titer to *Salmonella typhi* challenge and a similar dose-dependent inhibition of the delayed-type hypersensitivity response (as measured by the tuberculin reaction test). This suggests that cypermethrin affects both humoral and cell-mediated immune functions.

Miticides such as milbex,[39,46] herbicides such as paraquat,[47] diquat,[47] linuron,[48] and monum,[43] and the fungicides triphenyltin acetate,[49] zineb,[39] ziram, eptam, maneb,[41] and thiram[39] have all been shown by laboratory tests to depress various parts of the immune response in rats, mice, or guinea pigs. An in-depth review and summary of the immune effects of these and other pesticides and environmental contaminants can be found in Street[41] and Exon et al.[31]

WILDLIFE IMMUNOTOXICOLOGY STUDIES

Relatively few studies have examined the immune effects of environmental contaminants on wildlife species, either in the laboratory or in the field. Deer mice (*Peromyscus maniculatus*) have been exposed in laboratory studies to the plant growth regulators chlorocholine chloride and glyphosine and to the OC compounds in Arochlor 1254 (a mixture of PCBs).[50-52] White blood cell counts, hemolysin titers, and splenic plaque forming cells (PFCs; the number of spleen cells capable of producing antibodies) were decreased by the plant growth regulators, which also caused an increased viremia following challenge with Venezuelan equine encephalitis virus as compared to control animals. The PCB mixture decreased antibody titers.

Three wild bird species that have been domesticated and adapted to the laboratory environment have been used in immunotoxicology testing: (1) the mallard (*Anas platyrhynchos*); (2) the bobwhite (*Colinus virginianus*); and (3) the Japanese quail (*Coturnix coturnix japonica*). Several studies of immunotoxic effects also have been undertaken using the domestic chicken (*Gallus gallus*) as the test animal. In mallards, exposure to dieldrin-treated feed increased mortality due to duck viral hepatitis as compared to control animals.[33] Oral exposure to fuel oil and oil-dispersant mixtures decreased resistance of mallards to *Pasteurella multocida* infections but had no effect on splenic PFC numbers.[53] Ducks exposed to lead shot by oral gavage had decreased antibody titers to sheep red blood cells, but there was no measurable effect on lymphocyte proliferation.[54] Subchronic exposure of mallards to selenium-contaminated drinking water significantly impaired the delayed-type hypersensitivity response to tuberculin (a measure of cell-mediated immune function), had no effect on antibody production or cellular phagocytosis,[55] and increased susceptibility of 19-day-old ducklings to duck viral hepatitis.[56]

Bobwhite quail exposed to carbaryl-treated feed had higher mortality rates from *Histomonas meleagridis* infections than did control birds.[57] DDT exposure had no effect on antibody production in laboratory-reared Japanese quail.[58] Laboratory exposure of chickens to mercury, PCP, tri-o-tolyl phosphate, lead, TCDD, and PCBs affected many aspects of the immune response.[59-64] These tests in chickens as well as an even larger number of studies relative to immunological effects of ingestion of aflatoxins[65-68] are valuable in establishing methods for studying toxicological effects on the immune systems of other bird species.

To date, no reports of immunotoxicological tests on free-ranging wildlife have been published. Two studies utilized mallards maintained in seminatural outdoor enclosures. One studied the effect of lead shot on the mallards' ability to produce antibodies to sheep red blood cells,[77] and the other examined the effect of selenium-treated artificial streams on mallard antibody production, phagocytosis, white cell counts, and juvenile survival.[56]

Weeks and Warinner[69,70] collected fish (spot and hogchoker) from a polluted river near the Chesapeake Bay and demonstrated a decreased phagocytic response of macrophages. Wild-caught perch (*Perca fluviatilis*) from a cadmium-contaminated river exhibited increased numbers of lymphocytes.[71] Other studies have used field-caught fish and invertebrates (earthworms, polychaetes, and even amoeba) in laboratory tests of immunotoxicity.[72-74]

In order to conduct field tests of immune function of wildlife, it is necessary to have available a set of assays that includes at least one measure each of cell-mediated immunity, antibody production, and nonspecific immune functions. Initial positive screening tests could be followed by more in-depth laboratory studies if interest warrants, similar to the tier testing scheme in human toxicology[75] or described by Weeks et al.[76] In the laboratory, the author and her co-workers have developed such a set of assays that can be used in birds. Nonspecific immune function is measured by total and differential white blood cell counts, the ability of macrophages to phagocytose opsonized yeast cells, in vivo carbon clearance assays, and the chromium-release natural killer cell assay. Cell-mediated immunity is assayed using the lymphocyte migration inhibition test, and the tuberculin delayed-type hypersensitivity test or dermal (wing petagyum) phytohemagglutinin (PHA) reaction. A graft-vs-host response using embryonated eggs is under development. Quantification of antibody titers and PFC response to sheep or rabbit red blood cells is used to determine humoral immune status. Lymphocyte proliferation assays also are included. Most of these assays are described in Fairbrother and Fowles[55] and have been well characterized for mallards. We have now used the tests with bobwhite, avocets (*Americana recurvirostra*), blacknecked stilts (*Himantopus mexicanus*), Canada geese (*Branta canadensis*), and starlings (*Sturnus vulgarus*) in the laboratory and will soon be using them with free-ranging avocets and starlings.

CONCLUSION

Our knowledge of and ability to examine the effects of chemicals on the immune system has increased dramatically over the past 10 years, with the advent of new, sophisticated techniques for studying cellular and molecular mechanisms. Basic research in comparative immunology needs to continue and should be supported to expand our knowledge of the complexities involved in all parts of the immune system of different classes of animals. Additionally, immunotoxicological studies, with either free-living or captive wildlife, will provide a better understanding of which environmental contaminants are potential immunotoxicants.

Specific recommendations for future research in wildlife immunotoxicology include continuing to develop a battery of laboratory immune function tests to screen chemicals for immunomodulatory effects in birds, particularly the mallard. Development of these tests should be coupled with the assessment of immunotoxic effects of selected pesticides and toxicants. These tests then need to be adapted for field monitoring and their validity tested in actual field situations (e.g., monitoring birds and small mammals that live in or around hazardous waste sites or on agricultural drain water receiving impoundments). Geographic information systems could be used to look for correlations in space and time of pesticide use or chemical disposal and disease outbreaks in free-ranging wildlife.

ACKNOWLEDGMENTS

The information in this document has been funded wholly or in part by the U.S. Environmental Protection Agency. It has been subjected to the agency's peer and administrative review and it has been approved for publication. Mention of trade names or commercial products does not constitute endorsement or recommendation for use.

REFERENCES

1. Roitt, I., J. Brostoff and D. Male, *Immunology*, Gower Medical, London, 1985.
2. Miller, J., and H. Masur, AIDS-related infections, *Sci. Am.*, August, 50–57, 1990.
3. Davis, J. W., L. H. Karstad and D. O. Trainer, *Infectious Diseases of Wild Mammals*, Iowa State University Press, Ames, 1981.
4. Davis, J. W., and R. C. Anderson, *Parasitic Diseases of Wild Mammals*, Iowa State University Press, Ames, 1971.
5. Friend, M., *Field Guide to Wildlife Diseases, Vol. 1, General Field Procedures and Diseases of Migratory Birds*, Resource Publ. No. 167, U.S. Fish and Wildlife Service, Washington, DC, 1987.
6. Wobeser, G. A., *Diseases of Wild Waterfowl*, Plenum Press, New York, 1981.
7. Friend, M., Waterfowl management and waterfowl disease: independent or cause and effect relationships?, *Trans. North Am. Wildl. Nat. Res. Cong.*, 46, 94–102, 1981.
8. Darwin, C., *The Origin of Species*, Mentor edition, New American Library, New York, 1859, pp. 79–80.
9. Leopold, A., *Game Management*, Charles Scribner's Sons, New York, 1933, pp. 324–341.
10. Andrewartha, H. G., and L. C. Birch, *The Distribution and Abundance of Animals*, University of Chicago Press, Chicago, 1954, pp. 482–485.
11. Holmes, J. C., Evolutionary relationships between parasitic helminths and their hosts, in *Coevolution*, D. J. Futuyma and M. Slatkin, Eds., Sinauer Associates, Sunderland, MA, 1982, pp. 161–185.
12. Lack, D., *Natural Regulation of Animal Numbers*, Oxford University Press, Oxford, U.K., 1954, pp. 162–178.
13. Errington, P. L., *Of Predation and Life*, Iowa State University Press, Ames, 1967, pp. 176–178.
14. Freeland, W. J., Parasitism and behavioral dominance among male mice, *Science*, 213, 461–462, 1981.

15. Freeland, W. J., Pathogens and the evolution of primate sociality, *Biotropica*, 8, 12–24, 1976.
16. Hamilton, W. D., and M. Zuk, Heritable true fitness and bright birds: a role for parasites?, *Science*, 218, 384-386, 1982.
17. Ricklefs, R. E., *Ecology*, Chiron Press, Newton, MA, 1973, pp. 536–537.
18. Anderson, R. M., The regulation of host population growth by parasitic species, *Parasitology*, 76, 119–157, 1978.
19. Anderson, R. M., Parasite pathogenicity and the depression of host population equilibria, *Nature*, 279, 150–152, 1979.
20. Anderson, R. M., The persistence of direct life cycle infectious diseases within populations of host, in *Lectures on Mathematics in the Life Sciences*, Vol. 12, S. Evin, Ed., American Mathematical Society, Providence, RI, 1979, pp. 1-66.
21. Anderson, R. M., The influence of parasitic infection on the dynamics of host population growth, in *Population Dynamics*, R. M. Anderson, B. D. Turner and L. R. Taylor, Eds., Blackwell Scientific, Oxford, 1979, pp. 245-281.
22. Anderson, R. M., Depression of host population abundance by direct life cycle macroparasites, *J. Theor. Biol.*, 82, 283-311, 1980.
23. Anderson, R. M., Transmission dynamics and control of infectious disease agents, in *Population Biology of Infectious Diseases*, R. M. Anderson and R. M. May, Eds., Dahlem Konferenzen, Springer-Verlag, Berlin, 1982, pp. 149–176.
24. Anderson, R. M., H. C. Jackson, R. M. May and A. M. Smith, Population dynamics of fox rabies in Europe, *Nature*, 289, 765-771, 1981.
25. Anderson, R. M., and R. M. May, Regulation and stability of host parasite population interaction. I. Regulatory processes, *J. Anim. Ecol.*, 47, 219–249, 1978.
26. Anderson, R. M., and R. M. May, Population biology of infectious diseases. I., *Nature*, 280, 361-367, 1979.
27. Anderson, R. M., and R. M. May, Infectious diseases and population cycles of forest insects, *Science*, 210, 658-661, 1980.
28. May, R. M., and R. M. Anderson, Regulation and stability of host parasite population interactions. II. Destabilizing processes, *J. Anim. Ecol.*, 47, 249–268, 1978.
29. May, R. M., and R. M. Anderson, Population biology of infectious diseases. II, *Nature*, 280, 455-461, 1979.
30. Yuill, T. M., Myxomatosis and fibromatosis, in *Infectious Diseases of Wild Mammals*, J. W. Davis, L. H. Karstad and D. O. Trainer, Eds., Iowa State University Press, Ames, 1981.
31. Exon, J. H., N. I. Kerkvliet and P. A. Talcott, Immunotoxicity of carcinogenic pesticides and related chemicals, *Environ. Carcinogen. Rev., (J. Environ. Sci. Health)* C5(1), 73–120, 1987.
32. Koller, L. D., and J. G. Vos, Immunologic effects of metals, in *Immunologic Considerations in Toxicology*, R. P. Sharma, Ed., CRC Press, Boca Raton, FL, 1981, pp. 67–79.
33. Friend, M., and D. O. Trainer, Experimental dieldrin-duck hepatitis virus interaction studies, *J. Wildl. Manage.*, 38, 896-902, 1974.
34. Loose, L. D., K. A. Pittman, K. Benitz and J. B. Silkworth, PCB and HCB induced immunosuppression in mice, *Toxicol. Appl. Pharmacol.*, 41, 201-202, 1977.
35. Crocker, J. F. S., R. L. Ozere, S. H. Safe, S. C. Digout, K. R. Rozee and O. Hutzinger, Lethal interaction of ubiquitous insecticide carriers with virus, *Science*, 192, 1351–1353, 1976.
36. Radhakrishnan, C. V., N. P. Thompson and D. J. Forester, Susceptibility of chickens fed pp-DDT to histomoniasis, *Bull. Environ. Contam. Toxicol.*, 8, 147–152, 1972.
37. Glick, B., Antibody-mediated immunity in the presence of mirex and DDT, *Poultry Sci.*, 53, 1476–1484, 1974.
38. Rodica, G., and M. Stefania, Effects of some insecticides on the bursa of Fabricius in chicks, *Arch. Exp. Veterinaer. Med.*, 27, 723-728, 1973.
39. Perelygin, V. M., M. B. Shpirt, O. A. Aripov and V. I. Ershova, Effects of some pesticides on immunological reactivity, *Gig. Sanit.*, 36, 29, 1971.

40. Kaminski, E. J., C. A. Fischer, G. L. Kennedy and J. C. Calandra, Response of pulmonary macrophages to lead, *Br. J. Exp. Pathol.*, 58, 9–12, 1977.
41. Street, J. C., Pesticides and the immune system, in *Immunologic Considerations in Toxicology*, R. P. Sharma, Ed., CRC Press, Boca Raton, FL, 1981, pp. 31–66.
42. Fan, A., J. C. Street and R. M. Nelson, Immunosuppression in mice administered methyl parathion and carbofuran by diet, *Toxicol. Appl. Pharmacol.*, 45, 235, 1978.
43. Nikolaev, A. I., L. A. Ponomareva, I. S. Geller, M. I. Rozgon and F. S. Garipova, Immunodepressive action of some poisonous chemicals, *Farmakol. Toksikol. (Moscow)*, 35, 352–355, 1972.
44. Siwicki, A. K., M. Cossarini-Dunier, M. Studnicka and A. Demael, In vivo effects of the organophosphorus insecticide trichlorphon on immune response of carp (*Cyprinus carpio*). II. Effect of high doses of trichlorphon on nonspecific immune response, *Ecotoxicol. Environ. Safety*, 19, 99–105, 1990.
45. Desi, I., L. Varga, I. Dobronyl and G. Szklenarik, Immunotoxicological investigation of the effects of a pesticide; cypermethrin, *Arch. Toxicol., Suppl.*, 8, 305–309, 1985.
46. Aripdzhanov, T. M., Experimental study of the effects of anthio and milbex on immunologic reactivity and certain autoimmune processes of the body, *Gig. Sanit.*, 38, 101, 1973.
47. Styles, J. A., Studies on the effects of paraquat and diquat on cells in culture. Viability of macrophages and fibroblasts incubated with paraquat and diquat, *Br. J. Exp. Pathol.*, 55, 71–77, 1974.
48. Dinoeva, S. K., Dynamics of changes in the immune structure of lymphatic follicles of the spleen following pesticide poisoning, *Gig. Sanit.*, 39, 85, 1974.
49. Verschuuren, H. G., E. J. Ruitenberg, F. Peetoom, P. W. Helleman and E. F. van Esch, Influence of triphenyltin acetate on lymphatic tissue and immune response in guinea pigs, *Toxicol. Appl. Pharmacol.*, 16, 400–410, 1970.
50. Olson, L. J., and R. D. Hinsdill, Influence of feeding chlorocholine chloride and glyphosine on selected immune parameters in deer mice, *Peromyscus maniculatus*, *Toxicology*, 30, 103–114, 1984.
51. Fairbrother, A., T. M. Yuill and L. J. Olson, Effect of ingestion of chlorocholine chloride and cyclophosphamide on Venezuelan equine encephalitis virus infections on deer mice (*Peromyscus maniculatus*), *Toxicology*, 31, 67–71, 1984.
52. Porter, W. P., R. D. Hinsdill, A. Fairbrother, L. J. Olson, J. Jaeger, T. M. Yuill, S. Bisgaard, W. G. Hunter and K. Nolan, Toxicant-disease-environment interactions associated with suppression of immune system, growth, and reproduction, *Science*, 224, 1014–1017, 1984.
53. Rocke, T. E., T. M. Yuill and R. D. Hinsdill, Oil and related toxicant effects on mallard immune defenses, *Environ. Res.*, 33, 343–352, 1984.
54. Trust, K. A., M. Miller, J. K. Ringelman and I. A. Orme, Effects of ingested lead on antibody production in mallards, *J. Wildl. Dis.*, 26, 316–322, 1990.
55. Fairbrother, A., and J. Fowles, Subchronic effects of sodium selenite and selenomethionine on several immune functions in mallards, *Arch. Environ. Contam. Toxicol.*, 19, 836–844, 1992.
56. Whiteley, P., and T. M. Yuill, Effects of Selenium on Mallard Duck Reproduction and Immune Function, EPA/600/3-89/078, PB 120 692/AS, ERL-COR-574, U.S. Environmental Protection Agency, Environmental Research Laboratory, Corvallis, OR, 1989.
57. Zeakes, S. J., M. F. Hansen and R. J. Robel, Increased susceptibility of bobwhites (*Colinus virgianus*) to *Histomonas meleagridis* after exposure to Sevin insecticide, *Avian Dis.*, 25(4), 981–987, 1982.
58. Bryan, T. E., R. P. Gildersleeve and R. P. Wiard, Exposure of Japanese quail embryos to o,p'-DDT has long-term effects on reproductive behaviors, hematology, and feather morphology, *Teratology*, 39, 525–535, 1989.
59. Bridger, M. A., and J. P. Thaxton, Humoral immunity in the chicken as affected by mercury, *Arch. Environ. Contam. Toxicol.*, 12(1), 45–49, 1983.

60. Prescott, C. A., B. N. Wilkie, B. Hunter and R. J. Julian, Influence of a purified grade of pentachlorophenol in the immune response of chickens, *Am. J. Vet. Res.*, 43, 481–487, 1982.
61. Foil, L. D., H. W. Chambers, R. S. Stinson and B. Glick, Immunological agents of tri-o-tolyl phosphate-induced delayed neurotoxicity in chickens, *Toxicol. Appl. Pharmacol.*, 56, 259–264, 1980.
62. Vengis, V. E., and C. J. Mare, Lead poisoning in chickens and the effect of lead on interferon and antibody production, *Can. J. Comp. Med.*, 38, 328–335, 1974.
63. Nikolaides, E., B. Brunstrom and L. Dencker, Effects of the TCDD congeners 3,3′,4,4′-tetrachlorobiphenyl and 3,3′,4,4′-tetrachloro-azoxybenzene on lymphoid development in the bursa of Fabricius of the chick embryo, *Toxicol. Appl. Pharmacol.*, 92(2), 315–323, 1988.
64. Harris, S. J., O. H. C. Cecil, J. Bitman and R. J. Lillie, Antibody response and reduction in bursa of Fabricius and spleen weights of progeny of chickens fed PCBs, *Poultry Sci.*, 55, 1933–1940, 1987.
65. Kadian, S. K., D. P. Monga and M. C. Goel, Effect of aflatoxin B1 on the delayed type hypersensitivity and phagocytic activity of reticuloendothelial system in chickens, *Mycopathologia*, 104, 33–36, 1988.
66. Chang, C. F., W. E. Huff and P. B. Hamilton, A leukocytopenia induced in chickens by dietary ochratoxin A, *Poultry Sci.*, 58, 555–558, 1979.
67. Boonchuvit, B., and P. B. Hamilton, Interaction of aflatoxin and paratyphoid infections in broiler chickens, *Poultry Sci.*, 54(5), 1567–1573, 1975.
68. Giambrone, J. J., U. L. Diener, N. D. Davis, V. S. Panagais and F. J. Hoerr, Effects of purified aflatoxin on broiler chickens, *Poultry Sci.*, 64, 852–858, 1985.
69. Weeks, B. A., and J. E. Warinner, Effects of toxic chemicals on macrophage phagocytosis in two estuarine fishes, *Mar. Environ. Res.*, 1, 327–335, 1984.
70. Weeks, B. A., and J. E. Warinner, Functional evaluation of macrophages on fish from a polluted estuary, *Vet. Immunol. Immunopathol.*, 12, 313–320, 1986.
71. Sjobeck, M. L., C. Haux, A. Larsson and G. Lithner, Biochemical and hematological studies on perch, *Perca fluviatilis*, from the cadmium-contaminated river Emån, *Ecotoxicol. Environ. Safety*, 8, 303–312, 1984.
72. Goven, A. J., B. J. Venables, L. C. Fitzpatrick and E. L. Cooper, An invertebrate model for analyzing effects of environmental xenobiotics on immunity, *Clin. Col.*, 5, 150–154, 1988.
73. Anderson, R. S., C. S. Giam and L. E. Ray, Effect of hexachlorobenzene and pentachlorophenol on cellular and humoral immune parameters in *Glycera dibranchiata*, *Mar. Environ. Res.*, 14, 317–326, 1984.
74. Cyr, R. J., and R. L. Bernstein, Morphological changes and depressed phagocytic efficiency in dictyostelium amobae treated with toxic concentrations of cadmium, *Environ. Res.*, 35, 66–78, 1984.
75. Luster, M. I., A. E. Munson, P. T. Thomas, M. P. Holsapple, J. D. Fenters, K. L. White, L. D. Lauer, D. R. Germolec, G. J. Rosenthal and J. H. Dean, Development of a testing battery to assess chemical-induced immunotoxicity: National Toxicology Program's guidelines for immunotoxicity evaluation in mice, *Fund. Appl. Toxicol.*, 10, 2–19, 1988.
76. Weeks, B. A., A. Fairbrother, A. Goven and D. P. Anderson, Immunological biomarkers to assess environmental stress, in *Biomarkers: Biochemical, Physiological, and Histological Markers of Anthropogenic Stress*, R. J. Huggett, Ed., Lewis Publishers, Chelsea, MI, 1992, pp. 211–234.
77. Rocke, T. E., personal communication.

26

Modeling the Effects of Pesticides on Wildlife: The Need to Evaluate Our Leaps of Faith and Pious Hopes

Christian E. Grue

At least initially, models describing the effects of pesticides on wildlife populations will depend largely on data collected from captive animals in controlled laboratory conditions. Critical to the use of these data is knowledge of the extent to which they reflect or predict effects in the field. To date, most assessments of the hazards of pesticides to wildlife have relied on (1) the assumption that effects observed under laboratory conditions also occur in the field (leaps of faith), and (2) the conclusions at the end of manuscripts that observed effects may ultimately reduce the survival or reproduction of free-living individuals or populations (pious hopes[1]). How valid are these leaps of faith and pious hopes?

As noted in Chapter 21, one could argue that the relatively few instances of recurring adverse effects of pesticides on wildlife populations indicate that ecological risk assessments based largely on laboratory data[2] have been effective in protecting nontarget species. Hill[3] described the two most commonly used standardized toxicity tests (acute, LD_{50} and subacute, LC_{50}) and how the resultant data and statistics can be used to initially evaluate the direct toxic hazard of a particular pesticide to wildlife relative to that of another chemical tested in the same manner.

Conversely, one could argue that for the vast majority of chemicals in use, little is known about their direct and indirect effects on free-living individuals, and that an increasing number of studies have reported pesticide-related mortality in the field or sublethal effects in captive and wild individuals that may impair reproduction and survival. Wildlife mortality associated with the use of pesticides has been reviewed by Grue and co-workers.[4,5] Reviews by Peakall,[1] Grue et al.,[6] Bennett,[7] and Fairbrother[8] have described the many ways pesticides can directly affect the physiology and behavior of wildlife following sublethal exposure to these chemicals. Unfortunately, few studies have assessed the effects of sublethal exposure to pesticides currently in use on the reproduction and survival of free-living wildlife,[6–8] and an even smaller number of studies have assessed the effects of pesticides on habitat quality (availability of cover and food).[4,5]

Our lack of knowledge of the actual hazards of pesticides to nontarget wildlife is not surprising given the complexity of even a basic conceptual model (Figure 1), and the potential number of pesticide/wildlife exposure scenarios. Furthermore, many factors may affect the vulnerability of wildlife to pesticides, including species sensitivity; physiological condition; behavioral traits; the timing, rate, and frequency of applications; and the chemical formulation.[4,9,10] Sensitivity may vary among chemicals and species.[11,12] Route of exposure, sex, age, diet, and environmental conditions can also alter an individual's sensitivity to pesticide exposure. Body size appears to be an important

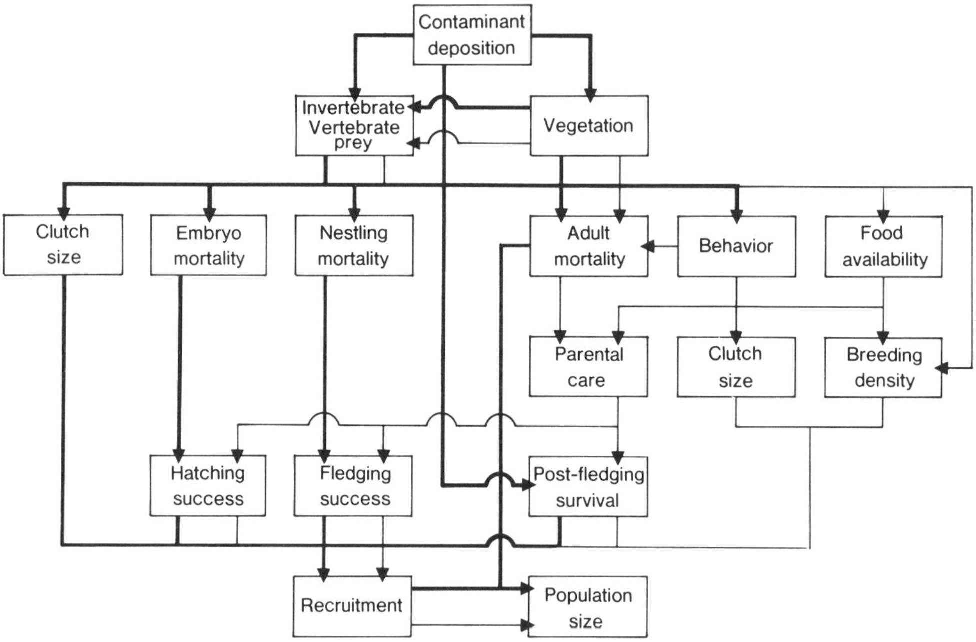

FIGURE 1. Potential direct and indirect effects of environmental contaminants on birds. Bold lines denote direct effects; light lines represent indirect effects. Schematic may be adapted for other terrestrial vertebrates by substituting the appropriate terminology. (Reprinted from Grue et al.[5] With permission).

factor governing the vulnerability of wildlife to some pesticides. For example, small birds and mammals may be more susceptible than their larger counterparts to the effects of organophosphorus pesticides because their high metabolic rates require them to ingest large quantities of contaminated food per unit body weight or make them less tolerant of pesticide-induced anorexia.[6,7] Poor physiological condition could increase these effects. Behavioral traits, particularly foraging habits, frequently increase the vulnerability of nontarget individuals and species to pesticides by increasing their exposure to chemicals. Behaviors that congregate wildlife in treated habitats appear to be associated with many of the largest incidents of pesticide-induced mortality of wildlife, particularly birds.[4] The potential hazards appear to be greatest for species whose foraging habits include the treated substrate or target organisms. Examples include the vulnerability of waterfowl to applications of insecticides (organophosphates and carbamates) to turf[13] or emergent small grains,[14,15] and the mortality of black-billed magpies (*Pica pica*) following the use of the organophosphorus insecticide famphur to control warbles on cattle in the Pacific Northwest.[16]

The direct toxic effects of pesticides on the availability of food or cover may also have significant indirect effects on wildlife species, but have not been included in most hazard assessments.[2] The reproductive strategies of many species of wildlife have evolved to synchronize reproductive activity with prey abundance. Although the indirect effects of pesticides on wildlife are poorly known, several studies have documented significant declines in nontarget insect populations following insecticide applications.[4,17]

The development of conceptual and predictive models for assessing the effects of pesticides on wildlife populations and their subsequent validation will provide a framework for guiding future research to fill data gaps, and force us to evaluate our leaps of faith and pious hopes. The importance of the development of conceptual models cannot be overemphasized. As noted recently by Grue et al.[6] and Di Giulio et al.,[18] there appear to have been many "missed opportunities" to collect needed data that could have been exploited if researchers had a better appreciation of data gaps or greater flexibility in study design. This appears to be particularly true for data on biomarkers, such as plasma and brain cholinesterase,[6] and a variety of in vitro techniques.[8,18]

What data are needed to evaluate the extent to which results from the laboratory can be extended to the field, and what approaches can be utilized to fill these data gaps? Data on exposure to pesticides and its toxicity need to be collected and compared among captive and free-living aminals.[19] Data on the biological implications of observed direct and indirect effects on individuals and populations[20] must also be obtained. The latter poses the greatest challenge. Pesticides are known to affect captive animals in a variety of ways, many of which appear sublethal under controlled environmental conditions, but may actually impair survival or reproduction in free-living wildlife.[6,7] As Heinz[21] recently asked, "How lethal are sublethal effects?" Finally, we must know the potential biases associated with the collection of these data.[22]

Achieving a full understanding of the biological implications of pesticide use will not be easy. A combination of complementary laboratory and field studies will be necessary. These studies will need to focus on developing the necessary tools for monitoring pesticide exposure, documenting effects, and identifying the underlying mechanisms. The studies by DeWeese and co-workers,[23] Powell,[24] Busby and co-workers,[25 and related papers] provide good examples of relatively comprehensive assessments in the field; the latter combined several laboratory, controlled field, and field studies in assessing the effects of aerial applications of the organophosphorus insecticide fenitrothion on breeding white-throated sparrows (*Zonotrichia albicolis*) in New Brunswick forests. Undoubtedly, the most comprehensive study of the overall ecological effects of pesticide use on a wildlife species conducted to date is that described by Potts[26] and Rands[27] on the effects of herbicides and insecticides on the survival of the grey partridge (*Perdix perdix*) in Great Britain. An equally comprehensive study is currently underway in Great Britain, sponsored by the Ministry of Agriculture, Fisheries, and Food. The Boxworth Project[28] will determine the long-term (1983 to 1993) ecological effects of three different cereal crop management regimens (including pesticide programs) on animal and plant species. Data comparable to that collected in these studies are necessary to effectively model the effects of pesticides on wildlife populations.

In conclusion, models provide a basis for understanding systems, a means for identifying data needs, and ultimately, a means for making predictions. The development of both conceptual and predictive models will be critical to the advancement of wildlife toxicology and ecological risk assessments for pesticides.[29] As noted by Emlen,[30] models offer many opportunities to advance the science of wildlife toxicology, but can just as easily "lead us wildly astray." Models should not be construed as an end in themselves, but as one of the many tools for improving our understanding of how pesticides affect wildlife populations. Objective, critical evaluations are necessary at each step of the modeling process, not only in terms of the conceptual framework, but also in terms of the quality of the data upon which the predictive capabilities of the models are built. To this end, it is essential that we critically examine our leaps of faith and pious hopes.

ACKNOWLEDGMENTS

The author is an employee of the U.S. Fish and Wildlife Service. The Washington Cooperative Fish and Wildlife Research Unit is jointly supported by the U.S. Fish and Wildlife Service, University of Washington, Washington Department of Ecology, Washington Department of Fisheries, Washington Department of Natural Resources, Washington Department of Wildlife, and the Wildlife Management Institute.

REFERENCES

1. Peakall, D. B., Behavioral responses of birds to pesticides and other contaminants, *Residue Rev.*, 96, 45–77, 1985.
2. Urban, D. J., and N. J. Cook, Hazard Evaluation Division Standard Evaluation Procedure—Ecological Risk Assessment, EPA 540/9-85-001, U.S. Environmental Protection Agency, Washington, DC, 1986.
3. Hill, E. F., Acute and subacute toxicology in evaluation of pesticide hazard to avian wildlife, in *Wildlife Toxicology and Population Modeling: Integrated Studies of Agroecosystems*, R. J. Kendall and T. E. Lacher, Jr., Eds., Lewis Publishers, Chelsea, MI, 1993.
4. Grue, C. E., W. J. Fleming, D. G. Busby and E. F. Hill, Assessing hazards of organophosphate pesticides to wildlife, *Trans. North Am. Wildl. Nat. Res. Conf., 48, 200–220, 1983.*
5. Grue, C. E., L. R. DeWeese, P. Mineau, G. A. Swanson, J. R. Foster, P. M. Arnold, J. N. Huckins, P. J. Sheehan, W. K. Marshall and A. P. Ludden, Potential impacts of agricultural chemicals on waterfowl and other wildlife inhabiting prairie wetlands: an evaluation of research needs and approaches, *Trans. North Am. Nat. Res. Conf.*, 51, 357–383, 1986.
6. Grue, C. E., A. D. M. Hart and P. Mineau, Biological consequences of depressed cholinesterase activity in wildlife, in *Cholinesterase-Inhibiting Insecticides: Impacts on Wildlife and the Environment*, P. Mineau, Ed., Elsevier, Amsterdam, 1991.
7. Bennett, R. S., Do behavioral responses to pesticide exposure affect wildlife population parameters?, in *Wildlife Toxicology and Population Modeling: Integrated Studies of Agroecosystems*, R. J. Kendall and T. E. Lacher, Jr., Eds., Lewis Publishers, Chelsea, MI, 1993.
8. Fairbrother, A., Immunotoxicology of captive and wild birds, in *Wildlife Toxicology and Population Modeling: Integrated Studies of Agroecosystems*, R. J. Kendall and T. E. Lacher, Jr., Eds., Lewis Publishers, Chelsea, MI, 1993.
9. Buttiker, W., Ecological effects of insect control on bird populations, in *The Ecological Effects of Biological and Chemical Control of Undesirable Plants and Animals*, D. J. Kuenen, Ed., E. J. Brill, Leiden, The Netherlands, 1961, pp. 48–60.
10. Tucker, R. K., and J. S. Leitzke, Comparative toxicology of insecticides for vertebrate wildlife and fish, *Pharmacol. Ther.*, 6, 167–220, 1979.
11. Schafer, E. W., Jr., W. A. Bowles, Jr., and J. Hurlbut, The acute oral toxicity, repellency, and hazard potential of 998 chemicals to one or more species of wild and domestic birds, *Arch. Environ. Contam. Toxicol.*, 12, 355–382, 1983.
12. Hudson R. H., R. K. Tucker and M. A. Haegele, *Handbook of Toxicity of Pesticides to Wildlife*, 2nd ed., Resource Publ. 153, U.S. Fish and Wildlife Service, Washington, DC, 1984.
13. Kendall, R. J., L. W. Brewer, R. Hitchcock and J. R. Mayer, American widgeon mortality associated with turf application of diazinon AG500, *J. Wildl. Dis.*, in press.
14. White, D. H., C. A. Mitchell, E. J. Kolbe and J. M. Williams, Parathion poisoning of wild geese in Texas, *J. Wildl. Dis.*, 18, 389–391, 1982.
15. White, D. H., C. A. Mitchell, L. D. Wynn, E. L. Flickinger and E. J. Kolbe, Organophosphate insecticide poisoning of Canada geese in the Texas panhandle, *J. Field Ornithol.*, 53, 22–27, 1982.
16. Henny, C. J., L. J. Blus, E. J. Kolbe and R. E. Fitzner, Organophosphate insecticide (fam-

phur) topically applied to cattle kills magpies and hawks, *J. Wildl. Manage.*, 49, 648–658, 1985.
17. Grue, C. E., M. W. Tome, T. A. Messmer, D. B. Henry, G. A. Swanson and L. R. DeWeese, Agricultural chemicals and prairie pothole wetlands: meeting the needs of the resource and the farmer—U.S. perspective, *Trans. North Am. Wildl. Nat. Res. Conf.*, 54, 43–58, 1989.
18. Di Giulio, R. T., E. P. Gallagher and C. Habig, The role of in vitro studies in ecological hazard assessments, in *Wildlife Toxicology and Population Modeling: Integrated Studies of Agroecosystems*, R. J. Kendall and T. E. Lacher, Jr., Eds., Lewis Publishers, Chelsea, MI, 1993.
19. Grue, C. E., G. V. N. Powell and C. H. Gorsuch, Assessing effects of organophosphates on songbirds: comparison of a captive and free-living population, *J. Wildl. Manage.*, 46, 766–768, 1982.
20. Stromborg, K. L., C. E. Grue, J. D. Nichols, G. R. Hepp, J. E. Hines and H. C. Bourne, Postfledging survival of European starlings exposed as nestlings to an organophosphorus insecticide, *Ecology*, 69, 590–601, 1988.
21. Heinz, G. H., How lethal are sublethal effects?, *Environ. Toxicol. Chem.*, 8, 463–464, 1989.
22. Mineau, P., and D. B. Peakall, An evaluation of avian impact assessment techniques following broad-scale forest insecticide sprays, *Environ. Toxicol. Chem.*, 6, 781–791, 1987.
23. DeWeese, L. R., C. J. Henny, R. L. Floyd, K. A. Bobal and A. W. Shultz, Response of Breeding Birds to Aerial Sprays of Trichlorfon (Dylox) and Carbaryl (Sevin-4-ol) in Montana Forests, Spec. Sci. Rep. Wildl. 224, U.S. Fish and Wildlife Service, Washington, DC, 1979.
24. Powell, G. V. N., Reproduction by an altricial songbird, the red-winged blackbird, in fields treated with the organophosphate insecticide fenthion, *J. Appl. Ecol.*, 21, 83–95, 1984.
25. Busby, D. G., L. M. White and P. A. Pearce, Effects of aerial spraying of fenitrothion on breeding white-throated sparrows, *J. Appl. Ecol.*, 27, 743–755, 1990.
26. Potts, G. R., Population dynamics of the grey partridge: overall effects of herbicides and insecticides on chick survival rates, *Proc. Int. Congr. Game Biol.*, 13, 203–211, 1977.
27. Rands, M. R. W., Pesticide use on cereals and the survival of grey partridge chicks: a field experiment, *J. Appl. Ecol.*, 22, 49–54, 1985.
28. Greig-Smith, P. W., The Boxworth experience: effects of pesticides on the fauna and flora of cereal fields, in *The Ecology of Temperate Cereal Fields*, L. G. Firbank, N. Carter, J. F. Darbyshire and G.R. Potts, Eds., Blackwell Scientific, Oxford, 1991, pp. 333–371.
29. Kendall, R. J., Farming with agrochemicals: the response of wildlife, *Environ. Sci. Technol.*, 26, 238–247, 1992.
30. Emlen, J. M., Terrestrial population models for ecological risk assessment: a state of the art review, *Environ. Toxicol. Chem.*, 8, 831–842, 1989.

27

Monitoring Data as a Source of Information for Modelers

Sarah Gerould

Data collection for models that examine the effects of pesticides and other contaminants on birds can take many forms. When repeated measurements are taken at the same locations over time, data collection falls into the category of monitoring. Avian monitoring can be used in answering specific questions involving single species or specific environmental conditions, or it can be oriented toward general questions involving status and trends of populations and/or contaminant residues. Likewise, the scale of monitoring programs can extend geographically, from local to global, or temporally, from a single season to multiple decades. These scaling factors must be taken into consideration when designing studies, interpreting data and using the data in models because ecological parameters are often particular to and measurable only at specific scales. For example, O'Connor (Chapter 29) points out that in small-scale population monitoring programs, the evaluation of emigration is necessary but nearly impossible. Because of biological variations among species, scale must be consciously considered when formulating the goals and applications of a monitoring program. For example, most large-scale monitoring programs are used for assessing status and trends, and for triggering small-scale, site-specific studies on particular issues.

The goals of monitoring programs are reflected in what is measured by the programs. Monitoring programs designed to measure status and trends in bird populations are used to draw inferences about changes in bird populations or communities and to assess the overall impact of stressors such as pesticide use, hunting, pathogens or loss of physical habitat. The May Waterfowl Breeding Population and Habitat Survey provides population data to the Fall Flight Forecast to regulate decisions on hunting and other management programs.[1,2] O'Connor (Chapter 29) discusses the usefulness of population monitoring for determining the dependence of various parameters on avian density, a necessary but difficult to evaluate interaction used in ecological modeling.

Monitoring the effects of pesticides on birds can take the form of incident monitoring programs. Incident monitoring programs have mechanisms and procedures in place to investigate causes of mortality in birds found dead in the environment. Some comprehensive programs collect data by actively searching for dead birds in targeted pesticide-application areas, but most programs rely on the chance discovery of bird kills by pesticide applicators, land owners, or the general public. Design of programs involves issues such as public awareness and the role of law enforcement in prosecuting cases in which pesticide use has resulted in bird kills. The incident monitoring program in Great Britain is used in connection with regulatory screening as a fail-safe mechanism for required laboratory testing (see Chapter 30).

Residue monitoring can be used to describe trends and assess needs for follow-up studies. In the following chapter Steffeck describes how the National Contaminant Biomonitoring Program measured concentrations of certain organochlorines in tissues of waterfowl, starlings and fish, with the goal of assessing national trends in contaminant residues in fish and wildlife. Residue monitoring programs often trigger detailed follow-up studies. For example, when data showed organochlorine concentrations in animals from the Lower Gila River to be among the highest in the nation, studies were initiated to determine the source and fate of contamination.[3]

The interpretation of contaminant residue information in bird tissue is limited by difficulties in extrapolating between laboratory toxicity and exposure information, and field residue data. This difficulty is compounded because (1) laboratory data linking exposure to toxicity show extensive species to species variability (Chapter 31); (2) degradation, volatility, and metabolism of pesticides alters the concentration in the environment and increases the uncertainties in estimates of pesticide exposure in the field (Chapter 15); and (3) pharmacokinetic modeling and empirical data linking tissue residues to effects are limited.

REFERENCES

1. Reynolds, R. E., R. J. Blohm, F. A. Johnson and J. B. Bortner, 1990, Status of Waterfowl and Fall Flight Forecast, Canadian Wildlife Service and U.S. Fish and Wildlife Service, Laurel, MD, 1990.
2. U.S. Fish and Wildlife Service, SEIS 88: Final Supplemental Environmental Impact Statement: Issuance of Annual Regulations Permitting the Sport Hunting of Migratory Birds, U.S. Fish and Wildlife Service, Washington, DC, 1988.
3. Kepner, W. G., Organochlorine Contaminant Investigation of the Lower Gila River, Arizona, U.S. Fish and Wildlife Service, Phoenix, 1987.

28

The Role of Monitoring in Assessing Pesticide Effects on Avian Species and Their Habitats

Donald W. Steffeck

ABSTRACT

Pesticides are used in large quantities in the United States and throughout the world to control a variety of plants and animals. Pesticides have been used by man for centuries; however, the use of commercial, synthetic chemicals has expanded dramatically in the last 50 years. The kinds of pesticides used in agricultural and, more recently, in urban areas has been dynamic. Impacts to nontarget species have been documented, but relatively few long-term monitoring programs have been instituted. The National Pesticide Monitoring Program, initiated in the mid-1960s, has been the only nationally oriented program to evaluate pesticide effects on birds. Monitoring for pesticide effects should involve an integrated approach that collects multiple types of information to assess potential impacts. Information collected should include tissue residue concentrations, biological effects, and chemical/physical environmental data. Assessment of monitoring data needs to be supplemented with site-specific studies to ascertain cause and effect relationships. Monitoring information needs to be understandable, accessible, and comparable. Data generated must be available in a data management system for use by other groups, including the public. Monitoring initiatives should consider data users' needs, such as modeling and regulatory use, when designing new or updated programs.

KEY WORDS

monitoring, pesticides, avian, modeling, National Pesticide Monitoring Program

INTRODUCTION

Monitoring programs are completed for a variety of specific purposes. These programs can be vehicles for providing managers responsible for regulatory and policy decisions with information on pesticide exposure and effects. Because technical scientific journals contain vast amounts of site-specific or technical data, time constraints can force decisionmakers to become segregated from the research community. It behooves both management and scientists to facilitate communication. Providing managers with science-based recommendations, even if the data sets are not yet definitive, can assist in ensuring that decisions consider all relevant information. Long-term monitoring programs can summarize large data sets and be informational conduits, thereby serving as intermediate steps between the research community and managers.

DEFINITION OF MONITORING

According to *Webster's New Collegiate Dictionary*, monitoring as related to environmental conditions, is defined as "to watch, observe, or check especially for a special purpose." Environmental monitoring can be distinguished from general research studies, for this discussion, as consisting of a series of repeated measurements usually at the same locations over time. In general, three major types of monitoring can be identified: (1) site-specific studies, in which a particular species and/or process is investigated at a particular location; (2) compliance monitoring, whereby measurements are made at a discharge point or location to document compliance with a permit or other imposed guidelines; and (3) long-term monitoring to assess the status and trends of local, regional, or national environmental parameters. Most of the following discussion is an assessment of long-term monitoring.

A SUMMARY OF PESTICIDE USE IN THE UNITED STATES

Pesticides to control unwanted plants and animals have been in use worldwide for centuries. Initial pest control measures used ingredients found in the environment. For instance, the Greeks used sulfur as a fumigant in 1000 B.C.; the Romans used hellebore to control rodents and insects in 100 B.C.; the Chinese used arsenic to control garden insects in the year 900; and in the early 19th century the Persians used pyrethrum, from a type of chrysanthemum, to control lice.[1,2] By the late 1800s the scientific use of pesticides had begun. The application of pesticides at specific times was matched with pest damage or vulnerability.

The widespread use of synthetic pesticides was initiated immediately following World War II. The organochlorine (OC) class of pesticides, specifically dichlorodiphenyltrichloroethane, or DDT, was first used experimentally in the United States in 1942.[1] DDT was used very successfully by the military to eradicate lice and mosquitoes. OC pesticides disrupt the function of the nervous system, but the exact mode of action is unknown. The initial successes of relatively cheap and plentiful petrochemical compounds such as the OC pesticides ushered in an era of unquestioned reliance on these compounds.[2] The Patuxent Wildlife Research Center of the U.S. Fish and Wildlife Service began to study the impact of OC pesticides to birds and other nontarget organisms in the mid-1940s. Because the additives used to assist delivery of pesticides can influence toxicity, technical grade pesticides consisting of the active ingredient are generally used to measure biotic effects. By 1962, with the publication of *Silent Spring* by Rachel Carson,[3] the serious associated impact on nontarget organisms, especially birds, was being documented. The continued documentation of the widespread impact of OC pesticides to birds[4] and other organisms eventually lead to the banning of DDT in 1972, and the subsequent removal or reduction in use of other OC pesticide use in the United States. However, because OC compounds are extremely stable in the environment and highly soluble in lipids, these compounds are still widely found in the tissues of fish and birds.[5,6]

A second class of synthetic pesticides, the organophosphates (OPs), was first discovered in 1938[1] and came into commercial production in the 1950s. As the use of OC pesticides was curtailed, the use of OP pesticides increased, and these pesticides are now one of the largest groups of pesticides in use.[2] OP pesticides inhibit acetylcholinesterase, an enzyme that breaks down acetylcholine (ACh). ACh transmits nerve impulses across the synapse between nerve cells. In the absence of acetylcholinesterase, the nerve continues to send impulses resulting in cell damage and death.[7]

An additional group of pesticides, the carbamates, was developed in the 1950s and subsequently have become widely used. The carbamates are also acetylcholinesterase inhibitors, but are metabolized more rapidly than OPs. Both of the newer groups of pesticides vary widely in toxicity by specific pesticide and species. However, both OP and carbamate pesticides have been responsible for numerous avian die-offs.[8-10]

The development of pesticides has continued, with recent trends toward more specialized pesticides. In addition to insecticides and herbicides, which have been used extensively for decades, organism-specific pesticides that have been developed include fungicides, bactericides, nematicides, molluscides, rodenticides, avicides, miticides, piscicides, and algicides.[1,11] In 1976, in response to continued concerns regarding the unilateral use of synthetic pesticides, the integrated pest management approach was initiated at an international conference on entomology.[1] Integrated pest management encourages the use of a combination of procedures to control pests and results in reducing the pesticide load in the environment. This approach is a departure from recent standard agricultural practices which were based on regularly scheduled applications of pesticides to prevent the occurrence of pests.

The current use of pesticides in the United States is relatively extensive. In 1982, the U.S. Environmental Protection Agency (EPA) registered about 1400 pesticide active ingredients. The total amount of pesticides used in the United States in 1980 was estimated at 544 million kg of active ingredient, of which 72% was for agricultural use, 21% for industrial and government use, and 7% for home and garden use.[12] More than 64.7 million hectare-treatments are estimated to be used on agricultural crops and forests each year. Over 100 different OP and carbamate pesticides are registered as the active ingredients in thousands of products. These two groups of pesticides constitute a major portion of all pesticides used today.[13] The two most heavily used pesticides in agriculture are Alachlor [2-chloro-N-(2,6-diethylphenyl)-N-methoxymethyl)-acetamide] and Atrazine [6-chloro-N-ethyl-N′-(1-methylethyl)-1,3,5-triazine-2,4-diamine], both herbicides, which constitute about 25% of national agricultural pesticide use.[14]

Pesticides have been responsible for increasing agricultural productivity and suppression of pest-related disease. However, the impact of pesticides on nontarget organisms, including birds, is a documented problem. Impact on birds occurs through acute toxicity from direct ingestion of pesticides, and through both chronic and indirect effects on survival and reproduction. Consideration of pesticide effects must account for direct on-site impacts and for off-site movement of pesticides through runoff and recharge of surface waters from pesticide-contaminated groundwater. The effects of increasing urban uses of pesticides must be considered in addition to more traditional agricultural uses. For instance, pesticides from urban sources such as lawn care, garden use, and termite control, have been found in urban watersheds.[15]

EXISTING AVIAN MONITORING PROGRAMS RELATED TO PESTICIDES

Several long-term bird population monitoring or survey programs are conducted by the U.S. Fish and Wildlife Service (FWS) and other organizations for breeding birds, waterfowl, mourning doves (*Zenaidura macroura*), American woodcock (*Philohela minor*), and wintering birds.[16-20] Several regional programs or surveys of pesticide accumulation have been conducted. Pintail wings were sampled in California,[21] and woodcock wings have been evaluated in an 11-state area;[22-25] in Australia, pesticide residue levels in wings of 12 species of waterfowl were surveyed in 1977 and 1978.[26] Active regional monitoring programs are also conducted for the Chesapeake Bay, the Great Lakes, and Puget Sound. Additionally, several states conduct incident monitoring programs for bird kills.

Nevertheless, only one monitoring program has evaluated pesticides related to birds on a national level. This program, the National Pesticide Monitoring Program, was established as a multiagency effort in the mid-1960s to track environmental concentrations of persistent OC pesticides. The program was established in response to public concern for pesticide effects to nontarget species following the 1962 publication of *Silent Spring*.[3] Each agency in the program periodically collected and analyzed samples from various media involving nationwide networks of stations. The U.S. Fish and Wildlife Service established and operated networks for avian wildlife, starlings and waterfowl, and freshwater fish; it was the only agency that continued to collect data into the 1980s.

The avian portion of the National Pesticide Monitoring Program involved the collection and analysis of starling and waterfowl tissue for 14 OC compounds.[6,27]. The collections were initiated in 1965 and continued at 2- to 3-year intervals until 1986. In 1984, polychlorinated biphenyls (PCBs) were added to the program. Starlings were collected from a network of 129 sites located throughout the contiguous United States. Waterfowl wings were collected from hunters from representative locations throughout the contiguous 48 states. The most recent analyses for starlings indicate the following: OC chemicals continue to be present in tissues, but at generally lower concentrations than previously reported; there is geographic variation, especially for DDE and PCBs; and occurrence frequencies for most OC compounds were stable (1982 vs 1979).

The National Pesticide Monitoring Program documented that OC pesticides have bioaccumulated in several species of birds throughout the United States. Several general hotspot areas for specific compounds have been identified, such as elevated concentrations of DDE in the southwestern United States, and PCBs in the Great Lakes region. Environmental levels and frequencies of occurrence for these compounds have been established and trends identified. This information is valuable and has been used to evaluate environmental regulations for OC chemicals. However, the U.S. Fish and Wildlife Service discontinued the avian contaminant monitoring program in 1989 because of fiscal constraints and the stabilization of status and trends. The National Pesticide Monitoring Program depended exclusively on tissue residue data. The OP and carbamate pesticides generally do not bioaccumulate as parent compounds in bird tissues. Therefore, the program was limited to addressing exposure to OC chemicals and did not evaluate biological effects.

Several national monitoring programs that include biological components have been established or proposed since 1986.

The National Oceanic and Atmospheric Administration's (NOAA) National Status and Trends Program was established in 1984. It consists of about 200 coastal and estuarine stations throughout the United States. Chemical contaminants measured are chlorinated pesticides, PCBs, polychlorinated aromatic hydrocarbons (PAHs), and 12 trace elements. Sediment, mussels, and fish are sampled to determine residues. Some biological assessment information is also collected.[28]

The National Water Quality Assessment Program is conducted by the U.S. Geological Survey. This program completed a pilot phase, and the full program began October 1990. Information for a wide range of contaminants, including pesticides, will be collected from 60 combined surface and groundwater basins from locations throughout the United States. Only 20 basins will be intensively studied at any one time.[29]

Sampling information will include hydrological data, chemical data for water and sediment, and biological tissue residue and aquatic community data for benthic invertebrates and fish. The addition of a related avian component may be considered in the future.

The Environmental Monitoring and Assessment Program is being developed by the U.S. EPA. This program will design and implement a number of networks to assess environmental quality and regulatory success on a regional and national basis. A grid system will be used to assess landscape characteristics, and subsampling for evaluating specific issues will occur for eight ecosystems. Sampling will be done using ecological indicators under a statistically based design (U.S. EPA draft report of July 20, 1989, "Environmental Monitoring and Assessment Program Conceptual Overview and Issues"). Birds are being considered as indicator species and agricultural pesticides will be assessed.

The U.S. FWS has decided to revise and expand its contaminant monitoring program to fully address contaminant issues affecting its trust resources. More extensive geographic coverage and a suite of bioassessment methods will be incorporated into the revised program relative to the National Pesticide Monitoring Program.

Except for the National Status and Trends Program, these new generation monitoring programs contain the assessment of causes and effects of contaminants as a project goal. This is in contrast to former long-term programs that generally emphasized reporting of status and trends for environmental variables.

AN INTEGRATED APPROACH TO INFORMATION COLLECTION AND ANALYSIS

Monitoring programs can be an effective tool in assessing potential pesticide effects on avian species. Because no single method can fully assess pesticide exposure or effects to birds, an integrated approach is warranted. Tissue residue data can be useful in determining the amount of pesticide bioaccumulated by an organism and in identifying temporal and geographical trends of persistent compounds.[30] Because of the impact on nontarget species, many of the pesticides currently used are designed to rapidly degrade or metabolize. Hence, tissue residue data are an ineffective measurement of pesticide exposure for most of the modern pesticides. However, additional data that measures biological effects such as ecological surveys, organism health or biomarkers, and bioassay/toxicity test information, collected in a coordinated manner with abiotic samples, will provide a more useful and relevant database.

Ecological survey data, which can include data such as population levels, species diversity, or other community parameters, have been used to assess the health of aquatic environments.[31-34] Because pesticides can adversely affect bird survival, behavior, and nesting success,[35] changes in bird populations can be monitored to assess pesticide impacts. For instance, documented and potential pesticide impacts to birds have been identified in wetlands,[36] forests,[37] agricultural areas,[38-40] and urban areas.[41] Monitoring of bird populations in relation to pesticide effects is not currently done on a national basis. Specific objectives must be considered in developing any monitoring program, such as whether national or regional geographic evaluations will be assessed, or whether individual indicator species or communities will be evaluated.[42]

Measurement of organism health or biomarkers can give an indication of exposure and biological effect prior to mortality and resultant population changes. Thus, assessment of these data can provide an anticipatory element to a monitoring program thereby acting as an "early warning system." Organism health or biomarker method development has been expanding in recent years and includes a wide range of bioassessment techniques. It has been shown that OC pesticides decrease the thickness of eggshells, and that shell thickness can be used to evaluate potential impacts.[43] The changes in eggshell ultrastructure and eggshell breaking strength can also be used to evaluate eggshell quality.[44] OP and carbamate insecticides are cholinesterase (ChE) inhibitors that have been responsible for avian die-offs. Evaluating the inhibition of ChE can document the cause of mortality, but nonlethal exposure can also be monitored for exposure of birds to these specific pesticides.[8] Immunoassays have been developed for many classes of pesticides and have been used to screen environmental samples for the presence of pesticides.[45] A variety of additional organism health or biomarker methods are in the process of being more fully developed and include measures of chromosome damage, tumor incidence rates, animal behavior, ChE inhibition, bioenergetics status, serum chemistries, and numerous enzyme induction protocols.

Bioassay or toxicity test information has been extensively used to identify the toxicity of pesticides to birds and other wildlife.[46] These tests have also documented pesticide effects regarding embryotoxicity and teratogenicity,[47] the effects of cold stress,[48] and technical vs granular pesticide formulations.[49] However, these tests are generally administered in a laboratory setting. In order to monitor pesticide effects in so-called real-life situations, exposure in field situations can provide a more realistic evaluation of effects. The innovative combination of field and laboratory experiments could allow a better understanding of potential effects.[50] For example, in situ exposures of pesticides to model or indicator species under semicontrolled conditions can provide a link between laboratory studies and observed conditions in the field.[8,51]

The use of supplemental assessment methods to monitor pesticide effects on avian species should also be considered. Because birds are mobile and can evade exposure to specific pesticide events, it is advantageous to consider supplemental or alternative methods of determining effects of pesticides on avian species. The compiling and evaluation of bird mortalities, known as incident monitoring, may provide additional insights into the safety or hazards of a pesticide. Wildlife incident monitoring is conducted by a number of federal and state agencies and a private group; however, no national synthesis is being done at this time.[52] In cases in which a pesticide is suspected of having a high avian mortality potential, it is appropriate to target monitor a pesticide during actual use. This involves designing and implementing pre- or post-registration field studies to ascertain potential effects on avian species. Pesticides that are either new or awaiting re-registration, demonstrating potential negative effects, may be subject to targeted monitoring to document their effects in more detail.[52] The collection of abiotic parameters, such as pesticide residues in water, sediment, soil, etc., and hydrological/atmospheric transport data, in a coordinated manner with biological data will assist in understanding potential effects.

The collection and interpretation of any of the aforementioned data collection methodologies will provide useful and needed information. All of these techniques have positive and negative aspects, however, and no single method can unilaterally monitor pesticide impacts. An effective monitoring program should integrate several abiotic and biotic sampling methodologies from all of the aforementioned sampling categories. In order to fully identify cause and effect, a monitoring program should be coordinated with site-specific special studies. Bioassessment and analytical methods development should also be fully integrated into a monitoring program to ensure that new methods and issues are considered and evaluated.

USE OF MONITORING INFORMATION

The ultimate goal of monitoring programs is the protection of the environment, living resources, and/or human health. Properly designed and coordinated monitoring programs can provide the information necessary to manage natural resources by documenting existing conditions and, if data are consistently collected over time, identify trends in these conditions. Monitoring results can also include the following uses: setting standards such as pesticide use, hunting regulations, and human consumption; evaluating the effectiveness of contaminant abatement programs such as "best management practices;" acting as an early warning system for future problems; and enhancing knowledge of aquatic and terrestrial ecosystems.[53]

The costs of monitoring programs are shared by federal, state, and local agencies, with compliance monitoring financed by the private sector. The Bureau of Economic Analysis of the U.S. Department of Commerce estimates that in 1985, $451 million was spent nationally on water regulation and monitoring. However, total spending for water pollution abatement programs was approximately $25.4 billion. Hence, about $0.02 were spent on monitoring for every dollar spent on water pollution abatement. Most of the funding for monitoring was spent on compliance monitoring.[53]

Monitoring programs have been criticized for failing to provide adequate information for environmental management decisions.[53] To justify its costs, a monitoring program must have a clear set of relevant objectives, be technically sound, and provide usable information to decisionmakers. Monitoring strategies need to be clearly identified and to address issues of concern to the users of the monitoring information. Such considerations will ensure that monitoring provides valuable and useful information.

The information collected by a monitoring program must be summarized and presented in an understandable manner for natural resource managers and public policy decisionmakers. The completion of technical reports and publication in peer-reviewed scientific publications is also mandatory as a last check for quality control/quality assurance. However, if reporting of monitoring data is limited to publication in the scientific literature, this information will not be used to its fullest extent for evaluating and mitigating environmental impacts. Thus, summaries of findings, in nontechnical terms, should be an additional product of monitoring programs. In addition to providing understandable information to agency and public policy decisionmakers, a monitoring program should include a feedback mechanism for decisionmakers to provide recommendations for program improvement.

A monitoring program is a step in assessing the status and trends of the effects of pesticides on avian populations. To ascertain more specific cause and effect information, special study information will be required from a variety of sources. A monitoring program can be used as a framework to organize the various information sources regarding pesticide effects by developing a database, coordinating with ancillary databases, providing data collection criteria to assure compatibility of data, and acting as a data source for related information needs. A monitoring program needs to maintain continued coordination with agency, academic, private sector, and public information group professionals to remain relevant and contemporary.

Pesticide monitoring data are used to evaluate a range of pesticide-related effects. For instance, the effects of pesticides on nontarget organisms could influence pesticide use on agricultural lands, state and federal wildlife management areas, and urban areas such as lawns and golf courses.[41] State and federal regulators of pesticides could also use monitoring information to make recommendations affecting how, where, and when pesticides may be used. Public health officials can use monitoring information to evaluate consumption of various foodstuffs. There are currently a number of pesticide and other contaminant-related consumption advisories for fish and wildlife throughout the United States. The U.S. FWS, EPA, and individual states could use pesticide monitoring information to assess impact on avian populations and endangered species. The widespread use of pesticides in agricultural and urban areas requires additional information to fully assess acute and chronic impacts to migratory and nonmigratory birds. Ecological Risk Assessments evaluate both individual and population-level responses to xenobiotic chemicals including pesticides. Monitoring information will also be used to evaluate and address concerns related to review of Clean Water Act permits; review of various federal, state, and local projects; assessment of natural resource damages through authorities granted by Superfund and the Clean Water Act; and the evaluation of nonpoint source pollution and proposed mitigation alternatives.

In order to respond to these and other important issues in a timely manner, a proactive, integrated approach to monitoring is necessary. Models can be used as a tool to help understand existing conditions; direct research to important data needs; and predict future conditions.[54,55] Modeling of population ecology and wildlife toxicology should be an integral part of monitoring programs that address pesticide effects. This is especially true of avian populations that migrate, because exposures can take place over a wide area and specific data for all variables will be hard to obtain. If a central database that organizes monitoring and related site-specific information is included as part of a monitoring program, data for modeling efforts could be enhanced.

It is important that wildlife toxicology and population ecology issues be communicated to policy and regulatory managers. Monitoring programs can be a way to convey this information to managers and the general public. In order for monitoring programs to be most useful, researchers must participate in the development of monitoring efforts and provide specific recommendations for methodologies to be used. For monitoring programs to be fully successful, the research community should make simple, straightforward suggestions regarding the best methods currently available for use in monitoring. Natural resource management decisions will be made without the benefit of science if the scientific community does not provide usable, understandable information. Monitoring programs can be a way of collecting, coordinating, and transferring information, thereby serving as an intermediate step from pesticide and site-specific data to the more generalized, conclusional information needed by managers and the public.

CONCLUSIONS

Pesticide use is widespread and dynamic, and can adversely affect avian populations.

Only one national long-term monitoring program has directly assessed pesticide impacts to avian species. This program assessed OC pesticides commonly used in the mid-1960s by collecting tissue residue data. Methodologies for the program remained consistent and were not altered.

A new or expanded contaminant-related monitoring program should integrate and coordinate a variety of abiotic and biotic assessment methodologies to assess potential pesticide impacts to avian populations, and should include well-defined goals and objectives.

Monitoring programs should include a centralized information management system that coordinates with other monitoring and data generating entities. Modeling of population ecology and wildlife toxicology information should be an integral part of monitoring programs to assist in understanding existing conditions, identifying data gaps, and predicting future conditions.

A monitoring program should provide summarized, understandable information to agency and public policy decisionmakers and allow for feedback by the decisionmakers.

REFERENCES

1. Ware, G. W., *Pesticides: Theory and Applications*, W. H. Freeman, San Francisco, 1983.
2. Best, C., Natural pest controls, *Garbage*, 1, 40–49, 1989.
3. Carson, R., *Silent Spring*, Houghton-Mifflin, Boston, 1962.
4. Jefferies, D. J., The effects of organochlorine insecticides and their metabolites on breeding birds, *J. Reprod. Fert. Suppl.*, 19, 337-352, 1973.
5. Schmitt, C. J., J. L. Zajicek and P. H. Peterman, National contaminant biomonitoring program: residues of organochlorine chemicals in U.S. freshwater fish, 1976–1984, *Arch. Environ. Contam. Toxicol.*, 19, 748–782, 1990.
6. Bunck, C. M., R. M. Prouty and A. J. Krynitsky, Residues of organochlorine pesticides and polychlorobiphenyls in starlings (*Sturnus vulgaris*) from the continental United States, 1982, *Environ. Monitor. Assess.*, 8, 59–75, 1987.
7. Davies, D. R., Neurotoxicity of organophosphorus compounds, in *Handbook of Experimental Pharmacology, Volume XV, Cholinesterase and Anticholinesterase Agents*, G. D. Koelle, Ed., Springer-Verlag, Berlin, 1963, pp. 860–882.
8. Hill, E. F., and W. J. Fleming, Anticholinesterase poisoning of birds: field monitoring and diagnosis of acute poisoning, *Environ. Toxicol. Chem.*, 1, 27–38, 1982.
9. Eisler, R., Diazinon Hazards to Fish, Wildlife, and Invertebrates: A Synoptic Review. Biol. Rep. 85 (1.9), U.S. Fish and Wildlife Service, Washington, DC, 1986.
10. Eisler, R., Carbofuran Hazards to Fish, Wildlife, and Invertebrates: A Synoptic Review, Biol. Rep. 85 (1.3), U.S. Fish and Wildlife Service, Washington, DC, 1985.
11. Bigger, J. W., and J. N. Seiber, Fate of Pesticides in the Environment, Publ. 3320. University of California, Division of Agriculture and Natural Resources, Oakland, CA, 1987.
12. Ware, G. W., Fundamentals of Pesticides: A Self-Instruction Guide, Thompson Publications, Fresno, CA, 1982.
13. Smith, G. J., Pesticide Use and Toxicology in Relation to Wildlife: Organophosphorus and Carbamate Compounds, Resource Publ. 170, U.S. Fish and Wildlife Service, Washington, DC, 1987.
14. Gianessi, L. P., A National Pesticide Usage Database, Report to the Office of Standards and Regulations, U.S. Environmental Protection Agency, Washington, DC, 1986.
15. North Carolina Department of Natural Resources and Community Development, Investigation of the Effects and Uses of Biocides and Related Compounds in North Carolina, Division of Environmental Management, Raleigh, NC, 1983.
16. Droege, S., and J. R. Sauer, North American Breeding Bird Survey Annual Summary, 1988, Biol. Report 89 (3), U.S. Fish and Wildlife Service, Washington, DC, 1989.
17. U.S. Fish and Wildlife Service and Canadian Wildlife Service, Status of Waterfowl and Fall Flight Forecast, U.S. Fish and Wildlife Service, Washington, DC, 1989.
18. Dolton, D. D., Mourning Dove Breeding Population Status, 1990, U.S. Fish and Wildlife Service, Washington, DC, 1990.
19. Bertner, J. B., American Woodcock Harvest and Breeding Population Status, 1990, U.S. Fish and Wildlife Service, Washington, DC, 1990.
20. Lebaron, G. S., The 89th Christmas bird count, *Am. Birds*, 43, 551–560, 1989.
21. Ohlendorf, H. M., and M. R. Miller, Organochlorine contaminants in California waterfowl, *J. Wildl. Manage.*, 48, 867–877, 1984.
22. Stickel, L. F., M. A. McLane and J. F. Krietzer, Organochlorines and mercury in woodcock: a progress report, in *Proc. 4th Am. Woodcock Workshop*, Higgins Lake, MI, October 13–15, 1971.

23. McLane, M. R., E. H. Dustman, E. R. Clark and D. L. Hughes, Organochlorine insecticide and polychlorinated biphenyl residues in woodcock wings, 1971-1972, *Pestic. Monit. J.*, 12, 22-25, 1978.
24. McLane, M. A. R., D. L. Hughes and G. H. Heinz, Changes in levels of organochlorines in woodcock wings from 1971 to 1975, *Environ. Monit. Assess.*, 4, 105-111, 1984.
25. McLane, M. A. R., L. F. Stickel, E. R. Clark and D. L. Hughes, Organochlorine residue in woodcock wings, 11 states—1970-1971, *Pestic. Monit. J.*, 7, 100-103, 1973.
26. Olsen, P., H. Settle and R. Swift, Organochlorine residues in wings of ducks in southeastern Australia, *Aust. Wildl. Res.*, 7, 139-148, 1980.
27. Prouty, R. M., and C. M. Bunck, Organochlorine residues in adult mallard and black duck wings, 1981-82, *Environ. Monitor. Assess.*, 6, 49-57, 1986.
28. A Summary of Selected Data on Chemical Contaminants in Sediments Collected during 1984, 1985, 1986, and 1987, NOAA Tech. Memo. NOS OMA 44, National Oceanic and Atmopsheric Administraiton, Rockville, MD, 1988.
29. Hirsch, R. M., W. M. Alley and W. G. Wilber, Concepts for a National Water Quality Assessment Program, U.S. Geological Survey Circ. 1021, Denver, CO, 1988.
30. Wang, M. P., B. M. Braune and W. K. Marshall, The Use of Wing Parts for Monitoring Environmental Residues, Tech. Rep. Ser. No. 63, Canadian Wildlife Service, Ottawa, 1989.
31. Gammon, J. R., The Use of Community Parameters Derived from Electrofishing Catches of River Fish as Indicators of Environmental Quality, in *Seminar on Water Quality Management Tradeoffs*, Rep. No. EPA-905/9-80-009, U.S. Environmental Protection Agency, Washington, DC, 1980.
32. Karr, J. R., Assessment of biotic integrity using fish communities, *Fisheries*, 6, 21-27, 1981.
33. Karr, J. R., P. R. Yant, K. D. Fausch and I. J. Schlosser, Spacial and temporal variability of the index of biotic integrity in three midwestern streams, *Trans. Am. Fish. Soc.*, 116, 1-11, 1987.
34. Plafkin, J. L., M. T. Barbour, K. D. Porter, S. K. Gross and R. M. Hughes, Rapid Bioassessment Protocols for Use in Streams and Rivers: Benthic Macroinvertebrates and Fish, EPA 444/4-89-001, U.S. Environmental Protection Agency, Washington, DC, 1989.
35. Grue, C. E., and B. K. Shipley, Interpreting population estimates of birds following pesticide applications—behavior of male starlings exposed to an organophosphorus pesticide, in *Estimating Numbers of Terrestrial Birds*, C. J. Ralph and J. M. Scott, Eds., Allen Press, Lawrence, KS, 1981, pp. 292-296.
36. Grue, C. E., L. R. De Weese, P. Mineau, G. A. Swanson, J. R. Foster, P. M. Arnold, J. N. Huckins, P. J. Sheehan, W. K. Marshall and A. P. Ludden, Potential impacts of agricultural chemicals on waterfowl and other wildlife inhabiting prairie wetlands: an evaluation of research needs and approaches, *Trans. North Am. Wildl. Natl. Res. Conf.*, 51, 357-383, 1987.
37. Peakall, D. B., and J. R. Bart, Impacts of aerial application of insecticides on forest birds, *Can. Res. Counc. Crit. Rev. Environ. Contr.*, 13, 117-165, 1983.
38. Henry, C. J., L. J. Blus, E. J. Kolbe and R. E. Fitzner, Organophosphate insecticide (Famphur) topically applied to cattle kills magpies and hawks, *J. Wildl. Manage.*, 49, 648-658, 1985.
39. Henry, C. J., E. J. Kolbe, E. F. Hill and L. J. Blus, Case histories of bald eagles and other raptors killed by organophosphorus insecticides topically applied to livestock, *J. Wildl. Dis.*, 23, 292-295, 1987.
40. Stromborg, K. L., Modern pesticides and bobwhite populations, in *Proc. 2nd Natl. Bobwhite Quail Symposium*, Stillwater, OK, 1982, pp. 69-73.
41. Stone, W. B., and P. B. Gradoni, Wildlife Mortality Related to Use of the Pesticide Diazinon, Report of the New York State Department of Environmental Conservation, Delmar, 1985.
42. Steele, B. B., R. L. Bayn and C. V. Grant, Environmental monitoring using populations of birds and small mammals: analysis of sampling effort, *Biol. Conserv.*, 30, 157-172, 1984.

43. Bunck, C. M., J. W. Spann, O. H. Pattee and W. J. Fleming, Changes in eggshell thickness during incubations implication: for evaluating the impact of organochlorine contaminants on productivity, *Bull. Environ. Contam. Toxicol.*, 35, 173–182, 1985.
44. Henry, C. J., and J. K. Bennett, Comparison of breaking strength and shell thickness as evaluators of white-faced ibis eggshell quality, *Environ. Toxicol. Chem.*, 9, 797–805, 1990.
45. Versar, Inc., Pesticide monitoring in the regulatory process: a cooperative effort, in *Proc. Interagency Pesticide Monitoring Workshop*, Arlington, VA, June 6–7, 1989, pp. 1–53.
46. Hudons, R. H., R. K. Tucker and M. A. Haegele, Handbook of Toxicity of Pesticides to Wildlife, Resource Publ. 153, U.S. Fish and Wildlife Service, Washington, DC, 1984.
47. Hoffman, D. J., and P. H. Albers, Evaluation of potential embryotoxicity and teratogenicity of 42 herbicides, insecticides, and petroleum contaminants to mallard eggs, *Arch. Environ. Contam. Toxicol.*, 13, 15–27, 1984.
48. Maguire, C. C., and B. A. Williams, Cold stress and acute organophosphorus exposure: interaction effects on juvenile northern bobwhite, *Arch. Environ. Contam. Toxicol.*, 16, 477–481, 1987.
49. Hill, E. F., and M. B. Camardese, Toxicity of anticholinesterase insecticides to birds: technical grade versus granular formulations, *Ecotoxicol. Environ. Safety*, 8, 551–563, 1984.
50. Kendall, R. J., Wildlife toxicology, *Environ. Sci. Technol.*, 16, 448–453, 1982.
51. Grue, C. E., G. V. N. Powell and C. H. Gorsuch, Assessing effects of organophosphates on songbirds: comparison of a captive and a free-living population, *J. Wildl. Manage.*, 46, 766–768, 1982.
52. Pesticides and Birds: Improving Impact Assessment. Report of Avian Effects Dialogue Group, The Conservation Foundation, Washington, DC, 1989.
53. *Managing Troubled Waters: The Role of Marine Environmental Monitoring, Report of the Committee on a Systems Assessment of Marine Environment Monitoring, Marine Board*, National Research Council/National Academy Press, Washington, DC, 1989.
54. Tipton, A. R., R. J. Kendall, J. F. Coyle and P. F. Scanlon, A model of the impact of methyl parathion spraying on a quail population, *Bull. Environ. Contam. Toxicol.*, 25, 586–593, 1980.
55. Emlen, J. M., Terrestrial population models for ecological risk assessment: a state-of-the-art review, *Environ. Toxicol. Chem.*, 8, 831–842, 1989.

29

Population Patterns and Process Parameters—Issues in Integrating Monitoring and Models

Raymond J. O'Connor

ABSTRACT

Much of the existing information on population processes in birds comes from intensive single-species studies and is therefore potentially biased by those features predisposing these species to detailed study. Analysis of national data sets for Britain and North America show that density-dependent processes are probably regularly present in bird species, although density dependence is technically difficult to demonstrate. Most bird species thus possess the capacity to recover from environmental insults, though at species-specific rates. Variation in habitat quality is an important source of density-dependent regulation, with the best habitats holding core populations whose offspring can recolonize poorer habitats. Much evidence exists to show that habitat saturation is not uncommon, resulting in a reservoir of birds that are either not breeding at all or breeding at less than maximum rates. Spatial and scale dynamics may be more important in the maintenance of regional populations than is reflected in most population studies, but choice of scale is operationally a difficult issue for bird population models. Most bird population models to date have been "bookkeeping" models tracking the propagation of cohorts in line with demographic parameters. Few have extended to include submodels describing the consequences of environmental impacts.

KEY WORDS

density-dependence, demography, metapopulations, habitat use, biogeography

INTRODUCTION

The integration of ecotoxicological indicators with population ecology will eventually necessitate a synthesis of fields ranging from tissue culture to population dynamics. However, the wide variety of population processes—reproduction, mortality, movement, and their respective variation with density—gives the typical population study system a variability less frequently encountered in physiological studies, considering the more precise techniques used there. Moreover, many population processes are scale dependent,[1-3] manifesting themselves in different ways in studies of different scale. For example, movements in and out of a study plot typically average out in studies considered on a regional scale. These processes have major implications for the design of population monitoring studies, and therefore for modeling of such systems.

This chapter focuses heavily on the lessons to be learned from large-scale population monitoring programs, principally in Britain and North America, but also draws on information from other types of population studies. The bulk of our information on population dynamics processes in birds has emerged from a broad spectrum of species-oriented studies.[4] However, such a basis generates biases in our understanding of dynamics in two ways. First, researchers tend to assume that the patterns observed in each species are characteristic of the species rather than being a manifestation of underlying flexibility of life history pattern. For example, much of Lack's[4] interpretation of the population dynamics of the great tit (*Parus major*) studies in Wytham Woods, Oxford, was predicated on the assumption that the species was adapted to rich oak woods, with conifers a very poor secondary habitat. Yet studies of the same species in The Netherlands show substantially greater benefit accruing to populations nesting in coniferous woodland. Such differences, also apparent in the work of Dhondt,[5] indicate that species biology can alter in different environments. Moreover, the author's analysis of national data for the great tit in Britain[6] showed that while many of the patterns apparent nationally have paralleled those determined from the Oxford study, other features noted in the studies outside Britain (for example, greater incidence of second clutches) were also apparent for Britain as a whole, although not typical in the Oxford study.

A second bias arising from single species studies on single study plots relates to scale issues. Most study plots are relatively small, the territories encompassed ranging from a few dozen to at most a few hundred pairs of the study species. Under such conditions there is a bias against studying the effect of emigration and immigration from the study plot. With small plots, researchers may be confident that their searching and recording effort is such that unbanded birds appearing in the population are indeed immigrants, but they can rarely be sure that birds disappearing from the study have died or have emigrated (cf. the various studies in Newton[7]). Moreover, the larger the study plot, the harder it is to determine such effects.

In recent years considerably greater emphasis has been placed on the analysis of national monitoring programs.[8-15] These studies have been developed partly because with the passage of time the national monitoring schemes involved have accrued sufficient data to warrant studies of population trends. In addition, however, these analyses have proven suitable for the investigation of population processes difficult to investigate within the confines of single sites.[10,15-18] These studies provide much information on large-scale dynamics, but lack the resolution of the single plot studies. A third approach based on cross-sectional studies has recently been adopted by a number of workers.[19-23] These studies have focused especially on the impact of forest fragmentation on the distribution of bird populations and on how birds have responded to these changes. Although not designed as monitoring studies, nor intended to follow changes in bird numbers over sustained periods in the manner of the single site studies, they nevertheless reveal much information about the nature of population dynamics in birds.

This chapter provides an overview of the findings in these various types of studies so as to indicate the types of patterns of change recorded in bird populations, the processes underlying these, and their implications for the choice and determination of parameters for population models connecting to ecotoxicological parameters.

BASIS OF POPULATION MODELS

The population of birds present at a single site must follow the equation

$$P_{n+1} = P_n + B + D + I + E$$

where P_n and P_{n+1} are the population sizes of birds (measured as birds, pairs, females, or other consistent measure of population size) in years n and $n+1$ on the site being studied; B is the number of birds (pairs, etc., as appropriate) born on the plot in the intervening period; D is the number of birds dying; and E and I are the number of birds emigrating from or immigrating to the study plot. The equation can be written explicitly to show the annual change in population as

$$dP = B + D + I + E$$

where dP is the year to year change in population level. All terms are to be measured in consistent units, typically pairs. However, many population models work with only the female component of the population. Differences between the female- and bird-based models may arise in the case of polygamous species. Bird population models usually assume yearly time intervals, as so few birds breed other than on an annual cycle. The main exceptions are parrots (Psittacidae) and zebra finches (Estrildidae) and their relatives in Australia,[24] the common quelea in Africa,[25] and various oceanic seabirds.[26] A few temperate zone species, however, have been shown recently to skip an occasional breeding year.[7]

The population balance equation specifies that the population of birds on the study site will rise or fall according to the balance between births and deaths and net movement in and out of the site. Hence, to detect the effects of some new chemical introduced into the environment, for example, one ideally wants an equilibrium population ($P_n = P_{n+1}$ in the long term), preferably with low variance about the equilibrium level. Failing this, one needs information on the time course of the population on some control site matched to the observation site on all relevant variables. One can then standardize the values P_n and P_{n+1} to correct for global changes (such as weather, resource base, etc.) assigning the residual effects to that of the environmental effect (such as introduction of a new compound). In general, we assume that the offending compound will impact either the birth or death rate of the population, with the effects being made manifest as a change in population size. Changes in population size over time are therefore being treated here as a secondary measure of changes in the underlying population processes.[27] An alternative approach is to target the primary population processes (birth and death rates) for our monitoring by gathering information on changes in these processes. Because such primary processes are directly impacted by the foreign compound, they are potentially more valuable in the information their changes supply.[27] Note that in either approach one is assuming no net loss or gain of individuals from the study site through movement. This assumption is often made by population ecologists because of the difficulty in tracking birds dispersing any distance from the study site. A growing body of evidence, discussed later in this chapter, indicates that this assumption is highly suspect.

How valid is this approach of population balance equations? A major problem lies in the equation's assumption that birth and death processes and movement rates are independent of population size;[4,28] i.e., in general we need to write the equation for population change above as

$$dP = B(P) + D(P)(P) + I(P) + E(P)$$

where the variable P in each of the process terms indicates the sensitivity of the process to the prevailing population density. Note that the change in population is dependent on density in a manner determined by the combination of the individual density dependencies of the four population process terms.

Figure 1 presents empirical evidence as to density dependence in the dusky flycatcher *Empidonax oberholseri*. Density dependence implies a regulatory process in population dynamics, as exemplified in Figure 1: at low population densities the population can increase relatively readily; nearer the equilibrium population increases are smaller and tend to reverse once above the equilibrium; and at very high densities population decrease is more frequent than is any continuing increase. The net effect is for the population to vary about an equilibrium value with very large excursions increasingly severely checked. Where density-dependent regulation is prevalent, it is brought about by patterns of compensatory mortality (discussed further later). Lack[4] and others have placed considerable emphasis on population regulation through density dependence, but the ideas have not been extensively developed for bird population census work. This is partly because the analysis of density dependence is demanding in terms of data[29-31] and because of unsatisfactory assumptions by various methods of analysis.[32,33] Table 1 summarizes an analysis of density dependence for British birds, based on national Common Bird Census data for Britain. The table exemplifies the difficulties of density dependence analysis: only two species (the mallard *Anas platyrhynchos* and great tit) could be proven to be regulated by density dependence, at least with the analytical method used.[34] Yet nearly half of the species showed some evidence suggesting that density-dependent regulation was operating in their populations, such that a more sensitive analytical technique might have given statistically reliable results. (The problem arises because the test developed by Varley and Gradwell is extremely conservative as to type I errors, though with a consequent increased probability of a type II

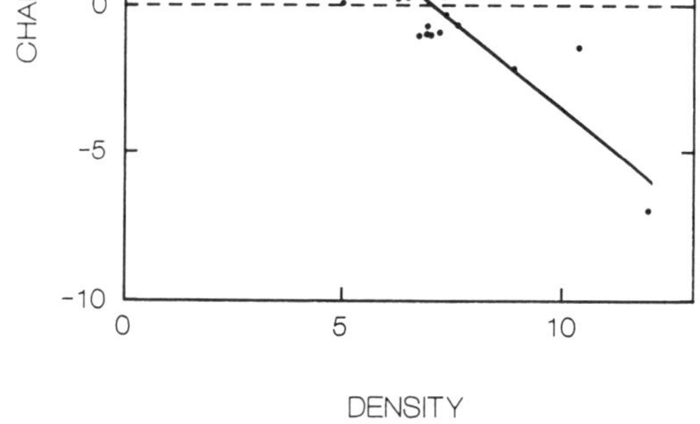

FIGURE 1. The relationship of population change between successive years in the dusky flycatcher (*Empidonax oberholseri*) to population levels in the initial year. The underlying density dependence suggested by the regression line shown was confirmed by the double regression test of Varley and Gradwell[34] with t test values of −6.87 and −4.86 ($p < 0.05$) for the 21 years 1966 to 1988. The density data used are the Breeding Birds Survey data for the contiguous United States.

Table 1. Incidence of Density Dependence in British Breeding Birds

	No. of Species	Percentage
Confirmed as density dependence	2	2.4
Probable density dependence	10	23.8
Confirmed density independence	20	47.6
Uncertain	10	23.8

Note: Information based on 16 years of Common Birds Census data and use of the Varley-Gradwell double regression methods.

error). Greenwood and Baillie[35] have recently reported statistically rigorous confirmation of density dependence for some of these species based on longer runs of Common Bird Census data. For North America, the author recently reported evidence showing that some 50% of 34 species studied show evidence of density-dependent processes in their population dynamics.[15] Of some 865 populations examined within individual states, only 14 (1.6%) had patterns inconsistent with density dependence. Taken together, these analyses indicate a likelihood of substantial involvement of density-dependent processes in bird populations in Britain and in North America, as originally suggested by Lack.[28]

Why is density dependence such a problem for monitoring environmental impacts? The difficulty arises because of the nonlinearity implicit in density dependence of birth and death rates or of immigration and emigration rates. Because birth rates fall or death rates rise with increased population densities, the per capita gains and losses change with density. The larger population, of course, experiences more deaths, even if unregulated, but with density dependence present there are relatively more deaths at the higher densities. Hence, relatively fewer survivors enter the breeding population the following year, thereby decreasing the ability of the population to grow further. The reverse is true of birth rates. Now consider a scenario in which reproduction and mortality, both density regulated, occur at separate times of the year, starting with deaths. At high populations relatively more birds die; the breeding population for the following year is reduced, and because the population is reduced, each of the survivors produces more offspring than they would have produced at the higher densities. The effect is to introduce additional recruitment that compensates for the mortality effect, buffering the population level against change. A similar effect can operate with density dependence in each of the other sources of mortality. For example, when the U.S. FWS deliberately altered hunting regulations for mallard so as to change the pattern of mortality falling on the hunted population, the net effects on mallards in North America was negligible because of the compensatory mortality effect brought about by density dependence.[36] Krebs[37] presents a simple population model for the great tit in Wytham Woods, Oxford, showing how large-scale population excursions quickly decay to an equilibrium value. Dobson[38] has extended this analysis to several dozen British species and has shown that most species have density-dependent dynamics that promote either damped or oscillatory return to equilibrium following a population excursion. It follows that for many, but not all, species the effects of an environmental insult will die away, although the time scale will vary.

If a population is depressed by an environmental factor, its return time will be similar to that of a propagule of initial immigrants to a remote island. MacArthur and Wilson[39] modeled the dynamical behavior of such propagules and asked if demographic factors might affect how quickly such a population might establish itself. This question is analogous to the reestablishment of a population reduced to low levels by environmental perturbation. If the per capita birth rate is b and the per capita death rate is m, the intrinsic rate of natural increase of the species r is the algebraic difference between b and m. MacArthur and Wilson's calculations showed that the population would most quickly reach equilibrium levels when the ratio r/b is large. Clearly, large values of this ratio can be achieved either by increasing r or by decreasing b, but a decrease in b may itself decrease r, depending on the value of m. MacArthur and Wilson concluded from their analysis that the best colonizers will be those with large r values obtained through low death rates rather than through high birth rates. However, MacArthur and Wilson's analysis is based on deterministic dynamics and gives little weight to stochastic events. Richter-Dyn and Goel[40] developed a stochastic model incorporating realistic dependence of b and m on population size. Their model suggested that for most populations with reasonable birth rates, establishment of an equilibrium population occurs above a critical value around 10 to 15 individuals. With smaller groups, the probability of the population surviving long enough to cross that critical threshold is approximately $1 - (m/b)^p$, where p is the initial population size. Hence, either a high birth rate or a low mortality rate will allow the population to recover to the threshold level above which the establishment is fairly certain. Taken with the r/b criterion for larger numbers, this suggests that maximization of birth rate is the optimal strategy, although contradictory statements appear later in this chapter in relation to Goodman's[41] work.

Because the effects of population perturbations die away, ecotoxicological indicators of those perturbations also die away; i.e. they are essentially transients within the system. Such indicators may themselves have physiological half-lives that influence the probability of following the time course of perturbation. Assuming that this problem can be addressed by using indicators of sufficiently long half-life, we can consider how the indicators might be influenced by the demographics of the impacted population, particularly the implications for the interpretation of monitoring results. Rabinovich[42] has empirically investigated the sensitivity of populations to changes in birth and death rates. He showed that data for 18 species carefully chosen for high quality of population information in fact conformed to the Richter-Dyn-Goel model, with response times linearly related to the r/b ratio and to fecundity in the way expected. Similarly, the success of species establishing themselves in Britain for the first time has been shown to be proportional to their annual egg production.[43] Furthermore, and more importantly because of its analogy to the perturbation of established populations of concern here, the recovery times of bird populations perturbed by an extremely severe winter in Britain were directly correlated with their reproductive outputs.[44] These findings for bird populations have the important implication that different types of environmental insult will be detectable differentially for different species. The effects of a pulse-type perturbation will quickly disappear in a species of high reproductive output, but will be longer lasting in a species with low output, more so when scaled with respect to r/b. On the other hand, a perturbation that reduces the death rate, and therefore the ratio of r/b, will be much more persistent in the population.

AVIAN HABITAT DYNAMICS

Considerable insight into the implications of density dependence for monitoring the population dynamics of affected species comes from considering two models of avian habitat dynamics. The models are usually described in terms of the distribution of birds across different habitat, but it can apply equally well to spatial patches. The simplest version of the theory is proposed by Brown.[45] Consider a series of habitat patches of different quality, such that patch A is better than patch B and patch B is better than patch C. In this context the quality can be set either by higher reproductive success or by reduced mortality. The following discussion can be cast in terms of reproductive success alone without loss of generality. As long as there are territorial vacancies available in patch A, newly arriving individuals can settle into these, and patch A is the only patch occupied. As the number of individuals to be accommodated in an area increases, however, not all individuals can live in patch A, and some must choose a patch elsewhere. Patch B is therefore chosen, because this offers the next highest rate of reproductive success. When all vacancies in patch B have been exhausted in turn, patch C is occupied, and so on in sequence. Thus, the population sequentially fills the spectrum or hierarchy of habitats of decreasing quality. At some point it is possible that the next available habitat is of such poor quality that a bird without a territory may not breed there but may become a "floater" awaiting the creation of a vacancy in a patch of acceptable quality due to the chance death of a territory holder. In the extreme, it may miss a breeding season altogether. Such floaters are often young birds that will do better the following year as experienced adults. In this simple model, the best habitats or patches are always occupied and the poorest acceptable patches are occupied only in years of high population density. Monitoring numbers in patches of the best type, therefore, shows nothing of the variation in overall population numbers, except when population levels fall to below the level that can be accommodated in patch A alone. Hence, although one might logically have expected to obtain the best indicator of population levels by monitoring the best habitat, that habitat is actually the one least sensitive to population changes in the area.

The Brown model[45] allows for no interaction between territory holders within a given patch other than for the maintenance of territorial boundaries; i.e., all birds have essentially the same fixed territory size (assumed of fixed quality) and therefore equal opportunities of reproductive success. Territory sizes are not subject to density-dependent adjustment. Nevertheless, the Brown model allows for density dependence in reproduction by the population as a whole: the greater the regional population level, the greater the proportion of individuals nesting in patches B, C, and so on, each of which is of successively poorer quality. Because a territory in patch C produces fewer offspring than a territory in patch B, and one in B produces fewer offspring than in patch A, the more birds that are distributed into patches of type B and type C, the lower the average reproduction in the population. The per capita reproductive output thus declines with population growth, checking further population expansion. The Brown model thus allows density-dependent regulation of the regional population, even though this regulation would not be observed on a study site located solely within a single patch type.

Fretwell and Lucas[46] developed a model that subsumes the ideas of Brown but allows more flexibility. In particular, Fretwell and Lucas postulate that density dependence may arise within patches. Consider the first bird being excluded from patch A. If the reproductive success it will experience in a patch of type B is markedly below that of a patch A territory, it can be expected to compete more fiercely for a type A territory. If so, the costs to the individuals already on patch A territories of yielding a little may be smaller than the costs of defense against a persistent intruder unwilling to pay the penalty of moving to a type B patch. If so, average territory size in patch A will reduce slightly, probably leading to a slightly reduced success as compared with a larger territory.[47] The scenario may be repeated with further individuals seeking to breed in patch A, but the process is not unlimited. A limit will be reached when the average reproductive success reached by individuals in crowded patch A is the same as that to be enjoyed by the first individual to enter the empty patch B. At this time fitness returns in both patches are equal, and the birds should be indifferent as to which they nest in. As patch B fills up, however, it may in turn bring about density-dependent reduction of breeding success there, slowly at first, but more steeply at higher levels of population density, and there may be benefit in nesting instead in the type A patch. If so, both patches will continue to experience further crowding as new individuals enter both in such a way as to keep the overall fitness the same. As both patches decline in quality, however, the third patch (type C) becomes viable and in turn fills up. The birds thus distribute themselves over a range of patch types in such a way that all individuals enjoy equal reproductive success but that success falls with total population density.

The Fretwell-Lucas model displays density dependence within an individual patch type, and this in turn ensures that the regional population is also subject to density dependence. Clearly, some hybrids of the Brown and Fretwell-Lucas models are possible, depending on the extent to which conditions in individual patch types are conducive or otherwise to invasion by latecomers. Knapton and Krebs[48] demonstrated that habitats may be subject to far greater crowding when settlement is synchronous than when the incoming birds attempt to fill the area sequentially. In the latter case, late-arriving birds are excluded from the habitat at territory sizes well above those tolerated when settlement is synchronous. Because environmental changes can dramatically alter the breeding season of birds,[11] either compressing or spreading out the season, such synchronicity is potentially a confounding factor in the population dynamics of species being monitored as indicators of impacts.

The models just described are essentially theoretical but much evidence is available to demonstrate their appropriateness to avian populations. This evidence covers the following points.

Within Individual Habitats Reproductive Success Decreases with Density

This was the original thesis of Lack,[4,28] and has been apparent in a variety of single site studies since then.[7,38] The effect has also been apparent on a larger scale: in Britain many of the species whose dynamics have been investigated at a national level have shown decreased clutch size or hatching and fledging success in years of high population density.[16,17,49,50] The effect has also been documented at a continent-wide level in North American populations of the dickcissel (*Spiza americana*).[51]

More Birds Use Inferior Habitat at High Densities

Evidence for this conclusion comes largely from large-scale regional and national studies. European wrens (*Troglodytes troglodytes*) expanded into new habitats as their population recovered from an eightfold reduction brought about by an excessively severe winter.[52] Eurasian kestrels (*Falco tinnunculus*) used successively poorer habitats in which they laid progressively smaller clutches with lower rearing success as their population expanded in Britain,[17] and the same pattern of hatching success has been observed in mistle thrushes (*Turdus viscivorus*).[49] The effect has also been apparent within North American bird populations. Dickcissels in Kansas concentrated into patches of good habitat until all available patches had been filled, and less suitable patches of vegetation (shorter or less dense in cover) were used only at highest densities.[53] Robinson[54] has shown that isolated woodlots in Illinois now become such peripheral habitat for wood thrushes (*Hylocichla mustelina*) that they cannot produce enough young to compensate for the impact of the cowbird (*Molothrus ater*). Van Horne[55] reviewed the evidence for the use of "sink" habitats by birds and cautions that it occurs sufficiently frequently to be a problem in censusing bird populations. Pulliam[56] has recently shown that very substantial populations may persist (through immigration of birds from prime habitat) in "sink" habitats unable to sustain themselves by their own reproductive output.

High Density Populations Often Contain Floaters

Much anecdotal evidence exists to show that breeding season populations typically contain individuals without territories that move in to take over vacancies that arise when the territory holder dies. However, relatively little systematic investigation of this phenomenon has taken place, though Smith[57] considers the adaptive value of the behavior. Four studies, however, document removal of birds from the population and the subsequent fate of the territories involved.[58-61] Edwards' experiment was expressly designed to investigate the consequences for pesticide impact monitoring of the presence of such species. Birds were mist-netted from the population early in the breeding season (but after territorial boundaries had been delineated) and the fate of the territory was subsequently followed. The proportion of territories refilled with newly arriving species varied substantially between species, and the speed of recolonization also varied. Nevertheless, a substantial number of vacancies were rapidly filled, such that monitoring based on birds that were not individually identified would have detected little or no change in overt population level. For example, had the original territory holders been killed by exposure to a contaminant, their replacement by floaters could conceal their deaths!

The fate of these floaters has not been experimentally studied, but O'Connor and Fuller[50] have considered the options available to floaters other than simply waiting in the hope of territory openings becoming available. These options include the following.

Movement Outside the Normal Range

Some evidence exists to suggest that starlings (*Sturnus vulgaris*) in Britain move farther north in years of high population density.[50] These northern populations then persist so long as the adults (which are relatively faithful to their breeding areas) survive or as long as further recruits arrive from the south. If the southern population is reduced (for example, by winter mortality during a severe winter), then in the following spring surviving juveniles first fill vacancies in the southern population, so that the northern population declines in numbers. If reproductive output continues at high levels, juveniles from later generations move north again. Zimmerman[53] has shown a similar phenomenon in dickcissels in the United States, with variation between northern and southern populations as numbers vary.

Sequential Use of Territories

Parr[62] has described this behavior of golden plover (*Pluvialis africaria*) breeding on moorland in Britain, where suitable breeding habitat was in short supply. Parr found that golden plovers formed pairs and waited in the vicinity of occupied territories. When the pair actually breeding left the site, often following clutch failure, the waiting pair moved in and occupied the territory and themselves started a breeding attempt, in some cases successfully. Parr found cases of even third replacements breeding, principally when the second pair also lost their eggs. Replacements were more frequent when the earlier pairs failed in a nesting attempt than when they were successful, although not exclusively so. Where populations with this breeding system are impacted by a contaminant causing the death of breeding pairs, this successional replacement process would hide the impact of the contaminant unless the pairs were individually banded.

"Guest Breeding"

Much evidence exists to indicate that territorial song and other displays are the primary processes by which territorial defense is achieved. In at least one study, great tits breeding at high density exploited this system to breed successfully within the boundaries of an existing territory by not engaging in territorial display. Eyckerman[63] found cases of pairs occupying nest holes within a territory defended by another pair, but going about their breeding activities without engaging in territorial challenges to the owner of the territory. The effect was to permit them to breed as subdominants within a territory. Such cases of breeding at high density would, of course, be a problem for censusing oriented toward counting birds in territorial song. The extent to which such behavior occurs in birds in general is not well understood.[57,64]

Polygamy

The Orians-Verner-Willson hypothesis explains polygamy in terms of the heterogeneous distribution of breeding resources. As a result, some territory owners hold extremely rich territories while others are rather poor in quality. Those birds with the richest resources may be able to support two or more females on their resource base, even though the male contribution to each breeding attempt therein may be reduced. Only males with resources above the polygamy threshold can adopt this strategy, and the threshold becomes higher for each successive female added to the harem. If breeding densities are high, polygamous breeding might be preferred by females over nonbreeding, provided the right set of circumstances prevailed. Direct evidence is available for a density-dependent component to this process in the dickcissel.[53] Such a possibility again has implications for monitoring environmental impacts.

MOVEMENT

One of the most difficult issues in bird censusing is addressing the issue of movements of birds between sites. Because a significant movement by birds takes the birds beyond the bounds of individual study plots, such birds are rarely tracked by researchers. As a result, surprisingly little is known about these birds.[62,65] In Britain, where there is a high density of birdbanders throughout the country and a dense population of people sufficiently informed to return bands found on dead birds, increased movement throughout the country has been documented in birds in years of high population densities.[6] It seems likely that such behavior is far more common than is evidenced by this reporting in the literature. Taylor et al.[66] have shown that patterns of variation in densities across sites of different overall densities are consistent with a push-pull model of animal movements.[67] Their data essentially consist of phenomenological evidence: when one plots the variance of bird populations against the mean, one frequently finds a linear relationship when both variables are logarithmically transformed. Thus, larger populations have more variance, but the increase progresses very systematically. Were the change purely one of scaling, the slope of such plots would be 2. In principle, however, spatial variance (i.e., variation in abundance across sites at a given time) may vary with mean density with exponents other than 2, either above or below that value. In the same way, one can plot the temporal variance (i.e., variation in abundance at a

Table 2. A Preliminary Analysis of Spatiotemporal Patterns of Variance in British Birds

	Spatial Variance		
Temporal Variance	**Decreasing**	**Constant**	**Increasing**
Decreasing	20	0	0
Constant	58	3	0
Increasing	4	0	0

Note: Each species was categorized according to whether each of its spatial and temporal variances decreased or increased or was independent of the corresponding mean density (see text for details). Density data were obtained as index values from the British Trust for Ornithology's Common Bird Census.

given site) against the mean abundance (at the site) and obtain similarly linear double logarithmic plots. Again, the neutral scaling exponent is 2, but bird populations may display exponents either above or below that value.

Combining these two ideas allows the possibility of nine cells corresponding to a particular biological phenomenon.[68] Most importantly, some of these categories correspond to patterns discussed in this chapter. For example, under the Fretwell-Lucas model, individual sites of different habitat type become increasingly similar as the population builds up and fills the less preferred habitats. The effect is to reduce the variance of the population at higher densities, i.e., to yield a Taylor variance-mean exponent below 2. Again, the Fretwell-Lucas model predicts that a patch of prime habitat will nearly always be filled, so that temporal variation in its numbers is low, while a similarly sized patch of poorer habitat will be occupied only in years of high overall density, thus leading to an inverse correlation of temporal stability with mean abundance at the plot. This pattern is therefore associated with a Taylor variance-mean exponent less than 2. Applying these ideas systematically to a sample of British bird species (Table 2) shows the relative distribution of these combinations over a wide range of species. The remarkable feature of the table is that no species show temporal exponents greater than 2 (which would correspond to outbreaks of various types[68]). Breeding birds in Britain do not "explode" in the way that insect populations do. Approximately 24% of the species fall into the category corresponding to Fretwell-Lucas type dynamics (upper left cell). Substantially more (68%) fall into the category of increased temporal stability at higher densities but neutral spatial dynamics. Such a category corresponds approximately to what might be termed "diffusion dynamics," possibly best approximated by the Brown model[45] described previously. In other words, in different years birds may settle into various areas in proportion to their resources, but consistently good sites are occupied most frequently. The phenomenon has been described in some detail by the author,[69] particularly as a strategy exploited by migrant species. In fact, there are a disproportionate number of migrants represented in the cell of Table 2 (30 vs 15%), in agreement with this. The lower left cell of Table 2 corresponds approximately to the Knapton-Krebs[48] dynamics: if many birds return simultaneously to a site in a given year, settlement densities are substantially higher than normal, greatly increasing the density of that plot relative to others and generating a correlation between increased variance and higher mean densities. Rather few species (5%) show this behavior. The remaining cell in Table 2, of neutral spatial and temporal stability, is occupied by three raptorial species (European sparrowhawk [*Accipiter nisus*], common buzzard [*Buteo buteo*], and Eurasian kestrel). All three species are relatively long-lived birds with strong territorial behavior, so it is perhaps not surprising that they should fall into this category. Nevertheless, several other raptors are found elsewhere in the table.

Implicit in the concepts encapsulated in Table 2 is the idea of substantially greater mobility of bird populations than is generally taken into account in population dynamic studies. Hanski[70,71] argued that this is true of animal populations in general. His model leads to the conclusion that animal populations are significantly affected by processes leading to random extinction of local breeding populations and recolonization of others. His model predicts that most species will be either very common or very rare. Preliminary analysis of data for Irish and British breeding populations indicates that this is indeed the case. On the other hand, Brown[72] argues instead that large-scale distributions are set deterministically by issues of resource utilization, with generalist species able to utilize a wide range of resources and therefore spread abundantly into many areas. In contrast, specialists can find only a restricted distribution of their resources and are limited thereby. However, this cannot be the complete explanation because birds in Britain show the correlation between habitat breadth and abundance predicted by Brown's model,[73] yet also conform to the Hanski model.[82] This issue is probably scale dependent, for a pattern of extinction and recolonization at a local level could still average out to relative stability at a more regional level, corresponding to a metapopulation.[74] If so, it implies that the monitoring programs operating at the metapopulation level or above offer the best hope of detecting the impact of environmental changes on bird populations, while studies at individual sites would be subject to a background of stochastic variation in population numbers, making it difficult to interpret localized impacts of contaminants, at least in the absence of individually marked birds.

INCORPORATING POPULATION PROCESSES INTO MODELS

Emlen[75] critically reviews the issues involved in developing terrestrial population models for ecological risk assessment. He suggests that demographic models should be subdivided into two basic parts: a bookkeeping submodel that translates demographic parameters into a description of the population dynamics resulting, and a group of submodels describing the impacts of the physical, chemical, and biotic environment on these demographic parameters. In general, the bookkeeping models for bird populations will necessitate the use of projection matrices such as the Leslie matrix because survivorship and reproductive effort in birds is typically age dependent.[7,27] One of the most important decisions to make about such bookkeeping relates to the spatial scale of that submodel. Leslie matrices have been routinely used in studying the changes in populations of particular study sites, but the boundaries of these sites are often arbitrary, chosen for the convenience of the researcher rather than coinciding with a demographic boundary. The evidence described above in connection with the Fretwell-Lucas model of avian habitat dynamics shows quite convincingly that birds will move relatively freely between different habitats or patches in response to population pressure. Additionally, the work of Hanski[70] and of Villard et al.[74] shows that these processes are likely to be apparent at a phenomenological level. Hence, although at first sight it would seem sensible to track the population of birds within a contaminated area, the reality is that both the birds affected and the contamination can spread significantly beyond the boundaries of that area. The choice of a suitable scale for such monitoring is therefore one of the most difficult issues for avian population modelers on which to decide.[78] Two approaches have already emerged as holding promise for the study of these spatial issues. The approach taken by Taylor et al.[66] summarizes the spatial patterning of individual species in a small number of phenomenological constants, which then can be applied to the dynamics of individual sites. The alternative approach, adopted by Fahrig and Paloheimo,[76] creates an explicit transition matrix to track changes between sites. Further development of this latter model is severely hindered by the paucity of data on short distance dispersal of birds from sites on which they are studied. Nor do we have any idea of the resolution required of spatial data either for Taylor's analysis or for Fahrig and Paloheimo's model. A sensitivity analysis of these models even at their present state would be informative in identifying what precision of data need to be gathered for effective use within the models.

Both theory[45,46] and empirical data indicate that density dependence is widespread, though not universal within bird populations. Hence, any projection matrix used to track the dynamics of bird populations should allow for the operation of density dependence. Although the proximate factors behind density dependence can vary substantially between species, they must in general impact either reproduction (i.e., from egg to fledgling), juvenile and immature survival (i.e., from fledgling to adulthood), or adult survival. The author noted previously that the effects of changes in birth rate or death rate had different effects on the time course of recovery of populations from environmental perturbations. It follows that changes in the patterning of density dependence between reproduction and mortality will significantly affect the recovery trajectories. Fowler[77] showed that species of varying size were differentially sensitive to density dependencies of different types, and this idea has been subsequently developed by Emlen and Pikitch.[78] They conducted an extended sensitivity analysis of simple network population models containing density-dependent factors and found that species of different body size were differentially sensitive to changes in the various parameters of the model. For small species, population size was critically set by the level of juvenile survival, with a 10% change in that parameter having a 30 to 42% change in population level. Small species were secondarily affected by changes in adult survival, with a 10% survival resulting in a 16 to 26% change in population level. For large species, on the other hand, immature and subadult survivorship was critical, with a 10% change altering population level by 46 to 65%. Adult fecundity was the next most important element, but only weakly: a 10% change in fecundity resulted in only a 7% change in population level, indicating the degree of buffering of the population. Intermediate-sized species proved to be most resistant to alterations in the model parameter values. Adult survival was most important, but its impact was only about one fifth that of early survival on the small species.

Emlen and Pikitch also examined the frequency with which "pseudo-extinctions" would occur as input parameters altered. A "pseudo-extinction" was defined as a case of the population level falling below 20% of that for standard input values. This measures the propensity of the model for taking the population toward extinction, extinctions themselves being such rare events that their occurrence provides rather poor information about model behavior. Using this measure, small species were found to be most influenced by changes in early survival, with less influence from late survival and fecundity having only a minor effect. For a large species, pseudo-extinction frequency was highly sensitive to changes in adult mortality, but additionally increased with variability in survival. For intermediate-sized species, the results were similar to those for population density, except that both immature and adult survival had critical influence on pseudo-extinction frequency. Overall, these intermediate size species were relatively well buffered against environmental impact. These results make it clear that different levels of resolution of the various input parameters will be needed with species of different sizes. For large species, greater resolution is needed as to adult mortality than to any other parameter. In contrast, for a small species both postfledging and adult survival need to be well defined, and fecundity may also need to be considered.

It is also informative to consider how variance in these parameters may arise naturally. Recent research has shown that the net contribution of individuals to future generations is far more skewed than has generally been realized, often with a very small number of birds accounting for the majority of future recruits to bird populations.[7] When such populations are subjected to environmentally driven variation, however, in populations of any size, these demographic differences (which differ between individuals but which have to be averaged, at higher population levels, across increasing numbers of individuals) paled in comparison to the environmental variability that impacts all individuals.[41] In the context of environmental impacts, therefore, population variability and the associated risk of extinction are most susceptible to environmental variation. Under such conditions the chances of extinction of the population can be reduced by the inclusion of "hot spots" in the range, places where mean growth rate in the population is consistently positive. Such sites are likely to be areas of prime habitat. In this context it is ironic that the U.S. EPA requirements for pesticide registration trials require that field sites be located in areas of high density. The intention is to study the "worst case scenario." However, if the site chosen is a "hot spot," any tendency toward local extinction induced by the chemical's impact will be buffered, according to Goodman's[41] arguments. On the other hand, if the site is a "sink habitat,"[56] changes in numbers on the site will be far more influenced by conditions in prime habitat elsewhere than on the test site. The dilemma perhaps illustrates in an extreme form the difficulty of choice facing bird population modelers.

In general, models of bird populations work on annual time steps, given the discreteness of avian breeding seasons. Much of the discussion here has dealt with the bookkeeping component of Emlen's[75] dichotomy. With short-lived ecotoxicological indicators, however, it is possible to contemplate giving a minimal role to such annual turnover bookkeeping and to place the emphasis of the model on the environmental submodels of Emlen's dichotomy. Tipton et al.[79] described one such model, in which a transition matrix model was used to track the impact of methyl parathion spraying on a bird population. Here, the time periods were chosen to provide several steps (matrix classes) along the decay curve of a physiological indicator, with the transition from class to class modified by mortality and loss of indicator. Such approaches may well be the most productive to use in short-term trials, though the problems induced by behavioral patterns (e.g., incursions by floaters) will remain an issue.

No discussion of modeling of environmental impacts on bird populations would be complete without mentioning the work of Potts.[80,81] Although Emlen and Pikitch[78] consider species-specific models as too dedicated to be useful in the broader regulatory context, Potts' study of the grey partridge *Perdix perdix* in England is exemplary. This particular species has undergone massive declines worldwide. By a careful autecological study, Potts has been able to show that the species has been impacted in England by three factors:

1. habitat changes brought about by hedgerow removal and field enlargement in the interests of improved cereal production, using large machines in large fields;
2. increased predation by more numerous predators associated with a decline in the number of gamekeepers (staff employees to tend to game management activities on English estates); and
3. an insecticidal side effect of fungicides used in cereals (the main brood-rearing habitat of the partridge), such that insects favored by partridge chicks were being selectively killed.

Potts[80] developed a relatively simple population model that incorporated simple monitoring of these factors, yet yielded very high predictability of partridge numbers. The spatial and temporal scale issues already cited, and the differential sensitivity of different sized birds, require that we consider the biology of potential target species before constructing the model. Because these considerations constrain the generalities sought by Emlen and Pikitch, it remains sensible to ask whether a relatively small increment of research effort may not yield a species-specific but high-quality model of the type developed by Potts.

ACKNOWLEDGMENTS

I thank Randall P. Boone for the analysis of the dusky flycatcher data used here and thank Sam Droege of the Patuxent Wildlife Research Center for providing a copy of the Breeding Birds Survey results on which that analysis was based. Ray Pierotti kindly let me have advanced sight of his manuscript. I thank Shirley Moulton for secretarial support in the preparation of this manuscript. Analysis of the Common Birds Census data was originally conducted while I was Director of the British Trust for Ornithology; the CBC and my post were funded by the Nature Conservancy Council, whose support is gratefully acknowledged. The analysis of North American data was conducted as part of a project funded jointly by American Cyanamid, CIBA-Geigy, Dupont, FMC Corporation, Mobay Corporation, and Rhone-Poulenc.

REFERENCES

1. Wiens, J. A., Scale problems in avian censusing, in *Estimating Numbers of Terrestrial Birds* (*Studies in Avian Biology No. 6*), C. J. Ralph and J. M. Scott, Eds., Cooper Ornithological Society, Lawrence, KS, 1981, pp. 513–521.
2. Wiens, J. A., *The Ecology of Bird Communities, Vol. 1, Foundations and Patterns*, Cambridge University Press, Cambridge, U.K., 1989.
3. Maurer, B. A., Avian community dynamics in desert grasslands: observational scale and hierarchical structure, *Ecol. Monogr.*, 55, 295–312, 1985.
4. Lack, D., *Population Studies of Birds*, Clarendon Press, Oxford, 1966.
5. Dhondt, A. A., The regulation of numbers in Belgian population of Great Tits, in *Proc. Adv. Study Institute Dynamics of Numbers in Populations*, PUDOC, Warsaw, 1971, pp. 532–547.
6. O'Connor, R. J., Pattern and process in Great Tit (*Parus major*) populations in Britain, *Ardea*, 68, 165–183, 1980.
7. Newton, I., Ed. *Lifetime Reproduction in Birds*, Academic Press, London, 1989.
8. Robbins, C. S., Effects of forest fragmentation on breeding bird populations in the Piedmont of the Mid-Atlantic region, *Atlantic Nat.*, 33, 31–36, 1980.
9. O'Connor, R., and C. J. Mead, The stock dove in Britain, 1930–80, *Br. Birds*, 77, 181–201, 1984.
10. O'Connor, R. J., Long-term monitoring of British bird populations, *Ornis Fenn.*, 62, 73–79, 1985.
11. O'Connor, R. J., and M. Shrubb, *Farming and Birds*, Cambridge University Press, Cambridge, U.K., 1986.
12. Robbins, C. S., D. Bystrak and P. H. Geissler, The Breeding Birds Survey: Its First 15 Years, 1965–1979, Resource Publ. 157, U.S. Fish and Wildlife Service, Washington, DC, 1986.

13. Robbins, C. S., J. R. Sauer, R. Greenberg and S. Droege, Population declines in North American birds that migrate to the Neotropics, *Proc. Natl. Acad. Sci. U.S.A.*, 86, 7658–7662, 1989.
14. Baillie, S. R., Integrated population monitoring of breeding birds in Britain and Ireland, *Ibis*, 132, 151–166, 1990.
15. O'Connor, R. J., Population variation in relation to migrancy status in some North American birds, in *Ecology and Conservation of Neotropical Migrant Landbirds*, J. M. Hagan, III and D. W. Johnston, Eds., Smithsonian Institution Press, Washington, DC, 1992, pp. 64–74.
16. O'Connor, R. J., Population regulation in the yellowhammer *Emberiza citrinella* in Britain, in *Bird Census Work and Nature Conservation*, H. Oelke, Ed., Dachverstandes Deutscher Avifaunisten, Gottingen, 1981, pp. 190–200.
17. O'Connor, R. J., Habitat occupancy and regulation of clutch size in the European Kestrel *Falco tinnunculus*, *Bird Study*, 29, 17–26, 1982.
18. Williamson, K., and L. A. Batten, Ecological implications of the Common Birds Census, *Pol. Ecol. Stud.*, 3, 237–244, 1977.
19. Whitcomb, R. F., J. F. Lynch, P. A. Opler and C. S. Robbins, Island biogeography and conservation: strategy and limitations, *Science*, 193, 1030–1032, 1976.
20. Whitcomb, R. F., C. S. Robbins, J. F. Lynch, B. L. Whitcomb, M. K. Klimkiewicz and D. Bystrak, Effects of forest fragmentation on avifauna of the eastern deciduous forest, in *Forest Island Dynamics in Man-Dominated Landscapes*, R. L. Burgess and D. M. Sharpe, Eds., Springer-Verlag, New York, 1981, pp. 125–205.
21. Robbins, C. S., Effects of forest fragmentation on bird populations, in *Management of North-Central and Northeastern Forests for Non-Game Birds: Workshop Proceedings*, R. M. DeGraaf and K. E. Evans, Eds., Gen. Tech. Rep. NC-51, U.S. Forest Service, St. Paul, MN, 1979, pp. 198–212.
22. McClellan, C. H., A. P. Dobson, D. S. Wilcove and J. F. Lynch, Effects of forest fragmentation on New- and Old-World bird communities: empirical observations and theoretical implications, in *Wildlife 2000: Modeling Habitat Relationships of Terrestrial Vertebrates*, J. Verner, M. L. Morrison and C. J. Ralph, Eds., University of Wisconsin Press, Madison, 1986, pp. 305–313.
23. Fuller, R. F., *Bird Habitats in Britain*, T. & A. D. Poyser, Calton, England, 1982.
24. Immelman, K., *Australian Finches*, Angus & Robertson, Sydney, 1982.
25. Ward, P., The migration patterns of *Quelea quelea* in Africa, *Ibis*, 113, 275–297, 1973.
26. Schreiber, R. W., and N. P. Ashmole, Seabird breeding seasons on Christmas Island, Pacific Ocean, *Ibis*, 112, 363–394, 1970.
27. Ricklefs, R. E., Fecundity, mortality, and avian demography, in *Breeding Biology of Birds*, D. S. Farner, ed., National Academy of Sciences, Washington, DC, 1973, pp. 336–347.
28. Lack, D., *The Natural Regulation of Animal Populations*, Clarendon Press, Oxford, 1954.
29. Royama, T., Population persistence and density dependence, *Ecol. Monogr.*, 47, 1–35, 1977.
30. Reddingius, J., Models for testing: a secondary note, *Oecologia*, 83, 50–52, 1990.
31. Solow, A. R., Testing for density dependence, *Oecologia*, 83, 47–49, 1990.
32. Hassel, M. P., Insect natural enemies as regulating factors, *J. Anim. Ecol.*, 54, 323–334, 1985.
33. Vickery, W. L., and T. D. Nudds, Detection of density-dependent effects in annual duck censuses, *Ecology*, 65, 96–104, 1984.
34. Varley, G. C., and G. R. Gradwell, Predatory insects as density dependent mortality factors, *Proc. 16th Int. Congr. Zool.*, 1, 240, 1963.
35. Greenwood, J. J. D., and S. R. Baillie, Population studies at the BTO: the second 50 years, *Ibis*, in press.
36. Nichols, J. D., M. J. Conroy, D. R. Anderson and K. P. Burnham, Compensatory mortality in waterfowl populations: a review of the evidence and implications for research and management, *Trans. North Am. Wildl. Nat. Res. Conf.*, 49, 535–554, 1984.

37. Krebs, J. R., Regulation of numbers in the Great Tit (Aves: Passeriformes), *J. Zool. Lond.*, 162, 317-333, 1970.
38. Dobson, A. P., Mortality Rates of British Birds, D.Phil. thesis, Oxford University, Oxford, U.K., 1981.
39. MacArthur, R. H., and E. O. Wilson, *The Theory of Island Biogeography*, Princeton University Press, Princeton, NJ, 1967.
40. Richter-Dyn, N., and N. S. Goel, On the extinction of a colonizing species, *Theor. Pop. Biol.*, 3, 406-433, 1972.
41. Goodman, D., How do any species persist? Lessons for conservation biology, *Conserv. Biol.*, 1, 59-62, 1987.
42. Rabinovich, J. E., Demographic strategies in animal population: a regression analysis, in *Tropical Ecological Systems: Trends in Terrestrial and Aquatic Research*, F. B. Golley and E. Medina, Eds., Springer-Verlag, New York, 1975, pp. 19-40.
43. O'Connor, R. J., Biological characteristics of invaders amongst bird species in Britain, in *Quantitative Aspects of the Ecology of Biological Invasions*, M. Williamson and H. Kornberg, Eds., The Royal Society, London, 1986, pp. 81-96.
44. O'Connor, R. J., Comparisons between migrant and non-migrant birds in Britain, in *Animal Migration*, D. J. Aidley, Ed., Society for Experimental Biology Seminar Series, No. 13, Cambridge University Press, Cambridge, U.K., 1981, pp. 167-195.
45. Brown, J. L., The buffer effect and productivity in tit populations, *Am. Nat.*, 103, 347-354, 1969.
46. Fretwell, S. D., and H. L. Lucas, A territorial behaviour and other factors influencing habitat distribution in birds, *Acta Biotheor.*, 19, 16-36, 1969.
47. Schoener, T. W., Theory of feeding strategies, *Annu. Rev. Ecol. Syst.*, 2, 369-404, 1971.
48. Knapton, K. R., and J. R. Krebs, Settlement patterns, territory size, and breeding density in the song sparrow (*Melospiza melodia*), *Can. J. Zool.*, 52, 1413-1420, 1974.
49. O'Connor, R. J., Dynamical aspects of avian habitat use, in *Wildlife 2000: Modeling Habitat Relationships of Terrestrial Vertebrates*, J. Verner, M. L. Morrison and C. J. Ralph, Eds., University of Wisconsin Press, 1986, Madison, pp. 235-240.
50. O'Connor, R. J., and R. J. Fuller, Bird population responses to habitat, in *Bird Census and Atlas Studies: Proc. 8th Int. Bird Census Conf.*, K. Taylor, R. J. Fuller and P. C. Lack, Eds., British Trust for Ornithology, Tring, England, 1985, pp. 197-212.
51. Fretwell, S. D., Is the Dickcissel a threatened species?, *Am. Birds*, 31, 923-932, 1977.
52. Williamson, K., Habitat preferences of the Wren on English farmland, *Bird Study*, 16, 53-59, 1969.
53. Zimmerman, J. L., The territory and its density dependent effect in *Spiza americana*, *Auk*, 88, 591-612, 1971.
54. Robinson, S. K., Population dynamics of breeding neotropical migrants in a fragmented Illinois landscape, in *Ecology and Conservation of Neotropical Migrant Landbirds*, J. M. Hagan, III and D. W. Johnston, Eds., Smithsonian Institution Press, Washington, DC, 1992, pp. 408-418.
55. Van Horne, B., Density as a misleading indicator of habitat quality, *J. Wildl. Manage.*, 47, 893-901, 1983.
56. Pulliam, R., Sources, sinks and population regulation, *Am. Nat.*, 132, 652-661, 1988.
57. Smith, S. M., The "underworld" in a territorial sparrow: adaptive strategy for floaters, *Am. Nat.*, 112, 571-582, 1978.
58. Stewart, R. E., and A. W. Aldrich, Removal and repopulation of breeding birds in a spruce-fir forest community, *Auk*, 68:471-482, 1951.
59. Morse, D. H., The foraging of warblers isolated on the small islands, *Ecology*, 52, 216-228, 1971.
60. Hensley, M. M., and J. D. Cope, Further data on removal and repopulation of breeding birds in a spruce-fir forest community, *Auk*, 68, 483-493, 1951.
61. Edwards, P. J., Re-invasion by some farmland bird species following capture and removal, *Pol. Ecol. Stud.*, 3, 53-70, 1977.
62. Parr, R., Sequential breeding by Golden Plovers, *Br. Birds*, 72, 499-503, 1979.

63. Eyckerman, R., Some observations on the behaviour of intruding great tits *Parus major* and on the success of their breeding attempts in a high density breeding season, *Gerfaut*, 64, 29-60, 1974.
64. Dhont, A. A., and J. Schillemans, Reproductive success of the Great Tit in relation to its territorial status, *Anim. Behav.*, 31, 902-912, 1983.
65. Greenwood, P. J., Mating systems, philopatry, and dispersal in birds and mammals, *Anim. Behav.*, 28, 1140-1162, 1980.
66. Taylor, L. R., I. P. Woiwod and J. N. Perry, Variance and the large scale spatial stability of aphids, moths and birds, *J. Anim. Ecol.*, 49, 831-854, 1980.
67. Taylor, R. A. J., and L. R. Taylor, A behavioural model for the evolution of spatial dynamics, in *Population Dynamics*, R. M. Anderson, B. D. Turner and L. R. Taylor, Eds., Blackwell Scientific, Oxford, 1979, pp. 1-28.
68. McArdle, B. H., K. J. Gaston and J. H. Lawton, Variation in the size of animal populations: patterns, problems and artefacts, *J. Anim. Ecol.*, 59, 439-454, 1990.
69. O'Connor, R. J., Behavioural regulation of bird populations: a review of habitat use in relation to migration and residency, in *Behavioural Ecology: Ecological Consequences of Adaptive Behaviour*, BES Symp. No. 25, R. M. Sibly and R. H. Smith, Eds., Blackwell Scientific, Oxford, 1985, pp. 105-142.
70. Hanski, J., Dynamics of regional distribution: the core and satellite hypothesis, *Oikos*, 38, 210-221, 1982.
71. Hanski, J., Single-species metapopulation dynamics: concepts, models, and observations, *Biol. J. Linn. Soc.*, 42, 17-38, 1991.
72. Brown, J. H., On the relationship between abundance and distribution of species, *Am. Nat.*, 124, 255-279, 1984.
73. O'Connor, R. J., Organization of avian assemblages—the influence of intraspecific habitat dynamics, in *Organisation of Communities—Past and Present*, BES Symp. No. 26, J. Gee and P. R. Giller, Eds., Blackwell Scientific, Oxford, 1987, pp. 163-183.
74. Villard, M. A., K. Freemark and G. Merriam, Metapopulation theory and neotropical migrant birds in temperate forests: an empirical investigation, in *Ecology and Conservation of Neotropical Migrant Landbirds*, J. M. Hagan, III and D.W. Johnston, Eds., Smithsonian Institution Press, Washington, D.C., 1992, pp. 474-482.
75. Emlen, J. M., Terrestrial population models for ecological risk assessment: a state-of-the-art review, *Environ. Toxicol. Chem.*, 8, 831-842, 1989.
76. Fahrig, L., and J. Paloheimo, Determinants of local population size in patchy habitats, *Theor. Pop. Biol.*, 34, 194-213, 1988.
77. Fowler, C. W., Density dependence as related to life history strategy, *Ecology*, 62, 602-610, 1981.
78. Emlen, J. M., and E. K. Pikitch, Animal population dynamics: identification of critical components, *Ecol. Model.*, 44, 253-273, 1989.
79. Tipton, A. R., R. J. Kendall, J. F. Coyle and P. F. Scanlon, A model of the impact of methyl parathion spraying on a quail population, *Bull. Environ. Contam. Toxicol.*, 25, 586-593, 1980.
80. Potts, G. R., The effects of modern agriculture, nest predation, and game management on the population ecology of partridges *Perdix perdix* and *Alectoris rufa*, *Adv. Ecol. Res.*, 11, 2-79, 1980.
81. Potts, G. R., *The Partridge—Pesticides, Predation, and Conservation*, Collins New Naturalist, London, 1986.
82. O'Connor, R. J., unpublished data.

30

Understanding the Impact of Pesticides on Wild Birds by Monitoring Incidents of Poisoning

Peter W. Greig-Smith

ABSTRACT

There are several possible approaches to monitoring the effects of pesticides on wild birds. This chapter is concerned with reactive investigations into incidents of mortality suspected to be caused by pesticide poisoning. It addresses the principles of incident investigations, and the role of monitoring in the assessment of pesticide effects in the United Kingdom.

The results of incident investigations can be used as (1) a trigger for further, more specific study; (2) a relative indicator of the scale of problems; or (3) an absolute measure of population impacts. Each of these entails a set of rather different limitations. Common to all, however, is uncertainty in knowing the likelihood of detecting effects that occur in correctly diagnosing the cause of birds' deaths and in judging the implications for populations.

In the United Kingdom there is a long-running scheme for investigation of suspected poisoning incidents, including reported mortality of birds. Some results from the scheme are reviewed to demonstrate its value in identifying pesticide problems, chiefly through interpretation as in (1) and (2), and to consider how it might be extended for use as an absolute index as in (3).

Requirements for monitoring to determine the population-level significance of incidents are demanding. It is important to assess what proportion of incidents is reported, to identify sublethal effects as well as mortality, and to be able to relate the effects to population dynamics in the light of "normal" levels of mortality. Currently, this cannot be achieved. For regulatory purposes in the United Kingdom, incident data are used to distinguish cases of misuse or abuse from approved use. Decisions on the unacceptability of mortality due to approved uses are ultimately subjective, and are generally made well below levels that would have any serious population implications for the species concerned.

KEY WORDS

monitoring, wildlife mortality, field investigation, diagnosis of poisoning, pesticide registration

INTRODUCTION

"Monitoring" can take many forms. For the purpose of assessing non-target effects of pesticides on birds, monitoring approaches fall into three categories:

1. Population surveys, which divide into two types. First, there are those dealing with changes in the density of species, usually over a long time scale; e.g., the Christmas Bird Counts in the United States[1] and the Common Bird Census and other programs run by the British Trust for Ornithology in the United Kingdom.[2] There are also several specific surveys of particular species, such as herons.[3] Second, some programs are concerned directly with the levels of contaminants in birds, chiefly of persistent pollutants such as organochlorine (OC) insecticides. For example, the Institute of Terrestrial Ecology and Nature Conservancy Council in the United Kingdom have carried out regular assessments of OC pesticide residues in four species of predatory birds,[4-6] and there are several contaminant monitoring programs in the United States.[7] Also, specific periodic surveys of OC residues (DDT and metabolites, and dieldrin) in woodpigeons (*Columba palumbus*) have been carried out on behalf of the United Kingdom Working Party on Pesticide Residues[8] to identify changes since the withdrawal of these compounds from use.
2. Surveillance of pesticide usage, to reveal any effects occurring as a result of known applications of products, under realistic farm conditions. This has formed part of the data requirements for regulatory approval of certain products, when it is considered preferable to monitor restricted commercial usage than to conduct experimental trials.[9,10]
3. Investigations into reported incidents of suspected animal poisoning. This is inevitably reactive, and is usually concerned with identifying the involvement of an unknown pesticide, rather than examining the consequences of a known event.

This chapter is concerned only with the last of these approaches, and seeks to address the opportunities for using the results of incident monitoring to assess the impact of pesticide poisoning on wild bird populations. Information from the Wildlife Incident Investigation Scheme in the United Kingdom[11,12] is used to illustrate generalizations about the advantages and limitations of incident data.

UNITED KINGDOM WILDLIFE INCIDENT INVESTIGATION SCHEME

The scheme has been operated since the early 1960s, primarily to provide surveillance for potential environmental side effects of newly approved products, allowing action to be taken if necessary to prevent further cases of animal poisoning. In recent years, greater emphasis has been given to cases revealing evidence of pesticide misuse (careless or accidental failure to adhere to correct practice) and abuse (deliberate, illegal poisoning of animals).[12]

Investigations involve a combination of field visits by biologists, examination of carcasses by veterinarians to determine the cause of death (including disease, injury, and starvation as well as poisoning), and laboratory studies to detect residues of pesticides (Table 1) or effects on the activity of brain acetylcholinesterase.[13]

The scheme is directed chiefly at identifying short-term effects of nonpersistent pesticides, and is also selective, concentrating on insecticides and vertebrate control agents, which are most likely to cause poisoning of wildlife. Although analysis for residues of persistent OC compounds or their metabolites is sometimes carried out, it is usually not possible to identify the source of contamination, unless it is from a highly concentrated toxic bait. The significance of gradual accumulation of persistent chemicals must be assessed by a different approach[6] than that described here.

Table 1. Investigation Procedures Available within the U.K. Wildlife Incident Investigation Scheme

Field information

Details supplied by the notifier (number of casualties, circumstances of death)

Visit by a field biologist (collection of samples, local pesticide use)

Autopsy of carcasses

Determination of the cause of death (injury, starvation, disease, poisoning, etc.)

Tests for disease

Identification of bacterial and viral pathogens, and parasites

Biochemical assays

Measurement of brain AChE inhibition and reactivation if organophosphorus or carbamate pesticides are suspected

Residue analysis

Multiresidue methods

- Organophosphorus compounds (GC with flame photometric detector; confirmation by GC/MS or TLC with esterase inhibition detection)
- Organochlorine compounds (GC with electron capture detection; confirmation by GC/MS)
- Carbamate compounds (GC with nitrogen-selective detector; confirmation by TFAA derivatization)
- Synthetic pyrethroids (GC with electron-capture detection; confirmation by GC/MS)
- Anticoagulant rodenticides (reversed-phase HPLC with fluorescence detection; confirmation by pH switching or normal phase HPLC)

Specific analyses

- Strychnine (HPLC with UV detection; confirmation by GC with nitrogen selective detector)
- Alphachloralose (GC of trimethylsilyl derivative with electron-capture detector; confirmation by HFBI derivatization)
- Paraquat (HPLC)
- Metaldehyde (GC of acetaldehyde released by acid, with flame ionization detector)

These incident investigations represent one response to public reporting of suspected animal poisoning, in which there are opportunities for detailed scientific study. In addition, there are several telephone response programs that do not involve investigations, but provide advice or referral of calls about suspected cases affecting animals. These deal with substantial numbers of cases in the United Kingdom (ca. 400 per year) and France (ca. 1500 per year).[14,15] Few other countries, with the exception of The Netherlands,[16] have well-developed national schemes to investigate incidents of bird poisoning.

For many years, incidents caused by the deliberate abuse of pesticides have outweighed those due to approved use or misuse (which may be difficult to distinguish from each other). On average, 75% of all incidents concerning birds and mammals, in which pesticide involvement is confirmed, are due to abuse.[17] This means that only about 20 incidents involving wild birds, in which normal pesticide use was the apparent cause of mortality, are reported in the United Kingdom each year.

The numbers of individual bird victims are also small. As Table 2 shows, a large majority of incidents was reported as a result of less than 10 sick or dead birds being found. For birds of prey, whose relatively small populations may be most vulnerable to large-scale mortality, there were no incidents involving more than eight birds. Only for gamebirds, pigeons, and songbirds were there cases of pesticide poisoning in which more than 100 birds were found dead. In contrast, a number of incidents in North America have involved much greater mortality.[18]

Table 3 lists the large-scale incidents (defined as those in which more than 50 birds were reported dead) in England and Wales between 1980 and 1989. Some of these occurred as a result of new pesticide uses, or were unexpected cases in which birds were exposed to the chemical by a route that had not been anticipated during preregistration testing. However, a substantial number were due to evident abuse or the cause was undetermined. Only 16 cases of large-scale mortality in 10 years were apparently due to the approved use of the products concerned. At first sight, this seems to indicate either a remarkable success of preregistration assessment in eliminating potential problems, or considerable inefficiency in detecting the incidents that occur. These possibilities are discussed here by considering the ways in which incident data are used or could be used.

USES OF INCIDENT DATA

There are three distinct ways in which the results of incident monitoring can be used:

1. as a trigger for further, more specific study of the problems that are suggested by incidents of a particular type;
2. as a relative index, showing variations in the occurrence of poisoning cases; and
3. as a measure of the severity of a pesticide's impact on populations of nontarget species.

Each entails a different set of assumptions, and is affected by a different set of limitations, which are discussed in turn. Common to all, however, is uncertainty in knowing the likelihood of detecting mortality when it occurs, in correctly diagnosing the cause of death, and in judging the implications of the observed mortality for population stability in the short or long term.

Incidents as a Trigger for Further Study

The requirements for this use of incident data are not very demanding. Most critical is to identify thresholds of concern that signal the need for a reappraisal of existing data on the toxicity or hazard of a product, or the need for additional data. These may be established through detailed knowledge of the species found victims of poisoning, if there have been independent research studies on their population ecology. More often there is insufficient knowledge to immediately assess the significance of mortality observed in an incident, and thresholds are based on subjective decisions about the seriousness of mortality, erring on the side of safety. In the case of rare species, such as many birds of prey,[19] a single death may be enough to act as a trigger. Precedents offered by experience with similar pesticide uses may be helpful; for example, insecticidal seed treatments against cereal pests periodically cause poisoning incidents, and a newly introduced product may require further investigation if its early use led to incidents more extreme than these "normal" cases.

A single incident can be the trigger for more study, particularly if it involves the deaths of many birds. However, a series of separate incidents, each unimportant in

Table 2. Numbers of Birds Found Dead in Suspected Poisoning Incidents in England and Wales, 1980–1989.

		Number of Birds Found Dead		
		1–10	**11–100**	**100+**
Birds of prey (including owls)	+	152	0	0
	–	176	0	0
Waterbirds (ducks geese, herons)	+	19	19	0
	–	74	15	0
Gulls & plovers	+	13	4	0
	–	28	26	6
Gamebirds (pheasants, partridges)	+	35	18	3
	–	44	13	3
Pigeons & doves	+	58	36	4
	–	55	36	5
Corvids (crows, magpies, jays)	+	129	24	0
	–	70	9	0
Songbirds (finches, thrushes, and others)	+	36	33	4
	–	68	33	1
All incidents[a]		820 (73%)	274 (24%)	24 (2%)

Note: Information is shown separately for those cases in which pesticide involvement was confirmed (+) and those attributable to other causes, or unknown (–). Values are the numbers of incidents investigated.

[a]Birds from more than one category may be involved in a single incident.

itself, could equally be a cause for concern. Cooke[20] discussed the difficulties in establishing fixed thresholds. Sometimes the incidents that act as triggers are due to routes of exposure and mechanisms of action that were anticipated during preregistration risk assessment, but were not expected to occur on such a large scale. In other cases, they represent unexpected effects, either due to environmental conditions that modify the expected patterns of exposure or susceptibility of wildlife, or to novel routes of exposure.

A recent example of an unexpected effect occurred in August 1989, when an onion field in Kent, England, was sprayed with the herbicide sodium monochloroacetate to control broadleaved weeds. Within 24 hours, over 300 small birds (predominantly greenfinches [*Carduelis chloris*], but also linnets [*Acanthis cannabina*], goldfinches [*Carduelis carduelis*], and house sparrows [*Passer domesticus*]) were found dead or dying within adjacent gardens. Although it seemed unlikely that the birds could have been exposed to sufficient spray to poison them (the toxicity of sodium monochloroacetate to vertebrates is low[21]), residues were detected in the crop contents, brain, liver, muscle, and intestines of a pooled sample of 11 birds. It was concluded that these birds had probably received exposure by drinking from puddles of the spray formed on very dry ground, which occurs only in unusual circumstances.

A further example, with major regulatory implications, is the poisoning of wild geese by organophosphorus seed dressings that occurred in the United Kingdom in the mid-1970s.[22] Following the withdrawal of dieldrin from use as a seed treatment against wheat bulb fly, two compounds were assessed as potential replacements, involving

Table 3. Incidents of Bird Poisoning in England and Wales, 1980–1989[a]

Date	Casualties	Pesticide Detected	Source
Seed Treatments			
Nov 1983	200 woodpigeons	Chlorfenvinphos	Treated wheat seed
Dec 1983	62 feral pigeons	Chlorfenvinphos	Treated wheat seed
Oct 1986	30–60 greylag geese	Carbophenothion	Unknown
Mar 1987	140 woodpigeons	Bendiocarb	Treated pea seed
Mar 1987	200+ woodpigeons	Bendiocarb	Treated pea seed
Insecticide Sprays			
Sep 1982	50 pheasants	Disulfoton	Aerial application to brussel sprouts
Mar 1984	55 brent geese	Triazophos + dimethoate	Applied as a mixture to winter wheat
Mar 1987	200+ woodpigeons	Omethoate	Spray on winter corn
Mar 1987	50–60 brent geese	Triazophos	Application to winter wheat
Herbicide Spray			
Aug 1989	300 songbirds (mainly greenfinches)	Sodium monochloroacetate	Application to onion field
Source Unknown			
Jun 1983	60–70 mallard and domestic ducks	Diazinon	—
Aug 1983	95 songbirds, including greenfinches and linnets	Chlorpyrifos	—
May 1984	70 rooks + 1 starling	Demeton-S-methyl	—
Dec 1985	50+ starlings	Triazophos	—
Feb 1987	50 woodpigeons	Demeton-S-methyl	—
Nov 1989	60 feral pigeons	Dimethoate	—

Notes: 1. In addition, there were 12 incidents attributed to deliberate abuse of pesticides, involving endrin (2), dimethoate (1), triazophos (1), mevinphos (6), alphachloralose (1), and aldrin (1).

2. 31 other incidents were reported involving deaths of 50 to 370 birds, but no evidence of pesticide poisoning was established.

3. Species mentioned above are woodpigeon (*Columba palumbus*), feral pigeon (*Columba livia*), greylag goose (*Anser anser*), pheasant (*Phasianus colchicus*), mallard (*Anas platyrhynchos*), greenfinch (*Carduelis chloris*), linnet (*Acanthis cannabina*), brent goose (*Branta bernicla*), rook (*Corvus frugilegus*), and starling (*Sturnus vulgaris*).

[a]More than 50 birds were found dead, and the cause was attributed to the approved use of pesticides.

evaluation of the possible risks to wild birds. Chlorfenvinphos was found to be particularly toxic to pigeons (LD_{50} 13.5 to 25.8 mg/kg[23]) and therefore carbophenothion was preferentially used in order to prevent possible poisoning of pigeons. Once in commercial use, a series of incidents was caused by exposure of geese to carbophenothion, involving the deaths of about 1500 greylag geese (*Anser anser*) and pink-footed geese (*Anser brachyrhynchos*) between 1971 and 1975.[24] The regulatory response was to restrict the use of carbophenothion to areas in which geese would not be at risk.

This example illustrates three points. First, preliminary assessment suggested that carbophenothion was of acceptably low toxicity to geese (the birds most likely to feed heavily on treated seed) based on toxicity tests using Canada geese (*Branta canadensis*) (LD_{50} value of 29 to 35 mg/kg[25]). However, further laboratory studies to explore why the field hazard was greater than had been anticipated, showed that *Anser* geese are more susceptible to carbophenothion than are *Branta* geese.[26] This demonstrates the potential importance of differences between species in susceptibility. Second, the level of mortality of pink-footed geese was estimated to be at least 540 birds in 1975, representing about 1% of the world's population; this was considered to be unacceptable, and the 1% level has since been proposed as a possible general criterion on which to judge the seriousness of direct pesticide effects on bird populations.[20] Third, restriction of carbophenothion use was accompanied by a switch to chlorfenvinphos for wheat bulb fly control, despite the fact that it was expected to cause an increased frequency of pigeon poisoning incidents (and has done so; Figure 1).[27] Nonetheless, a clear decision was made to accept this additional mortality of a very abundant resident pest species so as to avoid adverse effects on an uncommon species of conservation interest.

Incidents as a Relative Index of Problems

The number of incidents reported inevitably varies with time and place. In order to judge whether patterns in numbers represent a trend, and whether regional differences are real, investigations must be consistent.

For a particular species, it may be informative to examine the proportion of the analyses carried out for a specific type of pesticide that are positive. This should help to remove the potentially confounding variations due to differences in the proportions of incidents reported, and variation in the investigation effort. An example is given in Figure 2, showing that the occurrence of poisoning of predatory birds by alphachloralose apparently doubled in 1989 and 1990 from its previous level. However, when the data are expressed as the percentage of analyses that revealed the presence of residues, this trend is shown to be false. For other birds over the same period, there was no evidence of a trend in alphachloralose poisoning when examined both ways (Figure 2).

It is generally not justifiable to make detailed comparisons of incidents involving different species or different pesticides because the rates of occurrence are likely to be inherently biased. However, useful comparisons are possible for related groups of species or chemicals if the raw numbers are adjusted to take account of other causes of death and patterns of usage. This approach was used by Greig-Smith[27] to review poisoning of wild birds by seed treatments, pellets, and granules. Table 4 lists similar data covering all pesticide uses over the period 1984 to 1989.

Variation in the numbers of incidents in which pesticides were detected appears to show a preponderance of poisoning cases that affect birds of prey and corvids (Table 4). However, most of these were caused by deliberate abuse, and the incidents attributed to normal use greatly affected only pigeons and doves. When the figures are weighted by the total numbers of incidents, or just those in which pesticides were implicated, the pattern shows a more even level of poisoning among waterbirds, gamebirds, pigeons, and doves. If deliberate abuse cases are excluded, these three groups of birds were affected to about the same extent, which was twice the frequency of poisoning of other birds.

A second approach to weighting data of this kind is to use information on the relative quantities or frequency of use of different pesticide products available from periodic surveys.[28] This provides information about usage on different crops, and is a

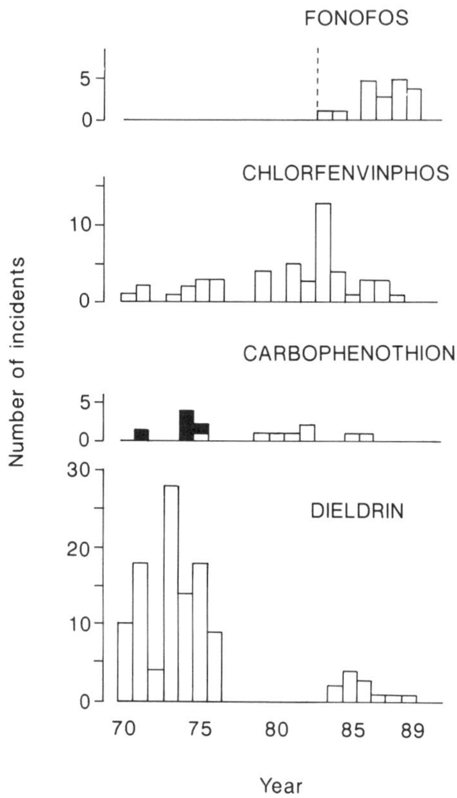

FIGURE 1. Trends in the occurrence of wildlife poisoning due to insecticidal seed treatments in England and Wales, from 1970 to 1989. Most incidents involved seed-eating birds, but dieldrin cases in the late 1980s all concerned predatory species. Shading indicates major incidents involving geese in Scotland.[22] Fonofos was approved as a seed treatment in 1982; some subsequent incidents may be the result of spray formulations.

powerful means of placing into context the numbers of poisoning cases identified. Table 5 shows that several widely used chemicals were involved in disproportionately few incidents, whereas some of those that caused most incidents have limited use. This approach is particularly valuable for helping to indicate a relative lack of problems for a chemical in widespread commercial use.

Assessing the Implications of Mortality for Bird Populations

Incident investigations are heavily biased toward cases of animal mortality and completely fail to identify indirect effects of pesticides. They are unlikely to be initiated as a result of sublethal poisoning effects on birds (although there are cases in which cats or dogs fall ill after exposure to pesticides, but subsequently recover; see Greig-Smith et al.[29]).

Even in cases of substantial mortality, the implications for local populations are rarely clear. This reflects uncertainty in the likelihood of detecting incidents that occur, correctly identifying pesticide involvement, and relating the scale of mortality to population processes.

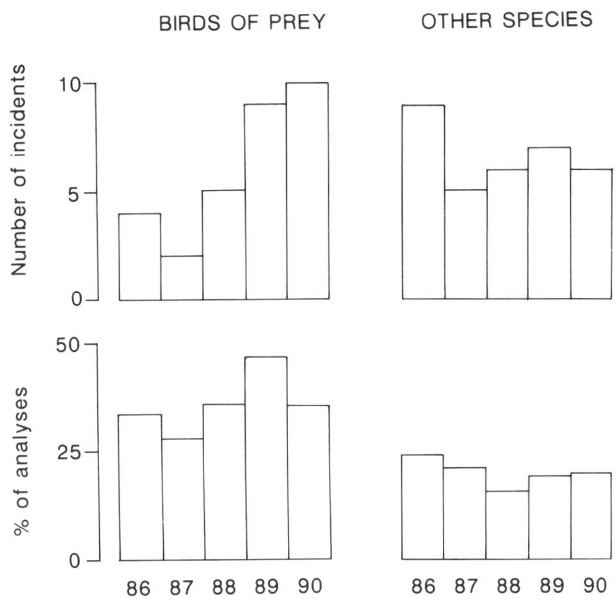

FIGURE 2. Recent trends in poisoning of predators and other birds by alphachloralose in England and Wales, shown as the number of incidents and the percentage of analyses in which residues were detected in birds.

What Proportion of Incidents are Discovered?

The success or failure of the scheme in identifying incidents depends on the combined effect of several probabilities, which are summarized in Table 6. The probability that birds receiving lethal exposure will die on or near to a treated field (P1) depends on the nature of the chemical and its toxic action (rapid acute effects are more likely to produce dead bodies at the site than delayed or chronic toxicity) and on the species. (Large, very mobile animals may be more likely to leave the area before succumbing, although in some cases, exposure may be more harmful to small-bodied species.) Sublethal effects on behavior may strongly bias the likelihood of affected birds being observed.[30]

After death, the probability of dead bodies being found by people visiting the site (P2) is a function of many aspects, including the size and conspicuousness (e.g., plumage patterns) of the birds; the numbers that die or are observed behaving oddly; the rate at which they are removed by scavengers; and the accessibility of the site to people who might report the incident. Finally, the likelihood that the incident will be notified to the scheme (P3) depends on the size of the impact (i.e., how many birds die) and the conservation interest of the species concerned, as well as on public knowledge of the arrangements available to deal with such cases.

The combination of these factors is clearly complex, and many are virtually unmeasurable, although preliminary attempts have been made to assess the likely scale of some biases.[31] There are certain circumstances in which the overall probability may be close to one; for example, pesticide problems that affect large-bodied flocking birds such as swans or geese are almost certain to be detected if there is rapid mortality in farmland near human habitation. In contrast, regular poisoning of substantial numbers of territorial songbirds, particularly if they reach non-field habitats before dying,

Table 4. Summary of Wildlife Incidents Involving Birds in England and Wales, 1984–1989 (The relative occurrence of poisoning due to the approved use of pesticides in different groups of birds is shown in this table.)

	Total No. of Incidents	No. Due to Pesticidal Poisoning:				Approved Use as % of		
		Total	Approved Use	Misuse	Abuse	Total No. of Incidents	Total No. Due to Pesticide Poisoning	Total No. of Incidents minus Abuse Cases
Birds of prey (including owls)	212	101	16	1	84	7.6	15.8	12.5
Waterbirds	65	22	12	4	6	18.5	54.6	20.3
Gulls and plovers	41	11	3	2	6	7.3	27.3	8.6
Gamebirds	56	24	10	3	11	17.9	41.7	22.2
Pigeons and doves	124	52	29	4	19	18.8	55.8	27.6
Corvids	154	100	7	2	91	4.6	7.0	11.1
Songbirds	99	32	8	4	20	8.1	25.0	10.1
Total	646	285	77	17	191	11.9	27.0	16.9

Note: The category "approved use" includes some possible cases of misuse or abuse that could not be firmly determined.

Table 5. Numbers of Incidents Involving Poisoning of Wild Birds by Selected Insecticides/Molluscicides in England and Wales (1984–1989) in Relation to Use of Pesticides[a]

	Total Incidents	Incidents Due to Use or Misuse	Occurrences in Usage Surveys	Area Treated (10^3 ha)	100 × (Incidents Due to Use or Misuse)/(Occurrences in Usage Surveys)
Organochlorine Insecticides					
DDT	32	8	233	31.1	3.4
Dieldrin	19	8	10	2.1	80.0
Gamma-HCH	0	0	272	89.2	0
Endosulfan	0	0	232	10.7	0
Contact Organophosphorus Insecticides					
Chlorfenvinphos	13	12	107	36.3	11.2
Chlorpyrifos	0	0	340	35.6	0
Fenitrothion	0	0	63	15.1	0
Fonofos	17	14	3	6.0	466.7
Malathion	1	0	222	2.3	0
Pirimiphos-methyl	1	0	194	13.2	0
Triazophos	10	8	158	123.4	5.1
Systemic Organophosphorus Insecticides					
Demeton-S-methyl	5	3	1114	413.2	0.3
Dimethoate	3	2	491	99.0	0.4
Disulfoton	0	0	96	170.6	0
Omethoate	2	2	18	39.7	11.1
Phorate	4	4	59	35.3	6.8
Carbamate Insecticides/Molluscicides					
Aldicarb	2	2	203	90.2	1.0
Bendiocarb	4	4	16	16.7	25.0
Carbofuran	6	6	118	588.7	5.1
Methiocarb	1	1	608	149.0	0.2
Other Compounds					
Metaldehyde	6	4	192	93.6	2.1

[a]Data from Sly, M. J. A., 1986.[28]

Table 6. Factors Affecting the Probability that Birds Dying as a Result of Poisoning by Pesticides Will Be Reported to an Incident Investigation Scheme

P1	(Probability of birds dying on or near to the site of application)	Rapidity of onset of toxic effects Mobility of the species
P2	(Probability of carcasses being found)	Body size Conspicuousness of plumage Number of birds that die Scavenger pressure Frequency of observers visiting site
P3	(Probability of events being reported to the scheme)	Scale of mortality Conservation status of species involved Awareness of investigation scheme

is likely to pass unnoticed.[32] Reactive investigations of reported incidents are not a good way to address such problems.

Diagnosis of Pesticide Involvement

A range of techniques is available to identify whether a pesticide was the cause of death. These include field information about the circumstances of the incident, results of an autopsy, tests for disease, biochemical assays to detect brain acetylcholinesterase depression, and chemical residue analysis of gut contents or tissue samples (Table 1). Not all this information is available in every case, and the quality of the sample material is often poor, having been taken from carcasses lying undiscovered for many days.

If large residues of a pesticide or its metabolites are detected in tissues or in gut contents and are confirmed by a different type of analysis, the chemical can usually be regarded as the cause of death. However, the presence of small residues is not straightforward to interpret, for animals may be exposed to a chemical without necessarily being affected seriously. For birds, there is no background of data from laboratory dosing of wild species that would indicate the residue levels expected in a lethally poisoned animal, as has been accumulated for honeybees.[46] Only in the case of anticoagulant rodenticide poisoning is it possible to make a semiquantitative assessment of residue levels, as a result of research into long-term storage of rodenticides in the liver of quail,[33] and recent experimental dosing of barn owls (*Tyto alba*).[34] This allows a distinction to be made between the low levels (<0.05 mg/kg) that persist for many months after sublethal exposure, and higher levels (>1.0 mg/kg) that indicate recent, usually lethal, exposure (Figure 3).[46] Otherwise, the concentration of a chemical in a bird's gut content permits only a very crude calculation of the dose (usually an underestimate) which can be compared to lethal concentrations determined in laboratory tests (LD_{50} or LC_{50}).

Apart from these difficulties, it should be recognized that the presence of chemical residues constitutes firm, positive evidence of exposure, but failure to detect residues cannot be interpreted as definite proof that the animal was not poisoned. There may be categories of pesticides for which analyses are not available, and many products may degrade rapidly or be metabolized or eliminated from the body, so that they would be hard to detect in incident samples, which are rarely fresh.

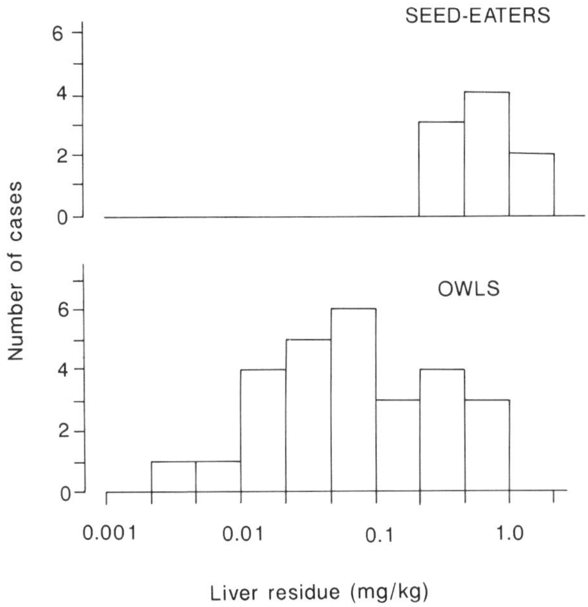

FIGURE 3. Residues of anticoagulant rodenticides in the livers of seed-eating birds and barn owls found dead in suspected poisoning incidents, 1984 to 1989. The low residue levels in many owls represent sublethal poisoning due to predation on contaminated prey. Seed-eating species (pigeons, pheasants) that consumed rodenticide baits directly had high residues indicative of acute death.

A further area of uncertainty concerns the fact that investigations deal only with dead animals. This means that the distribution of effects on survivors is unknown, although it may represent an important part of the hazard, including reproductive impairment or cases of marginally sublethal poisoning.[30] One solution would be to collect a sample of live-trapped animals at the site of an incident, if this can be done rapidly enough to avoid serious loss of the chemical or recovery from its effects.[35] In addition, it would often aid the interpretation of laboratory studies, particularly cholinesterase assays,[36] if parallel "control" samples of birds could be obtained from a matched site where they were not exposed to chemicals. However, this can be difficult to achieve, and is not justifiable for incidents in which the number of casualties is very small (Table 2).

Interpretation of Population Consequences

Clearly, the information obtained in the course of most incident investigations is inadequate to permit a detailed assessment of effects on the local population dynamics of birds. The limitations on what is feasible are such that this level of detail could never reliably be achieved. Instead, the following simple approaches have been used in particular cases.

1. For a few species, there may be local studies that provide an opportunity to measure the significance of mortality. For example, Cooke[37] conducted an annual winter survey of woodpigeons (*Columba palumbus*) roosting at a site in eastern England (Woodwalton Fen), surrounded by agricultural land on which insecticidal seed dressings were used. The number of poisoning incidents occurring at the site each winter (in which dead birds were found, or live birds were observed showing symptoms of intoxication) was compared to the index of winter roosting numbers. The lack of an obvious correlation

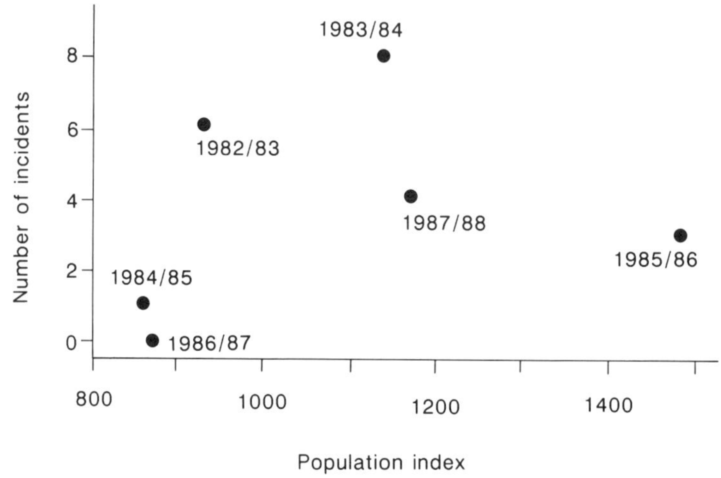

FIGURE 4. Numbers of local poisoning incidents involving woodpigeons, in relation to an index of winter population size, in six winters.[37]

(Figure 4) indicated that there was probably little impact on local population size. This result can be extrapolated very roughly by comparing the scale of mortality observed at Woodwalton Fen with that reported in other incidents, taking account of the time of year and other factors. On that basis, the poisoning in 1987 of 200 and 140 woodpigeons by a new use of bendiocarb as a pea seed treatment (Table 3) appeared high. Accordingly, that use of the chemical was rapidly withdrawn following those incidents.

2. The United Kingdom population of barn owls (*Tyto alba*) has been of particular concern to the Hawk Trust, which compiled information about the decline of this species during the period 1970 to 1985. A summary of the causes of death of barn owls[38] revealed that only 3% was attributed to chemical poisoning, compared to 52% caused by collisions with traffic and 16% due to other accidental injuries. Although this information was not based on detailed scientific investigations of casualties, and there is an untested possibility that some road traffic casualties could have been the result of sublethal debilitation after exposure to pesticides, the result implies that poisoning is a minor contributor to population changes. This approach provides a means of roughly assessing, at the national or regional level, whether pesticides are likely to influence population trends.
3. Some species of national conservation interest are the subject of regular censuses to determine their status. For rare species such as the red kite (*Milvus milvus*) and osprey (*Pandion haliaetus*), this may entail checks on individual breeding pairs. These circumstances provide unusually good opportunities both to detect poisoning incidents and to assess their importance. In 1989, there were 10 known incidents involving red kites, in which 11 adult and 2 nestling birds are known to have been poisoned, from a national population of only about 60 pairs. For such a small population, any mortality is likely to affect the numbers and breeding distribution of the species. The consequences may not be so extreme for other birds of conservation interest, but detailed awareness of their national status gives a baseline against which to assess the deaths of individuals.

For the future, it may be possible to use demographic models of wild bird populations to determine approximate thresholds of mortality which would be likely to affect local populations. This level could then be related to the occurrence of incidents, taking into account where possible the biases attending the discovery and reporting of casualties of the species. At present, there is sufficient information to attempt this only for very few of the species liable to be poisoned by pesticides.

The woodpigeon has been the subject of detailed population studies in the United Kingdom[39–41] because of its pest status in the 1950s and 1960s. This work involved an assessment of the effect on population density of intensive shooting, which is analogous to mortality caused by pesticide poisoning.

Woodpigeon populations have an annual cycle, in which a peak is reached after breeding (August to September), followed by high winter mortality of juveniles (over 50% dying before the following February). Murton[41] concluded that year to year variations in juvenile survival were dependent on food supply (cereal grain and clover), acting through competition, and resulting in local population regulation at a level appropriate to the availability of food. Winter density was therefore independent of the previous population level, and the subsequent breeding population was unrelated to winter numbers. (These studies were carried out in the 1960s, and more recent changes in farming practice have led to different patterns of population regulation.[42])

Murton's work demonstrated that even heavy killing of woodpigeons would have little effect on population density, particularly if it occurred early in the winter, when it would merely replace natural mortality, and could even increase the spring population by reducing competition among the survivors.[39] For high density populations of woodpigeons (150 birds per acre in winter) it would be necessary to kill at least 130 birds per acre in order to influence population stability, and even at low density (50 birds per acre), mortality would need to exceed 45 birds per acre. It is unlikely that even the largest incidents (Table 2) caused mortality on this scale.

Nevertheless, pesticide-related mortality could cause short-term, local changes in density, particularly when treated seed is sown in spring, at the time of minimum population density. The effects depend on whether birds are exposed before they have dispersed from their winter flocks (when all the local population is liable to be at risk, being attracted to a good food source, but surviving birds have the opportunity to redistribute before the main breeding season), or after dispersal (when mortality may be restricted to fewer birds, but they are unlikely to be replaced). Which of these is considered most serious is subjective.

It is difficult to establish a threshold level of mortality that might be used to assess individual incidents, even for this well-studied species. Because of the great year to year variation in density, criteria should be set as a proportion of local numbers, not a fixed number of birds. Also, the consequences of mortality will depend on the availability of food and the movements of the local population, which are rarely known.

Another farmland species that has been intensively studied in the United Kingdom is the grey partridge *Perdix perdix*.[43] Long-term research on the relationship of partridge population parameters to insect food supplies has resulted in a predictive model that successfully accounts for a very high proportion of variability in chick survival from place to place and year to year.[44] Although this work has been concerned almost exclusively with indirect effects of pesticides, in principle it could be used to interpret mortality caused by direct poisoning. However, Potts and Aebischer[44] point out that analysis of this type is reliably applicable only to bird populations at moderate or high densities (>5 pairs per square kilometer); a large proportion of British farmland species (about 60%) usually occur at lower densities than this.[45] Further, it is these species for which local mortality is probably most serious, but which are least likely to be revealed by the incident scheme.

Thus, even the best available population data for British farmland birds are not currently suitable for establishing operationally useful threshold criteria to help evaluate mortality in reported incidents.

CONCLUSION

Monitoring the effects of pesticides through reactive incident investigations has a number of advantages (Table 7). Perhaps most important of these are the opportunities it provides to cover extreme or unusual conditions that would be difficult or undesirable to simulate experimentally in the course of pre-registration testing of new products.[10] However, there are also shortcomings (Table 7) that severely limit the interpretation of incident investigations and their implications for bird population stability. Decisions made about the unacceptability of wildlife mortality due to approved uses are ultimately subjective, and are generally made well below the levels at which mortality would have any serious population implications for the species concerned.

Because of these advantages and disadvantages, incident monitoring must be seen as a complement to experimental testing. One of its main purposes is to act as a "trigger" for specific research or testing that was not conducted earlier in the period of pre-registration assessment of products. This does not imply that monitoring only identifies work that should have been carried out at an earlier stage. On the contrary, there are aspects of hazard assessment that are addressed most effectively by in-use monitoring, particularly for protected species. However, these are limited to certain birds and certain types of pesticide products. Monitoring cannot be used as a general method of identifying all adverse effects, and it is particularly weak as a means of establishing a lack of such effects. Decisions about how much emphasis to place on monitoring should be made flexibly, depending on the unique characteristics of each pesticide use.

ACKNOWLEDGMENTS

I am grateful to my colleagues in MAFF and SOAFD (Scottish Office Agriculture and Fisheries Department) who have contributed to the compilation of data from the Wildlife Incident Scheme, and to P. M. Brown, M. R. Fletcher, and A. D. M. Hart for advice and comments.

Table 7. Advantages and Limitations of Incident Monitoring for Revealing Effects of Pesticides on Wild Bird Populations

Advantages
Potential to reveal hazards not identified during preregistration testing
Provides coverage of use in a wide range of conditions
Tests safety under realistic farming practice
Less expensive than extensive field trials
Helps to minimize the extent of experimental testing, particularly for rare species
Limitations
Takes no account of indirect effects or long-term accumulation of persistent chemicals, and rarely reveals sublethal effects
Great uncertainty in knowing the proportion of incidents that are reported
Difficulty in identifying the cause of death in many incidents
Little opportunity to collect appropriate control data
Weak power to demonstrate an absence of effects

REFERENCES

1. Ralph, C. J. and J. M. Scott, Eds., *Estimating Numbers of Terrestrial Birds, Studies in Avian Biology, No. 6*, Cooper Ornithological Society, Lawrence, KS, 1981.
2. Marchant, J. H., R. Hudson, S. P. Carter and P. Whittington, *Population Trends in British Breeding Birds*, British Trust for Ornithology, Tring, U.K., 1990.
3. Reynolds, C. M., The Heronry Census: 1972-77 population changes and a review, *Bird Study*, 26, 7-12, 1979
4. Cooke, A. S., A. A. Bell and M. B. Haas, *Predatory Birds, Pesticides and Pollution*, Institute of Terrestrial Ecology, Cambridge, U.K., 1982.
5. Newton, I., *Population Ecology of Raptors*, T. & A. D. Poyser, Berkhamsted, U.K., 1979.
6. Newton, I., Determination of critical pollutant levels in wild populations, with examples from organochlorine insecticides in birds of prey, *Environ. Pollut.*, 55, 29-40, 1988.
7. Steffeck, D., The role of monitoring in assessing pesticide effects on avian species and their habitats, in *Wildlife Toxicology and Population Modeling: Integrated Studies of Agroecosystems*, R. J. Kendall and T. E. Lacher, Jr., Eds., Lewis Publishers, Chelsea, MI, 1993.
8. MAFF, Report of the Working Party on Pesticide Residues: 1985-1988, HMSO, London, 1989.
9. Hardy, A. R., Ecotoxicology of pesticides: the laboratory and field evaluation of the environmental hazard presented by new pesticides, in *Toxicology of Pesticides: Experimental, Clinical and Regulatory Aspects*, NATO ASI Ser., Vol. H13, L. G. Costa, Ed., Springer-Verlag, Berlin, 1987, pp. 185-196.
10. Greig-Smith, P. W., Intensive study versus extensive monitoring in pesticide field trials, in *Pesticide Effects on Terrestrial Wildlife*, L. Somerville and C. H. Walker, Eds., Taylor & Francis, London, 1990, pp. 217-239.
11. Hardy, A. R., M. R. Fletcher and P. I. Stanley, Pesticides and wildlife: twenty years of vertebrate wildlife incident investigations by MAFF, *State Vet. J.*, 40, 182-192, 1986.
12. Greig-Smith, P. W., Wildlife hazards from the use, misuse and abuse of pesticides, *Aspects Appl. Biol.*, 17, 247-256, 1988.
13. Greig-Smith, P. W., Use of cholinesterase measurements in surveillance of wildlife poisoning in the UK, in *Cholinesterase Inhibiting Insecticides. Their Impact on Wildlife and the Environment*, P. Mineau, Ed., Elsevier, Amsterdam, 1991, pp. 127-150.
14. Lorgue, G., K. Nahas, G. Keck and M. Rampaud, Intoxication of domestic and wild animals by anticoagulant rodenticides—A synthesis of data from the French national veterinary antipoison center, in *Proc. 12th Vert. Pest. Conf.*, 1986, pp. 82-87.
15. Tempowski, J. H., and G. N. Volans, A prospective survey of veterinary enquiries to the National Poisons Information Service, Proc. Symp. Animal Poisoning, British Agrochemical Association, Peterborough, 1989.
16. Spierenburg, T. J., P. E. F. Zoun and T. Smit, Poisoning of wild birds by pesticides, in *Wild Bird Mortality in the Netherlands, 1975-1989*, A. J. Baars and H. J. Over, Eds., Central Veterinary Institute, Lelystad, The Netherlands, 1989.
17. Greig-Smith, P. W., M. R. Fletcher, K. Hunter, M. P. Quick, A. D. Ruthven and I. C. Shaw, Investigations of Suspected Poisoning of Animals by Pesticides in Great Britain in 1987, Report of the Advisory Committee on Pesticides, MAFF, London, 1989.
18. Smith, G. J., Pesticide Use and Toxicology in Relation to Wildlife: Organophosphorus and Carbamate Compounds, Resource Publ. 170, Fish and Wildlife Service, Washington, DC, 1987.
19. Cadbury, C. J., G. Elliott and P. Harbard, Birds of prey conservation in the UK, *RSPB Conserv. Rev.*, 2, 9-16, 1988.
20. Cooke, A. S., The acceptability of pesticide-induced damage to wildlife, in *Pesticide Effects on Terrestrial Wildlife*, L. Somerville and C. H. Walker, Eds., Taylor & Francis, London, 1990, pp. 291-306.
21. Worthing, C. M., *The Pesticide Manual*, 8th ed., Commonwealth Agricultural Bureaux, London, 1987.
22. Stanley, P. I., and P. J. Bunyan, Hazards to wintering geese and other wildlife from the use

of dieldrin, chlorfenvinphos and carbophenothion as wheat seed treatments. *Proc. R. Soc. London Ser. B*, 205, 31–45, 1979.

23. Bunyan, P. J., D. M. Jennings and F. J. S. Jones, Organophosphorus poisoning: a comparative study of the toxicity of chlorfenvinphos (2-chloro-1-(2′,4′-dichlorophenyl)-vinyl diethyl phosphate) to the pigeon, the pheasant and the Japanese quail, *Pestic. Sci.*, 2, 148–151, 1971.
24. Hamilton, G. A., K. Hunter, A. S. Ritchie, A. D. Ruthven, P. M. Brown and P. I. Stanley, Poisoning of wild geese by carbophenothion-treated winter wheat, *Pestic. Sci.*, 7, 175–183, 1976.
25. Jennings, D. M., P. J. Bunyan, P. M. Brown, P. I. Stanley and F. J. S. Jones, Organophosphorus poisoning: a comparative study of the toxicity of carbophenothion to the Canada goose, the pigeon and the Japanese quail, *Pestic. Sci.* 6, 245–257, 1975.
26. Westlake, G. E., P. J. Bunyan and P. I. Stanley, Variation in the response of plasma enzyme activities in avian species dosed with carbophenothion, *Ecotoxicol. Environ. Safety*, 2, 151–159, 1978.
27. Greig-Smith, P. W., Hazards to wildlife from pesticide seed treatments, in *Application to Seeds and Soil*, Monograph No. 39, T. J. Martin, Ed., British Crop Protection Council, Croydon, 1988, pp. 127–134.
28. Sly, J. M. A., Review of Usage of Pesticides in Agriculture, Horticulture and Animal Husbandry in England and Wales 1980–1983, MAFF, London, 1986.
29. Greig-Smith, P. W., M. R. Fletcher, K. Hunter, M. P. Quick, A. D. Ruthven and I. C. Shaw, Pesticide Poisoning of Animals 1988: Investigations of Suspected Incidents in Great Britain, Report of the Environmental Panel of the Advisory Committee on Pesticides, MAFF, London, 1990.
30. Mineau, P., and D. B. Peakall, An evaluation of avian impact assessment techniques following broad-scale forest insecticide sprays, *Environ. Toxicol. Chem.*, 6, 781–791, 1987.
31. Hart, A. D. M., The assessment of pesticide hazards to birds: the problem of variable effects, *Ibis*, 132, 192–204, 1990.
32. Mineau, P., Avian mortality in agro-ecosystems. I. The case against granular insecticides in Canada, in *Field Methods for the Study of Environmental Effects of Pesticides*, Monograph No. 40. M. P. Greaves, B. D. Smith and P. W. Greig-Smith, Eds., British Crop Protection Council, Croydon, 1988, pp. 3–12.
33. Huckle, K. R., P. A. Warburton, S. Forbes and C. J. Logan, Studies on the fate of flocoumafen in the Japanese quail (*Coturnix coturnix japonica*), *Xenobiotica*, 19, 51–62, 1989.
34. Newton, I., I. Wyllie and P. Freestone, Rodenticides in British barn owls, *Environ. Pollut.*, 68, 101–117, 1990.
35. Hill, E. F., and W. J. Fleming, Anticholinesterase poisoning of birds: field monitoring and diagnosis of acute poisoning. *Environ. Toxicol. Chem.*, 1, 27–38, 1982.
36. Hill, E. F., Brain cholinesterase activity of apparently normal wild birds, *J. Wildl. Dis.*, 24, 51–61, 1988.
37. Cooke, A. S., Poisoning of woodpigeons on Woodwalton Fen, in *Field Methods for the Study of Environmental Effects of Pesticides, Monograph No. 40*, M. P. Greaves, B. D. Smith and P. W. Greig-Smith, Eds., British Crop Protection Council, Croydon, 1988, pp. 297–301.
38. Shawyer, C. R., *The Barn Owl in the British Isles*, The Hawk Trust, London, 1987.
39. Murton, R. K., Natural and artificial population control in the wood-pigeon, *Ann. Appl. Biol.*, 55, 177–192, 1965.
40. Murton, R. K., A. J. Isaacson and N. J. Westwood, The relationships between woodpigeons and their clover food supply and the mechanism of population control, *J. Appl. Ecol.*, 3, 55–96, 1966.
41. Murton, R. K., N. J. Westwood and A. J. Isaacson, A study of wood-pigeon shooting: the exploitation of a natural animal population, *J. Appl. Ecol.*, 11, 61–81, 1974.
42. Inglis, I. R., A. J. Isaacson, R. J. P. Thearle and N. J. Westwood, The effects of changing agricultural practice upon woodpigeon *Columba palumbus* numbers, *Ibis*, 132, 262–272, 1990.

43. Potts, G. R., *The Partridge: Pesticides, Predation and Conservation*, Collins, London, 1986.
44. Potts, G. R., and N. J. Aebischer, Control of population size in birds: the grey partridge as a case study, in *Towards a More Exact Ecology*, P. J. Grubb and J. B. Whittaker, Eds., Blackwell Scientific, Oxford, 1989.
45. O'Connor, R. J., and M. Shrubb, *Farming and Birds*, Cambridge University Press, Cambridge, U.K., 1986.
46. Greig-Smith, P. W., unpublished data.

31

Comparative Avian Toxicology: What Is Its Role in Predicting and Monitoring the Effects of Agricultural Pesticides

Edward W. Schafer, Jr.

ABSTRACT

Determining the comparative toxicology of pesticides and other chemical agents for wild bird species (quail, ducks, etc.) is an emerging area of environmental concern. In North America, excluding Central America and Mexico, over 700 wild bird species have been identified as common or occasional residents or visitors. Nevertheless, toxicological effects of most agricultural, industrial, or environmental chemicals are poorly characterized for almost all of these species. In fact, in-depth comparative toxicological studies of chemicals have been reported on less than ten wild bird species (including game birds). These species represent about 1% of all North American species. Where toxicological characterization has occurred, generally it has been limited to testing specific pesticides or groups of pesticides (organophosphates, carbamates, organochlorines, etc.) and not the broader array of potentially or purposefully introduced compounds of environmental concern. Additional research designed to better define the effects of these chemicals on live, intact birds and other warm-blooded animals is unlikely in the future due to increasing restrictions on the use of animals in research. Thus, wildlife toxicologists will have to make the most of previously gathered data and hone their predictive skills. This task will not be easy, and may, for some chemicals and many bird species, be impossible.

Besides these factors, specific bird species of concern to regulatory agencies and industry are usually limited in numbers or availability. In order to provide a means of assessing risk for species of limited availability, regulators and wildlife toxicologists have been seeking ways of extrapolating laboratory test results using available or abundant species, to other species, and/or to field situations. The concepts of surrogate, indicator, and sentinel species have been developed and used with limited success in the field of wildlife toxicology. Variations in chemical intoxication among and within family, genus, and species are large and chemical effects vary widely; associated environmental parameters also vary in quality and quantity. Given the current state of laboratory data and the status of existing databases that can be used to predict laboratory effects, predictions describing potential effects on untested species will have to be tempered by many qualifiers.

In the field, the situation is more complex because available databases are limited and inaccurate. Developing accurate data in the field is difficult and expensive even for a single bird species in a well-defined geographical area with a single chemical. The difficulty of performing a similar study involving several species and multiple sites is magnified many times by orders of magnitude. These difficulties can be overcome by persistence and innovative thinking on the part of regulators and industry. Close cooperation toward meeting common goals is an essential part of this effort. The lead taken by the Avian Effects Dialogue Group and other industry- and regulator-sponsored groups should result in advances in the knowledge and understanding of the complex issues currently facing wildlife toxicologists.

KEY WORDS

toxicology, pesticide, bird, monitoring, prediction

INTRODUCTION

North American Bird Populations

Approximately 650 species of wild birds are known to breed on the North American continent (excluding Mexico and Central America), and another 50 species are regular or occasional visitors.[1] The total of 700 species represents 75 bird families. Estimates of the total number of birds in North America at any given time vary considerably according to seasonal factors; however, although peak numbers are inaccurately known, they are often given as about 20 billion,[1] roughly 50 to 60 birds per human inhabitant. Estimated populations of up to 200 million waterfowl (ducks and geese) and 800 million upland game birds (pheasant, quail, etc.) are present in North America.[2,3] The waterfowl and gamebird species encompass five separate families but probably constitute no more than 5% of the avian populations of North America. The passerines (27 families) probably constitute more than 50% of the total postbreeding North American bird population.[4]

The number of individual birds within a given species or family vary considerably. Some species that are considered in danger of extinction may number less than 100 individuals, whereas an abundant species such as the European starling (*Sturnus vulgaris*) may number as many as 200 to 300 million individuals shortly after the breeding season.[5] Within passerines, the subfamily Icterinae (blackbirds and orioles) probably numbers about 2 billion in the postbreeding season, or up to 10% of the total North American population of birds.[2,5] Of the 13 most abundant species in North America, 5 are Icterines: the red-winged blackbird (*Agelaius phoniceus*), Western meadowlark (*Sturnella neglecta*), common grackle (*Quiscalus quiscala*), Eastern meadowlark (*Sturnella magna*), and brownheaded cowbird (*Molothrus ater*).[6] Other numerous passerines, besides starlings, are the house sparrow (*Passer domesticus*), horned lark (*Eremophila alpestris*), American robin (*Turdis migratorius*), common crow (*Corvus brachyrhynchos*), barn swallow (*Hirundo rustica*), and cliff swallow (*Hirundo pyrrhonota*).[6] Thus, of the 13 most abundant North American species, 12 are passerines. Only one other order, Columbiformes, contains a species on the most abundant list, the mourning dove (*Zenaida macroura*).[6] The northern bobwhite (*Colinus virginianus*), order Galliformes, ranks in the top 20 species, and the mallard (*Anas platyrhynchos*), order Anseriformes, in the top 50 species.[2]

Potential Exposure

All species of birds that inhabit or visit North America are exposed to agricultural, industrial, or other introduced chemicals, as well as naturally occurring chemicals in their food, water, and air. Birds are also ubiquitous throughout North America, feeding and breeding in a diversity of habitats year-round or seasonally. They are relatively large, numerous, readily observable, and usually identifiable as individuals. Birds are also aesthetically and economically important to the human population of North America. Over 125 million Americans are involved in observing or harvesting these animals each year.[7] For these reasons and many others, measuring the numbers, diversity, and incidental mortality of these species has been proposed as a means of monitoring or predicting environmental health, adverse environmental impacts, and species survivability, particularly as it relates to introduced chemical agents, mostly pesticides.[8,9] A number of specific concepts have been proposed to monitor or predict risk. Through 1990, these concepts have been elucidated at meetings of the Avian Effects Dialogue Group, a group convened by The Conservation Foundation at the request of industry and the U.S. Environmental Protection Agency (EPA) to provide guidance in the assessment of pesticide risk to birds.[10]

Species Concepts

The concepts of surrogate, indicator, and sentinel species have been used or described in various ways to allow for the testing or monitoring of small numbers of birds or bird species, and to extrapolate the observed effects to other species in the avian community or to other effects with some degree of confidence.[4,10-13] The definitions that follow delineate these concepts for purposes of discussion.

Surrogate species is an alternative species selected to represent a single species of concern. The surrogate species is usually selected because the species of concern is not readily available for study due to its small number, isolated distribution, legal status, or the physical damage that could occur during capture or holding. Based on physiological and behavioral parameters, the surrogate is assumed to represent the species of concern, to which it normally has close taxonomic associations. For use in the laboratory, a surrogate species must be tractable and available in numbers either through captive breeding or wild-capture techniques. In addition, surrogates used in the field should be readily observable and occupy a similar ecological niche as the species of concern. Thus, the response of the surrogate to applied chemical stress is assumed to be similar in the laboratory and field within a degree of accuracy that is acceptable to the investigator. This definition of surrogate species is more limited than is commonly used in many ecotoxicological studies.

Indicator species refers to a species selected to represent a larger group of species of concern. The indicator species is usually selected because those species of concern are too numerous to study individually, are not readily available for study, or cannot be studied without physical damage. The types of similarities to the species of concern often depend upon the question being posed by the study; physiological, behavioral, or taxonomic similarities may not all be necessary. Primary factors for the use of indicator species are often economics and availability. In the laboratory, an indicator species must be tractable and available in numbers through captive breeding or wild-capture techniques. Similar constraints exist for indicator species used in the field. In addition, the indicator species should have similar exposure potential and occupy a similar ecological niche as the species of concern. Responses of indicator species to applied chemical stress in the laboratory and the field are assumed to be similar to the species of concern; however, the ability to extrapolate results is often limited by the size and composition of the group that contains the species of concern. Indicator species are often proposed for study because of their potential to validate generalized laboratory studies. This definition of indicator species can be broadly applied, but contrary to its use in many ecological studies, its use in this case does not imply ecological monitoring.

Sentinel species is defined as a species of interest selected as a signal of impending potential harm or change to other species or the environment. The sentinel species is normally selected because of its availability, observability, and sensitivity to the toxicological issue of concern. The taxonomic similarity of the sentinel to other species of concern is often irrelevant, as is its similarity of exposure. In most cases, sentinel species are the most toxicologically sensitive and pose the greatest risk of exposure. Sentinel species are not used often in the laboratory investigation of the comparative effects of chemicals on birds, but are generally limited to field use. They may either occur naturally or be specifically placed on site; such species should be readily observable.

The scope of biological extrapolation among birds can be narrow (within a family) or broad (among families or across several orders). It can extend from confined laboratory studies to simulated field or pen studies, field investigations, and often monitoring activities. The scope of toxicological extrapolation can be equally diverse, ranging from acute, subacute, or chronic toxicity, to gross effects on reproduction, to actual or projected effects on individuals or populations.[9] Thus, risk assessment is an immense undertaking, but it has very important and long-term rewards if it can be reliably accomplished. Federal, state, and local government agencies, universities, private industry, and the general public all have a stake in this activity. After decades of study and discussion, the future direction and application of these concepts seem to be slowly progressing.

History

The domestic chicken (*Gallus domesticus*) was probably the first bird routinely used as an indicator for the effects of chemicals on other birds, primarily because of its availability, known husbandry requirements, and economics. It was used initially to determine the acute toxicity of chemicals to birds and continues to be used today to evaluate the veterinary importance of animal drugs for poultry. Finches, canaries, or other small songbirds have been used by miners for hundreds of years as sentinels to detect the presence of toxic gases in mines.

In the 1940s and 1950s, the northern bobwhite and the ring-necked pheasant (*Phasianus colchicus*) replaced the chicken as indicators of bird toxicity, as researchers assumed the responses of these semidomesticated wild birds were more closely related to those of other gallinaceous gamebirds than were those of the chicken. In the 1960s, the coturnix (*Coturnix coturnix*) and the mallard were added to the test species commonly used in the laboratory as indicators of avian toxicity.[4]

In the 1970s, four more species were added to the list of avian test subjects for which relatively large databases had accumulated: the European starling, red-winged blackbird, house sparrow, and house finch (*Carpodacus mexicanis*).[14-16] Since then, no significant published additions have been made to avian toxicology databases, although additional information has accumulated in the files of the U.S. EPA. Pesticide producers have contributed to this database by meeting the data requirements specified by the EPA under its regulatory authority over the registration and reregistration of pesticides. Primary EPA test species are northern bobwhite and mallard, which are used as indicator species of general avian toxicity in order to predict risks of environmental concern. Most other avian toxicology databases that have been assembled are specific to individual species or chemicals, or combine data from numerous sources and methods.

Comparative Avian Toxicology

Published studies covering the comparative acute toxicity of chemicals to birds are limited but available. Tucker and Haegele[17] summarized their laboratory studies, designed to determine whether the indicator species concept was valid, using 6 bird species (from 4 families) and 16 pesticides.[17] By using a ranking procedure, they showed that no significant differences in sensitivity existed among the species tested, but that large variation existed among species with respect to an individual pesticide. However, there was a significant difference in sensitivity between the chukar partridge (*Alectoris chukar*) and the house sparrow near the $p = 0.05$ level. As a result of their study, the authors concluded that if it was necessary to know the toxicity of a specific pesticide on a bird species, then "... the species itself should be used ... to avoid the need for extrapolation." Additional data from these authors and Hudson et al.[18] are also available in a handbook describing the acute effects of a large number of pesticides to many additional animal species.

In a 1967 study, data gathered by this author and others suggested that red-winged blackbirds were more sensitive than starlings to intoxication from 22 carbamates.[14] In 1972, using 7 species covering 4 families and 148 chemicals from a variety of sources, it was concluded that the ring-necked pheasant was significantly less sensitive than other species, and the house finch and the red-winged blackbird were significantly more sensitive than the remaining species, including mallards and pheasants.[15] Also in 1972, the results of comparative testing of 369 chemicals showed that red-winged blackbirds were significantly more sensitive to intoxication than starlings.[16] In 1973 a significant correlation between the oral or dermal acute toxicity of 17 chemicals to red-winged blackbirds, house sparrows, and red-billed quelea (*Quelea quelea*) was demonstrated.[19] In 1979, a study involving the toxicity of 36 pesticides from 3 different classes on 6 bird species covering 4 families showed that starlings were significantly less sensitive to the pesticides than the remaining 5 species. Again, red-winged blackbirds and sparrows were the most sensitive species tested. However, the authors concluded that "... the basis for the present use of indicator bird species for estimating or predicting hazards to other species of wild birds is not well established".[4] At the Second Annual Meeting of the Society of Environmental Toxicology and Chemistry (SETAC) in 1981, a presentation described the acute toxicity of 63 monosubstituted anilines and pyridines to 4 bird species (2 families) and 2 species of mammals.[20] This presentation, while not published outside the proceedings, showed that although generalized structure activity relationships exist for many of the tested chemicals, there is considerable variation between species. Definitive structure activity relationships between species were also questioned. In 1982, a comparative study of the acute toxicity and repellency of three chemicals to five African bird species (same family) showed that toxicological responses were similar only within individual chemicals.[21] In 1983 an acute toxicity study involving 130 chemicals and 3 species representing 3 families of birds concluded that red-winged blackbirds were significantly more sensitive than coturnix and starlings, and that the latter were not different.[22] Repellency (a physiological or behavioral measure) was not correlated with toxicity.

Only a limited number of studies have been reported on the comparative subacute intoxication of birds by chemical agents. Two early papers discussed the relationships between four species of young birds representing two families. Toxicological sensitivity was correlated with body size and chemical structure, but not species. However, the authors reported that "inconsistencies in the relative sensitivity of the four species . . . suggest that other species would be much more sensitive to different chemicals".[23,24] In a third related study, four pesticides were tested on four species (two families) and showed that passerines were considerably more sensitive to intoxication than members of Galliformes. This sensitivity was also thought to be a function of body weight.[25]

DISCUSSION

Literature sources indicate that it is not unusual for toxicity levels to differ by two to three orders of magnitude among orders of animals. Particularly true when data are collected from a number of unrelated sources, such differences can be due to test methodology; variations among family, species, and individual responses to chemical agents; and data interpretation.[26] Thus, although most environmental toxicologists agree on the need for concepts that will allow for the broad extrapolation of data to assist in providing accurate risk assessments, there are questions as to how accurately these data can be predicted.

In order to graphically illustrate the multitude of problems that occur when one tries to extrapolate data based on generalizations to specific chemicals or species, the data on acute avian toxicity collected by the Denver Wildlife Research Center was re-examined. These data are characterized by a broad spectrum of chemicals (more than 2500) and involve 45 species of birds from 15 families. It was gathered by the same individuals over a 20-year period using similar methods and the same facility. From this set of data, information was extracted for chemicals with acute toxicity data on at least ten bird species covering at least five families.[4,18,22] Where more than one data point (LD_{50}) was available for the same species, the lowest (most toxic) value for physically mature birds was used. Three specific areas not covered by other authors were targeted: toxicity distribution for individual chemicals among species or families (i.e., what is the range and distribution of toxicity for a given chemical); the location of recognized indicator, surrogate, or sentinel species on the toxicity spectrums of selected chemicals (i.e., how well do these species predict toxicity on a chemical-by-chemical basis); and the toxicity distribution among species or families of closely related chemical analogues (i.e., how well can data from a given chemical be used to predict toxicity for a closely related chemical across species or families).

Species and Family Toxicity Distribution

Figures 1 through 5 show the toxicity distributions for five chemicals representing four different chemical classes. They are examples of the species/family distribution of toxicity as well as the magnitude variation encountered in such a distribution. The figures display the information for both species and families for which the toxicity values (LD_{50}) fall within an artificially established order of magnitude range (i.e., 1 to 10 mg/kg, 11 to 100 mg/kg, etc.). The chemicals shown are 4-aminopyridine (Figure 1), aprocarb (Figure 2), fenthion (Figure 3), methiocarb (Figure 4), and Starlicide (Figure 5). These figures show that all of the toxicity distributions for five chemicals ranged over two to three orders of magnitude. This information is similar to that elucidated by Tucker and Leitzke[26] and shows the tremendous range of values that can occur even when using data sets with limited variables. Although the actual ranges of toxicity are overstated by this presentation, Figure 6 shows that the actual minimum range (represented by 4-aminopyridine) is slightly more than a single order of magni-

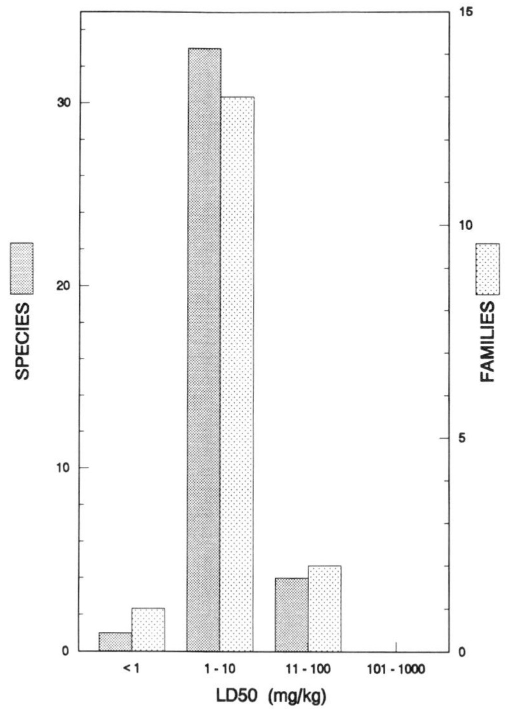

FIGURE 1. Acute oral toxicity of 4-aminopyridine to 13 families of birds.

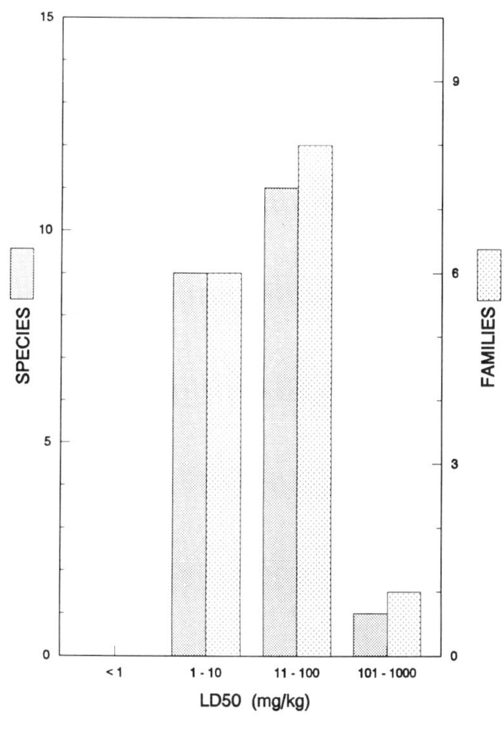

FIGURE 2. Acute oral toxicity of aprocarb to 11 families of birds.

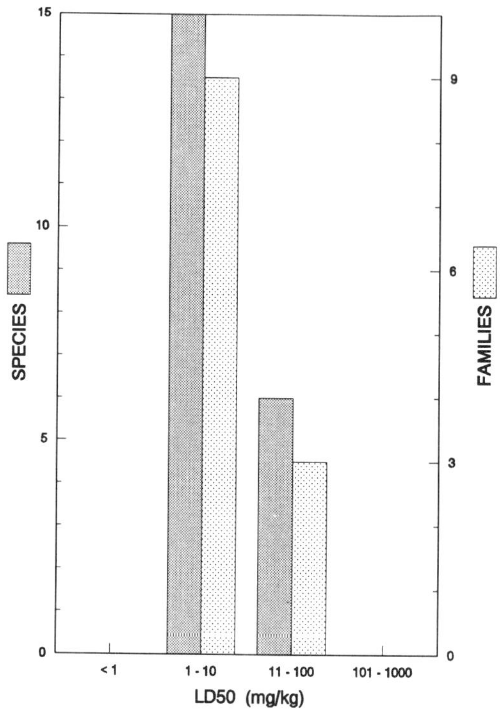

FIGURE 3. Acute oral toxicity of fenthion to 10 families of birds.

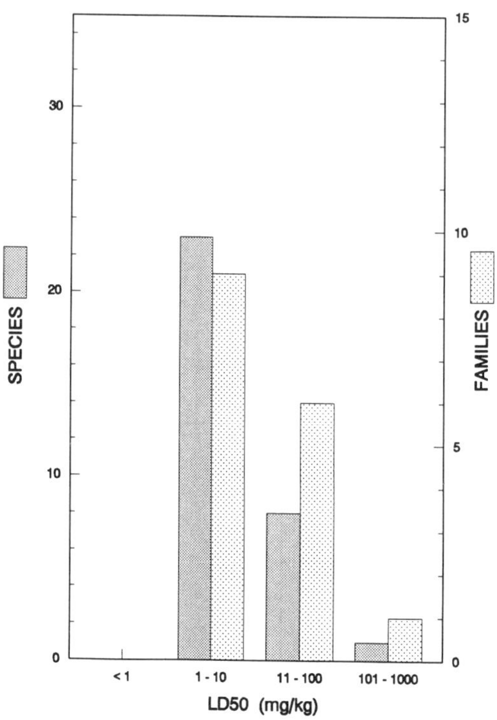

FIGURE 4. Acute oral toxicity of methiocarb to 13 families of birds.

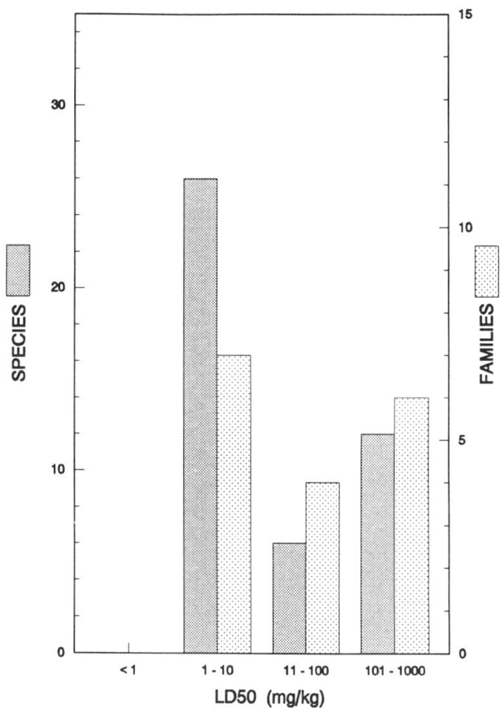

FIGURE 5. Acute oral toxicity of Starlicide to 15 families of birds.

tude, and the maximum (Starlicide) is very close to three orders of magnitude. Accepting such variation and applying those data to risk assessments could therefore result in either over- or underestimating risk by factors of up to 1000. Such estimation errors are not trivial, even if extrapolation factors are used in the risk assessment process to accommodate such variation.

Data associated with Figures 1 through 5 provide good examples of the differences between bird responses to different chemicals. The chemicals represented by these data are unique in that a single source existed for data on more than ten species of wild birds. Most pesticides have similar data for only two to five species, and often those data are generated by more than one laboratory.

Predictions Based on Sensitivity

Environmental toxicologists have been concerned about doing a better job of assessing risk by producing data on specific species that are very sensitive or very insensitive to chemical intoxication. Using moderately small to large data sets covering a number of chemicals and species, researchers have shown that some species exhibited a predictable sensitivity to chemical intoxication, and that such predictability could be used for risk assessments of specific chemicals. Such broad-scale generalizations assume, however, that predictability is not related in any meaningful way to chemical structure. They further assume that the structures represented in the sample used to establish a significant effect also represent other nontested chemical structures.

The fallacy of these assumptions can be demonstrated by examining Figure 6. This figure displays a comparison of the toxicity ranges for the five chemicals covered in the first five figures plus two additional chemicals for which bird toxicity data on at least

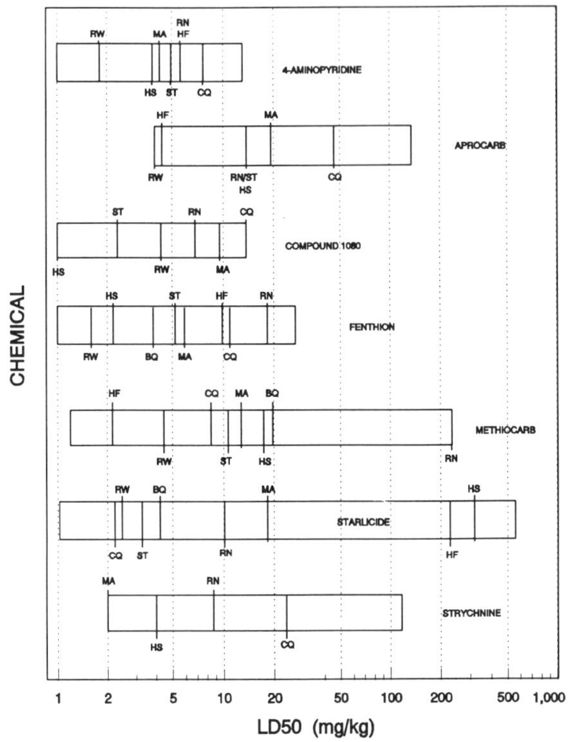

FIGURE 6. Location of selected species on a bar representing the toxicity range for seven chemicals. (See text for species abbreviations.)

ten species were available. The bar for each chemical represents the defined avian toxicity range for that chemical; lines and corresponding initials show the locations in this range of the toxicity values for eight avian species most often used as indicator species in North America. The red-winged blackbird (RW), house sparrow (HS), and house finch (HF) generally have been considered "sensitive" species, and the starling (ST), ring-necked pheasant (RN), and coturnix (CQ) as relatively "insensitive" species. The northern bobwhite (BQ) and mallard (MA), used in most EPA-mandated tests as indicators of avian toxicity, are intermediate in their sensitivity. Careful examination of the data for each chemical does not show a single case in which the three "sensitive" species are all more susceptible to intoxication than the four "insensitive" species. In fact, considerable overlap exists, and in some cases an "insensitive" species is the most impacted and a "sensitive" species is the least impacted. Not only is there a lack of pattern on a chemical by chemical basis, but the pattern changes dramatically among chemicals. If these data are representative of other potential data sets for other pesticides or industrial chemicals, it would be extremely difficult to justify the use of any one species as a surrogate for another or as an indicator of a larger group (e.g., coturnix for gallinaceous birds, red-winged blackbirds for passerines). Also, it would be difficult to select a sentinel species of known high sensitivity to chemical intoxication that can serve as a warning of impending chemical harm. The only way that this could be effectively accomplished is to accept that whatever predictions were going to be made could be off by two to three orders of magnitude in either direction. Thus, although general predictions can be made over large numbers of chemicals and a variety of chemical classes, if one examines the application of those predictions to individual chemicals, they do not hold up.

Predictions Based on Chemical Grouping

Figure 7 illustrates another data set covering 21 organophosphates, 9 carbamates, and 4 organochlorine pesticides that were tested by a single laboratory and investigator using 6 bird species, then ranked by order of sensitivity. Four of these species were included in the discussion of Figure 6, two being "sensitive" (red-winged blackbird and house sparrow) and two being "insensitive" (starling and coturnix). When these data were analyzed over all 35 chemicals (one chemical was tested that did not fall into any of the three classes), starlings were significantly less sensitive than all others to acute intoxication, and red-winged blackbirds (and probably house sparrows) were significantly more sensitive. If one considers each chemical group separately, these conclusions fail to hold for organochlorines. Red-winged blackbirds and starlings do not differ in their relative sensitivity to organochlorines; however, sparrows and coturnix do. The responses to carbamates and organophosphates match the overall response for all species.

Figure 8 presents the responses of starlings to the organochlorines in the previous data set. Starlings respond to organochlorines in a much different manner than they do to carbamates or organophosphates. Figure 9 shows data for the same chemicals for house sparrows; although sparrows appeared to respond similarly to organophosphates and carbamates but not organochlorines in Figure 7, they actually respond differently to all three chemical groups. Sparrows are highly responsive to organochlorines, moderately responsive to organophosphates, and much less responsive to

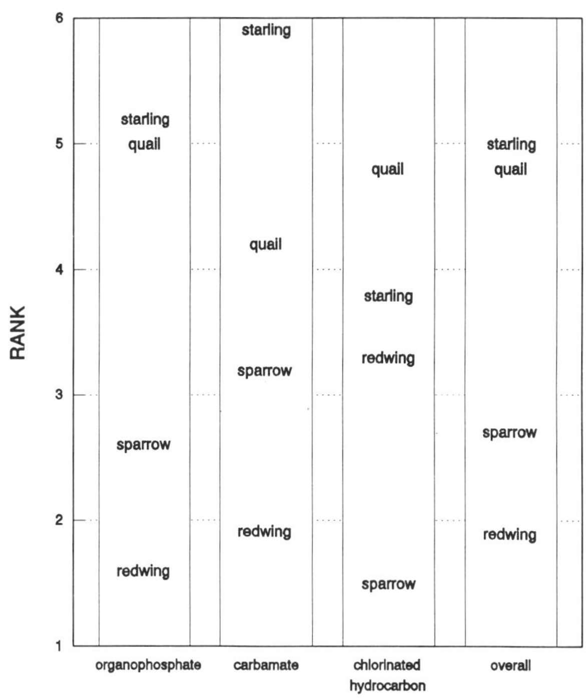

FIGURE 7. Comparison of ranking the acute oral toxicity sensitivity of four species of birds to three types of pesticides (1 = most toxic, 6 = least toxic).

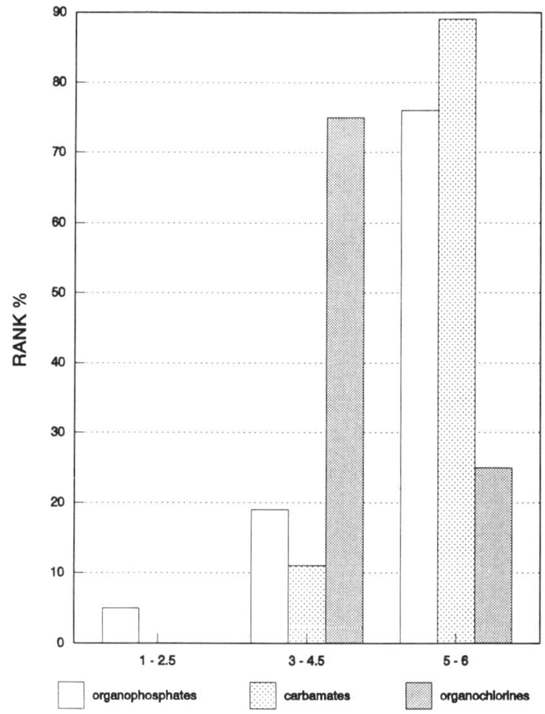

FIGURE 8. Distribution of the ranks of acute oral toxicity sensitivity of starlings to three types of pesticides.

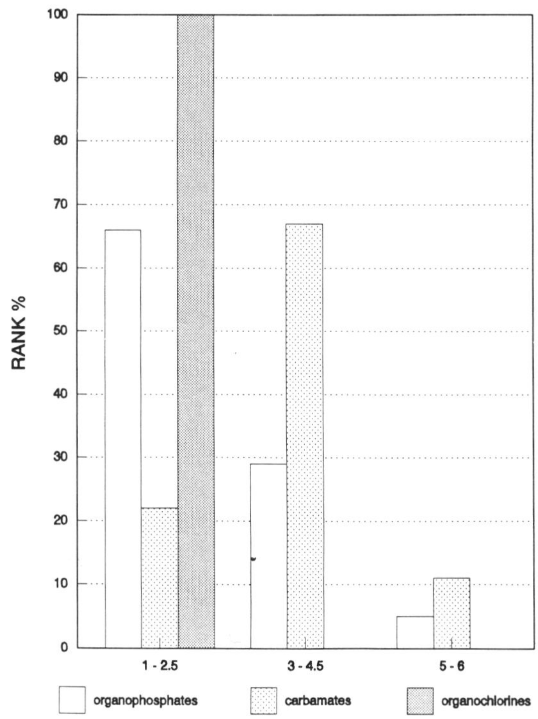

FIGURE 9. Distribution of the ranks of acute oral toxicity sensitivity of house sparrows to three types of pesticides.

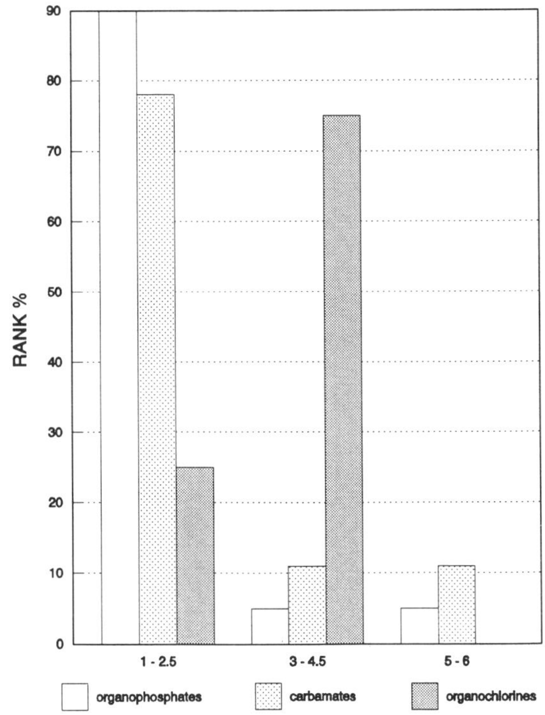

FIGURE 10. Distribution of the ranks of acute oral toxicity sensitivity of red-winged blackbirds to three types of pesticides.

carbamates. Figure 10 examines the same data set for red-winged blackbirds; however, the pattern is the reverse of that for starlings: red-winged blackbirds are highly sensitive to carbamates and organophosphates and only moderately responsive to organochlorines. Data on coturnix, although not presented, failed to show any difference in responses between any of the three chemical groups. Thus, practitioners using the concepts of indicator, surrogate, and sentinel species to predict results on individual or grouped species may have to accept a two to three order of magnitude difference in responses to individual chemicals. Compounding this lack of precision, they may also face an additional one to two order of magnitude difference in response if the chemical in question belongs to a group that the surrogate, indicator, or sentinel species responds to in a way not predicted by generalized sensitivity rankings.

Predictions Based on Chemical Structure

Many wildlife toxicologists would like to use the concepts of surrogate and indicator species to predict the toxicity of a particular chemical for which data are not available, based on the results for one or more closely related compounds. Such predictions are routinely made in the drug industry with a relatively high degree of accuracy because sufficient comparative data are available for the species of interest using closely related isomers. However, when the attempt is made to go from one wild species for which data exist to another wild species for which no data exist, results could be misleading. For example, in Figures 11 to 13, the author has extracted a limited amount of data from a much larger data set. In this study, 21 groups of 3 isomers of monosubstituted anilines or pyridines were tested for toxicity on 4 bird species. The study showed that starlings were the least sensitive, coturnix and house sparrows were similar in their

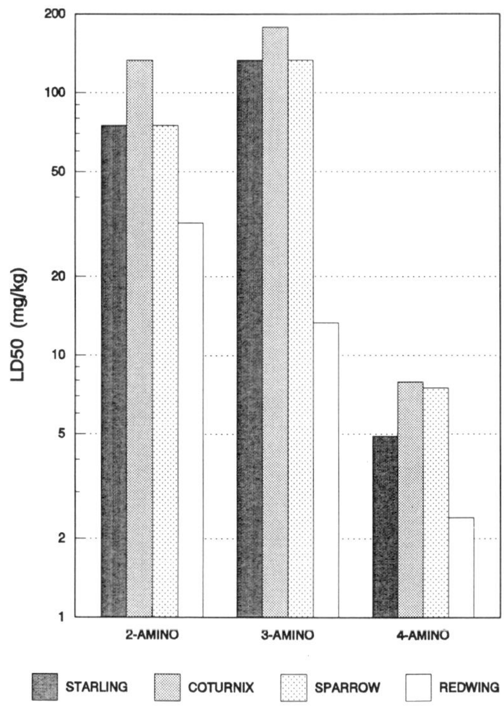

FIGURE 11. Acute oral toxicity of 2-, 3-, and 4-pyridineamines to four species of birds.

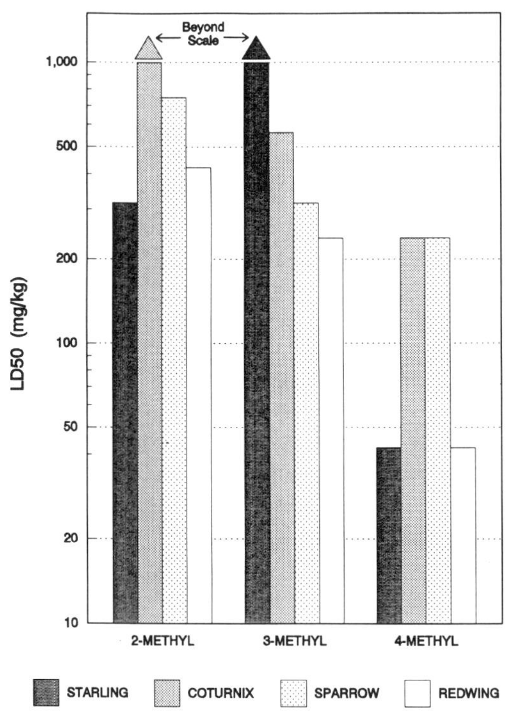

FIGURE 12. Acute oral toxicity of 2-, 3-, and 4-methylbenzenamines to four species of birds.

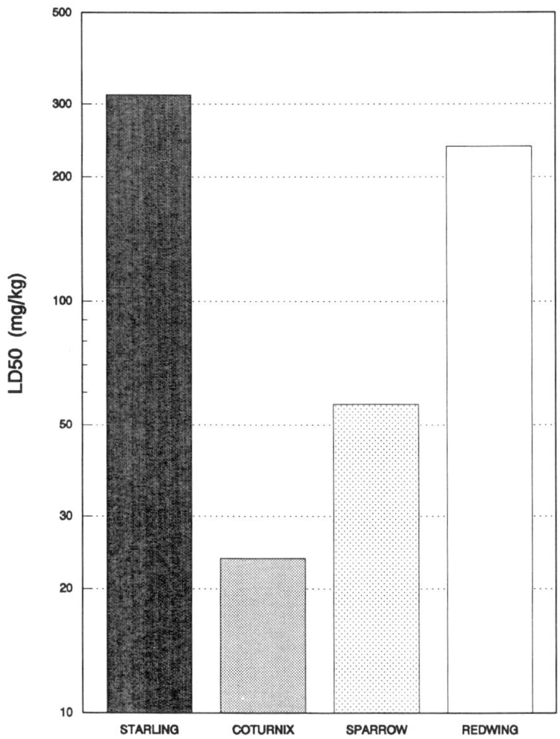

FIGURE 13. Acute oral toxicity of 4-aminobenzonitrile to four species of birds.

sensitivity to all the chemicals tested, and red-winged blackbirds were the most sensitive. It was also shown that for monosubstituted anilines across all species of birds, isomers were more toxic when proceeding structurally from the 2 to the 3 to the 4 isomer, and that the order of toxicity for monosubstituted pyridines increased from the 3 isomer to the 2 isomer to the 4 isomer. Although many of the data groups confirmed this general finding, there were numerous exceptions. For example, Figure 11 shows the results of testing three pyridineamine isomers. For these three chemicals, the 2-amino isomer is more toxic to three of the four species than the 3-amino isomer, but for one species (red-winged blackbird) the 3-amino isomer is more toxic than the 2-amino isomer. Although starlings were generally less sensitive to pyridineamine isomers than the three other species, coturnix are less sensitive than starlings; sparrows and starlings have a similar sensitivity compared to the more typical pairing of sparrows and coturnix. Although these differences are not large, they could be important if predictions of toxicity of a single isomer were being based on toxicity of another isomer. Figure 12 shows a similar situation of sensitivity/species reversals for methylbenzenamines. In this case, starlings are the most susceptible to intoxication from the 2-methyl isomer rather than the 3-methyl isomer, and starlings are similarly sensitive to intoxication by the 4-methyl isomer as are red-winged blackbirds. Figure 13 shows a similar reversal in species sensitivity to 4-aminobenzonitrile. In this case, if red-winged blackbirds were assumed to be the most sensitive species, toxicity estimates for coturnix and sparrows based on red-winged blackbird data could be off by at least one order of magnitude. Thus, the assumption about similar sensitivity of species to intoxication by isomers of chemicals within specific groups could be off by at least one order of magnitude.

RECOMMENDATIONS

At this time when producers, users, and regulators of chemicals are being asked to refine and improve their predictions of environmental risk, the tools available to improve the accuracy of models, to better define toxicological relationships, and to establish a basic understanding of the toxicological response of birds to introduced chemicals are further from their grasp. The accuracy of models is only as accurate as the information that goes into the models; if that information is incomplete or flawed, then the results will also be incomplete or flawed. Toxicologists need to address four questions that impact the future of comparative avian toxicology:

1. How good an estimate of potential environmental risk is acceptable and practical?
2. How can the scientific community access the limited amount of comparative avian toxicity data on wild bird species and use those data to support risk assessments, even though the data were not gathered under Good Laboratory Practice standards?
3. How can tests that were developed to generate comparative toxicity data be used for predictions on a chemical by chemical basis?
4. What do all these data mean when applied to the field and can laboratory tests be validated in the field?

These issues will become more critical with the public's increasing desire to be more informed on environmental issues and to minimize the environmental impact of chemicals. Perhaps it is time to qualify risk assessments by acknowledging that they give only crude, "ballpark" estimates and recognize the need for a type of directed post-registration monitoring to uncover "real" environmental risks.

REFERENCES

1. Robbins, C. S., B. Brun and H. S. Zim, *Birds of North America*, Golden Press, New York, 1966.
2. Robbins, C. S., D. Bystrack and P. H. Geissler, The Breeding Bird Survey: Its First Fifteen Years, 1965–1979. Resource Publ. No. 157, U.S. Fish and Wildlife Service, Washington, DC, 1986.
3. Benning, D. S., M. M. Smith and S. L. Rhoads, Waterfowl Status Report, 1973, Spec. Sci. Rep. Wildl. No. 188, U.S. Department of Interior, Washington, DC, 1975.
4. Schafer, E. W., Jr., and R. B. Brunton, Indicator bird species for toxicity determinations: is the technique usable in test method development?, in *Vertebrate Pest Control and Management Materials*, ASTM STP680, J. R. Beck, Ed., American Society for Testing and Materials, Philadelphia, 1979, pp. 157–168.
5. Meanley, B., and W. C. Royall, Nationwide estimates of blackbirds and starlings, in *Proc. 7th Bird Control Seminar*, Bowling Green State University, Bowling Green, OH, November 9 to 11, 1976, pp. 39–40, 1977.
6. Dolbeer, R. A., and R. A. Stehn, Population status of blackbirds and starlings in North America, 1966 to 1981, in *Proc. 1st Eastern Wildlife Damage Control Conference*, Ithaca, NY, September 27 to 30, 1983, pp. 51–61.
7. 1985 National Survey of Fishing, Hunting and Wildlife Associated Recreation, U.S. Department of Interior, Fish and Wildlife Service, Washington, DC, 1988.
8. Karr, J. R., Biological monitoring and environmental assessment: a conceptional framework, *Environ. Manage.*, 11, 249–256, 1987.
9. Lower, W. R., and R. J. Kendall, Sentinel species and sentinel bioassays, in *Use of Biomarkers for Assessment of Environmental Contaminant Exposure*, L. Shugart and J. McCarthy, Eds., CRC Press, Boca Raton, FL, 1990, pp. 309–331.
10. Avian Effects Dialogue Group, *Pesticides and Birds: Improving Impact Assessment*, The Conservation Foundation, Washington, DC, 1989.

11. *Proc. USEPA Workshop on Terrestrial Toxicology*, Society of Environmental Toxicology and Chemistry, Rockville, MD, March 25 to 27, 1981.
12. Glennon, J. P., Surrogate Species Workshop: Workshop Report, Lifesystems, Inc., Cleveland, July 27 to 28, 1981; July 20 to 21, 1982.
13. Glennon, J. P., Testing Triggers Workshop: Workshop Report, Lifesystems, Inc., Cleveland, July 22 to 23, September 1 to 2, 1982.
14. Schafer, E. W., Jr., R. I. Starr, D. J. Cunningham and T. J. DeCino, Substituted phenyl *N*-methyl carbamates as temporary immobilizing agents for birds, *J. Agric. Food Chem.*, 15, 287–289, 1967.
15. Schafer, E. W., Jr., and D. G. Cunningham, An Evaluation of 148 Compounds as Avian Immobilizing Agents, Spec. Sci. Rep. Wildl. No. 150, Fish and Wildlife Service, Washington, DC, 1972.
16. Schafer, E. W., Jr., The acute oral toxicity of 369 pesticidal, pharmaceutical and other chemicals to wild birds, *Toxicol. Appl. Pharmacol.*, 21, 315–330, 1972.
17. Tucker, R. K., and M. A. Haegele, Comparative acute oral toxicity of pesticides to six species of birds, *Toxicol. Appl. Pharmcol.*, 20, 57–63, 1971.
18. Hudson, R. H., R. K. Tucker and M. A. Haegele, *Handbook of Toxicity of Pesticides to Wildlife*, Resource Publ. No. 153, Fish and Wildlife Service, Washington, DC, 1984.
19. Schafer, E. W., Jr., R. B. Branton, N. F. Lockyer and J. W. DeGrazio, Comparative toxicity of 17 pesticides to the quelea, house sparrow and red-winged blackbird, *Toxicol. Appl. Pharmacol.*, 26, 156–157, 1973.
20. Schafer, E. W., Jr., and P. J. Savarie, The relationship of toxicity to structure of 63 monosubstituted anilines and pyridines tested in 6 animal species, in *Proc. 2nd Annu. Meet. the Society of Environmental Toxicology and Chemistry*, Arlington, VA, 1981, p. 84.
21. Shefte, N., R. Bruggers and E. W. Schafer, Jr., Repellency and toxicity of three bird control chemicals to four species of African pest birds, *J. Wildl. Manage.*, 46, 453–475, 1982.
22. Schafer, E. W., Jr., Acute oral toxicity, repellency and hazard potential of 998 chemicals to one or more species of wild and domestic birds, *Arch. Environ. Contam. Toxicol.*, 12, 355–382, 1983.
23. Heath, R. G., J. W. Spann, E. F. Hill and J. F. Kreitzen, Comparative Dietary Toxicities of Pesticides to Birds, Spec. Sci. Rep. Wildl. No. 152, U.S. Fish and Wildlife Service, Washington, DC, 1972.
24. Hill, E. F., R. G. Health, J. W. Spann and J. D. Williams, Lethal Dietary Toxicities of Environmental Pollutants to Birds, Spec. Sci. Rep. Wildl. No. 191, U.S. Fish and Wildlife Service, Washington, DC, 1975.
25. Hill, E. F., Toxicity of selected mosquito larvacides to some common avian species, *J. Wildl. Manage.*, 35, 757–762, 1971.
26. Tucker, R. K., and J. S. Leitzke, Comparative toxicology of insecticides for vertebrate wildlife and fish, *Pharmacol. Ther.*, 6, 167–220, 1979.

32

Monitoring Avian Populations for Pesticide Exposure and Effects: Uses and Cautions for Ecological Modelers

Sarah Gerould

ABSTRACT

Monitoring data can provide essential information in the creation and use of ecological models for assessing the effects of pesticides on avian populations. Information from monitoring studies can be used to formulate the original concept and structure for the model, and to help identify important parameters, estimate parameters and check model predictions. Care must be exercised in using monitoring information for validation of models because of hidden biases and unmeasured factors. Monitoring programs can help identify problems that may benefit from analysis using an ecological model. For each of these uses to be maximized, monitoring data must be available to modelers, and individuals involved in monitoring must be aware of how models can be used.

KEY WORDS

monitoring, models, avian, pesticides, biomarkers

INTRODUCTION

At each step in the creation of an ecological model for assessing the effects of pesticides on avian populations, monitoring data can provide essential information. Information from monitoring studies can be used to develop the original concept for the model, estimate parameters in the model, and check model predictions. For this information to be used, monitoring data must be available to modelers, and individuals involved in monitoring must be aware of how models can help them extract information from their data.

PROBLEM IDENTIFICATION

Models can also be useful in sorting out possible causes of declining populations documented by monitoring programs. Monitoring programs can identify problems that may benefit from analysis using an ecological model. Documentation of reductions in fledging success by population-monitoring programs and information showing substantial mortality gathered in incident-monitoring programs are examples of results indicating a possible population or community effect, but careful verification requires a modeling approach. Monitoring programs often trigger comprehensive investigations because they identify problems that require more extensive information.

The identification of sites with previously unknown problems is commonly cited as a goal of national monitoring programs. Although problem identification is often a direct benefit of monitoring programs, those interests may be more effectively served by a comprehensive one-time survey rather than repeated monitoring with a more limited suite of persistent chemicals. If trends in wildlife populations and health are the stated goal, however, repeated measurement of biomarkers and population indicators is essential.

MODEL STRUCTURE

Monitoring information can be used to establish those factors important to consider in structuring models. For instance, population monitoring can provide critical information on differential utilization of crops by different species. This information could be valuable in developing crop by crop, or chemical by chemical regulatory priorities, prioritization schemes that have been suggested to the U.S. Environmental Protection Agency (EPA) for regulation of pesticides. Monitoring of untreated sites can also provide a baseline for comparison with treated sites where birds are affected.

Population modeling studies have been useful in identifying variables of importance to population structure as well as factors that control population levels. For example, many population models include density-dependent parameters. O'Connor (Chapter 29) discussed the importance of considering density dependence in the monitoring of bird populations. In his analysis of species-oriented monitoring studies, he showed that 50% of the populations studied exhibited density dependence. Monitoring can provide data to determine whether populations are regulated in a density-dependent manner, but population monitoring must be carefully planned so that other issues affecting the importance of density dependence are considered when assessing the impact of pesticides on bird populations. Marginal habitat, variable utilization of habitat types, presence of nonbreeding members of the population, and other ecological and environmental factors may affect the importance of density dependence. In turn, density dependence mediates population-level effects of pesticides that cause mortality, affect reproduction, and alter suitability of habitat and food availability. Timing of both monitoring and pesticide application may be critical to understanding issues of density dependence, particularly when pairing, dispersal, and investigation of new habitats occur only at certain times of the year. Monitoring studies involving pesticide residues can also provide information about the avian exposure routes, such as ingestion of contaminated food and contact with contaminated environmental media.

MODEL PARAMETERIZATION

Monitoring information can help modelers determine which parameter values to include in a model. Monitoring of population cohorts can be used as input to Leslie matrix population models (Slade, Chapter 19) in order to predict population levels as cohorts move through time. Monitoring information on contaminant concentrations in biotic and abiotic media can be used in models such as those described by Bird (Chapter 16). Monitoring is a necessary element in many statistical models, and has been used effectively in management of waterfowl by the United States Fish and Wildlife Service (U.S. FWS).[1] Biomarkers can also provide potentially useful information for model parameterization in population models. Nonetheless, biomarkers are an indirect measure of effects on populations, and integration of biomarker information into population models demands care and creativity.

Monitoring programs are increasingly adopting biomarkers as a source of information on sublethal effects under field conditions. The availability and sophistication of tools for monitoring sublethal effects has expanded considerably in recent years, and there is also great potential for using such tools for monitoring exposure. For example, genetic damage has been used to indicate exposure in a variety of organisms (McBee, Chapter 36).

More work is needed in the area of interpretation in order to fully realize the potential usefulness of biomarkers. Even the interpretation of cholinesterase (ChE) inhibition, the most extensively used biomarker for carbamate and organophosphate pesticides, is not always straightforward. Although ChE inhibition has been used to quantify and relate different routes of exposure in controlled settings,[2] its use is not yet routine in large-scale monitoring programs other than as a forensic tool in documenting exposure and determining cause of death. One reason may be the physiological response typical of ChE inhibition. For most organophosphate pesticides, inhibition increases as the exposed animal metabolizes the pesticide to the most toxic form, and decreases as the animal recovers. Timing of sampling is critical, and it is impossible to determine from one or two measurements whether levels are rising or falling.[5] For this reason, strict correlations between pesticide exposure and field pesticide concentrations are not straightforward, especially when timing of application is uncontrolled or unknown. As Hooper has pointed out (see Chapter 5), additional information is also needed in relating biomarkers to lethality and nonlethal regulatory endpoints, such as reproduction. Despite the difficulties of using some biomarkers, they should not be overlooked in providing information for models that couple chemical fate with ecological effects. Ideally, terrestrial models will be developed that can relate fate and effects, similar to those available for aquatic systems.[3]

VALIDATION AND INTERPRETATION

In an earlier chapter, Barton cautioned that before a model can be used in a regulatory setting it must be validated. Monitoring programs can be used to provide a source of data for independent validation of model predictions. In the study proposed in this volume by Grue et al. (see Chapter 51), an agricultural wetland experiment site could be used for parameterizing the model. Long-term data from more than 30 years[1] of population monitoring could be used to test the model prediction for validation purposes. As Fairbrother pointed out in Chapter 25, ongoing comprehensive regional programs on the Chesapeake Bay and the Great Lakes as well as in other countries can also provide detailed information on populations and effects.

Care must be exercised in using monitoring information for validation of models. Many factors, most unmeasured, may influence the health or size of populations. Monitoring data may be biased unexpectedly by such factors as the choice of sites. The location and number of sites are often selected for reasons relating to budget, politics, and ease of sampling rather than biology. When using national status and trends programs as sources of information on geographic variability, the modeler must be cognizant of the method by which sites were chosen. In addition, Hallam (Chapter 49) cautions that if large-scale monitoring information is used to assess national status and trends, the regulator or national policymaker must be aware also that toxicology works at the level of the individual and its local environment, not the nation. For that reason, averaging must be done on the scale of the individual within its local environment.

Biases are prevalent in incident monitoring programs, which rely largely on uncontrolled data collection by the general public, and incorporate sociological biases that may overwhelm any effort at cross-chemical or cross-species comparisons. Thus, caution is required when interpreting data from incident monitoring programs developed for the purpose of validating predictions of risk, or when generating an index of relative susceptibility of bird populations. Cause of death must be carefully determined for each bird mortality, eliminating the possibility of factors such as disease and electrocution. Observer bias is an assumed part of incident monitoring programs. As Greig-Smith points out in Chapter 30, incident monitoring programs rely on (1) chance and timely encounter by the pesticide user or the public before the carcass is scavenged or too decomposed for analysis, (2) willingness of the observer to report mortality, and (3) knowledge of where to report it. Large, visible, and popular species are more likely to be observed and reported than smaller, obscure species. Incident monitoring, therefore, works best for comparing those pesticides that are available and toxic to such species. Legal pesticide use cannot be assumed. Despite these problems in comparing pesticides and species, incident monitoring programs have the advantage of generating information under realistic field conditions, and are a useful early warning system or final check for regulators of pesticides.

INTERFACE WITH USER GROUPS

The value of monitoring programs in triggering further investigation, whether ecological modeling, field studies, or forensic work, depends on the availability of monitoring information to those who can use it. Regulators of pesticides require information on the effects of pesticide use on avian populations. An example of a successful integration of population monitoring, modeling, and management is the use of population monitoring and modeling to set hunting regulations and develop national policy and programs for waterfowl management. The U.S. FWS uses the results of the May Waterfowl Breeding Population and Habitat Survey to determine allowable harvest levels.[1,4] Birds with declining populations are targeted for protection and more intensive management.

In order for models to be successfully integrated into the regulatory process, regulators of pesticides should be actively involved in the initial development of monitoring programs and associated ecological models. Modeling of pesticide fate in the environment is currently used as a basis for regulatory decisions. Ecological modeling has great potential to augment this regulatory process, especially if it is coupled with directed monitoring in field studies. These studies, already required for registration of some pesticides, can supply information needed by modelers on exposure and effects, and further reduce the uncertainties of model predictions.

REFERENCES

1. U.S. Fish and Wildlife Service, SEIS 88: Final Supplemental Environmental Impact Statement: Issuance of Annual Regulations Permitting the Sport Hunting of Migratory Birds, U.S. Fish and Wildlife Service, Washington, DC, 1988.
2. Driver, C. J., M. W. Ligotke, P. Van Voris, B. D. McVeety, B. J. Greenspan and D. Drown, Routes of uptake and their relative contribution to the toxicological response of northern

bobwhite (*Colinus virginianus*) to an organophosphate pesticide, *Environ. Toxicol. Chem.*, 10(1), 21–33, 1991.

3. Hallam, T. G., R. R. Lassiter, J. Li and W. McKinney, Toxicant-induced mortality in models of *Daphnia* populations, *Environ. Toxicol. Chem.*, 9, 597–621, 1990.
4. Reynolds, R. E., R. J. Blohm, F. A. Johnson and J. B. Bortner, 1990 Status of Waterfowl and Fall Flight Forecast, Canadian Wildlife Service and U.S. Fish and Wildlife Service, Laurel, MD, 1990.
5. Hill. E. F., personal communication.

33

Applying Life History, Food Web, Population Viability, and Genetic Models to Pesticide Studies

Thomas E. Lacher, Jr.

Previous chapters have discussed the use of a number of mathematical and matrix-based approaches to modeling biological processes in an ecotoxicological context.[1-4] The four chapters that follow have no single unifying theme; rather, they present additional approaches and amplify the breadth of perspectives available. Topics covered vary from simple demographic models to whole ecosystem models. Each approach demonstrates different yet important applications for modeling ecotoxicological processes.

Simple demographic models have computational advantages and are easily modified to include information on the effects of pesticides on survival or reproduction. They are the closest thing to "ready-made" models for ecotoxicology. Meyer and Boyce[5] use demographic exponential growth models (both age- and stage-structured) to examine the effects of a chemical perturbation on variance in per capita growth rates. Impacts that lower survival or reproduction will increase the variance in per capita growth rates, thereby increasing the probability of extinction. Long-lived species with low fecundity are more seriously affected when adult survivorship is depressed; short-lived species are more sensitive to effects on juvenile survivorship and fecundity. These results agree with previously published analyses[6] using a different modeling approach. These conclusions will allow wildlife toxicologists to begin to focus their research efforts on the appropriate parameters.

Belovsky[7] applied optimal foraging theory to assess possible ecological effects of a pesticide application. Research of this nature will clarify the effects of a xenobiotic chemical on the foraging behavior of a nontarget organism. As Belovsky showed, the indirect effects of a pesticide can be more important than the direct effects. DeAngelis[8] used food web analysis to extend the study of direct and indirect effects to the level of ecosystems. The various approaches applied in food web analysis serve to clarify a number of different problems related to the presence and propagation of toxicants through the ecosystem. By varying the level of the analysis, from conceptual to qualitative to quantitative, one obtains different insights on the effect of the toxicant on food web structure and function. The regulatory process frequently demands accurate information on the effects of a chemical on one or more species; for these kinds of results, a simulation model is required. The lack of generality of simulation models compromises their utility, however. DeAngelis argues that "hybrid" models that combine "generic food web models with detailed age-structured and spatially explicit representations for populations of particular interest" may be required.

Population viability (or vulnerability) analysis (PVA) is the "process for estimating species-, location-, and time-specific criteria for persistence" of a population.[9] A great deal has been published in recent years on the management of viable populations for conservation and the question of minimum viable populations.[10] Agricultural chemicals with real or potential hazards for wildlife populations must be a part of the management plan for any species threatened or in decline. Marcot[11] presents an overview of this process and its implications for ecotoxicology. The value of PVA lies in the incorporation of habitat, demographic, and genetic information into a risk analysis methodology.

All models presented to this point have been statistical, quantitative, or mathematical. McBee[12] presents a new laboratory model for the study of genetic effects. Cytogenetic assays are currently used with success to study the genetic effects of chemicals on mammals.[13] There has been little use of these techniques in avian genetic toxicology, due to the intractable nature of chromosomes in birds. Techniques employed in mammals, such as the direct examination of mitotic chromosomes, have limited applicability in birds. Other techniques (synaptonemal complex analysis, flow cytometric analysis, and DNA alkaline unwinding assay) show promise. McBee cautions that genetic toxicity studies address only immediate effects at the chromosomal and molecular levels. Long-term coordinated studies of genetics and demography will be required to demonstrate a link between induced mutations and reproductive success or population viability.

The diversity of approaches to modeling ecological and genetic processes can overwhelm the researcher not accustomed to models or quantitative methods. Toxicologists should take heart that within ecology there are divisions between empiricists and theoreticians.[14] No one can be the master of all trades; the best ecology is done through the combined efforts of theoretical and experimental scientists. The same will be true in toxicology. Experimental toxicologists must seek out good theoreticians with an interest in applied environmental science. As Paul Ehrlich[15] has stated, "Neglect of the theory ecologists have already developed is costing many billions of dollars annually and possibly mortgaging the future of our species." Facing such consequences, the rewards of a cooperative effort are great indeed.

REFERENCES

1. Bird, S. L., Fate and exposure modeling in terrestrial ecosystems: a process approach, in *Wildlife Toxicology and Population Modeling: Integrated Studies of Agroecosystems*, R. J. Kendall and T. E. Lacher, Jr., Eds., Lewis Publishers, Chelsea, MI, 1993.
2. Brisbin, I. L., Jr., Applications of a modified Richards sigmoid model to assess the uptake and effects of environmental contaminants upon birds, in *Wildlife Toxicology and Population Modeling: Integrated Studies of Agroecosystems*, R. J. Kendall and T. E. Lacher, Jr., Eds., Lewis Publishers, Chelsea, MI, 1993.
3. Chesser, R. K., K. B. Willis and N. E. Mathews, Impacts of toxicants on population dynamics and gene diversity in avian species, in *Wildlife Toxicology and Population Modeling: Integrated Studies of Agroecosystems*, R. J. Kendall and T. E. Lacher, Jr., Eds., Lewis Publishers, Chelsea, MI, 1993.
4. Slade, N. A., Models of structured populations: age and mass transition matrices, in *Wildlife Toxicology and Population Modeling: Integrated Studies of Agroecosystems*, R. J. Kendall and T. E. Lacher, Jr., Eds., Lewis Publishers, Chelsea, MI, 1993.
5. Meyer, J. S., and M. S. Boyce, Life historical consequences of pesticides and other insults to vital rates, in *Wildlife Toxicology and Population Modeling: Integrated Studies of Agroecosystems*, R. J. Kendall and T. E. Lacher, Jr., Eds., Lewis Publishers, Chelsea, MI, 1993.
6. Emlen, J. M., and E. K. Pikitch, Animal population dynamics: identification of critical components, *Ecol. Model.*, 44, 253–273, 1989.

7. Belovsky, G. E., Modeling avian foraging: implications for assessing the ecological effects of pesticides, in *Wildlife Toxicology and Population Modeling: Integrated Studies of Agroecosystems*, R. J. Kendall and T. E. Lacher, Jr., Eds., Lewis Publishers, Chelsea, MI, 1993.
8. DeAngelis, D. L., What food web analysis can contribute to wildlife toxicology, in *Wildlife Toxicology and Population Modeling: Integrated Studies of Agroecosystems*, R. J. Kendall and T. E. Lacher, Jr., Eds., Lewis Publishers, Chelsea, MI, 1993.
9. Soulé, M. E., Introduction, in *Viable Populations for Conservation*, M. E. Soulé, Ed., Cambridge University Press, London, 1987, pp. 1-10.
10. Soulé, M. E., *Viable Populations for Conservation*, Cambridge University Press, London, 1987.
11. Marcot, B. G., Analyzing and monitoring population viability, in *Wildlife Toxicology and Population Modeling: Integrated Studies of Agroecosystems*, R. J. Kendall and T. E. Lacher, Jr., Eds., Lewis Publishers, Chelsea, MI, 1993.
12. McBee, K. Adaptation of mutagenicity and clastogenicity assays for avian species, in *Wildlife Toxicology and Population Modeling: Integrated Studies of Agroecosystems*, R. J. Kendall and T. E. Lacher, Jr., Eds., Lewis Publishers, Chelsea, MI, 1993.
13. McBee, K., J. W. Bickham, K. W. Brown and K. C. Donnelly, Chromosomal aberrations in native small mammals (*Peromyscus leucopus* and *Sigmodon hispidis*) at a petrochemical waste disposal site. I. Standard karyology, *Arch. Environ. Contam. Toxicol.*, 16, 681-688, 1987.
14. Kareiva, P., Renewing the dialogue between theory and experiments in population ecology, in *Perspectives in Ecological Theory*, J. Roughgarden, R. M. May and S. A. Levin, Eds., Princeton University Press, Princeton, NJ, 1989, pp. 68-88.
15. Ehrlich, P. R., Discussion: ecology and resource management—is ecological theory any good in practice, in *Perspectives in Ecological Theory*, J. Roughgarden, R. M. May and S. A. Levin, Eds., Princeton University Press, Princeton, NJ, 1989, p. 306-318.

34

Life Historical Consequences of Pesticides and Other Insults to Vital Rates

Joseph S. Meyer and Mark S. Boyce

ABSTRACT

In simple demographic models, the per capita growth rate can be shown to be a concave function of most vital rates; i.e., age-specific fecundity or survival. As a consequence, for a given magnitude of variation in vital rates, the consequent variance in per capita growth rates will be greater when vital rates are low. Therefore, variance in per capita growth rates will be greater for populations in which survival or reproduction has been suppressed as a consequence of pesticides or other toxic substances. This will have further ramifications by decreasing long-term population growth rates in stochastic population projections, but should also enhance the rate of evolution for resistance. Yet, there will be negative ramifications for the probability of extinction in finite populations.

Applications of these principles are presented for populations of relatively short-lived species, including sage grouse (*Centrocercus urophasianus*), white-crowned sparrows (*Zonotrichia leucophrys*), and great tits (*Parus major*), and two long-lived species, the northern spotted owl (*Strix occidentalis caurina*) and the snail kite (*Rostrhamus sociabilis plumbeus*). In general, the concavity of growth rate effect is much greater for short-lived species, whereas variation in population growth rates approximates a linear function of vital rates for long-lived species.

The authors caution that such observations on demographic models may not apply under density dependence. Populations tightly regulated by density-dependent mechanisms may be more sensitive to nonlinearities and age-specificity of density dependence than to demographic structure.

KEY WORDS

birds, models, population, environmental variability

INTRODUCTION

Pesticides affect avian population growth rates by altering survivorship or fecundity. Although low concentrations of some toxicants actually have been shown to increase survival or fecundity,[1,2] high concentrations of a toxicant usually decrease survival and/or fecundity. Hence, effects of pesticides on avian population growth rates can be modeled to a first approximation using projection matrices in the same way other disturbances that decrease survivorship and/or fecundity (e.g., decreased food supply, disease, harvesting) have been modeled for management purposes.[3]

Emlen[3] recommended using projected mean population density and probability of extinction as indicators of toxicant effects on populations. Emlen and Pikitch[4] recently examined the sensitivity of population growth rates for several "model" avian species to decreases in survivorship and fecundity. They demonstrated that mean population densities and probabilities of extinction for long-lived species with relatively low fecundity are most sensitive to changes in adult survivorship, whereas mean population densities and probabilities of extinction for shorter-lived species with relatively high fecundity are most sensitive to changes in juvenile survivorship. Those results can be used to help decide the appropriate structure for a model of a toxicant's effects on an avian population and can help prioritize data gathering needs.

We argue that the population growth rate (λ) may often be of interest because it offers a more direct measure of the perturbation consequences of a toxicant. As such, population growth rate does not require assumptions about the behavior of the system over long time horizons as is necessary for population size and the probability of extinction. This chapter examines the effects of alterations in survivorship and fecundity schedules on the value of population growth rate for short- and long-lived avian species. We demonstrate that population growth rate often is a nonlinear function of fecundity or survivorship. The curvilinear nature of this relationship and the variability in estimates of the population growth rate depend on the structure of the model and assumptions regarding variability in the survivorship and fecundity schedules.

PROJECTION MODELS

Projection-matrix models are usually based on some form of the Leslie matrix, whose solution is given by the Euler equation

$$1 = \sum_{x=0}^{\Omega} \lambda^{-x} \cdot l_x \cdot b_x \tag{1}$$

where λ is population growth rate, l_x is probability of survival to age class x, b_x is fecundity of females in age class x, and Ω is oldest age class in the population. Population growth can also be expressed as the per capita rate of increase (r), where $r = \ln(\lambda)$. Letting s_i equal the probability of survival from age class i to age class $i + 1$, Equation 1 can be rewritten as

$$1 = \sum_{x=0}^{\Omega} \left[\lambda^{-x} \cdot \prod_{i=0}^{x-1} (s_i) \cdot b_x \right]. \tag{2}$$

Adult survivorship and fecundity in avian species are often independent of age.[5] Hence, for a species reproducing at age α, we can assume that s (adult survivorship) and b (fecundity) are constants for age classes $\geq \alpha$ and that $b = 0$ for age classes $< \alpha$. Equation 2 then simplifies to

$$1 = \sum_{x=\alpha}^{\Omega} \left[\lambda^{-x} \cdot \prod_{i=0}^{\alpha-1} (s_i)\, s^{x-\alpha} \cdot b \right]. \tag{3}$$

If we assume infinite longevity ($\Omega = \infty$) and no reproductive senescence, Equation 3 can be rewritten as[5]

$$0 = \lambda^{\alpha} - s \cdot \lambda^{\alpha-1} - \prod_{i=0}^{\alpha-1} (s_i) \cdot b. \tag{4}$$

We refer to this solution as a stage-structured model. If reproductive senescence occurs, or the birds have a finite longevity, or adult survivorship and fecundity are not constant, an iterative solution of Equation 3 must be used to estimate λ. We refer to this iterative solution as an age-structured model.

SENSITIVITY OF POPULATION GROWTH RATES

For species that breed at age 1 ($\alpha = 1$), Equation 4 can be rearranged to solve directly for the population growth rate

$$\lambda = s + s_0 b. \tag{5}$$

Thus, population growth rate is a linear function of survivorship and fecundity for species that reproduce at age 1, and are nonsenescent. By taking the partial derivative of λ with respect to survival and fecundity, we obtain the following sensitivities:

$$\partial \lambda / \partial s_0 = b,$$

$$\partial \lambda / \partial s = 1,$$

$$\partial \lambda / \partial b = s_0.$$

For this case, growth rate sensitivities thus depend on the magnitude of s_0 and b.

We can illustrate these sensitivities with data for the great tit (*Parus major*) population in Wytham Wood, England, for which $\alpha = 1$, $s = 0.38$, $b = 4.25$, and a value of $s_0 = 0.15$ produces $\lambda = 1.01$. Figure 1A shows that the decrease in λ for great tits is relatively large and equal for equal percentage decreases in s_0 or b (i.e., they have the same relative sensitivities, even though the absolute sensitivities differ), whereas the effect of a similar percentage decrease in s is smaller. These results are consistent with the simulations reported by Emlen and Pikitch[4] for "model" short-lived birds.

Because the second partial derivatives of λ with respect to survivorship and fecundity equal zero ($\partial^2\lambda/\partial s_0^2 = \partial^2\lambda/\partial s^2 = \partial^2\lambda/\partial b^2 = 0$; i.e., the derivatives of the above corresponding first derivatives) for species in which $\alpha = 1$ and no reproductive senescence occurs, there is no concavity or convexity in the population growth rate functions. However, r will always be a concave function (with respect to the abscissa) of survivorship and fecundity, because the second partial derivatives of $\ln(\lambda)$ with respect to survivorship and fecundity are negative for all s_0, s, and $b \geq 0$, as follows:

$$\partial^2 \ln \lambda / \partial s_0^2 = - b^2/(s + s_0 b)^2,$$

$$\partial^2 \ln \lambda / \partial s^2 = - 1 /(s + s_0 b)^2,$$

$$\partial^2 \ln \lambda / \partial b^2 = - s_0^2/(s + s_0 b)^2.$$

If reproductive senescence or a maximum age are imposed on the model, the population growth rate functions for species with $\alpha = 1$ become concave. Iterative solutions

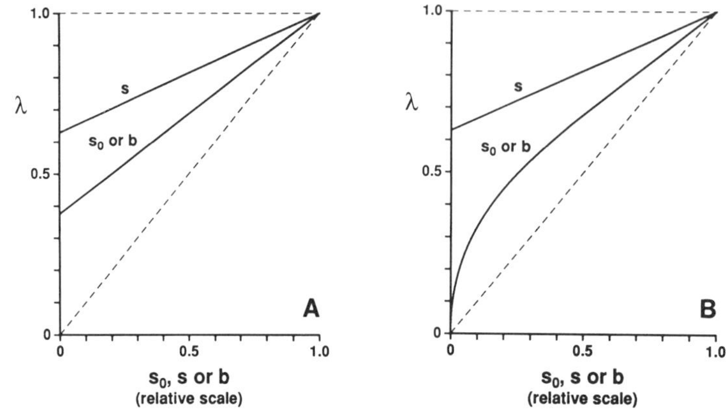

FIGURE 1. Population growth rate (λ) as a function of survivorship or fecundity for (A) a stage-structured model and (B) an age-structured model of a great tit population in Wytham Wood, England ($\alpha = 1$, $\Omega = 5$); 1.0 on the abscissa represents the empirical mean of survivorship or fecundity.

of Equation 3 for great tits with $\Omega = 5$ are shown in Figure 1B. Below approximately 50% of the original values of s_0 or b, values of λ predicted by the age-structured model (reproductive senescence or finite longevity) begin to diverge considerably from values of λ predicted by the stage-structured model (infinite longevity and no reproductive senescence) and decrease rapidly toward 0 as s_0 and b approach 0. The shape of the curve of r vs s_0 or b also becomes more concave. However, the curve of λ vs s remains linear and intersects the ordinate at the same value of λ as did the stage-structured model. Because the slope of λ vs s_0 or b increases rapidly at low values of s_0 or b, the sensitivity of λ to juvenile survivorship and fecundity becomes even greater in the age-structured model than it was in the stage-structured model.

Similar curves result for an age-structured model of a white-crowned sparrow population in California,[6] even when adult survivorship and fecundity are not constant with age (Table 1 and Figure 2). λ and r are also concave functions of survivorship and fecundity in an aquatic invertebrate, *Daphnia pulex*.[2,7] We conjecture that λ and r are generally concave functions of survivorship and fecundity whenever a species has finite longevity and reproductive senescence.

As a consequence of this concavity, age-structured models are more likely to predict population extinction during stochastic simulations of a given reduction in survivorship or fecundity than are stage-structured models. Moreover, the statistical uncertainty in λ and r will be greater generally at low values of survivorship or fecundity,

Table 1. Life Table for a White-Crowned Sparrow Population in California[6]

Age Class (x)	l_x	s_x	b_x
0	1.000	0.174	0
1	0.174	0.483	3.142
2	0.084	0.524	3.333
3	0.044	0.205	3.556
4	0.009	0.444	3.750
5	0.004	0	4.000

Note: Average values of l_x for two cohorts are presented, whereas values of b were estimated from all females studied.

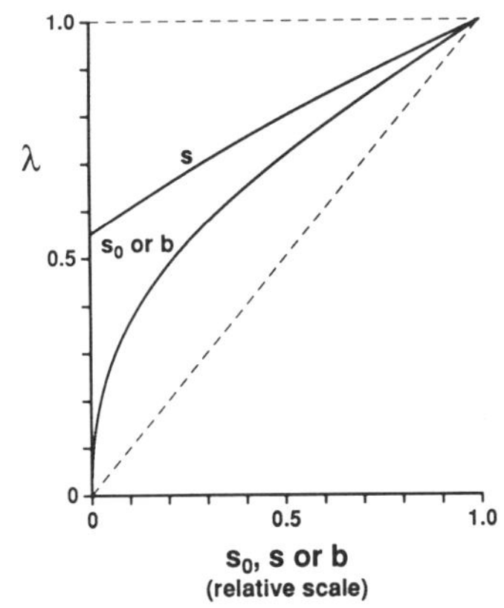

FIGURE 2. Population growth rate (λ) as a function of survivorship or fecundity for an age-structured model of a white-crowned sparrow population in California ($\alpha = 1$, $\Omega = 5$).

because a given variation in s_0, s, or b will be associated with larger variations in λ and r at low values of survivorship and fecundity than at high values (Figure 3). This assumes no change in variability of survivorship or reproduction due to the pesticide. However, if exposure to a toxicant increases the variability in survivorship or reproduction, an even greater variability in λ and r would be expected at low values of s_0, s, or b.

Associated with this increased variability of λ and r, we also expect an increased rate of evolution for resistance to the pesticide. Recall from Fisher's[8] fundamental theorem of natural selection that the rate of change in fitness is equal to the variance in fitness. It may also be relevant that the probability of extinction for nonresistant genotypes will be relatively high, but a genotype extinction argument is complicated by the fact that resistance usually will be polygenic.

For species that breed at age 2 ($\alpha = 2$), Equation 4 can be written as

$$0 = \lambda^2 - s \cdot \lambda - s_0 \cdot s_1 \cdot b \tag{6}$$

and solved by the quadratic equation

$$\lambda = (s + [s^2 + 4 \cdot s_0 \cdot s_1 \cdot b]^{1/2})/2. \tag{7}$$

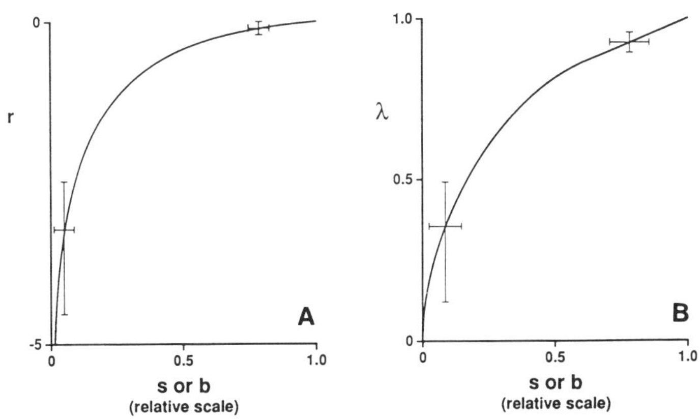

FIGURE 3. Variability in (A) r and (B) λ due to concavity for high and low mean values of survivorship or fecundity.

By taking the partial derivative of λ with respect to survival and fecundity, we obtain the following sensitivities:

$$\partial\lambda/\partial s_0 = s_1 \cdot b \,/\, u^{1/2},$$

$$\partial\lambda/\partial s_1 = s_0 \cdot b \,/\, u^{1/2},$$

$$\partial\lambda/\partial s = {}^1\!/\!_2 \cdot (1 + s/u^{1/2}),$$

$$\partial\lambda/\partial b = s_0 \cdot s_1 \,/\, u^{1/2},$$

where $u = s^2 + 4 \cdot s_0 \cdot s_1 \cdot b$ (always >0).

Thus, population growth rate is a nonlinear function of survivorship and fecundity for species that first reproduce at age 2 and are nonsenescent. Therefore, the magnitude of sensitivities will depend on the absolute values of s_0, s_1, s, and b. Interestingly, not all of the second partial derivatives have the same sign:

$$\partial^2\lambda/\partial s_0^2 = -2 \cdot s_1^2 \cdot b^2/u^{3/2},$$

$$\partial^2\lambda/\partial s_1^2 = -2 \cdot s_0^2 \cdot b^2/u^{3/2},$$

$$\partial^2\lambda/\partial s^2 = 2 \cdot s_0 \cdot s_1 \cdot b/u^{3/2},$$

$$\partial^2\lambda/\partial b^2 = -2 \cdot s_0^2 \cdot s_1^2/u^{3/2}.$$

Therefore, λ is a concave function of s_0, s_1, and b, because these second derivatives are negative for all s_0, s_1, and $b \geq 0$; whereas λ is a convex function of s, because its second derivative is positive for all $s \geq 0$. However, r will always be a concave function of survivorship or fecundity, because the second partial derivatives of $\ln(\lambda)$ with respect to survivorship and fecundity are always negative, as follows (see also Boyce[9]):

$$\partial^2\ln\lambda/\partial s_0^2 = -\frac{2 \cdot s_1^2 \cdot b^2}{u^{3/2}} \cdot \left[\frac{s + 2 \cdot u^{1/2}}{(s + u^{1/2})^2}\right],$$

$$\partial^2\ln\lambda/\partial s_1^2 = -\frac{2 \cdot s_0^2 \cdot b^2}{u^{3/2}} \cdot \left[\frac{s + 2 \cdot u^{1/2}}{(s + u^{1/2})^2}\right],$$

$$\partial^2\ln\lambda/\partial s^2 = -\frac{s}{u^{3/2}},$$

and

$$\partial^2\ln\lambda/\partial b^2 = -\frac{2 \cdot s_0^2 \cdot s_1^2}{u^{3/2}} \cdot \left[\frac{s + 2 \cdot u^{1/2}}{(s + u^{1/2})^2}\right].$$

We illustrate these sensitivities with data for a northern spotted owl population in northern California,[10] for which $\alpha = 2$, $s_0 = 0.29$, $s_1 = 0.903$, $s = 0.903$, and $b = 0.335$ ($\lambda = 0.99$). Figure 4 shows that the decrease in λ for northern spotted owls is relatively large for a given percentage decrease in s, whereas the effect is relatively small for a similar percentage decrease in s_0, s_1, or b (see also the sensitivity analysis of Noon and Biles[11]). These results are consistent with the simulations reported by Emlen and Pikitch[4] for "model" long-lived birds.

If senescence is imposed on the model for species with $\alpha = 2$, the magnitude of the effect depends on the oldest age class assumed for the population. Iterative solutions of Equation 3 for northern spotted owls with $\Omega = 15$ are shown in Figure 4B. The value of

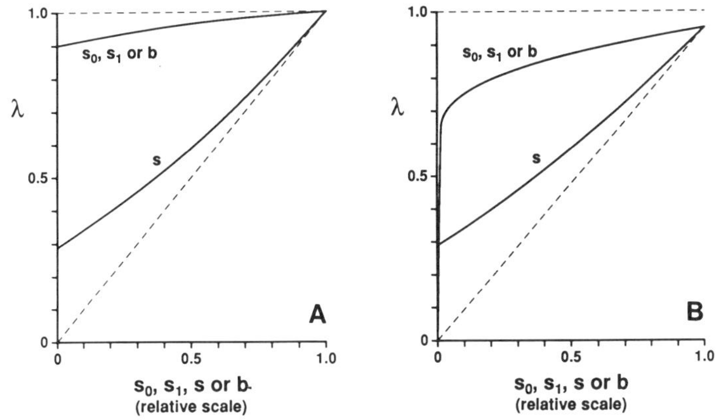

FIGURE 4. Population growth rate (λ) as a function of survivorship or fecundity for (A) a stage-structured model and (B) an age-structured model of a northern spotted owl population in northern California ($\alpha = 2$, $\Omega = 15$).

λ at the original survivorship and fecundity values is approximately 4% less than λ for the stage-structured model, so for this long-lived species there is little difference in the projection-matrix solutions of the two models at the original survivorship and fecundity values, if senescence does not occur before age 15.[12] For the age-structured model, the curve of λ vs s becomes slightly less convex, yet it intersects the ordinate at the same value as did the stage-structured model. The population growth rate functions for s_0, s_1, and b become more concave in the age-structured model.

Below approximately 50% of the original values of s_0, s_1, or b, values of λ predicted by the age-structured model begin to diverge considerably from values of λ predicted by the stage-structured model and decrease rapidly toward 0 as s_0 and b approach 0 (Figure 4B). Because the slope of λ vs s_0, s_1, or b increases rapidly at low values of s_0, s_1, or b, the sensitivity of λ to juvenile and subadult survivorship and to fecundity becomes even greater in the age-structured model. Additionally, the curve of λ vs s_0, s_1, or b intersects the curve of λ vs s at a low percentage of the original survivorship and fecundity values; however, because they intersect at such low values of survivorship and fecundity, this phenomenon probably has no practical consequence for modeling effects of pesticides on long-lived species.

Similar curves result for an age-structured model of a snail kite (formerly known as Everglade kite) population in Florida,[13] for which $\alpha = 3$, $\Omega = 25$, $s_0 = 0.578$, $s = s_1 = s_2 = \ldots = s_{24} = 0.900$ and $b = 0.367$ (Figure 5). At some value less than about 50% of observed survival and fecundity, population growth rates for long-lived species with $\alpha > 1$ in age-structured models appear to be more sensitive to a given percentage change in pre-reproductive survivorship or fecundity than they are to the same percentage change in adult survivorship.

Finally, we consider a sage grouse population in southeastern Wyoming whose survivorship is intermediate between our other examples (Table 2 and Figure 6). This is a particularly relevant species to consider because juveniles are especially sensitive to loss of insects in the diet due to pesticide application.[14] Because $\alpha = 1$ and this is an age-structured model, both the curve of λ vs s and the curve of λ vs s_0 or b are concave. Moreover, population growth rate is approximately equally sensitive to a given percentage change in survivorship and fecundity above 40% of the original survivorship and fecundity values. Below 20% of the original values, however, the sensitivity to changes in s_0 and b are much greater than the sensitivity to changes in s.

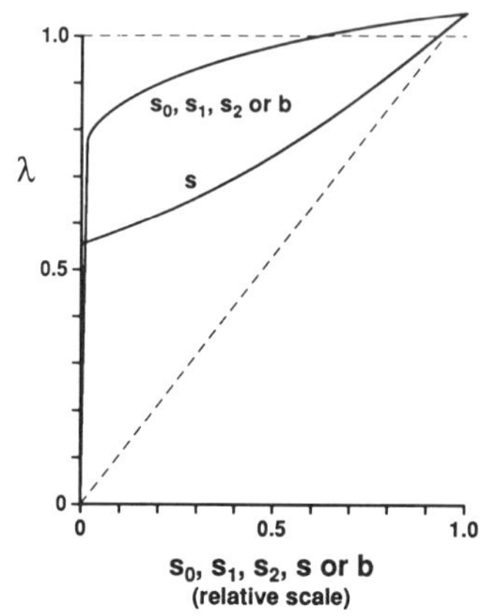

FIGURE 5. Population growth rate (λ) as a function of survivorship or fecundity for an age-structured model of a snail kite population in Florida ($\alpha = 3$, $\Omega = 25$).

Thus, when the effects of small changes in survivorship and/or fecundity on population growth rate are to be modeled, it appears that age- and stage-structured models provide similar results. Either the analysis by Emlen and Pikitch[4] or the first partial derivatives of λ presented above can be used as guides to the relative intensity of effort that should be expended on gathering survivorship and fecundity data. However, when the effects of large changes in survivorship and/or fecundity on population growth rate

Table 2. Life Table for a Sage Grouse Population in Southeastern Wyoming

Age Class (x)	l_x	s_x	b_x
0	1.000	0.185	1.88
1	0.093	0.500	3.75
2	0.046	0.500	3.75
3	0.023	0.500	3.75
4	0.012	0.500	3.75
5	0.006	0	3.75

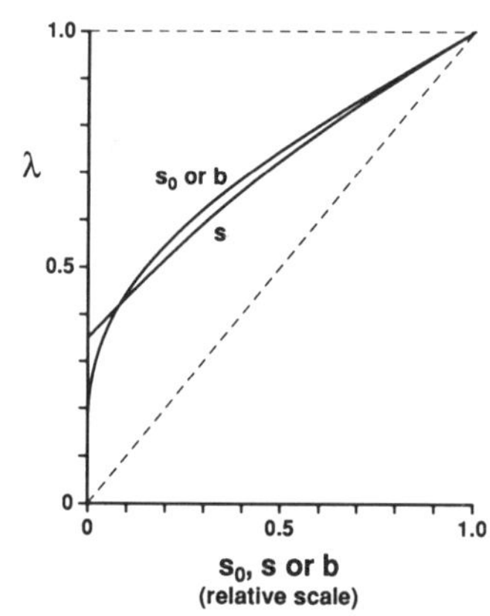

FIGURE 6. Population growth rate (λ) as a function of survivorship or fecundity for an age-structured model of a sage grouse population in Wyoming ($\alpha = 1$, $\Omega = 5$).

are to be modeled, the selection of model type (age-structured vs stage-structured) may markedly influence the results. Moreover, the assumed age of senescence will affect the outcome of an age-structured model. In age-structured models, the sensitivity of λ to changes in pre-reproductive survival and fecundity always becomes high in magnitude for low values of these parameters.

VARIABILITY IN POPULATION GROWTH RATES

For stage-structured models, the standard error of the estimate of λ, [$SE(\lambda)$], can be approximated as follows[5]:

$$SE(\lambda) = \{\sum_{\pi} [(\partial \lambda / \partial \pi)^2 \cdot SE(\pi)^2]\}^{1/2}. \quad (8)$$

where π is each of the survivorship and fecundity terms included in the model, and $SE(\pi)$ is standard error of the estimate of each survivorship and fecundity term. When $\alpha = 1$, Equation 8 becomes

$$SE(\lambda) = [b^2 \cdot SE(s_0)^2 + SE(s)^2 + s_0^2 \cdot SE(b)^2]^{1/2} \cdot \quad (9)$$

When $\alpha = 2$, Equation 8 becomes

$$SE(\lambda) = \{[s_1^2 \cdot b^2 \cdot SE(s_0)^2 + s_0^2 \cdot b^2 \cdot SE(s_1)^2 + (s + u^{1/2})^2 / 4 + s_0^2 \cdot s_1^2 \cdot SE(b)^2] / u\}^{1/2} \cdot \quad (10)$$

Short-lived birds often have high fecundity and reproduce at age 1. In general, we expect $b^2\, SE(s_0)^2 >> SE(s)^2$ and $b^2\, SE(s_0)^2 >> s_0^2\, SE(b)^2$ in Equation 9 for short-lived birds, except when $SE(s_0)$ is relatively small. In fact, because juvenile survivorship is usually more variable than adult survivorship or fecundity,[15] $SE(\lambda)$ will often be approximately equal to $b \cdot SE(s_0)$ for short-lived birds.

Assumptions about how the variability in survivorship and fecundity decreases as survivorship or fecundity decreases can affect $SE(\lambda)$. We used the following three methods for calculating $SE(s_i)$:

Method A: $SE(s_i) = [s_i\, (1 - s_i)/n]^{1/2}$,

Method B: $SE(s_i) = CV(s_i)\, s_i$,

Method C: $SE(s_i) = SE(s_i)_{initial}$,

where $s_i = s_0$, s_1, or s; n = sample size; and $CV(s_i) = SE(s_i)_{initial}/s_{i,initial}$. Method A assumes a binomial probability of survivorship[5] and is most appropriate for within-sample estimates of $SE(\lambda)$. Unfortunately, Method A is conceptually difficult to apply to among-years estimates of $SE(\lambda)$, because an initial, hypothetical sample size (n) must be computed as

$$n = s_i(1 - s_i)/SE(s_i)^2$$

where $SE(s_i)^2$ is the estimate of annual variance in each of the survivorship terms. Method B assumes that $SE(s_i)$ is a constant percentage of s_i. Method C assumes a constant $SE(s_i)$, regardless of the value of s_i.

Analogous to Methods B and C for $SE(s_i)$, we used the following two methods for calculating $SE(b)$:

$$\text{Method D:} \quad SE(b) = CV(b)\, b,$$

$$\text{Method E:} \quad SE(b) = SE(b)_{initial},$$

where $CV(b) = SE(b)_{\text{initial}}/b_{\text{initial}}$.

The sensitivity of $SE(\lambda)$ is illustrated with respect to survival and fecundity in Figure 7A, using the previous data for the great tit and estimates of $SE(s_0) = 0.05$, $SE(s) = 0.10$, and $SE(b) = 0.30$ for annual variability around the original values of survivorship and fecundity. For great tits, $SE(\lambda)$ is relatively insensitive to changes in s and $SE(s)$, regardless of the assumptions made about $SE(s)$. However, $SE(\lambda)$ decreases considerably as juvenile survivorship decreases, if either Method A or B is used to calculate $SE(s_0)$, because the term $b^2\, SE(s_0)^2$ in Equation 9 is large and decreases rapidly as $SE(s_0)$ decreases; $SE(\lambda)$ changes very little only if Method C is used to calculate $SE(s_0)$, because the term $b^2\, SE(s_0)^2$ in Equation 9 is large and does not decrease when $SE(s_0)$ is assumed to be constant. Estimates of $SE(\lambda)$ decrease considerably as fecundity decreases using either Method D or E for calculating SE(b), because changes in the term $s_0^2\, SE(b)^2$ in Equation 9 have little effect on $SE(\lambda)$; instead, the major cause of the decrease in $SE(\lambda)$ is the decrease in the term $b^2\, SE(s_0)^2$ as b decreases. Therefore, if variability in population growth rate is to be estimated relatively accurately for short-lived avian species, the majority of the research effort should be directed toward determining fecundity and how the variability in juvenile survivorship changes as the value of juvenile survivorship changes.

Because the value of s_0 for great tits was calculated from Equation 5 by assuming that $\lambda = 1.0$, we do not have an empirical estimate for $SE(s_0)$. The value $SE(s_0) = 0.05$ was selected because it produces $CV(s_0) = 33\%$, which is about the same as that found empirically for the northern spotted owl.[10] Because juvenile survivorship is thought to

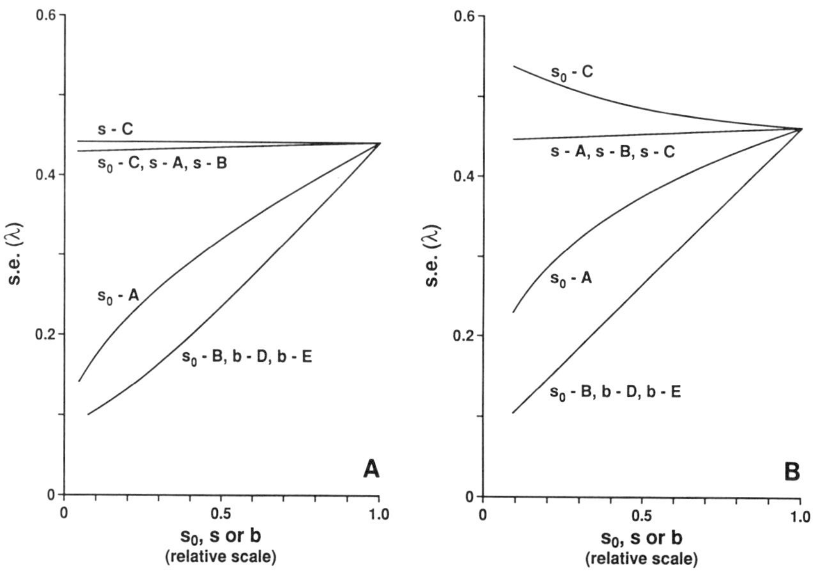

FIGURE 7. Annual variability in estimates of population growth rate [SE(λ)] as a function of survival or fecundity and various assumptions about variability in survivorship and fecundity, for (A) a stage-structured model and (B) an age-structured model of a great tit population in Wytham Wood, England ($\alpha = 1$, $\Omega = 5$). Curves are coded according to the variable that was decreased and the method used to compute its variability (see text for descriptions of methods used to compute standard errors of survivorship and fecundity); e.g., s_0-*B* represents $SE(\lambda)$ computed for a range of values of s_0, with $SE(s_0)$ computed by method B.

be highly variable, $SE(s_0) = 0.05$ may be an underestimate for great tits. A higher value of $SE(s_0) = 0.10$, for example, increases the values of $SE(\lambda)$ shown in Figure 7A by a factor of about 1.8; however, the qualitative trends in $SE(\lambda)$ do not change (results not shown).

It might seem surprising that an intensively studied bird species such as the great tit has such high variability in the estimate of $SE(\lambda)$ (e.g., $CV(\lambda) = 24\%$ at the original values of survivorship and fecundity when $SE(s_0) = 0.05$, and $CV(\lambda) = 45\%$ when $SE(s_0) = 0.10$). Because this variability is mostly caused by high annual variability in estimates of juvenile survivorship, it is unlikely that $CV(\lambda)$ could be decreased considerably simply by increasing sampling intensity and, thus, decreasing the variance in juvenile survivorship due to among-years sampling error. Instead, it appears that inherently high annual variability in juvenile survivorship (a characteristic of many avian species) will almost always produce relatively high estimates of $SE(\lambda)$ for short-lived birds. This suggests that stochastic models of exponential population growth for short-lived birds will almost always predict a high probability of extinction. Furthermore, this suggests that it will be difficult to conclude that λ of a pesticide-treated population of short-lived birds is significantly different than λ of a control population. In this example for great tits, assuming a pooled $SE(\lambda)_p = 0.25$, the difference between population growth rates for a treatment population, $\lambda_{treatment}$, and a control population, $\lambda_{control}$, would have to be

$$|\lambda_{treatment} - \lambda_{control}| > 1.96 \cdot [2 \cdot SE(\lambda)_p^2]^{1/2} = 0.69$$

to be significant at the $\alpha = 0.05$ level—a catastrophic decrease in λ. An even larger pooled $SE(\lambda)_p = 0.45$ would require an unlikely difference in λ of 1.25.

For long-lived birds whose age of first reproduction is 2, Equation 10 is more difficult to interpret. The sensitivity of $SE(\lambda)$ is illustrated with respect to decreases in survivorship and fecundity in Figure 8A, using the previous data for the northern spotted owl and estimates of $SE(s_0) = 0.11$, $SE(s_1) = 0.024$, $SE(s) = 0.024$ and $SE(b) = 0.046$ for within-population variability around the original values of survivorship

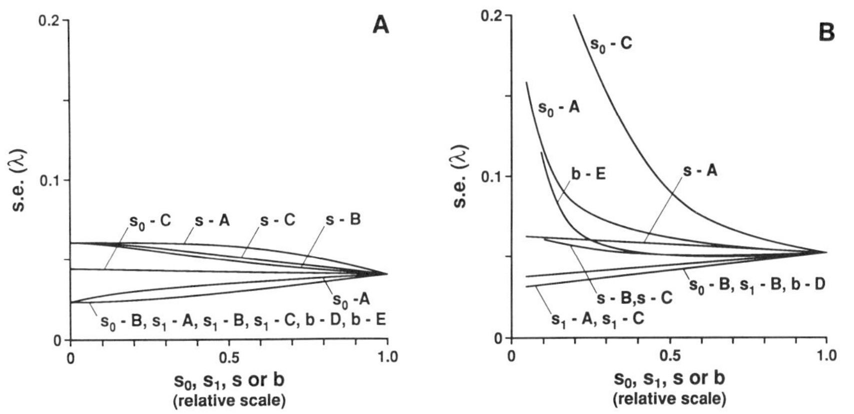

FIGURE 8. Within-population variability in estimates of population growth rate [SE(λ)] as a function of survivorship or fecundity and various assumptions about variability in survivorship and fecundity, for (A) a stage-structured model and (B) an age-structured model of a northern spotted owl population in northern California ($\alpha = 2$, $\Omega = 15$). Curves are coded according to the variable that was altered and the method used to compute its variability (see text for descriptions of methods used to compute standard errors of survivorship and fecundity); e.g., s_1-*C* represents $SE(\lambda)$ computed for decreasing values of s_1, with $SE(s_1)$ computed by method C.

and fecundity.[10] The same five methods described above were used to calculate $SE(s_i)$ and $SE(b)$ in simulations in which survivorship and fecundity were decreased below their original values. Because annual variability in survivorship and fecundity from 1985 to 1989 for this owl population[16] are similar to the pooled within-population variabilities used by Thomas et al.,[10] estimates of $SE(\lambda)$ for within-population variation and for annual variation are similar.

Surprisingly, $SE(\lambda)$ increased slightly as s decreased (Figure 8A), no matter which method was used to calculate $SE(s)$, because s dominates the value of the denominator in Equation 10 for this long-lived species. On the other hand, $SE(\lambda)$ remained approximately constant or decreased slightly as s_0, s_1, and b decreased. Variability in estimates of λ was much smaller for this population of northern spotted owls (CV(λ) = 5%) than was variability in estimates of λ for great tits, even though $CV(s_0) = 38\%$ for the owls. Hence, one would be much more likely to demonstrate that two estimates of population growth rate for northern spotted owls are significantly different. In this example, assuming a pooled $SE(\lambda)_p = 0.05$, the difference between population growth rates would only have to be

$$|\lambda_{treatment} - \lambda_{control}| > 1.96 \cdot [2 \cdot SE(\lambda)_p^2]^{1/2} = 0.14$$

to be significant at the $\alpha = 0.05$ level.

Monte Carlo simulations were used to estimate $SE(\lambda)$ for age-structured models to examine whether stochastic estimates of variability in age-structured models differed from the previous deterministic estimates of variability in stage-structured models. For northern spotted owls, 1000 random combinations of survivorship and fecundity were sufficient to obtain stable estimates of $SE(\lambda)$; whereas 5000 random combinations of survivorship and fecundity were necessary to obtain stable estimates of $SE(\lambda)$ for great tits.

To perform a Monte Carlo simulation, the authors calculated the nominal values of s_0, $SE(s_0)$, s_1, $SE(s_1)$, s, $SE(s)$, b, and $SE(b)$ by the same methods used to generate results of the stage-structured models in Figures 7A and 8A. Beta distributions were then generated using those survivorship and fecundity values (as recommended by Emlen and Pikitch[4]), and we randomly selected values of s_0, s_1, s, and b from the beta distributions and estimated λ. This process was repeated 1000 or 5000 times for each combination of nominal survivorship and fecundity values, and the standard deviation of the 1000 or 5000 estimates of λ was computed to estimate $SE(\lambda)$.

Under most assumptions about variability in survivorship and fecundity for great tits, $SE(\lambda)$ is approximately the same in Monte Carlo simulations of the age-structured model (Figure 7B) as it is in deterministic estimates of the stage-structured model (Figure 7A). When $SE(s_0)$ is computed by Method A (binomial probability of survivorship), however, $SE(\lambda)$ is approximately 10 to 20% greater in the age-structured model than in the stage-structured model (although $SE(\lambda)$ still decreases as s_0 decreases), and when $SE(s_0)$ is computed by Method C (constant $SE[s_0]$), $SE(\lambda)$ actually increases as s_0 decreases. As occurred with the stage-structured model, estimates of $SE(\lambda)$ increased by approximately 1.8-fold when we assumed $SE(s_0) = 0.10$ instead of 0.05 (results not shown).

Similarly, under most assumptions about variability in survivorship and fecundity for northern spotted owls, $SE(\lambda)$ is approximately the same in Monte Carlo simulations of the age-structured model (Figure 8B) as it is in deterministic estimates of the stage-structured model (Figure 8A). Here, however, $SE(\lambda)$ increases not only as s_0 decreases when $SE(s_0)$ is computed by Methods A and C, but $SE(\lambda)$ also increases as b decreases when SE(b) is computed by Method E [constant $SE(b)$]. At low values of s_0 and b, the

Monte Carlo estimates of $SE(\lambda)$ for the age-structured model are 2 to 5 times greater than the estimates of $SE(\lambda)$ for the stage-structured model.

For both species, these increases in $SE(\lambda)$ appear to be caused by the concavity in the λ vs s_0 and λ vs b curves in the stage-structured models, because concave functions contain inherently larger variability than linear functions (see discussion in section "Sensitivity of Population Growth Rates"). The reason $SE(\lambda)$ did not increase when method E was used to compute $SE(b)$ for great tits (although it did for northern spotted owls) is probably because $SE(b)$ contributes little to the value of $SE(\lambda)$ in short-lived birds, as we discussed above for the deterministic estimates of $SE(\lambda)$ in stage-structured populations. Method B ($SE[s_i]$ = a constant percentage of s_i) tends to produce the lowest values of $SE(s_i)$ at intermediate values of s_i (i.e., when s_i is not close to 0 or 1); hence, it tended to produce the lowest estimates of $SE(\lambda)$ for any given survivorship or fecundity variable.

Thus, the assumptions used to compute variability in survivorship or fecundity can drastically alter the estimates of variability in λ. As mentioned earlier, the binomial probability model for computing $SE(s_i)$ (Method A) may be appropriate for estimates of within-population variability, but it is questionable when applied to among-years variability. Computing $SE(s_i)$ or SE(b) as a constant percentage of s_i or b (Methods B and D) is probably a conservative assumption in most instances. However, if toxicant exposure tends to increase variability in survivorship and fecundity (a distinct possibility if differential resistance to toxicants occurs in the population), then assuming constant variability in survivorship and fecundity (Methods C and E) or some intermediate function may be more appropriate. However, the assumption of constant variability in survivorship or fecundity may be unreasonable at extremely low values of s_i or b.

CONCLUSIONS

The authors have presented deterministic computations for estimating population growth rates and their variability in stage-structured population projection models and in stochastic (Monte Carlo) simulations of age-structured population projection models. When the effects of small changes in survivorship and/or fecundity of avian species exposed to pesticides or other insults are modeled, age- and stage-structured models provide similar predictions of λ and $SE(\lambda)$. Sensitivities of λ and $SE(\lambda)$ with respect to small changes in survivorship and fecundity can be predicted using the equations presented here.

However, effects of larger decreases in survivorship or fecundity will depend upon assumptions about longevity and reproductive senescence in the population (stage- or age-structured models) and how the variability in survivorship and fecundity changes as survivorship or fecundity decreases. Strongly concave functions of λ vs survivorship or fecundity will probably require stochastic modeling to estimate λ and $SE(\lambda)$.

Results presented here may guide researchers in allocating effort in demographic studies of the effects of pesticides on avian species. They support generally the importance of juvenile survivorship and fecundity to population growth rates in short-lived birds and the importance of adult survivorship to population growth rates in long-lived birds, as proposed by others.[3,4,17,18] In addition, the authors show that to evaluate the effect of pesticides on avian populations it is essential to estimate the variance in survivorship and fecundity.

The demographic models that have been studied here are exponential growth models whose application can be relevant only for short time periods, e.g., immediately subsequent to the application of a pesticide. Models to predict the long-term population consequences of pesticides must incorporate density dependence and the ramifications of pesticide use to insect prey.

ACKNOWLEDGMENTS

We thank the Wyoming Water Research Center, the National Council for Air and Stream Improvement, and the Association of O&C Counties for supporting our research. Mark Boyce was supported by the National Science Foundation under RII-8610680. Thanks to C. M. Perrins of the Edward Grey Institute at Oxford for use of unpublished data on great tits, and to Barry Noon of the U.S. Forest Service's Redwood Science Laboratory for astute comments on the manuscript.

REFERENCES

1. Smyth, H. F., Jr., Sufficient challenge, *Food Cosmet. Toxicol.*, 5, 51–58, 1967.
2. Meyer, J. S., Variability and Bias in Estimating Cladoceran Population Growth Rates, Ph.D. thesis, University of Wyoming, Laramie, 1986.
3. Emlen, J. M., Terrestrial population models for ecological risk assessment: a state-of-the-art review, *Environ. Toxicol. Chem.*, 8, 831-842, 1989.
4. Emlen, J. M., and E. K. Pikitch, Animal population dynamics: identification of critical components, *Ecol. Model.*, 44, 253-273, 1989.
5. Lande, R., Demographic models of the northern spotted owl (*Strix occidentalis caurina*), *Oecologia*, 75, 601-607, 1988.
6. Baker, M. C., L. R. Mewaldt and R. M. Stewart, Demography of white-crowned sparrows (*Zonotrichia leucophrys nutalli*), *Ecology*, 62, 636-644, 1981.
7. Meyer, J. S., C. G. Ingersoll, L. L. McDonald and M. S. Boyce, Estimating uncertainty in population growth rates: jackknife vs. bootstrap techniques, *Ecology*, 67, 1156–1166, 1986.
8. Fisher, R. A., *The Genetical Theory of Natural Selection*, Dover, New York, 1930.
9. Boyce, M. S., Population growth with stochastic fluctuations in the life table, *Theor. Pop. Biol.*, 12, 366–373, 1977.
10. Thomas, J. W., E. D. Forsman, J. B. Lint, E. C. Meslow, B. R. Noon and J. Verner, A Conservation Strategy for the Northern Spotted Owl, Report of the Interagency Scientific Committee to Address the Conservation of the Northern Spotted Owl, U.S. Department of Agriculture–Forest Service/U.S. Department of the Interior–Bureau of Land Management/ U.S. Department of the Interior–Fish and Wildlife Service/U.S. Department of the Interior–National Park Service, Portland, OR, 1990.
11. Noon, B. R., and C. M. Biles, The mathematical demography of the spotted owl in the Pacific Northwest, *J. Wildl. Manage.*, 54, 18–27, 1990.
12. Boyce, M. S., A Review of the U.S. Forest Service's Viability Analysis for the Spotted Owl, National Council for Air and Stream Improvement, Corvallis, OR, 1987.
13. Nichols, J. D., G. L. Hensler and P. W. Sykes, Jr., Demography of the Everglade kite: implications for population management, *Ecol. Model.*, 9, 215–232, 1980.
14. Johnson, G. D., and M. S. Boyce, Feeding trials with insects in the diet of sage grouse chicks, *J. Wildl. Manage.*, 54, 89–91, 1990.
15. Charnov, E. L., Life history evolution in a "recruitment population": why are adult mortality rates constant?, *Oikos*, 47, 129–134, 1986.
16. Franklin, A. B., J. A. Blakesley and R. J. Gutierrez, Population Ecology of the Northern Spotted Owl (*Strix occidentalis caurina*) in Northern California: Preliminary Results, Unpublished report, Department of Wildlife, Humboldt State University, Arcata, CA, 1990.
17. Lebreton, J.-D., and J. Clobert, Bird population dynamics, management and conservation: the role of mathematical modelling, in *Bird Population Studies: Their Relevance to Conser-*

vation and Management, C. M. Perrins, J.-D. Lebreton, and G. J. M. Hirons, Eds., Oxford University Press, Oxford, U.K., 1991, pp. 105–124.

18. Young, H., A consideration of insecticide effects on hypothetical avian populations, *Ecology*, 49, 991–993, 1968.

35

What Food Web Analysis Can Contribute to Wildlife Toxicology

Donald L. DeAngelis

ABSTRACT

Food web analysis occupies a central place in ecology, as it provides the means for calculating the propagation of effects through the biotic components (species populations) of an ecosystem. Various approaches are used in food web analysis, including qualitative analysis, mathematical modeling, and computer simulation modeling. Qualitative methods may be useful in situations in which little quantitative information is available regarding strengths of food web interactions; however, there is a high degree of uncertainty in the predictions. Mathematical models are useful for obtaining general analytic results concerning system behavior. Concepts such as system resistance and resilience to disturbances have been developed from the study of such models and are applicable to problems of ecotoxicology. More specific and accurate predictions of indirect food web effects may require the use of computer simulation models. This chapter reviews both the modeling approaches and the ecological concepts related to the analysis of food webs and attempts to integrate these with the field of wildlife toxicology.

KEY WORDS

food webs, computer simulation models, qualitative analysis, mathematical models, individual-based models

INTRODUCTION

A central theme of ecological research is the understanding of the intricate web of relationships that connect all living things on earth. Among the most important of these relationships are the trophic or feeding interactions through which herbivores and carnivores are ultimately dependent on the energy fixed by the plant primary producers. The organization of primary producers, herbivores, and carnivores is called an ecological community or food web.

The simplest set of feeding linkages is a food chain. Colinvaux[1] gave an example of the terrestrial food chain:

pine trees → aphids → spiders → titmice → hawks

Because most animals eat more than one type of prey and most prey are eaten by more than one type of consumer, most chains are actually parts of more complex food webs. The "food web" is one of the most useful concepts in ecology and has long been the focus of both theory and analysis.

It is useful, to begin with, to distinguish food web analysis from food web theory. Food web theory has the goal of providing explanations for the size (number of species), structure (number of trophic levels, degree of niche overlap, etc.), and stability (species coexistence vs displacement, constancy of populations vs severe fluctuations, etc.) of food webs. In particular, food web or community ecologists seek an understanding of how these properties are related to external conditions and the strength and form of trophic interactions between species populations. These objectives are sometimes referred to as "strategic" goals (see Figure 1).

Food web theory and ecosystem theory in general often have been criticized for not being especially useful in providing principles and methods for carrying out environmental assessments, such as those prescribed by the National Environmental Protection Act (NEPA).[2] These criticisms are valid, but it also should be kept in mind that general theory is seldom directly applicable to particular situations, with which assessments are almost always concerned. "The inability of ecological theory to predict precisely future population sizes, the rate at which a fauna will collapse following insularization of the habitat, or the response of an ecosystem to a complex series of insults does not necessarily represent a failure of the theory," stated Ehrlich.[3]

Food web theory attempts to establish general relationships regarding food web structure and dynamics; e.g., the lengths of food chains as functions of environmental conditions, the stability of food webs as a function of connectedness, and the niche overlap of species using similar resources. At best, such relationships may hold in a statistical sense across a large collection of food webs, but cannot be expected to provide predictions for a specific system.

STRATEGIC GOALS	STRATEGIC MODELS
DEVELOPING GENERAL THEORY	Strategic models are abstract and not specific to any particular system. Many aspects of a system are ignored to reveal the most important causal mechanisms.

TACTICAL GOALS	TACTICAL MODELS
ANALYZING PARTICULAR CASES	Tactical models are usually specific to a given system or a given process. 1. For strategic goals Tactical models can be used to test some aspects of general theory or deduce particular consequences of such theory. 2. For applied goals Tactical models can be used as predictive tools to assess impacts on particular systems.

FIGURE 1. Models with strategic and tactical objectives.

The goals of food web analysis tend to be more tactical and applied than those of food web theory. Food web analysis may involve using food web models of particular systems to try to determine such specific things as how a change in one part of a food web provokes changes in other parts of the web.

There are two possible aspects to such tactical analyses (see Figure 1). First, the theoretician may use such analyses to derive generalizations on which theory can be built. Thus, the analysis serves the overall strategic goal of theory. As an example, Bartell et al.[4] used a detailed aquatic food web simulation model to derive general results concerning top-down and bottom-up controls in food webs. Second, the applied ecologist, a wildlife ecologist, for example, may use such food web analysis to look at specific effects, such as the effects of a toxicant on a particular ecosystem.

Food web analysis, like the study of any complex system, comes in many forms of varying degrees of appropriateness for a particular type of problem. As Kelly et al.[5] pointed out, a toxicant can influence a population through different pathways of effects (Figure 2). One such pathway is that of effects of the toxicant on the individual members of the population that encounter the chemical directly. Alternatively, individual population members may encounter the toxicant after it has passed through the food web and perhaps been concentrated or modified chemically. An example of this is the concentration of fluorides from industrial emissions by mangrove trees;[6] the leaves of mangroves can accumulate high concentrations of fluorides that can then be exported by litterfall to the detrital-based food webs of estuaries.

Further, the consequences the toxicant has for a population may not even involve the chemical effects the toxicant has directly on the population members, but may

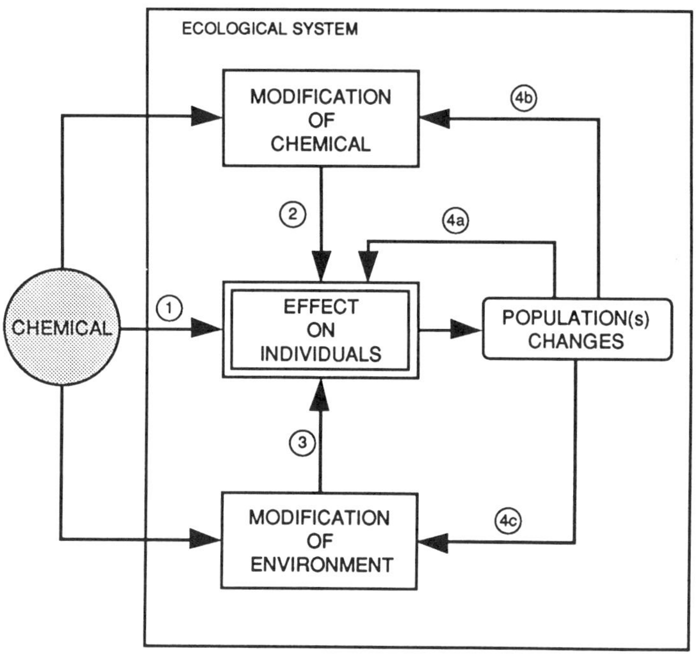

FIGURE 2. General ways by which a chemical toxicant affects an ecological system. There are effects on individuals through direct contact with the chemical. The toxicant may also affect the biotic environment of the individual, thus indirectly affecting individuals. The chemical toxicant may also be modified by its movement through the ecological system and this modified chemical can have different effects on the individual and environment. The effects on the individual cause changes at the population level that may have feedback effects on the individual. (From Kelly et al.[5] With permission.)

result instead from the chemical effects on species that are prey, predators, or competitors for the population in question. There may be some temptation to ignore effects that act indirectly through the food web with the justification that direct effects themselves are so difficult to estimate that it is a waste of effort to try to take into account indirect effects. It may be dangerous to ignore such indirect effects, however. The crucial roles of "keystone predators" in many systems has been pointed out by Paine.[7] A disturbance changing the abundance of such a predator will have enormous indirect consequences for its prey species, some of which will change in abundance relative to others. The analogous concept of a "keystone mutualist" was developed by Gilbert.[8] There are many obligate mutualisms in nature, including pollinator/angiosperm mutualisms. A disturbance to a keystone mutualist, or one that has a pivotal location in the web of interactions, may drastically affect many other species that depend on the mutualism.

This chapter attempts to review some of these methods of food web analysis. Examples of specific applications within the domain of food web modeling are presented.

GENERAL APPROACHES TO FOOD WEBS

One can approach any study of a given natural system, such as a food web, from a number of levels (Figure 3). At the most basic level is the formulation of a conceptual framework in which hypothesized causes and effects are explicitly stated. At a second level is the attempt to make whatever predictions are possible on the basis of qualitative information through formal techniques. This requires that the hypothesized causal relationships be carefully articulated, although the attachment of quantitative values

TYPES OF MODELS

KNOWLEDGE REQUIRED		PURPOSE
	CONCEPTUAL MODELS	
GENERAL KNOWLEDGE OF CASUAL RELATIONSHIPS		GENERAL UNDERSTANDING
	QUALITATIVE MODELS	
KNOWLEDGE OF CONNECTIONS AND FEEDBACKS		DERIVING SIMPLE GENERALIZATIONS
	ANALYTIC MODELS	
KNOWLEDGE OF FUNCTIONAL RELATIONSHIPS		ESTABLISHING A GENERAL BODY OF THEORY
	SIMULATION MODELS	
DETAILED QUANTITATIVE KNOWLEDGE OF PROCESSES		MAKING DETAILED PREDICTIONS FOR SPECIFIC SYSTEMS

FIGURE 3. A sequence of levels at which a system such as a food web can be studied.

may be lacking. Qualitative mathematical approaches are available to derive deductions from these relationships. At a third level is the representation of phenomena in terms of mathematical equations. These equations may be rather abstract, ignoring some of the complexity of particular ecological systems so that the properties of the system can be understood in a general sense that applies across a variety of similar systems. Some quantification is required here, but usually only at the level of specifying general functional relationships between components of the food web. At a fourth level is the attempt at making precise predictions for the specific system. This level is highly quantitative. An attempt is made to integrate all important and measurable aspects of the system into the model.

These four levels of approach can be illustrated in the study of food webs. I will refer to them as (1) conceptual modeling, (2) qualitative model analysis, (3) analytic or quantitative model analysis, and (4) simulation modeling. Figure 3 outlines these approaches. They differ from one another in their objectives and in the knowledge and database needed to use them. Brief surveys of each level of approach and how it is used are given below.

Conceptual Approaches

The main goal of food web analysis is to understand how various effects propagate through food webs. A variety of different chains of cause and effect converge on a given species in the web. For example, Levin[9] noted, "Chemicals may reduce an individual's capacity to resist infection, subtly but importantly changing the nature of a parasite-host relation. Similarly, they may shift competitive balances, affect pollination systems, or weaken a prey's ability to escape predation."

One ingenious way of helping to represent and classify environmental cause and effect chains converging on a particular species is the "envirogram" of Andrewartha and Birch.[10] An envirogram is a kind of dendrogram that traces a variety of causes through chains leading down to effects on a single species population. Figure 4 shows an envirogram for the European rabbit in Australia. Causes are separated by type (resources, predators, and malentities) and into those that are "distal" or indirectly impinging on the population and those that are at the "centrum," or directly impinging the population. For example, in the envirogram of the rabbit, landscape topography is a distal cause, but it influences the flow of rainwater and, therefore, affects vegetation, the direct food resource of the rabbit. This approach has obvious advantages for helping one catalog the various pathways by which a toxicant may affect a given species through chains of abiotic and biotic cause and effect. Although the focus of envirograms is on a particular species, food web influences are explicitly taken into account.

The structure of cause and effect in food chains is complicated by the occurrence of causal loops, or what Hutchinson[11] called "circular causality." Such cause and effect loops are known as feedback loops and have been described in ecological systems since the beginning of the science of ecology. The concept of feedback is awkward, though not impossible, to express satisfactorily in envirograms. In Figure 4 the "rabbit" appears as the object of all causal chains. It also occurs earlier in some of the chains, because the number of rabbits in earlier generations and the warrens built by them affect the current rabbit population.

Feedback effects are usefully conceptualized in diagrams such as the energy and nutrient flow diagrams used in the approach to ecological systems known as systems ecology. An example showing the complexity of even a simplified approximation of a complete food web is the diagram of carbon flows in a marsh ecosystem by Ulanowicz[12] (Figure 5). The "loops" may not be apparent at first because the return flows of carbon

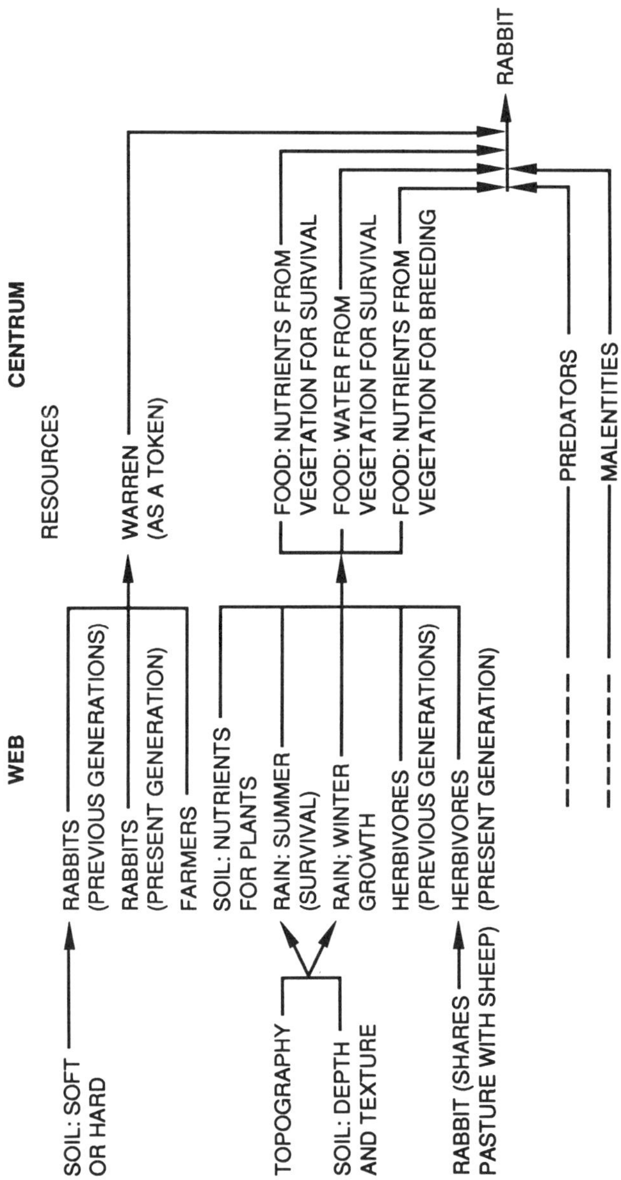

FIGURE 4. Part of an envirogram for a European rabbit in Australia. (From Andrewartha and Birch.[10] With permission.)

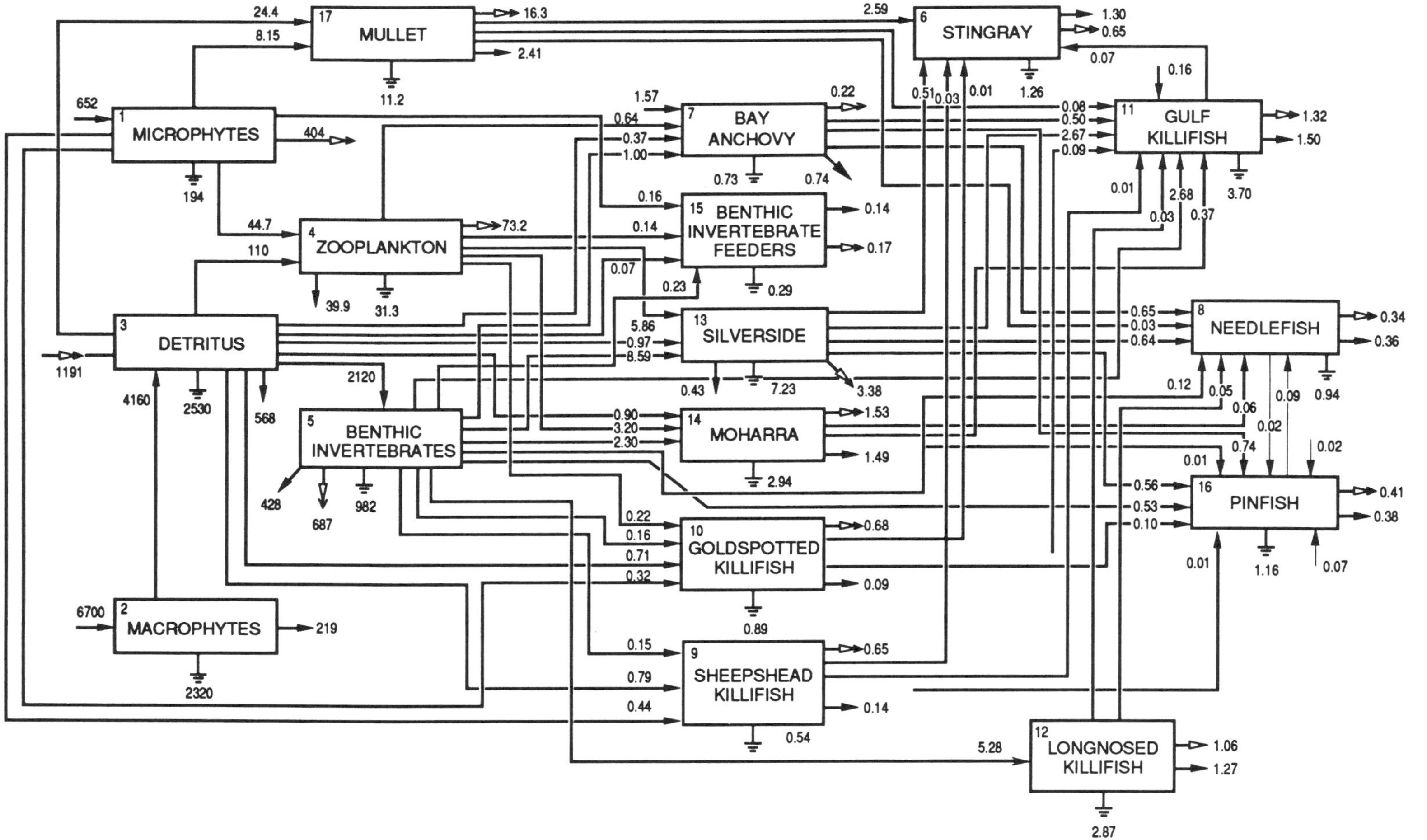

FIGURE 5. A schematic diagram of carbon flows (g m^{-2} day^{-1}) for taxa of a marsh food web, Crystal River, Florida. Flows to detritus are represented by arrow pointing downward. The double arrows represent losses to detritus and the ground symbols represent respiration losses. (From Ulanowicz,[12] based on M. Homer and W. M. Kemp, unpublished manuscript.)

to detritus are not drawn. The cause and effect relationships from a material flow viewpoint are obviously quite complex, as this diagram demonstrates the relative importance of the direct and indirect effects of one species on another in the food web. There are many more pathways that the indirect effects may follow, as opposed to the one pathway of the direct effect, so it cannot be assumed necessarily that the direct effect will be the most important.

While this material or energy flow diagram indicates some of the causal relationships in food webs, a variation on this representation, the signed digraph, is more general as it can represent any sort of positive and negative effect between species in a food web, not just material flows. Food web digraphs are generally used to represent three types of interactions between species: (1) one species positively affects another species, which is negatively affected by the relationship (predator-prey or parasite-host relationships); (2) both species affect each other negatively (competition); and (3) both affect each other positively (mutualism). A simple diagram of two competing autotrophs, a herbivore and a carnivore, are shown in Figure 6.

Qualitative Analysis

It is much easier to characterize the topological structure of connections within a food web, such as the feedback diagrams of Figure 5 and 6, than it is to assign quantitative values to these connections. This has led ecologists to ask whether information relevant to the assessment of toxicological and other effects on food webs could be derived from the qualitative characterization alone. A number of different approaches have been suggested and a few of these are mentioned.

The digraph representation of Figure 6 can be used to yield information about systems. One would like, for example, to be able to take such a diagram and make predictions concerning how a change in some parameter for one species (reflecting the impact of a toxicant on that species) would affect other species in the web. This can be done for such systems through a method of "loop analysis" developed by Levins.[13,14] An application by Briand and McCauley[15] will give an example of the approach.

Briand and McCauley[15] used loop analysis to study the problem of algae in lakes. A simple digraph of the particular system of interest is shown in Figure 7. This system consists of nutrients (N), inedible algae (IA), edible algae (EA), herbivorous grazers (H), zooplankton carnivorous on the herbivores (C), and planktivorous fish (P). There is self-damping (loss rates that increase with compartment size) only in the nutrients.

Suppose that the growth rate of one of the components is permanently changed due to some toxicant impact. Such a change will be called a "press disturbance." Bender et

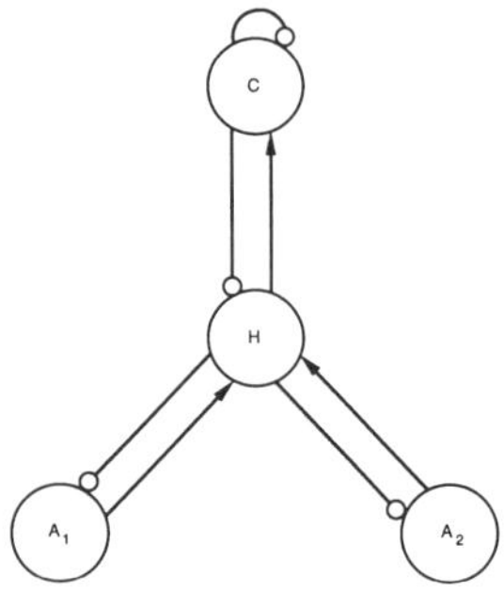

FIGURE 6. A simple digraph of two autotrophs, A_1 and A_2, a herbivore, H, and a carnivore, C. The arrows represent positive effects and the circles negative effects. The carnivore has self-regulation.

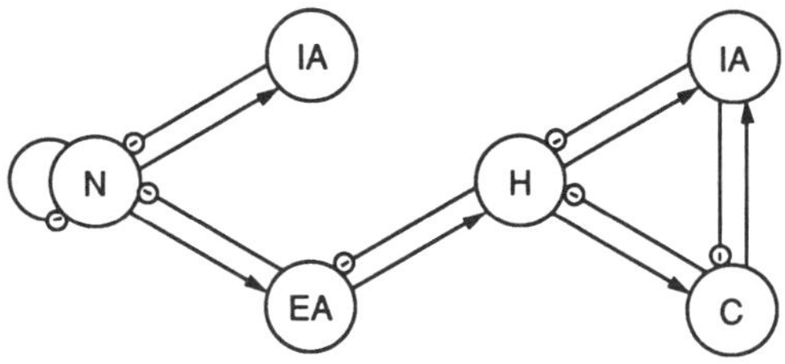

N - NUTRIENTS
IA - INEDIBLE ALGAE
EA - EDIBLE ALGAE
H - HERBIVORES
C - CARNIVORES
P - PLANKTIVORES

FIGURE 7. A digraph model of a lake planktonic community. There is self-regulation in the nutrient compartment. (From Puccia and Levins,[14] adapted from Briand and McCauley.[15])

al.[16] distinguished between two types of disturbances: pulse disturbances and press disturbances, corresponding roughly to acute and chronic inputs. A pulse disturbance is a relatively instantaneous alteration in the density of one or more species. One can imagine some event, such as the release of a toxicant, that has an immediate effect on mortality, but then quickly degrades and has no further toxic effects. A press disturbance involves the depression of one or more species populations for a sufficiently long time that all the remaining species in the food web have time to feel the effects and approach a new steady state. This also could involve the total removal of one or more of the species.

Briand and McCauley[15] used loop analysis to examine the direction (0, +, or -) of changes in any of the system components that would result from a change in any other component. These are listed in Figure 8, with some corrections by Puccia and Levins.[14] Consider a change in any of the components in the left-hand column. For example, an increase in nutrient (N) affects inedible algae (IA) positively and an increase in planktivorous fish (P) affects carnivorous zooplankton (c) and edible algae (EA) negatively but inedible algae positively. While these results give only the direction of change and not the magnitude, the advantage of the method is that it gives some quick general information.

		N	IA	EA	H	C	P
NUTRIENTS	N	0	+	0	0	0	0
INEDIBLE ALGAE	IA	-	0	0	-	+	-
EDIBLE ALGAE	EA	0	0	0	+	-	+
HERBIVORES	H	0	+	-	0	0	0
CARNIVORES	C	0	-	+	0	0	+
PLANKTIVORES	P	0	+	-	0	-	0

FIGURE 8. Qualitative predictions of changes in compartment sizes for lake planktonic community using digraph model. The left column represents compartments that are changed and the top row represents compartments that are affected (positively, +; negatively –; or 0 for no change). (From Puccia and Levins,[14] adapted from Briand and McCauley.[15])

There are, however, limitations on how much qualitative information of this form can be extracted from most situations. Yodzis[17] has examined the effects of press disturbances for 16 published food webs; for example, the Narragansett Bay, Rhode Island, food web (see Figure 9; for the species names, see Briand and McCauley[15]). Yodzis represented the food webs as digraphs, assuming that the autotrophs, but no other species, had self-regulating feedback. His goal was to attempt to predict the direction of effects (increase or decrease) that one species population would feel when another species population was given a press disturbance (a constant rate of addition of members of its population, which is elevated to a constant, higher level). Yodzis could have used the loop analysis approach outlined above, but chose to use another approach. He assigned quantitative values to the species interactions randomly within plausible ranges. To analyze the effects that a press disturbance on one species population has on another, Yodzis used a Monte Carlo approach, numerically solving the model for each of 100 different randomly chosen webs of interaction strengths. This statistical approach compensates for the fact that the actual interaction strengths are not known.

One would intuitively expect a great deal of qualitative regularity in the propagation of effects through a food web. For example, if a press disturbance causes a predator population to increase, we would expect a negative effect on the population of its prey, and this qualitative effect should be independent of precise parameter values in the food web. However, Yodzis' results did not show this. For disturbances to predators, in 52% of the predator-prey relationships studied in the 16 food webs, the direction of effect was not determined. (Yodzis' criterion for "not determined" was that in at least 5 of the 100 Monte Carlo samples, the population size deviated from the general trend, or increased rather than decreased in this case.) The reason for this is the prevalence of indirect effects. For example, even though an increase in the predator population may

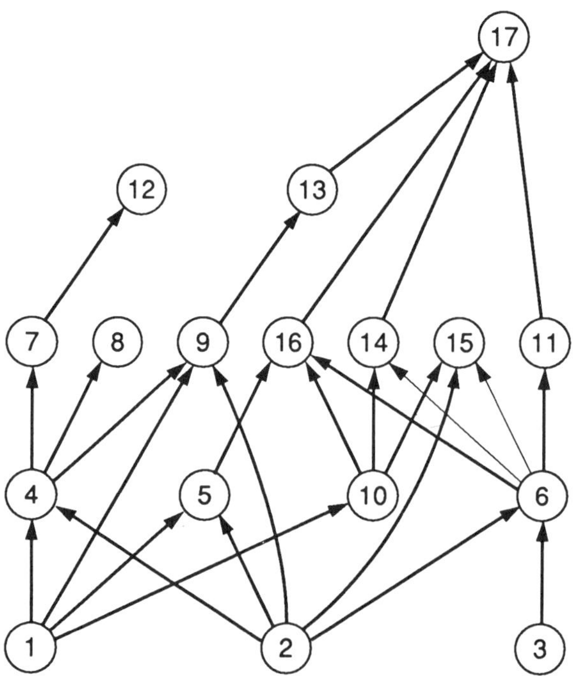

FIGURE 9. Food web model of Narragansett Bay. (From Yodzis.[17])

have a direct negative effect on the prey population, it may also affect competitors of the prey population negatively. Reduction in competitors would affect the prey population positively. Such indirect effects, Yodzis showed, can easily outweigh the direct effects.

The lesson from Yodzis'[17] study is that food webs may be so complex that in the absence of accurate knowledge concerning the interactions, it may be impossible to make even qualitative predictions with a strong degree of confidence. The problem may be even worse than Yodzis' analysis indicates, because the published food webs that he analyzed are great simplifications of the actual webs.

Quantitative Model Analysis

Mathematical model analysis refers here to the use of such models for predicting the quantitative effects of a disturbance, such as a toxicant loading, to one part of a food web on other parts of the web. This is a step toward greater precision than simply trying to determine whether another component increases or decreases. However, analytic models tend to be fairly simple and abstract in form. Their advantage is that they are easy to analyze mathematically. Most of the analytic models used in the study of food webs are descendants of the Lotka-Volterra equations, as described by May[18] and Pimm,[19] among others.

The use of analytic models allows a number of interesting properties of food webs to be quantified. Two of the most basic quantitative concepts for food webs are resistance and resilience.

Resistance and resilience concepts are fundamental to the quantitative study of ecosystems and food webs and have been used since the early development of systems ecology and mathematical ecology in the 1970s.[20,21] Resistance of a population is the tendency for that population not to be affected by a temporary or permanent toxicant loading applied to some part of the ecosystem. Resilience of a population is a measure of how fast a population that has been affected in some way by a loading can return to its previous state.

Resistance

When one speaks of resistance, it is first important to specify resistance to what sort of disturbance? Earlier, we introduced pulse and press disturbances. How do pulse and press disturbances affect a food web? The effects can be quite different. The point made by Bender et al.[16] was that experiments done to determine the interactions among species in a food web can give different information, depending on whether the experiments are pulse or press disturbances. Pulse disturbances generally give information only on direct effects between one population and another. To determine all of the indirect effects over a long term, press disturbances are necessary.

One study of the resistances for a press disturbance is the model study of Harrison and Fekete.[22] They studied a set of simple linear models for eight different ecosystems, each with six compartments (Figure 10), that were originally constructed by Webster et al.[20] Harrison and Fekete did not apply the disturbance directly to the compartment but to nutrient input fluxes or fluxes between different compartments. This changed the equilibrium values of the compartments.

In one model experiment, Harrison and Fekete[22] plotted changes in available nutrients, X_5, as a function of time following the imposition of the press disturbance that increased the loss rate of nutrients to the external environment by 5%. The magnitudes of change in X_5 through time are shown in Figure 11 for all eight of the ecosystem models.

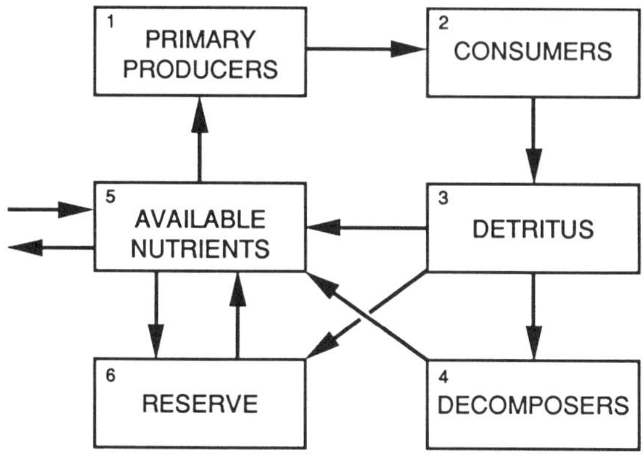

FIGURE 10. A diagram of nutrient cycling in a tundra food web, one of eight different ecosystem types analyzed by Webster et al.[20] and Harrison and Fekete.[22]

One striking result of Harrison and Fekete's analysis is the wide variation in time scales of the responses. The stream ecosystem reached its change in the steady-state value of X_5 = -4.75 within a few years, the lake system required over 200 years to approach the same value, and the grassland, tropical forest, temperate forest, and tundra systems required thousands of years to reach their ultimate changes. Note that the early change in the temperate forest was faster than those of the tundra, tropical forest, and grassland, but that its long-term change was slower than the others.

This example illustrates the temporal aspect in the concept of resistance. A pulse disturbance of only a few years would have seriously affected only the stream and lake ecosystems. The press disturbance ultimately (after thousands of years) had approximately the same effect on most of the ecosystems, causing final deviations in the nutrient compartment of X_5 = -4.76. The reason for this is that X_5 is determined entirely by the rate of outflow from compartment 5 in those ecosystems where this was

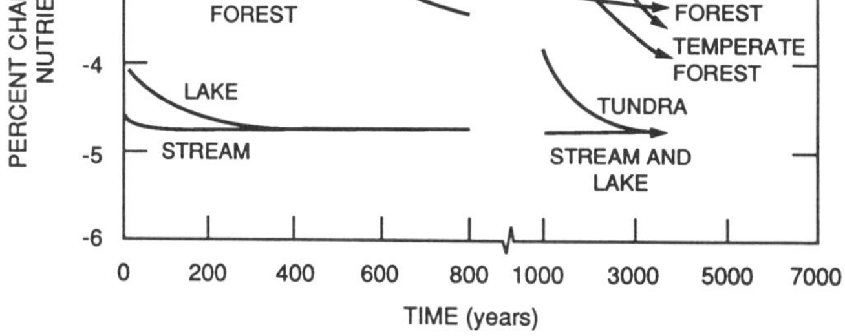

FIGURE 11. Changes in available nutrients as a function of time after imposition of a press disturbance on the loss rate of nutrients for eight different ecosystem types. (From Harrison and Fekete.[22])

the only outflow. The ecosystems that did not reach this value (stream and lake) had other outflows.

There are a number of elaborations on the concept of resistance. One concept related to resistance is the "transmission of variance" in a food web, which has been explored by Carpenter.[23] Carpenter was interested in the effects of variability in environmental factors (such as weather, nutrient supply, and flushing rate of the lake) on phytoplankton biomass. Not only will there be direct effects of these variabilities on phytoplankton, but there may also be indirect effects that result from the effects of the environmental variance on other trophic levels. Carpenter examined the indirect effect that may result because environmental variability causes variability in the recruitment of piscivorous fish. The variability of this recruitment could affect the numbers of piscivorous fish from year to year and could be transmitted down the food chain to the phytoplankton.

Carpenter[23] simulated these possible effects in a food web model. The piscivorous fish in the model was divided into ten age classes. When no variability in fish recruitment was assumed, the spectral analysis of phytoplankton primary production showed only white noise. However, strong variability in fish recruitment resulted in a strong periodicity (an apparent cycle of about seven years) in the phytoplankton. Thus, there were strong effects of variability in the top trophic level of the food web transmitted down the food chain to the phytoplankton.

Resilience

Resilience is a measure of how fast a population can recover or return to its former state following a pulse perturbation, which could be a short-term input of toxicant, for example. Recovery depends on two basic things: (1) how rapidly the toxicant is flushed from the system or degraded and (2) the intrinsic rate of biotic recovery. The potential for biotic recovery can be illustrated for the case of a single population very easily. Consider a population of size N governed by the logistic equation,

$$dN/dt = r(1 - N/K)N, \tag{1}$$

where r is the intrinsic growth rate, K is the environmental carrying capacity of the population, and dN/dt is the rate of change of N. The solution of this simple model is possible and from it one finds that the rate of recovery of the population is r.

Extension of the concept of resilience to large food webs is more difficult. Analytic solutions are not in general possible. However, extensive numerical analyses of food web models have made some generalizations possible. One such generalization is that the resilience of a food web is closely related to the turnover rate of either energy or nutrient in a food web, whichever is limiting production. Turnover rate is defined as the ratio of input of a substance to the amount of that substance stored in the system in steady state. Hence, if some nutrient, for example, phosphorus (P), is limiting production in an ecosystem, then the turnover rate of P is defined as the input rate of $P(I_p)$ divided by the total amount of P stored in the system (P_{tot}), or I_p/P_{tot}.

A large number of numerical studies have been done on model food webs, leading to some generalizations regarding food web resilience. Pimm and Lawton,[24] for example, showed that on average longer food chains will be less resilient than shorter chains. One can interpret this as resulting from the probability that given two food chains, the food chain with more compartments (trophic levels) is more likely to have at least one compartment with a smaller turnover rate than any in the shorter chain. A single compartment can decrease the resilience of the whole food chain.

Species Turnover

In recent years food web and community theory have moved away from "equilibrium theory," or the idea that populations have deterministic equilibrium points around which they fluctuate in response to environmental fluctuations. The equilibrium concept has been replaced by concepts that relate observed population dynamics to spatial scale. The degree of constancy of a population depends to some degree on the spatial scale at which one looks. On small spatial scales, there are continual extinctions and recolonizations; i.e., there is a high turnover rate of a species. The composition of a community thus is constantly changing. However, on larger spatial scales, turnover time is much slower. Even in the midst of continual turnover of species, there may be some characteristics that are preserved through time. The theory developed by MacArthur and Wilson[25] assumed that on a habitat island, turnover occurred but the total number of species remained about the same. However, Wiens [26] has pointed out that data indicate that turnover rates do not always balance and species numbers can fluctuate significantly in many systems.

Species turnover rates and how they scale with system size may have important implications in studying the effects of toxicants on food webs. A high proportion of the turnover in a community may involve uncommon "in and out" species. Thus, much species turnover will be ecologically trivial. Noise from this can mask deterministic long-term changes. The interpretation of local dynamics through measurement of local causes is likely to be difficult. This suggests that both local and regional level measurements may be necessary to discern systematic change.

Simulation Studies

Analysis of the type discussed in the preceding sections has value in a generic sense in that it can give insights to classes of problems. However, it usually cannot give answers to specific problems. Another approach to the problem of giving detailed information about specific problems has developed under the name of simulation modeling.

Swartzman and Kaluzny[27] wrote an introduction to simulation modeling in ecology, in which they described simulation models as "mechanistic" in that they are generally based on detailed mathematical descriptions of the mechanisms and processes that occur in food webs. For example, processes of energy production through photosynthesis, energy loss through respiration, and energy transfers from one trophic level to the next through consumption are modeled. Analogous modeling of water and nutrient flows and recycling through the detritus are also included when useful.

Simulation models usually attempt to incorporate all information useful to solving a problem, even though the model formulation may be highly complex. For example, Van Dyne[28] described the data that was incorporated into the simulation model of a shortgrass prairie: "(1) We measured what we called driving variables, such as precipitation, wind, temperature, and other meteorological phenomena; (2) we measured what we called system-state variables, such as the dynamics over time, of soil water, vegetation, animal populations, and many other variables; and (3) we measured a few of what we called rate processes, which were rates of change, such as decomposition, food-intake rates, and photosynthesis rates. During the period from 1968 to 1975, there were more than 16,000 time-series of data collected for different driving and state variable processes." Thus, with respect to the amount of information typically used, simulation models differ from mathematical models—the food web models of mathematical ecologists.[24,29] In mathematical models, the emphasis is on generality and compactness of the model and the ease with which it can be analyzed. Hence, mathematical models generally avoid detail. They may represent the interactions between predator and prey by a simple function that requires specifying only a few parameters, whereas a simulation model will attempt to describe the interaction in detailed terms of physiology and behavior, such as predator's foraging strategies and prey capture abilities and a prey's escape mechanisms. Time variations in processes, such as the seasonality of feeding and reproduction are also commonly included in simulation models but not mathematical models. Mathematical models are usually analyzed mathematically and simulation models are studied through computer simulation. Like all attempts to separate entities into dichotomous classes, this classification does not always work, as there is actually a spectrum of model types from the highly abstract to the highly detailed and specific.

Because simulation models have a mechanistic basis and are flexible in the amount of detail that can be included in describing processes, they are well suited for modeling the food web effects of toxicants. Simulation modeling is also necessary to represent accurately the fates of the toxicants; i.e., the way they are transferred through and modified by the food web.

Despite its advantages, simulation modeling also has some disadvantages. A large amount of information is necessary to implement a simulation model, such models are usually not highly general, and they may be awkward to use because of their complexity. This tradeoff between mathematical and simulation models means that circumstances may dictate the use of one type or the other.

An increasing amount of simulation modeling is being used to attempt to follow simultaneously the fate and effects of toxicants in food webs. No attempt is made here to survey this field, but an excellent illustrative example is provided. The example is the recent development of an integrated fate and effects model (IFEM) for estimation of risk in aquatic systems.[4] This integrated model is a synthesis of two models that were developed independently, the fates of aromatics model (FOAM)[30] and the standard water column model (SWACOM).[31]

The conceptual structure of FOAM is illustrated by Figure 12, which shows the main pathways of transfer of a polycyclic aromatic hydrocarbon (PAH) in a body of water. The particular version of the model developed in 1983 included, in addition to water and sediments, three functionally defined primary producer populations and six consumer types. The relevant processes of production, excretion, respiration, consumption, mortality, and decomposition of each of these biotic components were described mathematically. The transfer rates of PAHs from one biotic component to another depended on the biomasses and on rates of some of the biotic processes, such as respiration. Equations describing the dynamics of PAHs were formulated in terms of

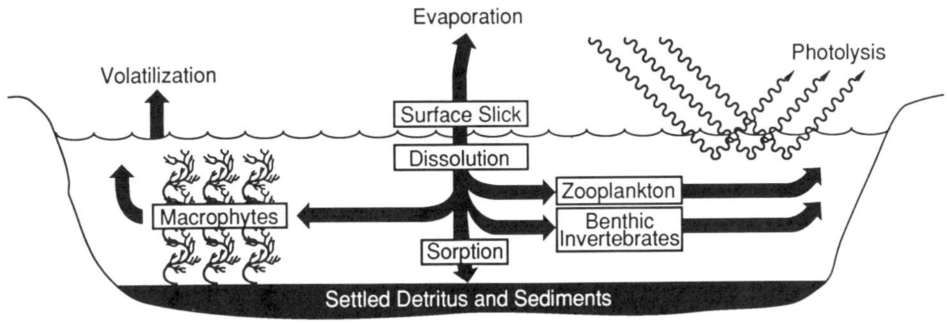

FIGURE 12. Conceptual structure of FOAM. (From Bartell et al.[32])

these transfer rates, so that the dynamics of the contaminant could be computed for any sort of acute or chronic input.

IFEM followed not only the movement of the toxicant in the system, but also the biotic effects, based on the SWACOM part of the model. SWACOM describes the pelagic portion of a northern lake with ten phytoplankters, five zooplankters, three planktivorous fish, and one piscivorous fish. The effects of toxicants on one species can precipitate effects on another species, both through transfer of the toxicant and alterations in the prey and predator densities of the other species.

Some of the results generated by SWACOM show the advantages of the integration of fates and effects in a single model. In one series of simulations, a toxicant was assumed to increase the mortality of fish and zooplankton by 10% and to reduce maximum photosynthesis of phytoplankton by 50%. When the toxicant stress was introduced early in the growing season, it depressed both the phytoplankton and the consumers. When the introduction was late in the season, however, it depressed the consumers but resulted in an increase in phytoplankton. This increase in phytoplankton results from trophic relationships that vary seasonally. The lesson is that complex responses in food webs can result from differential sensitivity of organisms to toxicants.

Food Web Simulations Combined with Individual-Based Modeling

Even assuming that models such as IFEM,[4] which follow the movement and effects of a toxicant through the food web, can be extended to a variety of food web types, this will not be sufficient to deal with the complex problem of predicting toxicant effects. The predictions that are needed are often even more detailed and specific than the usual food web simulation models can approach. A need exists to predict the consequences of toxicants for particular species populations in particular places. Food web simulation models usually lump species into functional groups. These models also usually have little or no spatial resolution or resolution of populations into life stages. However, the heterogeneous spatial dispersion of both toxicant and populations, and the occurrence within populations of life stages that may be particularly susceptible to toxicants are of crucial importance. The importance of these factors means that standard models such as IFEM may not give enough specific information on what the toxicant effects on a particular population will be.

Clearly, models are needed that can simulate the pathways by which toxicants move and the direct and indirect effects these toxicants have on functional groups in the food web. Furthermore, we must develop "hybrid" models that, in addition to these needs, can also simulate in detail the processes by which the toxicant may affect a given species population of interest in all of its spatial and life cycle detail. A model for simulating a wood stork brooding colony is now in development. This model describes the daily foraging and energetics of each individual adult wood stork and the growth of each individual nestling in a colony during the nesting season. The model is spatially explicit, with the foraging area around the colony divided into spatial cells that have different temporal characteristics. The fish and invertebrate populations that the wood storks depend on are modeled as one functional group, so the food web is quite simplified in that sense. However, these prey densities vary from cell to cell in space, depending on the yearly hydrologic pattern across the landscape.

Although the particular use of the wood stork model is to study the effects of different patterns of water level change on reproductive success, models of this type could be adapted to study possible uptake of some toxicant that is heterogeneously distributed in the system. Such a model, if it could be accurately parameterized, would take into account the variance in dosages that would accumulate across the members of the population.

CONCLUSIONS

Food web and ecosystem theory have been criticized because they have not provided useful contributions to environmental impact assessments. Ecosystem theory has been rich in concepts, such as diversity-stability relationships, community matrices, network analysis, and ecosystem resistance and resilience, some of which have been described above. Yet, as critics have pointed out, nearly all have been irrelevant to impact assessment.

This argument is essentially true and probably applies to the possible contributions of food web theory to the effects of toxicants on wildlife. One could reasonably expect that the attempts at developing general theory will not be useful in making predictions about specific systems. Many theoretical predictions have been tested successfully in laboratory systems, but these predictions apply only to relatively simple systems. Reliable food web theory that will apply with equal validity to larger systems outside of the laboratory can develop only as fast as large-scale controlled experiments capable of alternative hypotheses can be performed.

Despite this somewhat pessimistic prospect, methods of food web simulation modeling hold a fair amount of promise in estimating toxicant pathways and their biotic effects. It is highly important that ecosystem models integrating fate and effects of toxic substances (such as IFEM of Bartell and colleagues) be developed for all major ecosystem types and be available in user-friendly forms. These models may be relatively generic, with species aggregated into functional groups. Hybrid models that combine these generic food web models with detailed age structured and spatially explicit representations for populations of particular interest must also be developed to deal with particular situations.

ACKNOWLEDGMENT

I thank S. M. Bartell and G. W. Suter for their helpful comments. This work was sponsored by the Electric Power Research Institute under Contract No. RP2932-2 (DOE No. ERD-87-672) with the U.S. Department of Energy, under contract DE-AC05-84OR21400 with Martin Marietta Energy Systems, Inc.

REFERENCES

1. Colinvaux, P., *Ecology*, John Wiley & Sons, New York, 1986.
2. Suter, G. W., Ecosystem theory and NEPA assessment, *Bull. Ecol. Soc. Am.*, 62, 186–192, 1981.
3. Ehrlich, P. R., Discussion: ecology and resource management—is ecological theory any good in practice?, in *Perspectives in Ecological Theory*, J. Roughgarden, R. M. May and S. A. Levin, Eds., Princeton University Press, Princeton, NJ, 1989, pp. 306–318.
4. Bartell, S. M., A. L. Brenkert, R. V. O'Neill and R. H. Gardner, Temporal variations in regulation of production in a pelagic food web model, in *Complex Interactions in Lake Communities*, in S. R. Carpenter, Ed., Springer-Verlag, New York, 1988, pp. 101–119.
5. Kelly, J. R., T. W. Duke, M. A. Harwell and C. C. Harwell, An ecosystem perspective on potential impacts of drilling fluid discharges on seagrasses, *Environ. Manage.*, 11, 537–562, 1987.

6. Murray, F., Cycling of fluoride in a mangrove community near a fluoride emission source, *J. Appl. Ecol.*, 22, 277-285, 1985.
7. Paine, R. T., Food web complexity and species diversity, *Am. Nat.*, 100, 65-75, 1966.
8. Gilbert, L. E., Food web organization and the conservation of neotropical diversity, in *Conservation Ecology*, M. E. Soule and B. A. Wilcox, Eds., Sinauer Associates, Sunderland, MA, 1980, pp. 11-33.
9. Levin, S. A., Ed., *New Perspectives in Ecotoxicology*, ERC Rep. No. 14, Ecosystems Research Center, Cornell University, Ithaca, NY, 1982.
10. Andrewartha, H. G., and L. C. Birch, *The Ecological Web*, University of Chicago Press, Chicago, 1984, 506 pp.
11. Hutchinson, G. E., Circular causal systems in ecology, *Ann. N.Y. Acad. Sci.*, 50, 221-246, 1948.
12. Ulanowicz, R. E., *Growth and Development: Ecosystems Phenomenology*, Springer-Verlag, New York, 1986.
13. Levins, R., Qualitative analysis of partially specified systems, *Ann. N.Y. Acad.* Sci., 231, 123-138, 1974.
14. Puccia, C. J., and R. Levins, *Qualitative Modeling of Complex Systems*, Harvard University Press, Cambridge, MA, 1985.
15. Briand, F., and E. McCauley, Cybernetic mechanisms in lake plankton systems: how to control undesirable algae, *Nature*, 273, 228-230, 1978.
16. Bender, E. A., T. J. Case and M. E. Gilpin, Perturbation experiments in community ecology: theory and practice, *Ecology*, 65, 1-13, 1984.
17. Yodzis, P. P., The indeterminacy of ecological interactions as perceived through perturbation experiments, *Ecology*, 69, 508-515, 1988.
18. May, R. M., *Stability and Complexity in Model Ecosystems*, Princeton University Press, Princeton, NJ, 1973.
19. Pimm, S. L., *Food Webs*, Chapman & Hall, New York, 1982.
20. Webster, J. R., J. B. Waide and B. C. Patten, Nutrient recycling and the stability of ecosystems, in *Mineral Cycling in Southeastern Ecosystems*, CONF-740513, F. G. Howell, J. B. Gentry and M. H. Smith, Eds., National Technical Information Service, Springfield, VA, 1975, pp. 1-27.
21. Harwell, M. A., W. P. Cropper, Jr., and H. L. Ragsdale, Analysis of characteristics of a nutrient cycling model, *Ecol. Model.*, 12, 105-131, 1981.
22. Harrison, G. W., and S. Fekete, Resistance of nutrient cycling systems to perturbations of the flow rates, *Ecol. Model.*, 10, 227-241, 1980.
23. Carpenter, S. R., Transmission of variance through lake food webs, in *Complex Interactions in Lake Communities*, S. R. Carpenter, Ed., Springer-Verlag, New York, 1988, pp. 119-135.
24. Pimm, S. L., and J. H. Lawton, Number of trophic levels in ecological communities, *Nature*, 268, 329-331, 1977.
25. MacArthur, R. H., and E. O. Wilson, *The Theory of Island Biogeography*, Princeton University Press, Princeton, NJ, 1967.
26. Wiens, J. A., *The Ecology of Bird Communities, Vol. 2, Processes and Variations*, Cambridge University Press, Cambridge, U.K., 1989, 316 pp.
27. Swartzman, G. L., and S. P. Kaluzny, *Ecological Simulation Primer*, Macmillan, New York, 1987.
28. Van Dyne, G. M., Ed., *Ecosystem Concept in Natural Resource Management*, Academic Press, New York, 1969.
29. Goh, B.-S., *Management and Analysis of Biological Populations*, Elsevier, Amsterdam, 1980.
30. Bartell, S. M., R. H. Gardner and R. V. O'Neill, An Integrated Fates and Effects Model for Estimation of Risk in Aquatic Systems, Spec. Tech. Publ. 971, American Society for Testing and Materials, Philadelphia, 1988, pp. 261-274.
31. O'Neill, R. V., S. M. Bartell and R. H. Gardner, Patterns of toxicological effects in ecosystems: a modeling study, *Environ. Toxicol. Chem.*, 2, 451-461, 1983.
32. Bartell, S. M., R. H. Gardner, R. V. O'Neill and J. M. Giddings, Error analysis of predicted fate of anthracene in a simulated pond, *Environ. Toxicol. Chem.*, 2, 19-28, 1983.

36

Adaptation of Mutagenicity and Clastogenicity Assays for Avian Species

Karen McBee

ABSTRACT

Members of the class Aves have received considerable attention in studies of toxic effects of pesticides on nontarget organisms. However, one category of toxic effects for which there are few data from avian taxa is genetic toxicity. Several classes of pesticides (or their metabolites) such as carbamates, DDT, mercury pesticides, and some organophosphate insecticides have tested positively for mutagenicity or clastogenicity in bacterial, plant, or mammalian systems. It is, therefore, not unreasonable to assume that many pesticides may have similar mutagenic effects in exposed avian species. The paucity of information on mutagenic effects of pesticides in birds is most likely due to the somewhat intractable nature of bird chromosomes. Techniques developed for use in mammalian taxa show promise as tools to evaluate mutagenic and clastogenic effects at the molecular and chromosomal levels in exposed avian taxa. These include synaptonemal complex analysis of meiotic chromosomes and flow cytometric analysis of nuclear DNA content variation. Possible application of these tools along with conventional chromosome aberration assays to studies of mutagenic effects of pesticides in birds are discussed.

KEY WORDS

avian, clastogenicity, chromosome aberrations, flow cytometry, synaptonemal complex.

INTRODUCTION

It has long been recognized that many pesticides have the ability to alter the normal genome of a cell.[1-3] Alterations of genetic material may occur by one of three major means: (1) point mutations, (2) chromosome breaks, and (3) changes in chromosome number. Compounds that cause these alterations are broadly known as mutagens. Mutagens that specifically cause chromosome breaks are known as clastogens and those that cause changes in normal chromosome number are known as mitotic poisons. Alterations in somatic cell DNA may increase the probability of developing neoplasia and, through the induction of focal lesions, may be implicated in such broad physiological effects as heart disease, cataracts, and aging.[4,5] In germ cells, an increase in mutagenesis may increase the probability of spontaneous abortion, birth defects, heritable genetic defects, and infertility.[4] Therefore, an understanding of potential genetic hazard to exposed organisms is a critical component in the decision-making processes for pesticide development, use, and regulation.

A number of prokaryotic and eukaryotic systems have been used to document the mutagenicity and clastogenicity of several pesticides; however, very little is known about the genetic effects of pesticides in either avian laboratory models or wild bird populations. As an example, in a recent review of the effects of pesticides on avian populations no mention was made of genetic toxicity.[6] The present chapter briefly reviews the genetic toxicity of selected pesticides, discusses the organization of the avian genome in light of genetic toxicology studies, discusses genetic toxicity assays successfully applied to avian and other vertebrates, and suggests future areas for the development of avian genetic toxicity assessment systems.

MUTAGENICITY AND CLASTOGENICITY OF PESTICIDES

DDT and Its Derivatives

Moutschen-Dahmen et al.[1] reviewed the mutagenicity of several chlorinated, organophosphorus, and carbamate insecticides. The mutagenicity of DDT and its derivatives has been studied in several prokaryote and eukaryote assay systems, consistently giving negative results in prokaryotic systems and lower eukaryotic systems. However, several assays using mammalian cell cultures have yielded conflicting results. V-79 Chinese hamster cell lines exposed to DDT showed no increase in chromosome aberrations, whereas DDE tested in the same cell line showed a significant increase in aberrations.[7] Mahr and Miltenburger,[8] using Chinese hamster cell line B14 F28, found that DDA did not increase chromosome breaks, DDE induced slightly higher rates of chromosome breaks, and DDD and DDT both induced significant increases in gaps and breaks. The p,p′- isomers of DDT, DDD, and DDE resulted in nearly twice the level of chromosome damage seen with o,p′- isomers when tested with rat kangaroo cell cultures.[9]

Bone marrow cells of laboratory mice (*Mus musculus*) exposed intraperitoneally to DDT showed significantly increased numbers of chromosome deletions and gaps plus deletions at concentrations presumably encountered naturally in secondary consumers.[10] However, Legator et al.[11] found no significant increases in aberration frequencies in bone marrow cells of Osborne-Mendel rats dosed intraperitoneally or orally with DDT. Analysis of mouse meiotic cells in diakinesis showed increased frequencies of breaks and univalents after a 36-hour exposure to DDT.[12]

Dominant lethal assays are indirect indicators of major genetic alterations such as translocations and aneuploidy induction which result in inviable zygotes. Several researchers have tested DDT for dominant lethality in mice and rats[1] with conflicting results. Epstein and co-workers[13,14] found no evidence of induced dominant lethality in mice; however, Clark[12] found decreased numbers of fetal implants and increased frequencies of dead embryos. Cytogenetic examination of human workers occupationally exposed to DDT indicated no significant differences in chromosome aberrations when compared to unexposed workers.[15]

Other Chlorinated Insecticides

The genetic toxicity of aldrin and dieldrin is reviewed by Ashwood-Smith.[16] Aldrin, dieldrin, and endrin have all tested negatively in prokaryotic assays of gene mutation.[1]

Aldrin and dieldrin were shown to induce unscheduled DNA synthesis, a measure of damage repair, in human fibroblast cultures but not in primary rat hepatocytes.[17,18] Several human cell lines have shown increased frequencies of chromosome aberrations after exposure to aldrin or dieldrin.[19-21] Most in vivo assays of these compounds have been negative, including the micronucleus assay in Swiss albino mice,[22] dominant lethal mutation assay in mice,[14] and chromosome aberration assay in Chinese hamsters.[23] However, chromosome aberration frequencies were increased in bone marrow cells of mice and rats dosed at 19 and 38 mg/kg.[19]

Male rats exposed to heptachlor in their diet showed an increased number of chromosome abnormalities and produced positive results in the dominant lethal test over a three-generation period.[24] Endrin and methoxychlor tested negatively in prokaryotic systems for point mutations and in eukaryotic systems for DNA damage and unscheduled DNA synthesis.[2] Chlordecone (Kepone®) gave negative results for point mutations in the Ames Salmonella test,[18] but caused significant increases in chromosome aberrations in M3-1 cell cultures of the Chinese hamster.[25] DBCP (1,2-dibromo-3-chloropropane), a brominated organochlorine nematocide, showed mutagenic activity in the Ames *Salmonella* test after metabolic activation, resulted in a dominant lethal effect in rats, and caused increased sister chromatid exchange (SCE) frequencies in Chinese hamster cell cultures and increased chromosome aberration frequencies for in vivo rat bone marrow and spermatogonial cells and for in vitro Chinese hamster cells.[26] Kilacar® (bis-[parachlorophenyl]-cyclopropylmethanol), an insecticide-acaricide, tested as weakly mutagenic in prokaryotic systems, but showed a 300% increase in chromosome aberrations over controls in Syrian hamster bone marrow cells. A significant dominant lethal effect was not demonstrated.[27]

Organophosphorus Pesticides

Trichlorofon, malathion, vamidothion, demeton, dimethoate, metasystox-R, acephate, azinphos-methyl, disulfoton, methyl parathion, monocrotophos, and trichlorfon have all tested positively for point mutations in prokaryotic and/or eukaryotic systems.[1,2,28]

In vitro increases in sister chromatid exchange frequencies have been observed for dichlorvos (DDVP) in Chinese hamster cells and human lymphocytes; for malathion in human fetal lung fibroblasts; for methyl parathion in Chinese hamster cells, human Birkett lymphoma cells, and normal human lymphoid cell lines; for demeton, trichlorfon, dimethoate, malathion, methidathion, fenthion, and metasystox-R in Chinese hamster V79 cells;[1,29-31] and for acephate, demeton, disulfoton, methyl parathion, monocrotophos, and trichlorfon in Chinese hamster ovary cells.[2]

In vivo mammalian assays of micronucleus formation (MN) have been positive for dursban,[32] methyl parathion, and phorate.[33] In vivo mammalian bone marrow assays for chromosomal aberrations (CA) have been positive for dimethoate,[1] malathion,[34] methyl parathion, and phorate.[35]

Carbamate Pesticides

A number of carbamate pesticides have provided negative results for point mutations in prokaryotic systems.[1,2] The fungicide, benomyl, has tested positively in the mouse lymphoma point mutation assay.[2] The thiocarbamate herbicide, diallate, has tested positively for point mutations in prokaryote, mammal, and insect point mutation assays.[2] In vivo assays of chromosome damage with carbaryl have been negative,[1] but endodan caused increased chromosome aberrations in hamster bone marrow cells.[27]

AVIAN NUCLEAR GENOME ORGANIZATION

Knowledge of avian genetic toxicology, and avian genetics as a whole, lags far behind that of mammals. Only about 6% of the nearly 9000 described species of extant birds have been karyotyped, with few of these examined in detail.[36-39] In contrast, about 40% of known extant mammalian species have been karyotyped, with many species having undergone extensive genetic and cytogenetic examination.[37] As a result, we have a very limited understanding of the range of variability in patterns of structural organization of the avian genome. This lack of knowledge is due, in large part, to the somewhat intractable nature of bird chromosomes. Birds are characterized by large diploid ($2n$) numbers and large numbers of microchromosomes (Figure 1).[37,38] Diploid chromosome numbers range between 40 in the stone curlew (*Burhinus oedicnemus*) to over 100 in the Rallidae and Ramphastidae. Nearly 70% of birds studied have diploid numbers ranging between 76 and 84.[37,40]

With the exception of the family Accipitridae which has relatively few microchromosomes, bird karyotypes are characterized by several pairs of macrochromosomes and a uniquely large number of microchromosomes.[40] The first three pairs of macrochromosomes are typically very large, a middle series of macrochromosomes are intermediate in size, and the microchromosomes may be 1 μm or less in size. These microchromosomes may be barely resolvable with light microscopy and may number as many as 60. Because of the difficulty in accurately identifying and characterizing all these small elements it is not uncommon to see the diploid number of a bird reported as a range rather than a single number. Many investigations of chromosomal patterns in birds have concentrated only on the few macrochromosomes while essentially ignoring the microchromosomal complement. Chicken (*Gallus domesticus*) microchromosomes ap-

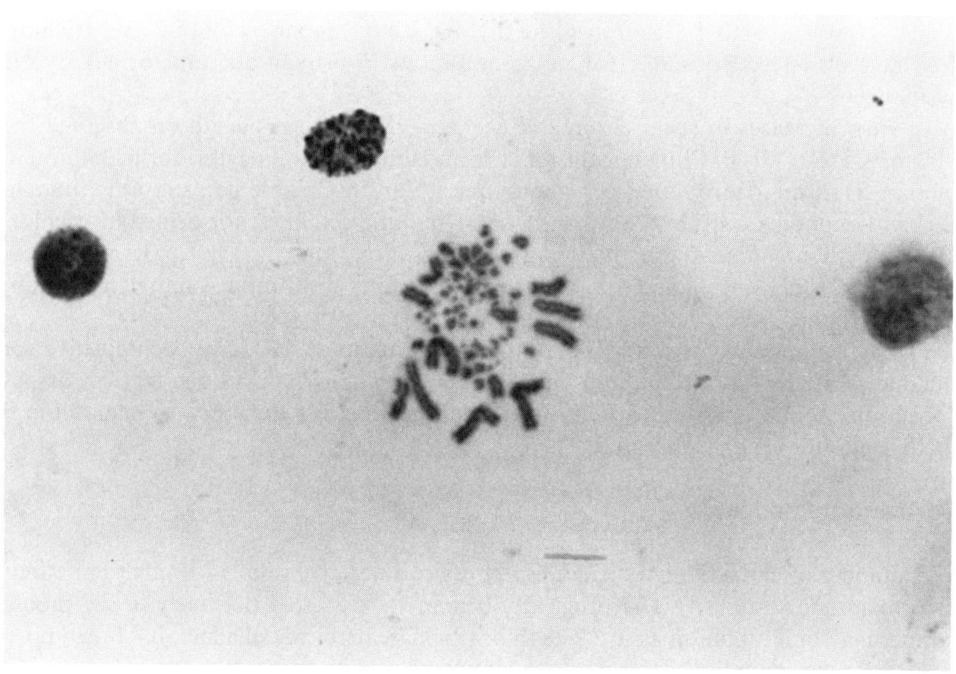

FIGURE 1. Standard bone marrow mitotic metaphase chromosome spread from a domestic chicken (*Gallus domesticus*).

parently undergo replication earlier in the S (synthesis) phase of the cell cycle than do macrochromosomes, but microchromosomes of the rock dove (*Columba livia*) undergo late replication.[37] However, the functional and evolutionary significance of maintaining these large numbers of microchromosomes is not understood.

The sex determination system in birds is also unusual in that it is the female that is heterogametic (ZZ males and ZW females). The Z chromosome is typically a medium-sized macrochromosome and comprises about 10% of the genome;[36] the W is usually much smaller, may be totally C-band positive for constitutive heterochromatin, and is not easily discernible from the many microchromosomes.[38]

There is no apparent pattern to C-band variation in avian autosomes. Positive C-banding for constitutive heterochromatin is visible in centromeric regions of some macrochromosomes but absent in others. In some species many microchromosomes will be C-band positive, whereas in other species microchromosomes appear devoid of heterochromatin.[38] Birds do not appear to have large blocks of heterochromatin forming short arms on chromosomes as do some mammals.

G-banding is a differential staining technique using a process of enzyme digestion followed by Giemsa staining. Areas of chromatin rich in adenine-thymine base pairs stain darkly, while areas relatively rich in guanine-cytosine base pairs do not stain. Unique banding patterns resulting from this process allow identification of homologous chromosome pairs, detection of chromosome polymorphisms, and identification of unique chromosome rearrangements for comparison of intra- and interspecies chromosomal variability. With the exception of white-throated sparrows and some juncos, little is known about intraspecies variability.[36] Considerable G-band pattern variability has been described for the few species studied.[38,41,42]

Analysis of meiotic chromosomes in birds has also been limited.[36,38] Studies of chromosomes during meiosis may be especially useful because they should provide information on a number of parameters unavailable from mitotic studies including orientation and extent of homologous and heterologous pairing of bivalents, chiasma frequency, more accurate determination of diploid number, and more accurate descriptions of microchromosome morphology.[36,43]

Although our knowledge of avian chromosomal morphology and variability is scant, information on genome organization at the molecular level is considerably greater. Most of this work, however, is based on investigations of only one species, the domestic chicken.[37,38]

The average haploid DNA content for birds is between 1.3 and 1.8 pg and appears relatively constant regardless of the diploid number.[37,43] This is about one half that of reptiles and only about one third the average amount for mammals.

Eukaryotes are characterized by both unique (single copy) DNA and repeated sequences of DNA. The repeated sequences may be classified as highly repetitive (HR; $> 10^5$ copies) and middle or moderately repetitive (MR; $< 10^5$ copies). Highly repetitive sequences are often associated with heterochromatin and may never be transcribed.[43] They have been hypothesized to play a role in chromosome recognition events and rearrangement events.[37]

Not only are birds characterized by relatively small amounts of DNA, but they also have relatively small amounts of repetitive DNA.[37,42,43] These two characteristics may be linked because taxa possessing large genomes also typically possess large amounts of repeated DNA.[37] The typical eukaryotic pattern for arrangement of unique and repetitive sequences is that first documented in the African clawed frog, *Xenopus laevis*. This pattern has about three fourths of the unique or single copy DNA interspersed with about three fourths of the repeated DNA. Single copy sequences of between 800 and 1000 base pairs (bp) alternate with short interspersed sequences of repetitive DNA about 300 bp in length.[37,43] In most cases, mammals also have this short-period interspersed DNA pattern. A second pattern of arrangement for repetitive and unique sequences is that found in *Drosophila*. The *Drosophila* genome is characterized by a long-period interspersed pattern in which repeated sequences of 5600 bp or more are alternated with unique sequences of 13,000 bp. This same long-period interspersed pattern has been documented in the domestic chicken and the domestic Muscovy duck (*Cairina moschata*).[37,41] In the chicken, MR sequences are 204 kb long and interspersed single-copy sequences are 5 kb or more. In approximately 60% of the genome, unique sequences are even longer.

The significance of the short-period interspersed pattern seen in mammals and most other eukaryotes vs the long-period interspersed pattern seen in birds and some insects is not understood. Long-period interspersed patterns appear to be associated with small genome size.[43] Shields[37,44] suggests a correlation with ecotypic specialization. Crain et al.[45] suggest that animals with short-period interspersed patterns retain greater potential for speciation and genetic divergence while species with long-period interspersed patterns have lost this potential.

Regardless of their potential causes or functional significance, the unique nature of several avian genome characteristics suggests that extrapolation of genetic toxicity data from a microbial or mammalian model to a presumed avian effect may not always be appropriate. Good avian models of genetic toxicity are needed.

AVIAN GENETIC TOXICITY STUDIES

Only one avian genetic toxicity assay system has been well developed; however, it has not been widely used. The chick embryo cytogenetic test (CECT), developed by Stephen E. Bloom, was first used to check for spontaneous chromosomal anomalies in chickens with Marek's disease[46] and for spontaneous mutations in several strains of domestic chickens.[47–49] The system uses either 3- to 4-day or 6- to 7-day chick embryos (Figure 2). Chromosome aberrations are examined in cells obtained from limb bud, allantoic sac, or blood. Embryos are exposed to test compounds by an "air-cell" method in which a 0.5 cm^2 portion of the shell overlying the air cell is removed, test solution is dropped onto the inner cell membrane with a micropipette, and the opening in the shell closed with adhesive tape. Test compound is then gradually absorbed through the inner shell membrane into the embryo.[50] Alternatively, test compound may be applied by the "window" method in which the compound is applied directly to the embryo. In this procedure, a hole is punched in the air cell, the egg is positioned horizontally, the shell is cut away directly above the embryo, and test solution is dropped by micropipette directly onto the embryo. The latter method is preferred when there is concern about absorption of test solution through the inner shell membrane.[49] A mitotic inhibitor such as colchicine is added 18 hours after dosing via the same route as the test solution at the rate of 20 μL of 2 μg/mL solution per egg. Chick embryos are removed from eggs 2 hours later and processed by the solid tissue method.[49,50] After removal, the embryo is placed in a hypotonic solution of 0.9% sodium citrate for 30 minutes at room temperature. The embryo is then fixed in ethanol:acetic acid (3:1) for a minimum of 4 hours. After fixation, appropriate tissues are removed and placed in 50% acetic acid. Tissue is macerated with a scalpel to form a cell suspension. The cell suspension is transferred to a clean, warm (47°C) microscope slide using a microhematocrit capillary tube (75×1.2 mm) equipped with a rubber bulb. Transfer takes place by drawing cell suspension into the tube, expelling a small amount onto the slide, and quickly drawing the suspension back into the tube, leaving a ring of cells approximately

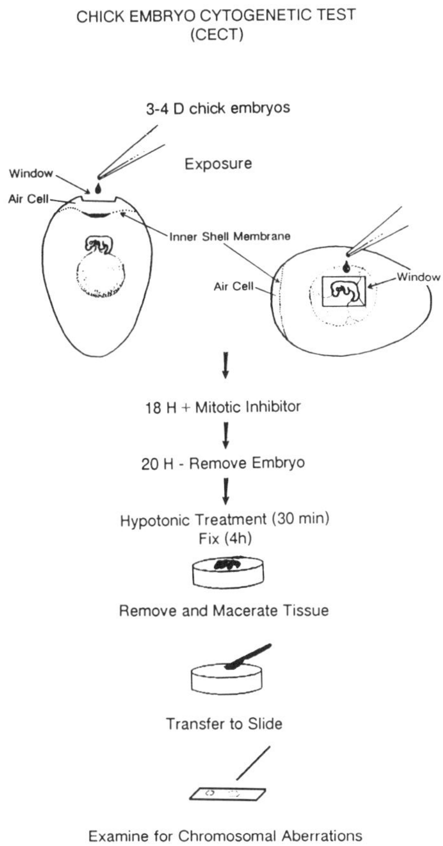

FIGURE 2. Diagrammatic representation of procedures involved in the CECT developed by S. E. Bloom.

1 cm in diameter on the slide. The process is repeated to make two to three rings of cells on each slide. Air dried slides are stained in 4% Giemsa in phosphate buffer for analysis of chromosome aberrations, or slides can be used in conventional G- and C-banding techniques.

Bloom and Hsu[51] developed techniques to document SCE in CECT. SCEs are reciprocal exchanges between chromatids of the same chromosome (Figure 3). The mechanism underlying SCE formation and its relationship to mutagenesis is not understood, but the frequency of SCE formation responds in a dose-related manner to exposure to many mutagens.[52] Alterations in the basic technique for chromosome aberration analysis in CECT involve dosing with bromodeoxyuridine (BrdU) 2 hours prior to dosing with test solution and use of a fluorescence-plus-Giemsa staining technique with Hoechst.[50] Aberrations and SCEs are scored in the first five pairs of macrochromosomes using a light microscope. These chromosomes account for approximately 50% of the DNA in a chick cell.[50]

The CECT system has been used to document chromosome aberrations and SCEs in chick embryos exposed to approximately 50 chemicals, including known mutagens, nonmutagens, dyes, aflatoxins, buffers, and culture media.[49,50] All nonmutagens were negative in the CECT. All known mutagens but three caused significant increases in SCE and one of those three, bleomycin, caused significant increases in chromosome aberrations.

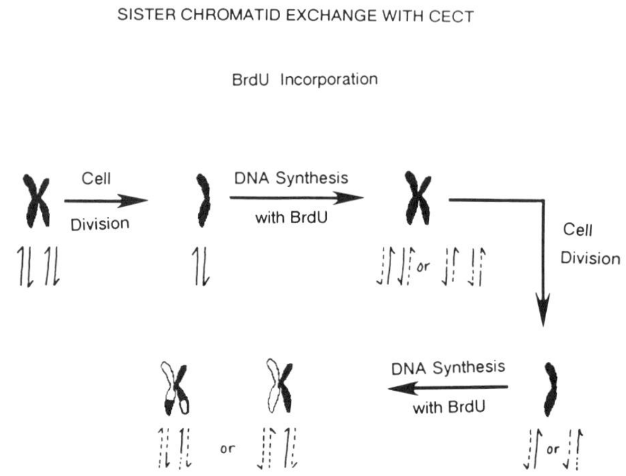

FIGURE 3. Illustration of the mechanism for visualization of SCEs. BrdU is incorporated via metaphase chromosomes prepared by CECT.

The CECT system has also been used to test the mutagenic and clastogenic potential of the fungicide Maneb (manganese ethylenebisdithiocarbamate) on 4-day embryos. Unincubated eggs were dipped for 30 seconds into five different aqueous concentrations of the pesticide plus controls, incubated for 4 days, and examined for SCEs and chromosome aberrations. Significantly increased levels of SCE were documented in embryos exposed to the two highest doses, but no clastogenic effects were observed at any dose level.[53]

The CECT system has also been applied in situ.[54,55] Unincubated herring gull (*Larus argentatus*) eggs were collected from five breeding colonies in polluted areas around the Great Lakes and from a colony in a relatively pristine area on the Atlantic Coast. After 6 days of incubation, embryos were processed for SCE analysis by Bloom's method. The average number of SCEs per chromosome for the colonies ranged from 0.069 to 0.101. The authors found no significant differences between any of the colonies, although levels of several contaminants were documented in eggs from the polluted colonies. Ellenton et al.[56] used extracts of egg homogenates from the six colonies in the Ames *Salmonella* assay to test for point mutations and in Chinese hamster ovary (CHO) cells to test for induction of chromosome aberrations and SCEs. All extracts were negative in the Ames test; however, all including controls were positive in the CHO cell culture assays. The authors concluded that herring gull eggs were unsuitable as biomonitors and recommended that research efforts to establish suitable avian genetic biomonitors be channeled elsewhere.[56]

POTENTIAL GENETIC TOXICITY ASSAYS

Mitotic Chromosomes

In spite of the relatively small amount of work done in avian cytogenetics, a number of protocols have been developed for extraction of mitotic metaphase chromosomes from several tissues or from cultured cells of birds. Shoffner et al.[57] reviews several of the older methods including the feather pulp technique and a short-term leukocyte culture technique.

Feather Pulp Technique

The feather pulp technique is promoted as rapid and easy to use. This method uses cells from the actively proliferating pulp at the base of the growing feather. Wing feathers from newly hatched birds provide the best material, although small body feathers from adult specimens can be used. A 0.05% solution of a mitotic inhibitor such as colchicine is injected either into the peritoneal cavity or directly into the wing vein 30 to 45 minutes prior to processing, allowing metaphase cells to accumulate. Recommended dose levels are 0.04 mL/30 g body weight for young birds and 1 mL/1.3 kg body weight for large adult birds.[58] After the mitotic inhibitor treatment, feathers are removed and the semisolid pulp is removed from the proximal end of the feather by squeezing out the pulp with a curved forceps. Pulp is placed immediately into a hypotonic solution of 0.45% sodium citrate and incubated at room temperature for 10 to 15 minutes. After hypotonic treatment, the pulp pellet is transferred to a 1:1 solution of glacial acetic acid:distilled water and fixed for a minimum of 30 minutes. Slides are prepared by removing a piece of pulp from the fixative, draining on a piece of bibulous paper and then smearing the pulp onto a clean, dry microscope slide in a circular motion. After the fixed pulp is smeared on the slide, the material is quickly covered with a coverslip and squashed by pushing downward on the slide with the thumb. The coverslip is then removed and slides can be stained and examined.[57]

Leukocyte Culture Technique

Shoffner et al.[57] describe a modification of a mammalian leukocyte culture technique that requires approximately 10 mL of blood per individual. Blood is collected in sterile, heparinized tubes either by cardiac puncture or from the wing vein. Leukocytes are isolated by centrifugation and plasma is removed and added to culture vials containing approximately 7 mL of Eagle's MEM (minimum essential medium). A mitogen such as phytohemagglutinin (PHA) is added to stimulate mitoses and cultures are incubated at 37.5°C for 72 hours. One drop of 0.025% colchicine is added to culture vials to arrest cells in metaphase 45 to 60 minutes before cell harvest. At the end of the colchicine treatment, contents of culture vials are centrifuged at 500 rpm for 5 to 10 minutes to form a cell pellet. The pellet is gently resuspended in approximately 5 mL of a hypotonic 0.45% sodium citrate solution, incubated for 15 minutes at room temperature, and recentrifuged. The hypotonic is decanted and the pellet fixed for 30 minutes in modified Carnoy's fixative (3:1 methanol:acetic acid). Slides are prepared by dropping the cell suspension onto clean, dry microscope slides and igniting the solution to facilitate chromosome spreading. Slides then can be air dried, stained, and examined.[57]

Bone Marrow Technique

An alternative method uses bone marrow from the tibia.[58] The tibia is removed, epiphyses clipped off, and the nonpink fatty distal portion of marrow removed and discarded. A narrow gauge needle is inserted into the marrow cavity and remaining marrow flushed from the cavity into a centrifuge tube with 5 mL of culture medium. Marrow is gently aspirated with a syringe and needle to disaggregate the cells. To this cell suspension, 0.1 mL of 0.001% colchicine is added. The suspension is incubated at 37°C for 37 to 42 minutes. After incubation, the suspension is centrifuged for approximately 2 minutes at 800 to 1000 rpm, supernatant decanted, the pellet gently resuspended in a hypotonic 0.75 *M* KCl solution, and incubated for an additional 20 minutes. The suspension is recentrifuged, hypotonic supernatant removed, and the pellet gently resuspended in modified Carnoy's fixative. Pellets are washed in three successive changes of Carnoy's fixative and slides are prepared by dropping suspension onto clean, dry microscope slides.[58] Hobart et al.[59] describe a similar bone marrow technique that does not include the first incubation in culture medium but uses an intraperitoneal injection of 0.05% colchicine at 0.3 to 0.4 mL/100 g body weight 20 minutes prior to sacrifice.

Fibroblast Cell Culture Technique

Shields[38] describes a cell culture method for establishment of a cell line that can be maintained for several weeks and harvested for chromosome extraction several times. Kidneys are removed from sacrificed individuals using sterile technique. Tissue is partially macerated by chopping with sterile scissors and scalpel and then dispersed with a collagen-specific protease such as dispase. Dispersed cells are cultured in Eagle's MEM supplemented with HEPES (2×10^{-2} *M*) and fortified with 20% fetal bovine serum (FBS).[60] Cultures are incubated at 37°C. At 1 hour prior to harvest, 10 μg/mL colchicine is added to culture flasks. Cells are harvested by trypsination, removed from culture flasks, and pelleted out by centrifugation. The pellet is resuspended in a hypotonic 0.075 *M* KCl solution for 12 to 15 minutes. Cell suspension is then fixed in modified Carnoy's fixative and dropped onto clean microscope slides as described for the bone marrow method.

Each of these methods has been successfully used in descriptive and systematic studies of avian chromosomes but have not been used in studies of avian genetic toxicology. Bone marrow and kidney fibroblast culture methods are disadvantageous in that both require killing specimens. However, both tend to yield larger numbers of analyzable metaphase spreads and better quality spreads[38,58] compared to the feather pulp and leukocyte culture methods. The latter two methods, however, can be used for sequential sampling from the same individuals or when study specimens cannot be killed for examination. The bone marrow method can be readily performed in the field and fixed cell suspensions frozen in liquid nitrogen and returned to the laboratory.

Both bone marrow and cell culture methods are conventional techniques in mammalian chromosomal aberration assays.[61] The bone marrow method has been successfully used in several in situ studies of the effects of environmental contaminants on populations of exposed small mammals.[62-65] However, problems associated with large diploid numbers and large numbers of microchromosomes will probably still confound the regular use of standard chromosome aberration assays in avian genetic toxicology. In most cases, only partial chromosomal complements (macrochromosomes) can be reliably scored for aberrations, and assays for aneuploidy will be almost impossible because of the variable number of microchromosomes visible in individual cells.

Meiotic Chromosomes

Studies of avian chromosomes during meiosis also have been limited in part because of high diploid number and large numbers of microchromosomes, and because wild birds are meiotically active for only a relatively short period of time.[36] Shields[66] describes a technique for extraction of chromosomes from testes. Testes are removed and placed in warm, distilled water (40°C) for 20 minutes. The tunica is removed and tubules dispersed in three washes of modified Carnoy's fixative at 4°C. Fixed cells are smeared onto albuminized slides, a coverslip is added, and the cells are squashed. Coverslips can be removed in fixative and slides air dried, stained, and analyzed.

More recently, Hale et al.[67] described a method for microspreading and settling of avian meiotic tissues for light and electron microscopic analysis of silver-stained synaptonemal complexes. The synaptonemal complex (SC) is a thin, ribbon-like proteinaceous structure that forms the longitudinal axis of chromosomal bivalents during pachynema of meiotic prophase I. Upon pairing, each chromosome of a meiotic bivalent forms an axis which becomes one half of the lateral elements of an SC. Silver preferentially binds to the two lateral elements of the SC rather than to the associated chromatin, giving a clear, linear representation of the paired chromosomes. Avian SCs are prepared according to the technique of Moses[68] for mammalian testes as modified by Greenbaum et al.[69] Testes are removed and washed in Eagle's MEM. The tunica is removed and tubules are immersed in 1 to 2 drops of MEM and macerated with curved forceps and scalpel. The resulting solution is drawn up into a 1-mL glass syringe. Contents are expressed and drawn back into the syringe three times, then centrifuged at 1000 rpm for 5 minutes. The pellet is resuspended in 1 to 1.5 mL MEM and placed on ice. Microspreading is accomplished by touching a plastic-coated microscope slide to the convex surface of a 0.5% (w/v) NaCl solution onto which the spermatocyte nuclei have been spread with a microhematocrit tube. Surface spread slides are fixed by immersion for 5 minutes in a 4% paraformaldehyde solution with 0.03% sodium dodecyl sulfate. Slides are then immersed in paraformaldehyde without sodium dodecyl sulfate for an additional 5 minutes, drained briefly, and quickly rinsed in 0.4% Photoflo® for 10 to 15 seconds. After drying, slides can be stained with silver nitrate following the method of Howell and Black.[70]

Synaptonemal complex analysis may be especially useful in studies of avian genetic toxicity for several reasons. Because chromosomes are completely paired during pachynema, only a haploid number of configurations must be examined. This technique yields large numbers of pachytene cells for examination. Also, because SC configurations are more elongate than mitotic preparations, morphology of microchromosomes can be more readily determined. Typically, studies of meiotic chromosome damage are carried out on cells in diakinesis/metaphase I. However, for damaged cells to persist through meiosis to these stages, chromosomes must undergo normal synapsis and chiasma formation. If damage is severe enough that these processes are disrupted, these cells would never been seen. Therefore, analyses of diakinesis/metaphase I may not give a realistic indication of structural damage in meiotic chromosomes. Because SC analysis is conducted on cells much earlier in meiosis, more and different types of structural damage may be recognized.[71]

Thus far, SC analysis in birds has been restricted to analysis of translocation and pericentric inversion heterozygotes in the domestic chicken.[72,73] Cawood and Breckon[71] and Allen et al.,[74] however, have shown that SC analysis can be an effective and rapid in vivo germ cell assay in Syrian hamsters (*Mesocricetus auratus*) exposed to X-rays and C57B1/6J mice and Chinese and Armenian hamsters exposed to mitomycin and cyclophosphamide. These mammalian studies suggest that SC analysis in avian species may be a feasible model of genetic toxicity.

Flow Cytometry

Measurement of variation in nuclear DNA content among populations of cells using flow cytometry (Figure 4) shows great utility as an assay of genetic toxicology because of the speed, precision, and ease with which large numbers of cells can be analyzed. The system is sensitive enough to distinguish among cells that differ by as little as 1 to 2% in DNA content.[75,76] Application of flow cytometry to analysis of chemically and physically induced chromosome damage has been suggested by several investiga-

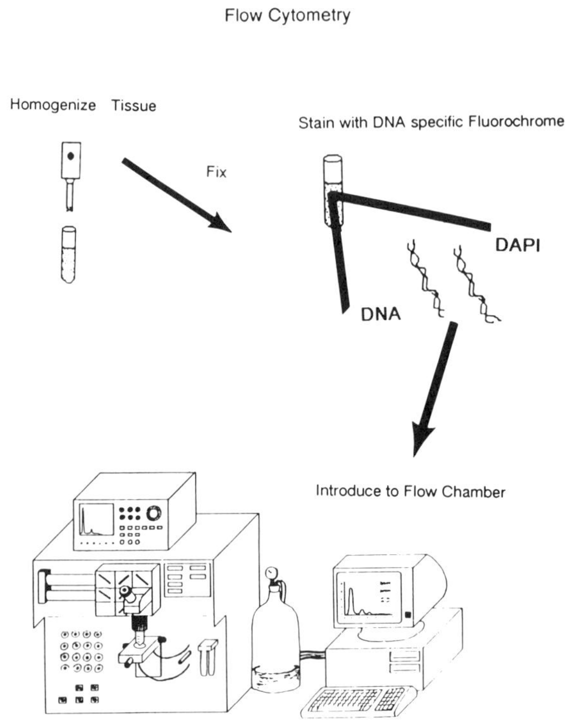

FIGURE 4. Illustration of the procedures of tissue disaggregation, fixation, and staining involved in flow cytometric analysis of nuclear DNA content.

tors.[75-77] Flow cytometry was first applied to analysis of chemically induced chromosome damage by Otto and Oldiges,[78] who reasoned that most structural chromosome aberrations that are not lethal or repaired in dividing cells will cause an unequal distribution of DNA to daughter cells, and that a growing cell population should exhibit an increased dispersion of nuclear DNA content over several cell cycles. They showed that the G_1 peak in somatic cells challenged with known mutagens had a typical broadened and flattened appearance and a higher coefficient of variation for mean DNA content when compared to untreated cells.

Spano et al.[79] demonstrated induction of cytogenetic damage in differentiating spermatogonia from testes of C57Bl/CnexC3H/Cne laboratory mice exposed to benzene. McBee and Bickham[80] used flow cytometry to document greater variation in DNA content in spleen cells of wild white-footed mice (*Peromyscus leucopus*) collected at a hazardous waste site compared to mice from a pristine reference site. Aquatic turtles (*Pseudemys scripta*) inhabiting seepage basins contaminated with low concentrations of radioactive and nonradioactive waste showed significantly greater variation in nuclear DNA content in blood cells as compared to turtles collected from a reference population.[81]

Few studies using flow cytometric measurement of nuclear DNA content have been performed on avian species. However, one study has used flow cytometric analysis of red blood cell DNA in mallards (*Anas platyrhychos*) exposed to environmental ^{137}Cs.[82] George et al.[82] released 100-day-old male mallards, previously unexposed to ^{137}Cs, on an abandoned nuclear reactor cooling tower reservoir. Blood samples periodically collected over 1 year from these animals and from two reference populations were analyzed for nuclear DNA content. Significant differences in coefficients of variation around mean DNA content between exposed individuals and reference individuals were detected after 9 months. Two individuals from the reservoir also showed evidence of aneuploidy after 9 months of exposure, but none of the reference animals showed the typical dual peak pattern indicative of aneuploidy. Nuclear DNA content has been determined for only a few other avian species such as domestic chicken, golden pheasant (*Chrysolophus pictus*), and Molluccan cockatoo (*Cacatua moluccensis*).[83] Interestingly, erythrocytes of the domestic chicken represent one of the widely accepted internal standards used in quantification of DNA content in animals with unknown DNA values;[84-86] therefore, rigorous methods for analysis of avian tissues are available. Heparinized blood is obtained by cardiac puncture or drawn from the wing vein, mixed with a citrate buffer, and can be stored for long periods of time in liquid nitrogen.[87] When ready for analysis, cell suspensions are thawed and mixed with a staining solution consisting of DAPI (4′,6-diamidino-2-phenylindole) and 0.04 *M* $MgCl_2$ in an 0.1 *M* tris-HCl buffer (pH 8.0),[81] or one consisting of propidium iodide and spermine tetrahydrochloride in trisodium citrate buffer (pH 7.6).[87]

Solid tissues such as spleen, bone marrow, or testis can be prepared by mincing with scissors and scalpel and shaking in a few milliliters of a pretreatment solution of citric acid and Tween® 20. The cell suspension is centrifuged and the pellet fixed in 70% ethanol. The fixed pellet can be stored refrigerated for several months. Fixed cells are prepared for analysis by centrifuging, resuspending in the pretreatment solution, and incubated for 10 minutes at room temperature. Approximately 5 vol of a staining solution containing DAPI and Na_2HPO_4 in deionized water are added to the fixed cells. Cells are ready for analysis after a 24-hour staining period.[88]

Flow cytometry may represent an excellent tool for analysis of effects in birds exposed to genetic toxins because the inherent problems of somatic chromosome analysis are overcome. Because bird erythrocytes are nucleated, adequate samples of blood can be drawn without killing or stressing animals and serial samples can be analyzed. Several solid tissues including spleen, liver, kidney, thymus, testes, and tumor cells can be examined,[87] allowing comparison of effects among different target organs and comparison of effects in somatic cells vs germ cells.

CONCLUSIONS

No models for avian genetic toxicity evaluation are currently available for wide use. Only one method, the chick embryo cytogenetic test, has been used with a number of test compounds. Several other methods for extraction of chromosomes have been developed. These methods are applicable to genetic toxicity studies but, because of unique features of avian chromosomal complements such as relatively high diploid numbers and large numbers of microchromosomes, direct examination of mitotic chromosomes for induced damage may not provide the most effective means of evaluating the impact of environmental mutagens on avian genomes. These unique chromosomal features along with unique features of avian genome organization at the molecular level emphasize the need for models of avian genetic toxicity rather than relying on extrapolation from mammalian or prokaryotic models.

Possible alternative models to mitotic chromosome studies include analysis of synaptonemal complex in prophase I meiotic cells and flow cytometric analysis of nuclear DNA content variation. Both of these systems have been effectively used with mammalian models in laboratory or field studies of genetic toxicity. Both also have been successfully applied to avian species, albeit not in a toxicological context. A third alternative not discussed here are molecular-level measures of genetic toxicity such as the DNA alkaline unwinding assay.[89] This assay system quantifies the amount of induced damage to the DNA molecule by determining ratios of single-stranded to double-stranded DNA in a sample. The technique has been successfully applied to fish and mammals[89,90] and should be readily adaptable to birds.

Finally, a critical assumption of any assay of genetic toxicity is that induced mutations ultimately have the potential to alter population parameters such as survivorship, population density, age structure, and reproductive potential. Mutagenic properties of many environmental contaminants are documented and altered demographic patterns have also been demonstrated in animals exposed to environmental contaminants;[91,92] however, coordinated studies of both genetic and demographic effects in populations exposed to xenobiotics such as pesticides have not been conducted. In order to have a thorough understanding of the effects of mutagenic contaminants on exposed populations, studies of genetic toxicity must not only address immediate effects at the chromosomal and molecular levels, but also must address the more long-term aspects of inheritance and reproductive success.

ACKNOWLEDGMENT

Support was provided by OCAST grant HN8-012 and AFOSR grant 89194. Monte L. Thies and Kathleen Thies provided helpful comments on earlier versions of this manuscript. Drawings are by Monte L. Thies.

REFERENCES

1. Moutschen-Dahmen, J., M. Moutschen-Dahmen and N. Degraeve, Mutagenicity, carcinogenicity, and teratogenicity of insecticides, in *Mutagenicity, Carcinogenicity, and Teratogenicity of Industrial Pollutants*, M. Kirsch-Volders, Ed., Plenum Press, New York, 1984, pp. 127–203.
2. Waters, M. D., S. S. Sandhu, V. F. Simmons, K. E. Mortelmans, A. D. Mitchell, T. A. Jorgenson, D. C. L. Jones, R. Valencia and N. E. Garrett, Study of pesticide genotoxicity, in *Genetic Toxicology: An Agricultural Perspective*, R. A. Fleck and A. Hollaender, Eds., Plenum Press, New York, 1982, pp. 275–326.
3. Epstein, S. S., and M. S. Legator, *The Mutagenicity of Pesticides: Concepts and Evaluation*, The MIT Press, Cambridge, MA, 1971.
4. Hsu, T. C., Introduction, in *Cytogenetic Assays of Environmental Mutagens*, T. C. Hsu, Ed., Allanheld, Osmun Publishers, Totowa, NJ, 1982, pp. 1–9.
5. Hartman, P. E., Mutagens: some possible health impacts beyond carcinogenesis, *Environ. Mutagen.*, 5, 139–152, 1983.
6. Risebrough, R. W., Pesticides and bird populations, in *Current Ornithology*, Vol. 3, R. F. Johnston, Ed., Plenum Press, New York, 1986, pp. 397–427.
7. Kelly-Garvert, F., and M. Legator, Cytogenetic and mutagenic effects of DDT and DDE in a Chinese hamster cell line, *Mutat. Res.*, 17, 223–229, 1973.
8. Mahr, U., and H. G. Miltenburger, The effects of insecticides on Chinese hamster cell cultures, *Mutat. Res.*, 40, 107–118, 1976.
9. Palmer, K. A., S. Green and M. S. Legator, Cytogenetic effects of DDT and derivatives of DDT in a cultured mammalian cell line, *Toxicol. Appl. Pharmacol.*, 22, 355–364, 1972.
10. Larsen, K. D., and S. M. Jalal, DDT induced chromosome mutations in mice-further testing, *Can. J. Genet. Cytol.*, 16, 491–497, 1974.
11. Legator, M. S., K. A. Palmer and I.-D. Adler, A collaborative study of in vivo cytogenetic analysis. I. Interpretation of slide preparations, *Toxicol. Appl. Pharmacol.*, 24, 337–350, 1973.

12. Clark, J., Mutagenicity of DDT in mice, *Drosophila melanogaster* and *Neurospora crassa*, *Aust. J. Biol. Sci.*, 27, 427–440, 1974.
13. Epstein, S., and H. Shafer, Chemical mutagens in the human environment, *Nature*, 219, 385–387, 1968.
14. Epstein S., E. Arnold, J. Andrea, W. Bass and Y. Bishop, Detection of chemical mutagens by the dominant lethal assay in the mouse, *Toxicol. Appl. Pharmacol.*, 23, 288–325, 1972.
15. Rabello, M., W. Becak, W. De Almeida, P. Pigati, M. Ungaro, T. Murata and C. Pereira, Cytogenetic study on individuals occupationally exposed to DDT, *Mutat. Res.*, 28, 449–454, 1975.
16. Ashwood-Smith, M., The genetic toxicology of aldrin and dieldrin, *Mutat. Res.*, 86, 137–154, 1981.
17. Ahmed, F., R. Hart and N. Lewis, Pesticide induced DNA damage and its repair in cultured human cells, *Mutat. Res.*, 42, 161–174, 1977.
18. Probst, G., R. McMahon, L. Hill, C. Thompson, J. Epp and S. Neal, Chemically induced unscheduled DNA synthesis in primary rat hepatocyte cultures, a comparison with bacterial mutagenicity using 218 compounds, *Environ. Mutat.*, 3, 11–32, 1981.
19. Georgian, L., The comparative cytogenetic effects of aldrin and phosphamidon, *Mutat. Res.*, 31, 103–108, 1975.
20. Majumdar, S., H. Kopelman and M. Schnitman, Dieldrin-induced chromosome damage in mouse bonemarrow and W1-38 human lung cells, *J. Hered.*, 67, 303–307, 1976.
21. Trepanier, G., F. Marchessault, J. Bansal and A. Chagonon, Cytological effects of insecticides on a human lymphoblastoid cell line, *In Vitro*, 13, 201, 1977.
22. Usha Rani, M., O. Reddy and P. Reddy, Mutagenicity studies involving aldrin, endosulfan, dimethoate, phosphamidon, carbaryl and ceresan, *Bull. Environ. Contam. Toxicol.*, 25, 277–282, 1980.
23. Bidwell, K., E. Weber, I. Nienhold, T. Conner and M. Legator, Comprehensive evaluation for mutagenic activity of dieldrin, *Mutat. Res.*, 31, 314, 1975.
24. Cerey, K., V. Izahovic and J. Ruttkay-Nedecka, Effect of heptachlor on dominant lethality and bone marrow in rats, *Mutat. Res.*, 21, 26, 1973.
25. Bale, S. S., Cytological effects of Kepone on Chinese hamster cells, *J. Hered.*, 74, 123–124, 1983.
26. Whorton, M. D., and D. E. Foliart, Mutagenicity, carcinogenicity and reproductive effects of dibromochloropropane (DBCP), *Mutat. Res.*, 123, 13–30, 1983.
27. Tzoneva, M., A. Kappas, V. Georgieva, R. Vachkova and V. Tziolas, On the genotoxicity of the pesticides Endodan and Kilocar in 6 different test systems, *Mutat. Res.*, 157, 13–22, 1985.
28. Pandita, T. K., Mutagenic studies on the insecticide Metasystox-R with different genetic systems, *Mutat. Res.*, 124, 97–2102, 1983.
29. Chen, H. H., J. L. Hsueh, S. R. Sirianni and C. C. Huang, Induction of sister-chromatid exchanges and cell cycle delay in cultured mammalian cells treated with eight organophosphorus pesticides, *Mutat. Res.*, 88, 307–316, 1981.
30. Chen, H. H., S. R. Sirianni and C. C. Huang, Sister-chromatid exchanges and cell-cycle delay in Chinese hamster V79 cells treated with 9 organophosphorus compounds (8 pesticides and 1 defoliant), *Mutat. Res.*, 103, 307–313, 1982.
31. Chen, H. H., S. R. Sirianni and C. C. Huang, Sister-chromatid exchanges in Chinese hamster cells treated with seventeen organophosphorus compounds in the presence of a metabolic activation system, *Environ. Mutagen.*, 4, 621–624, 1982.
32. Amer, S. M., and M. A. Fahmy, Cytogenetic effects of pesticides. I. Induction of micronuclei in mouse bone marrow by the insecticide Dursban, *Mutat. Res.*, 101, 247–255, 1982.
33. Grover, I. S., and P. K. Malhi, Genotoxic effects of some organophosphorus pesticides. I. Induction of micronuclei in bone marrow cells in rat, *Mutat. Res.*, 140, 131–134, 1985.
34. Dulout, F. N., M. C. Pastori and O. A. Olivero, Malathion-induced chromosomal aberrations in bone marrow cells of mice: dose-response relationships, *Mutat. Res.*, 122, 163–167, 1983.
35. Malhi, P. K., and I. S. Grover, Genotoxic effects of some organophosphorus pesticides. II.

In vivo chromosomal aberration bioassay in bone marrow cells in rat, *Mutat. Res.*, 188, 45-51, 1987.

36. Shields, G. F., Chromosomal variation, in *Avian Genetics: A Population and Ecological Approach*, F. Cooke and P.A. Buckley, Eds., Academic Press, London, 1987, pp. 79-104.
37. Shields, G. F., Organization of the avian genome, in *Perspectives in Ornithology. Essays Presented for the Centennial of the American Ornithologists Union*, A. H. Brush and G. A. Clark, Jr., Eds., Cambridge University Press, New York, 1983, pp. 271-290.
38. Shields, G. F., Bird chromosomes, in *Current Ornithology*, Vol. 1, R. F. Johnston, Ed., Plenum Press, New York, 1983, pp. 191-209.
39. Shields, G. F., Comparative avian cytogenetics: a review, *Condor*, 84, 45-58, 1982.
40. Sasaki, M., N. Takagi and C. Nishida, Current profiles of avian cytogenetics, with notes on chromosomal diagnosis of sex in birds, *Nucleus*, 27, 63-73, 1984.
41. Stock, A. D., F. E. Arrighi and K. Stefos, Chromosome homology in birds: banding patterns of the chromosomes of the domestic chicken, ring-necked dove, and domestic pigeon, *Cytogenet. Cell Genet.*, 13, 410-418, 1974.
42. Stock, A. D., and G. A. Mengden, Chromosome banding pattern conservatism in birds and nonhomology of chromosome banding patterns between birds, turtles, snakes, and amphibians, *Chromosoma*, *(Berlin)*, 50, 69-77, 1975.
43. Quinn, T. W., and B. N. White, Analysis of DNA sequence variation, in *Perspectives in Ornithology, Essays Presented for the Centennial of the American Ornithologists Union*, A. H. Brush and G. A. Clark, Jr., Eds., Cambridge University Press, New York, 1987, pp. 163-198.
44. Shields, G. F., and N. A. Straus, DNA-DNA hybridization studies of birds, *Evolution*, 29, 159-166, 1975.
45. Crain, W. R., E. H. Davidson and R. J. Britten, Contrasting patterns of DNA sequence arrangements in *Apis mellifera* (honeybee) and *Musca domestica* (housefly), *Chromosoma*, 59, 1-12, 1976.
46. Bloom, S. E., Marek's disease: chromosome studies of resistant and susceptible strains, *Avian. Dis.*, 14, 478-490, 1970.
47. Bloom, S. E., The origins and phenotypic effects of chromosome abnormalities in avian embryos, *Proc. 15th World's Poultry Congress, New Orleans*, McGregor and Warner, New York, 1974, pp. 316-320.
48. Bloom, S. E., Detection of normal and aberrant chromosomes in chicken embryos and in tumor cells, *Poultry Sci.*, 60, 1355-1361, 1981.
49. Bloom S. E., Chick embryos for detecting environmental mutagens, in *Chemical Mutagens, Principles and Methods in their Detection*, Vol. 5, A. Hollaender and F. J. de Serres, Eds., Plenum Press, New York, 1978, pp. 203-232.
50. Bloom, S. E., Detection of sister chromatid exchanges in vivo using avian embryos, in *Cytogenetic Assays of Environmental Mutagens*, T. C. Hsu, Ed., Allenheld, Osmun Publishers, Totowa, NJ, 1982, pp. 137-159.
51. Bloom, S. E., and T. C. Hsu, Differential fluorescence of sister chromatids in chicken embryos exposed to 5-bromodeoxyuridine, *Chromosoma*, 51, 261-267, 1975.
52. Brusick, D., *Genetic Toxicology*, Plenum Press, New York, 1980.
53. Arias, E., Sister-chromatid exchanges and chromosomal aberrations in chick embryos after treatment with the fungicide maneb., *Mutat. Res.*, 206, 271-273, 1988.
54. Ellenton, J. A., and M. F. McPherson, The herring gull as an indicator species for mutagenic activity in the Great Lakes, *Environ. Mutat.*, 3, 356, 1981.
55. Ellenton, J. A., and M. F. McPherson, Mutagenicity studies on herring gulls from different locations on the Great Lakes. I. Sister chromatid exchange rates in herring-gull embryos, *J. Toxicol. Environ. Health*, 12, 317-324, 1983.
56. Ellenton, J. A., M. F. McPherson and K. L. Maus, Mutagenicity studies on herring gulls from different locations on the Great Lakes. II. Mutagenic evaluation of extracts of herring-gull eggs in a battery of in vitro mammalian and microbial tests, *J. Toxicol. Environ. Health*, 12, 325-336, 1983.

57. Shoffner, R. N., A. Krishan, G. J. Haiden, R. K. Bammi and J. S. Otis, Avian chromosome methodology, *Poultry Sci.*, 46, 333-344, 1967.
58. Christidis, L., A rapid procedure for obtaining chromosome preparations from birds, *Auk*, 102, 892-893, 1985.
59. Hobart, H. L., S. J. Gunn and J. W. Bickham, Karyotypes of six species of North American blackbirds (Icteridae: Passeriformes), *Auk*, 99, 514-518, 1982.
60. Ryttman, H., and H. Tegelstrom, Chromosomal evolution in the family Phasianidae (Aves), *Hereditas*, 98, 71-75, 1983.
61. Preston, R. J., W. Au, M. A. Bender, J. G. Brewen, A. V. Carrano, J. A. Heddle, A. F. McFee, S. Wolff and J. S. Wassom, Mammalian in vivo and in vitro cytogenetic assays: a report of the USEPA Gene-Tox program, *Mutat. Res.*, 87, 143-188, 1981.
62. McBee, K., J. W. Bickham, K. W. Brown and K. C. Donnelly, Chromosomal aberrations in native small mammals (*Peromyscus leucopus* and *Sigmodon hispidus*) at a petrochemical waste disposal site. I. Standard karyology, *Arch. Environ. Contam. Toxicol.*, 16, 681-688, 1987.
63. McBee, K., Chromosomal aberrations in native small mammals (*Peromyscus leucopus*) at a waste disposal site. II. Inherited aberrations and fragile sites indicated by G-band analysis, *Environ. Toxicol. Chem.*, 10, 1321-1329, 1991.
64. Thompson, R., A Study of Chromosomal Aberrations in the Cotton Rat, *Sigmodon hispidus*, Exposed to Hazardous Waste, M.S. thesis, *University of Texas Health Science Center*, Houston, 1987.
65. Tice, R. R., B. G. Ormiston, R. Boucher, C. A. Luke and D. E. Paquette, Environmental biomonitoring with feral rodent species, in *Short-Term Bioassays in the Analysis of Complex Environmental Mixtures*, S. S. Sandhu, D. M. Demarine, M. J. Mass, M. M. Moore and J. L. Mumford, Eds., Plenum Press, New York, 1987, pp. 175-179.
66. Shields, G. F., Meiotic evidence for pericentric inversion polymorphism in Junco (Aves), *Can. J. Genet. Cytol.*, 18, 747-751, 1976.
67. Hale, D. W., E. J. Ryder, P. D. Sudman and I. F. Greenbaum, Application of synaptonemal complex techniques for determination of diploid number and chromosomal morphology of birds, *Auk*, 105, 776-779, 1988.
68. Moses, M. J., Synaptonemal complex karyotyping in spermatocytes of the Chinese hamster (*Cricetulus griseus*). I. Morphology of the autosomal complement in spread preparations, *Chromosoma*, 60, 99-125, 1977.
69. Greenbaum, I. F., D. W. Hale and K. P. Fuxa, The mechanism of autosomal synapsis and the substaging of zygonema and pachynema from deer mouse spermatocytes, *Chromosoma*, 93, 203-212, 1986.
70. Howell, W. M., and D. A. Black, Controlled silver staining of nucleolus organizer regions with a protective colloidal developer: a 1 step method, *Experientia*, 36, 1014-1015, 1980.
71. Cawood, A. H., and G. Breckon, Synaptonemal complexes as indicators of induced structural change in chromosomes after irradiation of spermatogonia, *Mutat. Res.*, 122, 149-154, 1983.
72. Kaelbling, M., and N. S. Fechheimer, Synaptonemal complex analysis of chromosome rearrangements in domestic fowl, *Gallus domesticus*, *Cytogenet. Cell Genet.*, 36, 567-572, 1983.
73. Kaelbling, M., and N. S. Fechheimer, Synaptonemal complex analysis of a pericentric inversion of chromosome 2 of domestic fowl, *Gallus domesticus*, *Cytogenet. Cell Genet.*, 39, 82-86, 1985.
74. Allen, J. W., G. K. DeWeese, J. B. Gibson, P. A. Poorman and M. J. Moses, Synaptonemal complex damage as a measure of chemical mutagen effects on mammalian germ cells, *Mutat. Res.*, 190, 12-24, 1987.
75. Deaven, L. L., Application of flow cytometry to cytogenetic testing of environmental mutagens, in *Cytogenetic Assays of Environmental Mutagens*, T. C. Hsu, Ed., Allenheld, Osmun Publishers, Totowa, NJ, 1982, pp. 325-351.
76. Steen, H. D., and T. Lindmo, Flow cytometry: a high-resolution instrument for everyone, *Science*, 204, 402-404, 1979.

77. Irons, R. D., and W. S. Stillman. 1985. Flow cytofluorometric analysis of blood and bone marrow, in *Toxicology of the Blood and Bone Marrow*, R. D. Irons, Ed., Raven Press, New York, 1985, pp. 101–118.
78. Otto, F. J., and H. Oldiges, Flow cytogenetic studies in chromosomes and whole cells for the detection of clastogenic effects, *Cytometry*, 1, 13–17, 1980.
79. Spano, M., F. Pacchierotti, R. Uccelli, R. Amendola and C. Bartoleschi, Cytotoxic effects of benzene on mouse germ cells determined by flow cytometry, *J. Toxicol. Environ. Health*, 26, 361–372, 1989.
80. McBee, K., and J. W. Bickham, Petrochemical-related DNA damage in wild rodents detected by flow cytometry, *Bull. Environ. Contam. Toxicol.*, 40, 343–349, 1988.
81. Bickham, J. W., B. G. Hanks, M. J. Smolen, T. Lamb and J. W. Gibbons, Flow cytometric analysis of the effects of low-level radiation exposure on natural populations of slider turtles (*Pseudemys scripta*), *Arch. Environ. Contam. Toxicol.*, 17, 837–841, 1988.
82. George, L. S., C. E. Dallas, I. L. Brisbin and D. L. Evans, Flow cytometric analysis of red blood cell DNA of mallards accumulating environmental ^{137}Cs over time, *Ecotoxicol. Environ. Safety*, 21, 337–347, 1991.
83. Tiersch, T. R., R. W. Chandler, S. S. Wachtel and S. Elias, Reference standards for flow cytometry and application in comparative studies of nuclear DNA content, *Cytometry*, 10, 706–710, 1989.
84. Vindelov, L. L., I. J. Christensen and N. I. Nissen, Standardization of high-resolution flow cytometric DNA analysis by the simultaneous use of chicken and trout red blood cells as internal reference standards, *Cytometry*, 3, 328–331, 1983.
85. Noguchi, P. D., and W. C. Browne, The use of chicken erythrocyte nuclei as a biological standard for flow microfluorometry, *J. Histochem. Cytochem.*, 26, 761–763, 1978.
86. Tannenbaum, E., M. Cassidy, O. Alabaster and C. Herman, Measurement of cellular DNA mass by flow microfluorometry with use of a biological internal standard, *J. Histochem. Cytochem.*, 26, 145–148, 1978.
87. Vindelov, L. L., I. J. Christensen and N. I. Nissen, A detergent-trypsin method for the preparation of nuclei for flow cytometric DNA analysis, *Cytometry*, 3, 323–327, 1983.
88. Gohde, W., J. Schumann and J. Zante, The use of DAPI in pulse cytophotometry, in *Pulse Cytophotometry*, D. Lutz, Ed., European Press, Ghent, 1978, pp. 229–232.
89. Shugart, L. R., Quantitation of chemically induced damage to DNA of aquatic organisms by alkaline unwinding assay, *Aquat. Toxicol.*, 13, 43–52, 1988.
90. Katner, P. M., and H. S. Schwartz, A fluorescence enhancement assay for cellular DNA damage, *Mol. Pharmacol.*, 22, 145–151, 1982.
91. French, N. R., B. G. Maza, H. O. Hill, A. P. Aschwanden and H. W. Kaaz, A population study of irradiated desert rodents, *Ecol. Monogr.*, 44, 45–72, 1974.
92. Rowley, M. H., J. J. Christian, D. K. Basu, M. A. Pawlikowski and B. Paigen, Use of small mammals (voles) to assess a hazardous waste site at Love Canal, Niagara Falls, New York, *Arch. Environ. Contam. Toxicol.*, 12, 383–397, 1983.

37

Analyzing and Monitoring Population Viability

Bruce G. Marcot

ABSTRACT

A viable population has adequate numbers and distribution to confer a high likelihood (95%) of persistence, "persistence" meaning at least a century or a millennium, according to the scientific literature. Viability of wildlife populations can be affected by demographic, genetic, and environmental factors. Demographic risks include variation in reproductive and survival rates among individuals and over time. Genetic risks include loss of genetic variation from inbreeding depression, genetic drift, founder effects, and genetic bottlenecks. Environmental risks comprise systematic and catastrophic loss of habitat, including effects of habitat fragmentation and isolation; changes in climate, environmental quality, and fires; and interactions with biological agents such as competitors, predators, disease, and parasites.

A procedure for assessing population viability risks entails: (1) identifying the current and future extent of suitable habitat, (2) relating habitat to capability for supporting populations, (3) identifying and analyzing major viability risk factors, and (4) summarizing viability factors within a risk analysis framework. An integrated program of research and monitoring can be used to reduce areas of uncertainty and test management hypotheses in an adaptive management framework. Specific evaluation criteria should be developed for assessing results of research and monitoring.

KEY WORDS

population viability, risk analysis, demography, environmental stochasticity, population viability analysis, (PVA)

INTRODUCTION

Declines in wildlife populations and their habitats have led to a series of legal challenges to wildlife management plans.[1] Questions have been raised as to the viability of small, isolated populations. This chapter summarizes the concerns for viability of wildlife populations and presents a framework for assessing and planning for their long-term persistence.

POPULATION VIABILITY CONCEPTS FOR MANAGEMENT

In the first half of the 1980s, the concept of "minimum viable population" was used to describe a density above which vitality and safety of a wildlife population are ensured and below which extinction is inevitable.[2,3] Initially, the concept of a viability threshold was aimed at identifying risks of and avoiding genetic erosion of a single, isolated population. Later, the threshold concept was also applied to avoiding demographic declines.[4-6] However, as other factors complicating viability of wildlife populations entered the analysis (Table 1), the concept of an easily identifiable minimum population size became problematic.[7]

A viable population has been defined more recently in terms of probability of persistence. A population is viable when it is capable of being self-sustaining (i.e., $\lambda \geq 1.0$) and having a very high likelihood (e.g., $\geq 95\%$) of avoiding a multitude of risks of extinction over the long term, such as 100, or even 1000, years.[8] Population sizes and distributions necessary to meet these criteria vary among species and circumstances, such as seen with viability analyses for the grizzly bear, *Ursus arctos horribilis*[9,10] and the Concho water snake, *Natrix harteri paucimaculata*.[11] Regardless of variation among species and locations, some of the central principles guiding management of viable populations are (1) managing for reproductive individuals is crucial, (2) both numbers and distribution of those individuals are important to viability, (3) continued existence, not merely presence at nonzero population sizes, is the goal, and (4) distribution must allow for interaction of individuals in the population.[12]

IDENTIFYING VIABILITY RISKS

A risk analysis approach has been used to help assess the degree of threat to wildlife populations under various management scenarios.[8,13-16] Of special concern is the isolation of small populations of species that have narrow geographic distributions and a high degree of resource specialization, and that number up to the mid-hundreds. However, viability analysis and planning should also apply to populations that are currently not in jeopardy; in the spirit of preventive maintenance, the aim should be to ensure their continued existence at viable levels.

Factors Causing Viability to be at Risk

Risk factors that can endanger the future vitality of a population can be categorized as demographic, genetic, and environmental (Table 1).

Demographic Risks

This category refers to the probability of a population decline resulting from variation in age-specific vital rates of survivorship and fecundity.[17] Two kinds of variations in vital rates can contribute to fluctuations and declines in populations. The first, environmental (population) stochasticity, is generally caused by factors extrinsic to the population that affect mean values of vital rates averaged over all members of the population. Mean values of vital rates can fluctuate over time because of variations of environmental conditions affecting the population more or less as a whole, such as seasonal or annual fluxes in prey availability or weather conditions. By incurring a few chance years or breeding pulses with low fecundity and survival rates, resultant population declines can increase risks of local extinction.[18,19] A second kind of variation,

Table 1. Factors that Can Cause Population Declines or Extinction

Demographic
- Environmental stochasticity: random variations in annual population means of birth and death rates
- Demographic stochasticity: random variations of birth and death rates among individuals at a given point in time

Genetic
- Inbreeding depression
- Genetic drift
- Founder effect
- Genetic bottlenecks

Environmental
- Loss of habitat
  - Systematic processes
    - Vegetation succession
    - Habitat fragmentation
    - Climatic changes
  - Catastrophic processes
    - Storms
    - Fires
    - Floods
    - Geologic events (e.g., volcanoes)
    - Droughts
    - Human-caused acute alterations of habitat conditions
- Biological interactions
  - Predators
  - Competitors
  - Disease
  - Parasites
  - Forest insect and disease outbreaks
  - Reduction in prey abundance or availability
- Pollutants and toxicants

Modified from Salwasser et al.[12]

demographic (individual) stochasticity, is intrinsic to the population and refers to variation in vital rates among individuals within a population at a given point in time.[20]

The greater the variance in demographic vital rates, the greater the variation in occupancy rates of habitats over time. Moreover, the finite rate of population change (λ) is sensitive to each vital rate parameter to varying degrees.[21] For example, with untruncated, stable stage-class projection matrices, which represent lack of senescence in fecundity rates, λ is much more sensitive to adult survivorship than to other vital rate parameters. Thus, one can expect that effects of demographic stochasticity on the finite rate of population change would vary considerably according to such individual and population parameters as longevity, reproductive senescence, and adult fecundity and survival rates.

Values for the mean and variances of vital rates, representing environmental and demographic stochasticity, are poorly known for most wild populations. The subjacent mechanisms and major causes of environmental (population) and demographic (individual) variations also are not well understood for most populations.

Environmental and demographic stochasticity are usually viewed as independent events, i.e., environmental factors that influence reproduction and survival, such as annual fluctuations in weather and in prey availability, are typically modeled as independent from the variation in number of progeny among individuals within a specific year. In other words, changes in the expected (population average) value of a vital rate parameter (environmental change) do not necessarily alter the shape (variance) of its distribution among individuals of the population.

Density dependence is often an important consideration in modeling environmental and demographic stochasticity of terrestrial vertebrate populations. The magnitude of demographic uncertainty is strongly density dependent in many populations. The environmentally induced component of demographic uncertainty is only moderately density dependent in general, and only slightly density dependent in a geographically structured population (metapopulation). Small, isolated populations generally do not exhibit a metapopulation structure, and thus might experience greater degrees of density dependence on environmental stochasticity than larger populations.

Genetic Risks

Another source of viability risk is genetic risks. This refers to adverse effects on population vital rates from loss of genetic variation within the population.[22-25] Specifically, loss of allelic diversity means that deleterious, recessive alleles can be expressed in greater proportion in homozygous conditions. Loss of allelic diversity can arise from several kinds of population dynamics.

Inbreeding depression occurs in small, isolated populations in which related individuals interbreed. Related individuals share a higher proportion of genes than do unrelated individuals. Conversely, unrelated individuals better represent the diversity of genes available in the general gene pool of the population. Inbreeding causes the loss of some alleles and fixation in homozygous genotypes of other recessive alleles, some deleterious, which are expressed phenotypically under homozygous conditions. This decreases genic diversity of the population. It is generally assumed that the loss of genetic diversity as from accumulated inbreeding results in lower survival, reproductive rates, and overall fitness of individuals.

Genetic drift causes increases in the proportion of homozygosity in a population. It is related in one sense to inbreeding depression, as it occurs in small, isolated populations. Genetic drift is the random loss of alleles (and their associated genes) among generations in populations from a "sampling effect." In small populations, not every type of gene can be passed on to the next generation. Some vanish as chance recombinations fail to statistically "select" the gene in the next generation.

Progressive loss of genetic variation is a concern because it renders the population less able to respond, with different phenotypes or behaviors, to changes in environmental conditions.[26] Overall, genetic effects become greater concerns with smaller population sizes. As populations get very small (10, 20, or 30 up to 100 to 120 breeding individuals), the probability of inbreeding or genetic drift increases nonlinearly, i.e., a linear decline in population size may trigger an exponential decline in genetic diversity, leading to a rapid erosion (simplification) of the genome, an increase in expression of deleterious genotypes, and a significant decline in reproductive fitness of individuals and their offspring.

Another source of genetic simplification of a population is the founder effect. This refers to the colonization of an area by a small number of individuals that represent only a small proportion of the overall genetic diversity of the source population. If the colonizers interbreed and remain reproductively isolated from their source, the gene pool of their new population is markedly simpler and, in short order, may contain a disproportionately high frequency of homozygosity, especially of deleterious alleles.

Founders and their progeny may incur greater degrees of inbreeding or genetic drift if the population remains small.

Related to the founder effect is the phenomenon of genetic bottlenecks. A genetic bottleneck is defined as the loss of a significant portion of a population's allelic diversity. This occurs when an isolated population declines in numbers to a point at which inbreeding or drift can operate to further simplify the genome. Factors affecting the severity of genetic bottlenecks include the original genetic diversity of the population, the size and demographic characteristics of the population, the type of social structure and mating systems as affecting rate of interchange of alleles, and particularly the duration of the bottleneck period. Some species, such as whooping cranes (*Grus americana*), have rebounded with surprising vigor after serious founder or bottleneck effects, whereas others, such as the Puerto Rican parrot (*Amazona vittata*), the black-footed ferret (*Mustela nigripes*), and the Florida panther (*Felis concolor coryi*), may be unable to recover from genetic bottlenecks and might be suffering irrevocable loss of genetic variation.

Environmental Risks

The third major category of viability risk factors is environmental risks, which can affect population viability through loss of habitat, biological interactions with other species, and direct threats to individuals from pollutants and toxicants. Loss of habitat is perhaps the most acute short-term threat to populations. Habitat can be lost from systematic, chronic events, including vegetation succession, habitat fragmentation, and climatic changes.

Habitat fragmentation is probably one of the most insidious causes of declines in habitat suitability for many wildlife species, although often there is little direct evidence of its effects on population vital rates, size, and trends. Fragmentation severs a contiguous population into partially or fully isolated subpopulations (i.e., a metapopulation, *sensu* Shaffer.[23]) The subpopulations then suffer greater threats from other risk factors, such as loss of genetic variation and demographic and environmental stochasticity.[27–32] Interactions among habitat fragmentation and population isolation, successful recolonization of vacant habitats, and demographic stability are poorly understood, but potentially can greatly affect population regulation.[33] (See Morrison et al.[34] Chapter 3, for recommendations on managing habitat of metapopulations to provide for viability.) Habitat fragmentation also increases the amount of forest edge in a landscape. Edge habitats provide introduction routes for predators and competitors[32] and can adversely change temperature and moisture regimes in forest interiors.

Habitat suitability also can decline from slow changes in climatic conditions.[35] In urban areas, increases in acid precipitation, decreases in air quality from smog and airborne particulates, and changes in temperature means and variances from increasing urbanization can contribute to increased physiological stress of vegetation[36] and capability of habitats to continue to produce resources necessary for maintaining prey and other resources.[37–40]

Habitat also can be lost from catastrophic, acute events, including storms, fires, floods, volcanoes and other geologic activities, outbreaks of forest insects and disease, droughts, and short-term, human-caused changes in habitat availability or conditions. Because of their stochastic nature, it is particularly difficult to predict the likelihoods and effects of catastrophic events. Ideally, the following parameters should be statistically evaluated to estimate probabilities of habitat loss from catastrophes: location, frequency, extent, duration, type of catastrophic event, and the specific effect on habitat conditions from each event type. Such data are seldom available. Variation in amount, distribution, connectivity, and quality of habitats over time, as from catastrophes and longer-term changes, can greatly influence likelihoods of population extinction.[41,42] One may also find lag effects on populations from catastrophes.[43] Wildfires often drastically change habitat conditions, and wildlife have been documented to shift behaviors and distributions accordingly.[44-47]

Another risk to viability of wildlife populations is from biological interactions. Starvation and predation potentially can affect population size.[48] Presence of predators and competitors can potentially change foraging habits and use of habitats, and ultimately affect reproductive fitness.

The influence of disease and parasites, which are other factors possibly affecting the viability of populations, is little known. Some theoretical population models[49,50] suggest these may be important factors for consideration.[51]

Modeling Population Viability

Population Viability Assessment Procedure

A general procedure for assessing population viability of northern spotted owls (*Strix occidentalis caurina*[13]) entailed four main steps. First, suitable habitat is defined and its extent and distribution quantified over time for each planning alternative being tested. In this step, it is important to highlight the extent and role of secondary habitat for nonreproductive, "floating" members of the population. Also, juxtaposition of habitat patches with other habitat patches and with areas not suitable for reproduction or dispersal can affect their degree of suitability.

In the case of the spotted owl, wildlife biologists on each of the 19 United States National Forests within the range of the subspecies were asked to provide specific descriptions of forest types (denoted by dominant overstory tree species) and structures (successional growth stages) in which reproductive spotted owls have been consistently located. The habitats of nonreproductive, floating members of the spotted owl population have not been described and could not be included in this exercise. Next, the habitat descriptions were used to query forest inventory databases on geographic information systems (GIS) of each National Forest. This produced maps and tallies of the total area of suitable spotted owl habitat within each National Forest.

Distribution was assessed by measuring on the maps three nearest-neighbor distances from each designated spotted owl habitat area (SOHA; a land management category depicting forests allocated for conservation of spotted owl pairs). SOHAs are intended to provide for nesting, foraging, and roosting of territorial spotted owl pairs, and vary by size and spacing among the planning alternatives considered. To determine the degree to which distribution of SOHAs provides for dispersal, nearest-neighbor distances among formally designated SOHAs and other areas that would provide spotted owl habitat over time (*de facto* SOHAs) were compared with a cumulative frequency distribution of dispersal distances of juvenile spotted owls, as determined from radiotelemetry studies. This relationship helped determine that spacing of habitats within 15 miles may provide for 60% of dispersing owls and spacing within 6 miles may provide for 80%. Specific models or empirical evidence relating dispersal success (individual survival) to distance traveled or to spacing between habitats were not available.

Second, capability of habitat to provide for reproductive individuals is estimated over time for each planning alternative. Estimates of habitat capability should account for how various degrees of habitat suitability and habitat distribution, notably fragmentation, provide for the needs of reproductive pairs. Also, occurrence and distribution of habitat providing for dispersing individuals, especially juveniles, should be accounted for in capability estimates.

For spotted owl planning, habitat capability was estimated by calculating the expected occupancy rate of SOHAs given the expected amount of suitable habitat within a home-range size area. An empirical relationship between occupancy rate and habitat amount was derived from data provided by National Forest monitoring studies of spotted owls, and was calculated as a posterior Bayesian probability of occupancy given habitat amount. This calculation crudely accounted for habitat distribution and fragmentation at the scale of a spotted owl pair's annual home range size.

Total habitat capability, expressed as number of potential spotted owl pairs, was then calculated for all habitat areas, summed by spotted owl population, and projected over time under each planning alternative. Total habitat capability was not corrected for expected dispersal success rates, as empirical evidence of such relationships is lacking.

Third, effects of viability risk factors on population persistence are estimated qualitatively or quantitatively. Finally, the separate influences on population persistence from each risk factor are combined in a risk analysis framework. Modeling risk factors and evaluating their composite effects on population viability constitutes a major part of a population viability analysis and is discussed next.

With the spotted owl case study, a rule set was established by which the probability of continued existence of a well-distributed population could be classified on a five-rank ordinal scale ranging from very high to very low. Each rank in the scale was associated with specific ranges of values of four major factors that could affect persistence of spotted owls:

- Demographic factors: $P(Nt > \min)$, the likelihood that environmental variation of population vital rates would keep the population above a specified minimum density (defined as a density resulting from spacing spotted owl pairs at the median juvenile spotted owl dispersal distance of 17 mi)
- Genetic factors: Ft, the value of a calculated inbreeding coefficient
- Habitat occupancy factors: $P(O|S)$, the probability that a designated SOHA of a given amount of suitable habitat would actually support a breeding pair
- Distribution factors: the mean of the first, second, and third nearest-neighbor distances from a designated SOHA to other SOHAs, as compared to dispersal distances of juvenile spotted owls

For example, to rank a planning alternative as providing a "very high" probability of long-term population persistence, it would have to provide for $P(Nt > \min) > 0.95$, $Ft < 0.05$, $P(O|S) > 0.95$, and nearest-neighbor distances <2 mi, for at least a century. No planning alternative met these criteria for all portions of the range of the northern spotted owls because of natural or human-induced fragmentation of habitats and isolation of some populations. The final spotted owl plan ultimately provided, over the long term (100 years) and for most of the northern spotted owl populations, for only a moderate probability of continued existence of well-distributed populations, which was seen by the decision-makers as a compromise between economics and biology.

Assessing Demographic Risks

Factors affecting demographic stability of a population include those that contribute to both environmental and individual variability in fecundity and survivorship rates. The most commonly used tools for assessing environmental variation are stage-based population models and stochastic life tables.[52-54] Persistence time of populations has been estimated by use of static and stochastic life tables. Static population trend, typically measured by calculating a mean and confidence interval of the finite rate of population change (λ), with eigenanalysis of population projection matrices, is another modeling tool that has been used.[55] Lande[55] used partial differentiation and Noon and Biles[21] used simulations to assess sensitivity effects on λ and on population persistence time from varying vital rates, particularly age of first reproduction, reproductive senescence and longevity, and stage class-specific survivorship and fecundity rates.

Spatially explicit models of habitat fragmentation and occupancy are a promising facet of viability analysis.[27,28,56-58] Such models can be used to simulate use of habitat patches by individuals within a home range area. Lamberson et al.[59] devised a patch colonization model to assess population persistence of black-footed ferrets. Folse et al.[60] created a prototype computer system that joins models of (1) foraging and movement behavior of mountain lions (*Felis concolor*), (2) suitability of mountain lion habitat, and (3) a GIS to predict effects of habitat patch patterns on persistence of mountain lion family units in a landscape.

In modeling persistence time of populations based on demographic trends and parameter variations, one constraint is the need for unbiased, accurate estimates of population structures and vital rates.[61-63] In addition, a rigorous statistical basis for establishing a sampling frame for such monitoring studies is necessary for sound estimates. Accurate estimation of age-class specific survivorship rates is also a problem in population studies.

Assessing Genetic Risks

Assessments to date of how loss of genetic variation can affect population persistence have generally applied basic theory on the rate of loss of genetic variation in small, isolated populations. For example, Barrowclough and Coats[64] estimated the size of effective population (N_e) of northern spotted owls by using a formula designed to account for fluctuations in overall population size, dispersal distance of juveniles, density of adult pairs, and other factors. They then related population size to potential degree of inbreeding, and assumed that cumulative inbreeding rates represented declines in reproductive fitness of individuals. The approach was also used in the viability analysis conducted by United States Department of Agriculture (USDA) Forest Service.[13] However, there are no standards for, and there is much controversy over, the correct formula for estimating N_e and procedures for estimating its components. Also, there is generally little empirical evidence about how inbreeding depression affects fitness of individuals. Until empirical studies are conducted for specific populations, mathematical analyses of loss of genetic variation from inbreeding and other factors should be treated as hypotheses.

Other genetic factors that potentially affect the viability of populations include genetic drift, founder effects, and effects of genetic bottlenecks. Lacy[66] addressed the general problem of assessing loss of genetic variation in a managed population by accounting for genetic drift, mutation, immigration, selection, and isolation.

Assessing Environmental Risks

Fire simulation models[67,68,73] may prove useful if the base empirical data on frequency, extent, location, and effect of fires can be parameterized. Similar models predicting likelihood of catastrophic habitat loss from storms, floods, droughts, other natural catastrophes, and human activities could add realism to prognostications of persistence times of populations.

Viability Risk Analysis

It is not understood how the various factors affecting viability (Table 1) interplay. Gilpin and Soulé[69] hypothesized that various components of population phenotype, environment, and population structure and fitness interact in complex feedback loops to intensify likelihoods of population decline. Most others have addressed viability components of populations singly.

There is virtually no empirical understanding of how separate viability risk factors interact for most populations. In assessing the viability of northern spotted owl populations, the USDA Forest Service[13] assumed that various viability risk factors operated independently. Estimates of their effect on population persistence time or habitat occupancy were combined into a single, qualitative statement of composite risk. Similar decision theory approaches to viability risk analysis[15] have also been used for assessing population conditions and persistence of grizzly bears,[70] zoo populations of tigers (*Panthera tigris*[71]), concho water snake,[11] and other species.

Managing for Viable Populations

In developing the best strategy for managing for viable populations, each of the viability risk factors discussed in this chapter should be evaluated at least qualitatively. To begin, evaluations need not demand quantitative analyses of field data or use of complicated statistics and simulation models. Of course, much can be learned from a rigorous approach, and nothing replaces basic knowledge gained from field ecology studies of populations and habitats.

Federal land management agencies, in particular the USDA Forest Service and the United States Department of the Interior (USDI) Bureau of Land Management, are charged with managing public land under multiple-use scenarios. Devising habitat management guidelines in the context of a multiple-use mandate entails understanding impacts from concomitant use of the land for other interests. A risk analysis framework that rigorously arrays and assesses impacts on population viability would likely prove most helpful for crafting resource management guidelines that maintain viability. Such an assessment would identify management activities that affect demographic, genetic, and environmental factors that in turn put viability at risk.

RESEARCH AND MONITORING NEEDS

Monitoring population responses should be an integral part of habitat management guidelines. Specifically, the extent and quality of suitable habitat should be tracked over time. Key behavioral attributes of individuals and population parameters also should be monitored. These may include turnover rates of individuals on territories, frequency of pair-bond dissolution, and declines in reproductive rates of adults and survival rates of adults and juveniles. Such parameters constitute an "early warning system" to signal an impending population decline much sooner than simply monitoring habitat occupancy. With the spotted owl case study, such early warning parameters are suggested by ongoing empirical demographic studies[72] and spatially explicit demographic simulation models.[21]

However, we need to understand how such early warning signals can be influenced by declines in available habitat. For example, occupancy rates of habitats and turnover rates and rates of juvenile mortality during dispersal can be artificially inflated by short-term packing of individuals displaced from disturbed habitats. If the rate of disturbance (as from timber harvest, fires, urbanization, etc.) of suitable habitat slows, then turnover, occupancy, and juvenile mortality rates may relax to lower, equilibrial values.

Questions remain to be answered, however, including is there any guarantee that the declining turnover will not be followed by a continuation of decline in occupancy, and ultimately by local extinction? Are there thresholds of habitat distribution below which the population cannot equilibrate? Even if turnover rates and other parameters theoretically can be used as an early warning system, do we actually know enough to interpret them? Turnover rates should be tracked in a monitoring program, but interpreted in light of other parameters. It may be obscured by short-term effects, just as is occupancy rate. Additional studies focused on movement dynamics in changing landscapes, and spatial models of habitat occupancy that incorporate realistic changes in habitat amount and distribution[59] will be vital tools for helping gauge presence and degree of such effects, and thus in helping us correctly interpret early warning signals.

Also, in a monitoring program, specific criteria for triggering reevaluation of management guidelines should be defined. These criteria can change over time as we learn more about population ecology of the species of interest, but they are essential for testing hypotheses associated with particular management guidelines, i.e., for closing the loop in the process of adapting management direction to new information.

ACKNOWLEDGMENT

I thank Thomas E. Lacher, Kent F. Premo, and Hal Salwasser for helpful review comments on the manuscript.

REFERENCES

1. Wilkinson, C. F., and H. M. Anderson, *Land and Resource Planning in the National Forests*, Island Press, Washington, DC, 1987.
2. Lacava, J., and J. Hughes, Determining minimum viable population levels, *Wildl. Soc. Bull.*, 12, 370–376, 1984.
3. Reed, J. M., P. D. Doerr and J. R. Walters, Determining minimum population sizes for birds and mammals, *Wildl. Soc. Bull.*, 14, 255–261, 1986.
4. Samson, F. B., F. Perez-Trejo, H. Salwasser, L. F. Ruggiero and M. L. Shaffer, On determining and managing minimum population size, *Wildl. Soc. Bull.*, 13, 425–433, 1985.
5. Shaffer, M. L., Determining minimum viable population sizes for the grizzly bear, *Int. Conf. Bear Res. Manage.*, 5, 133–139, 1983.
6. Shaffer, M. L., and F. B. Samson, Population size and extinction: a note on determining critical population sizes, *Am. Nat.*, 125, 144–152, 1985.
7. Conner, R. N., Wildlife populations: minimally viable or ecologically functional?, *Wildl. Soc. Bull.*, 16, 80–84, 1988.
8. Shaffer, M. L., Minimum population sizes for species conservation, *BioScience*, 31, 131–134, 1981.
9. Eberhardt, L. L., R. R. Knight and B. M. Blanchard, Monitoring grizzly bear population trends, *J. Wildl. Manage.*, 50, 613–618, 1986.
10. Knight, R.R., and L.L. Eberhardt, Population dynamics of Yellowstone grizzly bears, *Ecology*, 66, 323–334, 1985.
11. Soulé, M. E., Risk analysis for the concho water snake, *Endangered Species Update*, 6(10), 19, 22–25, 1989.
12. Salwasser, H., R. S. Holthausen and B. G. Marcot, Implications of laws and regulations for viable population planning in the National Forest system (updated 3 Feb 1989), in *The*

Management of Viable Populations: Theory, Applications, and Case Studies, B. A. Wilcox, P. F. Brussard, and B. G. Marcot, Eds., Center for Conservation Biology, Stanford, CA, 1986.

13. USDA Forest Service, Final Supplement to the Environmental Impact Statement for an Amendment to the Pacific Northwest Regional Guide, 2 vol., U.S. Government Printing Office, Washington, DC, 1988.
14. Marcot, B. G., and R. Holthausen, Analyzing population viability of the spotted owl in the Pacific Northwest, *North Am. Wildl. Nat. Resource Conf.*, 52, 333-347, 1987.
15. Marcot, B. G., Concepts of risk analysis as applied to viable population assessment and planning, in *The Management of Viable Populations: Theory, Applications, and Case Studies*, B. A. Wilcox, P. F. Brussard and B. G. Marcot, Eds., Center for Conservation Biology, Stanford, CA, 1986.
16. Shaffer, M. L., Population viability analysis, *Conserv. Biol.*, 4, 39-40, 1990.
17. Goodman, D., Consideration of stochastic demography in the design and management of biological reserves, *Nat. Resource Model.*, 1(2), 205-234, 1987.
18. Iwasa, Y., and H. Mochizuki, Probability of population extinction accompanying a temporary decrease of population size, *Res. Popul. Ecol.*, 30, 145-164, 1988.
19. Berger, J., Persistence of different-sized populations: an empirical assessment of rapid extinctions in bighorn sheep, *Conserv. Biol.*, 4, 91-98, 1990.
20. Hatch, S. A., Testing for individual variation in breeding success, *Auk*, 105, 193-194, 1988.
21. Noon, B. R., and C. M. Biles, The mathematical demography of the spotted owl in the Pacific Northwest, *J. Wildl. Manage.*, 54, 18-27, 1990.
22. Barrowclough, G. F., Gene flow, effective population sizes, and genetic variance components in birds, *Evolution*, 34, 789-798, 1980.
23. Shaffer, M. L., The metapopulation and species conservation: the special case of the northern spotted owl, in *Ecology and Management of the Spotted Owl in the Pacific Northwest*, R. J. Gutiérrez and A. B. Carey, Eds., USDA Forest Service Gen. Tech. Rep. PNW-185, Portland, OR, 1985, pp. 86-99.
24. Emigh, T., and E. Pollak, Fixation probabilities and effective population numbers in diploid populations with overlapping generations, *Theor. Pop. Biol.*, 15, 86-107, 1979.
25. Lande, R., Genetics and demography in biological conservation, *Science*, 241, 1455-1460, 1988.
26. Caswell, H., Phenotypic plasticity in life-history traits: demographic effects and evolutionary consequences, *Am. Zool.*, 23, 35-46, 1983.
27. Fahrig, L., and J. Paloheimo, Effect of spatial arrangement of habitat patches on local population size, *Ecology*, 69, 468-475, 1988.
28. Hastings, A., Spatial heterogeneity and ecological models, *Ecology*, 71, 426-428, 1990.
29. Hastings, A., and C. L. Wolin, Within-patch dynamics in a metapopulation, *Ecology*, 70, 1261-1266, 1989.
30. Pulliam, H. R., Sources, sinks, and population regulation, *Am. Nat.*, 132, 652-661, 1988.
31. Quinn, J. F., and A. Hastings, Extinction in subdivided habitats, *Conserv. Biol.*, 1, 198-209, 1987.
32. Wilcove, D. S., From fragmentation to extinction, *Nat. Areas J.*, 7, 23-29, 1987.
33. Hughes, T. P., Recruitment limitation, mortality, and population regulation in open systems: a case study, *Ecology*, 71, 12-20, 1990.
34. Morrison, M. L., B. G. Marcot and R. W. Mannan, *Principles of Wildlife-Habitat Relationships*, University of Wisconsin Press, Madison, WI, 1992.
35. Graham, R. W., The role of climatic change in the design of biological reserves: the paleoecological perspective for conservation biology, *Conserv. Biol.*, 2, 391-394, 1988.
36. Layser, E. F., Forestry and climatic change, *J. Forest.*, (November), 678-682, 1980.
37. Kricher, J. C., Diversity in two wintering bird communities: possible weather effects, *Auk*, 92, 766-772, 1975.
38. Meslow, E. C., and L. B. Keith, A correlation analysis of weather versus snowshoe hare population parameters, *J. Wildl. Manage.*, 35, 1-15, 1971.

39. Stinson, C. H., Weather-dependent foraging success and sibling aggression in red-tailed hawks in central Washington, *Condor*, 82, 76–80, 1980.
40. Botsford, L. W., T. C. Wainwright, J. T. Smith, S. Mastrup and D. F. Lott, Population dynamics of California quail related to meteorological conditions, *J. Wildl. Manage.*, 52, 469–477, 1988.
41. Lande, R., and S. H. Orzack, Extinction dynamics of age-structured populations in a fluctuating environment, *Proc. Natl. Acad. Sci. U.S.A.*, 85(19), 7418–7421, 1988.
42. Leigh, E. G., The average lifetime of a population in a varying environment, *J. Theor. Biol.*, 90, 213–239, 1981.
43. Knopf, F. L. and J. A. Sedgwick, Latent population responses of summer birds to a catastrophic, climatological event, *Condor*, 89, 869–873, 1987.
44. Apfelbaum, D. S., and A. Haney, Bird populations before and after wildfire in a Great Lakes pine forest, *Condor*, 83, 347–354, 1981.
45. Balda, R. P., and J. H. Overturf, The Effect of Forest Fire on Breeding Birds in Ponderosa Pine Forests, USDA Forest Service, Rocky Mountain Forest and Range Experiment Station, Tempe, AZ, 1977.
46. Emlen, J. T., Habitat selection by birds following a forest fire, *Ecology*, 51, 343–345, 1970.
47. Soulé, M. E., D. T. Boulger, A. C. Alberts, R. Sauvajot, J. Wright, M. Sorice and S. Hill, Reconstructed dynamics of rapid extinctions of chaparral-requiring birds in urban habitat islands, *Conserv. Biol.*, 2, 75–92, 1988.
48. McNamara, J. M., and A. I. Houston, Starvation and predation as factors limiting population size, *Ecology*, 68, 1515–1519, 1987.
49. May, R. M., Conservation and disease, *Conserv. Biol.*, 2, 28–30, 1988.
50. Scott, M. E., The impact of infection and disease on animal populations: implications for conservation biology, *Conserv. Biol.*, 2, 40–56, 1988.
51. Thorne, E. T., and E. S. Williams, Disease and endangered species: the black-footed ferret as a recent example, *Conserv. Biol.*, 2, 66–74, 1988.
52. Crouse, D. T., L. B. Crowder and H. Caswell, A stage-based population model for loggerhead sea turtles and implications for conservation, *Ecology*, 68, 1412–1423, 1987.
53. Eberhardt, L. L., Population projections from simple models, *J. Appl. Ecol.*, 24, 103–118, 1987.
54. Eberhardt, L. L., Using age structure data from changing populations, *J. Appl. Ecol.*, 25(2), 373–378, 1988.
55. Lande, R., Demographic models of the northern spotted owl (*Strix occidentalis caurina*), *Oecologia*, 75, 601–607, 1988.
56. Bovet, P., and S. Benhamou, Spatial analysis of animals' movements using a correlated random walk model, *J. Theor. Biol.*, 131(4), 419–433, 1988.
57. Horton, S. P., K. C. Peeler and A. B. Carey, Geographic information systems analyses of spotted owl ranges, Workshop on uses of geographical information systems in avian biology, in *Proc. 59th Annu. Meet. Cooper Ornithological Society*, Moscow, ID, 22–26 June 1989 (Abstr.).
58. Laymon, S. A., and R. H. Barrett, Developing and testing habitat-capability models: pitfalls and recommendations, in *Wildlife 2000: Modeling Habitat Relationships of Terrestrial Vertebrates*, J. Verner, M. L. Morrison and C. J. Ralph, Eds., University of Wisconsin Press, Madison, 1986, pp. 87–91.
59. Lamberson, R., R. McKelvey, B. R. Noon and C. Voss, Effects of Varying Dispersal Capabilities on the Population Dynamics of the Northern Spotted Owl—Preliminary Results, USDA Forest Service, Pacific Southwest Research Station, Arcata, CA, 1989.
60. Folse, L. J., H. E. Mueller and A. D. Whittaker, Object-oriented simulation and geographic information systems, *AI Appl. Nat. Resource Manage.*, 4(2), 41–47, 1990.
61. Burgman, M. A., H. R. Akcakaya and S. S. Loew, The use of extinction models for species conservation, *Biol. Conserv.*, 43(1), 9–25, 1988.
62. Chao, A., Estimating animal abundance with capture frequency data, *J. Wildl. Manage.*, 52, 295–300, 1988.

63. Eberhardt, L. L., and M. A. Simmons, Calibrating population indices by double sampling, *J. Wildl. Manage.*, 51, 665-674, 1987.
64. Barrowclough, G. F., and S. L. Coats, The demography and population genetics of owls, with special reference to the conservation of the spotted owl (*Strix occidentalis*), in *Ecology and Management of the Spotted Owl in the Pacific Northwest*, R. J. Gutiérrez and A. B. Carey, Eds., USDA Forest Service Gen. Tech. Rep. PNW-185, Portland, OR, 1985, pp. 74-85.
65. Salwasser, H., S. P. Mealey and K. Johnson, Wildlife population viability—a question of risk, *North Am. Wildl. Nat. Res. Conf.*, 49, 421-439, 1984.
66. Lacy, R. C., Loss of genetic diversity from managed populations: interacting effects of drift, mutation, immigration, selection, and population subdivision, *Conserv. Biol.*, 1, 143-158, 1987.
67. Keane, R. E., S. F. Arno and J. K. Brown, Simulating cumulative fire effects in Ponderosa pine/Douglas-fir forests, *Ecology*, 71, 189-203, 1990.
68. Potter, M. W., and S. R. Kessell, Predicting mosaics and wildlife diversity resulting from fire disturbance to a forest ecosystem, *Environ. Manage.*, 4, 247-254, 1980.
69. Gilpin, M. E., and M. E. Soulé, Minimum viable populations: processes of species extinction, in *Conservation Biology*, M. E. Soulé, Ed., Sinauer Associates, Sunderland, MA, 1988, pp. 19-34.
70. Maguire, L. A., An Analysis of Augmentation Strategies for Grizzly Populations: The Cabinet-Yak Ecosystem as an Example, Report to U.S. Forest Service, Contract No. 40-3187-4-1748, 1986.
71. Maguire, L. A., and R. C. Lacy, Allocating scarce resources for conservation of endangered subspecies: partitioning zoo space for tigers, *Conserv. Biol.*, 4, 157-166, 1990.
72. Franklin, A., J. P. Ward, R. J. Gutierrez and G. I. Gould, Density of northern spotted owls in northeast California, *J. Wildl. Manage.*, 54, 1-10, 1990.
73. Roberts, D., personal communication.

38

Theoretical Ecology and Research on Agroecosystems

Thomas E. Lacher, Jr.

ABSTRACT

Ecologists have long been averse to conducting research in agroecosystems, yet these habitats may be ideal for the study of several areas of theoretical importance in ecology. Agroecosystems are controlled and replicated, an important prerequisite for experimental field research. Indeed, important research has already been conducted in agricultural fields in the areas of population ecology and host-parasitoid interactions. Other areas in which new insights can be derived from the study of agroecosystems include mechanisms of biological control, population genetics, the ecology of colonizing species, the study of disturbance and nonequilibrial systems, and metapopulation analysis.

KEY WORDS

theoretical ecology, agroecosystems, modeling, population ecology, disturbance, nonequilibrial systems

INTRODUCTION

The agroecosystem is the primary forum for ecotoxicological research. Ecologists have traditionally emphasized the importance of studying undisturbed ecosystems. In reality, relatively little ecological research is done in wilderness regions. These areas are frequently difficult to access and have little infrastructure for support of field laboratories. Unfortunately, little wilderness remains and few research facilities are located in close proximity to a wilderness region. A perusal of the most important journals for the publication of ecological research, such as *Ecology* and *The American Naturalist*, reveals that most research is carried out in disturbed areas, second growth habitat, or small biological reserves or field stations that protect only a small complement of the native flora or fauna. More studies are carried out in the laboratory or in artificial enclosures than in undisturbed wilderness. Our knowledge of natural ecosystems is biased toward disturbed, fragmented habitats with many details derived from the laboratory rather than the field.

Given the above, it is curious that so many ecologists are averse to studying agricultural areas. The argument is frequently presented that these areas are "unnatural" and therefore intrinsically less interesting. All four chapters presented in the preceding section, as well as Chapters 15 and 20 by Craig Barber, deal with applications of ecological or genetic models to ecosystems that are disturbed or under stress. At a time when concern is increasing about the long-term effects of human activities on the global environment (including wilderness areas[1]) studies on the dynamics of disturbed ecosystems assume paramount importance. The study of agroecosystems can contribute a great deal to our knowledge of not only environmental impacts but basic ecological processes as well.

ADVANTAGES OF STUDYING AGROECOSYSTEMS

Agroecosystems cover vast areas of the earth's surface. In the United States an estimated 1.36 billion acres are suitable for cultivation—62% of the total land area. Of this, 413 million acres are used for agriculture (19%), 414 million acres are used for grazing (19%), 376 million acres are forested (17%—most is second growth on former agricultural land), and 153 million acres (7%) are developed for residential or other land use activities.[2] In contrast, 4% is preserved in the National Park System, considered one of the best in the world.[3] World wide, 25% of the land is considered arable, and 11% is currently under cultivation. The remaining 14% is either tropical forest (8%) or potentially fertile arid land (6%).[4] Agricultural habitats are abundant and easily accessible world wide. Because agricultural areas are generally privately owned, once permission is granted by land owners, fewer logistical difficulties are associated with conducting research.

Within a region, agricultural practices are relatively standardized. Agroecosystems are therefore controlled and replicated. They represent experiments being conducted on a large scale, yet few researchers currently study these systems from an ecological perspective. Areas of high agricultural productivity are not likely to be abandoned any time soon so that they will be available for long-term studies. Because these areas are of economic importance, funding to study agroecosystems may be more readily available than funds to study wilderness areas, which by design are not intended to generate income. Finally, many agricultural crops have extremely broad geographical distributions. Corn fields can be found nearly world wide; *Zea mays* is planted in temperate deciduous forests, a variety of different grasslands, semiarid regions at many latitudes, tropical savannas, and lowland tropical forests. There are numerous opportunities to study the comparative structure and function of simple early successional stage ecosystems that cover a broad geographical area, encompass a wide variety of different climatic regimes, and coexist with an equally wide variety of plants, invertebrates, and vertebrates. All of the preparation, planting, labor, and maintenance of the sites is done at no cost to the investigator.

IMPORTANT ECOLOGICAL RESEARCH QUESTIONS IN AGROECOSYSTEMS

Agricultural scientists have studied agricultural production systems for hundreds of years. They have generated a vast amount of information on the practical aspects of agriculture. Ecotoxicologists and wildlife toxicologists now study agroecosystems and the surrounding habitat to assess the potential impacts of agricultural practices on native species of plants and animals. The need to incorporate a more ecological perspective into production agriculture[5–8] and ecotoxicology[9,10] is widely recognized. Little research has addressed agroecosystems from the perspective of contemporary community ecology or evolutionary biology, yet some important research questions can be addressed in agricultural systems.[11] Because the results of much of this research also will have practical applications in production aspects of agriculture,[11] more funding could become available than currently exists for basic research.

Population Ecology of Insects

The study of insect population dynamics has always been an important area of research in the field of pest control.[12] Pests are insects that eat food crops and their dynamics are intrinsically the same as insects that consume wilderness plants. Excellent opportunities are available to conduct, under controlled situations, replicated studies of the dynamics of insect populations in early successional conditions. Populations can be subjected to a variety of manipulations and perturbations to examine the relative importance of sex ratio effects, dispersal behavior, and other aspects of the life histories of insects on their population dynamics.[13] Data will have applied (pest control) as well as basic (models of population growth of colonizing species) importance.

Mechanisms of Biological Control

A closely related area is the study of mechanisms of biological control. Pioneering research on the dynamics of host-parasitoid relationships has been done on insects in the context of biological control by Michael Hassell and Robert May;[14] substantial opportunities for empirical and theoretical research are available in agroecosystems to further research along these lines. Indeed, a great deal of pioneering work on parasitoid behavior has been done in crop ecosystems by Murdoch and colleagues.[15,16] Research on the importance of variation in plant chemical defenses within populations or within individuals in a population[17] is another area of investigation best conducted in agroecosystems. Many important crops are now genetically homogeneous in their chemical defenses against pest organisms. Utilizing multiple defenses in crop plants may prove to be one of the most effective means of pest control in agroecosystems[18] and will provide important information on the evolution of plant-animal interactions.

Population Genetics of Insects

The application of synthetic chemicals to agricultural fields has been one of the largest experiments in directional selection ever conducted. Insects have demonstrated a remarkable ability to evolve resistance to pesticides, as documented by a recent National Academy of Sciences report.[19] The same report outlines several areas of important theoretical and applied research in population genetics that can be conducted on these systems. Via[20] presents an overview of the advances in ecological genetics that have been obtained from the study of insects in agricultural systems, particularly in relation to host plant adaptation to herbivorous insects.

Ecology of Colonizing Species

Most agricultural crops are early successional stage species that grow best on disturbed areas. Farmers recreate early successional stage conditions every planting season in order to provide ideal growing conditions for these crops. A great deal of our understanding of plant demographics and community structure derives from the research of British ecologists on agroecosystems[6] and much fruitful research remains to be done in this area. Knowledge of the ecology of colonizing species will assist in the development of more sustainable agroecosystems and will also provide insights into the process of restoration of degraded habitats.

Disturbance, Nonequilibrial Systems, and Metapopulation Dynamics

Agroecosystems are disturbed on an annual basis and can be viewed as nonequilibrial systems. In addition, the chemical and biological control mechanisms used to limit the growth rates of insect pests maintain the populations of most invertebrate species inhabiting agricultural fields (both prey and their predators) at low densities. Interest is growing in the effect of disturbance on the maintenance of species diversity and in the interactions of species in nonequilibrial systems.[21] Agroecosystems may be a fruitful area to study these phenomena. Because agricultural fields are relatively simple systems, it may be more feasible to design studies on the dynamics of interacting species that exist well below their carrying capacities.

Agricultural fields exist in a complex mosaic with undisturbed habitats; as such they present an ideal situation for research on the dynamics of metapopulations. Metapopulations are defined as "systems of . . . local populations connected by dispersing individuals."[22] The *Biological Journal of the Linnean Society* devoted an entire issue to metapopulations in 1991 (Volume 42, Numbers 1 & 2, January/February). Harrison[23] discusses several types of metapopulation structures that would be well suited to the study of the regional dynamics of bird populations exposed to pesticides in habitat mosaics, including source-sink metapopulations, patchy populations, and nonequilibrial metapopulations. Metapopulation analysis could prove to be an important paradigm in avian ecotoxicology and its application in agricultural settings should result in important advances in basic theory.

Other Areas of Research

Other potentially rich areas of research have been discussed in other papers. These include the applications of optimal foraging theory in the research done by Belovsky[24] on grasshoppers and birds and the work on sources and sinks discussed by Bowers[25] and Pulliam.[26] The point is that ecologists have traditionally avoided research in agroecosystems, yet it is possible to do research of theoretical importance in agricultural areas. There is concern that the best ecological theoreticians will be averse to cooperate with ecotoxicologists because such involvement will detract from their ability to conduct basic research. Carefully designed proposals can accommodate the needs of applied researchers and provide opportunities to gather data that will contribute to the development of basic theory. Slobodkin[27] stated, "In minimal systems it may be possible to see meaning and derivations more clearly and to examine our criteria for theoretical quality." Although he was not discussing agroecosystems, the minimal nature of the agricultural field might facilitate our understanding of ecology and evolutionary biology in ways that we have been reluctant to accept.

REFERENCES

1. Peters, R. L., and J. D. S. Darling, The greenhouse effect and nature reserves, *BioScience*, 35, 707–717, 1985.
2. Moore, J. A., Science as a way of knowing—human ecology, *Am. Zool.*, 25, 483–637, 1985.
3. Frome, M., *The National Park Guide*, Prentice-Hall, New York, NY, 1989.
4. Miller, G. T., Jr., *Living in the Environment*, 4th ed., Wadsworth, Belmont, CA, 1985.
5. Coleman, D. C., Ecology, agroecosystems, and sustainable agriculture, *Ecology*, 70, 1590, 1989.
6. Jackson, W., and J. Piper, The necessary marriage between ecology and agriculture, *Ecology*, 70, 1591–1593, 1989.

7. Paul, E. A., and G. P. Robertson, Ecology and the agricultural sciences: a false dichotomy?, *Ecology*, 70, 1594-1597, 1989.
8. Elliott, E. T., and C. V. Cole, A perspective on agroecosystem science, *Ecology*, 70, 1597-1602, 1989.
9. Rosemarin, A., Ecotoxicology on the upswing—but where are the ecologists?, *Ambio*, 17, 359, 1988.
10. Kendall, R. J., and T. E. Lacher, Jr., Eds., *Wildlife Toxicology and Population Modeling: Integrated Studies of Agroecosystems*, Lewis Publishers, Chelsea, MI, 1993.
11. Murdoch, W. W., Diversity, complexity, stability and pest control, *J. Appl. Ecol.*, 12, 795-807, 1975.
12. Wallner, W. E., Factors affecting insect population dynamics: differences between outbreak and non-outbreak species, *Annu. Rev. Entomol.*, 32, 317-340, 1987.
13. Anderson, R. M., Discussion: ecology of pests and pathogens, in *Perspectives in Ecological Theory*, J. Roughgarden, R. M. May and S. A. Levin, Eds., Princeton University Press, Princeton, NJ, 1989, pp. 348-361.
14. Hassell, M. P., and R. M. May, The population biology of host-parasite and host-parasitoid associations, in *Perspectives in Ecological Theory*, J. Roughgarden, R. M. May and S.A. Levin, Eds., Princeton University Press, Princeton, NJ, 1989, pp. 319-347.
15. Murdoch, W. W., J. Chesson and P. L. Chesson, Biological control in theory and practice, *Am. Nat.*, 125, 344-366, 1985.
16. Murdoch, W. W., J. D. Reeve, C. B. Huffaker and C. E. Kennett, Biological control of olive scale and its relevance to ecological theory, *Am. Nat.*, 123, 371-392, 1984.
17. Whitham, T. G., A. G. Williams and A. M. Robinson, The variation principle: individual plants as temporal and spatial mosaics of resistance to rapidly evolving pests, in *A New Ecology*, P. W. Price, C. N. Slobodchikoff and W. S. Gaud, Eds., John Wiley & Sons, New York, 1984, pp. 15-51.
18. Pimental, D., and A. C. Bellotti, Parasite-host population systems and genetic stability, *Am. Nat.*, 110, 877-888, 1976.
19. National Academy of Sciences, *Pesticide Resistance: Strategies and Tactics for Management*, *NAS*, Washington, DC, 1986.
20. Via, S., Ecological genetics and host adaptation in herbivorous insects: the experimental study of evolution in natural and agricultural systems, *Annu. Rev. Entomol.*, 35, 421-446, 1990.
21. Petraitis, P. S., R. E. Latham and R. A. Niesenbaum, The maintenance of species diversity by disturbance, *Q. Rev. Biol.*, 64, 393-418, 1989.
22. Hanski, I., and M. Gilpin, Metapopulation dynamics: brief history and conceptual domain, *Biol. J. Linn. Soc.*, 42, 3-16, 1991.
23. Harrison, S., Local extinction in a metapopulation context: an empirical evaluation, *Biol. J. Linn. Soc.*, 42, 72-88, 1991.
24. Belovsky, G. E., Modeling avian foraging: implications for assessing the ecological effects of pesticides, in *Wildlife Toxicology and Population Modeling: Integrated Studies of Agroecosystrems*, R. J. Kendall and T. E. Lacher, Jr., Eds., Lewis Publishers, Chelsea, MI, 1993.
25. Bowers, M. A., Use of space and habitats by individuals and populations: dynamics and risk assessment, in *Wildlife Toxicology and Population Modeling: Integrated Studies of Agroecosystems*, R. J. Kendall and T. E. Lacher, Jr., Eds., Lewis Publishers, Chelsea, MI, 1993.
26. Pulliam, H. R., Sources, sinks, and population regulation, *Am. Nat.*, 132, 652-661, 1988.
27. Slobodkin, L. B., The role of minimalism in art and science, *Am. Nat.*, 127, 257-265, 1986.

39

Experimental Design, Statistical Analysis, and Demographic Models

Michael R. Willig

ABSTRACT

The detection of individual and population-level responses of wildlife to exposure to biocides frequently requires a quantitative approach during the design and analysis stages of experimentation. Limited familiarity with a particular arsenal of statistical tools or models often constrains experimental design and compromises subsequent analytical options. Exposure to a broader spectrum of approaches and designs, as summarized hereafter, may provide the impetus to restructure experimental protocols in a fashion that may more appropriately define the quantitative responses by wildlife to biocides in either a laboratory or field setting.

INTRODUCTION

Like environmental sciences in general, the field of avian wildlife toxicology is increasingly characterized by experiments that require a variety of different types of quantitative analyses. Perhaps the most difficult task that a scientist faces after deciding which questions to address[1] comprises the design of experiments that are feasible to conduct and likely to provide data capable of falsifying hypotheses. In many situations, knowledge of available statistical procedures and their limitations is among the more important factors guiding experimental design. Nothing is more frustrating to the environmental scientist or to the biostatistician than to discover that an elaborate, time consuming, and costly set of experiments cannot be analyzed in a very powerful fashion because adequate statistical advice was not sought prior to the collection of data. In a recent review of experiences as an editor for *Ecology*, Fowler[2] enumerated ten common statistical errors in ecological research and remarked that "over half of the papers that involved a statistical analysis had one or more errors" and "many of these errors involved relatively basic statistical knowledge." In many such cases, triage is unfortunately all that can be accomplished with the data, and discrimination among competing hypotheses remains as a task for future investigation. The chapters within this section provide both a general overview and a modicum of advice to those interested in the experimental design and analysis of data relevant to avian wildlife toxicology. The chapters by Dixon and Garrett[3] and Lacher and Willig[4] consider issues in univariate and multivariate data analysis, whereas the subsequent chapters by Skalski and Smith[5] and Sauer et al.[6] provide statistical interfaces to modeling demographic processes in wildlife populations.

EXPERIMENTAL DESIGN AND STATISTICAL ANALYSIS

No attempt has been made to be broadly inclusive of the entire field of biometry, in part because a number of excellent intermediate and advanced texts already do so.[7-15] Rather, the intent of the contributions by Dixon and Garrett[3] and Lacher and Willig[4] is to review and develop, within the context of ecotoxicological studies, a selection of useful statistical approaches and points of view concerning experimental design.

Dixon and Garrett[3] review the interrelated concepts of inference space, pseudoreplication, and statistical power as they apply to field experiments. This nicely sets the stage for a more detailed elucidation of the theory and application of bioequivalence tests: statistical procedures that directly evaluate if a particular treatment has no effect. Although bioequivalence procedures have not received widespread coverage in ecological and wildlife literature, they show much promise in applications concerning the detection of analytes and the effects of biocides on wildlife.

The chapter by Lacher and Willig[4] focuses on a variety of topics in multivariate statistics of interest to the wildlife toxicologists. In particular, many questions concerning the effects of biocides on fitness components or reproduction are inherently multifaceted in nature, and thus appropriately analyzed by multivariable procedures. In addition to reviewing the use of multivariate analysis of variance, principal components analysis, and discriminant analysis, they provide a protocol that may be useful in the analysis of multivariable data, one which avoids Rao's paradox and identifies the contributions of particular variables to multivariate differences.

DEMOGRAPHIC INTERFACES

Quantitative demographic models are of heuristic and predictive value to those interested in questions concerning avian wildlife toxicology. Demographic models can aid in answering questions, as well as in formulating them. To this end, estimating population parameters from field data is an important step in demographic modeling.

Sauer et al.[6] detail statistical issues relevant to modeling demographic trends by focusing on the rate of population increase and sources of its variation. Distinction between chronic and acute effects of biocides and the use of statistical approaches to evaluate the existence and significance of changepoints (alterations in population trajectories) in avian populations are particularly noteworthy contributions. The effect of a stressor (in their example, severe winters rather than biocide applications) on Eastern bluebirds provides a concrete example of the kind of applications that trend analysis can contribute to the field of avian wildlife toxicology.

Finally, the contribution by Skalski and Smith[5] focuses on risk assessment to avian populations from biocides by elaborating two approaches based upon mean temporal abundance. The direct quantification approach, based upon replication of control and treatment conditions, as well as randomization, constitutes the "notochord" of an effective experimental design, whether considering single or multiple year censuses. Demographic models offer the advantage of identifying the components of fitness (fecundity and survivorship) and incorporating their interaction with the environment and biocides. The theory and application of such procedures to field studies based upon radiotelemetry and mark-recapture are considered in detail. Finally, epidemiological approaches to statistical analysis, which are based upon individual birds rather than study plots as replicates, provide a powerful tool for detecting the effects of biocides on wildlife demography.

Together, these four contributions[3-6] provide a glimpse of the opportunities available to the avian wildlife toxicologist. The design of experiments and the selection of subsequent analyses are really different faces of the same coin. A well-designed study is able to clearly define the question, place it in the context of contemporary science, and produce data amenable to quantitative analysis. It is our hope that the following four contributions will hasten that process.

REFERENCES

1. Platt, J. R., Strong inference: certain systematic methods of scientific thinking may produce much more rapid progress than others, *Science*, 146, 347–353, 1964.
2. Fowler, N., The ten most common statistical errors, *ESA Bulletin*, 67, 161–164, 1990.
3. Dixon, P. M., and K. A. Garrett, Statistical issues for field experimenters, in *Wildlife Toxicology and Population Modeling: Integrated Studies of Agroecosystems*, R. J. Kendall and T. E. Lacher, Jr., Eds., Lewis Publishers, Chelsea, MI, 1993.
4. Lacher, T. E., Jr., and M. R. Willig, Univariate and multivariate approaches to the analysis of ecotoxicological data, in *Wildlife Toxicology and Population Modeling: Integrated Studies of Agroecosystems*, R. J. Kendall and T. E. Lacher, Jr., Eds., Lewis Publishers, Chelsea, MI, 1993.
5. Skalski, J. R., and S. G. Smith, Risk assessment in avian toxicology using experimental and epidemiological approaches, in *Wildlife Toxicology and Population Modeling: Integrated Studies of Agroecosystems*, R. J. Kendall and T. E. Lacher, Jr., Eds., Lewis Publishers, Chelsea, MI, 1993.
6. Sauer, J. R., R. J. Barker and P. H. Geissler, Statistical aspects of modeling population changes from population size data, in *Wildlife Toxicology and Population Modeling: Integrated Studies of Agroecosystems*, R. J. Kendall and T. E. Lacher, Jr., Eds., Lewis Publishers, Chelsea, MI, 1993.
7. Conover, W. J., *Practical Nonparametric Statistics*, 2nd ed., John Wiley & Sons, New York, 1980.
8. Hollander, M., and D. A. Wolfe, *Nonparametric Statistical Methods*, John Wiley & Sons, New York, 1973.
9. Kleinbaum, D. G., and L. L. Kupper, *Applied Regression Analysis and Other Multivariable Methods*, Duxbury Press, North Scituate, MA, 1978.
10. Siegel, S., *Nonparametric Statistics for the Behavioral Sciences*, McGraw-Hill, New York, 1956.
11. Snedecor, G. W., and W. G. Cochran, *Statistical Methods*, 7th ed., Iowa State University Press, Ames, 1980.
12. Sokal, R. R., and F. J. Rohlf, *Biometry*, 2nd ed., W.H. Freeman, San Francisco, 1981.
13. Steel, R. G. D., and J. H. Torrie, *Principles and Procedures of Statistics*, McGraw-Hill, New York, 1960.
14. Winer, B. J., *Statistical Principles in Experimental Design*, 2nd ed., McGraw-Hill, New York, 1971.
15. Zar, J. H., *Biostatistical Analysis*, 1st ed., Prentice-Hall, Englewood Cliffs, NJ, 1974.

40

Univariate and Multivariate Approaches to the Analysis of Ecotoxicological Data

Thomas E. Lacher, Jr. and Michael R. Willig

ABSTRACT

The possible effects of agricultural chemicals on the reproductive biology of non-target species are a major concern of environmental groups, government agencies, and the general public. Particular concern focuses on the effects of agricultural chemicals on avian species. Current regulatory guidelines recommend that a number of variables related to reproductive success be measured in avian field studies. These include both original and derived variables. The traditional statistical testing procedure (univariate analysis of variance on all variables) is not an appropriate design when comparing many closely related and intercorrelated variables. By definition α (= 0.05) of the variables, on average, will be significant due to chance alone when the variables are independent and no treatment effect is real. When characters are correlated, one cannot assess the overall error rate, reducing statistical decision-making to a state of futility. Multivariate procedures are more appropriate in these cases, because the algorithms incorporate intercorrelation into the analysis. We discuss the utility of three kinds of multivariate procedures (multivariate analysis of variance, discriminant function analysis, and principal component analysis) for the evaluation of treatment effects on avian reproductive success. We present guidelines for the use and interpretation of these procedures, using examples based upon the analysis of morphological data, and evaluate their advantages and disadvantages compared to traditional univariate procedures.

KEY WORDS

data analysis, multivariate statistics, ecotoxicology, discriminant function analysis, principal component analysis

INTRODUCTION

The study of natural systems is based upon a long history of careful observation and induction. Modern environmental science has placed a strong emphasis on the hypothetico-deductive approach, a trend noted by several contemporary philosophers of science.[1,2] Hypothesis testing in ecotoxicology demands rigor in the collection of data, and these data generally are quantitative. Accuracy and precision are important because variability at a number of levels is inherent to these systems. This results in a concomitant dependence on the application of statistical methodology to resolve quantitative hypotheses in ecotoxicological research.

A broad spectrum of research questions characterizes the emerging field of ecotoxicology. A convenient classification of this spectrum considers those questions that are variable specific in nature to be distinct from those that examine a suite of variables that together estimate a composite characteristic which may have biological significance. Examples of variable-specific questions include: (1) the effect of different exposures, species, age groups, or sexes on the uptake of a particular chemical compound; (2) the effect of different chemical vehicles (such as aerosols vs granules) on the uptake of a particular toxicant; (3) the detection of a toxicant (or a degradation product) in the tissues of potential target species; and (4) the effect of various doses of a chemical compound on the mortality of organisms. In each of these examples, the focus is on a single dependent variable or on a limited set of dependent variables in their own right. In contrast, composite characteristics estimated by a suite of variables include: (1) the behavior of organisms, in which behavior is described by several variables such as flying, resting, foraging, and aggression; (2) physiological or biochemical status as measured by heart rate, blood pressure, and brain or serum acetylcholinesterase activity; and (3) reproduction as estimated by clutch size, and hatching or fledgling success. Within each of these suites, variables generally are related to each other in varying degrees; this relationship is expressed quantitatively as a correlation. Multivariate procedures are especially efficacious when dealing with intercorrelated variables. When research protocols involve the measurement of several variables of interest taken on the same experimental unit, investigators are confronted with a choice between univariate and multivariate statistical approaches.

Multivariate statistical procedures are not new, and procedures such as multivariate analysis of variance (MANOVA) have been used since the 1930s.[3] They have been used by systematists for many years,[4,5] and have become increasingly common in phytosociology[6-10] as well. James and McCulloch[11] review the application and interpretation of multivariate procedures in ecology and systematics. Multivariate procedures are not yet commonly used in ecotoxicology even though they have much to offer. This chapter reviews several multivariate procedures that should prove useful to ecotoxicologists, and discuss their advantages and disadvantages in relation to univariate methods. We illustrate the utility of a multivariate analysis of biological data with examples from systematics and evolutionary biology. This review should assist researchers and regulatory agencies in the design of appropriate guidelines for future research protocols.

MULTIVARIATE STATISTICAL PROCEDURES FOR ECOTOXICOLOGY

There are several distinctions between univariate and multivariate statistical procedures. Univariate procedures evaluate hypotheses concerning a single dependent variable, whereas multivariate procedures test hypotheses based upon several dependent variables acting simultaneously. Multivariate analogues pertain to many univariate designs. The differences between these analyses are highlighted in numerous good discussions suitable for the statistically naive as well as for the more advanced user.[3,12-18] The authors illustrate the major differences in assumptions and inference by comparing univariate (ANOVA) and multivariate (MANOVA) analyses of variance.

Hypothesis Tests and Assumptions

ANOVA tests the null hypothesis that the means of a single dependent variable are equal in k groups. The test of significance is based upon the F statistic. MANOVA tests

Table 1. Assumptions of ANOVA and MANOVA[3]

ANOVA	MANOVA
Random sampling	Random sampling
Statistical independence of observations	Statistical independence of observations
Samples in each group drawn from a population that is normally distributed	Samples in each group drawn from a population that exhibits multivariate normality
Equality of within-group variances (homoscedasticity)	Equality of within-group covariance matrices

the null hypothesis that the mean weighted linear combinations of p variables in multidimensional space (group centroids) are equal in the k groups.[3] No single test criterion is universally accepted, but several are commonly used, such as Wilks's lambda criterion (Λ), Pillai's trace criterion (V), Roy's greatest characteristic root criterion (Θ), and the Lawley-Hotelling trace criterion (τ).[15] Fortunately, results from these statistics are frequently similar,[19] particularly if departures from the null hypotheses are slight.[20] MANOVA is generally robust with respect to moderate violation of its basic assumptions, but each test criterion performs somewhat differently in relation to the major assumptions,[3] and each criterion is differentially sensitive to the distribution of discriminating variance of the variables (trace).[15] The major assumptions of univariate and multivariate procedures are analogous (Table 1). The major differences are MANOVA requirements for multivariate normality (univariate normality for all variables does not guarantee multivariate normality) and for equality of variance-covariance matrices (again, not guaranteed by the homogeneity of all variances in the separate ANOVAs). The equality of variance-covariance matrices actually requires two conditions: within-group variances must be statistically equal for all variables, and the correlation for any two variables must be statistically equal in all groups involved in the analysis.[3]

When to Choose MANOVA in Ecotoxicology

The decision to choose MANOVA over ANOVA will depend, to a large degree, on the nature of the hypothesis and selection of variables. For example, if the major interest in a study is to evaluate the levels of six different chemical products in the bodies of collected birds from several exposure and control areas, then it may be appropriate to analyze the levels of each of the chemicals with separate ANOVAs. Remediation may require knowledge of which chemicals are actually in higher concentration in tissues from exposed vs control sites. Even though all six measurements were taken on the same experimental unit, and may well be intercorrelated, the real interest will likely be in the levels of each of the six chemicals across the sites, and not in the overall differences of some new composite variable. However, even in this case caution must be taken to avoid an excessively high type I error because the same null hypothesis (no difference among sites) will be tested for each variable. In this case, some correction of α (probability of incorrectly rejecting a true null hypothesis) will be necessary, either using the Bonferroni method or the Dunn-Šidák method.[21] An exception to this would be a case in which, because of potentially grave effects on human health or wildlife, it is critically important not to underestimate any potentially significant effects. In such a case, conservative tests might be inappropriate. Moreover, under some circumstances, a multivariate approach may be more powerful than a series of univariate tests (i.e., a MANOVA can be significant when all of the separate ANOVAs were nonsignificant) and may represent the most efficacious approach for identifying areas of concern.

If ANOVA seems to be the method of choice, then a series of steps should be followed in subsequent analyses (Figure 1). These include tests of the critical assumptions, such as randomness of sampling, normality, and homoscedasticity. Based upon the results of these tests, a decision must be made to use parametric procedures or an equivalent nonparametric analysis.

Both parametric and nonparametric methodologies have provisions for the analysis of *a priori* comparisons, a more powerful option for tests of significance that is rarely employed. It is particularly applicable in designs in ecotoxicology, where the groups consist of a control and several levels of application of a pesticide, or a control and several modes of application of the same pesticide. Investigators can test a restricted set of orthogonal or nonorthogonal comparisons at a fixed experiment-wise error rate, usually set at $\alpha = 0.05$. When the design does not allow for *a priori* comparisons (rarely) or all pairwise comparisons are of interest, then one of a number of *a posteriori* or multiple comparison procedures can be used. Statistical power can be substantially reduced, however, especially when a large number of comparisons is involved. Concise discussions of these procedures are readily encountered in a variety of sources.[21,22] Collyer and Enns[23] provide a readable and useful guide to ANOVA, including *a priori* and *a posteriori* analyses as performed in commonly used statistical packages (BMDP, SAS, and SPSS-X).

If the objective of the study is to evaluate the effect of a chemical exposure on a number of variables measuring clutch, nesting, and fledgling success, then MANOVA

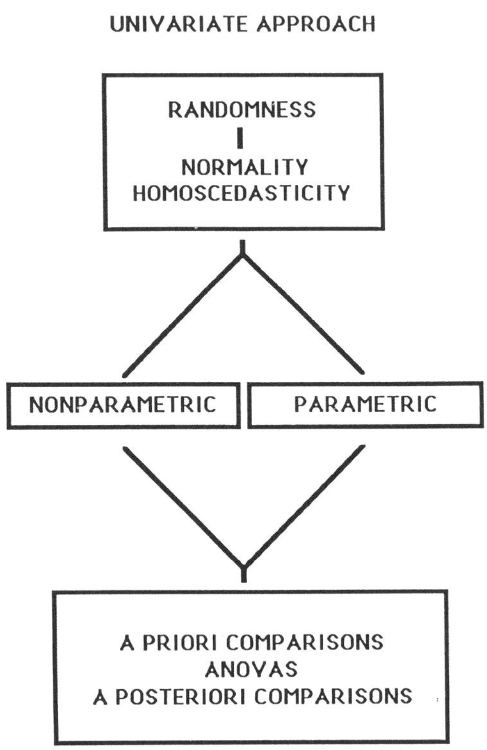

FIGURE 1. Pathway of procedures and decisions that should be followed when carrying out an ANOVA.

would be a better choice as these variables likely will be intercorrelated and no single variable is of concern by itself (Figure 2). (If all of the dependent variables are orthogonal, then the MANOVA approach simply yields the sum of the F ratios of the individual ANOVAs.[15]) MANOVA derives a number of composite variables that may have biological significance as a measure of overall reproductive success. A significant MANOVA among sites would mean the sites differed in overall reproductive success. The analysis can take several directions, depending upon results of initial test assumptions (Figure 3).

As discussed above, the critical assumptions for MANOVA are similar to those of ANOVA (Table 1). No commonly available tests are universally applied to assess multivariate normality; in any case, MANOVA is considered robust with respect to deviations from this assumption. Violations of the assumption of the homogeneity of within-group variance-covariance matrices can be more severe. All four of the test statistics discussed earlier are robust with respect to unequal matrices if sample sizes are equal and large, and the number of variables is small. The effect on the number of type I errors when these conditions do not apply and variance-covariance matrices are unequal is difficult to predict.[3] Because this assumption is frequently violated, precautions with regard to sample sizes should be incorporated into experimental design.

Whenever high communality (intercorrelation) characterizes a character suite, the pooled within-group variance-covariance matrix may be singular, and a MANOVA cannot be performed. (This occurs when one of the variables is a linear combination of the other variables.) Two solutions are reasonable. One or more variables suspected of being redundant can be deleted from the character suite and the reduced data set analyzed as before. Alternatively, the original character correlation matrix can be subjected to principal component analysis (PCA). Each experimental unit can be char-

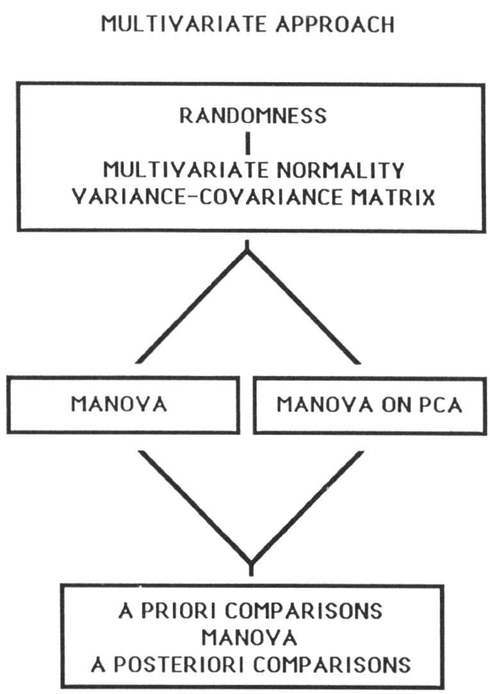

FIGURE 2. Pathway of procedures and decisions that should be followed when carrying out a MANOVA.

CORRELATION MATRIX

VAR	1	2	3	4	5	6	7	8	9	10	11
1	-										
2		-									
3	**		-								
4				-							
5			*	**	-						
6	**		**	**		-					
7				**	**	**	-				
8	**		**	**		**	**	-			
9			*	**	**	**		**	-		
10	**		**	**		**	**	**	**	–	
11			**			**		**	*	**	-
VAR	1	2	3	4	5	6	7	8	9	10	11

FIGURE 3. The character correlation matrix from a study on the effects of a pesticide application on starling reproduction. Characters are original and derived measures of various aspects of reproduction. Of 55 correlations, 31 were statistically significant.

acterized by a new set of uncorrelated variables (principal component axes) in which individual component scores represent each individual in a reduced number of dimensions but interindividual distances (= differences) are preserved. Principal component scores can then be subjected to MANOVA as suggested by Willig and Hollander;[24] an example of its application is presented in the section on case studies.

The same consideration should be given in MANOVA designs to the potential use of *a priori* and *a posteriori* comparisons with knowledge of their effects on experimentwise error rates.[25,26]

A researcher often is interested in evaluating the relative importance of the original dependent variables in contributing to group differences. Some authors have recommended performing univariate ANOVAs on all dependent variables if the MANOVA omnibus test is significant[12,15] under the concept of protection levels. Interpretation still can be misleading because of intercorrelation, and the table of pooled within-group correlation matrices should be presented along with such an analysis.[14] It is safer to restrict the evaluation of the relative importance of dependent variables to a more qualitative interpretation of the standardized discriminant function coefficients and the factor loadings. Standardized discriminant function coefficients are analogous to beta values in a multiple regression equation and present many of the same problems of interpretation. The loadings are correlations between the original set of dependent variables and the derived variables (discriminant function axes). Large correlations indicate that a given variable is important in accounting for group differences.[14]

MANOVA vs Discriminant Function Analysis

MANOVA and discriminant function analysis are two closely related multivariate procedures that are frequently confused and misapplied. The major distinction is that MANOVA is a multivariate test that assesses group differences based upon multiple dependent variables while holding the experiment-wise error rate constant. Examination of the derived, synthetic variables (also referred to as multiple criterion variables) can reveal complex biological effects not readily apparent from the examination of the individual dependent variables. The emphasis in discriminant analysis is to derive a synthetic variable, the discriminant function, that maximally distinguishes among groups already assumed or known to be different, and to use this criterion as a basis for predicting group membership. Two basic questions are involved: can we adequately discriminate among groups, and which linear combination of dependent variables is the best discriminator?

MANOVA will generally be the procedure of choice in ecotoxicology. Predicting group membership does not seem to be a priority in most ecotoxicological field studies, where the primary question is determining the effect of levels of a pesticide application on a suite of dependent variables. An exception may be a study in which data are available on a suite of reproductive or physiological variables from two aquatic sites: one pristine and one polluted by a chemical spill. Concern exists over the extent of the impact, and a sample of organisms is taken downstream from the spill. A discriminant analysis is conducted on the polluted and pristine sites as *a priori* groups, with the downstream cases assigned as unknowns; use of the predicted group memberships of the cases from the unknown site would assess the geographic scale of impact on these organisms resulting from the spill. If most unknown cases are assigned to the polluted group, it would indicate that the pollution had already moved downstream. Remediation would be suggested for sites with cases classified with the polluted group but not a priority for sites classified with the control.

CASE STUDIES

Adequacy of ANOVA vs MANOVA

Willig et al.[27] presented a critical analysis of univariate and multivariate approaches in the analysis of geographic variation in the morphology of bat populations. This research is analogous to an ecotoxicological study of variation in reproductive success in passerine populations across several levels of pesticide application. In morphometrics, proponents of a univariate approach conduct a set of analyses on each of a suite of variables. ANOVA is computationally simple and easy to understand, but it embodies several disadvantages. When several variables are compared, ANOVA is an inappropriate model of group difference. Significance is variable dependent, and group-specific contrasts (*a priori* or *a posteriori*) may not converge across individual variables. More importantly, morphological characters can be highly intercorrelated. The multivariate model for statistical analysis would seem most appropriate. However, some systematists attempt to avoid the problem by establishing a number of variables in advance that must be significant in order to consider the treatment groups distinct. This number is called the criterion variable. The overall error rate (OER) can be calculated if all variables are independent by the formula:

$$OER = 1 - \left[\prod_{i=1}^{c} (1 - \alpha_i) \right]$$

where α_i is the level of significance for character i in the ANOVA for a particular species and c equals the total number of variables in the analysis. However, when intercorrelation is present this formula is incorrect, and the decision becomes arbitrary.

MANOVA is a more appropriate model in this case because it incorporates the structure of the intercharacter correlation matrix, gives a single statement of group differences, and is frequently more powerful than a suite of ANOVAs. Disadvantages include the computational difficulty of MANOVA, the complicating factor of Rao's Paradox,[28] and the lack of association of particular characters with group differences. Also, singularity of the error matrix may prevent the analysis in some situations.

The question to resolve is whether univariate results emulate multivariate results. More specifically, are inferences drawn from a suite of univariate comparisons the same as those drawn from a single multivariate comparison? Willig and colleagues[27] tested the reliability of the arbitrary selection of univariate criterion values (percentage of significant ANOVAs) based upon a study of sexual and geographic variation in 12 cranial and 10 postcranial characters analyzed from both ANOVA and MANOVA perspectives. Comparison of results revealed little concordance between the results of the two analyses. In one case, none of the 10 postcranial characters were significant, yet the MANOVA was significant. At the other extreme, 11 of 12 cranial characters were significant based upon ANOVA, but nonsignificant using the multivariate approach. Essentially all possible combinations of significant and nonsignificant ANOVAs and MANOVAs were observed in their data. When a given percentage of significant ANOVAs was selected as a criterion value, the ANOVA approach yielded less than satisfactory results. Using the MANOVA outcome as the "correct" conclusion, no criterion value had a >85% success rate. The best success was obtained when the criterion was between 17 and 30% of the conducted comparisons. This occurred because very low criterion values led to exaggerated conclusions concerning group differences (a kind of type I error), whereas high criterion values increase the probability of "type II error." MANOVA is clearly the preferred statistical methodology for the morphological data examined by Willig et al.[27] In fact, any data set with a highly intercorrelated character set will best be analyzed by MANOVA. The authors expect that this will commonly be the case in many ecotoxicological studies as well.

The MANOVA/PCA Approach and the Question of Rao's Paradox

When ecotoxicological research involves the measurement of suites of characters, the MANOVA approach is preferred over the univariate approach. The MANOVA approach is not without difficulties. Hypotheses of group equality cannot be evaluated via MANOVA if the error matrix is singular, and high character intercorrelations can lead to the presence of Rao's Paradox.[28,29] Rao's Paradox is a situation in which some (or many) univariate analyses detect group differences, yet the MANOVA yields nonsignificance. The only way to completely avoid Rao's Paradox is to analyze characters that are orthogonal, an unlikely situation. A solution is to utilize principal component analysis, a data reduction procedure that constructs a set of orthogonal axes that may be considered independent characters for the ANOVA.[13,17,18] Willig and Hollander[24] analyzed the previously discussed data set on Brazilian bats to evaluate this approach. The character correlation matrix was subjected to a PCA, and a new set of variables were derived (the principal component axes). These orthogonal axes are linear combinations of the original character set. The principal component scores of each case (an individual bat in this analysis) were then used as the dependent variables.

If the disparity between the ANOVA and MANOVA results was attributable to Rao's paradox, then the MANOVAs run on principal component scores should have consistently higher significance levels than the MANOVAs run on the original characters. The results strongly supported this hypothesis. In nine cases in which Rao's Paradox may be important, the MANOVA/PCA procedure resulted in lower probability levels than the original MANOVA. Clearly, character correlation can decrease the power of the MANOVA.

When analyzing ecotoxicological data via MANOVA, investigators must be aware of this effect. Most commonly used packages of statistical procedures (e.g., SAS, SPSS-X, BMDP) will give ANOVA results as part of the MANOVA output.[25,30,31] If many ANOVAs are significant, but the MANOVA is not, then the original character matrix should be subjected to a PCA. The MANOVA should then be conducted on the principal component scores in order to evaluate the potential effects of Rao's Paradox. Willig and Hollander[24] recommend adhering to the MANOVA/PCA results for drawing conclusions. Investigators must be careful to use PCA and not factor analysis for this approach. These two methods use different algorithms and are frequently included in the same chapter in manuals for statistical software packages.

A Method for Assessing Variables of Importance

An important attribute of MANOVA is the generation of a new composite variable that may better summarize the biological differences among groups. A disadvantage is that it is more difficult to assess the importance of the original variables in the discrimination. Willig and Hollander[32] present a model that assesses the importance of individual variables in the discrimination of groups and measures the degree to which a group of differences are consistent. The model was developed to deal with research questions in systematics, but applications to ecotoxicology are shown here.

The original research question posed by their study was to evaluate which variables were important in accounting for sexual dimorphism in bat populations and determine the degree to which this sexual dimorphism is constrained by phylogeny. Sexes of each species for a given locality were compared using a stepwise discriminant analysis.[25] The initial analysis finds the best linear combination of variables for separating the sexes. The loadings of the original variables on the discriminant function axis are frequently used to assess their importance, but some variables important in discrimination can receive low loadings if they have a high communality with a more "important" variable.

In Willig and Hollander's[32] method, the importance of the original variables is determined in a series of steps. First, the original variables are correlated with the discriminant function scores for each individual. Second, this correlation coefficient is then squared to provide a measure of the proportion of the variation in the discriminant function scores that are attributable to variation in the original variable. This gives the importance of the discrimination of the original variable. Third, the polarity of the axes are standardized to facilitate the comparisons, such that males have high loadings and females low loadings. Therefore, the sign of the correlation indicates the direction of the effect; a positive correlation coefficient would mean that males are larger than females for that variable. A clear analogy to ecotoxicology would be in the comparison of reproductive effects on control and treatment fields after the application of a pesticide.

A study of sexual dimorphism will involve the calculation of only one discriminant axis. The method can be extended to scenarios which have more than two groups and more than one axis. Consider the formula:

$$I_i = \sum_{j=1}^{n-1} (\lambda_i)^2 (r_{i,j})^2$$

where I_i is the importance of variable i; n is the number of *a priori* groups; λ^2 is the proportion of variation accounted for by discriminant function i; and $r^2_{i,j}$ is the proportion of variation among scores for discriminant axis j accounted for by variation in variable i. The I values are scaled such that:

$$n - 1 > I_i > 0 \text{ or } 1 > I_i/(n - 1) > 0$$

Bar diagrams (Figure 4) provide a convenient way of viewing the I values. A comparison of the graphs for the same species from different areas in the bat study by Willig and Hollander provided a means of assessing character-specific phylogenetic con-

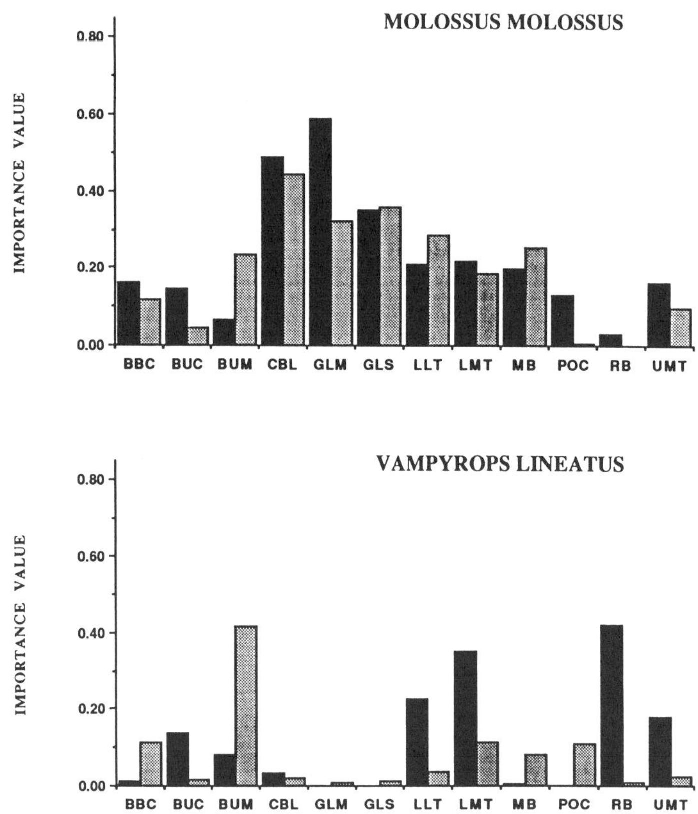

FIGURE 4. Bar diagram in which the importance value of a particular variable (x-axis) is represented by the height of the bars.[32] In ecotoxicology, the height of the bars may represent the effects of a pesticide on several variables related to reproductive success of a passerine. For *Molossus molossus*, significant concordance between two sites (solid, arid thorn scrub; shaded, savanna) in the expression of secondary sexual dimorphism is illustrated by the similar shapes of the distributions and a significant correlation between importance values. For *Vampyrops lineatus*, a lack of concordance is illustrated by dissimilar frequency distributions between sites and the absence of a significant correlation.

straints on the expression of secondary sexual dimorphism (see Figure 4 for discussion). This has a potentially important application in wildlife toxicology. For instance, bar graphs could be prepared from studies conducted with the same experimental design and on the same target species (for example, starlings), but at a number of different sites. Comparison of such graphs will reveal whether the same biological variables are affected by the chemical under a range of environmental conditions. This will determine whether the chemical has a specific physiological effect across a wide geographic area or affects the variables in a habitat-specific fashion. These results will assist in developing management plans for the timing and method of application in order to minimize deleterious impacts.

A MODEL RESEARCH PROTOCOL

Based upon the above scenarios, we envisage a model protocol to direct analyses in studies for which the MANOVA option is appropriate. MANOVA is performed on the original data, being careful to use the appropriate experimental design. A common error is to fail to use repeated measures designs when they are required. The design should also incorporate *a priori* comparisons if appropriate. In many cases, nonorthogonal contrasts should be seriously considered.[25] Rather than allowing the statistics to totally constrain the ecotoxicological questions of interest, the questions themselves should drive the experimental design.[26] For example, when a single control site exists along with several sites that differ in exposure to a pesticide, nonorthogonal contrasts of each exposed site to the control may constitute the most useful approach to hypothesis testing. If, on the other hand, a dose-response to different levels of a pesticide application is suspected, then polynomial contrasts are most appropriate. Finally, if sites are expected to respond to pesticide application as a step-cline or in a threshold-like fashion, then nonorthogonal comparison of adjacent sites may be the *a priori* design of choice. Fortunately, these designs are available in many statistical packages (e.g., SPSS-X).

As part of the MANOVA output, the battery of ANOVAs on the original characters should be requested. The MANOVA and ANOVA results should be compared in order to detect the possibility of Rao's Paradox. If evidence of the Paradox is found, then the original character correlation matrix should be subjected to PCA, and a MANOVA conducted on the principal component scores of the cases as the new variable set. If the MANOVA/PCA results confirm the suspicion of Rao's Paradox, then they should be used to draw inferences concerning group differences. As a final step, the original variables should be analyzed with stepwise discriminant function analysis and followed by the calculation of *I* values according to the method of Willig and Hollander.[32] Bar diagrams (Figure 4) will facilitate interpretation of which original variables are most important in distinguishing among groups.

SUMMARY

Although univariate analysis of variance is widely used in ecotoxicological research, many studies are more appropriately analyzed via multivariate techniques, such as multivariate analysis of variance, discriminant function analysis, and PCA. We illustrated the applications of multivariate approaches to ecotoxicology using a number of case studies.

All of the case studies presented deal with multivariate approaches to resolving questions in systematics. The authors chose examples from outside the realm of ecotoxicology, largely to emphasize the importance of intellectual interchange across disciplines. A substantial body of knowledge, theory, and methodology exists outside of the discipline of toxicology and has immediate and direct application to research on the effects of toxic substances on wildlife. Cooperative research endeavors involving scientists from basic and applied disciplines hold great promise not only in ecotoxicology, but in other areas of environmental science as well.

ACKNOWLEDGMENTS

We thank the Society of Environmental Toxicology and Chemistry and R. J. Kendall for the organization of the conference at which this chapter was presented. We are especially grateful to the corporations and federal agencies that helped to finance both the conference and the publication of this book. Susanne Swan of TIWET deserves special credit for her herculean efforts in organizing the conference. We also acknowledge Robert Owen, Randall Colbert, and especially Robert Hollander, the coauthors on the papers discussed in the case study sections of this chapter. Without their cooperation and consent this manuscript would not have reached fruition.

REFERENCES

1. Popper, K. R., *The Logic of Scientific Discovery*, Harper & Row, New York, 1968.
2. Kuhn, T. S., *The Structure of Scientific Revolutions*, 2nd ed., University of Chicago Press, Chicago, 1970.
3. Bray, J. H., and S. E. Maxwell, *Multivariate Analysis of Variance*, SAGE Publications, Beverly Hills, CA, 1985.
4. Sneath, P. H. A., and R. R. Sokal, *Numerical Taxonomy*, W. H. Freeman, San Francisco, 1973.
5. Neff, N. A., and L. F. Marcus, *A Survey of Multivariate Methods for Systematics*, privately published, 1980.
6. Gauch, H. G., Jr., *Multivariate Analysis in Community Ecology*, Cambridge University Press, New York, 1982.
7. Pielou, E. C., *The Interpretation of Ecological Data*, John Wiley & Sons, New York, 1984.
8. Digby, P. G. N., and R. A. Kempton, *Multivariate Analysis of Ecological Communities*, Chapman and Hall, New York, 1987.
9. Greig-Smith, P., *Quantitative Plant Ecology*, 3rd ed., Blackwell Scientific, Oxford, 1987.
10. Ludwig, J. A., and J. F. Reynolds, *Statistical Ecology: A Primer on Methods and Computing*, John Wiley & Sons, New York, 1988.
11. James, F. C., and C. E. McCulloch, Multivariate analysis in ecology and systematics: panacea or Pandora's Box?, *Annu. Rev. Ecol. Syst.*, 21, 129–166, 1990.
12. Cooley, W. W., and P. R. Lohnes, *Multivariate Data Analysis*, John Wiley & Sons, New York, 1971.
13. Harris, R. J., *A Primer of Multivariate Statistics, Academic Press*, New York, 1975.
14. Tabachnick, B. G., and L. S. Fidell, *Using Multivariate Statistics*, Harper & Row, New York, 1983.
15. Barker, H. R., and B. M. Barker, *Multivariate Analysis of Variance (MANOVA): A Practical Guide to its Use in Scientific Decision Making*, The University of Alabama Press, Tuscaloosa, 1984.
16. Manly, B. F. J., *Multivariate Statistical Methods: a Primer*, Chapman & Hall, New York, 1986.
17. Stevens, J., *Applied Multivariate Statistics for the Social Sciences*, Lawrence Erlbaum Associates, Hillsdale, NJ, 1986.
18. Cliff, N., *Analyzing Multivariate Data*, Harcourt Brace Jovanovich, San Diego, CA, 1987.

19. Pillai, K. C. S., and K. Jayachardran, Power comparisons of tests of equality of two multivariate hypotheses based on four criteria, *Biometrika*, 51, 313-326, 1967.
20. Morrison, D. F., *Multivariate Statistical Methods*, 2nd. ed., McGraw-Hill, New York, 1976.
21. Sokal, R. R., and F. J. Rohlf, *Biometry*, 2nd ed., W. H. Freeman, San Francisco, 1981.
22. Day, R. W., and G. P. Quinn, Comparisons of treatments after an analysis of variance in ecology, *Ecol. Monogr.*, 59, 433-463, 1989.
23. Collyer, C. E., and J. T. Enns, *Analysis of Variance, The Basic Designs*, Nelson-Hall, Chicago, 1987.
24. Willig, M. R., and R. R. Hollander, Multivariate morphometrics and Rao's paradox: the influence of character correlations in systematics, unpublished manuscript.
25. SPSS, Inc., *SPSS-X User's Guide*, McGraw-Hill, New York, 1983.
26. Hand, D. J., and C. C. Taylor, *Multivariate Analysis of Variance and Repeated Measures, A Practical Approach for Behavioral Scientists*, Chapman & Hall, New York, 1987.
27. Willig, M. R., R. D. Owen and R. L. Colbert, Assessment of morphological variation in natural populations: the inadequacy of the univariate approach, *Syst. Zool.*, 35, 195–203, 1986.
28. Corruccini, C. R., Univariate versus multivariate morphometric variation: an alternative viewpoint, *Syst. Zool.*, 36, 397-398, 1987.
29. Rao, C. R., Covariance adjustment and related problems in multivariate analysis, in *Multivariate Analysis*, P. R. Krishnaiah, Ed., Academic Press, New York, 1966, pp. 87–103.
30. *SAS User's Guide: Statistics*, SAS Institute, Inc., Cary, NC, 1985.
31. BMDP, *BMDP Statistical Software*, University of California Press, Berkeley, 1983.
32. Willig, M. R., and R. R. Hollander, Sexual dimorphism and phylogenetic constraints in bats: a multivariate approach, unpublished manuscript.

41

Statistical Issues for Field Experimenters

Philip M. Dixon and Karen A. Garrett

ABSTRACT

This chapter reviews four frequently misunderstood statistical aspects of the design and analysis of field experiments: generality of conclusions, pseudoreplication, statistical power, and testing for no effect. The choice of design determines the appropriate inference space (degree of generality of the conclusions) and the appropriate error terms to test hypotheses about treatments. Pseudoreplication, one form of use of inappropriate error terms, is common in analyses of field experiments, but it can be avoided by applying replicate treatments or restricting the generality of one's conclusions.

Statistical power is the ability of an experiment to detect differences between treatments. Field experiments frequently have low power because of few replicates and large residual variation. Power may be increased by changing the design, reducing the residual variation, or increasing the number of replicates.

Finally, bioequivalence tests are discussed, which are appropriate to test that a particular treatment has no effect. This goal is the reverse of the goal of a traditional hypothesis test, which is to show that a treatment has an effect. Because bioequivalence tests are not widely used, the authors demonstrate their use and interpretation with an example.

KEY WORDS

statistical analysis, experimental design, bioequivalence tests, power, sample size, pseudoreplication

INTRODUCTION

This chapter discusses four interrelated statistical topics applicable to the design and analysis of field experiments: (1) defining an inference space, (2) pseudoreplication, (3) statistical power, and (4) testing for bioequivalence. In the authors' experience, misunderstanding one or more of these issues causes most of the statistical mistakes made in the design and analysis of field experiments.

First, correctly defining the inference space is a necessary initial step in designing an experiment. The choice of inference space determines whether the conclusions from an experiment are to be applied generally or restricted to specific situations. Second, pseudoreplication occurs when treatments are applied to larger experimental units, such as a field, but "replicates" are collected from some smaller observational unit, such as a bird. Pseudoreplication usually results in an underestimate of the residual variance and an overestimate of the precision of the experiment. Third, statistical power is a concern in most field experiments; i.e., will a particular experiment be adequate to convincingly demonstrate interesting differences? Fourth, a traditional hypothesis test evaluates whether there is sufficient evidence to conclude that a treated group differs from some control group. This is the opposite of what is intended in most toxicological experiments, where the goal is to conclude that the treated group is the same, within some limit, as the control group. Bioequivalence tests, originally used in clinical statistics, were designed to demonstrate that there is no difference between treatments.

Our focus is on applications in wildlife toxicology, especially field experiments, to assess the consequences of pesticide use.[1,2] For illustration, a hypothetical experiment is used to determine whether quail eggs near fields receiving applications of a new pesticide have thinner eggshells than those on unsprayed fields. However, the principles discussed here (and often abused by experimenters) are relevant to field experiments in many disciplines.[3,4]

INFERENCE SPACE

Conclusions from an experiment make generalizations from the specific items used in the experiment (e.g., the quail on the experimental field) to a more general group (e.g., birds). The inference space of an experiment is the group to which the experimental conclusions can be correctly applied. This group may be small if conclusions are made only about animals in the experiment, or it may be large if conclusions are made about all related animals.

The design of the experiment determines whether the desired conclusions can be drawn. Some examples of assumptions that an experimenter may make include:

1. These quail eggs are representative of all quail eggs in the area, including those that evaded discovery.
2. These quail, though all from one geographic area, are representative of all quail.
3. These quail, though all of the same species, are representative of many or all species of birds.

Some assumptions can be defended on biological grounds; others can be justified by evidence from the experiment. For example, if geographical variation (assumption 2) is a concern, the experiment may be repeated in different areas.

If general results are desired, but an experiment provides only specific conclusions, then the experiment lacks what Campbell and Stanley[5] have called external validity. External validity determines whether conclusions from an experiment can be reliably generalized to other places or other populations. Experiments may also lack internal validity, which assesses whether "the treatment makes a difference in . . . [a] specific experimental instance."[5] The combination of experimental design and assumptions determines the external validity of the conclusions. If none of the assumptions listed previously are acceptable, then the conclusions should be limited to the quail used in the experiment. Then, the only issue is the internal validity of the experiment.

PSEUDOREPLICATION

One of the most frequent threats to external validity is pseudoreplication, a common occurrence in field experiments. Replication, the application of a treatment to more than one subject, is one of R. A. Fisher's three principles of experimental design[4,6] and is essential in a field experiment. Pseudoreplication occurs when repeated observations of a subject are substituted for replicated applications of a treatment on different subjects. In general, if it seems that the number of replicates can be increased indefinitely by splitting samples into increasingly smaller units, the experimenter is probably collecting pseudoreplicates, not true replicates.

Hurlbert[7] found that pseudoreplication was common in analyses of ecological field experiments. He surveyed 156 papers that he considered representative of the ecological literature of the late 1970s. Of the manipulative experiments, in which treatments are applied by the experimenter, 27% involved pseudoreplication, and of the studies applying inferential statistics, to generalize from one experiment to the class of all similar experiments, 48% involved pseudoreplication.

As an example of how easily one falls into a pseudoreplicated analysis, consider some possible experiments to determine the effect of a pesticide on quail eggshell thickness. Suppose an experiment is set up using two agricultural fields: the pesticide is applied to one field and not the other. The desired inference space is all agricultural fields, but only one field has received the pesticide application. If every egg in both fields is collected, the average eggshell thickness on each field is known exactly. Even if the treatment has absolutely no effect, it will appear to cause some difference between eggshell thickness, because any two sets of eggs will have a different mean thickness. The observed difference may be simply the difference inherent in the eggs of the two different fields, which could have been measured even in the absence of any treatment.

On the other hand, if the experiment is designed with many fields, thus dealing with the desired inference space, it is best to take more than one egg from each field because this increases the precision with which the eggshell thickness on each field is measured.[8] If one egg per field is measured, the residual variation includes components due to among-field variability and among-egg (within field) variability. If ten eggs per field are measured, the variation in average eggshell thickness includes the same among-field variability as before, but only one tenth the previous among-egg variability. If the among-egg variability is large, subsampling greatly increases the precision of the experiment, although it does not change the degrees of freedom available to test the treatment effect.

In some cases, constraints on an experimenter make an ideal set of replicates impossible. Pseudoreplication may be unavoidable, but the conclusions from such an experiment are not as strong as those from a replicated experiment. The typical pseudoreplicated experiment shows only that two fields differ, not that the difference was caused by a particular treatment. As Hurlbert[7] has written, "What is objectionable is when the tentative conclusions derived from unreplicated treatments are given an unmerited veneer of rigor by the erroneous application of inferential statistics."

STATISTICAL POWER

Another common misinterpretation in field experiments occurs when the usual statistical tests are used to conclude that a treatment has no effect. Suppose a statistical analysis is performed, and the usual null hypothesis (that the treatment has no effect) is not rejected. It would be wrong to conclude that the treatment has no effect just

Table 1. Coefficients of Variation for Replicate Measurements of Indices of Population Size[13]

Method	Coefficient of Variation
Roadside counts (gamebirds)	0.60–0.80
Pellet counts (deer)	1.00–1.40
Foot transects (deer, open habitat)	0.30–0.50
Foot transects (deer, wooded)	1.00–2.00

because the null hypothesis is not rejected. This raises the issue of statistical power, the ability of an experiment to detect a difference between treatments.

Two types of mistakes can be made when testing a statistical hypothesis. More attention is usually given to α, the probability of concluding that there is a significant effect when, in fact, there is none. β, the probability of missing an existing significant effect, may be of equal importance, however. There is a trade-off in designing a test: if a smaller α level is selected, β increases. The power of a test is $1 - \beta$, or the probability of correctly detecting an effect. Power is defined here in terms of hypothesis testing, but it can also be thought of in terms of estimation: the greater the power of an experiment, the narrower the confidence intervals for estimates.

The power of an experiment can be increased by any or all of the following:

1. increasing the number of replicates;
2. reducing the residual variance (e.g., by making more precise measurements); and
3. increasing the α level.

Power is also influenced by the choice of experimental design (e.g., a completely randomized design or paired plots) and the true difference between the treatments. An experiment is more likely to detect a difference if the true difference is large. This aspect of an experiment may or may not be susceptible to the artistry of an experimenter. Further discussions of power, its use in experimental design, and details of power calculations can be found in papers by Toft and Shea,[9] textbooks by Cox[10] and Cohen,[11] and computer manuals.[12]

In general, field experiments have low power because the typical number of replicates is small and the typical variability between fields or sites is large. Eberhardt[13] compiled some estimates of the variability in indices of population size (Table 1). These numbers are coefficients of variation (CV), the ratio of the standard deviation (SD) to the mean. Note that the least variable of these had a CV of 30%, while the most variable had an SD larger than the mean.

The large variability inherent in ecological studies means that more replicates are needed to detect an effect. Let us consider how many replicates are needed to achieve a given power. To calculate this power, we must estimate the residual variation, estimate the true difference, and specify the desired power, experimental design, and α level.

A common choice of power criterion is that an experiment should be large enough to have 80% power to detect an effect equal to 20% of the control value. Such values are frequently used in EPA[1] and FDA[14] calculations. Table 2 presents an example in which the response is a count, specifically population size, which is assumed to have Poisson variation; i.e., the mean of the count is equal to the variance. Also assume that there is no extra among-field variation and the treatments are assigned to fields in a completely randomized design. If the average abundance is 100 animals, then the smallest design to detect a change of 20 animals (i.e., a 20% effect) with 80% power has four treated and four control fields. If the average abundance is lower, then more fields must be

Table 2. Number of Replicate Fields Necessary to Detect a 20% Change in Number of Animals with At Least 80% Power

		No. of Replicate Fields		
Density	σ	**CRD Design**	**Paired Design $r = 0$**	**Paired Design $r = 0.5$**
100	10.0	4	5	4
50	7.1	7	8	5
25	5.0	13	14	8

Note: Three experimental designs are considered: (1) completely randomized design (treatments randomly assigned to fields), (2) paired plots design with no correlation between pairs ($r = 0$), and (3) paired plots design with moderate correlation between pairs ($r = 0.5$). One-sided *t*-tests with $\alpha = 0.05$. Residual variance assumed equal to the density, except for design 3, where residual variance = $(1 - r) * \sigma$.

used (Table 2). Even a design with four treated and four control fields is larger than the typical ecological field experiment.

To show the dependence of power and sample size on the experimental design, the minimum sample sizes were recalculated for a paired plot design under two assumptions: (1) moderate correlation between paired fields, and (2) no correlation between paired fields (Table 2). Pairing is appropriate when it reduces the residual variation because the paired fields are correlated. Fewer replicates are needed for an appropriate paired design to attain the same power as a completely randomized design.

However, the benefit of pairing depends on the correlation between plots. Pairing does increase the power when nearby plots are similar, but it also reduces the number of degrees of freedom in the error. If pairing does not reduce the residual variation, the power of a paired design is smaller than that of a completely randomized design because the residual variation is estimated with fewer degrees of freedom. Hence, the sample sizes needed to obtain a certain power are larger (Table 2). If the pairing is successful, however, the smaller residual variation is more important than the loss of degrees of freedom. Unfortunately, very little data are available on the scales of spatial variation that could be used to determine the optimal design.

Increasing the number of replicates, subsampling to reduce residual variation, and using more powerful designs are three common ways to increase the power of an experiment. Another way to increase power is to increase the number of control plots, while keeping the number of treated plots constant. This approach is especially useful when applying treatments is costly but sampling plots is relatively inexpensive. For example, in a field experiment to test whether aerial application of potassium fertilizer reduced cesium levels in vegetation and deer, applying treatments to large (approximately 2000 ha) plots would be expensive, but sampling deer and vegetation would be relatively cheap.

For this experiment, we recommended that the number of control plots be larger than the number of treated plots. Power curves are plotted for four designs with three treated and from three to nine control plots (Figure 1). Each curve exhibits the typical features of a power curve. When there is no effect, the power is equal to the α level. As the true effect size increases, the power of the experiment increases to an asymptote of 1 at extremely large effect sizes. Adding control plots increases the power because it increases the error degrees of freedom. For example, if the effect size is 2.0, the design with nine control plots has a power of 0.87, while the design with three control plots has a power of 0.64.

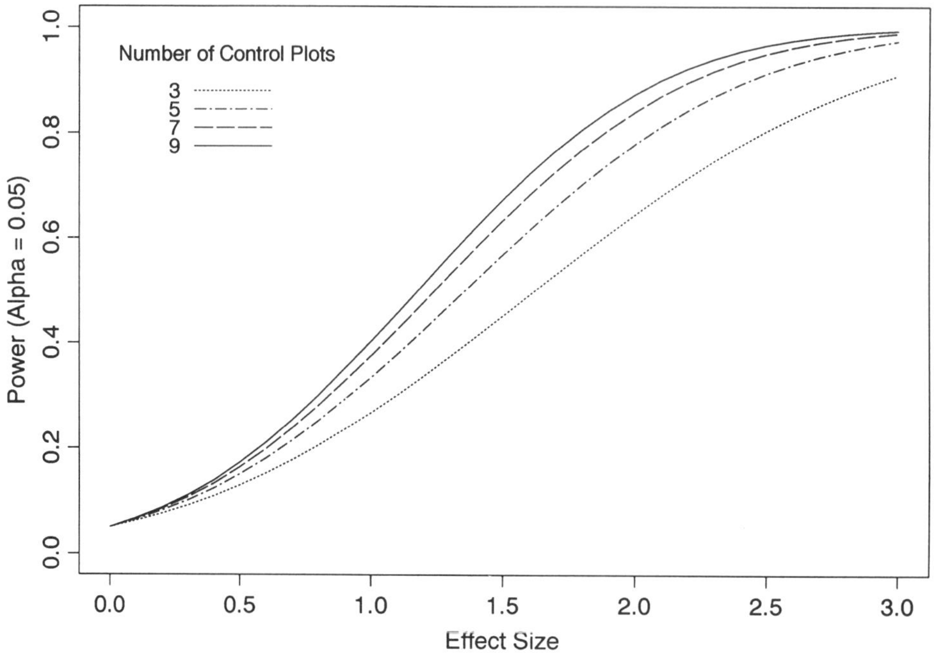

FIGURE 1. Power curves for one-way *t*-test with three treated plots and three or more control plots. Each curve gives the statistical power of that design as a function of the effect size (difference in mean divided by SD).

Adding control plots to increase power can only be used for some types of responses because it assumes that the among-replicate variation is the same in the treated and control plots. It is not appropriate for responses such as mortality rates or frequencies of developmental abnormalities. These types of responses are rare and often undetected in the absence of the treatment.[15] Hence, the variation in the response among control groups is low and not representative of variation in the treated group.

BIOEQUIVALENCE

The previous discussion has concentrated on the most common statistical application—demonstrating that a treatment does lead to a difference in some response. Many field experiments in wildlife toxicology have the opposite goal—demonstrating that a treatment does not cause a response. A nonsignificant hypothesis test is not sufficient evidence that a treatment has no effect. As discussed previously, a nonsignificant result may be due to a small difference between treatments or a large variability in the responses. Considering statistical power provides one way to design an experiment to be precise enough to detect effects of toxicological importance, but it is an indirect way to reach the conclusion that a treatment has no effect.

An experiment may also be too precise. A large experiment could reject the null hypothesis even if the effect of the treatment is extremely small. Depending on the aims of a study, a small treatment effect may be biologically the same as no effect. In this case the results are statistically, but not biologically significant. Again, the traditional hypothesis test does not answer the appropriate question.

Bioequivalence tests provide a direct way to test whether a treatment has no effect.[14] When a generic drug is developed to mimic a brand name drug, the performance of the generic must be shown not to differ from the brand name drug.[16] Our discussion of bioequivalence tests is more detailed than the discussion of the previous topics because they have not been used in the ecological or toxicological literature and few field biologists seem to know about them.

Because the difference between the effects of two drugs is rarely exactly zero, we must specify the limits within which we will consider the two drugs to be equivalent. These limits are usually symmetric. If Δ is the true difference between the drugs, and C_l and C_u are the lower and upper limits to the equivalence region, the bioequivalence test is

$$H_o\colon \Delta \leq C_l \text{ or } \Delta \geq C_u \text{ vs}$$

$$H_a\colon C_l < \Delta < C_u$$

The difference between standard statistical tests and bioequivalence tests is graphically illustrated in Figure 2. A standard statistical test of the hypothesis that $\Delta = 0$ rejects that hypothesis whenever the observed difference lies suitably far from zero (Figure 2a). A less precise experiment (one with more variation or fewer replicates) has a smaller rejection region; the null hypothesis is rejected for fewer observed differences. The two-way bioequivalence test rejects the null hypothesis whenever the observed difference lies suitably close to zero (Figure 2b). A less precise experiment will reject the null hypothesis less frequently. If the experiment is sufficiently imprecise, the

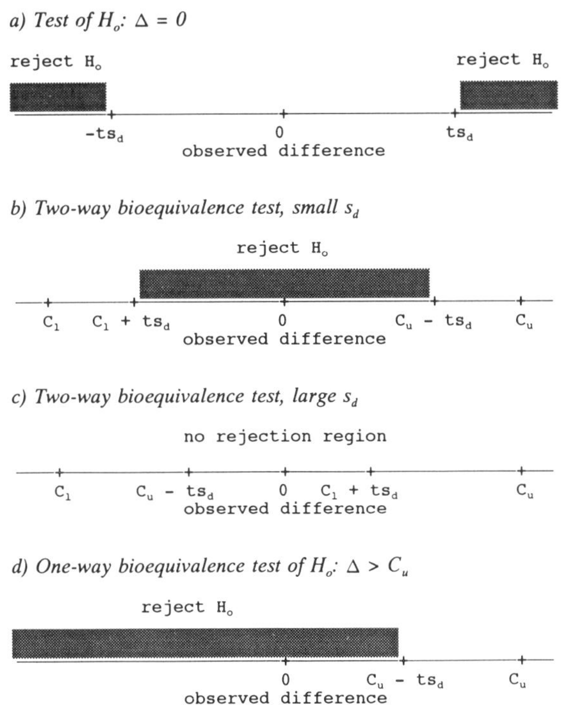

FIGURE 2. Comparison of rejection regions for standard tests of difference = 0 and bioequivalence tests. The rejection region shows the possible experimental outcomes (differences between treatments) that result in rejecting the null hypothesis. (a) Standard hypothesis test of true difference = 0. (b) Two-way bioequivalence test with small residual variation. (c) Two-way bioequivalence test with large residual variation. Note that lower and upper bounds to the rejection region cross, so there is no rejection region. (d) One-way bioequivalence test.

rejection region can shrink to the point that no experimental results will reject the null hypothesis (Figure 2c).

Bioequivalence tests are relatively new, and no consensus exists among statisticians as to the best test.[17-19] Two simple methods developed by Metzler[20] and Anderson and Hauck[21] assume a two-group comparison trial with independent, normally distributed responses, with mean μ_c in the control group, mean μ_t in the treated group, and common unknown variance. The unknown variance is estimated by the pooled SD with $(n_1 + n_2 - 2)$ degrees of freedom, just as in a standard test of the difference between two means.

The Metzler[20] protocol for an $\alpha = 0.05$ test is to use the 0.025 and 0.975 quantiles of the central t-distribution to compute a 95% confidence interval (CI) around the observed mean difference, just as one would calculate a typical CI around a mean. The null hypothesis is not rejected if the CI includes either C_u or C_l, the upper or lower equivalence limits. Alternatively, this test rejects the null hypothesis if the observed mean difference falls within a critical region defined by the C_l and C_u values, the pooled SD, and the appropriate t critical values (Figure 2b).

Anderson and Hauck[21,22] developed a slightly different test (Table 3) using the idea that the rejection region for a bioequivalence test is the intersection of two regions, one based on a test that the true difference is greater than C_l and one based on a test that the true difference is less than C_u. Their test has better statistical properties than the Metzler test, but is more difficult to calculate, requiring interpolation in tables of the cumulative t-distribution[23] or an algorithm to approximate the cumulative t-distribution.

A one-way bioequivalence may be appropriate for some experiments in wildlife toxicology. A one-way test is only concerned with differences between groups in one direction, rather than being concerned with any difference between groups. For example, we may not be interested in whether the field application of a pesticide reduces mortality. We would only be concerned that the pesticide does not increase mortality. Let Δ be the true difference in mortality between control and application groups and C_u be the largest increase in mortality that we can tolerate. The desired hypothesis test is

$$H_o\colon \Delta \geq C_u \text{ vs}$$

$$H_a\colon \Delta < C_u$$

One-way bioequivalence tests follow from the same principles used to construct two-way tests. The $\alpha = 0.05$ Metzler-like test procedure is to reject H_o if the observed difference exceeds the critical value x_c, which equals $C_u - t\, s_d$, where t is the 0.975 percentile of a central t-distribution and s_d is the pooled standard error of the differ-

Table 3. Calculations for Anderson and Hauck's Test of Bioequivalence

Bioequivalence Criterion	Calculated Values of			
	d_o	T_l	T_u	$P[T_l < T < T_u]$
$-0.05 < \Delta < 0.05$	1.58	−3.16	0.00	0.490
$-0.10 < \Delta < 0.10$	3.16	−4.74	−1.58	0.076
$-0.25 < \Delta < 0.25$	7.91	−9.48	−6.32	<0.001

Note: See text for details and formulae. For the example described in the text, the bioequivalence interval is symmetric, so the value of T_o is not affected by the choice of bioequivalence criterion. Hence, $T_o = d/s_d = 1.58$ for all three bioequivalence criteria.

ence. For a one-way test, the Anderson and Hauck procedure is similar to the one-way Metzler test, although again, it is necessary to evaluate the cumulative t-distribution. If the responses are proportions, another one-way bioequivalence test, based on a chi-square test, has been developed by Dunnett and Gent.[24]

The rejection region for a one-way bioequivalence test is shown in Figure 2d. Note that as the precision of the experiment increases, the size of the rejection region increases and that a one-way bioequivalence test will always accept the null hypothesis if the observed difference exceeds the critical difference. Also, if the residual variation is large, then the null hypothesis (that the pesticide causes increased mortality) can be accepted even if no difference in mortality is observed.

Examples of Bioequivalence Tests

Because bioequivalence tests have not been used in the ecological and toxicological literature, the calculation and interpretation of bioequivalence tests will be demonstrated using the eggshell thickness experiment example. Consider the following artificial results. The average shell thickness was 0.50 mm on five control fields and 0.45 mm on five treated fields. The SD among fields is 0.05 mm on both treated and control fields, so s_d, the standard error of the difference between fields, is 0.0316. In general, s_d is given by

$$S_d = SD\sqrt{\frac{1}{n_t} + \frac{1}{n_c}}$$

where SD is the standard deviation, and n_t and n_c are the number of treated and control fields. We will assume that the average thickness is normally distributed. If we use a traditional statistical approach, a standard t-test does not reject the null hypothesis ($t = 1.581$, $p = 0.15$).

To use a bioequivalence test, one must specify what range of differences are considered to be equivalent. Usually, this will be determined by the investigator's knowledge of the subject or specified by a regulatory agency. In this example, the calculations will be demonstrated using three increasingly liberal criteria for bioequivalence. These represent three different alternative hypotheses: (a) $-0.05<\Delta<0.05$, (b) $-0.10<\Delta<0.10$, and (c) $-0.25<\Delta<0.25$. These limits correspond to differences of $\pm10\%$, $\pm20\%$, and $\pm50\%$ of the control thickness, but they are arbitrarily chosen. In actual use, one would choose and use one criterion of equivalence.

Using the Metzler[20] procedure, a 95% CI around the mean difference is

$$d \pm t\, s_d$$

$$(0.50 - 0.45) \pm 2.305 \times 0.0316$$

where d is the observed difference, t is the 0.975 quantile of a t-distribution with 8 degrees of freedom (d.f.) (= 2.305), and s_d is the pooled standard error of the difference. The endpoints of the stringent bioequivalent criteria (0.05 and 0.1) are inside the 95% CI (0.123, -0.023), so the null hypothesis of differences between groups is not rejected. When the third criterion is used, its endpoint, 0.25, lies outside the 95% CI and the null hypothesis is rejected. If $\pm50\%$ is used as the criterion of equivalence, it can be concluded that the shell thicknesses are equivalent.

To use the Anderson and Hauck[21] procedure, first calculate two quantities: d_0, which standardizes the region of bioequivalence, and T_0, which standardizes the observed difference:

$$d_0 = \frac{C_u - C_l}{2S_d}$$

$$T_0 = \frac{d - (C_u + C_l)/2}{S_d}$$

where d is the observed difference, s_d is the standard error of the difference, and c_l and c_u are the lower and upper bounds of the bioequivalence region. Then calculate the probability that a random variable with a t-distribution with the appropriate degrees of freedom lies between two values, T_l and T_u, defined by:

$$T_l = -|T_0| - d_0$$

$$T_u = |T_0| - d_0$$

This probability is calculated from tables or approximations to the cumulative t-distribution.

In the eggshell thickness example used above (d), the observed difference is 0.05 mm, and s_d, the standard error of the difference, is 0.0316 mm. To apply Anderson and Hauck's test,[21] we must first specify the upper and lower limits of the equivalence region. Based on biological criteria, we might decide that differences in thickness between -0.05 mm and 0.05 mm had no biological effect. Hence, $c_u = 0.05$ mm and $c_l = -0.05$ mm. Values of d_0, T_0, T_l, and T_u calculated using the previous equations are given in Table 3. The appropriate t-distribution has $n_t + n_c - 2 = 8$ d.f. The significance level of the bioequivalence test is the proportion of the t-distribution with 8 d.f. that lies between -3.16 and 0. From a table of the cumulative t-distribution[23] or a computer program,[12] the cumulative probabilities associated with -3.16 and 0 are 0.01 and 0.50. The significance level is the difference, 0.49. Because this is large, we cannot conclude that the treatment has no effect on eggshell thickness, if a strict criterion for equivalence is used.

Calculations for the other two bioequivalence criteria are presented in Table 3. The two groups are equivalent if the ±50% criteria is used, but there is insufficient evidence ($p > 0.05$) to support equivalence if the stricter ±10% or ±20% criteria are used.

CONCLUSIONS

Designing and analyzing a useful field experiment is not easy. In our experience, most of the statistical mistakes made in the design and analysis of field experiments occur because one or more of the four issues discussed here—inference space, pseudoreplication, power, and bioequivalence—were misunderstood. It is hoped that awareness of the common problems will reduce their frequency.

However, statistical criteria are only one component in the design of a successful experiment. Statistical issues should always complement toxicological and ecological issues, not dictate the approach. To echo a well-stated conclusion of Moore and McLaughlin, ". . . the experimental models, sampling layout, and field methodology should be consistent with the researcher's intuitive understanding of ecosystem behavior."[25]

ACKNOWLEDGMENTS

We thank Gene Schupp, Joe Pechmann, and Kay Kirkman for reading and commenting on the manuscript. This work was supported by contract DE-AC09-76SROO-819 between the U.S. Department of Energy and the University of Georgia's Savannah River Ecology Laboratory.

REFERENCES

1. Fite, E. C., L. W. Turner, N. J. Cook and C. Stunkard, *Guidance Document for Conducting Terrestrial Field Studies*, EPA 540-09-88-109, U.S. Environmental Protection Agency, Washington, DC, 1988.
2. Avian Effects Dialogue Group, *Pesticides and Birds: Improving Impact Assessment*, The Conservation Foundation, Washington, DC, 1989.
3. Green, R. H., *Sampling Design and Statistical Methods for Environmental Biologists*, John Wiley & Sons, New York, 1979.
4. Federer, W. T., Principles of statistical design with special reference to experiment and treatment design, in *Statistics: An Appraisal*, H. A. David and H. T. David, Eds., Iowa State University Press, Ames, 1984, pp. 77-104.
5. Campbell, D. T., and J. C. Stanley, *Experimental and Quasi-experimental Designs for Research*, Rand McNally, Chicago, 1963.
6. Fisher, R. A., *Design of Experiments*, Oliver & Boyd, London, 1935.
7. Hurlbert, S. H., Pseudoreplication and the design of ecological field experiments, *Ecol. Monogr.*, 54, 187-211, 1984.
8. Christensen, R., *Plain Answers to Complex Questions*, Springer-Verlag, New York, 1987.
9. Toft, C. A., and P. J. Shea, Detecting community-wide patterns: estimating power strengthens statistical inference, *Am. Nat.*, 122, 618-625, 1983.
10. Cox, D. R., *Planning of Experiments*, John Wiley & Sons, New York, 1958.
11. Cohen, J., *Statistical Power Analysis for the Behavioral Sciences*, LEA Press, Hillsdale, NJ, 1988.
12. Lenth, R., *PowerPack User's Guide*, Russ Lenth, Ames, IA, 1987.
13. Eberhardt, L. L., Appraising variability in population studies, *J. Wildl. Manage.*, 42, 207-238, 1978.
14. Westlake, W. J., Bioavailability and bioequivalence of pharmaceutical formulations, in *Biopharmaceutical Statistics for Drug Development*, K. E. Peace, Ed., Marcel Dekker, New York, 1988, pp. 329-352.
15. Heinz, G. H., E. F. Hill, W. H. Stickel and L. F. Stickel, *Environmental Contaminant Studies by the Patuxent Wildlife Research Center*, ASTM STP 693, American Society for Testing and Materials, Philadelphia, 1979, pp. 8-35.
16. *The Bioavailability Protocol Guideline for ANDA and NDA Submission*, U.S. Food and Drug Administration, Washington, DC, 1977.
17. Westlake, W. J., Response, *Biometrics*, 37, 591-593, 1981.
18. Kirkwood, T. B. L., Bioequivalence testing—a need to rethink, *Biometrics*, 37, 589-594, 1981.
19. Rocke, D., On testing for bioequivalence, *Biometrics*, 40, 225-230, 1984.
20. Metzler, C. M., Bioavailability—a problem in equivalence, *Biometrics* 30, 309-317, 1974.
21. Anderson, S., and W. W. Hauck, A new procedure for testing equivalence in comparative bioavailability and other clinical trials, *Commun. Stat. Theory Methods*, 12, 2663-2692, 1983.

22. Hauck, W. W., and S. Anderson, A new statistical procedure for testing equivalence in two-group comparative bioavailability studies, *J. Pharm. Biopharm.*, 12, 83–91, 1984.
23. Fisher, R. A., and F. Yates, *Statistical Tables for Biological, Agricultural and Medical Research*, 6th ed., Hafner Publishing, New York, 1963.
24. Dunnett, C. W., and M. Gent, Significance testing to establish equivalence between treatments with special reference to data in the form of 2×2 tables, *Biometrics*, 33, 593–602, 1977.
25. Moore, S. F., and D. B. McLaughlin, Design of field experiments to determine the ecological effects of petroleum in intertidal ecosystems, *Water Res.*, 12, 1091–1099, 1978.

42

Statistical Aspects of Modeling Population Change from Population Size Data

John R. Sauer, Richard J. Barker, and Paul H. Geissler

ABSTRACT

Accurate modeling of population dynamics of widespread, migratory bird species from estimates of survival and recruitment is often difficult. Surveys that provide yearly indices of population size can be used to estimate an average finite rate of increase (λ). The analogies between the estimation of population trends from survey data and the estimation of rate of increase are reviewed. Statistical issues in modeling trends (and therefore λ) include sampling biases of sex and age classes and detection probabilities. Detection of a changepoint in time, when population dynamics change, can be difficult because of autocorrelation, sampling biases, or environmental variation in population trends. The authors present an example of detecting a change in a population associated with a known environmental change.

KEY WORDS

experimental design, statistical analysis, demographics, finite rate of increase, surveys, indices

INTRODUCTION

Modeling of animal population dynamics has traditionally been used to provide insights into the effects of changes in survival and fecundity rates on the intrinsic rate of population increase. However, models have not provided precise predictions of the observed dynamics of real populations. For example, Sykes[1] summarized his view of Leslie matrices by stating: "In demographic applications, the model has been found to give predictions of future populations which might most charitably be described as poor." Attempts to directly model population changes of mallards (*Anas platyrhynchos*) with discrete-time population models have proven unsuccessful, possibly because of changes in age structure and error in the estimation of survival and recruitment rates.[2,3] For mallards, a predictive model is used to estimate fall flight based on spring survey data and other information.[2] Further modeling is generally not used in management, primarily because uncertainty over whether hunter harvest is additive or compensatory to other sources of mortality has made development of models for optimal harvesting impossible. Recently, simulation models have been used to examine the dynamics of mid-continent mallards.[3,4] Because of our inability to develop accurate models for bird populations, survey data have traditionally been used as basic information for management decisions on migratory bird populations in North America.[2] Temple and Weins,[5] in a recent review of the use of bird populations as environmental indicators, made a distinction between primary population parameters such as birth, death, and dispersal, and secondary population parameters such as population size. They reviewed the biases that can arise when secondary population parameters are used to infer changes in primary population parameters such as birth and death rates. Secondary population parameters may not change in a predictable way in response to changes in birth or death rates because of the action of density-dependent survival or reproduction, migration, or compensatory mortality. However, they also suggest that basic information on changes in population size may have value as a summary of the often complex relationships among primary population parameters.

Information on changes in numbers of individuals within animal populations can be used to estimate intrinsic rates of increase. Caughley[6] wrote that intrinsic rates of increase "measures (sic) a population's general well-being, describing the average reaction of all members of the population to the collective action of all environmental influences." He defined three intrinsic rates of increase, (1) r_m, a theoretical maximum rate of increase, (2) r_s, a rate of increase for an observed life table, and (3) $\bar{\mathbf{r}}$, an observed exponential increase. Of these, r_m is only observed under optimal conditions and is not of particular relevance to this discussion. The observed life table rate, r_s is defined as demographic vigor and can be directly related through stable population theory to survival and fecundity rates. Field data can always be used to estimate $\bar{\mathbf{r}}$, but it is not directly related to survival and fecundity rates from a stable population. Therefore, interest exists in an evaluation of the conditions under which $\bar{\mathbf{r}}$ estimated from field data is related to stable population birth and death rates and also in an evaluation of some practical consequences of estimating $\bar{\mathbf{r}}$ from field data.

Caswell[7] suggested using intrinsic rate of increase as a summary statistic for life table response experiments. He developed methods for examining changes in intrinsic rates of increase from changes in population projection matrices in response to experimental treatments. A reasonable alternative design for such experiments might involve direct estimation of $\bar{\mathbf{r}}$ from the experimental populations, especially in the case of many migratory bird populations for which the exact structure of the population transition matrix is unknown and vital rates are difficult to estimate. Unfortunately, population size estimates are often extremely variable, reflecting sampling error, environmental effects, and changes in survival and recruitment rates over time.[3] Consequently, $\bar{\mathbf{r}}$ estimated from field data cannot be equated with r_s. Even when r_s is estimable from field data, decomposition of $\bar{\mathbf{r}}$ into birth and death rates is generally not possible because of identifiability constraints. One cannot assess the importance of density-dependent survival or fecundity in an age-specific population without information on both population size and the parameter of interest.

This chapter discusses the estimation of finite rate of increase (λ, which equals $\exp[\bar{\mathbf{r}}]$) from time-series data. After describing the design of a typical survey, some of the sampling constraints upon the analyses are reviewed, including biases in sampling age/sex classes and variation in detection probabilities. A method of estimating the composite finite rate of increase and a test for determining whether λ changed at some specified time are presented. The standard statistical technique for the estimation of unknown times at which λ changed (changepoints) is also reviewed. These techniques are illustrated with population survey data from the North American Breeding Bird Survey (BBS) for eastern bluebirds (*Sialia sialis*) in Tennessee, which contains a known intervention associated with severe weather in 1978.[8] Finally, we assess the ability of statistical techniques to correctly estimate known times at which λ changed in simulated population data given reasonable assumptions regarding probabilities of detection and autocorrelation (correlation among the counts in the time series).

Estimating Finite Rate of Increase from Estimates and Indices of Population Size

Most extensive surveys do not directly estimate population size. Instead, an index of population size is observed. These indices (C_i) are related to the population size (N_i) in year i by a detection probability (p_i), or

$$C_i = p_i N_i. \quad (1)$$

Much of the biological sampling literature in recent years has been devoted to designs for estimating p_i from survey data.[9,10] Only infrequently is p_i directly estimated in extensive surveys (e.g., Waterfowl Breeding Surveys[2]). In roadside surveys that consist of a series of point counts (such as the BBS,[11] the Mourning Dove Call-Count Survey,[12] and the American Woodcock Singing-Ground Survey[13]) it is assumed that p_i equals p and population changes are estimated from C_i. Bart and Schoultz[14] found that detection probabilities can be small and variable in roadside surveys, introducing the potential for both bias and inefficiency in estimation of λ.

The authors discuss analytical methods for estimating λ from two survey designs: (1) surveys that estimate N_i directly for a region, and (2) roadside surveys that rely on counts $C_{i,j}$ (j specifies the roadside survey route) within regions. In the first case, p_i is not an issue and λ can be directly estimated from N_i. However, detection probabilities, $p_{i,j}$, where subscripts denote years and routes, always exist in the second case, and estimation of λ must incorporate the potential for variation in $p_{i,j}$. The following section illustrates some of the sources of variation in $p_{i,j}$ using examples from roadside surveys.

Sources of Variation in Detection Probabilities

Count data from roadside surveys of birds are notoriously variable,[14] and much of the variation in the counts is due to variation in detection probabilities. Any environmental feature that can influence the counts contributes to this variability.

On roadside survey routes (routes are the sampling units that form the replicates), detection probabilities vary temporally and geographically, and several environmental features can affect detection probabilities both within and among replicates over time. For example, differences in the counting ability of observers cause detection probabilities to differ within a route over time (when observers change) and among routes (when different observers count the replicates). Also, bird species tend to differ in detection probabilities. Some of the important features that influence detection probabilities from count data are documented in this section.

Variation among Years

Within-year sampling effects change detection rates over time within a route. Routes are scheduled to be surveyed at the time of maximum breeding activity of the species. Breeding phenologies can differ from year to year, and these differences tend to be regional rather than local. Hence, surveys of all routes in a region may be poorly timed. Furthermore, because observers often have a limited number of days to run the route, surveys are sometimes conducted during marginal weather conditions. It is often unclear if low counts in a year represent changes in bird populations or year-specific sampling effects, especially if the year-specific effects occur near or during the time of surveys (e.g., severe spring weather and purple martin (*Progne subis*) populations.[15])

Variation in counts among years may occur because of differences in observers. While some observers will conduct the yearly surveys of a route for many years, many observers only survey routes for 1 or 2 years. For the American Woodcock Singing-Ground Survey, for example, each observer conducted an average of 3.95 yearly surveys during the period 1970 to 1988.[13] Although attempts are made to maintain the quality of observers, observer changes can have striking effects on the counts from a roadside survey. Observers tend to differ in knowledge of birds and ability to survey birds, and the actual survey locations tend to differ among observers. As an example, data are presented from a call-count route for mourning dove (*Zenaida macroura*) from Delaware (Figure 1). Without information on observers, the population seems to be declining. However, adding observer as a covariate in the analysis indicates an increasing population. In our analysis of American woodcock (*Scolopax minor*), we found that significant observer differences existed, and we incorporated them into our within-route trend models.[13] Finally, field studies we conducted several years ago with several competent observers on the same BBS route indicated that many differences existed among counts of the same species because of different observers.[16]

Population size may be associated with variation among routes and over time. Gates[17] suggested that ring-necked pheasant (*Phasianus colchicus*) males tended to increase calling frequency at high population densities. This suggests that detection probabilities increase in larger populations, which would tend to overemphasize changes in populations. It is interesting to note that ring-necked pheasant and northern bobwhite (*Colinus virginianus*) tend to show rapid changes in levels of population indices;[18] this may be due to this nonlinear response of calling males to changes in population density.

Variation among Routes

There is evidence that $p_{i,j}$ differs among routes both within and among regions. Clear regional differences, generally associated with habitat structure, exist in how birds are perceived by observers. Mourning dove call-counts provide an example of this, because birds are recorded separately as either heard or seen on the route. In forested regions in

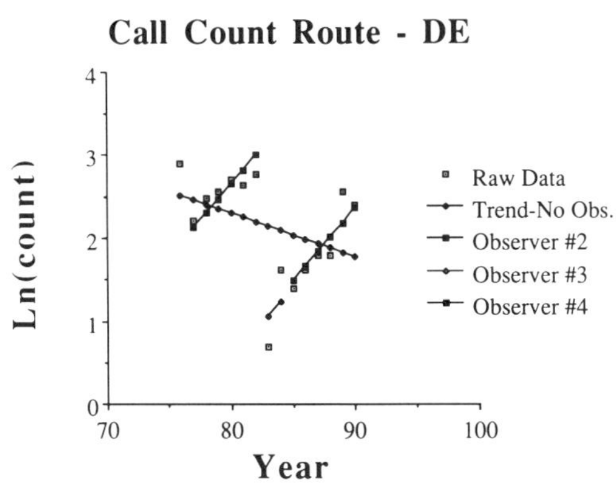

FIGURE 1. Count data from a single mourning dove call-count survey route in Delaware for the interval 1976 to 1990. Different observers are denoted by symbols. Increases occur within observers, but observer changes create the perception of a decline in counts.

the eastern United States, doves are more often heard than seen on call-count survey routes. In prairie habitats in the central United States, more doves are seen than heard.[12] Habitat differences exist even on adjacent routes, and can cause detection probabilities to differ within regions. Bart and Schoultz[14] documented that detection probabilities can differ between routes with high and low bird density, and large differences in route densities exist both within and among regions for most surveys. Finally, observer differences can also cause variation in detection probabilities among routes.

Variation among Species

Detection probabilities differ among species of birds. Although it is preferable to directly estimate detection probabilities, the effort involved with most statistical methods of estimating $p_{i,j}$ makes it impractical for roadside surveys. The authors have experimented with estimating detection probabilities from multiple runs of a single survey route using statistical methods for estimating p from binomial trials.[19] A single observer repeatedly surveyed a 30-stop BBS style route for several years, and detection probabilities are presented here for several species from the 1987 runs.[29] The estimates of p from these analyses are generally imprecise, but provide some indication of the relative magnitude of p (Figure 2). The median p from this analysis was 0.50, and values ranged from 0.18 (for northern bobwhite) to 0.95 (for downy woodpecker, *Picoides pubescens*). For several species with low p (for example, chimney swift, *Chaetura pelagica*, and northern bobwhite) it is apparent (although difficult to quantify) that either N or p changed during the time of the replicate counts.

Bias Associated with Sampling of Age/Sex Classes

Although some surveys provide an estimate of the total number of individuals in a population, roadside surveys only collect information on individuals seen and heard during the counts. These surveys are timed to occur before most young have fledged and fledglings are not counted; hence, no information exists on age structure of the populations. Additionally, a great deal of variation exists in how species are perceived on the survey routes. Birds are perceived by sight, call, song, or other noise (e.g.,

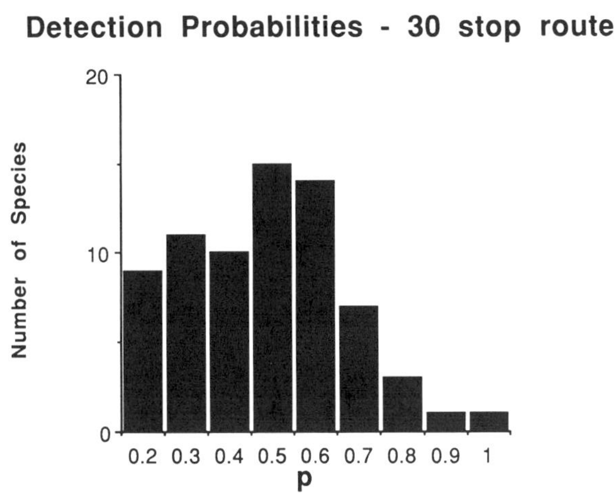

FIGURE 2. Detection probabilities (P) grouped in intervals of 0.1 by number of species from a 30-stop BBS style route.

drumming) on BBS routes, but the Mourning Dove and American Woodcock surveys rely almost entirely on male displays and songs such as coos and peenting or flight noises. A strong bias toward counting only males occurs in most species that are identified primarily by song, while it is unclear how sexes are represented in chip notes or sightings. One BBS observer has noted how he perceived each species recorded on his BBS routes,[30] and the authors have been analyzing the patterns of species identification methods among routes and between years. Here is a brief summary of two years (1988 and 1989) of data collected on one BBS route (Oldtown route in Maryland). Over the two years, 69.5% of birds were perceived by song, and results for species with N greater than 9 individuals indicate that most species are primarily identified by song (Table 1). Many species not identified by song are perceived by calls; hence individuals are not observed. It is unlikely that all age and sex classes of most species are adequately sampled, and likely that a bias toward observation of adult males exists.

METHODS

Identifying Population Finite Rates of Increase from Survey Data

Single Surveys

Caughley[6] suggests that $\bar{\mathbf{r}}$ can be calculated from a series of estimates or indices of population size by regression of the natural logs of counts on year. In our estimation of population λ, we use the multiplicative model:

$$C_i = \lambda^i \prod_{j=1}^{k} b_j^{o_j}, \tag{2}$$

where λ is the estimate of finite rate of increase (or exp($\bar{\mathbf{r}}$)), i indexes year, and the o_j terms are indicator variables for the k observers, which take on values of 1 for years when observer j conducted the survey and 0 for other years. The b_j terms are coefficients that describe differences in counts related to observer j. The observer covariables are used to absorb variability and eliminate bias associated with changing observers. Note that this is the model used by Caughley,[6] but generalized to allow observer differences in counting birds.

In practice, Equation 2 is estimated by taking the log of C_i (plus a constant to allow use of counts of 0) and regressing it on year (i) and the observer covariables. The resulting slope estimate is back-transformed[20] to provide an estimate of λ which is easier to interpret than $\bar{\mathbf{r}}$ in terms of population trends. It is assumed that detection probabilities are constant for counts from an observer on a route, but the presence of the b_j parameters allow detection probabilities to differ among observers.

Because the number of animals in any year is a function of the number in the prior year, it is likely that autocorrelation exists within time replicated survey data; hence, the estimated variances of slope parameter estimates are likely to be too small. Methods for testing for autocorrelation exist, and remedial methods for estimating variances of autocorrelated data are readily available,[21] although they are based on specific assumptions about the nature of the autocorrelation.

Table 1. Proportion (*P*) of Total Number of Identifications (*N*) Made by Song, Listed by Species, Over Two Surveys (1988 and 1989) of the Oldtown BBS Route

Species	*P*	*N*
Yellow-billed cuckoo	0.625	16
Pileated woodpecker	1.000	13
Red-bellied woodpecker	1.000	10
Eastern wood-pewee	0.941	17
Blue jay	0.000	14
American crow	0.000	46
Eastern tufted titmouse	0.700	30
White-breasted nuthatch	0.000	12
Carolina wren	0.750	12
Gray catbird	0.583	12
American robin	0.684	19
Wood thrush	0.905	42
Red-eyed vireo	0.982	57
Pine warbler	1.000	25
Hooded warbler	0.800	10
Ovenbird	1.000	17
Common grackle	0.000	19
Brown-headed cowbird	0.500	4
Scarlet tanager	0.935	31
Northern cardinal	0.870	23
Indigo bunting	0.838	37
Rufous-sided towhee	0.619	42
Chipping sparrow	0.778	27
Field sparrow	1.000	11

Data collected by D. Bystrak.[30]

Replicated Surveys

For roadside surveys, each route is considered a permanent monitoring site, and routes are randomly located within states and provinces. The authors note, but do not discuss, the issue of bias associated with the location of roadside routes.[22] Because of the possibility of p varying among routes and the sampling design containing multiple surveys over time at permanent locations, we estimate λ for each route using Equation 2 with the counts, observers, and slopes indexed by route. A regional weighted average λ is estimated as:

$$\bar{\lambda} = \frac{\sum_{j=1}^{n} w_j \lambda_j}{\sum_{j=1}^{n} w_j} \qquad (3)$$

where λ_j is the slope estimate and w_j is the weighting factor for route j ($j = 1, \ldots, n$). We currently weight slope estimates by marginal mean count for the route and the inverse of an estimate of the variance, which is related to changes in observers and number of years a route was surveyed.[23,24] These combined estimates are calculated for the smallest region of interest (i.e., a physiographic region within a state), and λ for larger regions is estimated as weighted averages (by land area) of the $\bar{\lambda}$ of component subregions. Because the assumption must be made that each route λ is an independent estimate of a regional λ, it is appropriate to estimate the variance among the route λ. The weighted regional average λ is bootstrapped[24] to obtain an estimate of the variance.

Determining Changes in Finite Rate of Increase in a Population

It is reasonable to assume that toxins and other environmental features can have either short-term (acute) or long-term (chronic) effects on bird populations. Oil spill mortality or severe winter weather are examples of acute effects, whereas chronic effects could result from long-term changes in survival or reproduction from yearly applications of pesticides. Both acute and chronic effects cause interventions in the population time series, but the structure of the change is modeled differently. Acute effects cause breakpoints in populations, which we define as a point on either side of which λ does not necessarily change but the size of the population does. Chronic effects cause changepoints, in which the trajectory of the population changes. Statistical aspects of determining breakpoints are similar to those of determining changepoints, and in the discussion below some methods for evaluating changepoints are reviewed.

Examining Finite Rates of Increase for Changepoints

In population studies, there is interest in (1) locating changepoints without prior knowledge of possible influences and (2) determining if a predetermined point is a changepoint. If the location of the changepoint is known, the problem is relatively simple, and standard statistical methods can be used to evaluate the statistical significance of the change in λ at the changepoint, although autocorrelation provides a complication. However, if the changepoint is unknown, it is much more difficult (from a statistical standpoint) to detect.

Evaluating Significance of Changepoints at Known Points

Single Routes. If the changepoint is known, standard statistical methods can be used to test for differences in slope for the two periods.[21] Suppose there are L observations, with a changepoint occurring at time k $(1<k<L)$. Two dummy variables can be set up and indexed as time trends. This model can be compared to a reduced-parameter model that only contains a single time trend parameter using a standard F test.

Multiple Routes. Geissler[23] provided a multiple-route analogue for the procedure discussed above that does not use within-route variances. He fits the model ([2], see above) to each route, and estimates $\Sigma\, w_{j,c}\lambda_j$ (similar to the numerator of Equation 3), in which the weighting factor is evaluated at the changepoint c $(w_{j,c})$. Then, a model with two time coefficients for pre- and post-changepoint periods (λ' and λ'') and covariables is fit to data from each route, and the quantity $\Sigma\, w_{j,c}\lambda''_j$ is estimated with the new slope parameter. The ratio of these two estimators, or:

$$\hat{\rho} = \frac{\sum_{j=1}^{n} w_{j,c}\,\lambda''_j}{\sum_{j=1}^{n} w_{j,c}\,\lambda'_j}, \tag{4}$$

should be 1.0 if no slope change occurs. Bootstrapping[24] can be used to find var $(\hat{\rho})$, and a t test can be used to test the statistical significance of the null hypothesis $\rho = 1$.

Location of Changepoints Unknown

Many ecologists are interested in explaining population changes. To do this, they usually first identify points when populations appear to change, then develop explanations for the changes. Unfortunately, identification of location and significance of changepoints is very difficult in autocorrelated time series.

Single Routes. Hinkley[25] described maximum likelihood estimation of regression parameters and changepoints from two-phase regression. However, the asymptotic tests he presented for testing for the equality of the two slope estimates and the significance of the changepoint may not always be appropriate.[26]

Multiple Routes. For multiple routes, the Hinkley method can be used on each route to determine if a changepoint exists. Using the estimates of changepoints on each route, the distribution of changepoints by date can be examined for peaks or for deviation from a uniform distribution, which would indicate a consistency of changepoints among routes.

The Relevance of Changepoint Analysis of Survey Data

In the tests for determining changepoints, the task is to detect an underlying point at which population size estimates or indices change. However, many reasonable population dynamics models indicate that population size changes often do not coincide directly with changes in survival and fecundity rates. Time lags in the expression of changes in survival and fecundity, autocorrelation in the counts, and environmental variation all can cause the appearance of multiple changepoints.

Simulation of Population Size

Because of varying detection probabilities and autocorrelation often associated with population counts, it is unlikely that either standard statistical methods for evaluating the significance of known changepoints or the Hinkley[25] method for determining unknown changepoints will produce acceptable results from population data. To evaluate the validity of statistical methods for detecting population changes, the authors conducted simulations of populations with three structures:

1. Population size independent among years and independent of density. In this case, population in year $t + 1$ was generated as:

$$\mathbf{x}_{t+1} = \mathbf{A}\mathbf{E}(\mathbf{X}_t),$$

where $\mathbf{E}(\mathbf{X}_t)$ is the expected population vector at time t and $\mathbf{A}$ is a stochastic projection matrix of transition probabilities.

2. Population size dependent among years but independent of density. The population in year $t + 1$ was generated as:

$$\mathbf{x}_{t+1} = \mathbf{A}\mathbf{X}_t.$$

3. Population size dependent among years and also density dependent. Population in year $t + 1$ was generated as:

$$\mathbf{x}_{t+1} = Q\mathbf{AX}_t,$$

where **Q** equals $1/(1 + bN_t)$, N_t is the population size at time t, and b is a constant prescribing the effect of population size at N_t on the projection matrix.

In all simulations, the projection matrix **A** was

$$\begin{bmatrix} S_{am} & 0 & RS_{ym} \\ 0 & S_{af} & (1 - R)S_{yf} \\ 0 & 2PS_{af} & (1 - \mathrm{R})2PS_{yf} \end{bmatrix}$$

$$\text{and } \mathbf{X}'_t = [M_t\ F_t\ Y_t]$$

where S equals survival of subscripted age/sex group; P is productivity; R is recruitment; and the elements of $\mathbf{X}'_t$ are the numbers of adult males, adult females, and young at time t. This matrix is almost identical to that used by Martin et al.[2]

The authors simulated a chronic effect on the population by changing productivity after year 5 so that the population declined after the changepoint at a rate of 1%/year in one case and 5%/year in another. Parameter values used were

$$S_{am} = S_{af} = S_{ym} = S_{yf} = E(R) = 0.5$$

and $E(P)$ equals 1.0 before the intervention and 0.98 (1% annual decline) or 0.90 (5% annual decline) after. R and P were generated as random normal variables with coefficients of variation (CVs) of 0.05 and 0.10, respectively. The realized number of survivors from each age class each year was modeled as a binomial random variable with parameters N equals number in each age class in year t and p equals survival probability.

The authors simulated 500 data sets, each with 10 years of data, for independent, autocorrelated, and density-dependent time series (the three cases discussed above) with either no change in the population or a changepoint after year 5 followed by a 1 or 5% decline. In each simulation, several alternative counts were examined, each of which represented data that might be collected by a survey. These counts were (1) population size at year t, (2) number of adult males at year t, (3) an index of total population size at year t, and (4) the index of adult males at time t. The indexes were generated as a realization of a binomial random variable with parameters N, the population size from the simulation, and p, the probability of detecting the animals, where p equals 0.5. Estimates for models were computed assuming changepoints occurring either at a known point or changepoints unknown. For reasons discussed below, the unknown changepoint model was only used for complete counts and no changepoint. For each simulation we recorded the number of times the null hypothesis was rejected (H_0: simple linear model adequately explains variation in the data vs a two-slope parameter linear model alternative). For all tests, nominal α was 5%.

RESULTS

Performance of Models Estimating Location of Changepoint, When No Changepoint Existed in the Simulated Population

To measure the validity of tests for determining if changepoints existed and how often the test found a changepoint when no changepoint existed in the simulated data was determined first. If nominal α was 5%, the tests should only be significant in 5%

of the simulations. The Hinkley[25] method of determining unknown changepoints had an α of 20.6% for the independent populations, which increased to 59.4% for autocorrelated cases (Table 2). These large percentages of significant results indicate that this test is finding changepoints where they do not actually exist. The poor performance of this test is not unexpected, as Feder[26] and others have criticized the method. In autocorrelated populations, changepoints occur (at least in moderately short intervals) due to increases and declines induced by the autocorrelation. Because of these problems, the authors did not use the model in simulations that contained changepoints.

Performance of Models Estimating Significance of Known Changepoint, When No Changepoint Existed in the Simulated Population

For independent data (case 1), the linear model had an appropriate α of 5.6%. However, the α was much larger (31.8%) for autocorrelated data. This suggests that changes in the population associated with the autocorrelation produce changepoints at

Table 2. Simulation Results Demonstrating the Effects of Population Structure and Choice of Count Data on Statistical Tests for Evaluating the Existence and Location of Changepoints

	No Change in Population Changepoint %		Percent Change	
Population Structure	**Known**	**Unknown**	**1%**	**5%**
Independent Population				
Total				
Complete count	5.6	20.6	6.5	37.6
Incomplete count	—	—	4.2	32.6
Adult males				
Complete count	5.6	—	7.5	37.4
Incomplete count	—	—	5.6	30.4
Autocorrelated				
Total				
Complete count	31.8	59.4	38.2	59.0
Incomplete count	—	—	28.8	48.2
Adult males				
Complete count	28.6	—	28.4	45.4
Incomplete count	—	—	23.4	40.8
Density Dependent				
Total				
Complete count	32.6	64.8	31.4	55.8
Incomplete count	—	—	25.4	46.2
Adult males				
Complete count	26.6	—	25.8	48.0
Incomplete count	—	—	22.0	40.6

Note: Population structure was independent, autocorrelated, or density dependent. Count data were either total population or adult male only, for complete census at a point or an incomplete count. Table entries indicate the proportion of tests that rejected the hypothesis that no changepoint existed in the population for 500 simulations when no changepoint existed, changepoint existed followed by a 1%/year decline in the population, or changepoint existed followed by a 5%/year decline in the population.

random locations throughout the time interval, and the linear model analysis determined that 31.8% of these changepoints were at the 5-year point, which is much more than the 5% expected by chance.

Performance of Models Estimating Significance of Known Changepoint, When Changepoint Existed in the Simulated Population

For these simulations, the population declined at 1% or 5% after a changepoint at year 5. In independent counts, a 1% decline in population after the changepoint caused only small variation from the nominal variation level, suggesting that the analysis has little ability to detect changes of this magnitude. In the autocorrelated cases, rejection probability was much higher, and rejection probability decreased for both incomplete counts and biased samples. Unfortunately, this increased ability to detect changes is a consequence of the autocorrelation introducing additional changepoints, not an increased ability to detect the actual changepoint.

Population changes of 5%/year after the changepoint had much higher rejection probabilities for the independent case and smaller increases in rejection probabilities for autocorrelated cases. As with the 1% change, density-dependent and biased sample cases had lower power than nondensity-dependent populations with complete counts. In each case in which data were analyzed using the known changepoint model, parameter estimates were unbiased, but in the autocorrelated and density-dependent cases the average model based standard errors (root mean model based variance) were much less than the standard deviation of the parameter estimates. This suggests that statistical tests in these cases are likely to indicate significance more often than they should.

Consequences for Estimation of λ

These results suggest that with standard techniques, it is difficult to determine if a real change in λ is occurring in a short time series from a single population if the changepoint is unknown. Even populations with no autocorrelation can appear to have changes. Even when the point of change was specified, autocorrelation created the perception of change almost one third of the time in our simulated populations. These results are not unexpected, as it is well known that autocorrelation leads to unbiased but inefficient parameter estimates.[21] Unfortunately, they call into question much of the hypothesis testing from observation of time series using simple models.

An Example: Eastern Bluebirds in Tennessee

The eastern United States experienced severe winters in 1977 and 1978. The effects of these winters were well documented for many permanent resident and short-distance migrant species.[27] Pitts[8] examined the effects of the severe winters on eastern bluebirds in Tennessee. He delineated the part of the state that had 15 days or more of subfreezing weather in January and February of 1978, which he considered to be a threshold for increased mortality in wintering eastern bluebirds. Because there were several years of severe weather and additional evidence of population declines in the latter part of the 1960s, the methods described above were used in an analysis of BBS data from 1966 to 1989 to test for differences in λ between the affected (experimental) and unaffected (control) parts of the state, using a known changepoint of 1978.

BBS routes in Tennessee were categorized as either experimental, if they were in the northern part of the state and were subjected to severe weather, or control if they were in the southern part of the state. The λ by route were estimated using Equation 2, then the regional weighted average λ was calculated using Equation 3 for each stratum for each treatment. Overall weighted λ were calculated for each treatment by weighting the stratum λ by areas within the strata. Composite annual indices of abundance were calculated using the linear model method.[28] Using Equation 4, the significance of the changepoint in 1978 was evaluated for experimental and control parts of the state. Finally, the Hinkley[25] method was used to determine, for each route in experimental and control regions, the location of the changepoint in the data.

Population λ estimates suggest generally increasing populations for both experimental and control regions in Tennessee. The bootstrapped λ values were >1 for both experimental ($\lambda = 1.002$, $SE(\lambda) = 0.0171$, $p>0.8$) and control (1.011, 0.0077, $p>0.1$) areas. Average λ for both regions over the 1966 to 1989 interval are similar; however, examination of annual indices of abundance for the experimental region clearly indicates a minimum in 1978 (Figure 3A). The control region also shows a decline in 1978, but not a clear population minimum (Figure 3B). Using Equation 4, the test for a change in λ in 1978 was highly significant for the experimental region ($\rho = 1.16$, $t = 5.46$, $d.f. = 23$, $p<0.0004$), but not significant for the control region ($\rho = 1.12$, $t = 2.94$, $d.f. = 9$, $p>0.375$). Note that the estimate for the control region has many fewer routes because it is a smaller portion of the state and several routes did not contain sufficient years on either side of the changepoint to allow estimation of period-specific λ.

Using the Hinkley[25] test for changepoints, the authors produced a histogram of predicted changepoints for each treatment (Figure 4). Observer information was not used in these regressions. Because the changepoint is in the center of the interval, the mean of the route changepoints was not used as a measure of accuracy. It can be observed from the histogram that many of the experimental routes had changepoints at 1978 or 1979, while the control routes were not centered on 1978. This provides evidence that 1978 was indeed a year of consequence to the experimental population.

DISCUSSION

The authors have presented evidence that many difficulties exist in detecting changes in populations from population survey data. These problems are associated with either sampling or demographic issues.

Sampling problems include variation in detection probabilities and the bias in sampling of sex or age classes. A method of estimating has been presented that makes minimum assumptions on variation in detection probabilities, and the authors have shown that these methods can be used to evaluate the importance of changes in λ.

Simulation results suggest that demographic factors affect the populations changes observed during population surveys. Autocorrelation will cause population changes that could appear to be environmentally related, but are actually due to intrinsic features of the population. These results suggest that it might be difficult to examine population size data for changes, and then associate the changes with environmental changes. This problem is most apparent in the examination of short time-series of population data, as long time-series can provide better estimates of λ that will be less influenced by the autocorrelation. The possibility exists that replicate routes may provide independent estimates of change that can be used to evaluate changepoints, as has been shown for eastern bluebirds. Consistency among independent routes in a region may provide stronger evidence of environmental effects in a population.

In practice, annual population counts or indices often are the only population data available as input for population modeling or management, and the eastern bluebird

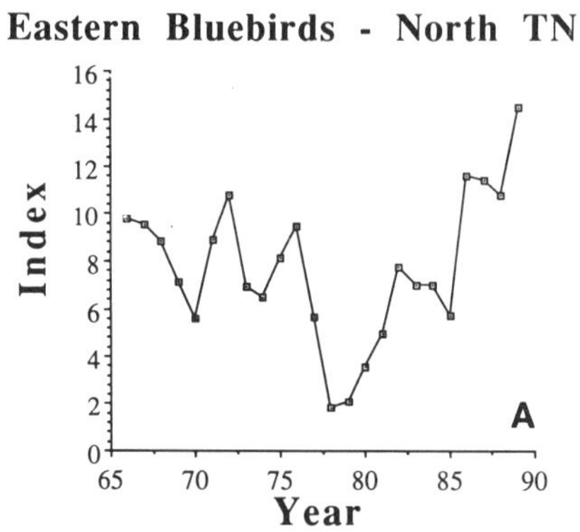

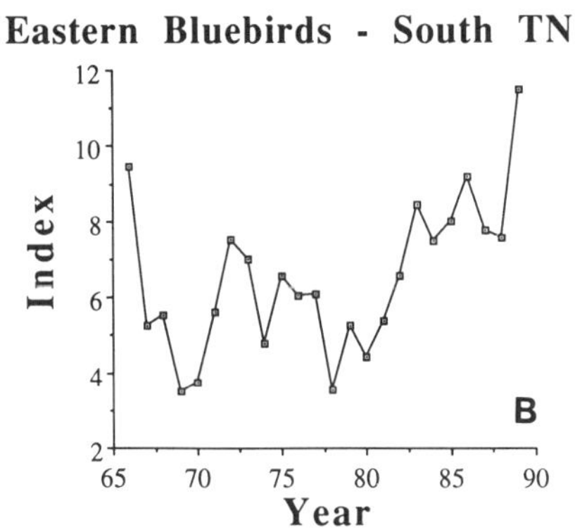

FIGURE 3. Composite population indices (the linear model method[28]) for (A) the experimental region in northern Tennessee and (B) the control region in southern Tennessee, as estimated from BBS data collected from 1966 to 1989.

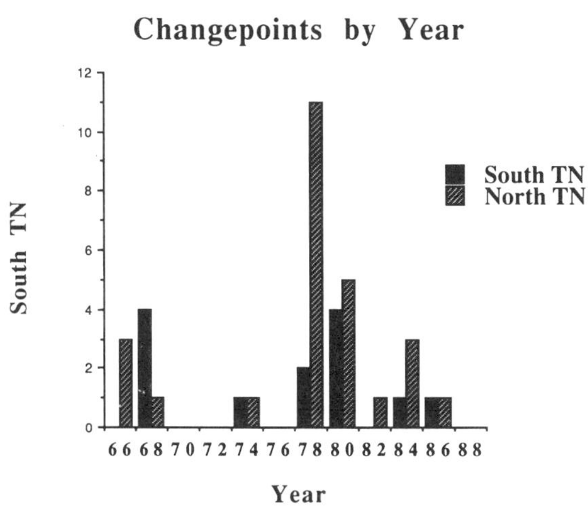

FIGURE 4. Changepoints estimated from BBS routes in experimental and control parts of Tennessee. Number of routes are presented that had changepoints in each 2-year interval, indexed by the first year in the interval.

example presented here demonstrates that there are environmental effects on populations that do occur over short periods of time and produce noticeable effects on population size. However, results presented in this chapter suggest that care is necessary in estimating population changes and interpreting patterns of change in counts over time. Short-term fluctuations in population time series may result from a variety of sources and may reflect only intrinsic attributes of autocorrelation within the time series. Therefore, simple analyses of associations of changes in populations with environmental factors always must be viewed with caution, because statistical tests may falsely indicate correlations when none exist.

ACKNOWLEDGMENTS

We thank G. W. Pendleton, J. D. Nichols, and W. Link for helpful discussions and review, and for assistance with programming. Diane Chu assisted with the eastern bluebird data analysis and with preparation of figures.

REFERENCES

1. Sykes, Z. M., Some stochastic versions of the matrix model for population dynamics, *JASA*, 64, 111-130, 1969.
2. Martin, F. W., R. S. Phospahala and J. D. Nichols, Assessment and population management of North American migratory birds, in *Environmental Biomonitoring, Assessment, Prediction, and Management—Certain Case Studies and Related Quantitative Issues,* J. Cairns, Jr., G. P. Patil and W. E. Waters, Eds., International Cooperative Publishing House, Fairland, MD, 1978, pp. 187-239.
3. Williams, B. K., and J. D. Nichols, Modeling and the management of migratory birds, *Nat. Resource Model.*, 4, 273-311, 1990.
4. Johnson, D. H., J. D. Nichols, M. J. Conroy and L. M. Cowardin, Some considerations in modeling the mallard life cycle, in *Waterfowl in Winter*, M. W. Weller, Ed., University of Minnesota Press, Minneapolis, 1988.
5. Temple, S. A., and J. A. Wiens, Bird populations and environmental changes: can birds be bio-indicators?, *Am. Birds*, 43, 260-270, 1989.
6. Caughley, G., *Analysis of Vertebrate Populations*, John Wiley & Sons, New York, 1977.
7. Caswell, H., Analysis of life table response experiments. I. Decomposition of effects on population growth rate, *Ecol. Model.*, 46, 221-237, 1989.
8. Pitts, T. D., Eastern bluebird population fluctuations in Tennessee during 1970-1979, *The Migrant*, 52, 29-37, 1981.
9. Pollock, K. H., and W. L. Kendall, Visibility bias in aerial surveys: a review of estimation procedures, *J. Wildl. Manage.*, 51, 502-510, 1987.
10. Lancia, R. A., J. D. Nichols and K. H. Pollock, Estimating the number of animals in wildlife populations, in *Wildlife Techniques Manual*, T. H. Bookhout, Ed., The Wildlife Society, Bethesda, MD, 1994.
11. Droege, S., and J. R. Sauer, North American Breeding Bird Survey Annual Summary 1988, Biol. Rep. 89, U.S. Fish and Wildlife Service, 1989, pp 1-16..
12. Dolton, D. D., Mourning dove status report, 1990, Office of Migratory Bird Management, Laurel, MD, 1990.
13. Sauer, J. R., and J. B. Bortner, Population trends from the American Woodcock singing-ground survey, 1970-1988, An evaluation, *J. Wildl. Manage.*, 55, 300-312, 1991.
14. Bart, J., and J. D. Schoultz, Reliability of singing bird surveys: changes in observer efficiency with avian density, *Auk*, 101, 307-318, 1984.
15. Sauer, J. R., M. K. Klimkiewicz and S. Droege, Population trends of the Purple Martin in North America, 1966-1986, *Nat. Soc. News*, 22, 1-2, 1987.

16. Sauer, J. R., S. Droege and D. Bystrak, Analysis of Observer Differences in the North American Breeding Bird Survey, Abstract, American Ornithologists Union Annual Meeting, Mississippi, 1987.
17. Gates, J. M., Crowing counts as indices to cock pheasant populations in Wisconsin, *J. Wildl. Manage.*, 30, 735–744, 1966.
18. Droege, S., and J. R. Sauer, Northern bobwhite, gray partridge, and ringed-necked pheasant population trends (1966–1988) from the North American Breeding Bird Survey, in *PERDIX V: Gray Partridge and Ring-Necked Pheasant Workshop*, K. E. Church, R. E. Warner and S. J. Brady, Eds. Kansas Department of Wildlife and Parks, Emporia, 1990.
19. Carroll, R. J., and F. Lombard, A note on N estimators for the binomial distribution, *JASA*, 80, 423–426, 1985.
20. Bradu, D., and Y. Mundlak, Estimation in lognormal linear models, *JASA*, 65, 198–211, 1970.
21. Draper, N. R., and H. Smith, *Applied Regression Analysis*, 2nd ed., John Wiley & Sons, New York, 1981.
22. Terborgh, J., *Where Have All the Birds Gone?*, Princeton University Press, Princeton, NJ, 1989.
23. Geissler, P. H., Estimation of animal population trends and annual indices from a survey of call-counts or other indications, in *Proc. Am. Stat. Assoc., Section on Surv. Res. Methods*, 1984, pp. 472–477.
24. Geissler, P. H., and J. R. Sauer, Topics in route regression, in *Survey Designs and Statistical Methods for the Estimation of Avian Population Trends*, J.R. Sauer and S. Droege, Eds., U.S. Fish and Wildlife Service, Washington, DC, 1990, pp. 54–57.
25. Hinkley, D. V., Inference in two-phase regression, *JASA*, 66, 736–743, 1971.
26. Feder, P. I., The log-likelihood ratio in segmented regression, *Ann. Stat.*, 3, 84–97, 1975.
27. Sauer, J. R., and S. Droege, Recent population trends of the Eastern Bluebird, *Wilson Bull.*, 102, 239–252, 1990.
28. Sauer, J. R., and P. H. Geissler, Estimation of annual indices from roadside surveys, in *Survey Designs and Statistical Methods for the Estimation of Avian Population Trends*, J.R. Sauer and S. Droege, Eds., U.S. Fish and Wildlife Service, Washington, DC, 1990, pp. 58–62.
29. Droege, S., personal communication.
30. Bystrak, D., personal communication.

43

Risk Assessment in Avian Toxicology Using Experimental and Epidemiological Approaches

John R. Skalski and Steven G. Smith

ABSTRACT

Assessment of environmental risks of pesticides to avian populations requires an integration of field sampling techniques, parameter estimation, measurement of population responses to environmental and chemical effects, and population modeling. Each of these phases of environmental risk assessment challenges our theoretical and biological skills and knowledge. Regulatory requirements, however, necessitate overcoming these limitations of existing methods and developing new statistical techniques tailored to assessing effects on wild populations. A framework for measuring risks to avian populations is presented along with recommendations on animal tagging studies to obtain estimates of needed model parameters. Experimental and epidemiological approaches to measuring population responses to pesticides are presented along with a critique of their relative merits in assessing environmental risks.

KEY WORDS

proportional hazards, survival analysis, Leslie matrix, abundance estimation, epidemiology

INTRODUCTION

In 1988, the United States Environmental Protection Agency (U.S. EPA) published a "Guidance Document on Terrestrial Studies." The guidelines provided general recommendations on the conduct of field toxicology studies, but failed to address the serious statistical issues of how to assess effects from multiplot experiments on free-ranging avian populations or how such results should be incorporated in risk assessment. With regard to licensing chemicals for environmental use, critical technical, philosophical, and regulatory issues remain about how best to measure effects and assess risks to populations.

The authors consider alternative endpoints that may be used for ecological risk assessment and the existing statistical methods that might be used to quantify the risk, and assign a level of statistical certainty to those assessments. Indeed, at the time the EPA Guidance Document was issued, statistical techniques to design and analyze multiplot survival or abundance experiments based on tagging studies did not exist. Without efficient statistical methods for design and estimation, very large field experiments (e.g., two treatments with ten replicate square-mile sites each) were considered necessary but cost prohibitive. Consequently, regulatory agencies and chemical industry representatives have recognized the need to develop new evaluation endpoints and techniques to assess avian effects.[1]

ASSESSING POPULATION RISKS

A first step in the statistical design of any field investigation is specification of biological and statistical goals of the study. Biological goals should specify the spatial, temporal, and taxon dimensions of the study, including: (1) pesticide, its use pattern, and concentrate range; (2) target avian species; (3) local agricultural practices; and (4) seasonal and annual dimensions of the investigation. To the above list of biological objectives must be added the response variable(s) or endpoint(s) to be measured in assessing risks to the population. Given the biological response, statistical goals of the study are then defined by an objective function. The objective function formalizes the biological and statistical objectives in terms of a mathematical expression that is subject to design optimization.

A common objective function is the precision of parameter estimates ($\hat{\theta}$) where

$$P\left(\left|\frac{\hat{\theta} - \theta}{\theta}\right| < \epsilon\right) = 1 - \alpha \tag{1}$$

θ is the true value and $(\hat{\theta} - \theta)/\theta$ measures the relative error in estimation. The precision expression (Equation 1) specifies that the relative error in estimation must be less than ϵ, $(1 - \alpha) \cdot 100\%$ of the time under similar sampling conditions. The other commonly used objective function is specification of the statistical power $(1 - \beta)$ of a test of the null hypothesis

$$H_o\colon \theta_C = \theta_T$$

against the alternative

$$H_a\colon \theta_C \neq \theta_T$$

when tested at significance level α.

In wildlife toxicology, no single endpoint has been universally acknowledged by scientific or regulatory communities as the best measure for assessing environmental risks. Therefore, begrudgingly, the task of selecting an endpoint often falls upon the wildlife biometrician when designing a field study. Emlen[2] reviewed alternative population endpoints useful in risk assessment and selected as favorable candidates (1) pseudoextinction (probability of a population falling below a prespecified density), and (2) temporal mean population density. Although the first of these endpoints is an abstraction that cannot be readily observed or quantified, a decline in mean temporal density is subject to experimental testing and verification. It is this endpoint of temporal mean density that will be this chapter's focus in risk assessment. A shift in mean density is an appealing endpoint because it is an unambiguous endpoint that integrates environmental, biological, and toxicological effects and their interactions. In other words, it is the sum total of all effects upon a population. Experimental procedures through the use of replication and randomization then can be used to assess the net consequences of a pesticide on population abundance.

Two statistical approaches to risk assessment based on a change in mean temporal abundance are addressed: (1) direct quantification and (2) parameter estimation for modeling population dynamics. Although direct measurements and tests of effects of pesticide applications on animal abundance can be undertaken, such investigations may not provide ecological and biological insights into the mechanisms of any observed changes. A mechanistic approach based on estimating and assessing effects on survival and fecundity, on the other hand, can provide insights into the nature of toxic effects, but risk assessment must then be model dependent. Obviously, each approach has its advantages; unfortunately, few investigations can afford both approaches.

Direct Quantification

A variety of tagging[3,4] and transect methods[5] exist to estimate population abundance or density at a study site. Surprisingly, statistical guidance on the design and analysis of multiplot mark-recapture experiments for testing effects on animal abundance is limited. Skalski[6] and Skalski and Robson[7] present tests of effects suitable for wildlife toxicology and provide statistical guidance on sample sizes, level of field replication, and effects of plot size and dispersal on the power of tests of hypotheses.

Crucial design elements are the randomization and replication of control and test conditions across the landscape in tests of hypotheses

$$H_o\colon \mu_{N_c} \leq \mu_{N_i} \tag{2}$$

against

$$H_a\colon \mu_{N_c} > \mu_{N_T}$$

where μ_{N_i} is the mean abundance for treatment $i(i = C, T)$. For readily biodegradable pesticides, a single year of testing provides a valid test of effects on mean temporal abundance. However, repeating the experiment over multiple years provides a test of treatment-by-year interactions and an assessment of carryover effects for the more environmentally persistent pesticides. Assessment of carryover effects should be based on repeated observations of the same populations over years, while treatment-by-year interactions should be assessed using yearly randomization.

Power of the test of hypotheses (Equation 2) will depend upon the spatial variance in animal abundance across sites (σ_N^2) and the average sampling error in estimating abundance among replicates

$$Var(\hat{\mu}_N) = \frac{\sigma_N^2 + \overline{Var\,(\hat{N}_i|N_i)}}{l} \tag{3}$$

where

l = number of replicate sites per treatment,

$\overline{Var(\hat{N}_i|N_i)}$ = average sampling error in estimating animal abundance across replicate sites.

Under ideal conditions in a homogeneous environment, animal abundance may be Poisson distributed such that $\sigma_N^2 = \mu_N$. Typically, σ_N^2 will be greater than μ_N because environmental patchiness causes a clumped or aggregated distribution of animals (i.e., $\sigma_N^2 > \mu_N$). In addition, for abundance studies, sampling error is a function of population size such that $\overline{Var(\hat{N}_i|N_i)} = \mu_N C$ where C is a proportionality constant dependent on the probabilities of capture during the population surveys. Consequently, the coefficient of variation (CV) of $\hat{\mu}_N$ based on Equation 3 can be expected to be of the magnitude

$$CV\,(\hat{\mu}_N) \geq \frac{\sqrt{\dfrac{\mu_N + \mu_N C}{l}}}{\mu_N} = \sqrt{\frac{1 + C}{\mu_N l}}. \tag{4}$$

Because variances σ_N^2 and $\overline{Var(\hat{N}_l|N_i)}$ are of order μ_N, field biologists have been reluctant to base assessments on animal surveys. The problems in estimating abundance caused by heterogeneity in capture probabilities among animals within a population[8] have also discouraged the use of mark-recapture methods in wildlife toxicology. Skalski,[6] however, showed intrapopulation heterogeneity is not necessarily a problem in tests of effects if the pattern of heterogeneity is similar across populations.

Parameter Estimation and Modeling

As an alternative to direct assessment, demographic models can be used to incorporate theoretical and empirical relationships between physical, biological, and toxicological conditions and survival and fecundity rates for purposes of assessing risks of pesticides on avian populations. Among the simplest yet most appropriate models for risk assessment are Leslie matrix models.[9] The general form of a projection matrix consists of age-specific survival and fecundity rates, which may be stationary or functions of biotic or abiotic factors over time. The projection matrix can be written as

$$\begin{bmatrix} f_1(\underset{\sim}{x},\underset{\sim}{y}) & f_2(\underset{\sim}{x},\underset{\sim}{y}) & \cdots & & f_k(\underset{\sim}{x},\underset{\sim}{y}) \\ S_1(\underset{\sim}{x},\underset{\sim}{y}) & & & & \\ & S_2(\underset{\sim}{x},\underset{\sim}{y}) & & & \\ & & \ddots & & \\ & & & S_{k-1}(\underset{\sim}{x},\underset{\sim}{y}) & 0 \end{bmatrix}_{K\times K} \tag{5}$$

where

$S_i(\underset{\sim}{x},\underset{\sim}{y})$ = survival rate for age class i(i = 1, . . ., K − 1) as a function of environmental covariates ($\underset{\sim}{x}$) and pesticide exposures ($\underset{\sim}{y}$), and

$f_i(\underset{\sim}{x},\underset{\sim}{y})$ = fecundity for age class i(i = 1, . . ., K) as a function of environmental covariates ($\underset{\sim}{x}$) and pesticide exposures ($\underset{\sim}{y}$).

Emlen and Pikitch[10] found adult mortality to be the critical parameter in determining temporal mean abundance and rate of pseudoextinction. For some small vertebrate species, juvenile mortality was the most crucial parameter determining the demographic success of the population.[10] The projection matrix also permits assessing the effects of toxicants on fecundity (i.e., f_i). Here, f_i refers to the net fecundity from egg laying through fledging to the season of projection in an avian population.

The constituents of the projection matrix (Equation 5) once again translate the exercise of risk assessment to that of accurately estimating demographic parameters and functional relationships between parameters and environmental exposure. Once the matrix is parameterized, deterministic or stochastic simulation studies can be used to assess the probable fate of a population (e.g., pseudoextinction) or, in the case of static conditions, used to estimate the finite rate of population increase (λ). The form of the Leslie matrix (Equation 5) assures at least one positive eigenvalue;[9] the largest positive eigenvalue is the finite rate of population increase. From the results of a specific investigation, hypotheses

$$H_o: \lambda \geq 1 \tag{6}$$

against

$$H_a: \lambda < 1$$

could be tested to assess the risk of population failure under stationary conditions. The authors are unaware of any research to propagate the sampling errors of S_i ($i = 1, \ldots, K - 1$) and f_i ($i = 1, \ldots,$ K) to the variance of λ. Using maximum likelihood techniques to derive estimates of $\hat{S}_i$ and $\hat{f}_i$, the test statistic

$$Z = \frac{\hat{\lambda} - 0}{\sqrt{V\hat{a}r\,(\hat{\lambda})}}$$

is asymptotically distributed as a standard normal deviate useful in testing hypotheses (Equation 6).

SURVIVAL STUDIES

Survival studies become the natural focus of avian field trials when risk assessment is to be based on demographic modeling. Age-specific survival rates in the Leslie matrix (Equation 5) must be related to environmental and exposure covariates through carefully designed field trials. Furthermore, net fecundity rates, besides being functions of clutch size, also depend upon survival of the hatch to recruitment into the trappable population. Hence, fecundity rates must include estimates of nesting success, nest abandonment rates, and hatchling and fledgling survival rates—all special applications of survival studies. Because of the obvious importance of survival studies in avian toxicology, the remainder of this chapter concerns itself with statistical and logistical aspects of survival analyses.

Statistical Concepts

Survival processes can be characterized by any of four equivalent expressions: a survivor function, a probability density function, a hazard function, or a survival probability. Let T be a non-negative random variable representing the death time of an individual from a homogeneous population. The survivor function for T is defined as the probability that T is at least as great as t; i.e.,

$$S(t) = P(T \geq t), \; 0 < t < \infty. \tag{7}$$

Note that if T is absolutely continuous, then $S(t) = 1 - F(t)$, where $F(t)$ is the cumulative density function with $S(0) = 1$ and $S(\infty) = 0$.

The probability density function (PDF) of T is

$$f(t) = \lim_{\Delta t \to 0^+} \frac{P(t \leq T < t + \Delta t)}{\Delta t} = -\frac{d}{dt}S(t). \tag{8}$$

By definition, a PDF has the properties $f(t) \geq 0$ and $\int_0^\infty f(t)dt = 1$. Also $S(t) = \int_t^\infty f(t)dt$.

The instantaneous rate of failure at time $T = t$ conditional on survival to time t is given by the hazard function and is defined as

$$\lambda(t) = \lim_{\Delta t \to 0^+} \frac{P(t \leq T < t + \Delta t | (T > + t))}{\Delta t} = \frac{f(t)}{S(t)}. \tag{9}$$

Note that from Equation 9, $\lambda(t) = -\, d/dt \log S(t)$ so that by integrating we obtain

$$S(t) = exp\left(- \int_0^t \lambda(u)du \right), \tag{10}$$

the survivor function. Using Equations 9 and 10, we see that the PDF can be written as

$$f(t) = \lambda(t)exp\left(- \int_0^t \lambda(u)du \right).$$

Survival probabilities are a fourth means of quantifying a failure time process. Here, the survivor function is partitioned into discrete time intervals and a vector of conditional survival probabilities ($S_1, S_1, \ldots, S_K$) is used to characterize the continuous process. The survival probability S_i is the probability a bird lives to time t_i given it was alive at time t_{i-1}. While the sampling occasions naturally lead to consideration of discrete time and survival probabilities rather than rates, there is still a continuous process at work. Consider, as before, the random variable T as the failure time of an individual from a homogeneous population. If the time of the initial marking of animals is denoted t_0 and the times of the subsequent resampling occasions are ($t_1, t_2, \ldots, t_K$), then the survival probabilities can be written in terms of the continuous survivor function or of the hazard function:

$$S_i = P(T > t_i | (T \geq t_{i-1})) = \frac{S(t_i)}{S(t_{i-1})} = exp\left(- \int_{t_{i-1}}^{t_i} \lambda(u)du \right). \tag{11}$$

Estimation

The four quantifications of survival—survivor function (Equation 7), PDF (Equation 8), hazard function (Equation 9), and survival probability (Equation 11)—can be used to characterize the survival process of a wild population. Survival probabilities are most commonly estimated in tagging studies, with the survivor function often the focus of radiotelemetry studies. The importance of the hazard function, as will be seen below, is in modeling the survival probabilities of the Leslie matrix as functions of environmental and pesticide exposure covariates.

Survivor Function

Kalbfleisch and Prentice[11] described methods of parameter estimation when a parametric distribution is assumed for death time. More commonly, however, a nonparametric method due to Kaplan and Meier[12] is used for estimation of the survivor function. The estimate produced by the method is known as the Kaplan-Meier or product-limit estimate.

The Kaplan-Meier estimate may be used to estimate the survivor function when the population is assumed to be homogeneous with regard to survival rates (i.e., no covariates affect survival rates). The estimate is based on the empirical survivor function (ESF); in other words, an estimate of the population survivor function. If there are no censored observations in a sample of size n birds followed through time, the ESF is defined as

$$\hat{S}(t) = \frac{\text{Number of observations} \geq t}{n}, \; t \geq 0.$$

If all death times are distinct, the ESF is a step function that decreases by $1/n$ just after each observed death time. In general, if there are d death times equal to t, the ESF decreases by d/n just after t.

If censored observations are found in the sample, the estimate must be altered somewhat. The Kaplan-Meier estimate is defined as follows: Let $t_1 < t_2 < \ldots < t_k$ be the distinct observed death times observed in a study with a sample of size n ($k \leq n$). The possibility of "ties" in the death times is allowed; i.e., more than one subject may die at each death time t_i, and we let d_i represent the number of deaths at time t_i ($i = 1,2, \ldots,k$). In addition to the death times, there are also censoring times M_i for individuals whose death times are not observed. The Kaplan-Meier estimate of the survivor function is defined as

$$\hat{S}(t) = \prod_{i:t_k < t} \frac{n_i - d_i}{n_i}$$

where n_i is the number of individuals at risk at time t_i, i.e., the number of individuals alive and uncensored just prior to t_i. As in the uncensored case, this is a step function that decreases by $(n_i - d_i)/n_i$ immediately after each death time t_i and is constant between t_i and t_{i+1}. Note that the Kaplan-Meier estimate reduces to the ESF if there is no censoring, because in that case, $n_1 = n$ and $n_i = n_{i-1} - d_{i-1}$. Variance estimates for $\hat{S}(t)$ are given in Kalbfleisch and Prentice[13] and Lawless.[14] Lawless also includes a section on estimation of the cumulative hazard function in the chapter on nonparametric estimation of the survivor function.

The Kaplan-Meier method is intended for situations in which the exact death or censoring times of every individual are known. Using radio transmitters and intense follow-up, some researchers have been able to pinpoint the days of death of moderately sized samples. Pollock et al.[15] reported on one study of a wild black duck population in which 50 female ducks were fitted with radios. After tagging, the location and status (alive, dead, or missing) of each bird were recorded daily from the date of release (8 November 1983 to 14 December 1983) until 15 February 1984. The authors reported that "diligent effort was made to locate each bird using roof-mounted antennas on trucks, strut-mounted antennas on fixed-wing aircraft, and hand-held antennas on foot and by boat."

The Kaplan-Meier method can still be applied if the sampling is not done daily; in other words, exact death times of each animal are not known. Pollock et al.[16] gave an example of the analysis of 52 bobwhite quail that were radiotagged and located weekly for 5 months. To calculate the estimate, however, it is necessary to assign a death time to each animal that dies. The animals that die in the interval between t_i and t_{i+1} could be assigned any death time therein, and the choice of assignment appears to be arbitrary. One solution is to assume that any deaths that occur during an intersampling interval occur uniformly throughout the interval. Note that similar assumptions must also be made for the censoring times of individuals censored within the interval (t_i, t_{i+1}). Depending on the use to which the Kaplan-Meier estimate is to be put, different assumptions about the death or censoring times may lead to different conclusions.

The Kaplan-Meier method, as originally devised, applies to a single group of subjects all entering the study at the same time, called "time zero." In epidemiological studies, patients often do not enter the study at the same calendar time, but all enter immediately after they have experienced a certain diagnosis or surgical procedure. Thus, time zero may refer to a different calendar time for each subject. However, in tag-release studies, there is no such "landmark" event in the life of a subject that "sets the clock." Rather, it is assumed that marked and unmarked animals experience the same hazards throughout the study periods, so time zero is assumed to be the same calendar time as for the animals already released. This assumption is implicit in the adaptation of Kaplan-Meier to "staggered entry" proposed by Pollock et al.[16] The solution simply involves the addition of the animals marked at time t_i to the number at risk at time t_{i+1}.

Survival Probabilities

The Mayfield method[17,18] provides estimates of daily survival when the objects studied can be viewed periodically. The method in recent years has been supported by statistical theory which provides variance estimates and special applications,[19-21] including tests of survival and estimation of functions of survival rates. The Mayfield method can be applied not only to studies of nest survival, its original intent, but also radiotelemetry data and egg and hatchling survival. The method assumes all individuals have equal survival and that the daily survival rate is constant through time. Hensler[19] illustrated the use of stratification in conjunction with the Mayfield method when survival processes are not homogeneous through time.

As an alternative to these classical avian survival techniques, White[22] provided a flexible estimation method permitting time-specific survival rates based on radiotelemetry methods. The advantage of White's approach is the ability to avoid assumption of constant survival rates inherent in the Mayfield method.

The previous Kaplan-Meier and Mayfield methods are inappropriate with tag-recapture sampling. Unlike radiotelemetry methods, passive tags may not be recovered because an animal dies and its body is not found or the animal survives but eludes recapture. In either event, the exact time of death is not known and recoveries are a complex function of both survival and capture probabilities. Consequently, other approaches to estimation must be used.

An array of statistical models have been developed for estimating survival rates based on the analysis of tag-release studies. Table 1 provides a key in the selection of an appropriate statistical analysis of avian tagging data. Among the options available, the use of release-recapture techniques of Cormack[23] and Burnham et al.[24] are likely to be most appropriate in avian field trials. For multiple release-recapture methods, expected numbers of recaptures of tagged birds in a sampling period is a function of release size (N_i, $i = 1, \ldots, l$) and period-specific survival (S_i, $i = 1, \ldots, k$) and capture probabilities (p_i, $i=1, \ldots, k$) (Table 2). Maximum likelihood estimates of survival and capture probabilities are calculated based on the capture histories of the marked birds. For this reason, uniquely tagged birds are required in the field studies.

SURVIVAL EFFECTS STUDIES

To be of value in risk assessment, the survival rates in a projection matrix (Equation 5) must be functions of environmental and exposure covariates. All previous survival

Table 1. Summary of Tag-Release Models That May Be Used in the Analysis of Avian Field Studies

Characteristics	Ref.
Tag recoveries upon death	
Tags from only exploited segment of population	25
Tags from all sources of mortality	26–29
Tag recoveries upon recapture	
Single release of tagged birds	24
Multiple release-recapture studies	
Single age class	30 23
Multiple age class	31 32

estimation techniques described, however, ignore explanatory variables in modeling the survival of tagged birds over time or across the landscape. In avian toxicology, on the other hand, characterization of shifts in baseline survival as a consequence of pesticide applications is the primary focus of the tagging study. The modeling of survival rates in avian toxicology is a recent advance in the statistical design and analysis of field trials.[33]

Two distinct approaches to the design of avian field trials are emerging to assess effects of pesticide exposure on avian populations and in subsequent risk assessment. The first of these approaches is based on multiplot field trials using replicate control and treatment sites in an experimental framework (see Kirk[34] on experimental designs). Plotwise covariates common to all individuals at a site are used to quantify the effects of environmental conditions and pesticide exposures on average survival rates of a population.[33] The alternative is to take an epidemiological approach in which the traits and exposure rates of individual birds are related to the fate of birds within a population. In the latter case, the birds within a population serve as replicates in an intrapopulation dose-response study.

Response Models

Survival probabilities need to be expressed as a function of concomitant variables such as pesticide dose to be useful in risk assessment and tests of effects on avian survival. This section briefly discusses some approaches to parameterizing avian survival rates.

Table 2. Expected Numbers of Recaptures of Tagged Birds in a Release-Recapture Study Consisting of $l = 4$ Releases and $k = 4$ Recapture Periods

Numbers Released	Recapture Periods			
	1	2	3	4
N_1	$N_1S_1p_1$	$N_1S_1S_2p_2$	$N_1S_1S_2S_3p_3$	$N_1S_1S_2S_3S_4p_4$
N_2		$N_2S_2p_2$	$N_2S_2S_3p_3$	$N_2S_2S_3S_4p_4$
N_3			$N_3S_3p_3$	$N_3S_3S_4p_4$
N_4				$N_4S_4p_4$

Logistic Model

In relating winter temperatures to the annual survival rates of the juvenile gray heron (*Ardea cinera*), North and Morgan[35] postulated a logistic function in which survival (S) was modeled as

$$S_i = \frac{\beta_2}{(1 + e^{\beta_0 + \beta_1 t_i})}$$

where t_i is the average temperature in the i-th year ($i = 1, \ldots, k$), β_0 is the intercept, β_1 is the slope, and β_2 is the asymptote (≤ 1). Typically, the asymptote is set at 1.

The same logistic model was used by Burnham et al.[24] when analyzing desert tortoise (*Gopherus agassizii*) survival probabilities as a function of carapace length and as a general approach to assessing dose-response relationships. The primary justification for using the logistic model is its mathematically tractable form and the properties of being approximately linear in its range and taking values between 0 and 1. The model often appears in the form of a linear logistic regression where

$$\ln\left(\frac{S_i}{1 - S_i}\right) = \beta_0 + \beta_1 x.$$

However, the biological basis of this response model is rarely if ever discussed in selection of its use. Use of the logistic model is usually a matter of convenience.

Exponential Model 1

When survival is to be modeled as a time varying function in capture-recapture theory, it is often presented as an exponential model where

$$S(t) = e^{-rt}$$

and where r is the instantaneous mortality rate. A reason for the use of the exponential model is that the instantaneous mortality rate can be partitioned into additive components where

$$S(t) = e^{-(r_m + r_H)t} = e^{-r_m t}\, e^{-r_H t}$$

implying multiplicative effects on the survival rate (r_m = instantaneous natural mortality rate; r_H = instantaneous harvest mortality rate).

If we assume lifetimes in a cohort follow an exponential distribution, the probability density function of the random variable T, where t denotes the observed lifetime of an individual is written as

$$f(t) = re^{-rt},\ t > 0,\ r > 0.$$

The expected or mean lifetime under the exponential model is

$$E(t) = \frac{1}{r}$$

with

$$Var(t) = \frac{1}{r^2}. \tag{11h}$$

Should the instantaneous mortality rate be a linear function of the treatment effects where

$$r = \alpha + \beta x$$

then adjustments to baseline survival, or $S(t)$, are multiplicative and time dependent. For some factor x

$$\begin{aligned} S(t|x) &= e^{-(\alpha + \beta x)t} \\ &= e^{-\alpha t} e^{-\beta x t} \\ &= S(t) e^{-\beta x t} \end{aligned}$$

with expected lifetime

$$E(t|x) = \frac{1}{\alpha + \beta x}.$$

Under treatment effects, the reciprocal of the expected lifetime of an individual with dose x, say μ_x, is a linear function of the reciprocal of the base lifetime (μ_0) where

$$\frac{1}{\mu_x} = \frac{1}{\mu_0} + \beta x.$$

Choice of this exponential survival model should include consideration of

- a constant hazard function, $\lambda(t)$, where

$$\lambda(t) = \frac{f(t)}{1 - F(t)} = r = \alpha + \beta x,$$

- treatment effects that are anticipated to have a multiplicative and time dependent effect on baseline survival, and
- treatment effects that are anticipated to have a linear effect on the reciprocal of the expected lifetime.

Exponential Model 2

An alternative formulation of the exponential model that has received extensive use in epidemiology[13,36] reparameterizes the instantaneous mortality coefficient where

$$r = e^{-(\alpha + \beta x)}$$

which assures $r>0$. Under this alternative parameterization, the survival to time t conditional on concomitant variable x can be written as

$$S(t|x) = e^{-(e^{-(\alpha + \beta x)}t)}$$

$$S(t|x) = S(t)^{e^{-\beta x}} \tag{12}$$

where $S(t) = e^{-(e^{-\alpha})t}$. In other words, the treatment effects are assumed to be constant through time and have an exponential effect on baseline survival. The expected lifetime with this model is

$$E(t|x) = \frac{1}{e^{-(\alpha + \beta x)}}$$

$$= e^{\alpha} \cdot e^{\beta x}$$

or relative to baseline life expectancy (μ_0)

$$\mu_x = \mu_0 e^{\beta x}$$

or similarly

$$\ln\mu_x = \ln\mu_0 + \beta_x .$$

Choice of this exponential regression model should include consideration of:

- a constant hazard function where
$$\lambda(t) = (e^{-\alpha})e^{-\beta x},$$
- treatment effects that are anticipated to have an exponential and time independent effect on baseline survival, and
- treatment effects that are anticipated to have a log-linear effect (i.e., multiplicative effect) on expected lifetime.

While these attributes have not been previously exploited in release-recapture theory, they have found their way into epidemiology under the concept of proportional hazards.

Both exponential models possess the properties of proportional hazards.[37,38] Implications of the properties include: (1) concomitant variables have a multiplicative effect on the hazard function, giving the model its name; (2) individuals with different concomitant variables have hazard functions that are proportional to one another; and (3) the ratio of hazard functions does not vary with time. Multiplicative changes in the hazard function are a reasonable assumption for environmental factors associated with plot, period effects, and chronic toxicity.

Under exponential model 2, the survivor function for individuals with covariate z is

$$S(t|z) = [S_0(t)]^{e(z\beta)}$$

where

$$S_0(t) = e\left[-\int_0^t \lambda_0(u)du\right].$$

Thus, the survivor function for an individual with covariate z is obtained by raising the baseline survivor function to a power. This relationship is also true for the conditional probability of surviving from an arbitrary time point t_1 until a second point t_2; i.e.,

$$P(t_1 < T \leq t_2|((T > t_1),z)) = [P_0(t_1 < T \leq t_2|(T > t_1))]^{e(z\beta)}. \tag{13}$$

This result is extremely important, for it permits the reparameterization of release-recapture models, taking advantage of the extensive literature on the use of the proportional hazard models in human health studies.

Multiplot Experiments

Replication and randomization of treatments to study sites are the quintessential elements in the design of any avian field experiment. The choice of experimental design (i.e., the restrictions used in randomizing treatments to sites) will depend on the environmental factors one wishes to control through blocking and the constraints imposed by the location and number of sites. Typically, the simplest design that adequately controls for extraneous sources of variability should be selected. Completely randomized or randomized block designs (paired design being a special case) will most likely be used for most studies. The treatment designs of avian field trials, on the other hand, may consist of a simple control and single type of pesticide application, dose response, carryover, or factorial treatment design. Choice of treatment design must be tempered, however, by the countereffect that as the number of treatments increases, the level of replication and subsequent statistical power decreases. In subsequent discussions, field studies to test and model survival probabilities will be assumed to be based on a completely randomized two-treatment design (i.e., control and one dose level). The extension of these design discussions to more elaborate experimental and treatment designs is straightforward.

Radiotelemetry Design

For illustration, equal numbers (balanced) of control and treatment plots will be employed in the field study. At each site, simultaneous survival estimation studies will be performed. At treatment sites, pesticide applications will be administered in the form of an intervention. Specifically, avian sampling will be performed for two or more periods prior to the pesticide spraying operation. Following the pesticide application, a minimum of two postapplication sampling periods will be performed. The sampling periods for all populations will be coincident. At each of n control and treatment sites, N_{ij} $(i = 1,2; j = 1,2, \ldots, n)$ radiotagged birds will be released at the initiation of the study. At k equally spaced intervals, the tagged populations will be queried to recover radiocollars from dead birds. After l sampling periods ($l < k$), pesticide applications will be applied to the n treatment sites. The basic model structure for this field design is represented by the expected numbers of radio tags recovered per period (Table 3).

An important distinction between the radiotelemetry model and the typical release-recapture model is the absence of the recovery/recapture probabilities (Table 2). Reparameterization of the survival rates takes into account plot, time, and treatment-specific effects using the exponential regression model (Equation 13). In general, let

$$S_{ijh} = S^{e^{(\beta_h + \tau_{ij}\eta_{ih})}} \tag{14}$$

where β_h = time period effect $(h = 2, \ldots, k)$
τ_{ij} = effect of plot j $(j = 1, \ldots, n)$ of ith $(i = 1,2)$ treatment, and
η_{ih} = effect of the pesticide in the i-th treatment $(i = 1,2)$ in the h period $(h = l + 1, \ldots, k)$.

The likelihood for the multiplot radiotelemetry survival study as a function of S_{ijh} can be written as

Table 3. Expected Recoveries of Radiotags as a Function of Release Numbers and Time-Specific Survival Rates for a Four-Period Survey Design

Treatment Type	Number Tagged	Expected Recoveries of Tags				Number Outstanding
		1	2	3	4	
Control	N_{11}	$N_{11}\ (1 - S_1)$	$N_{11}S_1\ (1 - S_2)$	$N_{11}S_1S_2\ (1 - S_3)$	$N_{11}S_1S_2S_3\ (1 - S_4)$	$N_{11}S_1S_2S_3S_4$
⋮						
Control	N_{1n}	$N_{1n}\ (1 - S_1)$	$N_{1n}S_1\ (1 - S_2)$	$N_{1n}S_1S_2\ (1 - S_3)$	$N_{1n}S_1S_2S_3\ (1 - S_4)$	$N_{1n}S_1S_2S_3S_4$
Treatment	N_{21}	$N_{21}\ (1 - S_1)$	$N_{21}S_1\ (1 - S_2)$	$N_{21}S_1S_2(1 - S'_3)$	$N_{21}S_1S_2S'_3\ (1 - S'_4)$	$N_{21}S_1S_2S'_3S'_4$
⋮						
Control	N_{2n}	$N_{2n}\ (1 - S_1)$	$N_{2n}S_1\ (1 - S_2)$	$N_{2n}S_1S_2\ (1 - S'_3)$	$N_{2n}S_1S_2S'_3\ (1 - S'_4)$	$N_{2n}S_1S_2S'_3S'_4$

Note: The pesticide is applied at treatment sites between periods 2 and 3.

$$L(R_{ijh}|S_{ijh},N_{ijh}) = \prod_{i=1}^{2}\prod_{j=1}^{n} \left[\binom{N_{ij}}{R_{ijl},\ldots,R_{ijk},N_{ij} - R_{ij}} \prod_{h=1}^{k} \left((1 - S_{ijh})^{R_{ijh}} S_{ijh}^{\left(N_{ij} - \sum_{g=1}^{h} R_{ijg}\right)} \right) \right]. \quad (15)$$

To assess treatment effects, the reparameterization of S_{ijh} (Equation 14) would be substituted into the likelihood (Equation 15). Newton-Raphson procedures can be used to iteratively solve for the parameters based on likelihood (Equation 15) and parameterization (Equation 14). Estimates of the treatment effects of the pesticide applications on survival would be obtained from the $\hat{\eta}_{2h}$ for $h = l + 1, \ldots, k$. Acute effects of the pesticides would be estimated by $\hat{\eta}_{2,l+1}$, while residual effects would be obtained from estimates of $\hat{\eta}_{2h}$ *for* $h = l + 2, \ldots, k$. Tests of effects would be based on analysis of deviance (ANODEV) under the null hypothesis

$$H_o\colon \eta_{2,l+1} \leq 0$$

against

$$H_a\colon \eta_{2,l+1} > 0$$

when only acute affects are considered.

To test for chronic effects, survival must be estimated for more than one period after the pesticide application. In radiotelemetry studies, estimates are available for each sampling period, so there must be at least two samples after application. (In release-recapture studies, estimates are not available for the final sampling period, so at least three postapplication samples are required if chronic effects are to be assessed.) The chronic effects can be tested in either a stepwise manner or by an overall test of effects for periods $l + 2, \ldots, k$ using ANODEV.

Although costs of radiotagging are considered small relative to the overall costs of performing a field trial, other technical issues are of concern. Primarily, radio transmitters may result in modified behavior and likely increased rates of mortality compared to nontagged animals. While the directions of the treatment comparisons should not be altered by the increased mortality, point estimates of the magnitude of the effect may be biased by any synergism between pesticide exposure and the stress of the radiotag. To a lesser extent, the same concerns also apply to tagging/banding studies.

Release-Recapture Designs

The costs and difficulties with radiotags can be minimized by simply banding the birds prior to release. The consequence of this choice is that the survival studies now need to incorporate an additional suite of parameters associated with time- and plot-specific capture probabilities. The basis of the release-recapture design at each plot is a series of banded bird releases and recapture periods. Multiple releases in addition to the recapture periods are essential to be able to separately estimate survival and capture probabilities. The expected numbers of recaptures or resightings are then functions of capture and survival probabilities (Table 2). However, the survey model takes into account more than simply the numbers of birds captured per period; rather, it uses the number of individuals in all 2^k–1 observable capture histories. For example, animals with capture history 1100 (i.e., 1 denotes capture, 0, escape) in a $k = 4$ period survey have the probability of occurrence as follows:

Table 4. Reparameterization of the Survival Probabilities at a Control and Treatment Plot for a Four-Period Study with Pesticide Application Following Period 2

	Radiotag Recovery Period			
Site	**1**	**2**	**3**	**4**
Control	$S^{e^{\tau_{1j}}}$	$S^{e^{\tau_{1j}} + \beta_2}$	$S^{e^{\tau_{1j}} + \beta_3}$	$S^{e^{\tau_{1j}} + \beta_4}$
Treatment	$S^{e^{\tau_{2j}}}$	$S^{e^{\tau_{2j}} + \beta_2}$	$S^{e^{\tau_{ij}} + \beta_3 + \eta_{23}}$	$S^{e^{\tau_{2j}} + \beta_4 + \eta_{24}}$

P(1100) = P(survived all four periods but captured only in periods 1 and 2)
+ P(survived first three periods and died before fourth period and recaptured in periods 1 and 2)
+ P(survived first two periods and died before third period and recaptured in periods 1 and 2)

$$= S_1S_2S_3S_4p_1p_2(1 - p_3)(1 - p_4) + S_1S_2S_3(1 - S_4)p_1p_2(1 - p_3) + S_1S_2(1 - S_3)p_1P_2$$

The resulting likelihood function for a single release-recapture study can be written as

$$L(R_{ij}|N_i, S_i, p_i) \alpha \prod_{i=1}^{k} S_i^{V_i} p_i^{a_i} (1 - p_i)^{V_i - a_i} w_i^{c_i} \tag{16}$$

where V_i = the number of previously marked animals known to be alive after recapture period i,
a_i = the number of previously marked animals captured in recapture period i,
c_i = the number of animals seen for the last time in recapture period i – 1, and

$$w_i = 1 - \sum_{j=1}^{k} \left\{ S_j p_j \cdot \prod_{h=1}^{j-1} S_j(1 - p_h) \right\}.$$

In constructing the likelihood for a multiplot toxicity trial, the likelihood (Equation 16) would be rewritten as product over the two treatments and n replicates such that

$$L \alpha \prod_{i=1}^{2} \prod_{j=1}^{n} \prod_{h=1}^{k} \left\{ S_{ijh}^{V_{ijh}} p_{ijh}^{a_{ijh}} (1 - p_{ijh})^{V_{ijh} - \alpha_{ijh}} w_{ijh}^{c_{ijh}} \right\}. \tag{17}$$

Explicit maximum likelihood estimates for survival (S) and capture (p) probabilities for the first K - 1 sampling periods are available in Cormack[23] for independent release-recapture studies. In estimating and testing effects of the pesticide application on the resultant survival rates of banded birds, the survival rates can be reparameterized (Equation 13) according to the scheme in Table 4. The dimensionality of the parameter space is reduced by this reparameterization scheme, and iterative procedures such as the Newton-Raphson method must be used. In the case of a field trial with five replicate control and treatment sites, each with four sampling periods, the fully parameterized model has 53 parameters which must be estimated simultaneously. This same design has 70 parameters with closed form estimators under the original parameterization using S_{ijh}. Inspection of the expected recaptures per period (Table 4) reveals the capture and survival probabilities in the last period are not separately estimable. The implication of this nonidentifiability is the need to have a minimum of two recapture periods following pesticide application to estimate acute mortality effects. At least three postpesticide application periods are necessary to estimate chronic effects.

As in the case of radiotelemetry studies, estimates of pesticide effects (η) and tests of hypotheses may be performed using maximum likelihood and ANODEV. Smith and Skalski[33] further discuss the modeling of avian field trials using proportional hazards models in conjunction with tagging and radiotelemetry approaches.

Applications of Multiplot Investigations

The above statistical theory has been applied to a variety of avian toxicology data sets. To this end, a computer program called SURPH (Survival with Proportional Hazards) was developed[33] to analyze release-recapture[23] and radiotelemetry[22] data based on the assumptions of proportional hazards[36] for avian field trials.

SURPH is a versatile computer program that permits estimation and tests of hypotheses concerning the effects of concomitant variables on animal survival rates. SURPH has wide applicability in the fields of wildlife ecology and wildlife toxicology, where concomitant variables may be related to changes in survival. The statistical software is capable of analyzing survival data from a single population or from multiplot field experiments. Any field study whose design would lend itself to ANOVA, regression, or covariance analysis (i.e., ANCOVA) can now use SURPH in the analysis of animal survival data. Specification of a design matrix analogous to that used in regression and ANOVA permits modeling of per period survival probabilities as a function of design factors (e.g., discrete period, plot, block, and treatment effects) and measured covariates associated with specific sampling periods or populations (e.g., dose-response, climatic, and habitat effects). By altering the design matrix, alternative survival models can be compared and tests of hypotheses performed.

To date, uses of SURPH to assess cohort-wide effects on survival have included: (1) analysis of batch releases of radiotagged bobwhite quail and cottontail rabbits in multitreatment, replicated plot experiments over time; (2) analysis of starling nesting box abandonment rates in multitreatment, replicated experiments in avian toxicology; and (3) analysis of release-recapture data in multitreatment, avian toxicology field trials. The advantage of this tag analysis is the ability not only to test for effects, but also to estimate the magnitude of the effects on survival and to model survival relationships. Applications have demonstrated the ability to detect treatment-by-year interactions, dose-response relationships, and the relative contribution of treatments and habitat characteristics to the spatial variation in survival rates. An important advantage of using the proportional hazards model is the ability to extract the hazard function directly from the survival analysis.

Epidemiological Studies

A so-called epidemiological approach to survival analysis is more appropriate for the analysis of blood chemistry data collected from individual birds tagged during avian field trials. The unique covariates of the individual birds, such as their age, condition, body size, and pesticide exposure (measured by cholinesterase), are then related to the subsequent fate of the birds. This epidemiological approach to avian toxicology is an alternative to the cohort-based technique of the previous section when individual pesticide exposure data can be obtained. This epidemiological approach has special applicability in the study of wide-ranging species in which study plots are inappropriate. Under these circumstances, use of individual covariates has advantages over cohort approaches to survival analysis, including the following: (1) allows a more realistic modeling of tagging studies by permitting heterogeneity in survival probabilities among individual animals within a population; (2) assesses survival relationships based on varying responses and covariate values of individual animals within a population; (3) uses individual birds as experimental units instead of replicate study plots in establishing dose-response models, thereby providing a cost-effective alternative to expensive multi-plot investigations; and (4) provides more detailed understanding of the exposure mechanism and interaction of behavioral, genetic, and physical traits of birds with pesticide effects.

An epidemiological approach can be either viewed as an economic alternative to multiplot field trials or used as one step in a tiered approach to field studies. Value of the approach lies primarily in the ability to relate exposure to the fate of individuals when individual exposure data are available. The challenge is in quantifying individual exposure, either through residence time or blood chemistry data.

Field Studies

The field studies incorporating epidemiological analyses can be based on either radiotelemetry or release-recapture techniques. Using radiotelemetry methods, a single release of radiotagged birds is sufficient for relating covariates to survival. All tagged birds are queried periodically to determine continued survival and the period when individual birds die. Individual attributes may be measured prior to release (e.g., sex, size, condition index, and blood chemistry) or following initial release (e.g., residence time and roosting habitat) for subsequent regression against survival.

In the case of tagging studies, multiple releases of birds in conjunction with multiple recapture periods are necessary to analyze the subsequent tagging data. With tagging studies, period-specific survival and capture probabilities are estimated along with regression coefficients relating concomitant variables to survival for all periods except the last.

Repeating the radiotelemetry or tagging studies at additional sites or over time is extremely valuable in assessing interactions between survival and habitat or climatic variables. The epidemiological analysis, besides using individual covariates, can incorporate indicator variables to indicate blocking factors such as site or year effects.

Statistical Model

As an illustration of analyzing the relationship between survival probabilities and concomitant variables, a radiotelemetry study with three survey periods will be considered. The study design will consist of a release of radiotagged birds during period 0 and queries during periods 1, 2, and 3. In practice, avian field trials would consist of multiple periods before and after a planned pesticide release. Furthermore, analysis models may include indicator variables for the replicate sites under control and test conditions. For simplicity, we will consider modeling only a single covariate with survival rates. The basic nature of the telemetry data is illustrated in Table 3. From such a design, three survival estimates can be computed between periods 0 to 1, 1 to 2, and 2 to 3. To complete the illustration, it will be assumed that blood cholinesterase of each bird is measured prior to release. An objective of the radiotelemetry study is to determine if survival may be related to this surrogate for pesticide exposure.

The probability that a bird survives between sampling periods will be modeled using an exponential proportional hazards model where

$$S^{e^{\alpha_j + \beta_j x_i}} = S_j^{e^{\beta_j x_i}}$$

and where

S = baseline survival
x_i = blood cholinesterase value for ith bird (i = 1, . . .,N),
α_j = effect of period j on survival, and
β_j = regression coefficient associated with a bird's blood cholinesterase and survival in period j.

This exponential regression model is used to parameterize the survival probabilities of birds in the study. The radiotelemetry model is based on a product of multiple Bernoulli trials. There is one multiple Bernoulli probability trial in the likelihood model for each bird released in the study. Each multinomial Bernoulli trial in turn is based on the likelihood of observing a bird with its traits having a particular capture history. The resulting statistical model is considerably more complex than the approaches used by Cormack[23] or Burnham et al.[24]. Iterative procedures are required to estimate the regression coefficients (e.g., β_j) associated with the relationship between survival and blood cholinesterase.

For a release of N birds followed through three query periods, the likelihood model can be written as

$$L = \prod_{i=1}^{N} \left\{ \left[1 - S_1^{e^{x_i\beta_1}}\right]^{1-w_i} \left[S_1^{e^{x_i\beta_1}}\left(1 - S_1^{e^{x_i\beta_2}}\right)\right]^{w_i(1-y_i)} \right.$$

$$\cdot \left[S_1^{e^{x_i\beta_1}} S_2^{e^{x_i\beta_2}} \left(1 - S_3^{e^{x_i\beta_3}}\right)\right]^{w_i y_i(1 - z_i)}$$

$$\left. \cdot \left[S_1^{e^{x_i\beta_1}} S_2^{e^{x_i\beta_2}} S_3^{e^{x_i\beta_3}}\right]^{w_i y_i z_i} \right\}$$

where $w_i = \begin{cases} 1 \text{ if } i\text{th animal survives period 1} \\ 0 \text{ otherwise} \end{cases}$

$y_i = \begin{cases} 1 \text{ if } i\text{th animal survives period 2} \\ 0 \text{ otherwise} \end{cases}$

$z_i = \begin{cases} 1 \text{ if } i\text{th animal survives period 3} \\ 0 \text{ otherwise} \end{cases}$

The likelihood can be equivalently written as

$$L = \prod_{i=1}^{N} \left[\prod_{h=1}^{t_j-1} S_k^{e^{x_i\beta_{jk}}}\right] \left[1 - S_j^{e^{x_i\beta_{ij}}}\right]^{I(t_j<3)} \tag{18}$$

where t_j is the period during which the bird dies (j = 1, . . .,N). The test of a regression relationship between blood cholinesterase values and survival would be based on a likelihood ratio test of the hypotheses

$$H_o\colon \beta_{11} = \beta_{12} = 0$$

$$H_a\colon \text{not all zero.}$$

The test statistic under the null hypothesis is asymmetrically chi-square distributed with two degrees of freedom. Single degree-of-freedom tests of regression relationships for each sampling period can also be constructed using likelihood ratio tests. It is important to note that tests of relationships can be performed using either single plot or multiplot field studies with these new survival models. Similar models can be developed with multiple concomitant variables and tag-recapture models. The likelihood model (Equation 18) illustrated provides a flexible approach to analyzing concomitant variables and assessing survival relationships in discrete time.

CONCLUSION

This chapter has attempted to provide a framework relating the statistical analysis of avian field trial data and environmental risk assessment. Population abundance or density estimation provides an integrated response variable for assessing effects on the long run mean abundance of avian species. Measures of bird abundance, to a degree, sacrifice demographic insights for a comprehensive measure of environmental and toxicological effects on a population.

The alternative is to model populations under potential assault as a function of fecundity and survival rates. The statistical analyses presented have focused on assessing and measuring effects on survival rates. For field studies to be useful in risk assessment, survival rates need to be modeled (i.e., $S(x)$) as a function of pesticide exposure and environmental covariates. Proportional hazard models used in human health risk assessment have been presented that can be incorporated into radiotelemetry and tagging models to quantify and test avian survival relationships. The relationship between survival and pesticide exposure can be developed using either multiplot field experiments with different rates of pesticide application or epidemiological methods using surrogates such as residence time or blood cholinesterase values. Choice of technique depends upon relative costs of the alternatives along with availability of cohort-wise or individual concomitant variables. The ability to assess risks on avian populations, however, depends upon using statistical methods tailored to the demands of avian field studies. Proper design and analysis of field studies can improve the interpretation of the field data and reduce overall costs through the use of more efficient estimation techniques.

REFERENCES

1. Avian Effects Dialogue Group, *Pesticides and Birds*, The Conservation Foundation, Washington, DC, 1989.
2. Emlen, J. M., Terrestrial population models for ecological risk assessment: a state-of-the-art review, *Environ. Toxicol. Chem.*, 8, 831–842, 1989.
3. Seber, G. A. F., *The Estimation of Animal Abundance*, Macmillan, New York, 1982.
4. Pollock, K. H., J. D. Nichols, C. Brownie and J. E. Hines, Statistical Inference for Capture-Recapture Experiments, *Wildl. Monogr.*, No. 107, 1990.
5. Burnham, K. P., D. R. Anderson and J. L. Laake, Estimation of Density from Line Transect Sampling of Biological Populations, *Wildl. Monogr.*, No. 72, 1980.
6. Skalski, J. R., Use of Capture Data to Quantify Change and Test for Effects on the Abundance of Wild Populations, Ph.D. dissertation, Cornell University, Ithaca, NY, 1985.

7. Skalski, J. R., and D. S. Robson, *Techniques for Wildlife Investigations: Design and Analysis of Capture Data*, Academic Press, San Diego, CA, 1992.
8. Otis, D. L., K. P. Burnham, G. C. White and D. R. Anderson, Statistical Inference from Capture Data on Closed Animal Populations, *Wildl. Monogr.*, No. 62, 1978.
9. Hansen, P. E., Leslie matrix models: a mathematical survey, in *Papers in Mathematical Ecology*, A. Csetenyi, Ed., Karl Marx University of Economics, Budapest, 1986.
10. Emlen, J. M., and E. K. Pikitch, Animal population dynamics: identification of critical components, *Ecol. Model.*, 44, 253–273, 1989.
11. Kalbfleisch, J. D., and R. L. Prentice, *The Statistical Analysis of Failure Time Data*, John Wiley & Sons, New York, 1980.
12. Kaplan, E. L., and P. Meier, Nonparametric estimation from incomplete observations, *JASA*, 53, 457–481, 1958.
13. Prentice, R. L., Exponential survival with censoring and explanatory variables, *Biometrika*, 60, 279–288, 1973.
14. Lawless, J. F., *Statistical Models and Methods for Lifetime Data*, John Wiley & Sons, New York, 1982.
15. Pollock, K. H., S. R. Winterstein and M. J. Conroy, Estimation and analysis of survival distributions for radio-tagged animals, *Biometrics*, 45, 99–109, 1989.
16. Pollock, K. H., S. R. Winterstein, C. M. Bunck and P. D. Curtis, Survival analysis in telemetry studies: the staggered entry design, *J. Wildl. Manage.*, 53, 7–15, 1989.
17. Mayfield, H., Nesting success calculated from exposure, *Wilson Bull.*, 73, 255–261, 1961.
18. Mayfield, H., Suggestions for calculating nest success, *Wilson Bull., 87, 456–466, 1975.*
19. Hensler, G. L., Estimation and comparison of functions of daily nest survival probabilities using the Mayfield method, in *Statistics in Ornithology*, B. J. T. Morgan and P. M. North, Eds.,Springer-Verlag, New York, 1985, pp. 289–301.
20. Bart, J., and D. Robson, Estimating survivorship when the subjects are visited periodically, *Ecology*, 63, 1078–1090, 1982.
21. Hensler, G. L., and J. D. Nichols, The Mayfield method of estimating nesting success: a model, estimators and simulation results, *Wilson Bull.*, 93, 42–53, 1981.
22. White, C. G., Numerical estimation of survival rates from band-recovery and biotelemetry data, *J. Wildl. Manage.*, 47, 716–728, 1983.
23. Cormack, R. M., Estimates of survival from the sighting of marked animals, *Biometrika*, 51, 429–438, 1964.
24. Burnham, K. P., D. R. Anderson, G. C. White, C. Brownie and K. H. Pollock, Design and Analysis Methods for Fish Survival Experiments Based on Release-Recapture, Monogr. 5, American Fisheries Society, 1987.
25. Brownie, C., D. R. Anderson, K. P. Burnham and D. S. Robson, Statistical Inferences from Band Recovery Data—A Handbook, Resource Publication No. 156, U.S. Fish and Wildlife Service, Washington, DC, 1988.
26. Cormack, R. M., Statistical appendix, in R. A. Fordham, Mortality and population change of Dominican gulls in Wellington, New Zealand, *J. Anim. Ecol.*, 39, 13–27, 1970.
27. Seber, G. A. F., The multi-sample single recapture census, *Biometrika*, 49, 330–349, 1962.
28. Seber, G. A. F., Estimating time-specific survival and reporting rates for adult birds from band returns, *Biometrika*, 57, 313–318, 1970.
29. Seber, G. A. F., Estimating survival rates from bird-band returns, *J. Wildl. Manage.*, 36, 405–413, 1972.
30. Clobert, J., J. D. Lebreton, M. Clobert-Gillet and H. Coquillart, The estimation of survival in bird populations by recaptures or sightings of marked individuals, in *Statistics in Ornithology*, B. J. T. Morgan and P. M. North, Eds., Springer-Verlag, New York, 1985, pp. 197–214.
31. Lakhani, K. H., Inherent difficulties in estimating age-specific bird survival rates from ring recoveries, in *Statistics in Ornithology*, B. J. T. Morgan and P. M. North, Eds., Springer-Verlag, New York, 1985, pp. 311–322.
32. Seber, G. A. F., Estimating age-specific bird survival rates for birds from bird-band returns when the reporting rate is constant, *Biometrika*, 58, 491–497, 1971.

33. Smith, S. G., and J. R. Skalski, Statistical design and analysis of avian field trials in environmental toxicology, in *Pesticides in Terrestrial and Aquatic Environments*, D. L. Weigman, Ed., Virginia Water Resource Research Center, Blacksburg, 1989, pp. 407–421.
34. Kirk, R. E., *Experimental Design: Procedures for the Behavioral Sciences*, Brooks/Cole, Belmont, CA, 1982.
35. North, P. M., and B. J. T. Morgan, Modelling heron survival using weather data, *Biometrics*, 35, 667–681, 1979.
36. Feigl, P., and M. Zelen, Estimation of exponential survival probabilities with concomitant information, *Biometrika*, 21, 826–838, 1965.
37. Cox, D. R., Regression models and life tables (with discussion), *J. Royal Stat. Soc. B.*, 34, 187–202, 1972.
38. Breslow, N. E., Analysis of survival data under the proportional hazards model, *Inst. Stat. Rev.*, 43, 45–48, 1975.

44

Statistical Approaches to Data Analysis in Wildlife Ecotoxicology

Michael R. Willig

ABSTRACT

The use of statistics in avian wildlife toxicology is mandated for a variety of reasons associated with identifying sources of variation and isolating the effects related to various treatment factors. Experimental design is essentially the adjustment of data collection methodologies to parallel available analytical techniques that provide resolution to hypotheses of biological interest. Increased focus on the use of *a priori* contrasts and the selection of *a posteriori* comparisons will increase the power of statistical analyses. Bayesian approaches to statistical analysis may become more common in ecotoxicological studies when true replication is not possible or the decision to terminate data collection is predicated upon the data already collected in an ongoing experiment. Multivariate statistical analyses can provide powerful approaches for evaluating hypotheses when they are applied with care, but their use as exploratory analyses has been neglected, especially in relation to construction of subsequent causal hypotheses for evaluation. Future research should examine attributes reflecting changes in the growth rate of populations, or parameters (e.g., age-specific survivorship and age-specific fertility) from which estimated population-level effects can be deduced with facility or be projected by modeling. Given the complexity of ecological systems and their response to stressors, rapid advancement in the field of avian wildlife toxicology will be predicated upon close cooperation between ecologists, biostatisticians, and environmental toxicologists.

KEY WORDS

multivariate statistics, Bayesian statistics, experimental design, population biology, ecological statistics

INTRODUCTION

The domain of topics considered by statistical approaches to data analysis precluded either a broad or comprehensive approach to the subject by the previous four chapters.[1-4] Rather than review or integrate the contents of those four contributions, the primary focus of this chapter is on useful quantitative approaches to wildlife ecotoxicological research which were not considered previously, and on previously considered quantitative approaches from a different perspective.

The use of statistical methodologies in avian toxicological studies is mandated for a variety of reasons associated with inherent sources of natural variation. In addition to responses to different levels of treatment factors, a variety of other sources of variation exist and impinge upon experimental results. In essence, the goal of experimental design and statistics is to identify sources of variation among groups and to distinguish between differences associated with treatment levels per se and those caused by other agents. Extraneous sources of variation should be experimentally controlled or applied to all levels of the treatments in an unbiased fashion. The ultimate goal in designing an experiment should revolve around the resolution of hypotheses. A single field experiment does not always address all hypotheses with equal power. As a result, one should be sure to identify critical questions and assign priority to their resolution before attempting to address peripheral questions within a single experimental design. In essence, one can view experimental design as the adjustment of data collection methodologies to parallel available analytical techniques that provide resolution to hypotheses of biological interest.

PARAMETER ESTIMATION

Measure of the uptake of biocides in target tissues is often required in ecotoxicological studies. Chemical assays frequently result in observations at or below the detection limits of the analytical methodologies. This results in data that are effectively left censored (i.e., data representing observations less than the mean are modified or deleted). Such left censored data often are used in descriptive and inferential statistical analyses after a constant (50% of the detection limit) has been substituted for points at the detection limits; alternatively, the observations below detection limits are excluded from consideration. Both approaches result in statistical biases in parameter estimations (means and variances). Recent work[5,6] suggests promise in the use of maximum likelihood estimation and regression of expected order statistics to circumvent biases in parameter estimation. More accurate parameter estimates are of interest in their own right because increased accuracy will enhance the predictive value of ecological models in which they are incorporated. Although the best statistical design and analysis may be performed on the correlates of fitness for a variety of wildlife taxa exposed to biocides, it is not axiomatic that individual effects will be translated into a demographic response by the population. Increased focus on population-level parameters should constitute a goal of wildlife toxicologists and regulatory agencies.

UNIVARIATE APPROACHES

In circumstances in which a particular dependent variable is of inherent interest, univariate analyses are clearly appropriate. In many experimental designs analysis of variance (ANOVA) is the most powerful approach for detecting group differences, but heteroscedasticity (unequal group variances) prevents its application with confidence. Transformations may be employed to produce homoscedasticity, but unless an effective transformation was selected in advance, levels of probability are only approximate. Even in situations in which ANOVA is applied, it may have less statistical power than recently developed alternatives. Rice and Gaines[7] describe a novel approach to analyzing data when variances are unequal; it circumvents these problems by evaluating a modified F statistic, which may be conceptualized as a weighted average of the Behrens-Fisher t statistic.[8] The approach is more powerful than ANOVA because it does not inflate small variance components in creating a pooled within-group mean square.[7]

A second limitation of the ANOVA approach is that it is a two-tailed test of mean differences. In certain ecotoxicological situations, such as those involving biocide dose response, ordered differences clearly constitute the alternative hypothesis (e.g., control <0.50 biocide application <0.75 biocide application). The application of ANOVA in this scenario results in a considerable loss of power, a situation straightforwardly analogous to conducting a two-tailed *t* test to evaluate a one-tailed hypothesis. Isotonic regression techniques, applicable to both parametric and nonparametric situations, are powerful alternatives to classical ANOVA approaches.[9] Operational difficulties of its application, especially when sample sizes are unequal, can be overcome through simulation to produce *p* values via microcomputer.[10]

A third limitation of ANOVA is that it does not identify which levels of a treatment factor differ; this facet of the analysis is relegated to the field of *a priori* (planned) and *a posteriori* (unplanned) comparisons.[11] Even though considerable thought should precede the design of experiments, *a priori* comparisons are not frequently conducted in the ecotoxicological literature, perhaps in part because *a posteriori* comparisons seem to address the same questions. This is unfortunate because *a priori* comparisons are more powerful than the ANOVA itself, whereas *a posteriori* comparisons are less powerful than the ANOVA. Two kinds of *a priori* comparisons may be of special interest in ecotoxicological research: those involving trend analysis and those in which a number of treatments are each compared to a single control. Situations in which different doses of a biocide are applied to agricultural fields are amenable to planned comparisons that evaluate if a parallel trend occurs in the dependent variable or else response thresholds exist.[12,13] In addition to the parametric techniques, nonparametric techniques are presented by Hollander and Wolfe [14] for similar questions that do not require interval scale values for the levels of the variables of interest.

In cases similar to those in which the effect of different biocides on acetylcholinesterase activity is being compared to levels in a control, nonorthogonal *a priori* contrasts would be appropriate. In these situations, the level of acetylcholinesterase in each biocide treatment is compared to the level in a control most effectively by the Dunn-Sidak test,[15] although some caution should be employed.[13]

In situations in which all possible comparisons between treatment levels are of interest, researchers provide little (if any) rationale for the selection of a particular *a posteriori* technique, making it difficult to judge the appropriateness of the selection. In an excellent review of unplanned comparisons, Day and Quinn[16] provide a detailed consideration of a large suite of statistical techniques and promote a strategy for their use in evaluating hypotheses. When making all possible pairwise comparisons, they in general conclude that the parametric Ryan's Q test or the nonparametric Joint-Rank Ryan's test should be used to evaluate group differences. They did not recommend the commonly used Duncan's multiple range test or the Student-Newman-Keuls (SNK) procedure, although they did consider Welsch's Step-Up procedure to be only slightly less powerful than Ryan's Q test.

BAYESIAN STATISTICS

Bayesian approaches to statistical analysis may be more appropriate than classical approaches in many ecotoxicological scenarios. This is especially true for situations in which true replication (*sensu* Reference 17) cannot occur for a variety of reasons (cost, agricultural practices, etc.) or for protocols in which the size of the final sample is at least in part determined by the kind of response reflected in data collected during an experiment.[18] In the latter case, probability levels from standard statistical tests may be misleading and biased toward overestimating the "significance" of differences among levels of treatment factors.[19,20] In the former case, parallels exist between studies of large-scale perturbations in ecology and studies in wildlife ecotoxicology. In particular, monitoring the response of nontarget species to the application of biocides in agricultural fields is expensive and often not easily replicated; if replicated, site selection frequently is not random within the inference space of interest to regulatory agencies. Nonetheless, unreplicated experiments such as these can be extremely informative[21,22] when analyses are pursued from a Bayesian perspective.[23,24]

Bayesian statistics analyze data to produce final probabilities (posterior probabilities) that are independent of the probabilities of data not obtained in the experiment (i.e., the probabilities of outcomes more extreme than those obtained in the experiment are not considered and do not exaggerate the strength of evidence against a null hypothesis). However, adherents to the Bayesian school of statistics, or subjectivists, do require the specification of the probability of truth of the hypothesis before, or at least separate from, the data generated to evaluate the question of interest. These prior probabilities must be subjectively chosen by the investigator and ought to reflect the biological insight of the investigator. The elegance of the approach is that the interpretation of data is formally related to the consumer's *a priori* view of the truth of the hypothesis, and this prior view must be specified for all to see. For the Bayesian statistician, defining the subjectivity of any test is an important element in understanding data analysis.

MULTIVARIATE APPROACHES

Multivariate analyses of data have become increasingly common in the ecological literature since their early application to questions in phytoecology[25] and numerical taxonomy[26] over three decades ago. Many comprehensive summaries of their application have appeared in the intervening years, with James and McCulloch[27] most recently classifying the plethora of techniques available, summarizing their objectives and limitations, and providing useful guidelines for their incorporation in ecological studies. All too often, multivariate inferences are made without due consideration to sampling design and methodological assumptions. Moreover, the utility of multivariate data analysis as an exploratory method that can suggest descriptive models or future hypotheses has been neglected in the ecological literature, in general, and ecotoxicological studies, in particular. Observations and previous experience with a system can directly lead to insights about cause and effect relationships and are the bases of the hypothetico-deductive method (Figure 1). When straightforward patterns are not easily discerned because multiple factors affect relationships among variables, multivariate approaches can be used as exploratory or heuristic devices capable of suggesting descriptive models that can, in turn, provide a framework for a causal model. In either case, controlled experiments in the field or laboratory or quasiexperimental designs involving time-series analysis or repeated measures can subsequently be evaluated by multivariable techniques.[27,28]

It is difficult, if not impossible, to provide effective directions for avoiding the "Scylla of oversimplification" that occurs when complex relationships among intercorrelated characters are evaluated via univariate techniques without risking exposure to the "Charybdis of assuming that patterns in data necessarily reflect factors in nature, that they have a common cause, or worse, that statistical methods alone have sorted

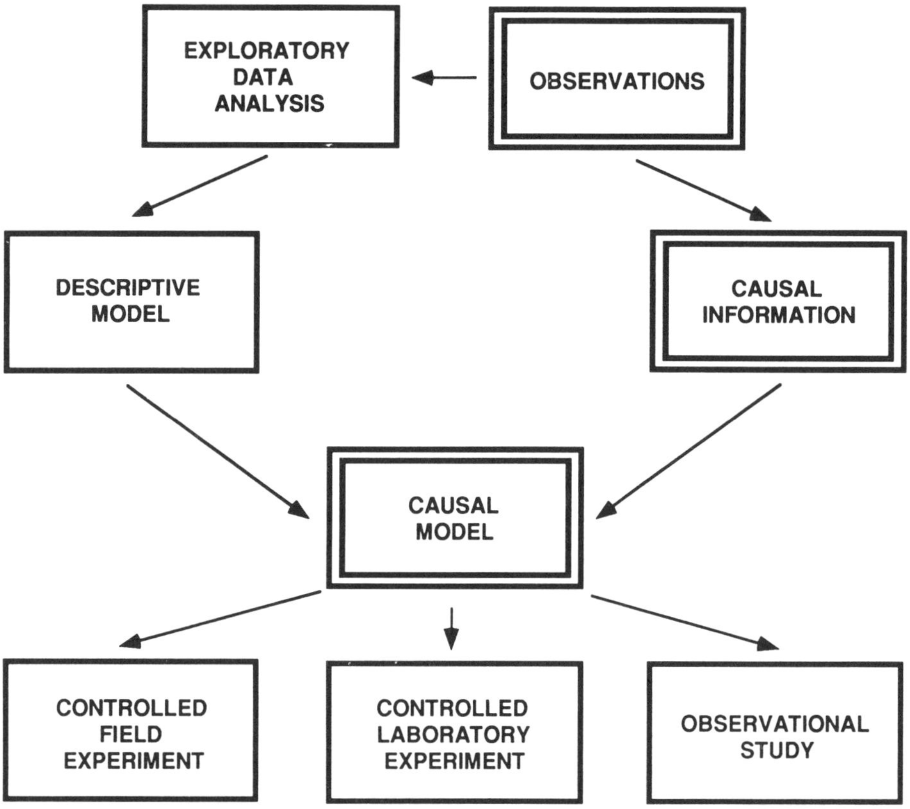

FIGURE 1. Research protocol emphasizing the role of multivariate statistics (applicable to issues surrounded by single boxes) in the formulation of models and in the evaluation of hypotheses.[27,28] Observation can directly yield insight into the cause and effect relationship between factors which then forms the basis for formulation of hypotheses in the hypothetico-deductive method (indicated by the linked double boxes). At other times, observations may be complex or confounded by a variety of variables, reducing the likelihood that inference of causal relationships will be directly obtained. In such circumstances, multivariate data analysis can act as a heuristic device to explore relationships among factors. These multivariate exploratory analyses produce a descriptive model which then gives rise to a causal model (indicated by the single boxes linking observations and causal model). Regardless of the nature of the study (field or laboratory), multivariate statistics may be used to evaluate specific hypotheses (controlled experiments) or detect patterns in the analysis of trends (observational studies).

out multiple causes" for the observed patterns.[27] Perhaps the best that can be expected is that the knowledge that both pitfalls exist provides a navigational tool for avoiding their consequences.

DEMOGRAPHIC INTERFACES

Effects at the level of the individual may be immediate or delayed, and may or may not have an effect upon population-level processes. Although the best statistical design and analysis may be performed on the correlates of fitness for a variety of wildlife taxa exposed to biocides, it is not axiomatic that individual effects will be translated into a demographic response by the population. Indeed, if the major concern surrounding the use of biocides is their effects on population parameters, then the research focus should examine attributes that can reflect changes in the growth rate of populations, or the parameters (e.g., age-specific survivorship and age-specific fertility) from which estimated population-level effects can be deduced with facility or be projected by modeling. The impact of biocides on population-level processes should receive consideration as an important component in regulatory decisions by governmental agencies.

It is well known, for example, that limited removal of individuals via hunting can have little or no effect on subsequent population levels. Moreover, the local effect of removing individuals from a population can be mediated by the immigration of individuals from surrounding habitats, resulting in the absence of a net change in the density of individuals over the long term. These same lessons from wildlife management can be applied to ecotoxicological research examining the effects of biocides on nontarget species rather than the impact of hunting on game animals. The extent to which ecological filters such as these operate will confound extrapolations from individual-level effects to those at the population level. Even the size and proximity of habitats surrounding agricultural fields can have an effect on population-level effects within the area targeted for biocide treatment.

Answers concerning the effects of biocides on wildlife populations cannot be obtained from a universally applicable general design. Only a clear statement of the biological question and a delineation of the level at which particular effects are to be manifested can yield a successful research design. The fields of ecology and wildlife management provide the general principles upon which effects can be predicted and the fields of statistics and modeling can provide a variety of technical approaches to analyzing data. It seems clear, given the complexity of ecological systems and their responses to stressors, that rapid advancement in the field of avian toxicology will depend upon close cooperation between ecologists, biostatisticians, and environmental toxicologists.

ACKNOWLEDGMENTS

Foremost, I would like to thank R. J. Kendall, T. E. Lacher, Jr., and the Society of Environmental Toxicology and Chemistry for the opportunity to attend the conference entitled "The Population Ecology and Wildlife Toxicology of Agricultural Pesticide Use: A Modeling Initiative for Avian Species;" the various federal agencies and corporations for making it a financial reality; and the participants themselves for creating a stimulating environment in which ideas could be discussed in a critical, yet friendly manner. Finally, the Department of Biological Sciences at Texas Tech University facilitated the gestation of the manuscript and its final preparation.

REFERENCES

1. Dixon, P. M., and K. A. Garrett, Statistical issues for field experimenters, in *Wildlife Toxicology and Population Modeling: Integrated Studies of Agroecosystems*, R. J. Kendall and T. E. Lacher, Jr., Eds., Lewis Publishers, Chelsea, MI, 1993.
2. Lacher, T. E., Jr., and M.R. Willig, Univariate and multivariate approaches to the analysis of ecotoxicological data, in *Wildlife Toxicology and Population Modeling: Integrated Studies of Agroecosystems*, R. J. Kendall and T. E. Lacher, Jr., Eds., Lewis Publishers, Chelsea, MI, 1993.
3. Skalski, J. R., and S. G. Smith, Risk assessment in avian toxicology using experimental and epidemiological approaches, in *Wildlife Toxicology and Population Modeling: Integrated Studies of Agroecosystems*, R. J. Kendall and T. E. Lacher, Jr., Eds., Lewis Publishers, Chelsea, MI, 1993.

4. Sauer, J. R., R. J. Barker and P. H. Geissler, Statistical aspects of modeling population changes from population size data, in *Wildlife Toxicology and Population Modeling: Integrated Studies of Agroecosystems*, R. J. Kendall and T. E. Lacher, Jr., Eds., Lewis Publishers, Chelsea, MI, 1993.
5. Newman, M. C., P. M. Dixon, B. B. Looney and J. E. Pinder, III, Estimating mean and variance for environmental samples with below detection limit observations, *Water Res. Bull.*, 4, 905–916, 1989.
6. Newman, M. C., and P. M. Dixon, UNCENSOR: a program to estimate means and standard deviations for data sets with below detection limit observations. *Am. Environ. Lab.*, 4/90, 27–30, 1990.
7. Rice, W. R., and S. D. Gaines, One-way analysis of variance with unequal variances, *Proc. Natl. Acad. Sci.*, U.S.A. 86, 8183–8184, 1989.
8. Behrens, W.-U., The comparison of means of independent normal distributions with different variances, *Biometrics*, 20, 16–27, 1964.
9. Barlow, R. E., D. J. Bartholomew, J. M Bremmer and H. D. Brunk, *Statistical Inference under Ordered Restrictions*, John Wiley & Sons, New York, 1972.
10. Gaines S. D., and W. R. Rice, Analysis of biological data when there are ordered expectations, *Am. Nat.*, 135, 310–317, 1990.
11. Sokal, R. R., and F. J. Rohlf, *Biometry*, 2nd ed., W. H. Freeman, San Francisco, 1981.
12. Mead, R., and K. J. Pike, A review of response surface methodology from a biometric viewpoint, *Biometrics*, 31, 803–853, 1975.
13. Keppel, G., *Design and Analysis, A Researcher's Handbook*, 2nd ed., Prentice-Hall, Englewood Cliffs, NJ, 1982.
14. Hollander, M., and D. A. Wolfe, *Nonparametric Statistical Methods*, John Wiley & Sons, New York, 1973.
15. Miller, R. G., *Simultaneous Statistical Inference*, 2nd ed., Springer-Verlag, Berlin, 1981.
16. Day, R. W., and G. P. Quinn, Comparison of treatments after an analysis of variance in ecology, *Ecology*, 59, 433–463, 1989.
17. Hurlbert, S. H., Pseudoreplication and the design of ecological field experiments, *Ecol. Monogr.*, 54, 187–211, 1984.
18. Berger, J. O., and D. A. Berry, Statistical analysis and the illusion of objectivity, *Am. Sci.*, 76, 159–165, 1988.
19. Edwards, W., H. Lindman and L. J. Savage, Bayesian statistical inference for psychological research, *Psychol. Rev.*, 70, 193–242, 1963.
20. Berger, J. O., *Statistical Decision Theory and Bayesian Analysis*, Springer-Verlag, Berlin, 1985.
21. Blau, G. E., and W. B. Neely, Mathematical model building with an application to determine the distribution of Dursban insecticide added to a simulated ecosystem, in *Advances in Ecological Research*, Vol. 9., A. MacFayden, Ed., Academic Press, London, 1990.
22. Frost, T. M., D. L. DeAngelis, S. M. Bartell, D. J. Hall and S. H. Hurlbert, Scale in the design and interpretation of aquatic community research, in *Complex Interactions in Lake Communities*, S.R. Carpenter, Ed., Springer-Verlag, New York, 1988, pp. 229–260.
23. Carpenter, S. R., Large-scale perturbations: opportunities for innovation, *Ecology*, 71, 2038–2043, 1990.
24. Reckhow, K. H., Bayesian inference in non-replicated ecological studies, *Ecology*, 71, 2053–2059, 1990.
25. Williams, W. R., and J. M. Lambert, Multivariate methods in plant ecology. I. Association analysis in plant communities, *J. Ecol.*, 47, 83–101, 1959.
26. Sokal, R. R., and C. Michener, A statistical method for evaluating systematic relationships, *Univ. Kan. Sci. Bull.*, 38, 1409–1438, 1958.
27. James, F. C., and C. E. McCulloch, Multivariate analysis in ecology and systematics: panacea or Pandora's Box?, *Annu. Rev. Ecol. Syst.*, 21, 129–166, 1990.
28. James, F. C., and C. E. McCulloch, Data analysis and the design of experiments in ornithology, in *Current Ornithology*, Vol. 2. R. F. Johnson, Ed., Plenum Press, New York, 1985, pp. 1–63.

45

The Future Application of Ecological Models in Environmental Risk Assessment

James A. Gagne

ABSTRACT

Population ecology theory and mathematical modeling offer theories and techniques that will be useful in the development of wildlife toxicology. However, integration of these theories and techniques into ecological risk assessment will be, at least, a long-term process. A major policy issue that currently limits the integration of population level models into ecological risk assessment is the very low level of "theoretical" mortality that regulators will accept. Despite this circumstance, mathematical models should be important products of future research that aims to reduce the uncertainties in our current evaluations of the toxicity of and the exposure to pesticides that form the basis for ecological risk assessment today.

KEY WORDS

population ecology, mathematical modeling, ecological risk assessment, pesticides

INTRODUCTION

My comments are written from the perspective of someone whose task is to obtain and maintain pesticide registrations. As such, when I think of environmental risk assessment, I am thinking about it as practiced by the United States Environmental Protection Agency's (U.S. EPA's) Office of Pesticide Programs, under the Federal Fungicide, Insecticide, and Rodenticide Act (FIFRA). Ed Fite outlines that procedure very well in Chapter 48.[1] It was pointed out in the discussion following this session at the 1990 Ecological Modeling Conference that environmental risk assessment is also done by other parts of the EPA under other laws, such as the Toxic Substance Control Act (TSCA). The data used in environmental risk assessments under FIFRA are more extensive than those used under TSCA. It would appear that there are policy variations in environmental risk assessment even within the EPA. To me, this is a significant discrepancy that merits further analysis and discussion.

As mentioned at the outset, the organizers of this conference felt that interactions between scientists in wildlife toxicology and population ecology would benefit both

0-87371-591-8/94/$0.00 + $.50

disciplines. Perhaps naively, it was also assumed that this benefit would help regulators to make decisions that are "better" in some sense. Conclusions and perspective gleaned from the cross-fertilization of wildlife toxicology and population ecology, however, cannot automatically be used in making regulatory decisions, for at least two reasons. First, the data and methods used by the EPA have been developed over a long period of time. Pesticides on the market have similar databases and have been evaluated with uniform methods. This first reason leads to my first conclusion: *Further integration of ecological modeling into environmental risk assessment will be a long-term process, requiring years, perhaps decades.* The second reason that the results from this conference cannot automatically be used in making "better" regulatory decisions is that the conference implicitly took a "bottom-up" view, instead of the "top-down" view correctly advocated by Lou Gross:[2]

> "This leads to suggested forms for a top-down approach. Here, one starts with the [risk] manager and attempts to ascertain levels of indifference to model outcomes. This provides a means to scale the amount of reductionism necessary to meet the goals of the manager, and doesn't waste effort dealing with minute details of biological processes if these are second or third order effects in the scheme of social and political constraints acting."

A brief review of Ed Fite's presentation will allow one to see the level of indifference to certain outcomes that are inherent in the current method. Recall that the quotient method compares the estimated environmental concentration (EEC) of a pesticide with the LD_{50} or LC_{50}. Typically, if the EEC is greater than one fifth the LC_{50} for terrestrial animals and one tenth the LC_{50} for aquatics, risk is assumed. The EPA has performed an analysis of the dose-response curves for terrestrial and aquatic organisms. The "average" slope of these curves is 5.78 probits per log cycle for terrestrial organisms and 9.95 probits per log cycle for aquatics. Using the one fifth and one tenth safety factors, these slopes, and a program to calculate probabilities from probits, one finds that the theoretical probability of mortality occurring at one fifth the LC_{50} for terrestrial organisms is 0.0000267, and the theoretical probability of mortality occurring at one tenth the LC_{50} for aquatic organisms is 0.00000000000000000000000012. For endangered species, EPA uses an additional twofold safety factor, resulting in theoretical probabilities of mortality of 0.00000000374 and 0.0000000000000000000000000000000000000125, respectively. These vanishingly small probabilities of mortality occurring lead me to my second conclusion: *It makes no sense to attempt to integrate population models into environmental risk assessments, because the acceptable level of theoretical mortality is essentially zero. Obviously, models for communities and ecosystems also do not merit consideration if the current triggers are followed.*

One may argue at this point that further discussion of the use of ecological models in environmental risk assessment is meaningless. I disagree. Although much of the theory that has been developed for populations is not currently germane, because of policy, it may become so in the future. Even more importantly, other classes of models discussed at this conference do have tremendous potential for use in improving risk assessments. These models will do what I will term "reducing the uncertainty" inherent in the data used in the current risk assessment method.

In Chapter 48, Fite[1] points out that there are several assumptions and uncertainties inherent in the current method. To paraphrase Fite, these are variability and lack of reproducibility in acute and subacute toxicity data; the bobwhite and the mallard do not capture the range of sensitivity of avian species; it is difficult to extrapolate from the laboratory to the field, especially parameters such as duration of exposure, age of the bird, etc., and residue levels in or on food items may not correlate well to actual exposure. This leads me to my third conclusion: *Research efforts aimed at reducing the uncertainties inherent in both the toxicity and exposure components of the hazard equation will pay large dividends. Wherever possible, the research should attempt to develop mathematical models that are carefully validated as part of the research effort. Also, these models should employ the top-down approach if at all possible.*

Examples of the sorts of models that I am talking about are models of the fate of pesticides in the food and environment of birds; models of the duration and magnitude of pesticide exposure that key bird species actually experience in the field; models to predict toxicity of the same pesticide across species or within the same species at different ages; and models to predict the toxicity of different pesticides to the same species. All these sorts of models were mentioned during the course of the 1990 conference. I also urge that this work be done outside the normal registration process by the EPA, universities, and industry, working in cooperation. Only after the research has been completed should there be an effort to change the testing requirements for registration.

REFERENCES

1. Fite, E., The Environmental Protection Agency's avian pesticide assessment model, in *Wildlife Toxicology and Population Modeling: Integrated Studies of Agroecosystems*, R. J. Kendall and T. E. Lacher, Jr., Eds., Lewis Publishers, Chelsea, MI, 1993.
2. Gross, L.J., Limitations of reductionist approach in ecological modeling: model evaluation, model complexity, and environmental policy, in *Wildlife Toxicology and Population Modeling: Integrated Studies of Agroecosystems*, R. J. Kendall and T. E. Lacher, Jr., Eds., Lewis Publishers, Chelsea, MI, 1993.

46

Population Models as Research Tools: An Empirical Perspective

Bill Williams and John Emlen

ABSTRACT

As environmental regulators have been asked to determine the probable impact of pesticides and other agricultural chemicals on populations of nontarget wildlife, the need for environmentally realistic population models has become more apparent. Although it is not yet practical to conduct field studies that determine the effects of chemicals on entire populations of wildlife, it is likely that efforts will be directed toward determining the impact of these chemicals on sub-populations or local populations of wildlife. This task is a difficult one, as the use of predictive models to simulate the effects of chemicals or other human-induced stressors on terrestrial populations will be impeded by the enormous variety of species and special conditions associated with sites of interest. It simply is not possible to construct models that deal appropriately with all species and circumstances of interest. Therefore, it would be useful, before building predictive population models, to determine which input (population) parameters are critical to the desired model output. Once these parameters are determined, they can be organized into hierarchical levels of importance vis a vis their relative impacts on population density or fitness. Use of such information will enable the construction of simpler and more generic, predictive terrestrial population models. This chapter proposes the consideration of models composed of two partly separable components: one comprising the purely mechanical descriptors of dynamics from given demographic parameter values, and the other describing the modulation of the demographic parameters by environmental factors such as changes in the physical environment, species interactions, pathogens, and xenobiotic chemicals. This approach allows the determination of the relative sensitivity of a specified population endpoint to each of the dynamic parameters that ultimately control population density and fitness.

A primary purpose of this chapter is to present some of the unifying concepts that link population ecologists (model developers) with toxicologists and field biologists charged with the collection, interpretation, and utilization of empirical data on the effects of chemicals on wildlife populations. The identification of critical parameters driving population changes will make models more reliable, indicate areas of priority in research, and enable researchers to glean the maximum amount of usable data from field research studies. In addition, the interaction of modelers and empirical researchers will increase the scientific return from both groups.

0-87371-591-8/94/$0.00 + $.50

KEY WORDS

population, models, survival, chemical effects

INTRODUCTION

Numerous models have been developed to characterize aquatic populations; fewer models have been developed to characterize terrestrial populations. Moreover, most of the latter models have been developed for purposes of wildlife management or harvesting, not for assessing or predicting the impact of chemicals. In order to be useful for avian risk assessments, terrestrial population models must have the following characteristics:

1. They must address endpoints that are of regulatory importance.
2. They must easily incorporate all relevant toxicological information.
3. They must use generally available population data.
4. They must have potential linkage to models of exposure to pesticides or toxicants.

The purpose of avian and mammalian population models is to evaluate or predict the impact of disturbances on a relevant population endpoint such as population density. In any regulatory scheme, human-induced change in population density will provide regulators with a quantitative means for assessing damage to the population of concern. Changes in density are brought about by changes in birth rate, death rate, immigration, and emigration, but these parameters do not, in themselves, constitute endpoints.

Other endpoints that could be used to evaluate the impact of a chemical on a population are relative abundances of age or size classes, the sex ratio, and alterations in population growth rate or alterations in body size structure. Endpoints must be clearly demarcated by spatial and temporal limits to avoid ambiguity and inconsistency.

Attempts to regulate numerous agricultural chemicals and pesticides based on potential risks to terrestrial wildlife have centered on die-offs and chemical contamination of various nontarget birds and mammals. Although impacts of chemicals on wildlife often can be temporally correlated with acute localized applications, it is usually difficult to relate adverse effects to a specific chemical exposure. Indirect methods of relating chemical exposure to effects have been developed for chemicals that do not persist in the environment, the food chain, or animal tissue. Methods such as determination of cholinesterase (ChE) depression depend on the fact that organophosphates (OPs) and carbamates are known to depress levels of circulating ChEs rather specifically. The relative successes of these new techniques, however, are overshadowed by the difficulties encountered when attempting to assess chemical risk on a broader geographical or regional scale. The interest in determining risk at these scales, however, highlights the need to develop predictive models that relate lethal and sublethal effects of chemicals on individuals to changes in population dynamics.

Research in risk assessment has increased the need to identify and evaluate extant population models and determine their suitability for use in terrestrial risk assessments.[1] The simplest approach would be to evaluate the adaptability of these models for incorporating the effects of xenobiotic chemicals on birth and death rates. Unfortunately, most such models are too simplistic to be used as reliable predictors of population effects, or are too site or species specific to be used in a general sense. In any event, a primary problem arises in the search for models for use as predictors of the effects of xenobiotic chemicals on populations of terrestrial wildlife: most extant models have been developed with a specific exposure scenario in mind and in addition do not adequately account even for the normal swings in population density that can be attributed to "natural" causes.

It is useful to divide population models into two parts. The first part describes population dynamics in terms of the demographic parameters; i.e., the bookkeeping formulae by which age- (or stage-) specific birth and death rates and density feedback patterns interrelate to drive population densities. The second part describes the relative impacts of environmental change, chemical toxicity, and the patterns of species interactions on the demographic parameters (Figure 1). In this arrangement, wherein the submodels relating environmental interactions are separate, the modeler is provided an opportunity to deal specifically with the bookkeeping portion of population dynamics to determine the sensitivity of the endpoint to alterations in each parameter (e.g., specific life history information) and thus which demographic parameters are critical to damage prediction. Input parameters that exert little influence on the desired model output can be sorted out and afforded a low priority without markedly affecting the model performance. For example, if population density were the output of interest and if a large percentage (80 to 90%) of the variance in population density were accounted for by juvenile mortality, the modeler would be justified in putting extra effort into defining the impact of external stresses (predation, pesticide-induced effects, etc.) on this parameter, giving less attention to others. Identification of the most influential parameters would justify the eventual extrapolation of the model to more generic conditions, because site and special scenarios play a relatively less important role when one parameter displays this level of importance in the model. In summary, the first step to development of practical risk assessment models for terrestrial populations is to

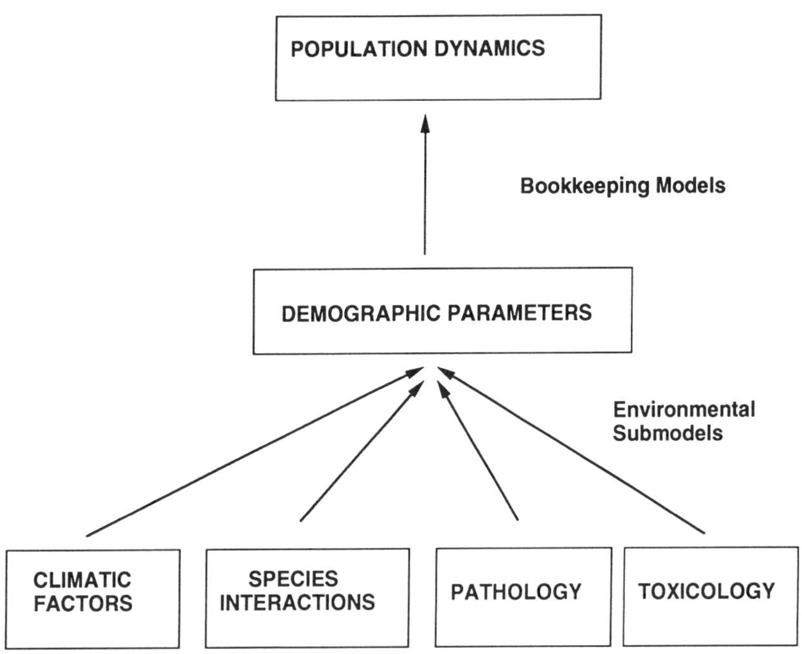

FIGURE 1. Population modeling can be divided into two parts: (1) demographic parameters representing bookkeeping parameters, and (2) relative impacts of external factors such as climate change, chemical insult, and species interactions.

determine the sensitivity of model endpoints to each of the interactive bookkeeping parameters.

The life histories of birds and mammals can be divided into three stages that for purposes of simplicity can be described as equally spaced in time: juvenile, late juvenile and early adult, and adult. In such cases, a Leslie matrix can be modified to incorporate density dependence and provide a simple and adequate model describing populations. Network models provide a more general approach, capable of incorporating complexities in life histories not often found in birds and mammals, while reducing to the Leslie matrix in simple cases.[2,3] In addition, network models easily incorporate density dependence information and can be extended to examine more complex heterogeneous environments and multispecies systems.

In many terrestrial wildlife populations, demographic parameters exhibit considerable natural variation through time (Figure 2). It is necessary in these cases to consider a modified output such as mean population density over time rather than population density. This endpoint acknowledges the significant variation in population density that arises from natural fluctuations in population numbers and is driven, at least in part, by age structure, variations in birth and death rates, immigration and emigration, and the impact of natural stresses on other parameters normally contributing to population density. Researchers must be prudent in their determination of population density, accounting for large normal variations in the population mean, $\bar{N}$, during any sampling scenario. It is necessary, in fact, to evaluate the population density over a time period that will provide a reasonable sample of population numbers sufficient to account for the natural variation in population density. Inappropriate "snapshot" determinations of population density can result in erroneous interpretation of the trend or direction of these changes. In Figure 2, successive samples of hypothetical population density (boxed points) suggest that the density has decreased between sampling years, and would likely result in some statement that the population density is decreasing. This short observation time frame, however, provides an unrealistic estimate of the actual long-range change in population density as indicated by the spread of N over the

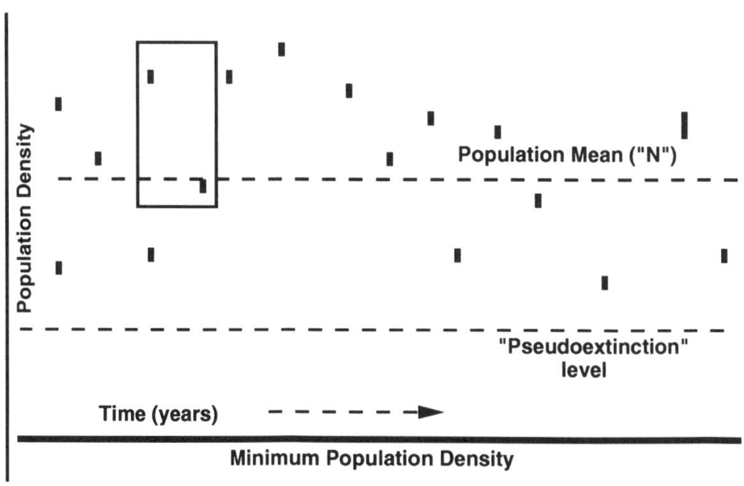

FIGURE 2. Hypothetical variation in population density over time indicates the importance of density change in relation to absolute numbers of animals. The importance of long-term observation of population trends is noted by the "decreased" population densities indicated in the box, while in fact these changes fall well within the natural fluctuation of population size.

selected period. Observation of N in this hypothetical case suggests, however, that these changes are well within the normal variation of population density caused by natural fluctuations. Attempts to appropriately account for the large natural fluctuations in demographic parameters of wildlife populations must incorporate sensitivity analysis of each input parameter and simulate the interrelationships of these input parameters in response to natural stresses.

Although complex in its approach, considerable theoretical work on the sensitivity of populations to changes in demographic parameters has been done[4,5] and applied to specific species by Ebert[6] and others. The works cited describe linear, extended, deterministic sensitivity analyses for network systems. These authors, however, were able to apply their concepts only to growth parameters. They did not incorporate the stochastic nature of the system needed to describe the interaction of parameters that is appropriate for realistic predictive modeling.

APPROACH

Birds have been used as the basis of many population models because of researchers' ability to relate population vitality to reproduction parameters such as clutch size, egg production, egg quality, and number of surviving hatchlings. These parameters can be quantified easily and provide a reasonable source of input parameters for models. In addition, many of the field studies being conducted in support of chemical testing and registration already use wildlife species of birds. Once again for purposes of discussion, birds are considered to have three life stages, all of roughly equal duration. Life history patterns for birds and mammals are also sufficiently related to body size so that these taxa can be split into small, medium, and large animals within which age-specific survival and fecundity patterns are reasonably consistent.[7] Generally, survival is low and fecundity correspondingly high in species with small individuals, whereas survival is high and fecundity is low in species with large individuals (Figure 3). Regardless of relative size adult mortality is usually lower than that of juveniles, and though increasing late in life, can be represented reasonably by a constant.[8-12] Fecundity usually rises with age, plateaus, and then falls off again later in life.[13-17] This course in fecundity can be described simply by attributing no reproduction to an early juvenile stage, some reproduction to the next stage (late juvenile and early adult), and higher reproduction to adults. A species grouping by size divided into these three life history categories can be used to relate mean birth and death rates, and the time variances associated with them to population endpoints (Figure 4). Although these categories may not easily accommodate all bird species, they provide a reasonable, practical first attempt to relate sensitivities within and between groups. In the approach suggested by Emlen and Pikitch,[18] life stage-specific survival and fecundity values were allowed to fluctuate randomly over time about "pivotal" values characteristic of each species type (small, medium, large) and each life history stage. The resulting population dynamics were thereby simulated and two endpoints of interest calculated: (1) mean population density over time N and (2) the number of times the population is predicted to drop below some finite percentage (e.g., 20%) of $\bar{N}$. The drops in N below the 20% level are termed "pseudoextinction events" and provide a useful means of determining the relative impact of a stress on the population. By repeating simulations and varying pivotal values and fluctuation variances one at a time, it is possible to ascertain how sensitive the above endpoints are to changes in each demographic parameter and their degree of fluctuation. The approach thereby provides an organizational tool that identifies the relative importance of factors affecting each demographic parameter in the overall

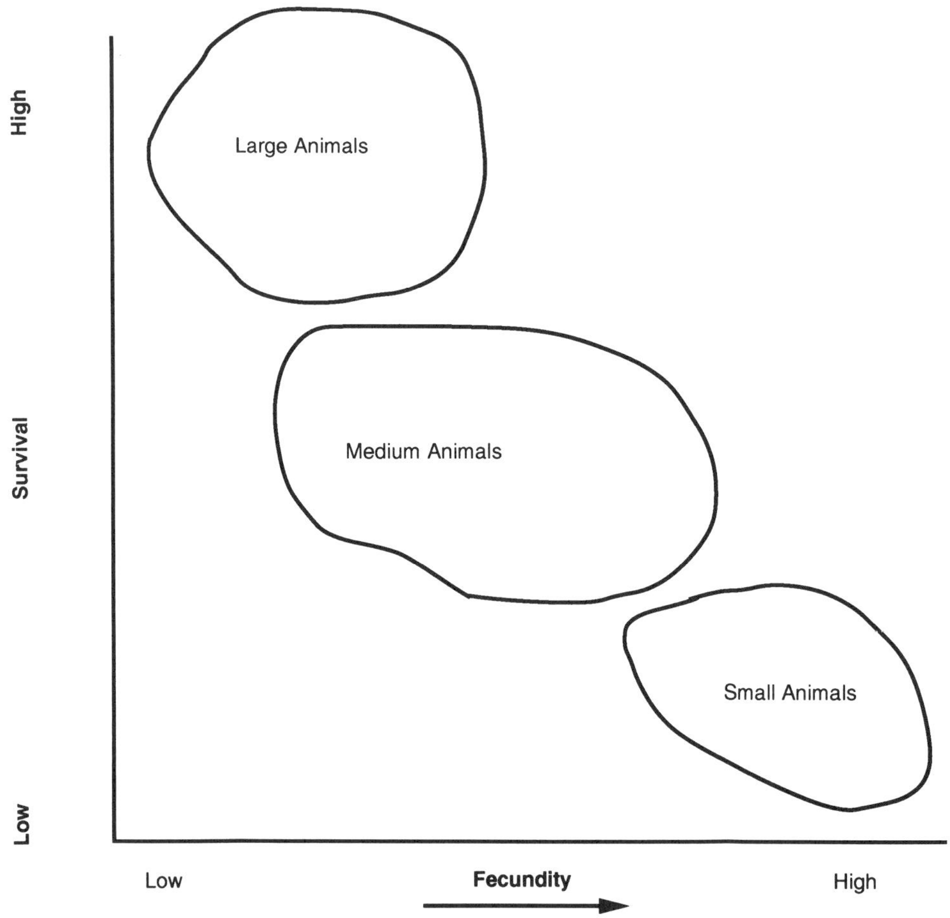

FIGURE 3. Theoretical grouping of animals by size suggests differences in reproductive and mortality strategies. The relation of fecundity to survivability is generally a function of size.

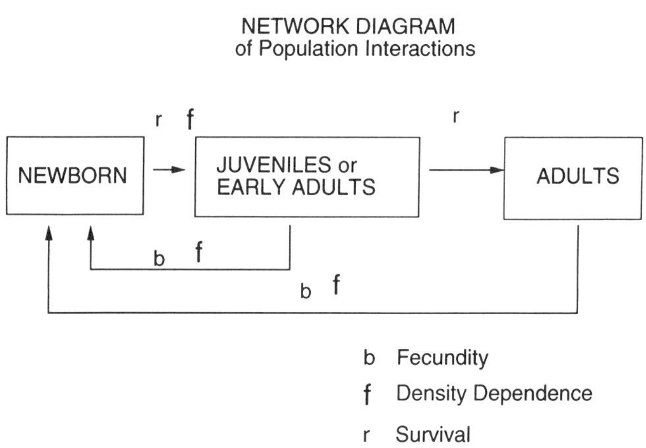

FIGURE 4. Relative life history stages of animals are used to organize sensitive population parameters. Each phase can be represented by an interaction function.

population model. In effect, this will provide a prioritized "shopping list" of data that need to be collected to fill each of the parameter cells. The researcher, working closely with the modeler, will be able to design and conduct laboratory and field experiments to provide data on a prioritized basis.

RESULTS

The results of the hierarchical evaluation presented above can be used to organize population research, and will constitute the basis for verification of models when empirical data are actually collected. They are summarized as follows.

Small Animals

1. For small animals, the most critical input parameter determining *N* is early survival. A 10% increase in survival can be expected to result in a 42% rise in *N*, while a 10% decrease in survival results in a 30% drop. The expected rise in *N* with a 20% increase in survival is 96%, while a 20% decrease in adult survival produces a 47% decrease in *N*.
2. The second most important (theoretical) critical parameter is adult survival. In this case, Emlen and Pikitch predict that a 10% increase in adult survival results in a 26% rise in *N*; a 10% drop in adult survival results in a 16% decrease in *N*.
3. Birth rates rank third in the Emlen and Pikitch analysis, resulting in a 20% increase in *N* with an input (birth rate) increase of 10%: a 17% drop results when birth rate falls 10%.

Medium Animals

1. The most important parameter appears to be adult survival, but the impact of changes in this variable results in only about 20% of the impact in small mammals.
2. No single demographic parameter stands out as the most critical determinate of population density.

Large Animals

In this case, the most significant parameter appears to be adult survival, with 10% rises and falls leading to 65% increases or 46% decreases in population mean.

DISCUSSION

It is apparent from the above synopsis of predicted effects and interactions of variations in critical population parameters that this tactic can be used to organize and prioritize work in the laboratory and in the field. This approach provides some insight into the ways researchers and modelers can team up to collect the most effective data for improving our understanding of population dynamics used to structure long-range research plans, prioritize research initiatives, and lend credibility to the process of developing research funding options. A practical use of this organizational tool has been illustrated by Emlen and Pikitch,[18] in which they relate a typical regulatory decision process that evaluates the impact of a specific toxic chemical (pesticide) on white-tailed deer, northern bobwhite, and a fox. In their example, the impact on the deer (large animal) is primarily on adult survival, while the greatest impact on mean population density for the northern bobwhite (small animal) is predicted to be related to changes in both first-year mortality and adult mortality. In addition, some information about the effect on fecundity on the bobwhite would be valuable. The impact on the fox or other medium animal is less clear, but illustrates the necessity of collecting additional information about fecundity and mortality.

SUMMARY

The usefulness of this approach as a research tool is seen in the ability to identify, prioritize, and evaluate the most immediately productive research questions. In concert with model development, and synchronized with model validation, this organizational approach provides empirical researchers, population modelers, and regulators with a means of directing research and planning future research needs.

REFERENCES

1. Emlen, J. M., Terrestrial population models for ecological risk assessment: a state of the art review, *Environ. Toxicol. Chem.*, 8, 831–842, 1989.
2. Lewis, E. R., Application of discrete and continuous network theory to linear population models, *Environ. Manage.*, 8, 375–442, 1976.
3. Hubbell, S. P., and P. A. Werner, On measuring the intrinsic rate of increase of populations with heterogeneous life histories, *Am. Nat.*, 113, 277–293, 1979.
4. Caswell, H., A general formula for the sensitivity of population growth rate to changes in life history parameters, *Theor. Pop. Biol.*, 14, 215–230, 1978.
5. Caswell, H., and P. A. Werner, Transient behavior and life history analysis of weasel (*Dipsacus sylvestris* Huds.), *Ecology*, 59, 53–66, 1978.
6. Ebert, T. A., Sensitivity of fitness to macroparameter changes: an analysis of survivorship and individual growth in sea urchin life histories, *Oecologia*, 65, 461–467, 1985.
7. Calder, W. A., *Size, Function, and Life History*, Harvard University Press, Cambridge, MA, 1984.
8. Deevey, E. S., Jr., Life tables for natural populations of animals, *Q. Rev. Biol.*, 22, 283–314, 1947.
9. Farner, D. S., Bird banding in the study of population dynamics, in *Recent Studies in Avian Biology*, A. Wolfson, Ed., University of Illinois Press, Urbana, 1955.
10. Ricklefs, R. W., Fecundity, mortality, and avian demography, in *Breeding Biology of Birds*, D.S. Farner, Ed., National Academy of Sciences, Washington, DC, 1973, pp. 336–347.
11. Botkin, D. B., and R. S. Miller, Mortality rates and survival of birds, *Am. Nat.*, 108, 181–192, 1974.
12. Charnov, E. L., Life history evolution in a "recruitment population:" why are adult mortality rates constant?, *Oikos*, 47, 129–134, 1986.
13. Coulson, J. C., and E. White, The effect of age on the breeding biology of the kittiwake, *Rissa tridactyla*, *Ibis*, 100, 40–51, 1958.
14. Perrins, C. M., Population fluctuations and clutch size in the great tit, *Parus major* (L.), *J. Anim. Ecol.*, 34, 601–647, 1965.
15. Keyfitz, N., *Introduction to the Mathematics of Populations*, Addison-Wesley, Reading, MA, 1968.
16. Simpson, C. D., Reproduction and population structure in the greater kudu in Rhodesia, *J. Wildl. Manage.*, 32, 149–161, 1968.
17. Drickamer, L. C., and B. M. Vestal, Patterns of reproduction in a laboratory colony of *Peromyscus*, *Evolution*, 20, 174–184, 1973.
18. Emlen, J. M. and E. K. Pikitch, Animal population dynamics: identification of critical components, *Ecol. Model.*, 44, 252–273, 1989.

47

Limitations of Reductionist Approaches in Ecological Modeling: Model Evaluation, Model Complexity, and Environmental Policy

Louis J. Gross

ABSTRACT

Population ecology has traditionally relied on mathematically simple, and usually tractable, models to mimic the basic dynamics of populations. Parameters in these models are not generally tied to physiology, but reflect aggregate properties of individual behavioral and physiological characteristics. The models are therefore biologically naive, with little direct relevance to practical problems of resource management or risk assessment, although they may be useful as general descriptors of possible population behavior. Models with higher precision and greater biological realism include many more details of population structuring (e.g., age, size, genetics, spatial, dominance, and satiation level), but thereby require many more parameters and assumptions about interactions. These complex models may work well for accurately predicting future population sizes and structures under alternative scenarios for a certain time frame for the few species in which we are willing to invest the requisite financial and research resources. However, the structure of the dynamical systems underlying these complex models is such that analysis of the general behavior of the model is either precluded or requires intensive numerical experimentation. These difficulties, coupled with the fact that the models typically view the habitat environmental variables (physical and biotic) as static, diminish the utility of these reductionist approaches to problems on larger spatial or longer temporal scales. What is needed for problems of environmental policy may well be a hybrid of reductionist and "top-down" approaches. Reductionism can suggest appropriate alternative forms for macrodescriptors of large-scale systems, which may then be coupled to methods that query policymakers as to the level of accuracy they deem necessary to differentiate between the effects of alternative decisions. Evaluation procedures for these models will be quite different from those appropriate for models used mainly for development of general theory.

KEY WORDS

mathematical models, reductionism, aggregation, scaling

INTRODUCTION

> It is a common fallacy to confuse scientists' models of reality with reality itself. A model is a map. A map is not the territory it describes.
>
> Richard Casement, in *Man Suddenly Sees to the Edge of the Universe*

Despite tremendously rapid growth in the development of mathematical models over the past several decades, in application to hosts of scientific disciplines, relatively little attention has been paid to the actual practice of testing and evaluating models. It may be reasonably argued that for most of the physical sciences, in which carefully controlled experiments often lead to observations with small sampling errors, questions of model testing should be relegated to philosophers. In these disciplines, there is no great difficulty in comparing model results with new experiments; rather the concern is whether the model results are consistent with alternative experiments. In the ecological sciences, however, appropriate experiments to test a model may be difficult, if not impossible to perform, and often the data obtained have so large an inherent variation that they do not allow discrimination between several different models. It is often extremely difficult to control all variables that may be affecting a particular behavior of interest unless the experiment is performed under laboratory conditions. How readily laboratory results can be subsequently extended to field conditions, for which there are many more independent and dependent variables, is typically open to much interpretation by the researcher. For these reasons, careful attention to evaluation procedures is essential, particularly if the models are to be used for policy considerations. Yet relatively little work has been done on the development of reasonable agreed-upon procedures for testing models, either for practical applications or in theory.[1]

The lack of substantive work on model evaluation is evident by perusing most of the texts available on mathematical modeling. It is not unusual for these texts to say so little on the subject that it is not even listed in the index (e.g., References 2 through 5), or the comments are limited to parameter estimation and statistical goodness-of-fit (e.g., References 6 and 7). Only a couple of texts illustrate concern for model testing; although some carefully include it in a list of the important attributes of modeling,[8] the general attitude seems to be that the issue is not central to the modeling process. One of the exceptions is the text by Mesterton-Gibbons,[9] which is infused with the importance of testing models, but proceeds mainly by case studies. This offers little in the way of a general approach to model testing, though the author clearly stresses that one important criterion for models being applied in decision-making is their flexibility. France and Thornley[10] devote a brief chapter to model evaluation with emphasis on the importance of testing throughout the modeling process and the difficulty of defining precise criteria for model evaluation.

An admitted difficulty faced by any modeler is the possibility of a conflict of interest in the testing of a model that the same modeler may well have spent a great deal of time and effort developing. Some authors[11,12] argue that criteria for model testing should be specified before model construction begins to ensure that bias does not creep into the evaluation process. In practice, this suggestion is rarely followed in ecology, and the literature is full of models that have never been tested in any substantive way. As Hall and DeAngelis[13] point out, "the testing of the adequacy of models vis-a-vis reality seems to us to be poorly developed and often deliberately and arrogantly ignored." As the quote from Casement indicates, models sometimes take on a life of their own, obscuring the purposes for which the model was ostensibly derived. This can lead to potential misuse of the model by individuals who merely pull a model off the shelf when needed without taking care that its use is compatible with the assumptions on which it was originally based.

The main objective in this chapter is to point out the dependence of model evaluation procedures on the problem to which the model is intended to be applied. The author suggests that useful classifications of models based on application and leading to different evaluation criteria are models for theory development; models for specific, carefully delineated systems; and models for policy decisions. The discussion includes ways in which this relates to other classification schemes (e.g., the generality-realism-precision continuum of Levins[14]), and point out some of the difficulties inherent in the application of highly reductionist models based upon the behavior of individuals. Another key point concerns the general inattention of the scientific community to the criteria that policymakers use in making decisions concerning biological systems, in particular the economic, social, and political implications.

MODEL EVALUATION

A wide variety of terms are used by different authors in discussions on model evaluation, including the term *evaluation* itself. Others include testing, accuracy, verification, validation, corroboration, desirability, domain of applicability, certification, realism, tuning, and curve-fitting. This plethora, and the very different meanings assigned to the terms by different researchers, contributes to the lack of agreement on what constitutes effective evaluation. Most authors point out that evaluation is coupled to the purposes for which the model is being constructed. The variety of reasons for constructing models thus leads to differing criteria for evaluation. Several terms are defined for the purpose of this chapter.

Verification. The model behaves as intended, in that the equations correctly represent the stated assumptions. The equations are self-consistent and dimensionally correct, their analysis is error free, and any computer coding has been carried out correctly.[11,15] France and Thornley[10] use the term *testing* synonymously with verification in this sense, as does the present author. In contrast, Jeffers [16] uses *verification* to indicate that a model behaves in ways that broadly fit expectations of how the modeled system behaves, while Shugart[17] uses it to indicate that the model has been investigated to ascertain if it can be made consistent with some set of observations. In the usage of Shugart,[17], verification therefore becomes specific to a particular data set, while in that of Jeffers,[16] it is specific to the modeler.

Validation. The model behavior is in agreement with the real system it represents with respect to the specific purposes for which the model was constructed. Inherent in this is a notion of both accuracy (that is how far apart in some metric the model behavior is from some components of the real system), as well as a domain of applicability (a prescribed set of conditions over which the model is intended to apply). A variety of measures of the domain of applicability have been introduced in theory,[18] but these seem to be of little use in real applications. The metric chosen to specify model accuracy would depend on the purposes for which the model was constructed. For example, if model output consists of a time-dependent vector variable $\mathbf{X}(t) = (X_1(t), \ldots, X_n(t))$ and the data to which the model is being compared were $(x_1(t_1), \ldots, x_n(t_1)), \ldots, (x_1(t_k), \ldots, x_n(t_k))$ for some time points $(t_1, \ldots, t_k)$, then a general form of a metric would be

$$\Psi\ [X,x] = \sum_{j=1}^{k} \sum_{i=1}^{n} f_{ij}(X_i(t_j) - x_i(t_j))$$

A typical choice for the f_{ij} functions would be a time-weighted least squares such as f_{ij} $(y) = w_i \omega_j y^2$. Here, the ω_j represents relative weights in time so that if emphasis were desired on more recent times, for example, the ω_j would be increasing functions of j. If comparisons were only desired at a single time, then all the weights ω_j would be set to zero, except at that time. The w_i represents relative weights of the different output variables. The above choice of $f_{ij}(y)$ assumes multiplicative effects of the time vs variable weights, though more complicated models could be chosen.

For example, it may be important for a model to determine the occurrence of rare events, one of which could be a population bottleneck; in this case the form of the f_{ij} would not be as above but depend only upon times t_j for which $x_i(t_j)$ is below some preassigned threshold. Such a circumstance may have occurred in the case of the passenger pigeon, due to the interaction between social effects and population size. In situations with limited data, investigators have utilized a variety of Monte Carlo schemes.[19] These involve resampling randomly from a data set split into a calibration component and one used for validation.

Calibration. This refers to the use of a data set to estimate parameters of a model, resulting in model behavior that is consistent with this data set. This is also called curve fitting or model tuning and involves many areas of statistics tied to parameter estimation. Also inherent is a notion of model accuracy, although only in the restricted sense of providing agreement to a particular data set.

Corroboration. The model is in agreement with a set of data independent of that used to construct and calibrate the model. This is one aspect of validation. It is quite different from the notion of corroboration of a scientific theory.[11] The issue of differences between models constructed to elaborate a theory vs those constructed as calculation tools or for prediction has been discussed by a number of authors.[13,20,21]

Evaluation. This term comprises validation plus attention to a variety of criteria, including appropriateness to objectives, utility, plausibility, elegance, simplicity, and flexibility.[10] Therefore, no set of simply objective criteria for evaluation exists, but a number with different weightings assigned to each through the preferences of the investigator. It is the areas emphasized in the evaluation criteria that serve to differentiate the modeler, who is often primarily interested in either theory development or a particular scientific question, from the manager or politician who is in general answerable to the public for decisions influenced by the model.

EVALUATION CRITERIA FOR DIFFERENT TYPES OF MODELS

One classification scheme for models considers where along the generality-realism-precision continuum they occur,[14] because no single model is capable of completely satisfying all three criteria. At the outset of a modeling project, a decision is made regarding where on this continuum the resulting model should be focused. This is intimately tied with the purposes for which the model is being constructed, and the evaluation criteria are chosen accordingly.

Models for theory development comprise the majority of the subject of mathematical ecology, with emphasis on generality, a slight amount of realism, and typically very little precision. Evaluation criteria here are tied more to biological reasonableness, rather than to biological reality, for the objective is to produce a theory for general patterns of nature.[20] Validation here takes the form of qualitative comparisons with nature rather than quantitative comparisons. Parameters in these models are often far removed from observable biology. Far too much effort in ecology has been wasted in trying to estimate model parameters such as the coefficients in Lotka-Volterra type models. If estimable at all, these coefficients would result from complex interactions of many physiological and population-scale processes. The models serve useful qualitative roles, but trying to squeeze them into a role for which they are not suited is a wasted effort. Calibration and corroboration are not appropriate for these models, for the objective is typically to investigate how model behavior varies qualitatively over what is viewed as a reasonable parameter space.

In a number of instances, however, models developed having mainly theoretical goals have proven to be useful in extensions to particular problems with close ties to observable biology. One example includes the McKendrick-von Foerster formulation of a population structured by age, size, or physiological state. This partial differential equation model is very general in form and serves as a basis for the theory of continuously structured populations. Yet it has been shown to be quite useful as a means to couple toxicological effects on individuals to population scale phenomena (see Chapter 49). Very simplistic epidemic models have been extended in a host of ways to analyze specific instances of disease spread, with varying success,[22] and simple host-parasite models have been extended to apply to several case studies of macroparasite infection and spread.[23] Other examples include the applications of very general reaction-diffusion models to many situations of animal movement.[24] Here, applications to particular field situations typically involve discretizations of the underlying partial differential equation or random walk model being approximated.[25] Thus, theoretical models not only are important for the development of general paradigms, but often lead to more realistic extensions closely coupled to biological data. This requires a step back from the generality of the original model, however, along with an attendant increase in model complexity and size of parameter space.

Models for specific systems (e.g., a given fish stock in a particular region, or a particular mammal population in a forest) are typically tied to data sets that, though often very limited, provide some guidance for choosing possible validation criteria as well as possible model formulations.[26] It makes little sense to construct a model with detailed social structuring in the population, if there are no data available to estimate the nature of this structuring. Population models of this sort are typically based around some method of fancy bookkeeping, particularly discrete age-structured models used for wildlife populations. The assumptions about the mortality and fecundity relationships in these models are typically the weakest components, and the most difficult to validate. Nevertheless, it is these components that detail the coupling between individuals and the population-scale effects of external environmental factors.[27] Evaluation criteria, therefore, may need to be more stringent in application to some model components than to others.

Of all applications of ecological modeling, the author believes the uses in policy analysis have the poorest record of results obtained for efforts expended. A graphical scheme to illustrate the situation is shown in Figure 1. The horizontal axis corresponds to a simple bounded set of possible decisions. The solid line portrays a modeling effort producing what might be considered superb results for a particular policy decision. In this case, any particular policy chosen results in an easily understood outcome that is completely deterministic. This outcome results from a biological analysis of the situation, with higher values corresponding to stronger negative effect on the population in

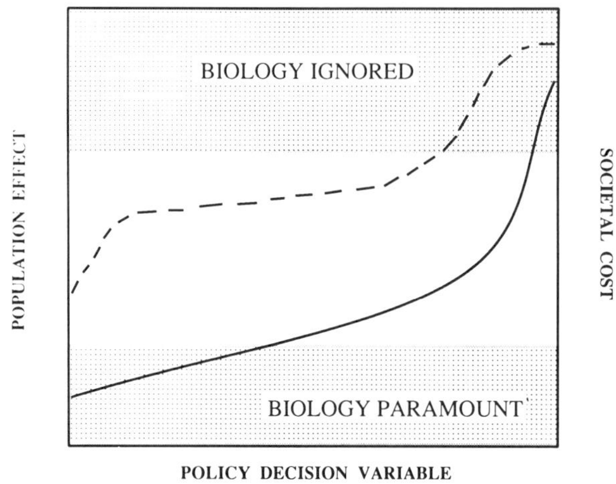

FIGURE 1. Scheme to illustrate the combination of social and political concerns (dashed curve) along with scientific criteria (solid curve) affecting a policy decision. Social cost and population effect are not on commensurate scales.

question. Thus, a rational decision may be made based upon some acceptable level of the outcome. Possible examples are population size remaining given a certain amount of habitat destruction, population size as a function of mean toxicant concentration per unit area of habitat, or fraction of suitable habitat remaining to a species after a given land-use scheme is implemented.

The difficulty with the preceding analysis is that typically quite different criteria, not included in the biological assessment of effects, are applied in making the policy decision. Figure 1 (dashed line) illustrates the effects of the policy decision on social factors such as loss or gain of jobs, expenditures to maintain the local economy, infrastructure or bureaucracy needed to carry out the decision, and the opportunity costs associated with these expenditures. These could just as well be political costs, such as the perceived political advantage to the decision-makers of carrying out a particular policy, including gain of financial support from certain constituents. The dashed curve corresponds to a particular societal cost function. The solid and dashed curves are not commensurate quantities; combining them in some way involves multiple-criteria optimization and construction of a utility function that makes assumptions about the effects of alternative combinations. Societal costs act to effectively constrain policy responses in many instances, in particular to the endpoints of the policy decision set.

Several possible scenarios are illustrated in Figure 1. The upper shaded region indicates a threshold such that if the social costs are within this area, these costs completely determine the policy choice with any possible population effects being ignored. Similarly, the lower shaded region illustrates a situation in which biological factors completely outweigh any political or social concerns, and policy choice is determined completely by biology. Cases corresponding to these situations are readily available. The recent controversy regarding the use of Alar on apples[28] illustrates that when direct effects of toxicants on humans are perceived to be large, policy decisions can be made based entirely upon social factors, independent of the biological data, which in this case indicated minuscule relative health effects on the population. As an alternative to this example of policy action being taken according to political and social factors, consider the case of the spread of resistance to antibiotics in bacterial populations endemic among humans. Here, despite the fact that many researchers feel that the widespread use of antibiotics in animal feed has fostered the development and spread of new bacterial strains with quite damaging human health effects, governmental agencies have taken very little action on the issue, except to request further studies.[29] This is one example of a common property resource overutilized for the benefit of one component of society but not rational from a total society viewpoint.[30] Although in general one may believe that an analysis of recent past policy decisions would indicate that it is situations involving direct effects on human health for which the biological impacts (the solid line in Figure 1) take precedence over the political, this may not always hold true. Certainly, direct human effects would more frequently take precedence relative to effects on other species.

Situations for which the social and political factors fall into the unshaded region are those in which our models have been least successful. In part this is due to the ignorance of social factors when structuring our models. What may appear to the biologists to be very large effects may well appear very small when viewed in a policy framework that takes account of social factors. The level of detail of our models must therefore take into account the level of indifference to details imposed not by biology, but by external social factors. Dealing with externalities is a standard difficulty in bioeconomics but with little agreement as to how to do so.[30].

Figure 1 is, of course highly simplified. The policy decision space is often not represented as a single variable, and multiple biological, social, and political criteria may be applied. The underlying view of Figure 1 assumes the world is deterministic, while often we can do no more than specify outcomes with certain probabilities, or in a mean sense. Probabilistic explanations introduce difficulties both in the analysis of models as well as in their explanation to policymakers. The view of Figure 1 is also static, ignoring both the dynamics of the system being modeled as well as the adaptive nature of policy decisions.

Key data are typically not available for many resource management problems. Uncertainties in extrapolation from limited data sets can lead to potentially large errors in predicting ecosystem response to external inputs such as toxicants.[31] In such situations, one goal of a management policy could be to explicitly attempt to force the system into very contrasting outcomes. A policy that leads to large deviations from a mean response of a system can be highly useful because it provides a method to evaluate the applicability of a model under very different circumstances. This process of adaptive management is a particularly useful means to quickly increase the accuracy and utility of models in situations with very limited data.[32,33] Another aspect of the information effects of policies is the potential for a policy to change the nature of currently operating social and political factors, primarily through education.

REDUCTIONISM AND ENVIRONMENTAL POLICY

A natural tendency in applying models derived for general theory to policy concerns, or to specific management situations, is to add complexity to increase potential precision and allow for calibration and corroboration, through the introduction of detailed submodels. Many of the detailed single species and multispecies models take this approach. It is one alternative to strictly empirically based models such as those for habitat indices, or statistical models coupled to a particular database. As noted above, this has led to useful results in several situations. It also requires either a much larger database or many more assumptions about particular model forms. These models have limitations, however, because they typically amalgamate all individuals within some class (age, size, physiological state, etc.) and assume uniformity within this class. Situations with strong neighborhood effects, or whose assumptions of uniformity within a class break down (e.g., population sizes are so low that a class consists of few individuals) are not readily handled by these extensions of theoretical models. Thus, there are a number of ongoing attempts to model populations as interacting individuals, tracking each separately in a simulation format.[34,35]

The new individual-based models have great appeal for a number of reasons. First, they allow explicit behavioral rules to be specified at the individual level and do not require ad hoc assumptions about the effect of these behaviors at the population scale. Second, such models are relatively easy to construct once the behavioral rules are specified, since essentially dynamic bookkeeping is done from that point. (It should be noted that efficient algorithms to handle this for simulations involving large numbers of individuals are not easily obtained.) Third, they naturally provide a means to handle individual interactions and neighborhood effects, and Monte Carlo methods provide for analysis of situations with small population sizes.

The problem is that these models quickly grow to require enormous numbers of assumptions about individual behavior for which the available evidence is limited. They are numerical, not qualitative, so determining the effects of a poorly understood assumption requires numerous simulations. Validation of the submodels is precluded without a vast amount of field observations of individual behavior. A saving grace of these models is that they are relatively easy to explain to managers since the key objects are individual organisms rather than more abstract population mortality and fecundity schedules. This implies a great future for these models in policy applications, because a key limitation in the use of models by managers may be the ability to understand, at some level, how the models are constructed. Understanding on the part of the manager may lead to more ready acceptance of the model results as well as feedback from the manager in the decision process.

The author would argue, however, that policy applications of these models are generally premature, mainly due to the enormous number of poorly understood assumptions from which the models are constructed. The parameter space may be so rich that essentially any result desired can be obtained through judicious choices of the parameters. Because the models require large Monte Carlo simulations, sensitivity analyses of these models are quite computer intensive. To determine how sensitive the model output is to a parameter, one must essentially conduct a Monte Carlo analysis on a Monte Carlo simulation by varying the parameter within some range and observing how model output is affected.

It is the author's belief that these highly reductionist models will be most useful in a policy sense if they provide means to suggest appropriate large scale macrodescriptors of systems. These macrodescriptors should be robust so that the outcomes are not highly sensitive to the details of individual behavioral assumptions. The best way to accomplish this is using a top-down approach. Here, one starts with the manager and attempts to ascertain levels of indifference to model outcomes. This approach provides a means to scale the amount of reductionism necessary to meet the goals of the manager, and reduces wasted effort dealing with minute details of biological processes, if these are second or third order effects in the scheme of acting social and political constraints.

SUMMARY

Models are not a panacea. In analyzing policy issues, models have two main uses: to allow investigation of system behavior under alternative assumptions when information is limited about key processes affecting the system; and (2) to provide a mechanism to predict system behavior under alternative management policies, possibly coupling this with criteria to determine optimal policy choice. Unless information is very detailed about a particular natural system, the investigation in the first use will be necessary before model results can be reliably applied to carry out the second use.

Just as modeling is an iterative process, so is social policy choice. The political process is often one of compromise that involves trade-offs for certain constituencies, made in an iterative manner. The objective is to couple this iteration with the iteration of the modeling process—not with a single model, or even a single class of models, but rather with models that are most effective at each stage of the policy decision process. For example, the choice of national land-use plans would require models on much larger, regional, or global scales than the implementation of that policy at smaller, local scales. Managing particular forest systems according to local guidelines, taking account of local wildlife, recreational use, and wood product production, would require finer detail in any model being utilized. When averaged over large regions, these local guidelines would conform to the large-scale policy, but may on a small scale be quite diverse.

Models used at these diverse spatial scales would be quite different in form. The detail necessary at smaller scales can come from reductionist approaches, including individual-based models that mimic the behavior of numerous individual organisms to allow prediction of population and community behavior. On larger scales, the hope is that these reductionist approaches will suggest the form for less detailed models of sufficient realism to be applied under very differing policy alternatives. There has been little experience to date in developing large-scale models in this manner, yet they are a critical component of regional and global planning. If their development is also closely coupled to input from those responsible for policy decisions, our detailed knowledge of local biology will have been successfully integrated with the constraints imposed by social factors to produce a rational method of environmental social choice.

REFERENCES

1. Shugart, H. H., and R. V. O'Neill, *Systems Ecology*, Dowden, Hutchison & Ross, Stroudsburg, PA, 1979.
2. Maki, D. P., and M. Thompson, *Mathematical Models and Applications*, Prentice-Hall, Englewood Cliffs, NJ, 1973.
3. Beltrami, E., *Mathematics for Dynamic Modeling*, Academic Press, Orlando, FL, 1987.
4. Rose, M. R., *Quantitative Ecological Theory: An Introduction to Basic Models*, Johns Hopkins University Press, Baltimore, 1987.
5. Jeffries, C., *Mathematical Modeling in Ecology: A Workbook for Students*, Birkhauser, Boston, 1989.
6. Giordano, F. R., and M. D. Weir, *A First Course in Mathematical Modeling*, Brooks/Cole, Monterey, CA, 1985.
7. Edelstein-Keshet, L., *Mathematical Models in Biology*, Random House, New York, 1988.
8. Roberts, F. S., *Discrete Mathematical Models*, Prentice-Hall, Englewood Cliffs, NJ, 1976.
9. Mesterton-Gibbons, M., *A Concrete Approach to Mathematical Modeling*, Addison-Wesley, Redwood City, CA, 1989.
10. France, J., and J. H. M. Thornley, *Mathematical Models in Agriculture*, Butterworths, London, 1984.

11. Caswell, H., The validation problem, in *Systems Analysis and Simulation in Ecology*, Vol. 4. B.C. Patten, Ed., Academic Press, New York, 1976, pp. 313-325.
12. Overton, W. S., A strategy of model construction, in *Ecosystem Modeling in Theory and Practice: An Introduction with Case Histories*, C. A. S. Hall and J. W. Day, Eds., John Wiley & Sons, New York, 1977, chap. 3.
13. Hall, C. A. S., and D. L. DeAngelis, Models in ecology: paradigms found or paradigms lost?, *Bull. Ecol. Soc. Am.*, 66, 339-346, 1985.
14. Levins, R., *Evolution in Changing Environments*, Princeton University Press, Princeton, NJ, 1968.
15. Levis, A. H., and I. S. Levis, Toward a conceptual framework for policy analysis modeling, in *System Theory Applications to Agricultural Modeling*, A. H. Levis and C. L. Quance, Eds., USDA Publ. ESCS-07, U.S. Department of Agriculture, Washington, DC, 1978.
16. Jeffers, J. N. R., *Modelling*, Chapman & Hall, London, 1982.
17. Shugart, H. H., *A Theory of Forest Dynamics: The Ecological Implications of Forest Succession Models*, Springer-Verlag, New York, 1984.
18. Mankin, J. B., R. V. O'Neill, H. H. Shugart and B. W. Rust, The importance of validation in ecosystem models, in *New Directions in the Analysis of Ecological Systems, Part 1*, Simulation Councils Proceedings Series, Vol. 5, G. S. Innis, Ed., Simulation Councils, LaJolla, CA, 1977, pp. 63-71.
19. Wegener, K. R., and R. F. Malone, A methodology for determining the predictive accuracy of mathematical models from a limited data base, in Analysis of Ecological Systems: State of the Art in Ecological Modeling, Developments in Environmental Modeling, Vol. 5, W. K. Lauenroth, G. V. Skogerboe and M. Flug, Eds., Elsevier, Amsterdam, 1983, pp. 75-80.
20. Levin, S. A., Mathematics, ecology, ornithology, *Auk*, 97, 422-425, 1981.
21. Loehle, C., Evaluation of theories and calculation tools in ecology, *Ecol. Model.*, 19, 239-247, 1983.
22. Aron, J. L., Simple versus complex epidemiological models, in *Applied Mathematical Ecology*, S.A. Levin, T. G. Hallam and L. J. Gross, Eds., Springer-Verlag, Berlin, 1989, pp. 176-192.
23. Dobson, A. P., The population biology of parasitic helminths in animal populations, in *Applied Mathematical Ecology*, S. A. Levin, T. G. Hallam and L. J. Gross, Eds., Springer-Verlag, Berlin, 1989, pp. 145-175.
24. Okubo, A., *Diffusion and Ecological Problems: Mathematical Models*, Springer-Verlag, Berlin, 1980.
25. Kareiva, P., Predator-prey dynamics in spatially structured populations: manipulating dispersal in a Coccinellid-aphid interaction, *Lect. Notes Biomath.*, 54, 368-389, 1984.
26. Cale, W. G., R. V. O'Neill and H. H. Shugart, Development and application of desirable ecological models, *Ecol. Model.*, 18, 171-186, 1983.
27. Emlen, J. M., Terrestrial population models for ecological risk assessment: a state-of-the-art review, *Environ. Tox. Chem.*, 8, 831-842, 1989.
28. Marshall, E., 1991, A is for apple, Alar, and . . . alarmist?, *Science*, 254, 20-22, 1991.
29. Wright, K., Bad news bacteria, *Science*, 249, 22-24, 1990.
30. McKelvey, R., Common property and the conservation of natural resources, in *Applied Mathematical Ecology*, S. A. Levin, T. G. Hallam and L. J. Gross, Eds., Springer-Verlag, Berlin, 1989, pp. 58-80.
31. O'Neill, R. V., R. H. Gardner, L. W. Barnthouse, G. W. Suter, S. G. Hildebrand and C. W. Gehrs, Ecosystem risk analysis: a new methodology, *Environ. Toxicol. Chem.*, 1, 167-177, 1982.
32. Ludwig, D., and R. Hilborn, Adaptive probing strategies for age-structured fish stocks, *Can. J. Fish. Aquat. Sci.*, 40, 559-569, 1983.
33. Walters, C., *Adaptive Management of Renewable Resources*, Macmillan, New York, 1986.
34. Huston, M., D. L. DeAngelis and W. M. Post, New computer models unify ecological theory, *Bioscience*, 38, 682-691, 1988.
35. DeAngelis, D. L., and L. J. Gross., *Individual-Based Models and Approaches in Ecology*, Chapman and Hall, New York, 1992.

48

The Environmental Protection Agency's Avian Pesticide Assessment Model

Ed Fite

ABSTRACT

The Federal Insecticide, Fungicide, Rodenticide Act (FIFRA), as amended, specifies that for registration or continued registration of a product, the United States Environmental Protection Agency (U.S. EPA) must determine that the pesticide will not cause unreasonable adverse effects on the environment. To determine this for nontarget wildlife species, the EPA has utilized a tiered assessment model that has remained relatively unchanged for the past 15 years. The model consists of a tiered approach that initially compares results of single species laboratory toxicity tests to environmental exposure estimates. If exposure estimates are below the threshold levels that elicit a biological response in laboratory studies, it is presumed that adverse ecological effects on avian species are minor. For pesticides that exceed the threshold levels, further testing under field conditions is needed to demonstrate that under typical conditions the likelihood of adverse effects is minimal.

The model remains relatively untested for chemicals that do not exceed the established criteria. While safety factors were applied in establishing threshold levels, uncertainties associated with extrapolating from a few laboratory toxicity tests to widely varying field conditions encompassing numerous species suggest the need for further refinement of the model.

KEY WORDS

wildlife, avian, pesticide assessment model, pesticide registration, field tests, laboratory toxicity tests

INTRODUCTION

Passage of the 1972 amendments to the FIFRA was part of a wave of environmental legislation that radically overhauled federal environmental regulatory authority. While federal regulation of pesticides began in 1910 and was significantly expanded in 1947, the 1972 amendments restructured the federal pesticide regulatory scheme and redefined its thrust. FIFRA was changed from a labeling law into an environmental protection statute with a primary purpose of ensuring that pesticide use was subject to a thorough environmental and human health hazard review.[1]

In response to the amended Act, and guided by administrative and judicial interpretations of the basic statutory standard, the U.S. EPA developed an avian pesticide assessment model that has remained relatively unchanged since 1975. The model consists of a series of tests arranged in a hierarchical or tiered system, progressing from basic laboratory toxicity tests to applied field studies. The initial evaluation of a pesticide compares the results of avian laboratory toxicity tests and estimated environmental concentrations with established screening criteria. In addition, when information (i.e., labeling or environmental chemistry data) indicates chemical persistence, repeat application, or the potential for bioaccumulation, chronic tests must be considered. If estimated environmental concentrations on avian ingestible items (food, grit) are below the established screening criteria, it is generally presumed that adverse ecological effects to nontarget avian species will be minor and no further testing is required. However, for those pesticides for which environmental concentrations exceed the criteria, it is presumed that adverse effects are significant and further laboratory and/or field studies are required. The purpose of the field study is either to refute the presumption of hazard or to provide some quantification of the effects that may occur.

The underlying basis for this model and rationale for its design were initially presented in the 1975 regulations.[1] Since then several publications[2,3] have addressed various aspects of the EPA assessment model, with Urban and Cook[4] giving a fairly thorough presentation. Nevertheless, it seems appropriate to reexamine the model in the context of what occured during the 1990 workshop and discuss some of its assumptions and uncertainties from a different perspective.

STATUTORY STANDARD

The avian pesticide assessment model used by the EPA is a product of FIFRA and the administrative and judicial interpretations of the basic statutory standard. While an in-depth review of FIFRA and guiding interpretations is beyond the scope of this chapter, a brief review of some of its major requirements is appropriate to aid in understanding the design of the model.

Under FIFRA, the basic environmental standard for regulatory determination is "unreasonable adverse effects on the environment." The term is defined in FIFRA to mean:

> Any unreasonable risk to man or the environment taking into account the economic, social, and environmental costs and benefits of use of any pesticide.

FIFRA, therefore, requires that the final determination be a risk/benefit balance. FIFRA further requires the EPA to establish regulatory criteria for registration, classification, and cancellation—criteria intended as screening mechanisms to tentatively determine which pesticides should be denied, registered, canceled, or restricted. The regulations provide that if a pesticide meets or exceeds these criteria a substantial question of safety exists and the burden of proof is on the registrant to convince the EPA that the hazard predicted by the acute or chronic screening criteria will not occur. In establishing criteria to prevent unreasonable adverse effects, FIFRA requires a margin of safety sufficiently large to ensure that risk of injury to nontarget wildlife is small to negligible and that laboratory data are sufficient to provide the basis for the criteria. FIFRA also requires that the EPA specify the data required to demonstrate that the hazard predicted by the acute or chronic screening criteria will not occur.[1,3]

Table 1. U.S. Environmental Protection Agency Testing Tiers for Avian Species

Test	When Needed
Tier 1	
Avian acute oral LD_{50} (upland gamebird or waterfowl)	All pesticides proposed for registration
Avian dietary LC_{50} (upland gamebird and waterfowl)	
Tier 2	
Avian reproduction studies (upland gamebird and waterfowl)	Labeling or environmental chemistry data indicate repeated or continued exposure or the potential for bioaccumulation
Special studies if needed	
Tier 3	
Simulated and actual field testing	EEC $\geq$ 1/5 LC_{50} or for granular products $LD_{50}/ft^2 > 1$

MODEL OVERVIEW

The avian pesticide assessment model is arranged sequentially in tiers, with each tier requiring more complex information than that of the previous tier (Table 1). The first two tiers involve toxicity and exposure data. Based on the premise that hazard is a function of toxicity and exposure, criteria have been established at each tier for guiding decisions on advancements to higher tiers (Table 1). It is assumed that if the toxicity and estimated exposure comparison is below the screening criteria, the potential to affect nontarget wildlife is low. However, if toxicity and exposure estimates equal or exceed the screening criteria a substantial question of safety is presumed to exist and additional data are needed to demonstrate the method of use or other factors mitigating the indicated hazard.

For discussion, the model can be separated into four parts: exposure estimates, toxicity data, criteria for comparing exposure and toxicity data, and the final tier, field tests (Figure 1).

EXPOSURE ESTIMATES

The model assumes that the major route of pesticide exposure to avian species is contaminated food sources. In the case of granular formulations, direct exposure through ingestion of granules is also considered. Residue estimates on avian food sources are based on information developed by Hoerger and Kenaga.[5] Based on their review of the existing literature and tolerance values published by the Department of Agriculture (USDA), they estimated "upper limit" and "typical" residue values for seven categories of vegetation that could be food sources for non-target wildlife species (Table 2). Residue values were estimated for immediately after and 6 weeks after application. The upper limit, considered the "worst" situation that may occur, was based on the higher residues from each reference, while the typical residue estimates were based on an average of the highest residues for all pesticide studied on the various plant categories. Kenaga[6] suggested that for insects, also a major source of food for many species and life stages of birds, residues are probably on the same order as those

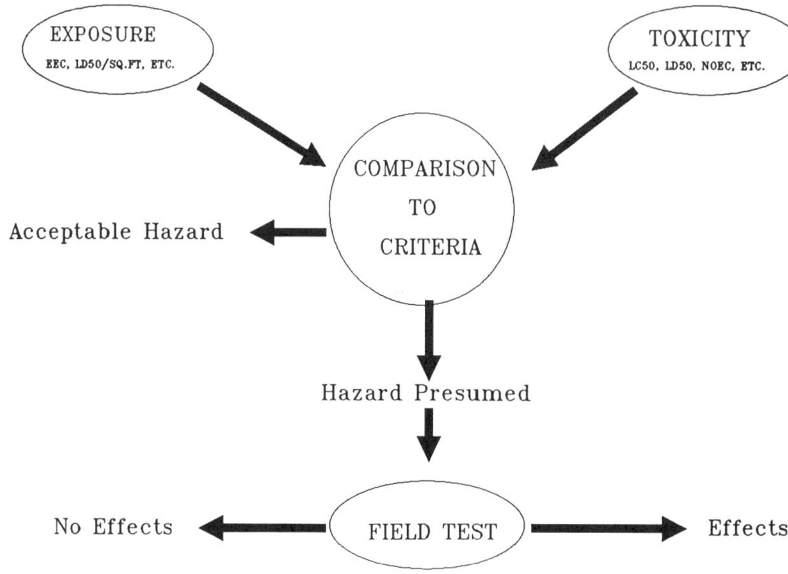

FIGURE 1. Avian pesticide assessment model.

on plants of similar surface area to mass ratios. For small insects "forage" estimates and for larger insects "pods containing seeds" estimates were recommended.

For pesticides applied as granular formulations, a different approach to estimating exposure is used due to the potential for direct ingestion of the formulated product. Both field studies and wildlife kill incident reports involving granular pesticide products have shown that direct consumption of granules occurs.[7] Granules may be ingested directly by birds when foraging for seed or grit at or below the soil surface on treated areas. Birds may also ingest granules adhered to the surface of invertebrate prey items as well as through ingestion of water or food sources contaminated with the pesticide. Because of these somewhat unique routes of exposure, toxicant per unit surface area adjusted for incorporation was developed as an indication of exposure of granular products to nontarget species. The major basis for using toxicant per unit area was a

Table 2. Maximum Expected Residues and Typical Residues of Pesticides on Differing Categories of Vegetation Types

	Residue (ppm) Based on a Pesticide Dosage of 1 lb/Acre			
	Immediately after Application		6 Weeks after Application	
Plant Category	Upper Limit	Typical Limit	Upper Limit	Typical Limit
Range grass	240	125	30	5
Grass	110	92	20	1–5
Leaves and leafy crops	125	35	20	<1
Forage crops	58	33	1.0	<1
Pods containing seeds	12	3	1.5	<1.0
Grain	10	3	1.5	<1.0
Fruit	7	1.5	1.5	<0.2

Source: From Hoerger and Kenaga.[5]

review article by DeWitt,[8] who suggested that effects on birds were correlated with the amount of toxicant applied per square foot. Results of acute oral tests are used for comparison because the mode of toxicant entry is more akin to an acute oral exposure than a dietary exposure.

TOXICITY TESTS

Avian toxicological data used in the assessment model are separated into acute oral, dietary, and chronic studies. Acute oral and dietary tests evaluate the quantity of pesticide required to affect avian species. Mortality is the major endpoint; however, other effects are also noted. Results are usually reported as the dosage or concentration that produces 50% mortality in test subjects (LD_{50} or LC_{50}). Chronic toxicity tests use sublethal pesticide concentrations over a longer exposure period, with the intent of evaluating effects on reproduction. Results usually are reported as no observable effect concentration and a lowest observable effect concentration.

Tier 1 tests (Table 1) are acute oral and dietary single species toxicity tests usually using either northern bobwhite or mallard as test species. Several reasons have been advanced for the selection of these species, including:

- demonstrated sensitivity to pesticides
- a dose-response to a variety of pesticides
- ecological significance
- aesthetic and economic importance
- ready availability for testing
- adaptability to laboratory conditions

However, as Tucker and Leitzke[9] point out, "If we were all quite honest about it, we probably admit in the final analysis that our choices of test species are based primarily upon convenience."

These avian toxicity tests follow standardized protocols. In the acute oral tests, several geometrically arranged dosages are administered in milligrams of active ingredient (a.i.) per kilogram of body weight to test birds. In the dietary test, the toxicant is mixed in the feed and fed in a geometric series in milligrams a.i. per kilogram of feed to the test animals for 5 days. The data are analyzed statistically (probit analysis). The results are usually a linear dose-response relationship that can be used to estimate the dosage in the acute oral test (LD_{50}) and the concentration in the dietary tests (LC_{50}) that in both cases produce 50% mortality in the test subjects. Generally, the 95% confidence limits and the slope are also reported with these estimates. These dose-response relationships can also be used to estimate dosages/concentrations that result in other percentages of mortality, such as 10% mortality in the dietary test (LC10), but they are less precise than the corresponding median lethal estimate.

Chronic studies (tier 2) are required when labeling directions or environmental chemistry data indicate that avian species may be subject to repeated or continuous exposure to the pesticide, that the pesticide may be stable in the environment, or that it may be stored or accumulated in plant or animal tissues. These tests expose usually a minimum of two avian species (mallard and bobwhite) to sublethal dietary concentrations for an extended period: a minimum of 10 weeks prior to onset of egg production and continuing through egg laying. Dosage levels (a minimum of two) for these tests are based on estimated field residue exposure concentrations. Data are recorded and evaluated for several reproduction parameters, body weight, food consumption, and toxicological symptoms or clinical observations that indicate potential deleterious sublethal effects influencing development and survival in the wild.

SCREENING CRITERIA

The screening criteria were developed to identify pesticides requiring additional scrutiny to determine whether they should be registered or, if registered, whether they should be classified for general or restricted use. Originally two sets of numerical criteria were established: criteria for classification and criteria for "rebuttable presumption against registration." These are used to separate pesticides into three categories: pesticides with acceptable hazard, pesticides with unacceptable hazard that may be mitigated by labeling, and pesticides with unacceptable hazard (Table 3). Although some of the criteria originally identified in the 1975 regulations have been modified in subsequent revisions of the regulations and redesignated criteria for "special review,"[3] they remain implicitly unchanged. As such, they are still appropriate for use in the assessment model to identify pesticides that must be closely scrutinized.

For pesticides in the category of acceptable hazard, it is presumed that an adequate safety margin exists and minimal impact to nontarget organisms will occur when the pesticide is applied by untrained users. The safety margins were established based on a review of avian dose-response data. From this it was estimated that a dose or exposure ten times lower than the LD_{50} or LC_{50} would yield a mortality of about 0.01% under typical slope conditions and 4.0% under minimum slope conditions. Further, a dose-response five times lower than the LD_{50} or LC_{50} would yield a mortality of about 0.1% under typical slope conditions and 10% under minimum slope conditions. Based on these figures, a safety factor of 5 was selected for nonendangered species and 10 for the endangered.[1] A further margin of safety is inherent in that these criteria use the maximum residues immediately after application.

For granular products the amount of toxicant per unit area is a more appropriate means to determine the hazard to nontarget wildlife. As already mentioned, DeWitt[8] related quantities of pesticide toxicant ingested by birds to quantities of toxicant deposited per unit area. Using several laboratory and field studies he concluded that losses of birds may be expected if the quantity of toxicant per square foot equals or exceeds the quantity causing death of northern bobwhite in short-term feeding tests. Tucker[10] reported that field kills of birds have occurred in many instances when the amount of pesticide applied per acre exceeded 50,000 mallard LD_{50} per acre; this is comparable to 1.1 LD_{50} values per square foot. Felthousen[11] proposed EPA criteria for granular products related to the amount of toxicant per square foot available to an animal on the assumption that wildlife mortality is likely if the quantity of toxicant per square foot equals or exceeds the LD_{50} or LC_{50}.

Chronic criteria are less explicit. The number of uncontrolled variables found in the toxicity studies as well as in the field conditions for pesticide use increases the uncertainty of establishing protective criteria. The initial criteria for evaluating a pesticide's potential chronic hazard is comparison of the no observable effect level (NOEL) found in the avian chronic test to maximum expected residues immediately following application. Pesticide levels that do not exceed the chronic criteria (NOEL) are presumed to present minimal chronic hazard to nontarget avian species.

For those pesticides for which estimated environmental concentrations exceed the criteria, field studies (tier 3) may be needed to negate or define the potential hazard. Several factors, or appropriate combinations of these factors, are also considered in addition to the criteria and are as follows:

Table 3. Screening Criteria Used to Identify Pesticides that Require Additional Scrutiny

			Unacceptable Hazard	
	Acceptable Hazard	Unacceptable Hazard that May Be Mitigated by Labeling	Non-Endangered Species	Endangered Species
Acute toxicity	$EEC < 1/5\ LC_{50}$	$1/5\ LC_{50} \leq EEC < LC_{50}$	$EEC \geq LC_{50}$	$EEC \geq 1/10\ LC_{50}$
Granular products	$LD_{50}/ft^2 < 1$	$LD_{50}/ft^2 \geq 1$		
Chronic toxicity	EEC < chronic NOEL		EEC ≥ chronic effect levels	EEC ≥ chronic effect levels

- The chemical properties of the pesticide (e.g., persistence, conversion to toxic metabolites, retention on food, and repellency);
- Intended use pattern (e.g., treated habitats, expected presence of endangered and other species, extent of use area, number of applications and treatment intervals);
- Margin between the estimated environmental concentration (EEC) and the criteria; and
- Dose-response relationship noted in laboratory tests, including dose-response slope, time of mortality, reproductive effects, signs of intoxication, abnormal behavior, and gross pathological changes.

The final determination of whether a field study will be requested is based on the weight of evidence factoring in all pertinent information. For some pesticides there may be sufficient data and information to conclude the pesticide presents sufficiently high hazard without the need for field data. An example of this is when the EEC is several multiples of the LD_{50} or LC_{50}.

FIELD STUDIES

The purposes of the field study are to either refute the assumption that hazard to wildlife will occur under conditions of actual use of the pesticide or to provide some quantification of the adverse effects that may occur. This twofold purpose exists because the FIFRA requirement to determine unreasonable adverse effects necessitates a risk-benefit analysis. Thus, if the presumption of unacceptable effects cannot be refuted, it is desirable to refine the risk component of the analysis by quantifying the adverse effects of actual pesticide through field studies.

The basic objective of field studies is to evaluate survival and/or reproductive success of nontarget avian species under actual conditions of pesticide use. Potential influences that pesticides can have on avian species include:

- Direct poisoning and death by ingestion, dermal exposure, and/or inhalation;
- Sublethal toxic effects indirectly causing death by reducing resistance to environmental stresses such as diseases, weather, or predators;
- Altered behavior such as abandonment of nests or young, change in parental care, or reduction in food consumption;
- Reduced food resources or alteration of habitat; or
- Lowered productivity through fewer eggs laid, reduced clutch size, or reduced egg fertility.

These effects can manifest themselves in avian species through reduced survival and lower reproductive success. If the results of a field study show that actual use of a pesticide does not affect survival or reproductive success or that only minor changes occur, it seems reasonable to conclude that the use of the chemical will not seriously impact wildlife. Further, if a field study quantifies survival and reproductive effects, one may be able to make reasonable projections on the meaning of the effects to a given nontarget species by using available natural history information about that species and basic theories of population dynamics.[12]

Field studies have evolved into two types, screening and definitive. The objective of the screening field study is to evaluate whether impact is occurring, while that of the definitive study is to estimate the magnitude of the impact. As with laboratory studies, field studies must be done with appropriate experimental designs and replication of treatments.

ASSUMPTIONS AND UNCERTAINTIES

The theoretical basis for the avian pesticide assessment model appears reasonable: hazard is a function of toxicity and exposure. By applying safety factors to account for the variability in the functional relationship, one can identify pesticides that require additional scrutiny. However, there are several uncertainties about the assumptions and data used in the model that need to be understood.

Toxicity Data

Extrapolating toxicity data to predict the potential hazard of pesticides to avian species has limitations. Laboratory toxicity tests in themselves have a relatively high uncertainty. Some insight into this uncertainty can be gained by examining a compendium of toxicity values for avian species,[13,14] often showing severalfold differences in reported toxicity values between tests for the same chemical and species and even larger differences in sensitivity between species. For example, Hill et al.[13] reported the LC_{50} of parathion for two separate tests to mallards as 76 ppm (95% confidence limit [CL] 61 to 93) and 275 (95% CL 183 to 373) over a threefold difference. For phosphamidon, Hill et al.[13] reported the LC_{50} for the bobwhite as 24 ppm (95% CL 10 to 37) and for the mallard as 712 ppm (95% CL 558 to 887), approximately a 30-fold difference. The model accounts for this variability to some extent in tier 1 when considering data from two dietary toxicity studies on two species by using the most sensitive to evaluate the potential for hazard.

Some of the variability in results may be attributed to differences in test conditions used between tests. The LC_{50} itself is only an estimate, with variability. The 95% CLs are usually reported, but do not give information about the precision of the median lethal dose or concentration estimate. These limits define an interval such that if all possible replicate 95% confidence intervals (CI) were determined under the same conditions, 95% of them would include the true LC_{50} of the population. The LC_{50} and its 95% CLs describe the distribution of the susceptibilities of the individual test organisms, but it gives no indication of the reproducibility or repeatability of the test. To obtain the precision of the estimated median lethal dose or concentration, replicated tests must be conducted.[15]

Further uncertainty results from extrapolating from the laboratory to the field. Laboratory tests are standardized in order to allow comparisons of results. However, these idealized test methods fail to show the effects of the variables that can greatly influence toxicity under field conditions. These variables include exposure duration, sex, age, nutritional status, diet, size, activity periods, seasonal variation in temperature and breeding conditions as well as other physiological and behavioral variables.[9] Unfortunately, while completed research identified these factors as potential influences on the sensitivity of species to toxicants under field conditions, the functional relationship between laboratory toxicity data and field effects remains poorly defined.

The model accounts for these uncertainties to some extent by requiring field studies for pesticides that exceed the hazard criteria. In contrast, pesticides that do not exceed the hazard criteria are assumed to have sufficient safety factors to minimize the likelihood of hazard and no additional data are generated. Given the poorly defined relationship between laboratory toxicity data and field effects, however, safety factors may be minimal. One fifth the LC_{50}, the initial hazard criterion, accounts for the variability only in sensitivity of test animals for a typical slope in a single test.[1] The variability in the estimated susceptibilities, the CLs, are not addressed, nor are the additional factors mentioned above. A random sample ($N = 10$) of CL taken from the toxicity compen-

Table 4. Exposure Safety Margins Estimated from the Difference between the Upper Limit and Typical Limit

Plant Category	Residue (ppm) Based on a Pesticide Dosage of 1 lb/Acre: Immediately after Application: Upper Limit	Typical Limit	Safety Factor
Range grass	240	125	1.9
Grass	110	92	1.2
Leaves and leafy crops	125	35	3.6
Forage crops	58	33	1.8
Pods containing seeds	12	3	4.0
Grain	10	3	3.3
Fruit	7	1.5	4.7

dium prepared by Smith[14] showed the average range of CLs to be 60% of the estimated LD_{50}, or an average variability of ~ ±33% of the estimated LD_{50} from the individual samples. This suggests that a dose-response five times lower than the LD_{50} could lead to mortalities greater than the 0.1 to 10% suggested in the 1975 regulation and, given the other biological and environmental variables, that further research is needed to refine these safety factors.

Exposure Estimates

The use of maximum residues immediately after application as the exposure measurement in the model may compensate to some extent for the uncertainties associated with toxicity extrapolation. Hoerger and Kenaga[5] emphasize that >95% of the actual residue values considered in the development of residues on avian food sources fell below the cited upper limit values. Further, they indicated the upper limits were weighted toward extreme residue levels, and for this reason more typical residue values were needed to represent the majority of pesticide-cultural practices, the typical limit values in Table 2. If the safety margins are assumed to be the difference between the upper limit and the typical limit, a safety factor can be calculated (Table 4). Depending on the vegetation type the upper limit overestimates the typical limit from 1.1 to 4.6 times.

Scant information is available to evaluate the adequacy of these safety factors of the screening criteria. For pesticides that have exceeded the screening criteria approximately 25 field studies have been completed and all have shown acute effects, several with relatively high numbers of dead birds (over 100). These results provide limited insight into the adequacy of the safety factors in that these pesticides exceeded the hazard criteria by relatively large margins.

As mentioned previously, data are limited for pesticides that do not exceed the hazard criteria. Fenitrothion, used at 0.20 lb a.i./acre for spruce budworm control, is one of the exceptions. Based on Hoerger and Kenaga,[5] fenitrothion used at 0.20 lb a.i./acre would result in upper limit residues ranging from 1.4 to 48 ppm, just exceeding the screening criteria of 31.6 ppm, one fifth the LC_{50} of fenitrothion to bobwhite quail. LC_{50} of fenitrothion to bobwhite quail is 158 ppm (95% CL 135 to 183) and to mallard is 2485 (95% CL 1693 to 3985).[13] Residues estimated on insects and seeds, major food sources for many small passerines in areas of spruce budworm control, range from approximately 2 to 12 ppm, well below the screening criteria. Yet, substantial bird mortalities, predominately small passerines, have been reported in several investigations of operational spraying of fenitrothion.[16]

Dimethoate use on alfalfa at 0.5 lb a.i./acre provides another example for the potential of false negatives from the model; i.e., concluding minimal effects when additional scrutiny is needed. At the above use rate based on Hoerger and Kenaga[5] the upper limit dimethoate residues range from approximately 3.5 to 120 ppm on avian food sources, with residue estimates on alfalfa, a forage crop, of 29 ppm. If the estimated residues on the alfalfa are assumed to provide an estimate of exposure, the screening criteria of 66 ppm, one fifth the LC_{50} for ring-necked pheasants of 332 ppm (95% CL 293 to 376),[13] is not exceeded. Nevertheless, following an application of dimethoate to alfalfa fields in Idaho, over 60 sage grouse were found dead within 11 days.[17]

While the model would identify these chemicals for further scrutiny if the maximum upper limit residues are used, if food preferences of species associated with these uses and the habitats are considered, errant indications of effects would be made. These instances with dimethoate and fenitrothion, coupled with no false positives from field studies (i.e., all field studies have shown acute effects), suggest that further refinement of the assessment model may be needed to assure that pesticides requiring additional scrutiny are identified.

In conclusion, uncertainties associated with the present avian pesticide assessment model must be considered for improvement in assessment methodologies to be made. Developing more sophisticated models may present no serious conceptual problem.[18] There is, however, a great need to better understand the functional relationship between laboratory toxicity data and field effects, as well as a need to identify and quantify the major variables, both biological and environmental, that influence the hazard of pesticides to avian species.

ACKNOWLEDGMENTS

I thank J. Akerman, L. Touart, R. Felthousen, and N. Cook for reviewing the manuscript.

REFERENCES

1. U.S. Environmental Protection Agency, Regulations for the Enforcement of FIFRA, Fed. Regist., 40 CFR (129) Part 162, 1975.
2. U.S. Environmental Protection Agency, Data Requirements for Pesticide Registration, Fed. Regist., 40 CFR Part 158, 1984.
3. U.S. Environmental Protection Agency, Special Review of Pesticides: Criteria and Procedures, Final Rule, *Fed. Regist.*, 40 CFR (229) Parts 154, 162, and 172, 1985.
4. Urban, D. J., and N. J. Cook, Hazard Evaluation Division, Standard Evaluation Procedure: Ecological Risk Assessment, EPA 540/9-95-001. U.S. Environmental Protection Agency, Washington, DC, 1986.
5. Hoerger, F. D., and E. E. Kenaga, Pesticide residues on plants correlation of representative data as a basis for estimation of their magnitude in the environment, in *Environmental Quality*, Vol. 1, F. Coulston and F. Korte, Ed., Academic Press, New York, 1972, pp. 9-28.
6. Kenaga, E. E., Factors to be considered in the evaluation of the toxicity of pesticides to birds in their environment, in *Environmental Quality and Safety*, Vol. 2, Academic Press, New York, 1973, pp. 166-181.

7. Flickenger, E., K. A. King, W. F. Stout and M. M. Mohn, Wildlife hazards from Furadan 3G applications to rice in Texas, *J. Wildl. Manage.*, 44, 190–197, 1980.
8. DeWitt, J. B., Methodology for determining toxicity of pesticides to wild vertebrates, *J. Appl. Ecol.*, 3(Suppl.), 275–278, 1966.
9. Tucker, R. K., and J. S. Leitzke, Comparative toxicology of insecticides for vertebrate wildlife and fish, *Pharmacol. Ther.*, 6, 167–220, 1979.
10. Tucker, R. K., Criteria Upon Which to Trigger Exercise of Conditional and Special Tests for Avian Wildlife, Ecological Effects Branch Criteria Paper, U.S. Environmental Protection Agency, Washington, DC, 1975.
11. Felthousen, R., Memorandum dated 9 September 1977 to Environmental Safety Section, Opp. U.S. EPA: Classification of Granulated Formulations, 1977.
12. Fite, E. C., L. W. Turner, J. J. Cook and C. Stunkard, Guidance Document for Conducting Terrestrial Field Studies, EPA 540/09-88/109, U.S. Environmental Protection Agency, Washington, DC, 1988.
13. Hill, E. F., R. G. Heath, J. W. Spann and J. D. Williams, Lethal Dietary Toxicities of Environmental Pollutants to Birds, Spec. Sc. Rep. Wildl., 191, *U.S. Fish and Wildlife Service, 1975.*
14. Smith, G. J., Pesticide Use and Toxicology in Relation to Wildlife: Organophosphorus and Carbamate Compounds, Resource Publ. 170, U.S. Fish and Wildlife Service, Washington, DC, 1987.
15. Stephan, C. E., Methods for calculating an LC_{50}, in *Aquatic Toxicology and Hazard Evaluation*, F. L. Mayer and J. L. Hamelink, Eds., ASTM STP 634, American Society for Testing and Materials, Philadelphia, 1977, pp. 65–84.
16. Pearce, P. A., and D. B. Peakall, The impact of fenitrothion on bird populations in New Brunswick, in *Fenitrothion: The Long-Term Effects of its Use in Forest Ecosystems*, National Research Council of Canada, Publ. No. 16073, Ottawa, 1977.
17. Blus, L. J., C. S. Staley, C. J. Henny, G. W. Pendleton, T. H. Craig, E. H. Craig and D. K. Halford, Effects of organophosphorus insecticides on sage grouse in southeastern Idaho, *J. Wildl. Manage.*, 53, 1139–1146, 1989.
18. Suter, G. W., Risk assessment models for terrestrial effects of pesticides, in *Pesticide Effects on Terrestrial Wildlife*, L. Somerville and C. H. Walker, Eds., Taylor & Francis, New York, 1990, pp. 335–346.

49

Individual-Based Mathematical Modeling Approaches in Ecotoxicology: A Promising Direction for Aquatic Population and Community Ecological Risk Assessment

Thomas G. Hallam and Ray R. Lassiter

ABSTRACT

Solutions to the many environmental problems facing mankind are presently beyond the realm of scientific experience and of a scale that makes their experimental study extremely difficult or impossible. This does not, however, obviate the need to quantify the risk associated with the environmental stresses affecting natural populations. Although theoretical developments are required for risk assessment, the current level of theoretical constructs for predicting effects of chemicals on systems is generally not sufficient to meet demands of present-day risk assessment. Modifications of population models from classical mathematical ecology generally have been unsuccessful in predicting effects; hence, future theoretical developments will probably come from a different basis.

One approach that holds promise in ecotoxicology is the use of physiologically based population models. This modeling approach focuses on the individual organism as the basal unit of the system. Herein, we present an assessment protocol consisting of a dynamic (ordinary differential equation) model of an individual organism, a population model represented by a system of hyperbolic partial differential equations, a chemical uptake model, and a quantitative structure activity relationship to determine effect due to chemical exposure.

The advantages and disadvantages of physiologically based population models are discussed. Extensions of the population models to study communities are also indicated. Applications of this approach are presently available only for aquatic communities.

KEY WORDS

mathematical modeling, ecotoxicology, aquatic population models, ecological risk assessment.

INTRODUCTION

For large numbers of the public, environmental problems no longer can be placed on the distant horizon. Answers to many environmental questions facing us are beyond the realm of current scientific experience and of a scale that makes their experimental study difficult, if not impossible. Nevertheless, it is still necessary to attempt to quantify risks associated with environmental and anthropogenic stresses affecting natural populations. Any remedy for the lack of empirical data must include development of theory for assessing risks in ecotoxicology. The current development level of theoretical constructs for predicting effects of chemicals on biological systems is generally not sufficient to meet the demands of present-day risk assessment.[1]

Relevant knowledge of physics and chemistry is at a practical level for ecotoxicological needs; biological understanding, even under unstressed conditions, is at a lower level. Because of this deficiency, the authors feel that one of the most pressing theoretical needs and an area with a most promising future is the theoretical investigation of the behavior of biological systems in a chemically stressed environment. Norton et al.[2] is an excellent state-of-the-art review of model approaches in risk assessment.

The future of ecological risk assessment lies in the development of theoretical constructs that are useful in unifying developments from many different academic disciplines. These constructs must be coupled interactively with appropriate empirical studies so realistic prognoses can be developed. Comprehensive theories that can impact the immense multidisciplinary area of ecotoxicology are needed. Mathematical models can be used to enhance our knowledge of chemically stressed systems.

The future for theoretical studies in ecotoxicology is assuredly strategically important. Nevertheless, the authors' experience indicates that progress will be hindered by intense difficulties related to theoretical developments, to acquisition of biological data for model parameterization and ecological data for model comparison, and to numerical computations.

Opinions about future directions of research must be viewed with caution. Frequently, the recommendation of the proposer is that the most promising direction of research is precisely the direction that proposer is following at the time. This chapter, which contains opinions and conclusions that are no exception to this rule, presents an idiosyncratic perspective on some possible directions for the application of mathematical models and their utility in ecotoxicology. In many senses, the logic presented is that used to derive the approach described in the Protocol section. To relate to the discussion of the future of theoretical approaches in ecotoxicology, it is useful to have the perspective of the authors toward mathematical modeling, toward the relationship of theoretical and empirical work, and toward the use of theoretical developments in the discipline of ecotoxicology.

A Modeling Perspective

Empiricists and theorists are each indispensable and should be complementary creatures. Scientific laws are empirically based with certain exceptions such as those laws derived from axiomatically based systems (such as mathematics). This position of mathematics, unique as it may be, can create difficulties in multidisciplinary endeavors in which the spectrum of model perceptibility can range from blatant misuse of models and mathematics to the sometimes indiscriminate discarding of fundamental information obtained from mathematical approaches.

Empirical models ideally should underlie theoretical models. Indeed, the most applicable models and those most likely to have an impact on science are those that have a sound underpinning in data. Conversely, a mathematical model that does not ultimately relate to data and appropriate levels of problem resolution is doomed to a barrage of skepticism.

It is generally accepted that theoretical and empirical ecology are independent subjects but interdependent approaches to study scientific problems.[3,4] These areas must nurture each other, even though these disciplines have their own methods of achieving understanding of nature. In the authors' opinion, the best approach for the multidisciplinary scientific studies required for ecotoxicology is an adaptive one, with strong feedback loops from each disciplinary component—such as chemistry, physics, toxicology, biology, and mathematics—to the others.

In many scientific arenas, in the ecological sciences generally, and in ecotoxicology specifically, there is often an insufficient database available to obtain scientifically justifiable conclusions relevant to pressing problems. Existing empirical models play a fundamental role, but such models often cannot extrapolate with the degree of accuracy needed for many environmental problems requiring reasonably accurate risk assessments. When a degree of predictability is desirable, theoretical constructs, based upon data and fundamental scientific influences, can enhance the predictive capability. The combination of the empirical and the theoretical approach will allow for developments additional to considerations based solely upon empirical or theoretical models.

While the need for empirical studies is accepted without question by most scientists, the need for mathematical modeling has not been universally accepted. Indeed, the fact that some mathematical models are useful in their own right is often overlooked. Models can serve many roles. They are useful organizational tools; they can be used to generate hypotheses to test or they may provide insight where it is not possible to do experimental work. Caswell,[5] discussing the relationship between theory and empiricism, has argued that subjecting every model to rigorous field testing makes no more sense than to demand that every experiment generate a significant theory. Again, an approach in which disciplines can operate independently but are integrated into a cooperative approach is desirable and certainly more useful than disciplines operating independently. It is, of course, a major task to assemble an environment and personnel capable of cooperating in an effective manner.

Testing and Utility of Mathematical Models

What does it mean to test a mathematical model? How good is a model? To test a model one must examine the model at the limits of applicability. There is, in the authors' opinion, a distinct difference in testing for the extent of application or the utility of a model and in "validating" the model. In terms of model validity, it is the opinion of the authors that a mathematical model cannot be validated. It can only be tested for consistency with the data and, of course, can be and must be invalidated in some situations. Models that are, by necessity, built on a finite set of hypotheses must be violated in an infinite dimensional nature.

If a model is always going to be wrong, when is a model useful? A model is successful when it accomplishes the goal for which it was created. The objective of the modeling effort, whether it is set by project managers or by the modeling group developing the model, must be precise and reasonable. The problem with vague objectives is they cannot be achieved. Realistic objective determination often leads to a minimax optimization problem in that the set of hypotheses tested must be maximal to ascertain the phenomenological behavior, while understanding and effort demands the number of considerations be kept to a minimum.

What is a set of hypotheses that realistically can be expected to be treated by a model? Perhaps in this response modelers should be more directly influenced by empirical scientists. A perspective often attributed by modelers to unsympathetic inquiring minds is the expectation that models should do "everything." Caswell[5] has indicated that by using rigorous scientific methods, usually conversation, he has determined that experimentalists seldom vary more than three factors in their experiments. He claims it is not reasonable to expect modelers to do more than empiricists in their efforts.

A caveat for modeling endeavors is that if the modeling goal is sufficiently specific it can be achieved. Models can be created to do whatever specific task is set. By imposing appropriate assumptions, it is not difficult to mimic a particular static or dynamic phenomenon. To construct a model to reproduce a particular phenomenon is not a great achievement, nor should this capability be unexpected. By contrast, it is a difficult task to start from "first principles," employ a process-oriented, mechanistic approach, and then mimic the evolution of generic or complex phenomena.

Modeling Approaches in Ecotoxicology

A spectrum of modeling approaches, ranging from the rudimentary to the system level, has been used to investigate ecotoxicological problems.[6] However, we lack success stories in ecotoxicology that involve mathematical modeling approaches in general and especially at the rudimentary level of population ecology, where nonmechanistic or statistical curve-fitting representations such as the logistic[7] or other sigmoid functions[8] are included in the models. There are several directions in which the difficulties arise. The authors believe that one of the difficulties occurs primarily because a fundamental aspect of toxicology, variation among individuals, is not adequately handled. The rudimentary level of aggregation is not often mechanistic and does not relate well to relevant biological data, which are frequently at the level of genetics or physiology of the individual organism. This lack of individual variation also hinders the application of toxicological data, which again is primarily at the level of the individual, and the determination of biological response. Concepts of toxicology such as tolerance or susceptibility are based upon the fact that different individuals may respond differently to the same stress. If appropriate individual variation is not present in the biological components of a population model, then assessment of effects of stress on a population cannot be extrapolated to field conditions in which variation determines survival.

The omission of variation is manifested in the utilization of rudimentary population models such as the logistic, in which no variation among individuals is present. Modifications of rudimentary population models or classical community models of mathematical ecology generally have not been successful in predicting effects. In fact, the problems are even more basic; it is difficult to estimate parameters or to obtain population-level dose-response representations for rudimentary models. Such exercises in risk assessment are often no more than curve-fitting statistical analyses, seldom capable of extrapolation to settings other than those under study. Hence, future developments will probably come from a different fount than the population and community models of classical ecology.

Many existing environmental models (the Exposure Analysis Modeling System [EXAMS])[9] have the capability to handle fate and transport problems. Descriptions of certain physical and chemical attributes of the environment are within tolerable levels because the primary formulations are mechanistic and process-oriented. Modeling treatments are generally deficient in their treatment of the biological components. Reasons for this deficiency include past approaches employing aggregation on an unjustified scale and insufficient mechanistic and process representation. Not only is this true for rudimentary approaches but also for many system-level approaches in ecotoxicology.

System-level approaches to problems in ecotoxicology purport to evolve from a "top-down" approach that formulates models at a population, community, or ecosystem level. Unfortunately, these approaches have often tried to address phenomena at the same level of hierarchy. Assumptions at the same level of resolution at which the problem of interest resides is inherently dangerous. In ecotoxicological investigations it is a mistake to ignore the fact that toxicants affect individuals. Toxicants do not affect populations or communities except through the impact on individuals that compose those populations or communities. Because of differences in individual tolerance to stress, it does not seem possible to aggregate and still obtain reasonable predictions of dose-response functions.

From the planning perspective, a system-level, top-down approach should always be employed. It may be possible to resolve problems in ecotoxicology by employing a system-level approach, but the authors think the best chance of obtaining significant results in the near future is to develop strategies at the system level, but place the resolution of the problem at the biological position of the chemical impact—the individual.

THE ROLE OF THE INDIVIDUAL ORGANISM IN POPULATION AND COMMUNITY RISK ASSESSMENT

Issues from the ecological and the toxicological components of ecotoxicology imply that individual variation should be a basal unit for theoretical developments in ecotoxicology. Demographic processes of birth, growth, and death govern population dynamics. Each of these processes is a function of individual characteristics such as physiology, genetics, and morphology. Again, the individual emerges, this time as an appropriate level of investigative focus for population ecology. Lomnicki[10] has argued for the same point because it is at the individual level that the evolutionary and environmental influences coalesce and are reflected.

Essentially three deterministic methods utilize individual information to study a population as a structured entity. One approach is the use of discrete matrix models such as Leslie or stage-structured models.[11] A second approach is the use of delay-differential equations.[12] Both methods can be obtained in connection with the extended McKendrick-von Foerster partial differential equation.[13] Because the McKendrick-von Foerster equation is capable of including the dynamics of an individual organism directly in the model formulation, it is an appropriately focused procedure for the study of population dynamics from an individual-based approach. The extended McKendrick-von Foerster equation is of the form

$$\rho_t + \rho_a + \sum_{i=1}^{n} (\rho S_i)_{m_{s_i}} = -\mu\rho$$

with ρ representing the population density function that depends upon the time t, a representing age, and the physiological variable m_{s_i}. The factors designated by S_i represent the growth rates of the physiological components of an individual. The factor μ is the mortality rate of the population and can be a function of variables such as age, size, or population density. This representation also requires definition of a birth process as a boundary condition and an initial distribution of the population.

All methods aggregate individual attributes to study the population. Of course, any population study must do some aggregation, even if the individual is followed or even if the McKendrick-von Foerster model is employed. However, the level of aggregation is fundamental to solution of the problem. Aggregation at the physiological level yields information about individuals. Hence, to study populations from a dynamic individual perspective, data from physiological levels are useful. This is consistent with the modeling principle that hypotheses should be imposed at the level below the focal level of problem formulation. The McKendrick-von Foerster approach allows incorporation of the physiological level of resolution.

In spite of the fact that the basic McKendrick-von Foerster model and its capabilities have been long known,[14-16] the study of population and community dynamics from the individual perspective is in its infancy. Despite their relative newness, individual-based approaches are supported by several principles recognized as important. The authors think that the extended McKendrick-von Foerster equation is sufficiently basic to be regarded as a foundational equation in ecology. The logistic equation, widely regarded and taught in ecology as dogma, has little to justify it as mechanically useful. We think the McKendrick-von Foerster equation is far more fundamental. In its extended form[17] this model tracks individuals within a population, allowing incorporation of individual dynamics in as complete detail as desired. In addition to the advantage of an individually tailored procedure to model a population, the individual-based approach is focused at the biological level, at which data are the most extensive. Kooijman alone[18,19] and with Metz[20] has pioneered the exploration of the role of the individual in ecotoxicology.

A disadvantage of the individual-based approach is that the inclusion of relevant biological detail often removes the problem from analytical mathematical tractability. Indeed, it is analytical tractability that has kept the classical models of mathematical ecology at the forefront of theoretical ecology since its golden age, the second quarter of this century. Fortunately, recent years have seen the advent of sufficient computing power to adequately study the individual-based models from a numerical viewpoint. A recent flurry of activity in the area of individual-based models has been prompted by the advent of powerful computers that have the capability of tracking thousands of individuals in populations.[21]

FUTURE NEEDS OF THEORETICAL ECOTOXICOLOGY

To describe the future of an area requires knowledge of present status and current needs of the area combined with knowledgeable extrapolation over the time frame of interest. In ecological risk assessment our present status generally is found wanting and our needs are great. In almost all related subject areas we have incomplete information, yet we must be able to address the needs of managers and regulators with a relatively sound scientific support system. A pressing need is to estimate uncertainty in a risk assessment. The authors believe no techniques exist that estimate uncertainty for biological systems to any reasonable degree, although progress is underway.[22] The foundation of uncertainty analysis, especially development of a sound theoretical base, is a prime area for future development.

An even more pressing need is biological predictability, even indications of biological reality, from mechanistic viewpoints that relate to biological data. Many population and community models are derived from nonmechanistic assumptions and, while much effort has been invested in the study of such models, it is difficult to rationalize that continued efforts will be productive in meeting future theoretical ecotoxicological needs. While granting that these models may serve as guides for our thoughts over a short time, they cannot be more than that. The longer time scales that are important for biological effects investigations indicate that a different methodology is required to relate to the biology that is relevant to chemical toxicity and tolerance.

AN ECOTOXICOLOGICAL PROTOCOL FOR INVESTIGATION OF BIOLOGICAL EFFECTS

To be useful for ecotoxicological problems, theoretical constructs must have characteristics that at least allow inclusion and representation of:

1. the relevant toxicological and ecological data
2. the dynamics of the chemical exposure
3. the chemical-biological interfaces
4. the dynamics of the ecological system
5. the determination of biological response to the stress

These characteristic markers can be used as indicators to assess the utility of a theoretical approach in ecotoxicology. Certainly any feasible approach ultimately must be able to incorporate portions of the vast literature and techniques of toxicology with those of population ecology.

The authors describe a protocol relating to these five aspects of a responsible model representation employed in an analysis of a chemically stressed population.[23,24]

Modeling Chemical Uptake in Individuals

To study the impact of chemical stress on a biological system, the authors have argued that in most cases this is best measured at the individual level, then extrapolated to the population and higher levels of biological organization. The degree of toxic chemical stress is related to the level of chemical in a compartment of the body or in a target organ of the individual. The amount of chemical in the organism is determined by uptake from the biotic and abiotic environments.

To adequately address problems of ecotoxicology it is necessary to consider the manner in which chemical exposure and the biological-chemical interfaces are formulated. Uptake models are available for aquatic organisms. For example, Food and Gill Exchange of Toxic Substances (FGETS)[25] allows fish to uptake chemical from both the food and the environmental pathways. FGETS, based upon thermodynamic potential, represents the chemical exchange between fish and the aqueous environment that occurs across gill membranes and across gut walls from ingestion of the contaminated resource. The general form of the model is

$$\frac{dB_T}{dt} = Sk_wC_w + C_FF - \frac{B_T}{W_TBCF}[Sk_w + Ek_E]$$

where B_t represents the total toxicant in the organism; C_W and C_F represent the concentrations of toxicant in the environment and in the food, respectively; F and E are the mass fluxes of food and feces, respectively; X is the bioconcentration factor (total concentration in the organism/C_W); k_E is the partition coefficient of chemical to excrement (C_E/C_A, where C_A and C_E are concentrations of the chemical in the aqueous portion of the organism and its feces, respectively); and k_W is the unit conductance. Hallam et al.[23] have also used a version of this model to consider uptake of chemical by *Daphnia*. An important aspect of this model approach is that the chemical-biological interface is developed in mechanistic detail so that uptake is not an arbitrary action. Composition, structure, and morphology of the organism are important in these uptake formulations. The body burden is a function of the total weight of the organism. For assessment purposes, therefore, it is necessary to track the weight of an individual as a dynamic variable.

A Dynamic Individual Model

For the model mentioned above, the weight of an organism is an important factor in mediating chemical uptake. Depending upon the mode of action of the particular exposure chemical, physiological characteristics other than weight may be of equal importance for the investigation of effects. We have previously considered effects of electroneutral, lipophilic chemicals on mortality in *Daphnia* and fish populations.[24] Because of the partitioning of lipophilic chemicals, their affinity toward organismal fat leads to bioconcentration of chemical in the lipid of a stressed organism. To investigate the effect of lipophilic chemical stressors, it is necessary to investigate dynamic lipid structure in individuals. Because of the fundamental role of lipids in reproduction, lipid storage seems to be an indispensable compartment for any study of population dynamics, not solely those concerned with the effects of lipophilic chemicals. This illustration demonstrates that the specific type of individual variation included in a population model is fundamental to the objective of the risk assessment problem. The determinants of variation employed in the model formulation must be consistent with the particular problem objectives because any factor that significantly affects the dose-response function is also an important determinant for the model formulation. For example, if a chemical stressor is significantly attracted to an organismal phase (such as lipid), then the variation maintained in the population study should include that phase representation in the individual model. If the chemical is known to bioaccumulate significantly in an organ of an individual organism and the concentration of chemical in that organ can have an effect on the health of the individual, it is important to have a mechanistic component reflecting this in the model.

The relationship between the type of chemical, its properties such as mode of action or partitioning capability, and the physiological and genetic characteristics of the individual are fundamental to the formulation of a dynamic individual model. To assess the effects of chemicals on a population, the chemical and the target organism must be considered a unique system. Here, *unique* may be defined as including equivalent classes of chemicals or organisms. If either the chemical or the organism is changed, the individual model must adequately reflect the new characteristics of the interaction. Within these constraints, it should be the goal of ecotoxicological research to develop an approach that is as generic as possible. An individual model based upon energetics has been the authors' main thrust and only the simplest mode of action for chemicals in the research has been considered.[23]

Determination of Effects of Chemicals on an Individual and on a Population

The determination of effects of a chemical on a population can be accomplished through internal model representations, or alternatively, they can be ascertained externally by an indicator triggered from within the model. The latter approach was chosen by the authors because it relates to the idea that a toxic stress is an external effect. At present only mortality effects have been addressed by Hallam and associates, utilizing a perspective different from the traditional for modeling death. Mortality due to chemical stress is not formulated directly but is assessed independently when the indicator, the concentration of chemical in the individual organism, is at lethal concentration for that particular chemical. This approach allows the use of some of the available relevant toxicological data developed essentially at the individual level.

One of the most useful methods for investigating individual toxicity effects is through quantitative structure activity relationships (QSARs). QSARs are mathematical expressions that relate biological endpoints or activity (such as mortality or a decrease in reproduction) to descriptors of chemical molecular properties or a sequence of chemical properties (such as partition coefficients or activity). In spite of the fact that the biological endpoint of a QSAR has little to do with the biological mechanisms that are affected by the toxicant stress, QSARs are useful methods to indicate the effects of chemicals on individuals.

For a study of population-level effects, information about the individual, and in particular, information about the distribution of the variables that are moderating the effect, must be extrapolated to the population level. One method is to assess the actual data associated with the variables. Unfortunately, in instances in which many compartmental variables determine an effect, empirical data frequently are not available. One reason for the lack of available data could be that it is not obtainable in the dynamic mode needed for risk assessment. For example, the variation in the lipid distribution in the *Daphnia* and fish populations[23,24] must be followed over time. Consequently, alternative methods of generating these data model those situations in which the distribution of the effect variable is obtained directly from the model, or they follow all individuals in the population explicitly in the model and accumulate properties to form the distribution in the population.

The authors' previous work has treated only lethal effects without regard for the progression of effects that lead up to lethality. An acute toxicity syndrome of fish, such as that observed by McKim et al.,[26] consists of progressive loss of schooling behavior, increasing hypoactivity and unresponsiveness to external stimuli, increased respiration and loss of equilibrium, followed by death. Inclusion of this type of compound effect requires additional formulations that relate ethological considerations with the physiological developments present in the current models. It is important that any representation of a toxicity syndrome, especially those on chronic scales, be mechanistically based. These mechanistic representations are probably more important at the community level than they are at the population level (e.g., because of the possibility of increased risk of predation caused by unusual behavior).

Sublethal effects relating to reproduction are important for population dynamics on longer time scales that may result from chronic toxic exposures. It is important to develop approaches to handle these stresses and to ascertain their impact on populations.

A Protocol for Determining Effects of Chemicals on Aquatic Populations

A generic procedure that can be used to determine the effects of chemicals on aquatic populations consists of the following steps:

1. Formulate a model that represents the individual from the perspective of the components affected by the chemical.
2. Formulate a method to determine effects of the chemical on the individual organism.
3. Formulate a model of a population that incorporates individual dynamics. An example of such a model is the extended McKendrick-von Foerster equation.
4. Determine the effects of the chemical exposure at the population level by accumulating individual effects.

This approach has proved feasible to investigate the effects of an electroneutral chemical on a population. Although this approach appears generic, consideration must be given to the interaction between the biology and the chemistry in cases of extension to chemicals with other modes of action. The individual model must reflect the important compartments of the individual that have affinity for the chemical under investigation. Clearly, from biological and chemical considerations, dynamics of the aqueous, blood, lipid, and protein phases are fundamental compartments for any individual model. However, other compartments may still be important for a particular chemical stressor. Representations of these other compartments must also be included in the individual model.

The biology and, consequently, the modeling and the numerical procedure, become more complicated above the level of the population. The authors have recently initiated studies of a predator-prey system consisting of *Daphnia* and fish. Each population in the system is physiologically structured and the predators have a size-structured range of prey in which they eat. These factors combine to make analysis of the system difficult due to numerous combinatorial factors that influence the dynamics. Of course, as with any large simulation, one must be concerned about the computational accuracy of the numerical procedures used in the analysis.

SUMMARY

Regulation of chemicals must account for risks of exposure to both humans and the environment. It is a tacit assumption that regulatory decisions should precede environmental release, a premise that demands a predictive capability based upon sound scientific knowledge. To be successful, risk assessment must have sound theoretical underpinnings that reflect bioassay testing and yield relevant applicable information for extrapolation to field studies. The long history of bioassay testing has resulted in the statistical analysis of bioassay data being firmly grounded.[27] However, straightforward and direct interpretation of bioassays to obtain an expected response of populations has not been possible. Regulatory actions intended to protect all life must currently depend upon the similarity of test species and test conditions to the situations to be protected by regulation. A common basis for bioassay results and field response is needed for ecotoxicology to progress significantly as a discipline.

Adequate assessment of the risk to biotic systems due to exposure of toxic chemicals must utilize basic theoretical constructs because, in many instances, it is not possible to do the empirical work necessary to answer ecotoxicological questions related to the longer time scales within which many biological effects operate.

The authors have argued that from both the ecological and toxicological perspectives the appropriate level of focus for ecotoxicological modeling is that of the individual organism. An advantage of this individual-based modeling approach is the database utilized is at appropriate levels for both biology and toxicology. Even though the approach is data intensive it is closer to the biology than classical mathematical models. Also indicated was that techniques for the individual-based approach are available, with a primary tool being the McKendrick-von Foerster model, a partial differential equation that can be used to represent the population. The use of this basic model for ecology requires the development of an individual model that adequately reflects the interaction of biology and chemistry.

While the authors feel that the individual-based approach is appropriate to ecotoxicology, there are pitfalls. It is a very data intensive procedure, and the necessary data often are not available for the species of interest. The numerical work is often complex and the computations profuse. Because of the complexity of the model, significant inroads into the full model analysis probably will not be through analytical mathematical studies. However, in spite of the complexities, analytical studies should be encouraged and probably will be necessary to adequately understand the models.

REFERENCES

1. Mittelman, A., *Ecosystem Model Selection Criteria,* Technical Resources, Inc., Rockville, MD, 1987.
2. Norton, S., M. McVey, J. Colt, J. Durda and R. Hegner, Review of Ecological Risk Assessment Methods, EPA/230/10-88/041, U.S. Environmental Protection Agency, Washington, DC, 1988.
3. Levin, S. A., The role of theoretical ecology in the description and understanding of populations in heterogeneous environments, *Am. Zool.,* 21, 865-875, 1981.
4. Kot, M., W. M. Schaffer, G. L. Truty, D. J. Graser and L. F. Olsen, Changing criteria for imposing order, *Ecol. Model.,* 43, 75-110, 1988.
5. Caswell, H., Theory and models in ecology: a different perspective, *Ecol. Model.,* 43, 34-44, 1988.
6. Moriarity, F., *Ecotoxicology,* Academic Press, New York, 1983.
7. Hallam, T. G., Population dynamics in a homogeneous environment, in *Mathematical Ecology: An Introduction. Biomathematics,* Vol. 17, T. G. Hallam and S. A. Levin, Eds., Springer-Verlag, Berlin, 1986, pp. 61-94.
8. Leberg, P. L., I. L. Brisbin, Jr., M. H. Smith and G. C. White, Factors affecting the analysis of growth patterns of large mammals, *J. Mammals,* 70, 275-283, 1989.
9. Burns, L. A., D. M. Cline and R. R. Lassiter, Exposure Analysis Modelling Systems (EXAMS) User Manual and Systems Documentation Report, EPA/600/3-82/023, U.S. Environmental Protection Agency, Athens, GA, 1981.
10. Lomnicki, A. *Population Ecology of Individuals, Monographs in Population Biology,* Vol. 25, Princeton University Press, Princeton, NJ, 1988.
11. Caswell, H., *Matrix Population Models: Construction, Analysis, and Interpretation*, Sinauer Associates, Sunderland, MA, 1989.
12. Nisbet, R. M., and W. S. Gurney, *Modeling Fluctuating Populations,* John Wiley & Sons, Chichester, U.K., 1982.
13. Metz, J. A. J., and O. Diekmann, *The Dynamics of Physiologically Structured Populations. Lecture Notes in Biomathematics, Vol. 68,* Springer-Verlag, Berlin, 1986.
14. Sinko, J. W., and W. Streifer, A new model for age-size structure of a population, *Ecology*, 48, 910-918, 1967.
15. Sinko, J. W., and W. Streifer, Applying models incorporating age-size structure of a population to *Daphnia, Ecology*, 50, 608-615, 1969.
16. Bell, G. I., and E. C. Anderson, Cell growth and division. I. A mathematical model with applications to cell volume distributions in mammalian suspension cultures, *Biophys. J.*, 7, 329-351, 1967.
17. Auslander, D. M., G. F. Oster and C. B. Huffaker, Dynamics of interacting populations, *J. Franklin Inst.*, 297, 345-374, 1977.
18. Kooijman, S. A. L. M., Population dynamics on basis of budgets, in *The Dynamics of Physiologically Structured Populations. Lecture Notes in Biomathematics,* Vol. 68, J. A. J. Metz and O. Diekmann, Eds., Springer-Verlag, Berlin, 1986, pp. 341-380.
19. Kooijman, S. A. L. M., The von Bertalanffy growth rate as a function of physiological parameters: a comparative analysis, in *Mathematical Ecology,* T. G. Hallam, L. J. Gross, and S. A. Levin, Eds., World Scientific, Singapore, 1988, pp. 3-45.
20. Kooijman, S. A. L. M., and J. A. J. Metz, On the dynamics of chemically stressed populations: the deduction of population consequences from the effects on individuals, *Ecotoxicol. Environ, Safety,* 8, 254-274, 1984.

21. Huston, M., D. L. DeAngelis and W. M. Post, New computer models unify ecological theory, *Bioscience*, 38, 682–691, 1988.
22. Barnthouse, L. W., G. W. Suter, S. M. Bartell, J. J. Beauchamp, R. H. Gardner, E. Linder, R. V. O'Neill and A. E. Rosen, User's Manual for Ecological Risk Assessment, Environ. Sci. Div. Publ. No. 2679, Oak Ridge National Laboratory, Oak Ridge, TN, 1986.
23. Hallam, T. G., R. R. Lassiter, J. Li and W. McKinney, Toxicant-induced mortality in models of *Daphnia* populations, *Environ. Toxicol. Chem.,* 9, 597–621, 1990.
24. Lassiter, R. R., and T. G. Hallam, Survival of the fattest: implications for acute effects of lipophilic chemicals on populations, *Environ. Toxicol. Chem.*, 9, 585–595, 1990.
25. Barber, M. C., L. A. Suarez and R. R. Lassiter, Kinetic exchange of nonpolar organic pollutants by fish, *Environ. Toxicol. Chem.,* 7, 545–558, 1988.
26. McKim, J. M., P. K. Schmieder, R. W. Carlson, E. F. Hunt and G. J. Neimi, Use of respiratory cardiovascular responses of rainbow trout (*Salmo gairdneri*) in identifying acute toxicity syndromes in fish. I. Pentachlorophenol, 2,4-dinitrophenol, tricaine methanesulfonate and 1-octonal, *Environ. Toxicol. Chem.,* 6, 295–312, 1987.
27. Finney, D. J., *Statistical Methods in Bioassay,* Macmillan, New York, 1978.

50

Integrating Ecological Modeling in Ecological Risk Assessment

James A. Gagne

ABSTRACT

A major goal of this conference was to encourage interaction and exchange of ideas between scientists in wildlife toxicology and population ecology. It was hoped that this interaction would help wildlife toxicologists design better experiments and draw more robust conclusions, by incorporating theory and techniques from population ecology and mathematical modeling. A longer term goal that was also implicit was that the process of ecological risk assessment ultimately would be improved. The four papers presented offer the diverse view of mathematical models and modeling that was evident throughout the conference. Chapters by Fite and Williams and Emlen are illustrative of how models are currently applied in ecological risk assessment, and how they can be used to guide efforts to refine the risk assessment process, respectively (Chapters 47 and 49). Gross and Hallam and Lassiter delve more deeply into how models are developed mathematically, used, and evaluated. Taken together the four chapters re-emphasize the crucial importance of explicitly stating the purpose for which a model is being developed, and its inherent limitations because of that purpose.

KEY WORDS

mathematical models, ecological modeling, ecological risk assessment

INTRODUCTION

The 1990 Ecological Modeling Conference was convened to encourage interactions and exchanges of ideas between scientists in the nascent discipline of wildlife toxicology and the mature discipline of population ecology. The organizers felt that the conference would prove beneficial to both disciplines. Wildlife toxicology is currently accumulating a large database, but is lacking in relevant theory. Population ecology has a rich assortment of theories and models that could provide a theoretical underpinning for wildlife toxicology. The viewpoint of the wildlife toxicologists at the workshop was, "Here are our concerns and what we can currently measure. How can we use popula-

0-87371-591-8/94/$0.00+$.50

tion ecology theory and models to help design better experiments and draw robust conclusions? What else should we be taking into account and measuring?"

The four chapters in this section are telling snapshots of the degree that ecological modeling is and can be integrated into environmental risk assessment. Fite[1] and Williams and Emlen[2] provide the perspectives of government officials charged with either deciding whether a pesticide will not cause adverse effects on the environment or performing research that refines the risk assessment process.

Hallam and Lassiter[3] and Gross[4] provide examples of how models are developed, used, and evaluated from the perspectives of academics. This author speaks from the perspective of someone in the regulated industry, attempting to work within the current system to obtain and maintain pesticide registrations.

Fite[1] presents an overview of the data used and the approach taken by the United States Environmental Protection Agency's (U.S. EPA's) Office of Pesticide Programs in estimating the risk a pesticide may pose to the environment, especially fish, aquatic invertebrates, and birds. A key consideration in the basic philosophy of the assessment is the legal dictum that the EPA must conclude that the pesticide will not cause unreasonable adverse effects on the environment. The general sequence of the assessment is that laboratory toxicity results for a pesticide are compared with estimates of the exposure to that pesticide received by an animal in the wild. If estimated exposure is substantially less than toxicity values, it is assumed that no adverse effects will occur. As estimated exposure approaches some proportion of the laboratory toxicity values (usually one fifth for terrestrial organisms and one tenth for aquatics), it is presumed that adverse effects are possible. Additional data, typically field tests, are usually required to rebut this presumption. This environmental risk assessment model has two limitations that must be kept in mind. First, it is not very well validated for the case in which no risk is presumed. Second, the numerous uncertainties inherent in its components are not well quantified. As examples, acute toxicity values are variable, the relationship between apparent toxicity in the laboratory to actual toxicity in the field is not well known, and the presence of pesticide residues may not relate directly to the exposure an animal actually receives. Although the model would likely benefit from additional data, a more complex but unvalidated model will not necessarily improve the accuracy of the risk assessment process.

Williams and Emlen[2] present a very fruitful area for cooperation between wildlife toxicologists and population ecologists: the identification of critical parameters in population models, and the subsequent attempts by experimenters to measure these parameters. In other words, what parameters are best suited for predicting the effects of a pesticide on population trajectories? A demographic approach is very useful because it employs parameters that are relatively easy to measure and have common sense appeal. By studying the sensitivity of the model to changes in the various demographic parameters, one can organize research efforts to measure variables that should affect population levels. This theme, that models can be used as tools for organizing experimental programs, was mentioned by several conference attendees. It is clearly one area in which ecological modeling in the larger sense can make a valuable contribution to the risk assessment process. Williams and Emlen do point out that in order for this contribution to be realized, additional efforts at communication must be made by population ecologists and wildlife toxicologists alike. This communication process will take time.

Gross[4] highlights both the utility and shortcomings of models. While models can provide valuable information and insights, they are often not formulated for the application in questions, nor are they adequately tested and evaluated. The evaluation process depends on the context in which the model is to be applied. Models may be classified as to context. One context is models for theory development, another is models for specific, delineated systems, and a third is models for decision-making. These three types of models require different evaluation criteria.

In an effort to standardize the evaluation process, Gross proposes the following definitions:

- *Verification*: The model behaves as intended. The equations are correct, they agree with the assumptions made, and the programming is correct,
- *Validation*: The model mimics the real system it represents for the specific purposes for which it was constructed. It accurately covers the range of conditions over which it is to be applied.
- *Calibration*: Data are used to determine parameters of the model, so that the model agrees with a particular data set.
- *Corroboration*: A set of data independent of that used to construct the model is in agreement with the model.
- *Evaluation:* Validation plus attention to other criteria, such as utility and plausibility, are assessed. These criteria often differ for the modeler and the decision-maker.

These definitions are then employed in the evaluation process for the three types of models listed previously. By far the most difficult models to evaluate are those for decision-making because criteria that are not included in the biological assessment are important in making a policy decision. One solution to this difficulty is to employ a top-down approach, in which the goals of the decision-maker, especially their levels of indifference to model outcomes, are considered before deciding how much detail about biological processes is needed in the model.

Hallam and Lassiter[3] discuss the development and application of models that employ information on individual organisms. This has great appeal because toxicological responses occur at the level of the individual. The basic mathematical structure is the McKendric-von Foerster partial differential equation, which can include information on the physiological level, and has a "dynamic individual perspective." These authors feel that this model can be regarded as a fundamental equation of ecology; its application has been hindered because it becomes analytically intractable when biological information is added. This difficulty can now be solved by a brute force approach using today's high-speed computers.

Hallam and Lassiter propose criteria for a protocol to evaluate biological responses. At minimum, this protocol should include relevant toxicological and ecological data, chemical and biological interfaces, dynamics of the ecological system, and determination of biological response to stress. This protocol is then applied to three situations: modeling chemical uptake in individuals, modeling energetics, and modeling the effects of chemicals on an individual and a population.

Lastly, Hallam and Lassiter propose a protocol with the following steps for modeling the effects of a chemical on an aquatic population: formulation of a model of the individual as affected by the chemical; formulation of a method to determine effects of the chemical on the individual; formulation of a model of a population that incorporates individual dynamics, such as the McKendric-von Foerster equation; and determination of effects of the chemical on the population by accumulating individual effects.

REFERENCES

1. Fite, E., The Environmental Protection Agency's avian pesticide assessment model, in *Wildlife Toxicology and Population Modeling: Integrated Studies of Agroecosystems*, R. J. Kendall and T. E. Lacher, Jr., Eds., Lewis Publishers, Chelsea, MI, 1993.
2. Williams, B., and J. Emlen, Population models as research tools: an empirical perspective, in *Wildlife Toxicology and Population Modeling: Integrated Studies of Agroecosystems*, R. J. Kendall and T. E. Lacher, Jr., Eds., Lewis Publishers, Chelsea, MI, 1993.
3. Hallam, T. G., and R. R. Lassiter, Individual-based mathematical modeling approaches in ecotoxicology: a promising direction for aquatic population and community ecological risk assessment, in *Wildlife Toxicology and Population Modeling: Integrated Studies of Agroecosystems*, R. J. Kendall and T. E. Lacher, Jr., Eds., Lewis Publishers, Chelsea, MI, 1993.
4. Gross, L. J., Limitations of reductionist approaches in ecological modeling: model evaluation, model complexity, and environmental policy, in *Wildlife Toxicology and Population Modeling: Integrated Studies of Agroecosystems*, R. J. Kendall and T. E. Lacher, Jr., Eds., Lewis Publishers, Chelsea, MI, 1993.

51

Potential Use of the Mallard Model to Predict Effects of Pesticides on Mallard Reproduction in the Prairie-Pothole Region of North Dakota

Christian E. Grue, Michael W. Tome, George A. Swanson, and Sarah Gerould

The development of conceptual and predictive models will undoubtedly have a major role in the advancement of wildlife toxicology and ecological risk assessments for pesticides. However, to date in only a few cases has modeling been applied to studies of the effects of pesticides on wildlife populations (for a recent review, see Reference 1). This is probably because the data necessary to model population or ecosystem effects have been collected in only a few studies. The most comprehensive study of the effects of pesticides on a wildlife species is summarized by Potts[2] and Rands.[3] Data these authors collected over a period of 8 years show a decline in survival rates of the grey partridge (*Perdix perdix*) chicks which the investigators attribute to the indirect effects of pesticides (i.e., pesticide-induced reductions in food). In comparison, most field studies on the effects of pesticides involve no more than two field seasons (for examples, see References in Grue et al.[4]).

Existing population models for wildlife species may, however, provide opportunities to incorporate data on pesticide effects. An affirmative answer to the following questions may be important in identifying existing models that may serve as early success stories in modeling the effects of pesticides on wildlife populations:

1. Does the model deal with a species of general interest?
2. Does the model deal with a habitat of concern, either in terms of ecological importance or potential for pesticide inputs?
3. Does the model simulate an operational, real-world setting?
4. Does the model have regional or national application?
5. Will the model assist in land management or regulatory decisions?
6. Do the methodologies exist to obtain needed data?
7. Is funding available to collect needed data?

The authors believe the model developed by Cowardin and co-workers[5] to assist in decisions in the management of the mallard (*Anas platyrhynchos*) within the prairie-pothole region of North Dakota provides such an opportunity.

The mallard model consists of four components: a waterfowl habitat database, a database on nesting success, regression models predicting the number of mallard pairs on four classes of wetland of different sizes, and a stochastic mallard productivity model.[5] The latter estimates the number of mallards fledged under specific conditions and the recruitment of young females into the fall population. The productivity model[5] contains several "determining factors," but does not currently include data on the direct or indirect effects of agricultural chemicals. Recent studies[6,7] indicate that some pesticides, particularly aerially applied insecticides, have a high probability of entering prairie wetlands and can significantly impair duckling survival either through direct toxicity or through pesticide-induced reductions in the aquatic invertebrates on which ducklings depend for food. The authors believe data on the direct and indirect effects of pesticides on mallard productivity could be collected within a 5-year period. This would greatly enhance the capabilities of the existing model.

The data that need to be collected and the approaches that could be used to obtain them are reviewed by Grue and coworkers.[6,8] Several research questions need to be answered:

1. What is the extent of contamination of prairie wetlands following operational application of agricultural chemicals?
2. What are the seasonal and cumulative effects of wetland contamination on waterfowl productivity?
3. How does the use of agricultural chemicals affect the quality of upland cover and nesting habitat?
4. To what extent are waterfowl exposed to pesticides in upland habitats?
5. What are the effects of this exposure on waterfowl productivity?

A conceptual model of the ways agricultural chemicals may affect the productivity of birds and other wildlife is provided by Grue and co-workers.[8] Chemical analyses and a variety of laboratory and in situ bioassays may be used to determine the extent of wetland contamination within the region. Answers to the remaining questions could be determined by conducting controlled field studies in enclosed (predator-proof fenced) wetland complexes or watersheds that simulate operational agricultural practices. The use of wing-clipped, pen-reared, wild strain, female mallards fitted with radio transmitters would facilitate studying the effects of selected chemicals on waterfowl reproduction (egg laying through class III ducklings) within these research areas.

The authors believe that the collection of these data and their incorporation into the existing mallard model will constitute an early success story for using models to predict the effects of pesticides on wildlife populations. More importantly, the resultant model could be used to more accurately predict the effects of land use practices and management decisions on waterfowl populations.

REFERENCES

1. Emlen, J. M., Terrestrial population models for ecological risk assessment: a state of the art review, *Environ. Toxicol. Chem.*, 8, 831–842, 1989.
2. Potts, G. R., Population dynamics of the grey partridge: overall effects of herbicides and insecticides on chick survival rates, *Proc. Int. Cong. Game Biol.*, 13, 203–211, 1977.
3. Rands, M. R. W., Pesticide use on cereals and the survival of grey partridge chicks: a field experiment, *J. Appl. Ecol.*, 22, 49–54, 1985.
4. Grue, C. E., A. D. M. Hart and P. Mineau, Biological consequences of depressed cholinesterase activity in wildlife, in *Cholinesterase-Inhibiting Insecticides: Impacts on Wildlife and the Environment*, P. Mineau, Ed., Elsevier, Amsterdam, 1991, pp. 151–209.
5. Cowardin, L. M., D. H. Johnson, T. L. Shaffer and D. W. Sparling, Applications of a Simulation Model to Decisions in Mallard Management, Fish Wildl. Tech. Rep. 17, Fish and Wildlife Service, Washington, DC, 1988.
6. Grue, C. E., M. W. Tome, T. A. Messmer, D. B. Henry, G. A. Swanson and L. R. DeWeese, Agricultural chemicals and prairie pothole wetlands: meeting the needs of the resource

and the farmer—U.S. perspective, *Trans. North Am. Wildl. Nat. Resource Conf.*, 54, 43–58, 1989.

7. Tome, M. W., C. E. Grue and L. R. DeWeese, Ethyl parathion in wetlands following aerial application to sunflowers in North Dakota, *Wildl. Soc. Bull.*, 19, 450–457, 1991.
8. Grue, C. E., L. R. DeWeese, P. Mineau, G. A. Swanson, J. R. Foster, P. M. Arnold, J. N. Huckins, P. J. Sheehan, W. K. Marshall and A. P. Ludden, Potential impacts of agricultural chemicals on waterfowl and other wildlife inhabiting prairie wetlands: an evaluation of research needs and approaches, *Trans. North Am. Nat. Resource Conf.*, 51, 357–383, 1986.

52

How Valuable are the Results of Models and Laboratory Studies when Extended to Field Situations?

Gary H. Heinz, Michael Brody, and Lawrence J. Blus

INTRODUCTION

This chapter summarizes the contributions of participants at the 1990 Ecological Modeling Conference during a 2-hour workshop, and attempts to represent fairly the diverse and sometimes conflicting perspectives that were presented. Although models and laboratory studies were supposed to share equal importance in this workshop, the discussion almost exclusively related to the use of models to assess the risk of pesticides to populations of birds. This greater emphasis on models vs laboratory data as a means to assess risk may reflect a difference in the perceived importance of laboratory studies and models or it may simply reflect a continuation of the direction the discussion took from an early stage. Although models dominated the discussion, laboratory data are the main line of information for predicting pesticide-induced harm to birds, and may remain so into the foreseeable future.

In some introductory remarks to the workshop, Heinz suggested that improvements are needed in our ability to extrapolate from laboratory results to what happens in nature when birds are exposed to pesticides. Each year, the United States Environmental Protection Agency (U.S. EPA) examines many pesticides for registration, reregistration, and other reasons. The examination of pesticides is carried out under the Federal Insecticide, Fungicide, and Rodenticide Act (FIFRA), which contains guidelines for different levels (tiers) of toxicity data, proceeding from laboratory tests to full-scale field tests. Given the enormous costs of field studies, most of the toxicological data collected on a pesticide will be of necessity laboratory data.

Even with laboratory data, there is often little more than LD_{50} and LC_{50} values for bobwhite and mallards and, perhaps, reproductive data for the same two species. Generally, such data are not put into what might be called a formal mathematical model, but are used in conjunction with estimates of pesticide exposure to determine whether an acceptable level of risk exists. In the future, it may be possible to use laboratory-derived data to build more formal mathematical models of risk.

Whether laboratory data are used in traditional ways or in models, their usefulness in understanding what will happen in the field needs to be improved. For example, interactions of different pesticides are very difficult to study in nature, but are more amenable to laboratory experiments. Likewise, the interactions of pesticides with natural environmental stresses such as temperature extremes, food shortages, and migration can be tested in the laboratory to expand the range of field conditions under which laboratory results apply. The more that is known about how variations in laboratory tests affect toxicity results, the more it will be possible to extrapolate results from the laboratory to the field.

Chemical formulations of pesticides can vary greatly. A larger body of laboratory data is needed to enable regulators to predict the possible harm these formulations may have on wild birds. Another important need is to know how the toxicity of technical grade pesticides mixed into laboratory diets may differ from the toxicity of the same pesticides when biologically incorporated into the foods of wild birds.

A presentation at the conference by Crystal Driver, Battelle Northwest, described the greater energy needs of wild vs captive birds. In the laboratory, this difference could be simulated to measure effects on toxicity. As a final example, some behaviors of laboratory test birds need to be studied carefully to learn how they could affect extrapolation of results to the field. Avoidance of pesticide-treated food is not uncommon in laboratory tests, yet birds have been killed by some of the same pesticides in nature. Why does the avoidance seen in the laboratory not always occur in the field?

It will never be possible in laboratory tests to perfectly mimic all the conditions that exist in nature. However, this should not be the goal of laboratory tests or those risk assessments based in part on laboratory tests. The goal should be to understand how the conditions of the laboratory tests bias our extrapolations to the field. When we better understand these biases, we will improvc our ability to correct or allow for them when assessing the risk a pesticide is likely to cause in free-living bird populations.

As mentioned earlier, most of what was said in this workshop dealt with the incorporation of field, not laboratory, data into models of population-level effects of pesticides. Before summarizing the many technical points brought out in this discussion, two coexisting themes that surfaced at this conference should be considered. The first theme was that regulatory agencies need practical means to measure and predict the hazards of pesticides to birds in order to make proper regulatory decisions. Even when these measures of hazard are in the form of population models, they would best be in a form that can be easily used in a regulatory process. In essence, this means they should be straightforward, simple models.

The second main theme of the conference sometimes seemed in conflict with the first; it can be expressed as the opinion that the best science be used to develop models of pesticide effects on birds at the population level. The reconciliation of these two themes was the struggle the conference participants faced for most of the workshop. This struggle took form in three phases throughout the workshop: problems, promise, and plans of action.

PROBLEMS

The first problem encountered was one of scale: what do we mean when we talk about a "population"? Is it sufficient to consider the population to be that occupying one farm where birds are being affected by a pesticide, or should it be the population at the county, state, or continental level? Even at a single site, individuals of different species may be parts of populations with vastly different geographic distributions. Those familiar with modeling were quick to point out that the model would have to be different at each level. In addition, one could not overlook the problems that migration would add to the model. Some argued that a small, discrete population would be best, but just as many others made a case that a large population would be preferred in order to be meaningful. In modeling, populations generally are not even defined by the total area they occupy, but by density or biomass of a species. It was even argued that there is no simple working definition of a "population."

Workshop participants repeatedly pointed out that different types of models are necessary for different questions. Some models are designed to predict what will happen in the future. Others can be used to explain what is happening at one time and place. Wildlife toxicologists may be confused in not knowing exactly what they want models to do for them. If the response is that we want models to measure pesticide impacts on birds, there is still confusion as to what "impacts" means. Should models be able to extrapolate results from one species to another? If so, very general models are needed.

There was also heavy criticism that we cannot begin to model the effects of pesticides on bird populations until we understand all the other major factors that affect bird populations in a given ecosystem. In other words, the nonpesticide effects must be modeled first. It was generally considered impossible among modelers to plug only pesticide exposure into a population model. Even with all factors entered, good models are the result of a lengthy interactive process of trial and error. The Potts model for the grey partridge (*Perdix perdix*), for example, includes 40 years of data. The Potts model succeeded because it included the indirect effects of herbicide use on insect abundance. Modeling indirect effects have not even been considered, which, as in the case of the grey partridge, may be more consequential to bird populations than direct toxic effects.

Finally, modeling pesticide effects on bird populations would be a very applied use of routine models, as contrasted to the state-of-the-art research for which modelers are most rewarded. Consequently, there is little research money for such work.

PROMISE

After much was said about the problems of adapting modeling to pesticide-induced changes in bird populations, the tide of the discussion slowly turned toward a more optimistic outlook—the promise that modeling holds for wildlife toxicology.

It was realized that models should not be the goal, but a tool to reach our goal of understanding how pesticides may affect bird populations. Models also can be a tool to understand natural variation in populations. Population dynamic models can show how large a study area is needed.

To resolve the question of how large a population should be modeled, it was suggested that a tiered approach may be needed to examine several different levels of models for populations at different geographic scales. Furthermore, the U.S. EPA is not required to address population effects of pesticides. Therefore, conducting initial studies with small populations seems the most productive alternative. As long as one defines the segment of the population that is of interest and recognizes the limits of the model to that scale of the population, modest size modeling is feasible. In addition, if a model only applies to a discrete, small area, regulatory officials may not be so tempted to apply it improperly everywhere. Models that address behavior and exposure issues, rather than population consequences, should be feasible very soon.

It was recognized that models of pesticide-induced population changes would not have to be 100% accurate. As long as the degree of uncertainty we are dealing with is known, we may still feel comfortable with a model. In avian toxicology one is also probably less concerned with predictions than with evidence that harm is occurring to a population at a given place and time. Investigators have well-defined populations, areas, and times in mind: for example, songbirds and gamebirds in farm fields and surrounding woods, primarily during the breeding season. It may be possible to define risk to populations as the difference in model projections when a model is run with and without pesticide effects.

More importantly, existing data are available to begin forming the building blocks of the population-level models. For example, the Tall Timbers Research Station in Florida has compiled an excellent database on northern bobwhite (*Colinus virginianus*) survival and reproduction over approximately 50 years. Survival of bobwhite dosed with methyl parathion has been followed for 5 years.

The success of the Potts grey partridge model shows that modeling of pesticide effects at the population level can be done. This model, applicable to a few square miles, contains only about six variables, including pesticide use. Although such models take time to perfect, useful information may be derived in a shorter period.

There are other promising species and situations in which useful models may be developed. One good example, discussed in the introductory remarks to the workshop, is the declining population of sage grouse (*Centrocercus urophasianus*) in southeastern Idaho. Extensive die-offs of sage grouse have been reported along the Snake River Plain, within and around cropland. Recent radiotelemetry studies of sage grouse using cropland indicate that at least 20% of the summer population was exposed to organophosphorus pesticides and 16% was seriously impacted by their use. To model population-level effects, one would have to know what proportion of the population used cropland. Included among the pesticides sprayed are parathion, methamidophos, dimethoate, disulfoton, aldicarb, and carbofuran. Grouse can be exposed to several of these pesticides over a relatively short period.

The sage grouse is a species that would lend itself well to population modeling. They are large gamebirds whose declining populations are of great concern. The sage grouse is a species that uses leks, where the breeding males display. Birds move out from these leks in summer, seeking more succulent vegetation. If close enough to cropland, they move into agricultural fields during the summer. Because leks can be selected for study that are varying distances from agricultural fields, different degrees of expected pesticide exposure can be built into a model.

Adult sage grouse can be trapped at the leks and equipped with radiotransmitters carrying mortality sensors. These grouse can be followed to their summer range and their time spent in agricultural fields monitored. Birds can be retrapped following normal operational sprayings of agricultural fields with carbamate and organophosphorus pesticides and their exposure measured using blood cholinesterase. Measures of hatching success of eggs, survival of young, and year-round survival of adults can be followed in radiotagged birds. A proportional hazard model can be developed to depict accurately the relationships between exposure of sage grouse to agricultural pesticides and population parameters such as recruitment and age-specific survival. The development of such a model would be a dynamic process over several years.

It was suggested that modeling of pesticide effects on bird populations appears difficult because it has not been done before. The same could have been said years ago about the complex field studies now being done on pesticide effects on birds.

Regardless of the apparent difficulty, population modeling will have to be attempted because pesticide pressures on wildlife resources demand it. Wildlife toxicologists have been using informal models for years; we will see the use of more powerful, formal models in the future.

PLANS OF ACTION

Woven throughout the workshop were suggestions to put into action plans for modeling pesticide effects at the level of avian populations. At the heart of these suggestions was the exhortation, "We need some success stories." To aid in this success, the authors derived a list of actions from comments made by the workshop participants.

1. We need to define clearly what questions we want to ask about pesticide effects on avian populations.
2. Modelers should tell wildlife toxicologists what needs to be looked at in risk assessment in order to develop good models.
3. Wildlife toxicologists need to give modelers enough grounding in wildlife toxicology so we can at least start, together, to develop models.
4. Regulatory scientists need to communicate to modelers and toxicologists what the regulatory constraints are in considering risk assessments of population-level effects.
5. We need to start analyzing data sets of field and laboratory findings to see how they work in models and to see how much uncertainty exists.
6. In the future, efforts to integrate population trends and modeling with pesticide use should be made part of the regulatory process.
7. From this conference, a nucleus should be formed of people who will collaborate to develop modeling as a tool to investigate population-level effects of pesticides on birds.

53

Research and Development Needs

Synthesis Committee*

ABSTRACT

The 1990 conference "Population Ecology and Wildlife Toxicology of Agricultural Pesticide Use: A Modeling Initiative for Avian Species" brought together population ecologists, ecological modelers, wildlife toxicologists, representatives of the regulatory agencies, and industry. A major goal of the conference was to forge ties between toxicologists and ecologists in order to improve the design and analysis of field research and to form cooperative ventures in ecotoxicology and theoretical ecology to enhance our ability to model toxicological processes and develop a predictive capability regarding the effects of chemicals on wildlife populations. The Synthesis Committee presents a list of recommendations for these two areas. Priorities in future data collection are provided, especially in the context of large-scale field studies. Next discussed are areas of ecological modeling and theoretical ecology that can contribute to the development of a better predictive capability in ecotoxicology.

KEY WORDS

environmental toxicology, ecological modeling, population ecology, research priorities, field studies, avian toxicology.

INTRODUCTION

Concern about the effect of agricultural chemicals on avian populations has been an important scientific and public issue during the past decade.[1] As a consequence, large-scale field trials for the screening and registration of agricultural chemicals have become common.[2] A driving force behind field trials is the need to understand the effects of chemicals on vital rates in animal populations in a natural setting so that the use of these chemicals can be appropriately regulated. Field trials generally are large scale but not necessarily long term, and merely documenting the magnitude of an effect in a 1- or 2-year study will rarely provide sufficient information as to the long-term viability of

*Synthesis Committee members in alphabetic order are James Fava, Roy F. Weston, Co.; James A. Gagne, American Cyanamid Company; Ronald J. Kendall, TIWET, Clemson University; Thomas E. Lacher, Jr., Archbold Tropical Research Center, Clemson University; Raymond J. O'Connor, Department of Wildlife, University of Maine; Suzanne Orenstein, Conservation Foundation; Rod Parrish, SETAC Foundation for Environmental Education; Bill A. Williams, ecological planning & toxicology, inc.

populations in the wild. Time lags are commonly observed in the dynamics of natural systems, and their presence often separates cause from effect.[3] The future role of field studies has therefore been questioned by some scientists and regulators.[4] Whether such field trials can establish any relationship between acute and chronic effects of agricultural chemicals on population viability over the long term is the principal concern.

One solution is to extend large-scale field studies over many years,[5-7] but this would be expensive and, alone, still will not answer many research questions. Population ecologists have complemented their field research protocols with the use of models to generate predictions about the behavior of populations over time. Models currently available in the literature can assist in the synthesis of field studies, and may direct wildlife toxicologists in the design of field studies that collect the most critical data for model development. This will in turn allow "theoretical toxicologists" to attain better predictive capabilities.

The purpose of the 1990 Kiawah Island conference that led to this book was to integrate concepts from wildlife biology, wildlife and environmental toxicology, population ecology, and ecological modeling to achieve a more unified approach to, and better interpretation of, toxicological field studies. The conference organizers sought to review the information available from large-scale field studies of the effect of pesticides on avian species and to explore the relevance of various models from the ecological literature to ecotoxicology. A long-term goal of ecotoxicologists is to model in detail the effects of chemicals on individuals, ultimately extending this capacity to model effects at the level of populations and communities. To do this, they must (1) determine how existing models can assist in this process, (2) explore the potential for development of heuristic and predictive models for ecotoxicology, and (3) determine the kinds of empirical studies needed to provide the necessary data as input to predictive models. Appropriately designed large-scale field studies will be useful in providing the basic data for this process.

WHAT LARGE-SCALE FIELD STUDIES PROVIDE

A frequent goal of large-scale field studies is to evaluate the effects of pesticides on avian species in the wild.[8-10] Field trials offer unusual opportunities to gather very detailed information on processes relating to the effects of chemical applications on individual birds; this can often be extended to evaluate the potential for longer term effects on avian populations. Field trials have the additional advantage of allowing some control over the field experiments rather than being natural experiments (based on observation alone); this facilitates the discovery of elusive cause and effect pathways.

Certain variables and processes have received emphasis in large-scale field studies. For example, field trials present the opportunity to obtain data on the fate and transport of pesticides under natural environmental conditions. Agricultural chemicals often show complex behavior in the variety of inorganic and biological matrices in the field, and the different degradation products can vary in toxicity. The integration of research in the field of environmental chemistry into field protocols is a high priority in ongoing and future field research.

Detailed information is also collected on the exposure of large numbers of individuals under field rather than laboratory conditions. The resultant data allow a number of different biochemical or physiological biomarkers of exposure to be evaluated as to their effectiveness in predicting other important sublethal and lethal biological effects such as altered behavior, depressed reproductive success, and death. These data can also lead to the formulation of dose-response relationships involving ecologically realistic routes of exposure (although routes of exposure are difficult to evaluate in the field). When this information is related to the work on environmental fate and transport of pesticides and metabolites, better estimates of risk can be obtained.

In his chapter in this volume, Lacher[11] presents definitions of lethal vs sublethal effects and direct vs indirect pathways. Sublethal effects are those that impair metabolism or reproductive success without causing mortality. Effects via indirect pathways are caused by chemicals affecting species other than that being studied; changes in their abundance or behavior then affect the study species, either lethally or sublethally. Few data have been collected on the indirect sublethal effects of pesticides on wildlife.[12,13] However, as several chapters in this volume have noted, these kinds of effects may be far more important over the long term. Techniques are available for monitoring indirect sublethal effects within the context of the large-scale field study. Particularly promising areas for study include research on the effects of agricultural chemicals on foraging ecology, habitat selection, and dispersal, and how these effects will alter reproductive success or susceptibility to predation and disease. The important point is that field trials allow ecotoxicologists to collect data that cannot be obtained under laboratory or simulated seminatural conditions.

The development of theory derived from observation and experiment and supported by modeling and simulation has assisted in the maturation of ecology as a science, and the same process will be required in ecotoxicology.[14] A great deal of basic information still needs to be collected within the context of large-scale field studies before we can begin to think about modeling ecotoxicological processes with any accuracy or precision. However, it is imperative that ecotoxicologists take advantage of the unique opportunities that large-scale field trials present. The data collected over the past several years is ready for use to assist in the development of theory in ecotoxicology.

FUTURE PRIORITIES IN DATA COLLECTION

Throughout the conference, participants stressed the need to make careful observations under field situations and to conduct carefully designed experiments. The collection of data is paramount to advancing any scientific discipline. The amount and kinds of data collected, however, should be guided in part by the need to resolve important theoretical questions and to provide information for the development of theories and models.

Discussion among the conference participants focused on areas in which data were lacking and specific areas requiring ecological modeling. Conferees Larry Brewer and Roy Hummel of TIWET suggested two pathways requiring data and amenable to simulation modeling. The first was an *exposure model* involving the movement of a chemical through the environment from application to avian exposure (Figure 1); field trials already provide a great deal of data relevant to exposure. The second, an *effects model*, would relate exposure to mortality, reproduction, and behavior. Ultimately, the goal would be some predictive capability on the ways in which exposure may affect population numbers and community structure. At this stage, relating exposure to mortality, reproduction, and behavior is a challenge; extending this to population- and community-level processes is still a distant prospect.

Peter Greig-Smith provided a list of topics or problems that came up during discussion that were related to data collection and modeling. The list overlaps to some degree with the suggestions of Brewer and Hummel (Table 1). Both Figure 1 and Table 1 merely represent sample "wish-lists" of avian ecotoxicologists, and many of the goals

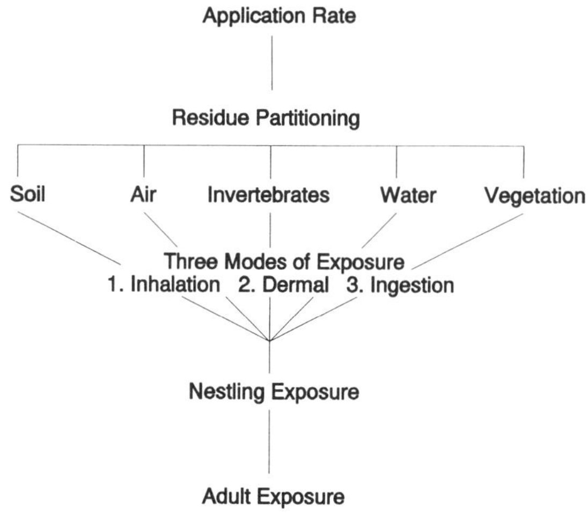

FIGURE 1. Schematic exposure model of chemical movement through the environment. The diagram represents the pathways and effects important in the exposure of birds to a pesticide. Toxicologists are working toward predictive models of exposure based on chemical application.

are unrealistic over the short term. The Synthesis Committee attempted to identify data that will be necessary both to improve experimental field studies and to use as inputs to various aspects of the modeling process.

There is a paucity of data relating avian mortality to exposure to pesticides (which vary widely in toxicity) under field situations. This lack compromises the development of theoretical models that extend information on exposure at the individual level to projections of changes in population numbers. We need carefully designed and controlled experiments under seminatural conditions to provide estimates of the relation between exposure and mortality for several age classes of a number of key indicator

Table 1. Areas in Avian Toxicology Needing Further Empirical Information and Potentially Amenable to Modeling

The fate and transport of pesticides
Routes of exposure in animals
The uptake, processing, and elimination of pesticides by animals, including bioaccumulation and detoxification
Dose-response relationships and toxic effects of biological importance
Interspecific extrapolations of toxicity and biomarkers
Demographic models of thresholds of concern (e.g., mortality, reproductive success, population growth rate)
Modeling assessment of field trial results over time (both direct and indirect effects)
Prediction of local consequences based upon the understanding of mechanisms at a broader level
Correlations of chemical use and environmental factors with population trends on a regional or national scale
Prediction of the effects of a total use pattern based upon knowledge of effects of individual chemicals
Effects of total use pattern on avian community ecology

species or appropriate animal models.[15] These experiments then need to be followed by well-designed field trials to verify that the controlled results can be extended to field situations as well.

Several biomarkers are currently used as indicators of avian exposure to pesticides. One of the most common biomarkers is acetylcholinesterase (AChE) depression, used to detect exposure to organophosphate and carbamate insecticides. More data are needed on the relationship between the degree of AChE depression and the probability of mortality for a variety of different species (ideally with different ecological characteristics) and a number of different compounds. The inability of most field trials to express the probability of mortality in the wild as a function of AChE depression is an example of one of the major unresolved issues in avian toxicology. It will be difficult to extend observations on effects at the individual level to predictions about long-term trends in population numbers without better empirically defined relationships between biomarkers of exposure and expected survivorship.

An important route of exposure of birds to pesticides is through the ingestion of chemically contaminated food. We lack information on the foraging behavior of many avian species. Critical studies for future ecotoxicological research include the evaluation of seasonal and temporal use of food resources as well as the patterns of spatial use of the habitat for foraging. The results of these studies will be especially applicable where agricultural fields are interspersed with natural habitats in a complex mosaic, as is often the case. Basic research on patterns of foraging, as well as more controlled testing of food selectivity of common species of birds, should be conducted as complements to ongoing field studies.

Although defining the relation between biomarkers (such as AChE depression) and mortality for the generalized individual in the population will be useful, field trials should strive to obtain information on the effects of a chemical application on survivorship and fecundity of explicitly defined age or stage classes. Age- or stage-structured data will greatly enhance the accuracy and precision of predictions of long-term trends in population numbers by enabling modelers to utilize matrix-based procedures that incorporate age class structure into the projections.

Ecologists know little about the degree to which the numbers of avian species are regulated by density-dependent or density-independent effects.[16–19] This is an area of basic ecological research, both experimental and theoretical, that has major implications for ecotoxicology. If an agricultural chemical causes incremental mortality early in an age class that generally is subjected to high levels of mortality attributable to density-dependent factors, the mortality attributable to pesticides may have little effect on population size. This is further complicated by the large variation in population response that can be observed. Many of the modeling approaches developed for the management of game or fish species are relevant here.[20–23] If we are ever to anticipate the long-term effects of age or stage class specific mortality in birds that is attributable to agricultural chemicals, then we will have to know a great deal more about the basic biology of the impacted avian species. We have only recently begun to realize the critical importance of basic ecological research to the evaluation of the risk to wildlife of toxic substances.[24,25]

We need to make better use of the information available on the long-term trends in avian population numbers. Clear distinctions must also be drawn regarding the scope of the population under study (i.e., "local" vs "regional"). Large databases are available through the Breeding Bird Survey, the Audubon Society's Christmas Bird Count, and other bird counts and surveys. One major retrospective study has already been completed.[19] This information will allow researchers to evaluate in general terms the numerical status of a population at the time of exposure in relation to long-term trends and will assist the researchers in understanding how vulnerable populations are to the effects of agricultural chemicals. Related to this, we will also require better information on the temporal and spatial variability,[26] not only in avian population density, but also in avian vital rates and development.[27,28] This is especially critical when comparing the effects of chemicals over large geographical regions.[29] This issue has also been addressed by Pulliam,[30] Bowers,[31] and O'Connor[17] in this book.

We need regional projects that compile existing databases of inventories of bird populations as well as inventories of land-use patterns.[32] The latter could be best accomplished by analyzing remote sensing imagery and aerial photography as input for a geographical information system.[7,33]

The previous discussion of priority needs for data is by no means exhaustive. The items presented do, however, follow a certain protocol that focuses on the manner in which effects at the level of the individual may eventually be extended to some general predictions about effects on populations at the local, regional, and national levels. We caution that results at the national and regional levels will be some time in coming.

PRIORITY APPROACHES IN MODELING

Toxicologists and ecologists gained different yet equally valuable insights from the conference. The ecologists gained an appreciation of the complexity of ecotoxicological research, especially of the constraints imposed on the research environment by the regulatory arena. Ecologists are to a large degree immune to the scientific and political decisions of regulators, and many gained a real appreciation of the difficulties faced by toxicologists. As a generalization, the toxicologists learned a great deal about the role of modeling in ecological research and were exposed to the great diversity of models available. At the end of the conference, all realized that many ecological models have relevance to ecotoxicology and some can be applied with little modification. Other, more predictive models will likely have to be developed anew. No single model or even class of models will provide quick answers.

As toxicologists begin to use modeling as a tool, they must be careful to express clearly the questions that they wish to answer and use rigor in the selection of variables and parameters.[20,21,34-36] Before theoretical toxicologists can begin to develop a modeling capability for pesticide effects in terrestrial ecosystems, a strong database on the effects of pesticides on wildlife populations under natural conditions is needed. The generation of specific predictions about the effects of a chemical via simulation will require the input of large amounts of data; simulation modeling can potentially yield very specific predictions, but only under a limited range of conditions. Several chapters in this volume present simulation approaches,[37,38] but ecotoxicologists need to avoid the temptation to concentrate their efforts on simulation alone at the cost of ignoring general unifying principles.[39] Not all theory needs to be testable; an equally important objective is the development of mathematical simplifications of ecotoxicological processes that suggest possible relationships and effects, rather than make specific predictions. The authors discuss several modeling approaches believed to contribute directly to the development of the discipline.

Fate and transport models[37] are useful for predicting the movement of chemicals and their products through the various matrices in the environment. These models are related more to chemistry than to ecology; they can be adapted to ecotoxicology, and research in this area should continue.

Other kinds of models are related to the behavior of generalized individuals. Hallam[40] has presented models of individual-level effects of certain classes of pesticides in this volume. The individual-based modeling approach presented by Hallam holds great promise in ecotoxicology as a method that provides a bridge between individual effects and population-level impacts. A large body of literature is available on models of foraging[41] and much of this literature has application in ecotoxicology, with little modification. Belovsky's chapter[42] discusses one such class of models, the application of linear programming techniques to the analysis of diets. His models of avian foraging include constraints imposed by pesticides. Belovsky notes that detailed ecological data are required to appropriately apply linear programming techniques, but when the basic foraging ecology of a species is well described, linear programming can prove to be a powerful technique for the management of pesticide use.

Pulliam[30,43] presented a group of landscape, or spatial geometry, models that are becoming more widely used in ecology[44] and seem to be of particular value in ecotoxicology. They will assist toxicologists in predicting exposure to species in landscape mosaics where species have differential use of habitats for foraging or breeding.[45,46] In fact, the Synthesis Committee see the wedding of landscape models with foraging models as an area in which the models are already well developed and can be applied to toxicology studies with little modification. Bowers's chapter[31] on habitat-specific vital rates is also of relevance when attempting to estimate the effects of agrochemicals on birds in spatially structured habitats. Consideration needs to be given to the appropriate spatial and temporal scale of models.[47] This will need to be done on a case by case and problem by problem basis.

Several recent studies have shown that birds can be exposed to organophosphate pesticides indirectly through the food chain.[10] The chapter by DeAngelis[38] on food-web analysis discusses an area in which models are well developed and can be applied to questions in ecotoxicology. Food-web theory has received several extensive reviews in recent years[48,49] and has facilitated the development of the more applied area of food-web analysis. One major objective of food-web analysis is to determine how changes or effects in one part of a web affect other segments of the web. This method shows promise for modeling the consequences of the application of a chemical when the basic nature of the affected food web is well defined.

The Committee also recommends exploratory research on several areas that were not represented at the conference but have potential application in ecotoxicological research. One topic that deserves investigation is the adaptive management approach of Walters.[50] At the heart of this technique is the deliberate manipulation of the ecological system under investigation away from the desired system state into a new parameter space. The response of the system then provides information on its dynamic properties and associated model parameters that would never be revealed by investigations centered on a homogeneous sample of sites. Such concepts have significant implications for the design of field trials intended to lead to model construction. A second, related technique acknowledges that pesticide use implicitly seeks the production of agricultural goods at lower cost, suggesting that we need to make more use of cost-benefit models based on economic rather than solely ecological theory. There is a growing literature on the costs and benefits of wildlife and wilderness upon which these analyses may be based.[51]

There was also interest in the potential of decision-support use models.[52] These models could be used to explore the consequences of alternative scenarios, in particular their associated financial costs. This would allow modelers to look at the trade-offs between adding complexity (for example, more detailed experiments, additional replicates, or longer studies) to a field trial, the improvement in knowledge (accuracy and/or precision) obtained therefrom, and the economic costs involved in doing so.[53] If the process is formalized into a decision-support system, subjective or qualitative weights can be assigned to given elements, allowing for the generation of numerical scores for each scenario. The regulatory agencies could assist in the development of weights, working in conjunction with universities and industry. This would facilitate decisions by regulatory agencies when confronted by complex data sets and highly mathematical models and simulations.

Adaptive management and decision-support use models by definition require that research be done on agroecosystems. Agroecosystems also offer many opportunities for research of fundamental importance to population and community ecologists. The line between "basic" and "applied" research has become less distinct; nevertheless, agroecosystems do allow ecologists and environmental toxicologists to perform collaborative research that can be of importance in management and theory.[54]

THE PET INITIATIVE

There was a strong sentiment at the close of the conference that in order to obtain the goals expressed during the summary sessions an organized and concerted initiative would be needed. The acronym PET (population ecology/toxicology) initiative was suggested by Dr. Kendall. The initiative as outlined had seven major objectives.

1. To publish a continuing newsletter on the PET initiative beginning in 1991. This newsletter is now published annually.
2. To finish the chapters from the conference and publish this book in 1993.
3. To establish scientific collaborations. One objective was to begin a scientific exchange program whereby ecologists and toxicologists could visit laboratories to exchange ideas. The costs of such a program would be shared by U.S. EPA and industry. A longer-range goal is the creation of joint research laboratories in which basic and applied scientists work together.
4. To design research projects that address heavily impacted and threatened habitats. Priority areas include agricultural fields (especially those areas in which agriculture and natural habitats exist as a mosaic), turf habitats, wetlands, coastal zones, and estuaries.
5. To use the annual SETAC meetings for special sessions focusing on the application of existing ecological models to ecotoxicological research and the development of new models with predictive capabilities. Several workshops were scheduled along these lines for the 1991, 1992, and 1993 meetings. We propose that these workshops continue on an annual basis.
6. To establish a regular channel of communication with U.S. EPAs Program Offices and Office of Research and Development so that research needs can be quickly expressed and funding priorities can be redirected for innovative initiatives. These efforts can be developed initially by the President and Board of Directors of SETAC, and the effort expanded at a later date to include research by ecologists, ecological modelers, wildlife biologists, and other scientists not affiliated with SETAC. An initial step should be to establish an Ecomodeling Synthesis Committee within SETAC that will set 5- and 10-year goals and objectives for research.
7. To reconvene in 5 years (1995) to update and review the work resulting from the first Kiawah conference. Our plan is to hold a second symposium and to publish a second book of symposium proceedings. Some of the original conference participants may be asked to return, but a number of new speakers will also be invited. We will seek out the best people active at that time in order to show how the field has progressed in 5 years. Rather than simply publishing a revised second edition, we will produce a new book, with some chapters and topics revised, and others presenting new ideas and information.

CONCLUSIONS

In order to appropriately manage the use of agricultural chemicals, we require cooperative ventures that unite basic researchers in population and community ecology, behavioral ecology, theoretical ecology, and ecological modeling; applied investigators in wildlife and environmental toxicology; and regulators in government. Toxicologists will better understand the effects of chemicals on fish and wildlife, ecologists will gain new insights on ecological patterns and processes, and regulators will be able to make better informed decisions. Such a meeting of minds, however, must pursue realistic and directed goals. Large endeavors, such as some field trials, will expend large amounts of money. There is no guarantee that they will yield meaningful results simply as a result of their size. Regardless of the techniques applied and the systems studied, our strongest recommendation is that clearly defined and meaningful questions be asked that are addressed by carefully designed experiments that parallel the development of theory.[55] No amount of theory and modeling alone can eliminate the need for the scientific method.

REFERENCES

1. Risebrough, R. W., Pesticides and bird populations, *Current Ornithology*, 3, 397–427, 1986.
2. Fite, E. C., L. W. Turner, N. J. Cook, C. Stunkard and R. M. Lee, Guidance Document for Conducting Terrestrial Field Studies, Ecological Effects Branch, Hazard Education Division, Office of Pesticide Programs, U.S. Environmental Protection Agency, Washington, DC, 1988.
3. Barbault, R., Ecological constraints and community dynamics: linking community patterns to organismal ecology. The case of tropical herpetofaunas, *Acta Oecol.*, 12, 139–163, 1991.
4. Avian Effects Dialogue Group, *Pesticides and Birds: Improving Impact Assessment*, The Conservation Foundation, Washington, DC, 1989.
5. Likens, G. E., Ed., *Long-Term Studies in Ecology*, Springer-Verlag, New York, 1987.
6. Magnuson, J. J., Long-term ecological research and the invisible present, *BioScience*, 40, 495–501, 1990.
7. O'Connor, R. J., Long-term bird population studies in the United States, *Ibis*, 133(Suppl. 1), 36–48, 1991.
8. Kendall, R. J., L. W. Brewer, T. E. Lacher, Jr., B. T. Marden and M. L. Whitten, The Use of Starling Nest Boxes for Field Reproductive Studies: Provisional Guidance Document and Technical Support Document, EPA/600/8-89/056, NTIS Publ. No. PB89 195 028/AS, National Technical Information Services, Springfield, VA, 1989.
9. Brewer, L. W., D. E. Dix, G. P. Cobb, III, M. J. Hooper, S. L. Tank, R. A. Craft, J. A. Buck, W. M. Buerger and C. M. Bens, Second Year Investigation of the Response of Selected Wildlife Populations to Planting Time Application of COUNTER^R 15-G Systemic Insecticide-Nematicide in an Iowa Cornfield Agroecosystem, report to American Cyanamid Company, Institute of Wildlife Toxicology, Western Washington University, Bellingham, 1989.
10. Tank, S. L., L. W. Brewer, G. P. Cobb, M. J. Hooper, R. A. Hummell, R. A. Craft, J. A. Buck, W. L. McKenzie and C. M. Bens, Third Year Investigation of Response of Selected Wildlife Populations to Planting Time Application of COUNTER^R 15-G Systemic Insecticide-Nematicide in an Iowa Corn Agroecosystem, report to American Cyanamid Company, Institute of Wildlife and Environmental Toxicology, Clemson University, Clemson, SC, 1990.
11. Lacher, T. E., Jr., Direct and indirect pathways in spatially structured habitats, in *Wildlife Toxicology and Population Modeling: Integrated Studies of Agroecosystems,* R. J. Kendall and T. E. Lacher, Jr., Eds., Lewis Publishers, Chelsea, MI, 1993.

12. Buerger, T. T., R. J. Kendall, B. S. Mueller, T. DeVos and B. A. Williams, Effects of methyl parathion on northern bobwhite survivability, *Environ. Toxicol. Chem.*, 10, 527–532, 1991.
13. Galindo, J., R. J. Kendall, C. J. Driver and T. E. Lacher, Jr., The effect of methyl parathion on the susceptibility of bobwhite quail (*Colinus virginianus*) to domestic cat predation, *Behav. Neural Biol.*, 43, 21–36, 1985.
14. Kendall, R. J., and T. E. Lacher, Jr., Ecological modeling, population ecology, and wildlife toxicology: a team approach to environmental toxicology, *Environ. Toxicol. Chem.*, 10, 297–299, 1991.
15. Morrison, M. L., Bird populations as indicators of environmental change, *Curr. Ornithol.*, 3, 429–451, 1986.
16. Greenwood, J. D., and S. R. Baillie, Effects of density-dependence and weather on population changes of English passerines using a non-experimental paradigm, *Ibis*, 133(Suppl. 1), 121–133, 1991.
17. O'Connor, R. J., Population patterns and process parameters: issues in integrating monitoring and models. In *Wildlife Toxicology and Population Modeling: Integrated Studies of Agroecosystems*, R. J. Kendall and T. E. Lacher, Jr., Eds., Lewis Publishers, Chelsea, MI, 1993.
18. Lack, D., *The Natural Regulation of Animal Numbers*, Clarendon Press, Oxford, 1954.
19. O'Connor, R. J., and R. B. Boone, A retrospective study of agricultural bird populations in North America, in *Proc. Symp. Ecological Indicators, Fort Lauderdale, Florida (October, 1990)*, Elsevier, Amsterdam, in press.
20. Potts, G. R., *The Partridge—Pesticides, Predation, and Conservation*, Collins, London, 1986.
21. Potts, G. R., The effects of modern agriculture, nest predation, and game management on the population ecology of partridges, *Perdix perdix* and *Alectoris rufa, Adv. Ecol. Res., 11, 2–79, 1980.*
22. May, R. M., Ed., *Exploitation of Marine Communities, Dahlem Conference*, Springer-Verlag, Berlin, 1984.
23. Getz, W. M., and R. G. Gaight, *Demographic Models of Fish, Forest and Mammal Resources*. Princeton University Press, Princeton, NJ, 1989.
24. Bascietto, J., D. Hinckley, J. Plafkin and M. Slimak, Ecotoxicity and ecological risk assessment, *Environ. Sci. Technol.*, 24, 10–15, 1990.
25. Rosemarin, A., Ecotoxicology on the upswing—but where are the ecologists?, *Ambio,* 17, 359, 1988.
26. Taylor, L. R., I. P. Woiwod and J. N. Perry, Variance and the large scale spatial stability of aphids, moths and birds, *J. Anim. Ecol.*, 49, 831–854, 1980.
27. Ricklefs, R. E., Comparative avian demography, *Curr. Ornithol.*, 1, 1–32, 1983.
28. O'Connor, R. J., *The Growth and Development of Birds*, John Wiley & Sons, Chichester, U.K., 1984.
29. O'Connor, R. J., and A. A. Whitman, Recent Population Changes in Some Bird Species that Use Gardens, Lawns, and Turf in the United States, Cyclostyl Contract Report to CIBA-GEIGY Corp., Greensboro, NC, 1990
30. Pulliam, H. R., Incorporating concepts from population and behavioral ecology into models of exposure to toxins and risk assessment, in *Wildlife Toxicology and Population Modeling: Integrated Studies of Agroecosystems*, R. J. Kendall and T. E. Lacher, Jr., Eds., Lewis Publishers, Chelsea, MI, 1993.
31. Bowers, M. A., Use of space and habitats by individuals and populations: dynamics and risk assessment, in *Wildlife Toxicology and Population Modeling: Integrated Studies of Agroecosystems,* R. J. Kendall and T. E. Lacher, Jr., Eds., Lewis Publishers, Chelsea, MI, 1993.
32. Devaul, H., A. A. Whitman, R. B. Boone, M. T. Jones, M. L. Hunter and R. J. O'Connor, An Analysis of Selected Extant Data for Birds in New England, report to Technical Resources Inc. for EPA contract #68-CO-0021/WB-94-04, 1990.
33. Hoekstra, T. W., and C. H. Flathers, Theoretical basis for integrating wildlife in renewable resource inventories, *J. Environ. Manage.*, 24, 95–110, 1987.

34. Emlen, J. M., and E. K. Pikitch, Animal population dynamics: identification of critical components, *Ecol. Model.*, 44, 253–273, 1989.
35. Dobson, A. P., Survival rates and their relationship to life-history traits in common British birds, *Curr. Ornithol.*, 7, 115–146, 1990.
36. May, R. M., Ed., *Theoretical Ecology: Principles and Applications*, 2nd ed., Sinauer Associates, Sunderland, MA, 1981.
37. Bird, S., Fate and exposure modeling in terrestrial ecosystems: a process approach, in *Wildlife Toxicology and Population Modeling: Integrated Studies of Agroecosystems*, R. J. Kendall and T. E. Lacher, Jr., Eds., Lewis Publishers, Chelsea, MI, 1993.
38. DeAngelis, D. L., What food web analysis can contribute to wildlife toxicology, in *Wildlife Toxicology and Population Modeling: Integrated Studies of Agroecosystems*, R. J. Kendall and T. E. Lacher, Jr., Eds., Lewis Publishers, Chelsea, MI, 1993.
39. Abrams, P. A., The contributions of ecological theory (review of J. Roughgarden, R. M. May and S. A. Levin, Eds., *Perspectives in Ecological Theory*, Princeton University Press, Princeton, NJ), *Trends Ecol. Evol.*, 4, 220–221, 1989.
40. Hallam, T., Individual based mathematical modeling approaches in ecotoxicology: a promising direction for aquatic population and community ecological risk assessment, in *Wildlife Toxicology and Population Modeling: Integrated Studies of Agroecosystems*, R. J. Kendall and T. E. Lacher, Jr., Eds., Lewis Publishers, Chelsea, MI, 1993.
41. Stephens, D. W., and J. R. Krebs, *Foraging Theory,* Princeton University Press, Princeton, NJ. 1986.
42. Belovsky, G. E., Modeling avian foraging: implications for assessing the ecological effects of pesticides, in *Wildlife Toxicology and Population Modeling: Integrated Studies of Agroecosystems*, R. J. Kendall and T. E. Lacher, Jr., Eds., Lewis Publishers, Chelsea, MI, 1993.
43. Pulliam, H. R., Sources, sinks and population regulation, *Amer. Nat.*, 132, 652–661, 1988.
44. Temple, S. A., Predicting impacts of habitat fragmentation on forest birds: a comparison of two models, in *Wildlife 2000: Modeling Habitat Relationships of Terrestrial Vertebrates*, J. Verner, M. L. Morrison, and C. J. Ralph, Eds., University of Wisconsin Press, Madison, 1986, pp. 301–304.
45. O'Connor, R. J., Behavioral regulation of bird populations: a review of habitat use in relation to migrancy and residency, in *Behavioural Ecology: Ecological Consequences of Adaptive Behaviour,* R. M. Sibly and R. H. Smith, Eds., Blackwell Scientific, Oxford, 1985, pp. 105–142.
46. O'Connor, R. J., Dynamical aspects of avian habitat use, in *Wildlife 2000: Modeling Habitat Relationships of Terrestrial Vertebrates*, J. Verner, M. L. Morrison and C. J. Ralph, Eds., University of Wisconsin Press, Madison, 1986, pp. 235–240.
47. Wiens, J. A., *The Ecology of Bird Communities,* Cambridge University Press, Cambridge, U.K., 1989.
48. Cohen, J. E., *Food Webs and Niche Space,* Princeton University Press, Princeton, NJ, 1978.
49. Pimm, S. L., *Food Webs,* Chapman & Hall, New York, 1982.
50. Walters, C., *Adaptive Management of Renewable Resources,* Macmillan, New York, 1986.
51. Bishop, R. C., S. R. Milliman, K. J. Boyle and B. L. Johnson, Benefit-cost analysis of fishery rehabilitation projects: a Great Lakes case study, *Ocean Shoreline Manage.,* 13, 253–274, 1990.
52. Halter, A. N. and G. W. Dean, *Decisions under Uncertainty, with Research Applications,* Southwestern Publications, Cincinnati, 1971.
53. Mumford, J. D., and G. A. Norton., Economics of decision making in pest management, *Annu. Rev. Entomol.*, 29, 157–174, 1984.
54. Lacher, T. E., Jr., Theoretical ecology and research on agroecosystems, in *Wildlife Toxicology and Population Modeling: Integrated Studies of Agroecosystems*, R. J. Kendall and T. E. Lacher, Jr., Eds., Lewis Publishers, Chelsea, MI, 1993.
55. Cairns, J., Jr., Paradigms flossed: the coming of age of environmental toxicology, *Environ. Toxicol. Chem.*, 11, 285–287, 1992.

Index